Planet Earth

SECOND EDITION
Revised Printing

John J. Renton
Department of Geology and Geography
West Virginia University

KENDALL/HUNT PUBLISHING COMPANY
4050 Westmark Drive   Dubuque, Iowa 52002

Copyright © 1994 by West Publishing Company

Copyright © 2002 by John J. Renton

Revised Printing 2006

ISBN 13: 978-0-7575-3378-5

Kendall/Hunt Publishing Company has the exclusive rights to reproduce this work, to prepare derivative works from this work, to publicly distribute this work, to publicly perform this work and to publicly display this work.

All rights reserved. No part of this publication may be reproduced, stored in a retrieval system, or transmitted, in any form or by any means, electronic, mechanical, photocopying, recording, or otherwise, without the prior written permission of Kendall/Hunt Publishing Company.

Printed in the United States of America

10 9 8 7 6 5

This Book Is Dedicated
to
My Wife and Family

CONTENTS IN BRIEF

Introduction:	An Overview of Earth	1
Chapter 1:	Earth and Its Place in Space	25
Chapter 2:	Plate Tectonics	47
Chapter 3:	Minerals	71
Chapter 4:	Volcanism	95
Chapter 5:	Igneous Rocks	139
Chapter 6:	Weathering	161
Chapter 7:	Soils	183
Chapter 8:	Mass Wasting	203
Chapter 9:	Streams	233
Chapter 10:	Glaciers	281
Chapter 11:	Deserts and the Wind	319
Chapter 12:	Ocean and Shorelines	349
Chapter 13:	Sedimentary Rocks	391
Chapter 14:	Groundwater	423
Chapter 15:	Rock Deformation and the Geologic Structures	451
Chapter 16:	Metamorphic Rocks	481
Chapter 17:	Mountain Building	499
Chapter 18:	Earthquakes and Seismology	521
Chapter 19:	Earth's Interior	547
Chapter 20:	Economic Geology and Energy	565
Chapter 21:	The Age of Earth	607
Appendix A:	Glossary	635
Appendix B:	Tables	663
Answers to Multiple Choice Questions		684
Index		685

CONTENTS

INTRODUCTION:
An Overview of Earth — 1

- Kinds of Rocks — 1
  - Igneous Rocks — 1
  - Sedimentary Rocks — 5
  - Metamorphic Rocks — 7
- The Internal Structure of Earth — 8
- The Surface of Earth — 12
  - The Ocean Basins — 12
  - The Continents — 17
- The Hydrosphere and the Atmosphere — 21
- The Rock Cycle — 21
- Conclusion — 22
- Concepts and Terms to Remember — 22
- Review Questions — 23

CHAPTER 1:
Earth and Its Place in Space — 25

- Introduction — 27
- The View for Earth — 27
  - The Contribution of the Ancient Greeks — 27
  - The Evolution of Modern Astronomy — 29
  - The Galaxies — 30
- Distances in Space — 30
- The Origin and Evolution of the Universe — 31
  - The Formation of Stars — 31
  - The Death of Stars — 31
- The Formation of Our Solar System — 32
- The Planets — 33
  - The Terrestrial Planets — 33
  - The Jovian Planets — 37
  - Pluto, the Maverick Planet — 40
- Asteroids — 41
- Meteors, Meteoroids, and Meteorites — 41
- Comets — 42
- Concepts and Terms to Remember — 44
- Review Questions — 44
- Thought Problems — 45
- For Your Notebook — 45

CHAPTER 2:
Plate Tectonics — 47

- Introduction — 49
- Historical Development — 49
  - F.B. Taylor — 50
  - H.B. Baker — 50
  - Alfred Wegener — 50
- Modern Developments — 52
  - Rock Magnetism — 52
  - Topography of the Ocean Bottom — 54
  - Seismic Investigation of Earth's Interior — 60
- The Principle of Plate Tectonics — 65
- Concepts and Terms to Remember — 67
- Review Questions — 67
- Thought Problems — 68
- For Your Notebook — 69

CHAPTER 3:
Minerals — 71

- Introduction — 73
- Minerals Defined — 73
- Chemical Composition of Minerals — 73
  - Compounds — 75
  - Chemical Bonding — 78
  - Crystal Structure — 80
- Mineral Identification — 80
  - Physical Properties — 80
  - Instrumental Identification Techniques — 85
- Mineral Classification — 85
  - Silicate Crystal Structures — 86
  - Nonsilicate Minerals — 91
- Concepts and Terms to Remember — 91
- Review Questions — 92
- Thought Problems — 93
- For Your Notebook — 93

CHAPTER 4:
Volcanism — 95

- Introduction — 97
- Distribution of Volcanoes — 99

Relationship to Plate Boundaries	99
Intraplate Volcanism	100
Ejected Materials	100
Gases	102
Liquids	103
Solids (Tephra)	106
Ash Flows	108
Mudflows or Lahars	108
Kinds of Volcanoes	109
Cinder Cones	109
Spatter Cones	109
Shield Volcanoes	110
Stratovolcanoes or Composite Volcanoes	111
Eruptive Intensity	112
Classification of Eruptive Intensity	114
Fissure Eruptions	116
Oceanic Ridges	116
Flood Basalts	117
Felsic Eruptions	118
Craters and Calderas	118
Other Volcanic Features	121
Hot Springs	121
Geysers	121
Fumaroles	122
Prediction of Volcanic Eruptions	123
Examples of Other Noteworthy Volcanoes	125
Mount Vesuvius	125
Santorini	126
Krakatoa	128
Mount St. Helens	128
The Hawaiian Islands	131
Sakurajima, Japan	136
Concepts and Terms to Remember	137
Review Questions	137
Thought Problems	138
For Your Notebook	138

CHAPTER 5:
Igneous Rocks — 139

Introduction	141
Melting and Crystallization of Solids	141
The Melting of Single Minerals of Constant Composition	142
The Melting of Mixtures of Minerals and Minerals of Variable Composition	142
Order of Mineral Crystallization	143
Rock Texture	146
The Rate of Cooling	146
Classification of Igneous Rocks	148
The Origin of Magmas and Plate Tectonics	150
Igneous Rock Bodies	154
Intrusive Rock Bodies	154
Concepts and Terms to Remember	158
Review Questions	158
Thought Problems	159
For Your Notebook	159

CHAPTER 6:
Weathering — 161

Introduction	163
Weathering Defined	164
Mechanical Weathering	164
Frost Action	164
Exfoliation	165
Spheroidal Weathering	167
Effect of Temperature	168
Plant Wedging	169
Chemical Weathering	170
Oxidation	170
Dissolution	171
Carbonation/Hydrolysis	173
Rates of Chemical Weathering	174
Temperature	174
Availability of Water	174
Particle Size	174
Composition	175
Metastable Materials	176
Regolith	176
Environmental Concerns	177
Trace Elements	177
Water Quality	179
The Effect of Mining on Water Quality	180
Concepts and Terms to Remember	180
Review Questions	180
Thought Problems	181
For Your Notebook	182

CHAPTER 7:
Soils — 183

Introduction	185
Soil Defined	185
Factors Controlling Soil Formation	185
Biological Factors	186
Physical Factors	187
Climate	188
Time	188
The Role of the Clay Minerals	188
Cation Adsorption	189
Cation Exchange Capacity	189

Soil Horizons	192
The O Horizon	192
The A Horizon	192
The E Horizon	192
The B Horizon	193
The C Horizon	194
Horizon Development	194
The Influence of Rainfall	194
The Effect of Water Volume and Chemistry	194
Tropical Regions	196
Types of Soil	196
Environmental Concerns	197
Concepts and Terms to Remember	201
Review Questions	201
Thought Problems	202
For Your Notebook	202

CHAPTER 8:
Mass Wasting   203

Introduction	205
Slope Stability	208
Factors Affecting Cohesion and Friction	208
Water Saturation	208
Ice Formation	209
Oversteepening of Slopes	210
Classification of Mass Wasting Processes	211
Creep	211
Earthflow	211
Landslides	213
Environmental Concerns	220
Causes of Slope Instability	220
The Effects of Earth Moving	222
Identification of Potentially Hazardous Sites	223
Methods of Slope Stabilization	224
Subsidence	226
Concepts and Terms to Remember	230
Review Questions	230
Thought Problems	231
For Your Notebook	231

CHAPTER 9:
Streams   233

Introduction	235
The Hydrologic Cycle	235
Drainage Systems	236
Stream Order	238
Stream Patterns	238
The Energy of a Stream	242
Gradient	244
Channel Texture	244
Water Volume	244
The Dynamics of Water Flow	245
The Processes of Erosion and Transportation	245
The Generation of Force	246
The Effect of Particle Size and Shape	246
Capacity and Load	247
The Development of Topographic Relief	249
Baselevel	249
Fluvial Landforms	249
Meandering	257
The Development of Grade	257
Climate Influences	260
Tectonic Influences	260
Human Influences	260
Alluvial Deposits	261
Channel Deposits	261
Floodplain Deposits	261
Deltas	264
The Mississippi Delta	265
Alluvial Fans	272
Environmental Concerns	274
Floods	274
Water Pollution	275
Concepts and Terms to Remember	276
Review Questions	277
Thought Problems	278
For Your Notebook	278

CHAPTER 10:
Glaciers   281

Introduction	283
The Distribution of Glaciers	283
The Origin of Glacial Ice	283
The Movement of Glacial Ice	286
Ice Thickness	286
Ice Temperature	287
Slope of the Bedrock and Ice Surface	287
The Glacial Budget	288
Processes of Glacial Erosion	289
Erosion by Alpine Glaciers	290
Erosion by Continental Glaciers	294
Glacial Deposition	297
Kinds of Glacial Deposits	297
Periglacial Regions	303
Mass Wasting	303
Weathering	303
Streams	305
The Wind	305

Understanding Continental Glaciation	305
The Great Ice Age	307
The Great Lakes	307
The Causes of Continental Glaciation	313
Celestial Causes	313
Volcanic Activity	313
Plate Tectonics	314
Environmental Concerns	315
Rising Sea Levels	315
Goiter	315
Concepts and Terms to Remember	316
Review Questions	317
Thought Problems	317
For Your Notebook	318

CHAPTER 11:

Deserts and the Wind — 319

Introduction	321
Relative Humidity	324
Kinds of Deserts	324
Rainshadow Deserts	324
Subtropical Deserts	325
Fog Deserts	326
Isolation Deserts	327
Erosion in the Desert	329
The Effectiveness of Water	329
The Arid Cycle of Erosion	331
Erosion by the Wind	337
Deflation	338
Abrasion	338
Deposition from the Wind	339
Loess	339
Dunes	340
Environmental Concerns	344
Overgrazing	344
Overcultivation	345
Fuel	345
Salinization	345
Concepts and Terms to Remember	346
Review Questions	347
Thought Problems	347
For Your Notebook	348

CHAPTER 12:

Ocean and Shorelines — 349

Introduction	351
Exploring the Ocean Bottom	351
Sonar	351
Submersibles	352
Deep-Sea Drilling	353
The Origin of the Ocean Basins	353
Cosmic Origins	353
Plate Tectonics	353
Supercontinent Cycles	356
Landforms on the Ocean Bottom	357
Oceanic Ridges	357
Abyssal Hills	358
Abyssal Plains	359
Deep-Sea Trenches	359
Shield Volcanoes	359
Continental Margins	359
Continental Trailing Edges	359
Continental Leading Edges	363
Movement of Ocean Water	364
Waves	364
Tides	365
Longshore Currents	367
Ocean Currents	367
Coastal Processes	374
Emergent or High-Energy Coastlines	375
Submergent or Low-Energy Coastlines	378
Reefs	382
Kinds of Reefs	382
Modification of Shoreline Processes	385
Environmental Concerns	385
Concepts and Terms to Remember	387
Review Questions	388
Thought Problems	389
For Your Notebook	389

CHAPTER 13:

Sedimentary Rocks — 391

Introduction	393
Classification of Sedimentary Rocks	394
Size	394
Composition	394
Sorting	394
Shape	395
Detrital Sedimentary Rocks	397
Breccia	397
Conglomerate	397
Sandstones	397
Siltstone	398
Shales and Mudstones	398
Nondetrital Sedimentary Rocks	398
Chemical Sedimentary Rocks	398
Biochemical Sedimentary Rocks	400
Evaporitic Sedimentary Rocks	402
Organic Sedimentary Rocks	402

Depositional Environments	403
Continental Deposits	403
Transitional Environments	405
Marine Environments	407
Lithification	409
Compaction	409
Cementation	409
Sedimentary Features	410
Beds	410
Ripple Marks	410
Cross-Bedding	410
Graded Bedding	411
Mud Cracks	413
Animal Trails	413
Interpretation of Ancient Depositional Environments	414
Walther's Law	417
Environmental Concerns (Sediments)	419
Sediment Pollution	419
Engineering Problems	419
Concepts and Terms to Remember	420
Review Questions	421
Thought Problems	421
For Your Notebook	422

CHAPTER 14:

Groundwater   423

Introduction	425
Porosity and Permeability	426
Porosity	426
Permeability	428
Aquifiers, Aquitards, and Aquicludes	428
The Water Table	430
Regional Water Tables	431
Perched or Hanging Water Tables	432
Water Wells	433
Confined Aquifers	433
Municipl Water Supplies	436
Karst Topography	437
Dissolution of Carbonate Rocks	438
Sinkholes	438
Surface Drainage	439
Subsurface Drainage	439
Caves and Caverns	440
Hot Springs, Fumaroles, and Geysers	443
Environmental Concerns	443
Groundwater Pollution	445
Dewatering	445
Saltwater Encroachment	446
Concepts and Terms to Remember	448
Review Questions	448
Thought Problems	449
For Your Notebook	449

CHAPTER 15:

Rock Deformation and the Geologic Structures   451

Introduction	453
Stress, Deformation, and the Strength of Materials	454
Kinds of Stress	454
Kinds of Deformation	455
The Strength of Rocks	458
Rock Deformation	458
Elastic Deformation	460
Plastic Deformation	460
Brittle Deformation	460
The Concept of Strike and Dip	461
Geologic Structures	461
Folds	461
Faults	464
Faulting After Initial Folding	468
Joints	469
Determining the Direction of Rock Movement	473
Rock Transport	473
Actual Direction of Fault Movements	474
Relative Direction of Fault Movements	475
Concepts and Terms to Remember	477
Review Questions	478
Thought Problems	478
For Your Notebook	479

CHAPTER 16:

Metamorphic Rocks   481

Introduction	483
Metamorphism	484
Metamorphic Textures and Classifications	484
Foliated Texture	485
Nonfoliated Texture	486
Kinds of Metamorphism	488
Contact Metamorphism	488
Dynamo-Thermal Metamorphism	488
Hydrothermal Metamorphism	489
Metamorphic Grade	491
Metamorphic Facies	491
Metamorphism and Plate Tectonics	492
Environmental Concerns	494
Concepts and Terms to Remember	495
Review Questions	496
Thought Problems	497
For Your Notebook	497

CHAPTER 17:

Mountain Building — 499

- Introduction — 501
- Definition of a Mountain — 501
- Kinds of Mountains — 502
    - Volcanic Mountains — 502
    - Domal and Block-Fault Mountains — 504
    - Foldbelt Mountains — 507
- Orogenic Styles — 511
    - Ocean-Continent Orogenesis — 511
    - Ocean-Island Arc-Continent Orogenesis — 514
    - Continent-Continent Orogensis — 515
- Concepts and Terms to Remember — 519
- Review Questions — 519
- Thought Problems — 520
- For Your Notebook — 520

CHAPTER 18:

Earthquakes and Seismology — 521

- Introduction — 523
- Distribution of Earthquakes — 523
- The Frequency and Location of Earthquakes — 524
    - Focus — 525
    - Epicenter — 525
- Measuring Earthquake Intensity and Magnitude — 526
- Intensity — 526
- Magnitude — 527
- Damage from Earthquakes — 530
    - Tsunami — 532
- Earthquake (Seismic) Waves — 533
    - Body Waves — 534
    - Surface Waves — 534
- The Seismograph — 335
- Locating Earthquakes from Seismic Data — 538
- WorldWide Seismic Network — 542
- Earthquake Prediction — 543
- Concepts and Terms to Remember — 544
- Review Questions — 545
- Thought Problems — 546
- For Your Notebook — 546

CHAPTER 19:

Earth's Interior — 547

- Introduction — 549
- Seismic Wave Velocities — 550
- The Crust-Mantle Boundary — 551
- The Mantle-Core Boundary — 552
- Isostasy — 555
    - Isostatic Balance — 558
- Crystal Loading and Unloading — 560
- Concepts and Terms to Remember — 562
- Review Questions — 562
- Thought Problems — 563
- For Your Notebook — 563

CHAPTER 20:

Economic Geology and Energy — 565

- Introduction — 567
- Classification of Natural Resources — 568
    - Ore, Protore, and Gangue — 568
    - Reserves and Resources — 568
- Metalliferous Deposits — 569
    - Concentrations of Metals — 569
    - Enrichment of Ore Bodies by Weathering — 571
    - Banded Iron Formations — 572
    - Alluvial and Eolian Concentrations — 574
- Nonmetallic Nonfuels — 577
- Fuels — 577
    - Coals — 578
    - Petroleum — 584
    - Oil Shales — 586
- The Future of Fossil Fuels — 586
- Nuclear Power — 588
    - Fission Reactors — 588
    - Fusion — 591
- Alternative Sources of Energy — 591
    - Geothermal Power — 591
    - Wind — 592
    - The Tides — 594
    - Hydropower — 594
    - Biomass — 595
    - Conservation — 597
    - Solar Energy — 598
- Environmental Concerns — 601
    - Coal Mining — 601
    - Environmental Problems — 602
    - Treating the Acid Mine Drainage Problem — 602
- Concepts and Terms to Remember — 604
- Review Questions — 605
- Thought Problems — 606
- For Your Notebook — 606

CHAPTER 21:
The Age of Earth — 607

Introduction — 609
Early Estimates of Earth's Age — 609
Dating Methods — 611
   Relative Dating — 611
   Absolute Dating — 619
   Dating by Sediment Accumulation,
     Ocean Salinity, and Earth's Heat — 619
   Radiometric Dating — 621
Rock Time Scales — 624
   The Modern Time Scale — 624
   Periods of the Paleozoic Era — 628
   Periods of the Mesozoic Era — 629
   Periods and Epochs of the
     Cenozoic Era — 631

Concepts and Terms to Remember — 632
Review Questions — 633
Thought Problems — 634
For Your Notebook — 634

APPENDIX A:
Glossary — 635

APPENDIX B:
Tables — 663

ANSWERS TO MULTIPLE CHOICE QUESTIONS — 684

INDEX — 685

PREFACE

At the beginning of every new semester, I tell my students that my primary objective is simply to have them wonder about Earth. My primary intent in writing this textbook is the same. I want you to wonder why the appearance of the land changes as you travel from place to place; whether stories that claim that the continents are drifting about the globe are true; why California experiences more severe earthquakes than any other state in the Union; and why most of the active volcanoes are located around the margin of the Pacific Ocean.

Assuming that I succeed in stirring your imagination, I want you to ask questions and seek answers to those questions. I want you to understand, appreciate, and *enjoy* geology. You will be amazed at how much more meaningful a landscape becomes when you understand how it formed. The Grand Canyon is simply a deep ravine until you understand the geologic processes that were responsible for its formation; then it literally becomes one of the wonders of the world. The seashore is simply a place to bask in the Sun until you understand the processes that are constantly reshaping it and why our feeble attempts to control the natural coastal processes are doomed to fail. Coastal wetlands are simply swamps until you understand how they form and appreciate the critical role they play in the marine ecosystem.

INTENDED AUDIENCE The most important consideration for a classroom instructor, or the author of a textbook, is the makeup of the audience. For more than 35 years, I have been teaching introductory physical geology to undergraduate students who come with a broad spectrum of backgrounds and interests. Many instructors would regard the great majority of these students as "nonscience," presumably referring to the fact that most are more interested in the arts or the humanities than in chemistry, physics, or mathematics. A serious misconception is that such students have neither the background nor the interest required to grasp basic scientific concepts. I strongly disagree. *All* students have the potential to learn such concepts regardless of whether they are interested in the sciences, the arts, or the humanities. Although some aspects of geology involve basic chemistry and physical parameters such as heat, pressure, and bond energies, most do not. For the most part, geology is an observational science and requires only that you observe and be interested enough to question how it all came to be. In the few instances where I felt it was necessary, I have used chemical symbols, formulas, and equations. If you consider them for what they are, namely, a shorthand way of presenting chemical information, they lose their seemingly formidable character. Besides, whenever I use them, I also indicate what the symbolism means in the hope that you will come to accept the chemical style of notation.

DESIGN OF THE TEXT From the initial overview of Earth to the concluding discussion of the exploitation of natural resources, this book is designed to methodically build your understanding of Earth. I have tried to present the material in a conversational style; I have always been of the opinion that a textbook should not be simply another lecture.

The book begins with an overview of Earth. In this chapter, I have attempted to present many of the topics and concepts that you will be learning throughout the semester without becoming too involved in the details. My intent is to introduce you to terms and concepts so that they will be familiar when you encounter them again later in the book. I also want you to see how the various topics fit together into a unified plan so that you can better understand how the topics of the individual chapters are interrelated.

UNIQUE FEATURES This textbook contains a number of features that I feel are unique. For example, I have introduced the topic of *plate tectonics* early in the text. Although plate tectonics is perhaps the most important single concept formulated in the history of the science of geology, most textbook authors place the chapter dealing with this topic late in their textbooks, apparently believing that you will not be able to understand plate tectonics until you have covered most of the other topics in the book. I obviously disagree. Not only do I feel that you *can* understand the basic precepts of the theory early in your studies, but I feel that it is *imperative* that you have an understanding of the theory from the very beginning of your study of Earth. The theory of plate tectonics has allowed geologists to explain and understand many geologic phenomena for the first time. By studying plate tectonics in the second chapter, you will be able to draw upon your understanding of the theory to better comprehend later topics.

Another unique feature of this book is an entire chapter devoted to *soils*. In most other textbooks, the topic of soils is little more than a small part of the chapter on weathering. Yet soil is one of the most distinctive materials on Earth's surface and, through its association with plants, affects all of our lives. In fact, without soils, terrestrial life as we know it could not exist. I have always felt that soils deserve a chapter dedicated to their study.

In Chapter 15, involving the discussion of rock deformation and the geologic structures, I have introduced the concept of the *strain ellipsoid*. Because it may be of more interest to students who intend to major in geology and who will use the concept in a structural geology course, I have included it in a box rather than in the text. The concept, however, is designed to assist *any* student to better understand the stress-strain relationship and the three-dimensional response of solids to stress. In this role, the concept of the strain ellipsoid is an important addition to the discussion of rock deformation.

The *environment* is a topic of critical importance to each of us. In order to make intelligent decisions about the causes and possible solutions of many environmental problems that we now face, and hopefully to prevent future problems, each of us must become more aware of the environment. To this end, I have included **Environmental Concerns** segments in most chapters. Rather than dealing with environmental geology as though it were an isolated entity unto itself, I have tried to show that an environmental impact is associated with nearly every area of geology. Although many environmental problems are the result of human activities, I believe that it is important to show how the environment is affected by natural processes such as weathering, volcanic eruptions, and floods. The individual treatments are not intended to be comprehensive, but rather to make you more aware of the environment and the problems that befall it. In this day when many college and university courses may not seem too relevant, I think it is important that there be no doubt about the place of geology in everyday life; the role geology plays in the environment certainly should demonstrate its relevance.

Another unique feature of the book is the combined opportunities for *self-testing*. One of my major concerns as a teacher is that you understand the concepts as they are presented. To this end, I have provided numerous opportunities for you to test your understanding of the material. Each chapter includes **Spot Reviews** that encourage you to stop and evaluate your comprehension of the material that you have just read. There is little reason to proceed unless you understand what you have already read.

The review materials at the end of each chapter are meant to enable you to evaluate your overall understanding of the chapter. They begin with an outline entitled **Concepts and Terms to Remember** that include the major concepts that have been presented in the chapter along with the associated key words. My purpose in including the outline was twofold. First I wanted to demonstrate how an outline can be used to organize material, a skill that I hope will help you to take better classroom notes. Secondly, I wanted the key terms to be placed in context rather than simply listed in alphabetical order where any interrelationship between term and concept or between terms is lost.

The outline is followed by two sets of **Review Questions.** The first set is a collection of *multiple-choice questions.* If you are enrolled in a typical large introductory class with a hundred or more students, this is the type of test question you are likely to see. When you use the multiple-choice questions to test your understanding of the material, keep in mind that it is as important to understand why three answers are *wrong* as it is to know why the one answer is *right.* You will find the answers to the multiple-choice questions after the appendixes.

The second set of review questions are *thought questions.* These are the kind of questions that you might expect if you are in a relatively small class where the instructor has the time to correct essay-type exams. These questions require you to synthesize and combine a number of different concepts that you have learned. Essay-type questions usually require more in-depth understanding of the material than is needed to answer objective-type questions where the answer is before you. In fact, individual questions may require more information than is presented in this book, or even in class, to formulate a complete answer. The purpose of such questions is to encourage you to go *beyond* this text and seek more information about certain topics.

A unique feature of the end-of-chapter materials is the **For Your Notebook** segment. I hope you are among the students who will take advantage of this segment. In my experience as a student of geology, and I still consider myself to be one, a great deal of what I have learned about geology and about Earth has come from field trips. There is a saying that the best geologist is the one who has seen the most rocks. The field is the place to see rocks.

The intent of the notebook is to involve you in the science of geology through "hands-on" experiences. The notebook is another attempt to show you that geology *is* relevant, that it touches your everyday life, and that it is around you wherever you go. The notebook segment encourages you to go on personal field trips, to *use* what you have learned by looking critically at the geology that surrounds you, and to keep notes on all that you see. In each segment, I suggest specific activities that will allow you to apply what you have learned in the chapter and list local sources of information that will make your field trip more successful.

As a teacher, perhaps the most exciting aspect of writing a textbook is the thought that I am able to touch the lives of individuals whom I will never know, that there is the possibility that what I have to say may somehow change their lives. Certainly, that is what has kept me in the introductory classroom for more than a quarter century. I consider this book to be an extension of my classroom. If you honestly desire to become, even temporarily, a student of geology and learn about Earth, this book was written for you.

ACKNOWLEDGEMENTS

The monumental task of writing a textbook cannot be accomplished without the professional services of individuals other than the author or authors. At this time, I would like to recognize those who were instrumental in the preparation of this textbook.

First, I would like to recognize the contributions of the individuals who reviewed the original chapter manuscripts. Every chapter in an introductory textbook is an area of specialization. Few authors are expert in *all* areas of geology, and yet, it is their responsibility to prepare a scientifically correct presentation of each topic. During the review process, each chapter is reviewed by an expert in the specific area of that chapter. The contributions of such reviewers assures the reader that the text of each chapter is both authoritative and correct. I thank all the reviewers who contributed their professional experiences to make this book possible:

Gary C. Allen   *University of New Orleans*
Richard Arnseth   *University of Tennessee*
R. Scott Babcock   *Western Washington University*
David Best   *Northern Arizona University*
Thomas Broadhead   *University of Tennessee-Knoxville*
Philip Brown   *University of Wisconsin-Madison*
F. Howard Campbell, III   *James Madison University*
Gaylen Carlson   *California State University-Fullerton*
Margie Chan   *University of Utah*
David Cok   *Adelphi University*
Charles Connor   *Florida International University*
Richard Conway   *Shoreline Community College*
Stewart Farrar   *Eastern Kentucky University*
Robert Filson   *Green River Community College*
James Firby   *University of Nevada-Reno*
Richard Flory   *California State University-Chico*
Richard Fluegeman   *Ball State University*
Kazuya Fujita   *Michigan State University*
Robert B. Furlong   *Wayne State University*
Norman Gray   *University of Connecticut*
Jack Green   *California State University-Long Beach*
Bryan Gregor   *Wright State University*
Jeff Grover   *Cuesta College*
Jack C. Hall   *University of North Carolina-Wilmington*
Bryce M. Hand   *Syracuse University*
Frank Hanna   *California State University-Northridge*
George Haselton   *Clemson University*
Barry Haskell   *Los Angeles Pierce College*
Jack Hyde   *Tacoma Community College*
Lance E. Kearns   *James Madison University*
James N. Kellogg   *University of South Carolina*
Albert Kudo   *University of New Mexico*

Martin B. Lagoe   *University of Texas at Austin*
Ian Lang   *University of Montana*
David Lumsden   *Memphis State University*
Erwin Mantei   *Southwest Missouri State University*
Richard Mauger   *East Carolina University*
Kevin McCartney   *University of Maine-Presque Isle*
George R. McCormick   *University of Iowa*
Eileen McLellan   *University of Maryland at College Park*
Nancy McMillan   *New Mexico State University*
Carelton Moore   *Arizona State University*
Alan Morris   *University of Texas at San Antonio*
David Nash   *University of Cincinnati*
John Nicholas   *University of Bridgeport*
Harold Pelton   *Seattle Central Community College*
James F. Peterson   *Southwest Texas State University*
Howard Reith   *North Shore Community College*
M. J. Richardson   *Texas A&M University*
Gary Rosenberg   *Indiana University-Purdue University*
Vernon Scott   *Oklahoma State University*
Douglas Sherman   *College of Lake County*
Charles Singler   *Youngstown State University*
Frederick Soster   *DePauw University*
Charles P. Thornton   *Pennsylvania State University*
Sam Upchurch   *University of Southern Florida*
Kenneth J. Van Dellen   *Macomb Community College*
William J. Wayne   *University of Nebraska*

I would like to acknowledge my colleagues who endured innumerable questions as I educated myself in the finer points of their specialty areas. I would especially like to thank Ron Harris and Joe Donovan who agreed to review the artwork for accuracy and authenticity and to Scott Morgan who prepared the glossary.

I want to express special thanks to Pat Lewis who was the copy editor for the final manuscript. As hard as I worked to choose the correct words and to construct the perfect sentence, she always found a better word and devised a better presentation. When I think of Pat's contributions to the book, I am reminded of the movie *Amadeus*. The music of Salieri was good, but Mozart made it memorable.

No instructor can write a textbook without recognizing the contributions of his or her students. Over my tenure as a teacher, more than 25,000 students have passed through my introductory geology course. Not all were inspired by my presentations, but the enthusiasm that many students showed and the questions they asked caused me to constantly strive to become a better teacher and, at the same time, made me a better student of geology. It has been said that teachers learn from their students. There is no doubt in my mind that the statement is true.

CHAPTER OUTLINE

INTRODUCTION

KINDS OF ROCKS
    Igneous Rocks
    Sedimentary Rocks
    Metamorphic Rocks

THE INTERNAL STRUCTURE OF EARTH

THE SURFACE OF EARTH
    The Ocean Basins
    The Continents

THE HYDROSPHERE AND THE ATMOSPHERE

THE ROCK CYCLE

CONCLUSION

INTRODUCTION

An Overview of Earth

This chapter is a brief summary of what you will be studying in the following chapters. Many of the topics introduced here will be the subject of later chapters and will be discussed in more detail at that time. The intent at this point is simply to introduce you to the basic structure of Earth, the kinds of rock of which Earth is made, and some of the more important geological processes. In particular, the intent is to impress upon you that the rocks that make up Earth and the processes that operate both on and within Earth are all part of a much larger plan.

KINDS OF ROCKS

Since the science of geology began more than two hundred years ago, our understanding of Earth has come mainly from the study of rocks exposed at the immediate surface in cliffs, stream channels, quarries, and excavations made for canals, railroads, and roadways. Occasionally, rocks have been available from mines and, within the past century, from wells drilled for oil and gas. Now, however, with the advent of advanced technology, we are drilling deep wells into Earth's crust from land-based and deep-ocean drilling platforms, using advanced seismic technology to study rocks deep below Earth's surface, analyzing rocks on the surfaces of distant planets, and using remote sensing to map their surfaces. By studying the minerals and other features found in these rocks and those brought to the surface from deep within Earth by volcanic eruptions, we can better understand how the terrestrial planets formed.

Rocks are consolidated mixtures of mineral grains. There are three different kinds of rocks: (1) *igneous*, (2) *sedimentary*, and (3) *metamorphic*.

Igneous Rocks

Igneous rocks form from the cooling and solidification of molten rock. Molten rock forms within Earth and rises because it is hotter and therefore less dense than the surrounding rock. Molten rock contained within Earth is called **magma**. Magma extruded onto Earth's surface is called **lava**.

When magmas cool and crystallize before reaching the surface, they create **intrusive** igneous rock bodies called **plutons** consisting of rocks such as granite. With time, erosion may strip away the overlying rocks, exposing these intrusive rocks at the surface. For example, intrusive rocks that originally formed deep within the North American continent are now exposed at the surface throughout the Sierra Nevada of California (Figure I.1).

FIGURE I.1 *A view of the Sierra Nevada, California. At one time the rocks now exposed throughout the Sierra Nevada were overlain by a range of continental arc volcanoes similar to the present-day Cascade Mountains. The volcanic mountains were removed by erosion, exposing the underlying granitic intrusions that now form the crests of the Sierra Nevada. (VU/ © Walt Anderson)*

The term *lava* is also used to describe the igneous rock that forms when molten lava cools by coming in contact with the atmosphere, water, or ice. The violent eruption of highly viscous magmas (those that flow with high internal resistance) produces a variety of rock fragments collectively called **tephra** or **pyroclastic materials**. Lava and rocks formed by the solidification of pyroclastic materials are referred to as **extrusive** igneous rocks or **volcanic rocks**. Volcanic rocks accumulate around a vent to create volcanic mountains such as those in the Cascade and Andes mountains along the eastern margin of the Pacific Ocean, in the Japanese Islands on the western margin of the Pacific, and in many mid-ocean islands including Hawaiian Islands (Figure I.2).

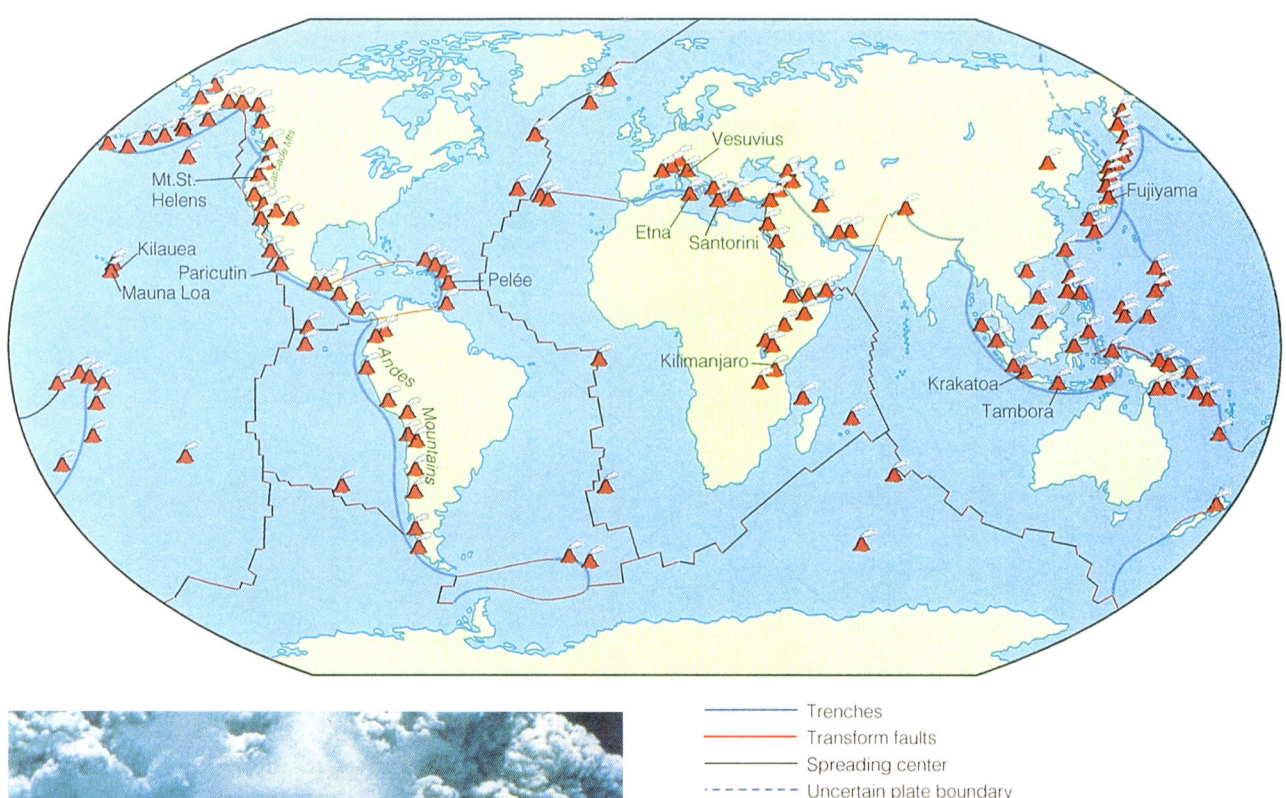

FIGURE I.2 (a) *Most of the active and dormant volcanoes of the world are associated with the edges of the lithospheric plates, especially along* zones of subduction. *Of these, the greatest percentage are located around the Pacific Ocean, the so-called* Ring of Fire.

(b) *The characteristic violent eruptive style of volcanoes associated with zones of subduction is illustrated by the eruption of* Mount Pinatubo *in the Philippines in 1991. (Courtesy of T.J. Casadevall/USGS)*

FIGURE I.3 *Few, if any, eruptions accompany the eruption of basaltic magmas. The eruption of Kilauea on the island of Hawaii in 1983 illustrates the typical quiet eruptive style of basaltic magmas. Kilauea has been in more or less continuous eruption since 1983. (Courtesy of USGS/HVO)*

The less viscous magmas (those that flow with less internal resistance) flow readily through fractures and erupt quietly at the surface with few if any explosions. The viscosity of magmas is largely determined by their composition, in particular, their silicon (Si) content. When low-viscosity magmas cool, they form a rock called **basalt**. Basalt is the most common rock type in the oceanic crust and in volcanic islands such as the Hawaiian Islands (Figure I.3).

Sedimentary Rocks

When rocks of any kind are exposed at Earth's surface, the mineral grains begin to be loosened, decomposed, or dissolved, a process called **weathering**. Rocks subjected to weathering are mechanically and/or chemically attacked, broken into smaller pieces, and decomposed (Figure I.4). The debris produced by *mechanical* and *chemical weathering* accumulates as a layer called **regolith**. Materials taken into solution during chemical weathering may reprecipitate later by either chemical or biochemical processes.

Once formed, the regolith is subjected to **mass wasting** that transports the weathered debris downslope to the valley floor (Figure I.5) where other agents of **erosion** including *streams, glaciers, waves,* or *wind* carry the debris away to basins or sites of **deposition**. Deposition takes place in a variety of environments. On land, these environments include floodplains, stream channels, lakes, ponds, and swamps. Eventually, however, nearly all materials stripped from the land will be deposited in the ocean (Figure I.6).

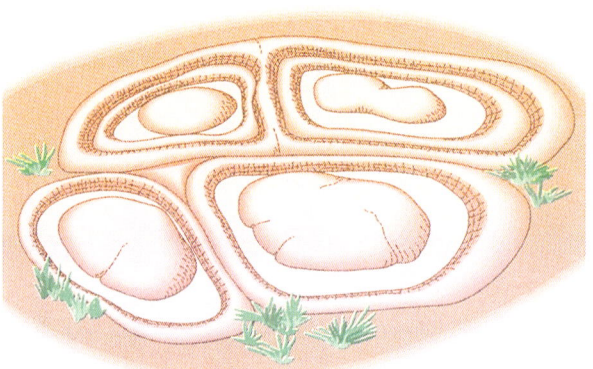

FIGURE I.4 *The end result of weathering is the rounding of rocks as sharp edges and corners are removed by a combination of processes. (Courtesy of E.N. Hinrichs/USGS)*

FIGURE I.5 *Many processes of mass wasting are slow and go on unnoticed. Only when a slope fails by a fast progression of mass wasting such as a landslide are we made aware of this most important of geologic processes. (Courtesy of J. T. McGill/USGS)*

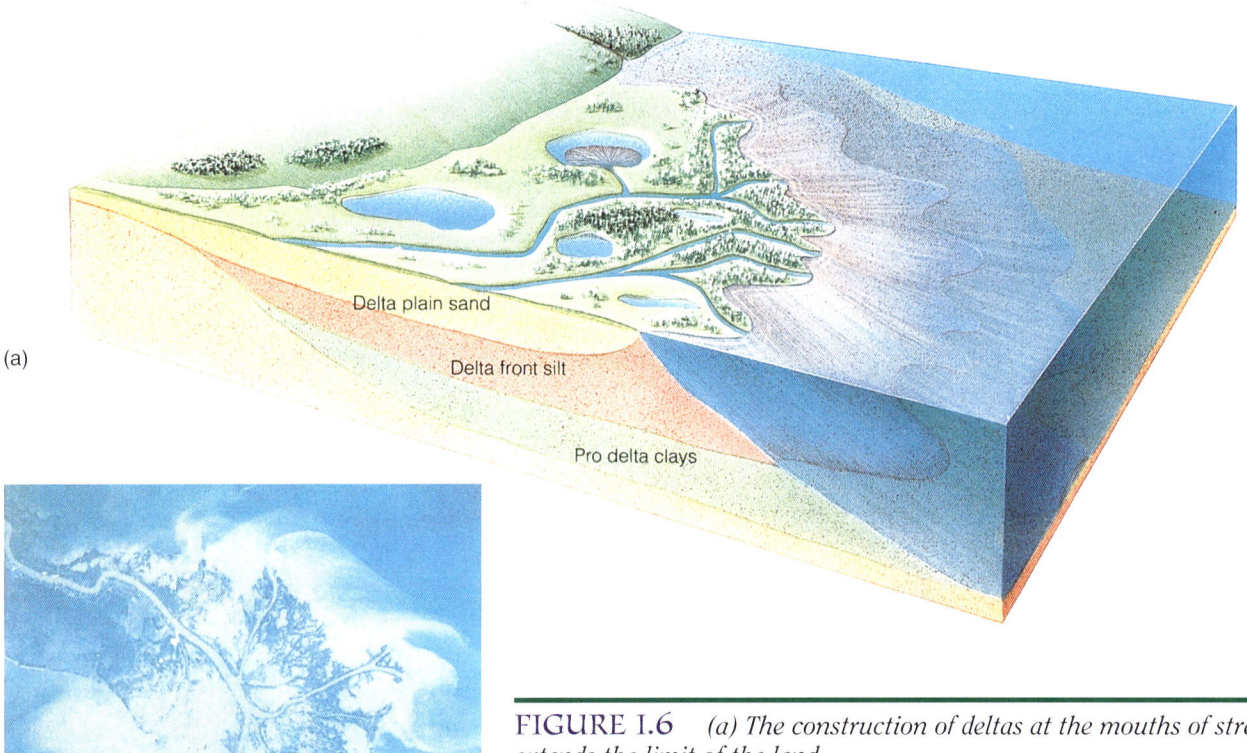

FIGURE I.6 *(a) The construction of deltas at the mouths of streams extends the limit of the land.*

(b) One of the major deltas of the world is the Mississippi delta which deposits material eroded from the southern half of North America into the Gulf of Mexico. Note the plume that forms at the active delta lobe of the Mississippi as the fine sediments carried by the lower density, fresh water of the river spread out over the more dense saline waters of the Gulf. (Courtesy of NASA)

Whether accumulated on land or in the ocean, each layer of sediment or precipitated minerals may be *buried* beneath other layers and undergo **lithification** (compaction and cementation) to form a **sedimentary rock**. Because the ocean floor is the major repository of materials stripped from the land, most sedimentary rocks are **marine** in origin. Sedimentary rocks made from materials accumulated on land are called **continental rocks**. The sedimentary rocks may eventually be uplifted by a mountain-building episode or exposed by a change in sea level and thus be subjected once again to the processes of erosion.

Shale, the most abundant sedimentary rock, forms from clay mineral-rich muds that accumulate on stream floodplains, in wetlands, and on the ocean bottom. The next most abundant sedimentary rock is **sandstone**, composed primarily of quartz grains derived from previously existing rocks. Because it is resistant to chemical attack, quartz survives long after other minerals have decomposed and accumulates as a residual by-product.

The third most abundant sedimentary rock, **limestone**, forms from calcium carbonate ($CaCO_3$) in solution. Certain organisms, both animal and plant, remove $CaCO_3$ from water to make shells and other protective coatings. When the organisms die, the remains of the shells and coatings accumulate and undergo lithification to form limestones. Other limestones are formed from $CaCO_3$ precipitated directly from solution without the intervention of organisms.

Metamorphic Rocks

Deep within Earth, rocks are subjected to conditions of temperature, pressure, and chemically active fluids that may not cause them to melt, but rather to *recrystallize* to form **metamorphic rocks**. Examples are the transformation of limestone into **marble** and shale into **slate**. Note that, in metamorphism, the recrystallization takes place while the rock is in the solid state; the original rock does not completely melt.

Although some metamorphic rocks form near the surface of earth, most metamorphic rocks form under conditions that exist only deep within the cores of developing mountains. Because the conditions of extreme metamorphism overlap those of igneous activity, it is not surprising that some rocks exhibit characteristics of both processes. Called **migmatite,** they form when metamorphic rocks are subjected to temperatures sufficiently high to cause partial melting but where the magma is unable to separate from the solid residue. Consequently, after cooling, the resultant migmatite records both igneous and metamorphic processes.

SPOT REVIEW

1. What are the three kinds of rock and how do they differ in composition and in their mode of formation?
2. What is the different between magma and lava?
3. What is the difference between weathering and erosion?

THE INTERNAL STRUCTURE OF EARTH

Our knowledge of Earth's interior is based upon various types of indirect evidence, mainly data generated from the study of earthquakes. The "picture" of Earth's interior the seismologists have generated shows a very thin **crust** and a thick **mantle** surrounding a central **core** (Figure I.7).

Materials in the crust mantle, and core have different **densities.** Density is mass per unit volume, usually measured in pounds per cubic foot (lb/ft^3) or grams per cubic centimeter (gm/cm^3). It is important to note that the density of the materials decreases systematically from the core to the crust (refer to Figure I.7), with the crust having the lowest density (162 to 181 lb/ft^3 or 2.6 to 2.9 gm/cm^3), the mantle having an intermediate density (355 lb/ft^3 or 5.7 gm/cm^3), and the core having the highest density (805 lb/ft^3 or 12.9 gm/cm^3). Note that because of the differences in density, each outer shell literally floats on the shell below.

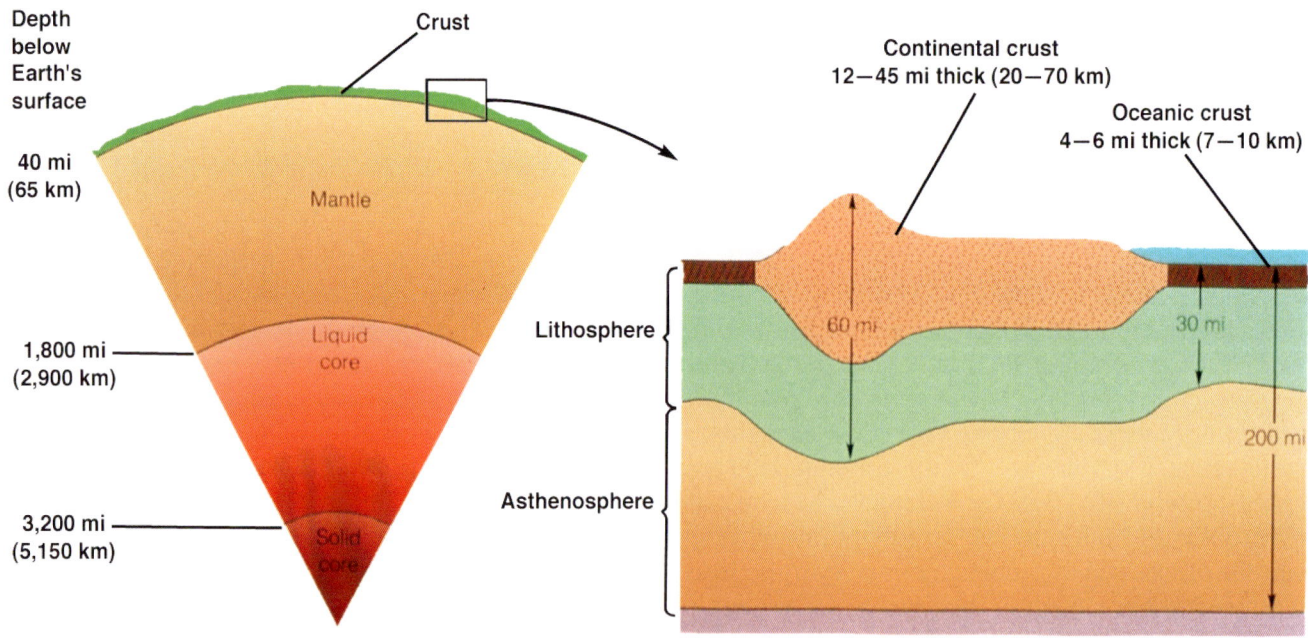

FIGURE I.7 *From our study of earthquakes, we know that Earth is made up of the core, mantle and crust, each differing in composition and density. The crust consists of two components, basaltic oceanic crust and granitic continental crust. The outermost portion of Earth, called the lithosphere, is composed of the crust and the brittle portion of the mantle, and overlies the asthenosphere, a plastic layer that plays an important role in plate tectonics.*

The core is probably composed of iron and nickel, not unlike the iron meteorites (one of the three main types of meteorites, which are pieces of Solar System debris that have penetrated the atmosphere and reached Earth's surface). Earthquake data indicate that the outer portion of the core is molten while the center of the core is either solid or a liquid with much greater resistance to flow than the enclosing molten portion.

The mantle makes up 82% of the volume of Earth and is solid throughout. However, a layer in the upper portion of the mantle called the **asthenosphere** is on the verge of melting. Some geologists estimate that as much as 10% of the asthenosphere may be molten although 1% or less is a more common estimate. In either case, the asthenosphere is considered to be a soft, weak, rocky mush that is actively convecting or moving in much the same manner as boiling water moves in a pot. Because the mobile asthenosphere underlies the brittle outer layers of Earth, it plays an important role in processes that affect Earth's surface.

The low-density outer layer of Earth is appropriately called the crust. The crust varies in thickness from 4 miles (7 km) to 45 miles (70 km) and consists of two parts. The thinner **oceanic crust** underlies the ocean basins while the thicker **continental crust** makes up the continents; the thickest part of the continental crust underlies the major mountain ranges. The oceanic crust is composed mostly of *basalt* while the continental crust is mainly made of *granitic* and *metamorphic rocks,* with a thin veneer of *sedimentary* rocks in many places.

Together, the crust and the outermost portion of the mantle above the asthenosphere form the **lithosphere** (refer to Figure I.7). Under the ocean basins, the lithosphere extends to depths of approximately 30 miles (50 km) while under major mountain ranges where the crust is the thickest, the base of the lithosphere may extend to depths of more than 60 miles (100 km).

An important discovery of the 1960s was that the lithosphere is broken into **plates** that fit together like a gigantic jigsaw puzzle. Driven by heat flow from within Earth, the plates move about Earth's surface carried on the underlying asthenosphere. At some plate margins, the plates are *diverging* or moving away from each other while at others, the plates *converge* or move toward each other (Figure I.8).

As plates move away from each other or *rift,* fractures extend down to the top of the asthenosphere. Molten rock, formed at the top of the asthenosphere, rises, cools, and solidifies to constantly form new lithospheric rocks.

Large continents can be torn apart by the **divergent plate** movements, resulting in smaller continents that move or "drift" about on Earth's surface. Beginning as a continental **rift zone** like the Rio Grande Rift in New Mexico and progressing to a **rift valley** such as the East African Rift Valley, continued

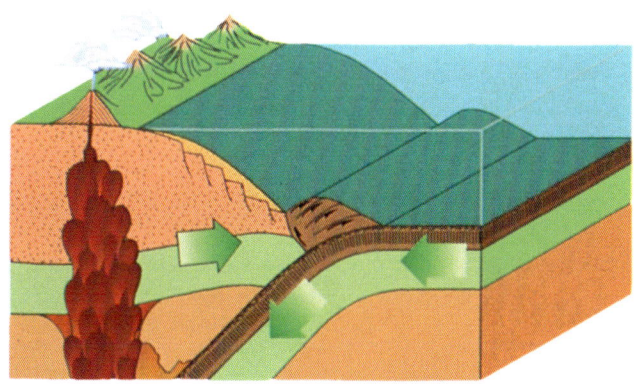

(a) Convergent Margin

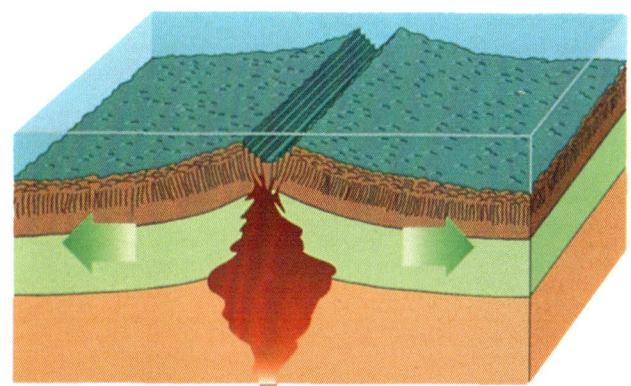

(b) Divergent Margin

FIGURE I.8   *The lithosphere is broken into plates that are moving to each other. (a) At convergent margins, the plates move toward each other and form zones of subduction. (b) At divergent margins, the plates move away from each other and form rift zones such as the oceanic ridges.*

movement of the plates results in the opening of **linear oceans** such as the Red Sea and the eventual formation of **ocean basins** such as the Atlantic and Indian oceans. The upwelling molten rock solidifies to form new lithospheric rocks and crust for the floor of the new ocean basin (Figure I.9).

The present continents were formed about 200 million years ago by the rifting of a huge continent called *Pangaea*. The Atlantic Ocean was created as the American continents and Europe and Africa moved away from each other at the rate of several centimeters per year. The Atlantic Ocean is still opening as the continents continue to drift apart.

Where the plates **converge**, the more dense oceanic lithospheric plate dives, or *subducts,* beneath the less dense continental lithospheric plate or beneath another oceanic plate. Water drawn down into the **zones of subduction** lowers the melting point of the mantle rocks and produces the molten rock that rises to Earth's surface to build volcanic mountain chains such as the Cascade Mountains and the Japanese Islands (Figure I.10). The compressive forces generated by the converging plates also create chains of folded mountains along the continental margin and generate earthquakes.

Over time, convergent plate movements cause continents to collide, forming still larger continents. For example, while the continents on opposite sides of the Atlantic continue to drift apart, Africa is moving northward toward Europe. At one time, Europe and Africa were separated by an ocean called the *Tethys Sea* (Figure I.11). Today all that remains of the Tethys Sea are the Mediterranean, Black, and Caspian seas. Geologically, the last vestiges of the Tethys Sea may disappear as Africa and Europe collide. The Alps are the result of the increasing convergence of the northern edge of Africa under the southern edge of Europe.

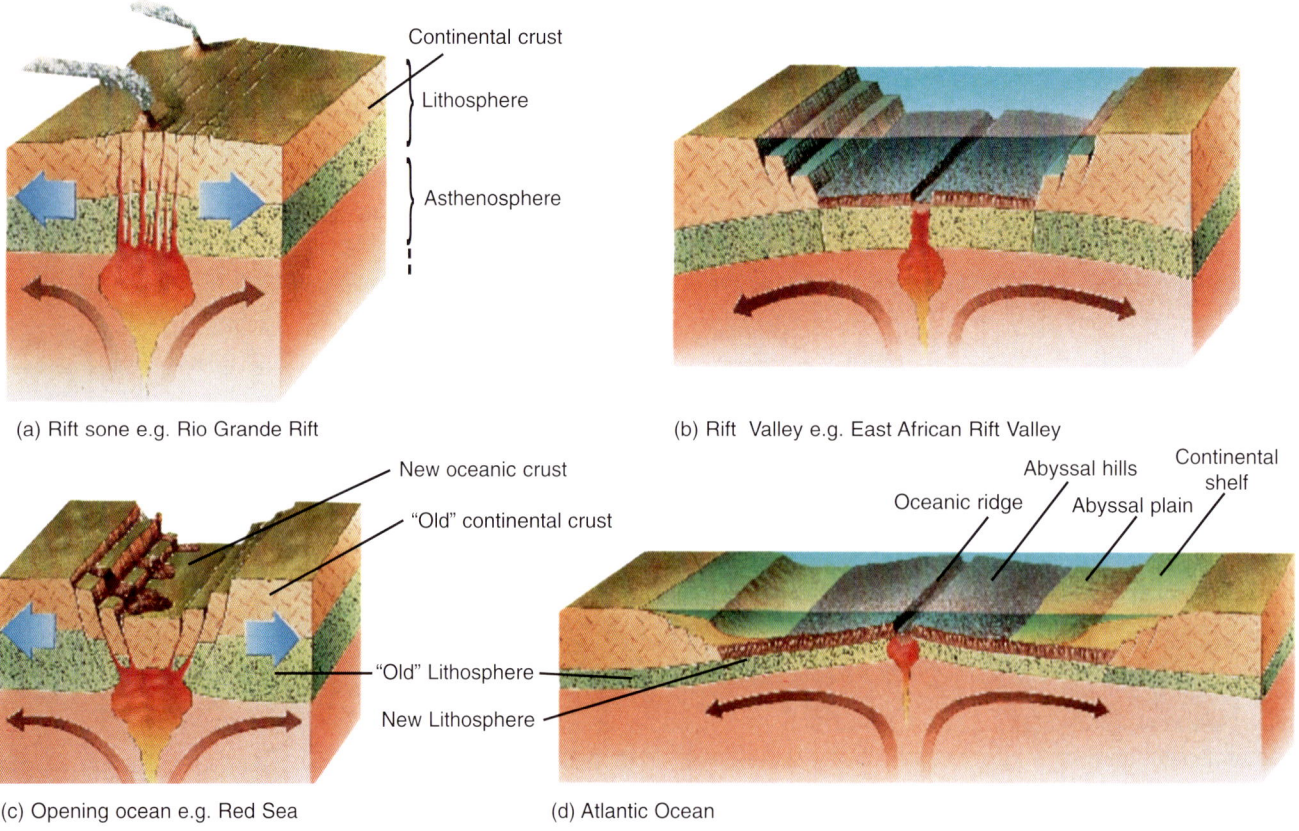

FIGURE I.9 *(a) The initial divergence of two plates invariably begins with the rifting of continental lithosphere, forming a rift zone. (b) In time, the rift zone develops into a rift valley, such as the East African Rift Valley, which in turn floods to form a linear ocean, (c) such as the Red Sea. (d) Eventually, the two continental masses completely separate to produce an ocean basin such as the Atlantic and Indian Oceans.*

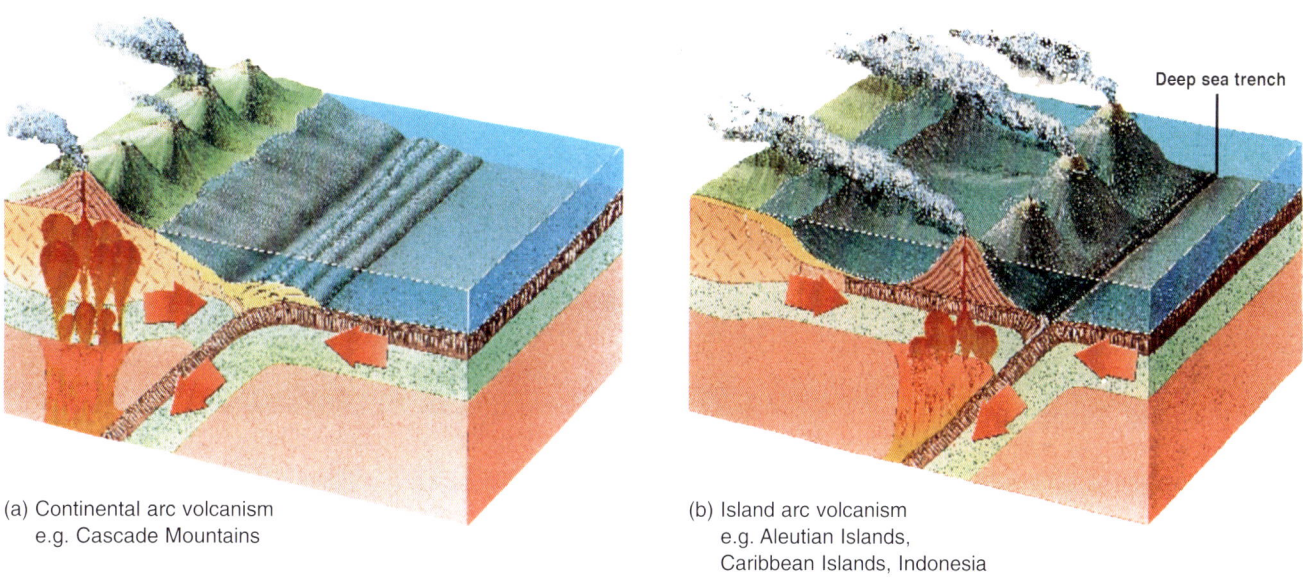

(a) Continental arc volcanism
   e.g. Cascade Mountains

(b) Island arc volcanism
   e.g. Aleutian Islands,
   Caribbean Islands, Indonesia

FIGURE I.10 *Volcanic mountain ranges form at zones of subduction. Depending on the location of the subduction relative to the continental margin, the volcanoes build either a continental arc (a) such as the Cascade Mountains or the Andes Mountains or they build an island arc (b) such as the Aleutian Islands.*

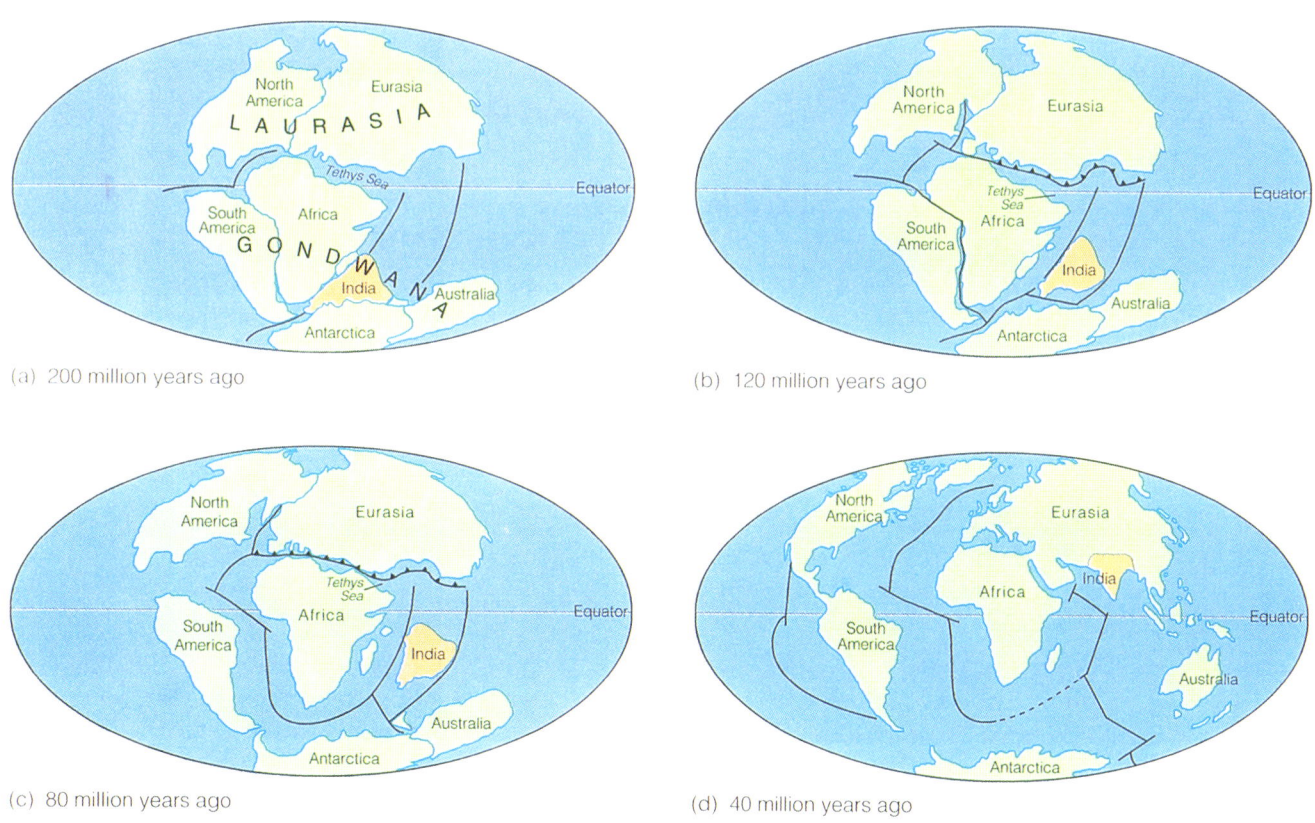

(a) 200 million years ago

(b) 120 million years ago

(c) 80 million years ago

(d) 40 million years ago

FIGURE I.11 *Until 200 million years ago, all the present continents were joined together into a supercontinent called Pangaea which consisted of a northern landmass called Laurasia and a southern land mass called Gondwana. Since then, the supercontinent broke into smaller pieces which drifted apart to become the present continents.*

The Himalaya Mountains are the result of the collision of India and Asia that began about 45 million years ago (Figure I.12). Although now slowing, earthquakes throughout the region indicate that the northward movement of India has not yet stopped. As a result of the crustal thickening caused by the collision, the Himalayas are still rising at the rate of about 2 inches (5 cm) per year.

The movement of the plates is the basis of **plate tectonics**, one of the most significant concepts to be proposed since the origin of the science of geology more than two hundred years ago. Since its introduction a little more than two decades ago, plate tectonics has explained many Earth processes including the "drifting" of continents, the formation of Earth's major mountains, the origin of molten rock, the origin of ore deposits, and many other geologic phenomena including the distribution and origin of volcanoes and earthquakes.

SPOT REVIEW

1. Prepare a chart comparing and contrasting the characteristics of the crust, mantle, and core.
2. In what way do the lithosphere and asthenosphere differ? How do they interact in plate tectonics?
3. How is plate tectonics involved in the creation of continents and ocean basins?

THE SURFACE OF EARTH

The two major components of Earth's surface are the *ocean basins* and the *continents*. Due to the lower density and greater thickness of the continental rocks, the continents rise (float) higher than the adjacent ocean basins (Figure I.13). This balance of the lower-density crust floating on the denser mantle, called **isostasy**, is analogous to wood blocks of different thickness and densities floating on water (Figure I.14). Each wood block sinks until the weight of water displaced is equal to the weight of the block. Thus, a thick block of low-density wood floats higher than a thin block of high-density wood. In similar fashion, the low-density granitic continental crust floats higher than the denser basaltic oceanic crust. As a result, the average elevation of the continents is about 0.5 miles (0.8 km) above sea level while the average depth of the ocean floor is 2.3 miles (3.7 km) below sea level. The variations in the elevation of Earth's solid surface relative to sea-level is shown in the hypsographic curve (Figure I.15). Note that the ocean basins are much deeper than the average elevation of the land. The compositional segregation of rocks into oceanic and continental crust is apparently unique to Earth. To our knowledge, none of the other planets have the equivalent of Earth's granitic continents and basaltic ocean basins.

The Ocean Basins

Oceans cover 71% of Earth's surface with the greatest area concentrated in the Southern Hemisphere

 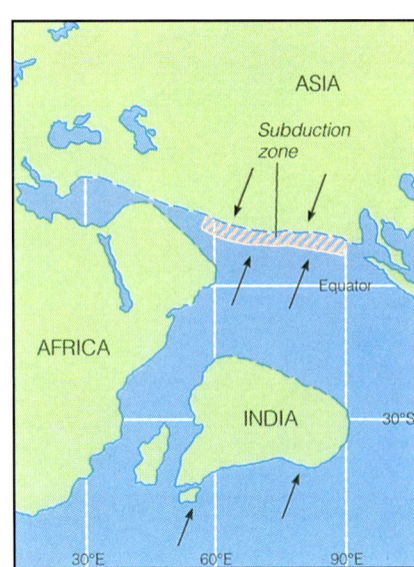 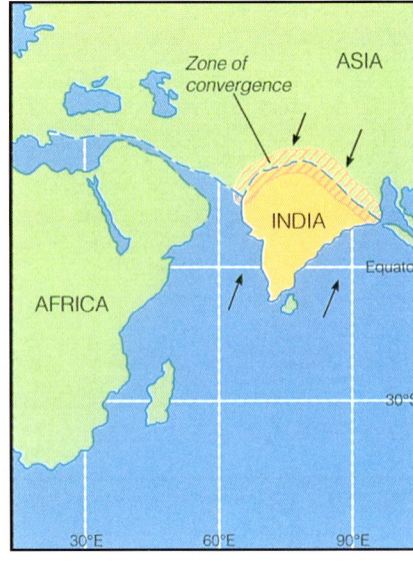

FIGURE I.12   *About 120 million years ago, a part of Gondwana that was to become India broke off and drifted northward. The collision of India with Asia about 45 million years ago resulted in the formation of the Himalaya.*

An Overview of Earth 13

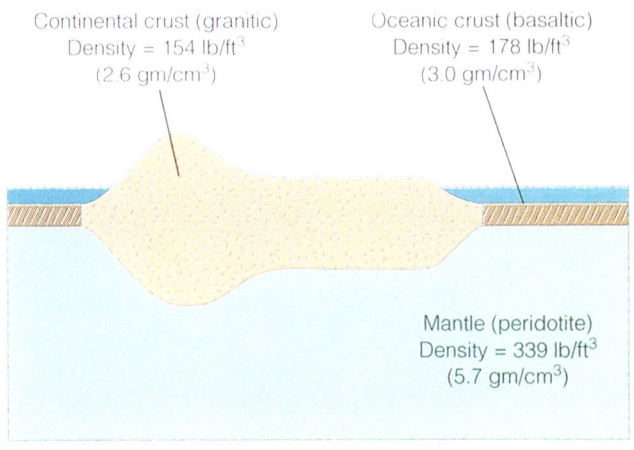

FIGURE I.13 *The crust, being the lowest density component of Earth, literally floats on the underlying, more dense mantle.*

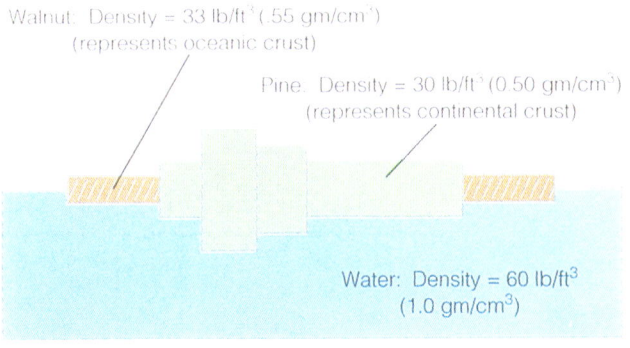

FIGURE I.14 *The relationship between the two components of the crust, the basaltic oceanic crust and the granitic continental crust, floating on the denser peridotitic mantle can be modeled using blocks of pine and walnut wood floating on water.*

(Figure I.16). The major features of the ocean basins include (1) *continental margins,* (2) *abyssal plains,* (3) *abyssal hills,* (4) *oceanic ridges,* (5) *seamounts,* and (6) *deep-sea trenches.*

The **continental margins** are the submerged portions of the continents and are transitional between the land and the deep-ocean basin. Continental margins are divided into three parts: (1) the *continental shelf,* (2) the *continental slope,* and (3) the *continental rise.* The **continental shelf** is the submerged portion of the continent immediately seaward of the shoreline. The width of the continental shelf is quite variable. Along much of the Pacific coast of North and South America, the shelf is only a few hundred feet (or meters) wide; in contrast, along the Atlantic coast the width averages about 100 miles (160 km) (Figure I.17). The average depth at the outer edge of the shelf is about 425 feet (130 m). For the most part, the angle of slope of the continental shelf is so gentle that, were it exposed to view, the surface would look nearly horizontal.

From the outer edge of the continental shelf, the **continental slope** descends more steeply and leads down to the continental rise, which grades imperceptibly into the abyssal plain. The **continental rise** consists primarily of land-derived sediments that have spilled over the edge of the continental shelf and accumulated at the base of the slope.

Except for the trenches, the **abyssal plain** is the deepest portion of the ocean basin with water depths averaging 12,500 feet (3,800 m). Within the abyssal plains, the rocks of the oceanic crust may be covered with a blanket of fine-grained sediment consisting of

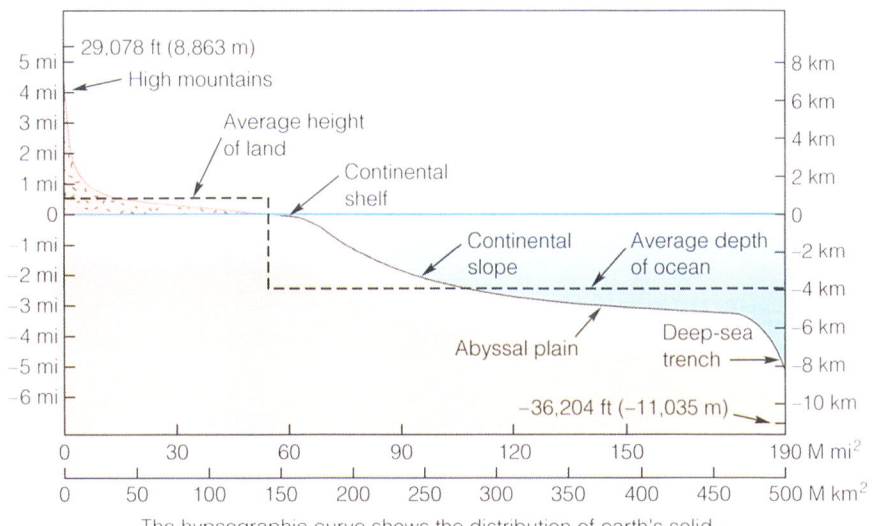

FIGURE I.15 *The hypsographic curve shows the distribution of Earth's solid surface at various elevations above depths below sea level.*

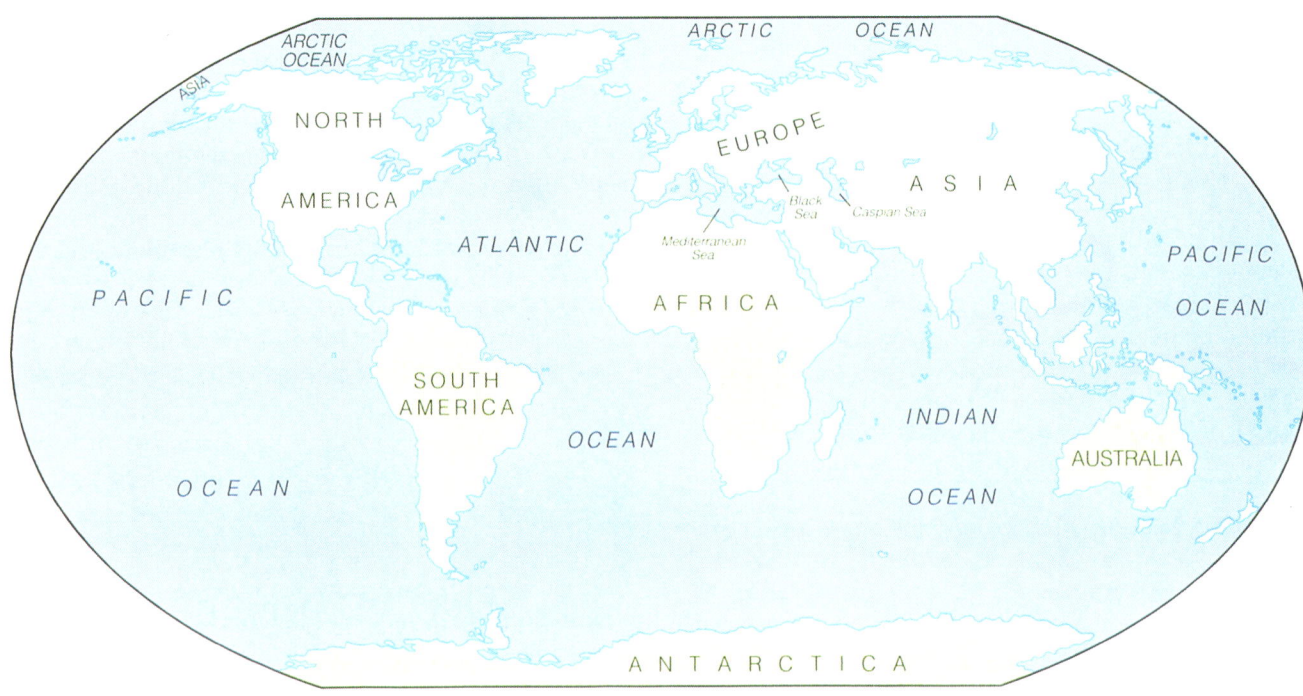

FIGURE I.16 *The oceans cover about 71% of Earth's surface with most of the ocean basins located in the Southern Hemisphere.*

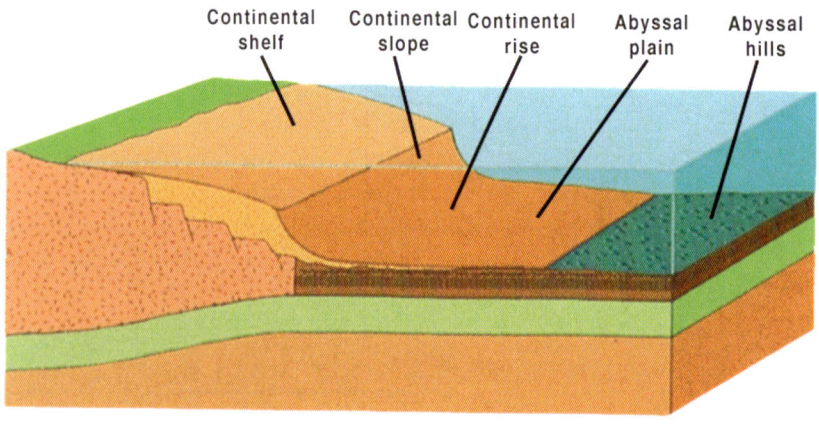

(a) Atlantic continental margin

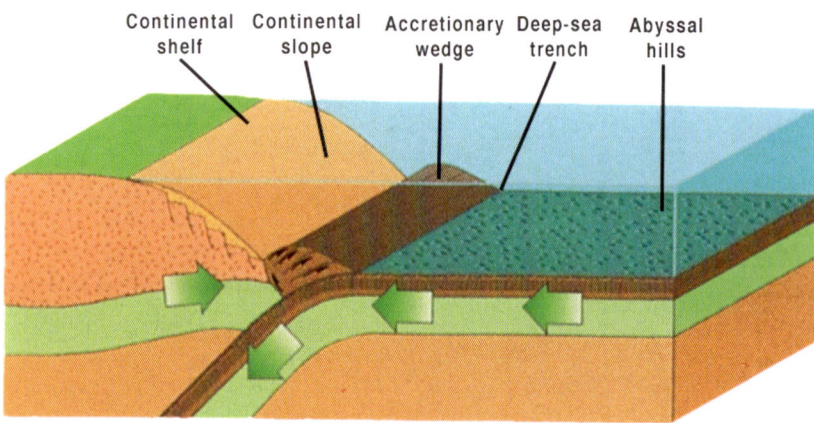

(b) Pacific continental margin

FIGURE I.17 *The structure of the continental margin depends on whether it adjoins an opening ocean or a closing ocean. (a) The Atlantic-type continental margin is sometimes referred to as a passive margin or a trailing edge. It adjoins an opening ocean and consists of a gently sloping continental shelf, averaging about 80 to 100 miles (128-160 km) wide that drops away to the abyssal plain. (b) The Pacific-type margin, sometimes called an active margin or leading edge, adjoins the zone of subduction of a closing ocean and drops steeply to the deep ocean bottom at the deep sea trench.*

material transported from the continental shelf by ocean currents, wind-borne dust carried from the land, and the microscopic remains of animals and plants that live and die in the surface layers of the ocean waters. Where they are covered with this blanket of sediment, the abyssal plains are the flattest features of Earth.

Portions of the abyssal ocean bottom adjoining the ocean ridges and areas deprived of land-derived sediments by the presence of deep-sea trenches are only thinly covered with the sediment provided by wind-borne dust and the microscopic remains of plants and animals. In these areas, the oceanic crust is exposed as an undulating and irregular surface called the **abyssal hills.**

The **oceanic ridges** are the dominant feature of the ocean bottom, forming as magma erupts and solidifies along the divergent plate margins. They interconnect throughout all the ocean basins, forming a continuous ridge complex more than 47,000 miles (75,000 km) long (refer to Figure I.9). Unlike the steep-sloped mountain ranges on land, the oceanic ridges are broad structures with gentle slopes, in places occupying about a third of the ocean floor. The oceanic ridges rise 5,000 to 10,000 feet (1,500-3,000 m) above the adjacent sea floor and may be more than 1,000 miles (1,600 km) wide. Although the oceanic ridges are for the most part submarine, at a few sites such as the island of Iceland, a portion of an oceanic ridge is exposed (Figure I.18).

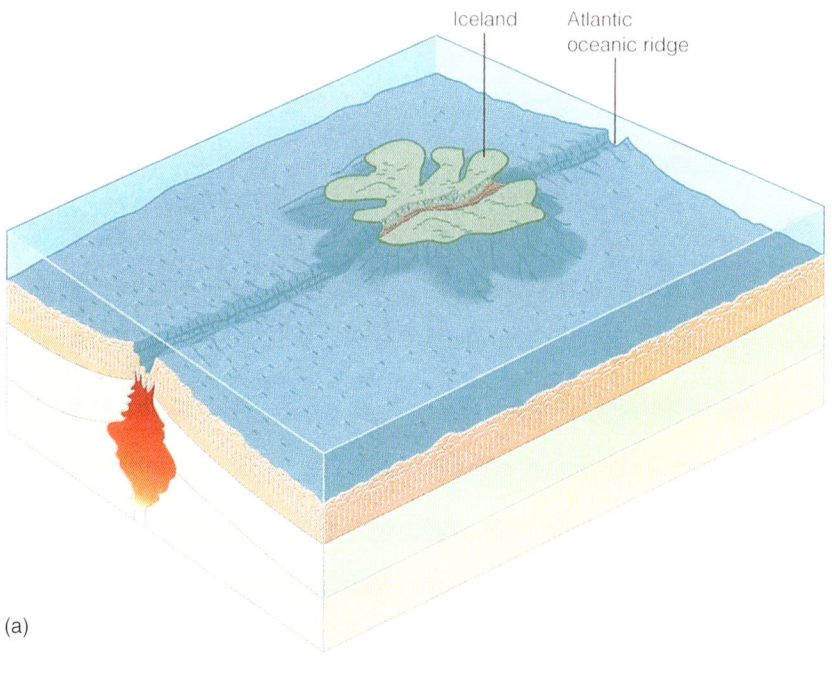

FIGURE I.18 *Most oceanic ridges are submarine, (a) an exception being the exposed portion of the North Atlantic Oceanic Ridge that is the island of Iceland. (b) The rift valley that parallels the summit of oceanic ridges is well exposed on Iceland. (VU/© Barry Griffiths)*

Thousands of conical submarine peaks called **seamounts** rise from the ocean bottom. Seamounts are volcanoes that are created on the ocean floor by the eruption of molten rock that originates from sources below the oceanic lithosphere. Because the cones build on the surface of a lithospheric plate that is constantly moving above a stationary source of molten rock, seamounts are often arranged in rows. Some peaks grow above sea level and become volcanic islands. An excellent example of such a chain of volcanic islands is the fiftieth state of the United States, Hawaii (Figure I.19).

The deepest portions of the ocean basin are the **deep-sea trenches**. Most deep-sea trenches are located around the margins of the Pacific Ocean. Although they vary is size, trenches may be up to 125 miles (200 km) wide and 1,500 miles (2,400 km) long and extend to 2 miles (3 km) below the surface of the adjoining deep-ocean floor. Trenches are located at sites where the oceanic lithosphere subducts into the mantle (Figure I.20).

SPOT REVIEW

What are the major features found on the ocean bottom and how are they formed?

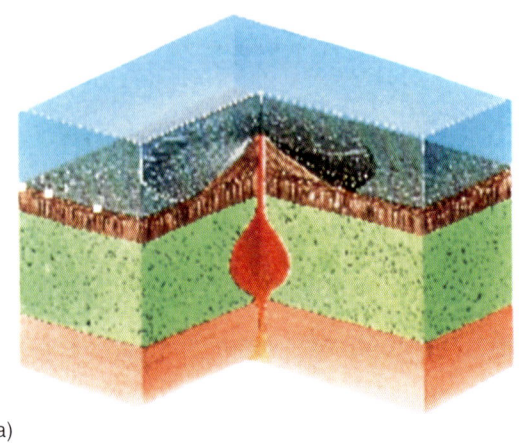
(a)

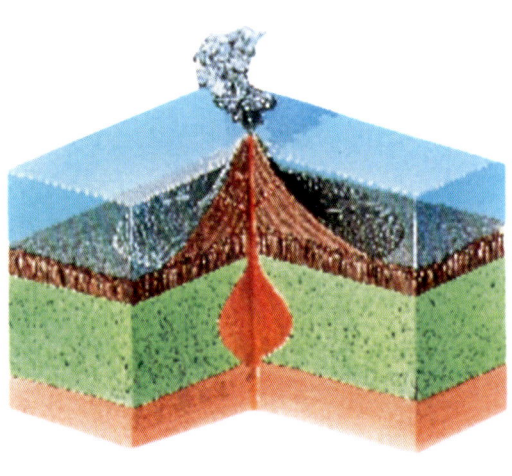
(b)

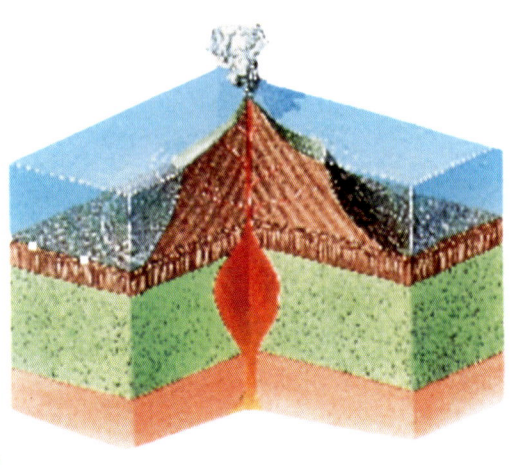
(c)

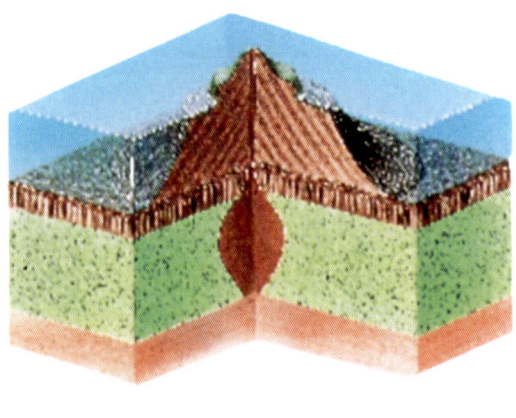
(d)

FIGURE I.19 *Volcanic islands such as the Hawaiian Islands are shield volcanoes that form over hot spots located beneath the oceanic lithosphere. Many do not build to the ocean's surface and remain submerged as seamounts. Others, however, may build above sealevel and become volcanic islands. Eventually, the volcanism on the island ceases as the island is carried off the hot spot by the moving oceanic plate.*

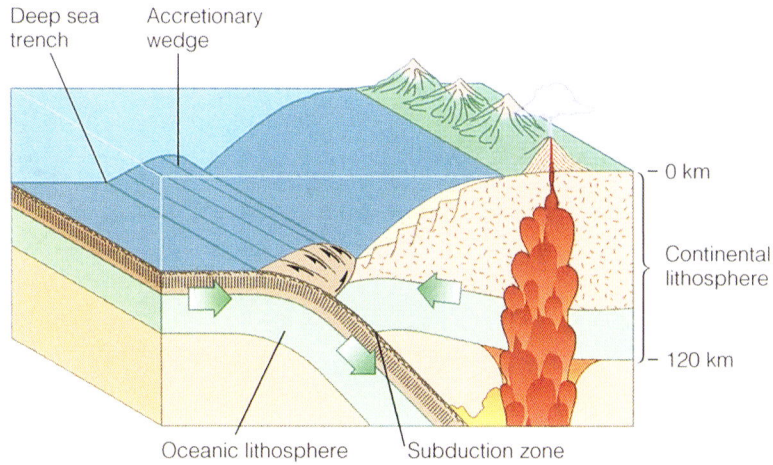

FIGURE I.20 *Deep-sea trenches form where the two lithospheric plates converge and the oceanic plate descends below the overriding plate along the zone of subduction.*

The Continents

The continents are part of the lithosphere and are made up mostly of masses of low-density granitic rocks. Although each continent is different, most follow a general pattern that features a central region of very old rocks, called the **craton**, surrounded and partially covered by a veneer of younger rocks that have been deformed by folding. The rocks of the craton are the oldest rocks of the continent, in some cases nearly 4 billion years old. Because they have not experienced major deformation for more than a half billion years, the cratonic rocks are very stable.

Inland from the deformed continental margins, much of the craton underlying the continental interior is covered by a few thousand feet of essentially horizontal sedimentary rocks, referred to as the **stable platform**. In certain areas, the surface covering of the craton may be removed by erosion, exposing the ancient rocks in an area called a **shield**. For the most part, shields are areas of low *relief* rising only a few hundred feet above sea level. The relief of an area is the difference in elevation between the hilltops and the valley floors. The shield of the North American continent, called the *Canadian Shield,* is centered over east-central Canada (Figure I.21). Shields also are found in the Scandinavian Peninsula, Siberia, Australia, Africa, and South America.

Landforms At first sight, the surface of a continent may seem to be a bewildering array of topographic features. Closer inspection, however, indicates that the various features can be subdivided into (1) *mountains,* (2) *hills,* (3) *plateaus,* and (4) *plains.*

Geologically, Earth's major **mountains** are long, linear zones where the rocks have been highly deformed as a result of horizontal, compressive forces in much the way an accordion folds when compressed. Although the term *mountain* usually implies high topographic relief, this is not always the case. Newly created mountains such as the Alps or the Himalayas do not show rugged, high elevations more than 26,000 feet (7,800 m) above sea level and exhibit steep slopes, but mountains such as the Appalachians that have undergone hundreds of millions of years of erosion are often quite low. When first formed, the Appalachians may have been as grand as the modern Alps, but they now rise to maximum elevations of only about 7,000 feet (2,100 m).

Distinguishing between mountains and hills is difficult because the distinction is arbitrary and often a matter of personal or local definition. A workable definition, however, might be that a **hill** is a rounded land surface that rises prominently above the surrounding land to elevations of less than 1,000 feet (300 m) while a mountain is a topographic feature higher than a hill; that is, it rises more than 1,000 feet above the surrounding terrain.

A **plateau** is a broad, flat region that has been uplifted significantly higher than the surrounding areas and usually consists of horizontally layered rocks. Subsequent to uplift, some plateaus have been subjected to deformation that has resulted in the faulting or folding of the rocks, creating an area of high relief. An example of such an area can be found in the state of Nevada where extension has broken the originally horizontal rocks into long, parallel mountain

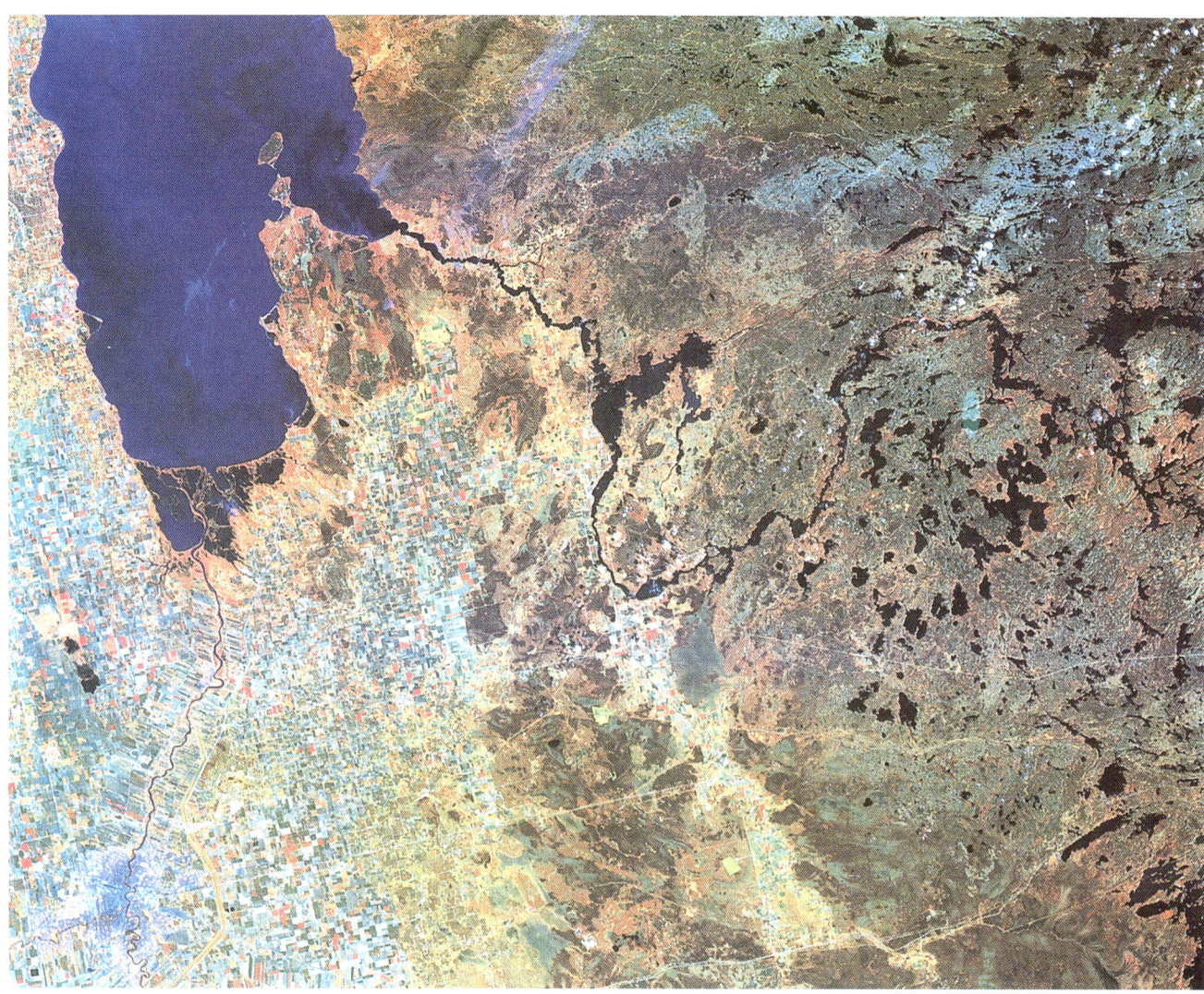

ridges and valleys (Figure I.22). In other areas, where the rocks have been uplifted with little or no subsequent deformation, except perhaps for the rocks at the edges of the plateau, the major topographic features may be deep canyons and scattered erosional features such as buttes and mesas. The best known example is the *Colorado Plateau* centered in the state of Utah and the impressive canyonlands of its southern margin, including Canyonlands National Park, Zion National Park, Bryce Canyon, Monument Valley, and one of Earth's most spectacular sights, the Grand Canyon in northern Arizona (Figure I.23).

Plains differ from plateaus in that they are generally flat or gently rolling areas. Although most plains are found at elevations of less than 1,000 feet (300 m), some plains such as the western Great Plains of North America are at higher elevations. Plains cover more than a third of the land surface of the United States and include both interior plains, such as those that make up the vast interior of the North American stable platform, and the coastal plains of the Atlantic and Gulf coasts.

SPOT REVIEW

1. What are the major structural features of the continents? How are they related to the lithospheric plates?
2. Describe the major surface features of the continents and explain how they differ.

An Overview of Earth   19

FIGURE I.21 *Shields are the cores of the continents and are composed of the most ancient rocks of the continental crust. Because these rocks have undergone hundreds of millions of years of erosion and no episodes of mountain building, the topography of a shield is commonly very subdued, as illustrated by the satellite photo of the Canadian Shield on page 18 (Landsat imagery courtesy f the Canada Centre for Remote Sensing, Energy, Mines, and Resources, Canada.)*

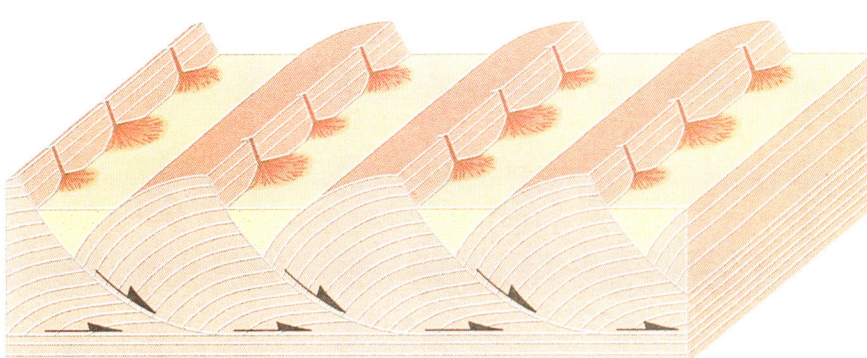

FIGURE I.22 *Where thick sequences of horizontal sedimentary rocks are subjected to tensional forces, they often break and rotate to form long parallel mountain ridges and valleys. An excellent example can be found in the Basin and Range Province that makes up most of the state of Nevada.*

(a) During Jurassic time, about 150 million years ago, much of western North America was a vast desert. The spectacular cliff in the photo shows the cross-beds that formed in the dunes of that desert. These fossilized dunes are now exposed in Zion National Park, Utah.

(b) One of the most spectacular examples of differential weathering of the horizontal beds of the Wasatch Formation in Bryce Canyon Amphitheater, Utah. The colors of the rocks within the canyon change depending on the angle of the Sun.

(c) Two of the best-known buttes in the United States are the Left and Right Mittens in Monument Valley on the Arizona-Utah border. The sandstone layer now exposed in the buttes once extended throughout the area. It was removed by stream erosion, leaving behind remnants that make up the mesas and buttes that characterize the valley. (Courtesy of E.D. McKee/USGS)

FIGURE I.23 *Some of the most scenic areas of the southwestern United States are found within the Colorado Plateau, as this collection of photos illustrates.*

FIGURE I.23 continued
(d) About 65 million years ago, a mountain-building episode affected western North America that was to have profound effects on the topography of the entire region. During this time, the Colorado Plateau began to rise. About 6 million years ago, as the drainage opened to the Gulf of California, the present course of the Colorado River was established, and it began the rapid downcutting that was to result in the formation of the Grand Canyon. (Courtesy of E.D. McKee/USGS)

THE HYDROSPHERE AND THE ATMOSPHERE

When people think of Earth, they automatically think of the solid part and often overlook two parts that are equally important, namely, the *hydrosphere* and the *atmosphere*. The **hydrosphere** includes all the water that exists on, above, and below Earth's surface. Most of the water (97%) is contained within the ocean basins that cover about 71% of Earth's surface. Most water outside the ocean basins is tied up in glacial *ice sheets* and *ice caps* (80%); the remaining 20% is contained underground as *groundwater*. The more obvious *streams* and *lakes* account for less than 1% of all nonoceanic water. For example, streams constitute a mere 0.005% of all water outside the ocean basins!

The **atmosphere**, the gaseous envelope that surrounds Earth, consists of *nitrogen* (78%), *oxygen* (21%), *argon* (0.9%), *carbon dioxide* (0.03%), and trace levels of various other gases including *water vapor*. More than half of the atmosphere's mass is below a height of 4 miles (6 km). Scientists do not agree upon an outer limit for the atmosphere. Some place the top of the atmosphere at a few hundred miles while others consider it to lie more than 6,000 miles (9,500 km) above Earth's surface. By controlling Earth's climate, the atmosphere indirectly controls many of the most important geologic processes responsible for shaping Earth's surface, including the rates of mass wasting, weathering, and erosion.

THE ROCK CYCLE

Our discussion thus far can be summarized in the **rock cycle**, which demonstrates graphically that the rocks and the processes that go on both within Earth and on its surface belong to cycles (Figure I.24). Nature is full of cycles. The seasons go through a cycle as Earth orbits the Sun. The daily cycles of day and night are the result of Earth rotating on its axis. The migrations of animals are a dramatic manifestation of the cycles of life. With neither beginning nor end, cycles are never-ending patterns of change.

The rock cycle consists of three individual rock cycles, one for each rock type, interconnected by the various internal and external processes. Depending on the pathway taken, one rock type may enter the cycle of another type. For example, rather than being forced *upward* onto land by a mountain-building episode and reentering the sedimentary cycle, a newly formed sedimentary rock may be thrust *downward* by subduction deep beneath the core of a rising chain of mountains. If the heat is not sufficiently intense to cause the rock to melt, it may enter the metamorphic rock cycle and recrystallize to form a different kind of rock, perhaps a marble or a slate depending upon its original composition. If, on the other hand, the temperature is sufficiently high, the rock will enter the igneous rock cycle, melt, and become magma. Igneous rocks or metamorphic rocks created deep within Earth may enter

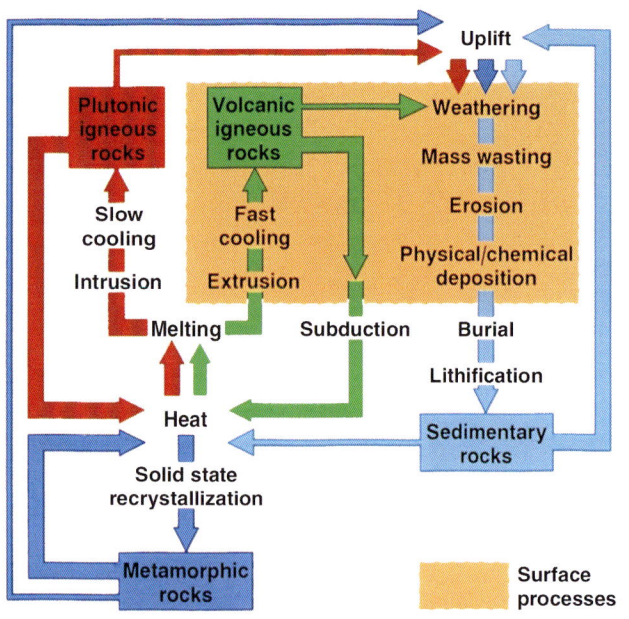

FIGURE I.24 *Each rock type has its own rock-cycle within which it may remain, undergo processes specific to its cycle, and produce another rock of the same type. However, each rock type may* exit *its cycle and enter the cycle of another rock type, be subjected to a different set of processes, and be converted to a different kind of rock.*

the sedimentary rock cycle by being uplifted during a mountain-building episode and exposed to weathering and erosion.

Note that except for the surface cooling of lava and pyroclastic rocks, the igneous and metamorphic rock cycles are the result of *interior processes* that may take place tens or hundreds of miles below the surface. Sedimentary rocks, lavas, and pyroclastic rocks are formed by *exterior processes* that act at or very near the surface of Earth.

The rock cycle we encounter most frequently in our everyday lives is the sedimentary rock cycle, simply because sedimentary rocks form primarily by surface or near-surface processes and, as a result, cover about 75% of Earth's land surface. We see the results of weathering and mass wasting as piles of rocks accumulated at the base of road cuts, in the development of soil, or in the debris left behind after a flood. Streams, the major agents of erosion, seem to be found almost everywhere, even in the driest desert. Many of us play on sites of deposition called beaches. The only portions of the sedimentary cycle that are beyond our everyday view are the processes of deep-sea sedimentation, burial, and lithification.

CONCLUSION

This textbook and the course in geology that you have just begun are going to be in large part a study of the rocks and processes contained in the rock cycle. At this point, you most likely have many questions. How does subduction come about? How are mountains created? What conditions cause rocks to recrystallize? How are bits and pieces of rocks and minerals put back together to make a sedimentary rock? The most fundamental requirement to learning is an inquiring mind. To learn, one needs only to ask "why?" and then seek the answer.

The intent of this chapter was to introduce you to some of the basic terminology of geology, to show you that there is a common thread that ties all the various topics together, *and* hopefully to spur your interest to learn more about Earth by asking "why?" The purpose of the remainder of this text is to satisfy your curiosity and provide some answers to your questions.

CONCEPTS AND TERMS TO REMEMBER

Kinds of rocks
  igneous rocks
    magma
    lava
    intrusive
    pluton
    granite
    tephra

pyroclastic
extrusive
volcanic
basalt
sedimentary rocks
  weathering
  regolith
  mass wasting

erosion
deposition
lithification
marine rocks
continental rocks
shale
sandstone
limestone

metamorphic rocks
  marble
  slate
  migmatite
Structure of Earth
  crust
  mantle
  core
  density
  asthenosphere
  oceanic crust
  continental crust
  plate tectonics
    lithosphere
    plate
    divergent plate
    rift zone
    rift valley
    linear ocean
    ocean basin
    convergent plate
    zone of subduction
Earth's surface
  isostasy
  hypsographic curve
  ocean basin features
    continental margin
    continental shelf
    continental slope
    continental rise
    abyssal plain
    abyssal hill
    oceanic ridge
    seamount
    deep-sea trench
  continental features
    craton
    stable platform
    shield
    landforms
      mountains
      hills
      plateaus
      plains
Hydrosphere
Atmosphere
Rock cycle

REVIEW QUESTIONS

1. What are the basic differences between igneous, metamorphic, and sedimentary rocks?
2. How can wooden blocks of varying thickness and density be used to model the relationship between the continental and oceanic lithosphere and the underlying asthenosphere?
3. Suggest a possible explanation why the folded Ural Mountains of Russia are located within the continent rather than along the margin of the continent.
4. By what mechanism(s) can one rock enter the cycle of another?
5. Describe the basic features found on the ocean bottom and explain how they differ.
6. Describe the basic continental landforms and explain how they are formed.

CHAPTER OUTLINE

INTRODUCTION

THE VIEW FROM EARTH
    The Contribution of the Ancient Greeks
    The Evolution of Modern Astronomy
    The Galaxies

DISTANCES IN SPACE

THE ORIGIN AND EVOLUTION OF THE UNIVERSE
    The Formation of Stars
    The Death of Stars

THE FORMATION OF OUR SOLAR SYSTEM

THE PLANETS
    The Terrestrial Planets
    The Jovian Planets
    Pluto, the Maverick Planet

ASTEROIDS METEORS, METEOROIDS, AND METEORITES

COMETS

CHAPTER 1

Earth and Its Place in Space

INTRODUCTION Traditionally, geology has been the study of Earth. Yet answers to many of our questions about the origin of Earth have come from combined observations of Earth, the Solar System, and the Universe. Accordingly, the science of geology has been extended to include other planets and distant stars as well. We scan the stars and planets in the hope of gaining insight into the formation of our Solar System. We send space probes to photograph distant planets looking for both similarities and differences that may shed some light on how they and Earth came to be as they are today. We look critically at Earth, at both its land and ocean basins and at the internal and external processes that constantly shape and modify our planet, and we wonder if perhaps other planets with similar characteristics orbit some distant stars. In this chapter, we will briefly look at the historical development of our understanding of the Universe and will examine the stars and the Solar System for what they tell us about Earth and its place in space.

THE VIEW FROM EARTH

The beginnings of astronomy are buried deep in antiquity. Archaeological evidence indicates quite clearly that ancient civilizations watched the sky and recorded the movements of the Sun, Moon, and planets. Civilizations as widely separated in time and space as the Egyptians (c. 1,000 B.C.), the Mayans of Mexico (third to tenth century A.D.), and the Incas of Peru (fourteenth to fifteenth century A.D.) used astronomical data to establish the orientations of their temples and cities. One well-known archaeological site, *Stonehenge* near Salisbury, England, was constructed sometime between 3,100 and 1,600 B.C. and may have been used for astronomical observations (Figure 1.1). Achievements such as these are yet another reason why terms such as "primitive," which imply a value judgment, should never be used to describe early cultures.

The Contribution of the Ancient Greeks

Although many cultures, including the Chinese, the Babylonians, and the Arabs, recorded the movements of the stars and the planets, none contributed more to the evolution of modern astronomy than the ancient Greeks. While earlier civilizations simply

Figure 1.1 *The archaeological site at Stonehenge may have been used by civilizations living more than 4,000 years ago to observe the heavens. (VU/© N. Pecnik)*

compiled data about the locations and movements of heavenly bodies, the early Greeks actually attempted to understand and explain their observations.

Pythagoras (c. 570–500 B.C.), whom you may remember for his contributions to geometry, was the first to suggest that Earth was a spherical body, similar to the Sun and the Moon; he believed that Earth

THE FIRST MEASUREMENT OF EARTH'S DIAMETER

Eratosthenes understood the significance of the observation that the noontime Sun produces shadows with different angles at different places. If the Earth were flat, the angles should be the same everywhere on Earth, but if the Earth were curved—that is, a sphere—the shadows should vary depending on the latitude. On a particular summer day, the Sun was directly above a certain water well in the town of Syene (located near the present city of Khartoum, Sudan) while at the same time at Alexandria in Egypt, the Sun was 7° south of the vertical. Assuming that the Sun's rays reached Earth following parallel paths, Eratosthenes constructed the drawing shown in Figure 1.B1. Knowing that the distance from Syene to Alexandria represented 7° of the 360° circumference of a sphere, Eratosthenes calculated the circumference of Earth by multiplying the distance from Syene to Alexandria by the ratio of 360/7. His calculated result was equivalent to 31,000 miles (49,910 km), quite close to the actual distance of 24,800 miles (39,903 km).

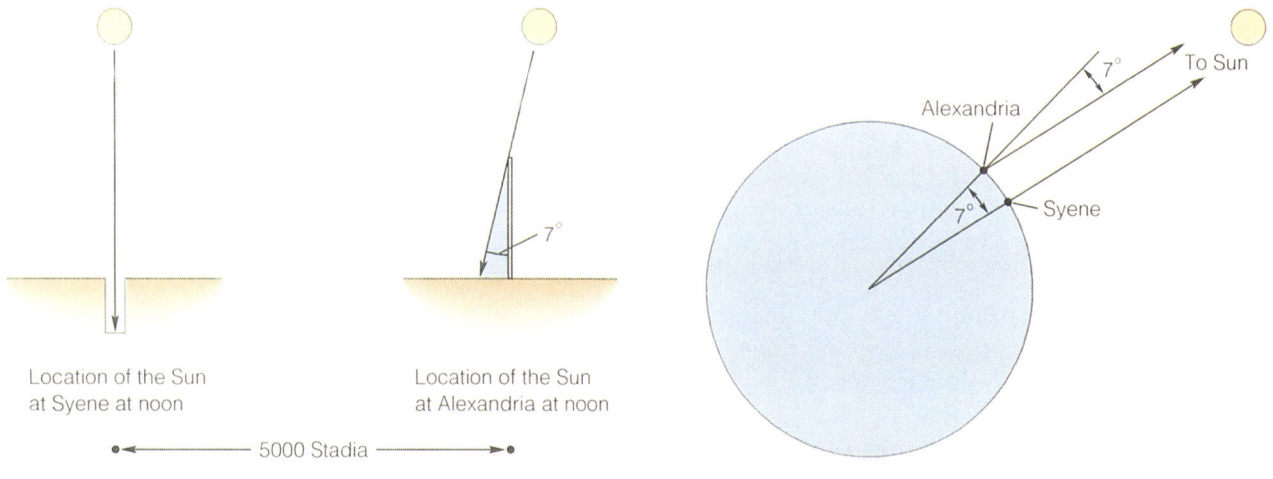

Figure 1.B1 Eratosthenes *knew that on a particular day, the sun shown directly down a water well at Syene while at Alexandria, 5,000 stadia to the north, it cast a shadow of 7°. He reasoned that the distance between Syene and Alexandria was 1/50 the circumference of Earth (7°/360°). Therefore, the circumference of Earth would be 50 times the distance between the two cities, or 250,000 stadia. With one stadium equal to about 655 feet (200 m), his calculated circumference of Earth was about 31,000 miles (49,910 km), approximately 25% larger than the known circumference of 24,777 miles (39,890 km). Considering the inaccuracies of the distance measurements at the time Eratosthenes made his calculations, the accuracy of his estimate of Earth's size is impressive.*

was the center of the Universe. The philosopher *Plato* (428–347 B.C.) accepted the idea of an Earth-centered Universe and maintained that all heavenly bodies were spherical and moved through the heavens following perfectly circular paths. *Aristarchus* (c. 310–230 B.C.) was the first to propose that the Sun was the center of the Universe, although at the time he could not demonstrate that Earth had any orbital motion. *Eratosthenes*, who lived about the same time as Aristarchus, is remembered for making the first realistic determination of the size of Earth, a truly spectacular accomplishment for the times.

Perhaps the most significant of the early Greek observers of the sky was *Ptolemy* (A.D. 100–200) who produced a compilation of all the astronomical data of his time. Ptolemy accepted an Earth-centered Universe and circular orbits for all celestial bodies. His work was to dominate astronomical thought for more than a thousand years. With the death of Ptolemy, the era of Greek dominance of astronomy came to an end, and the Sun-centered Solar System would have to wait until the fifteenth century when new ideas and curious minds would once again brighten the world of astronomy.

NEWTON'S LAWS AND EVERYDAY LIFE

Of all Newton's contributions, none touches our daily lives so much as his laws of motion. Newton's first law of motion, which formalizes the concept of *inertia*, states that, unless acted upon by an outside force, a body at rest tends to stay at rest and a body in motion tends to stay in motion in straight line. Consider yourself a passenger in the front seat of a car. A potential hazard looms ahead and the driver slams on the brakes. As the car rapidly slows, you continue to move forward at the original velocity of the car as your body "tends to remain in motion." Your forward motion will slow only when "acted upon by an outside force," presumably the restraining belt you are wearing. Otherwise, your forward motion will be stopped by the instrument panel or the windshield. Seatbelts in aircraft and cars are a direct application of Newton's first law of motion.

You are affected by the same law when an aircraft accelerates down the runway during takeoff. While the speed of the aircraft increases from a position of no motion, your body tends to "remain at rest" until acted upon by "an outside force," in this case the forward-moving seatback, which gives you the sensation of being thrown backward into your seat.

Another of Newton's laws states that for every action there is an equal and opposite reaction. Stationary objects are examples of Newton's laws at work. As you sit reading this book, your body exerts a downward force on the chair in which you are sitting, and at the same time, the chair exerts an equal upward force against your body. Because the two forces are equal and directed toward each other, neither you nor the chair experiences any motion. Consider, however, releasing a blown-up balloon without tying its neck. As the elasticity of the balloon causes its to regain the original smaller volume, the pressure applied to the gas rises within the balloon. As the gas exits from the balloon in one direction, the balloon moves in the opposite direction. The same principle applies to the operation of jet and rocket engines and to the recoil of a rifle as it is fired.

The Evolution of Modern Astronomy

As more accurate instruments were devised, discrepancies between the predicted and actual movements of the planets eventually led to a more precise understanding of the workings of the Universe. The long-held concepts of Ptolemy were challenged in the sixteenth century by *Nicolaus Copernicus* (1473–1543) who resurrected Aristarchus's idea that the Sun and not Earth is at the center of the Solar System. Because Copernicus's hypothesis explained the observed motions of the planets so well, it was only a few decades before his ideas were accepted and the Earth-centered Universe of Ptolemy was finally laid to rest.

Copernicus, like all the astronomers before him, had observed the heavens without the aid of magnification, but in 1609, *Galileo* (1564–1642) became the first to use a telescope to view the heavens, a significant contribution to astronomy. With the telescope, Galileo obtained the first clear view of distant objects such as the craters on the Moon, the moons of Jupiter, and the rings of Saturn. According to some accounts, he also used the telescope to view incoming commercial ships from afar and then played the financial market for his personal gain.

Isaac Newton (1642–1727) was born the year Galileo died. Newton's contributions to modern astronomy, physics, and mathematics include the basic rules of light and optics, the laws of force and motion, and the initial laws of gravity and planetary motion. What may be even more amazing is that Newton devised these rules during a two-year period (1665–1666) when he had retreated to the country to escape the bubonic plague then sweeping London. During this same period, he also invented calculus to enable him to formulate the rules.

Albert Einstein (1879–1955) extended our knowledge of gravity beyond Newton's work. Even though Einstein's contributions allow planetary movements to be calculated more precisely, Newton's laws are sufficiently accurate to solve most problems encountered today, even the launching of a spacecraft to distant planets.

The Galaxies

As we have seen, the early Greek astronomers placed either the Sun or Earth at the center of the Universe. Until well into the eighteenth century, astronomers thought that all stars belonged to a single huge star cluster or **galaxy,** which we now call the **Milky Way Galaxy.** During the eighteenth century, some astronomers proposed that hazy patches of light, which they called **nebulas,** could be other galaxies, each so far removed from our own that the individual stars could not be seen (Figure 1.2). With this new idea, our vision of the Universe expanded enormously. We now believe that the Universe is made up of billions of galaxies, each containing hundreds of billions of stars.

Figure 1.2 *Originally, astronomers thought the wisps of light they called* nebulas *were simply clouds of cosmic dust within the Milky Way Galaxy. When they realized that these patches of light were in fact other galaxies so distant from our own that the individual stars were not visible, the magnitude of the Universe increased dramatically. (© Royal Observatory, Edinburgh, Anglo-Australian Telescope Board)*

SPOT REVIEW

Starting with the ancient Greeks, discuss how our ideas about Earth, the Solar System, and the Universe have changed over time.

DISTANCES IN SPACE

As long as we deal with our own Solar System, units of measurement such as miles or kilometers are quite adequate. The distance from Earth to the Moon, for example, is 239,000 miles (384,000 km) while the distance from Earth to the Sun is 93 million miles (150 million km). Once we go beyond the confines of the Solar System, however, distances become so enormous that Earth-based units are far too small. A new distance unit is needed. The unit used is the **light-year.**

The light-year is defined as the distance that light will travel in one year. In a vacuum, light travels at a velocity of 186,000 miles (299,000 km) per second. Converted to miles, a single light-year equals 5.9 trillion miles (5.9 followed by 12 zeros) or 9.5 trillion kilometers. You can readily see why neither miles nor kilometers are appropriate for measuring space.

The distance from the Sun to the next nearest star is about 4 light-years. Note that this means that when you observe this star, you are seeing it as it was four years ago! The astronomers tell us that the Milky Way Galaxy is an average-sized galaxy with a diameter of 100,000 light-years. Our Sun is located toward the outer edge of the galaxy about 30,000 light-years from the center. Again, this means that when we observe stars in the center of our galaxy, we actually see the stars as they existed 30,000 years ago! The point is that the greater the distance we can look *into* space, the farther we are looking *back* in time! This is why astronomers anxiously awaited the orbiting Hubble Telescope (Figure 1.3). Far above the Earth's atmosphere, the Hubble Telescope is able to see farther into space, and therefore farther back in time, than will ever be possible using the largest Earth-bound telescope.

The nearest galaxy to Earth is about 2.5 million light-years away, which astronomers tell us is a typical distance between galaxies. As you look at the heavens and see those thousands of stars, recall that each is a Sun separated from us by distances so great that it has taken tens and even hundreds of thousands of years for the light to travel across the void of space. Also keep in mind that we only see a small,

Figure 1.3 *The* Hubble Space Telescope *is seen being deployed by the Space Shuttle* Discovery *on April 25, 1990. Although its optical capability was severely limited at the time of launch by faulty mirrors, this orbiting system has obtained other types of information heretofore unavailable. Repairs to the telescope's optic system were made in 1993. (Courtesy of NASA)*

perhaps a negligible, fraction of the stars that exist. In the great scheme of things, our Sun is one average-sized star out of many, and Earth is a relatively small planet of the Sun.

THE ORIGIN AND EVOLUTION OF THE UNIVERSE

In 1927, a Belgian astronomer, *Canon Georges Henri Lemaître,* theorized that all of the matter contained within the Universe was originally condensed into what he called the "primeval atom," a sphere that was perhaps the size of a golf ball! Individual atoms could not exist under the conditions that existed within the sphere. Instead, Lemaître hypothesized that the sphere was an incredibly condensed mass of the most fundamental, primitive subatomic particles from which atoms are derived. About 15 to 20 billion years ago, the ball exploded, and the matter expanded into space, an event called the **Big Bang.**

Immediately after the explosion, the expanded matter began to cool. After 0.01 seconds, the temperature had cooled to about 100 billion kelvins, and electrons began to form. Within a few minutes of the explosion, the simplest atoms, *hydrogen* and *helium*, began to form. Only a few percent of the heavier atoms, such as those that constitute minerals and rocks, were formed at this time. Most of the heavy atoms were formed later during the deaths of stars.

The Formation of Stars

About a million years after this cloud of matter rushed outward into space, eddy currents formed, and as they rotated faster, the hydrogen and helium atoms were contracted and compressed into a shrinking core. As the gas cloud collapsed, some of the gravitational energy was converted to heat, and temperatures within the core of the shrinking cloud began to rise. Eventually, the temperature and pressure within the core were so high that hydrogen began to combine chemically to form helium, giving off energy such as light and heat. It is in this way that **stars** are born. Within 5 to 10 billion years after the Big Bang, groups of stars, mutually attracted by the forces of gravity, formed into galaxies. Within the galaxies, stars continue to be born from clouds of hydrogen in interstellar space.

The Death of Stars

Once born, stars live by converting hydrogen to helium in a nuclear fire. As the supply of hydrogen dwindles, stars begin to die out like any other fire deprived of fuel. Depending upon their original mass, stars go through various stages as they die. Small stars with masses less than 0.4 times that of our Sun first shrink to **white dwarfs**, then cool to become a large cinder. Those with masses similar to our Sun's or up to three or four times as large expand tremendously to a **red giant** phase (Figure 1.4), during which helium begins to be chemically converted into carbon in the core. The very largest stars expand to become **supergiants**, until eventually, the outer materials of the expanding star are blown out into space to form a planetary nebula. Eventually, as the helium is consumed, the supergiant shrinks and ends its life as a white dwarf. Astronomers anticipate that the Sun will eventually enter a red giant phase and expand in diameter to engulf Mercury, Venus, and Earth.

The most massive stars expand to the red giant stage, collapse, and later explode into a **supernova** during which most heavy elements, such as those that make up minerals and rocks, are formed. An example of a supernova is the Crab Nebula (Figure 1.5). The debris from these explosions is thrown into interstellar space, providing the material from which new stars and planets are created.

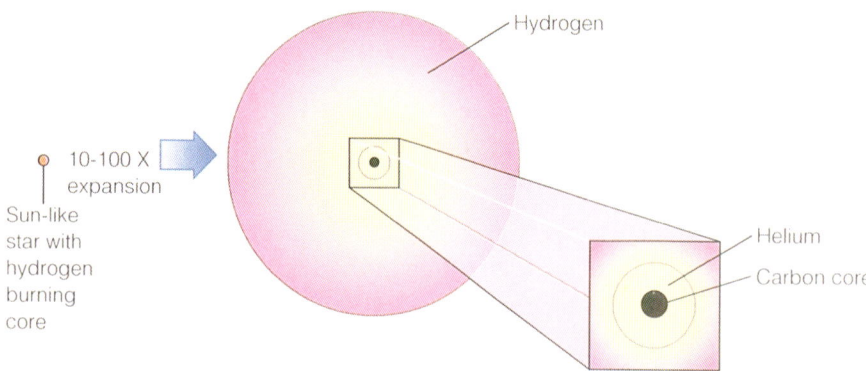

Figure 1.4 *The diameter of a red giant can be up to 100 times that of a star such as the Sun. As stars expand to the red giant phase, the element carbon, an essential component of all forms of life, begins to form within their cores.*

Figure 1.5 *All of the heavy elements, such as those making up minerals and rocks, form as massive stars die by exploding into a* supernova. *The debris from supernovas is the material from which other stars and their orbiting planets form. A well-known supernova, the* Crab nebula, *formed by the explosion of a star in the constellation Taurus on July 4, 1054. Chinese astronomers reported that for several weeks after its formation, the supernova was bright enough to be seen during the day and provided enough light to allow them to read by night.* (© California Institute of Technology 1959)

THE FORMATION OF OUR SOLAR SYSTEM

Our Sun was created from hydrogen, helium, and heavier atoms, along with bits of silicates, oxides, and carbon thrown into space by supernovas. The rotation, collapse, and contraction of such a cloud of gas and dust is thought to be initiated by the shockwave generated by a nearby supernova. In other words, the death of one star initiates the birth of another.

Our Sun was apparently born in this way in the outer reaches of the Milky Way Galaxy about 5 billion years ago. Over a period of a million or so years, as the rotational speed of the cloud or **solar nebula** increased, it flattened and contracted. Eventually, the concentration of matter in the center of the core became so great that gravity took over and the inner core collapsed. Pressure and heat generated in the core of the collapsing cloud started the fusion of hydrogen into helium, and the Sun was born. With the Sun now shining at the center, the remaining gas and dust orbited in the Sun's equatorial plane. The Sun emitted a stream of high-speed electrons and protons called the **solar wind** that drove the lighter elements into the cold outer reaches of the orbiting cloud while the denser materials remained behind.

Astronomers postulate that the orbital velocity of the disc-shaped cloud may have caused it to separate into rings not unlike those we see orbiting around the large outer planets (Figure 1.6). It should be noted that astronomers have seen what they interpret to be discs of rocky material orbiting other stars. Eddy currents within these rings caused small particles to aggregate into the **planetesimals**, which continued to grow larger. Once the mass of the planetesimals had become sufficiently large, they contracted under the force of gravity, creating **protoplanets**. Only internal compositional changes and subsequent separation into layers by density were needed to convert the protoplanets into **planets**.

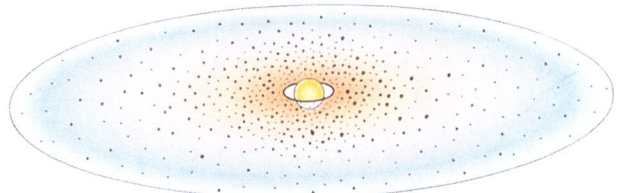
(a) New star forms at the center of a rotating cloud of space matter; density decreases away from the star.

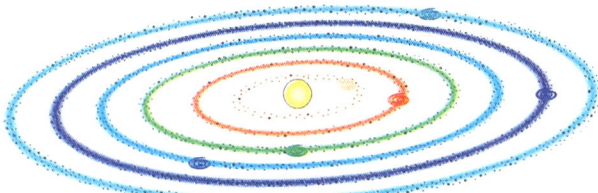
(b) Matter coalesces into concentric rings. Eddy currents form within rings and begin to condense matter into spherical forms (protoplanets).

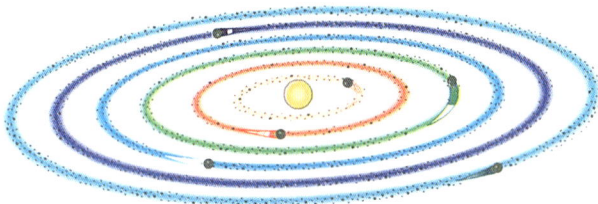
(c) Protoplanets continue to grow by condensing remaining matter within rings.

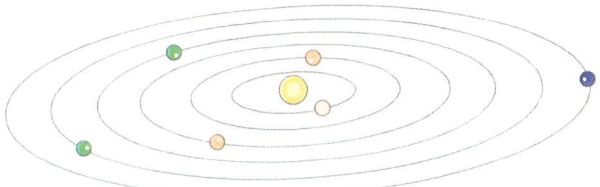
(d) Planets form by density and compositional stratification.

Figure 1.6 *Planets form from the orbiting cloud of gas and particles of ice and rock that surrounds a new-formed star. Materials within the cloud are separated as the* solar wind *forces the lower-density ices and gases to the outer reaches of the disc, leaving the denser materials closer to the star. With time, the bits of rock and ice accrete to form larger bodies called* planetesimals *which, with further accretion, increase in diameter and become structureless bodies called* protoplanets. *As temperatures within the protoplanets increase due to the crush of gravity, density segregation of materials takes place that eventually converts the protoplanet to a* planet.

SPOT REVIEW

1. Discuss ways in which the formation of stars, solar systems, and galaxies are similar.
2. In what ways do the deaths of stars contribute to the formation of new stars and the planets that surround them?

THE PLANETS

Nine known planets surround the Sun. With the exception of Pluto, the planets orbit the Sun within or nearly within a plane called the **ecliptic**, which is perpendicular to the Sun's rotational axis and extends outward from its equatorial plane (Figure 1.7).

Eight of the planets are grouped into two categories. The four planets closest to the Sun are known as the *terrestrial planets* because they share many characteristics with Earth (Latin *terra* = Earth). The next four planets are called the *Jovian planets* because they all resemble Jupiter. (Jove was the Roman name for Jupiter.) The ninth planet, Pluto, does not fit into the Jovian category; indeed, in size and density, it is comparable to the terrestrial planets. In fact, Pluto may consist of two smaller Moon-sized planets. Unlike the other planets, Pluto's highly elliptical orbit does not lie in the plane of the ecliptic.

The Terrestrial Planets

The **terrestrial (rocky) planets** include *Mercury, Venus, Earth,* and *Mars.* These planets are relatively small with diameters ranging from 3,000 to 8,000 miles (4,800–12,800 km). They have average densities of 5 gm/cm^3 and are composed primarily of the more refractory (heat-resistant) elements such as iron, magnesium, and silicon. They have few if any satellites or moons, rotate slowly on their axes, orbit the Sun at high velocity, and possess thin atmospheres.

Mercury

Most of what we know about the surface of **Mercury** comes from the 1974 flyby of the *Mariner 10* spacecraft, which recorded a rocky, cratered surface not unlike that of our Moon (Figure 1.8). Mercury is only 3,030 miles (4,875 km) in diameter. It's density of 5.43 gm/cm^3 indicates that it is composed of rocky materials similar to those of Earth. Besides craters,

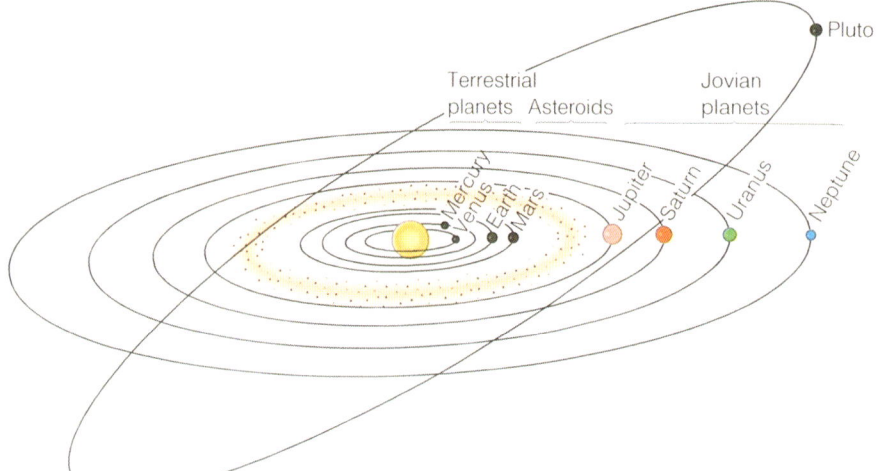

Figure 1.7 *With the exception of Pluto, all of the planets orbit the Sun in the same direction within or very near to an imaginary plane extending out from the Sun's equator called the* ecliptic.

Figure 1.8 *This view of* Mercury *was obtained by the Mariner 10 spacecraft on March 29, 1974. Due to its proximity to the Sun and the absence of an atmosphere, Mercury has the greatest range of surface temperatures of all the planets. (Courtesy of NASA)*

the most notable surface features are the traces of large fractures that break up the surface into what may be separate "plates."

Because Mercury is relatively close to the Sun and possesses no atmosphere to conduct heat from the sunlit side to the dark side, the temperatures swing from a scorching daytime high of 840°F or 450°C (lead melts at 626°F) to a frigid nighttime low of –238°F or –150°C (nitrogen freezes at –209°F). This day-night temperature range is the greatest observed on any of the planets.

Venus With a diameter of 7,516 miles (12,093 km) and a density of 5.24 gm/cm^3, **Venus** is similar in size and density to Earth. Enclosed in a continuous cloud cover, Venus has never been directly observed from Earth. The Sun's light reflects from the upper surface of Venus's atmosphere, causing the planet to appear as a very bright object in the night sky of Earth's Northern Hemisphere.

Venus was visited by 18 space vehicles between 1962 and 1985 including Russian landers that found a rock-strewn surface at the landing sites. Most recently, Venus was visited in 1990 by the *Galileo* spacecraft on its way to Jupiter and by the *Magellan* orbiter.

The surface temperatures on Venus, +860°F (+460°C), are hotter than on Mercury even though Venus is farther from the Sun. This is largely due to an excessive **greenhouse effect** caused by the high carbon dioxide content (96%) of the atmosphere, which absorbs and retains heat that otherwise would be radiated into space. The dense atmosphere on Venus also transfers heat efficiently from the sunlit side to the dark side; as a result, there is little difference between the daytime and nighttime temperatures. Making conditions even more inhospitable, the continuous cloud cover on Venus largely consists of sulfuric acid droplets, which undoubtedly contributed to the short life span of the Venusian landers and make the existence of life as we know it unlikely.

Radar images show that two-thirds of the Venusian surface is covered by rolling hills dotted with craters (Figure 1.9). The presence of volcanoes, mostly located around the equatorial region, indicates that sources of heat and magma exist beneath

Figure 1.9 *This image of Venus's northern hemisphere was produced at the JPL Multimission Image Processing Laboratory based on data collected by the Magellan, Pioneer, and Russian Venera spacecrafts (a). One of the volcanoes on Venus,* Maat Mons, *is seen in the three dimensional perspective (b). (Courtesy of NASA)*

the Venusian surface. The most recent data indicate that Venus may be volcanically active. One volcano, Monte Maxwell, is the highest point on the surface of the planet, towering nearly 40,000 feet (12,000 m) above the average surface elevation.

Earth Earth, the third planet from the Sun, has a diameter of about 8,000 miles (12,800 km) and a density of 5.52 gm/cm^3. During Earth's protoplanet stage, it underwent a process of compositional differentiation, which resulted in the formation of the **core, mantle,** and **crust** (Figure 1.10). During the gravitational collapse of the protoplanet, iron melted, mixed with a little nickel, and sank to the center of the planet forming the core. Earth's core is approximately 4,400 miles (7,000 km) in diameter with an inner portion of solid iron surrounded by an outer layer of molten iron. Movements in the liquid outer iron core are responsible for Earth's magnetic field.

As we discussed in earlier chapters, the core is surrounded by the less dense, magnesium- and iron-rich mantle. The upper mantle is composed of **peridotite**, made up primarily of the magnesium- and iron-rich silicate minerals **olivine** and **pyroxene**, while the lower mantle consists of a mixture of metallic oxides. The crust is made up of the lowest density materials. It consists of the relatively thin **oceanic crust**, composed mostly of the igneous rock **basalt**, and the thicker **continental crust**, composed primarily of igneous and

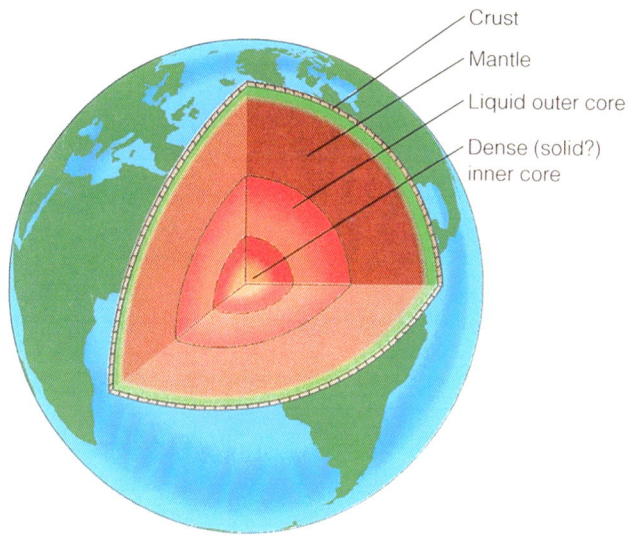

Distances from earth's center to surface of:
- Inner core: 900 mi.
- Outer core: 2200 mi.
- Mantle: 3900 mi.
- Crust: 3964 mi.

Figure 1.10 *Based on seismic data, we know that* Earth *consists of an innermost core of molten iron, the very center of which may be either solid or highly viscous, surrounded by* mantle *that accounts for most of Earth's volume, and a thin outer* crust.

metamorphic rocks. Because basalt is denser than the rocks that make up the continental crust, the oceanic crust forms basins on Earth's surface, now water filled, while the lower-density continental rocks rise above sea level as landmasses.

It seems unlikely that Earth was ever a homogeneous ball of molten rock, but it is certain that the young surface was racked with widespread volcanism. Some geologists believe that "oceans" of basaltic lavas may have existed at times. Enormous volumes of water vapor escaped into the primeval atmosphere from erupting volcanoes. Because Earth's surface was too hot for liquid water to exist, the water vapor collected in the atmosphere. Eventually, Earth's surface cooled to the point where the water in the atmosphere could condense and fall as rain. With the coming of the water, the major geologic processes of weathering, erosion, transportation, and deposition (as shown in the rock cycle on page 22 of the Introduction) were initiated. Ancient ocean basins filled with water that became salty as the soluble products released by weathering were removed from rocks by streams that washed the surface of the land.

Earth's atmosphere now consists primarily of nitrogen (78%) and oxygen (21%). Free oxygen was not abundant in the original atmosphere, but has been added mostly as a by-product of photosynthesis. Carbon dioxide is a small (0.03%) component of the atmosphere, but it is important because it plays a major role in controlling the temperature of the atmosphere via the greenhouse effect.

Mars

Of all the planets, **Mars,** the red planet, has most captured our imagination (Figure 1.11). Mars is the only other planet that has been seriously considered as a site for life as we know it. Tales of life on Mars date back to 1863 when two Italian astronomers, *Angelo Secchi* and *Giovanni Schiaparelli* drew maps of Mars showing what they described as *canali,* which in English means "channels" (Figure 1.12). Unfortunately, when their works were translated into English, the word *canali* was translated as "canals" rather than "channels." The difference is that channels can be cut by natural processes such as streams whereas the word *canal* implies a feature that is the work of some intelligent life. As a result of this error in translation, the idea spread—at least among the general populace—that life had been *proven* to exist on Mars!

We now know that canals do not exist on Mars although channels *do* exist. Observations made by

Figure 1.11 *This view of* Mars *show* Valles Marineris, *a water-carved valley twice as deep as the Grand Canyon and as long as the width of the United States from coast to coast. Note the other stream valleys that branch out from the main valley. The three dark red spots along the westernmost edge of the view are volcanoes. (Courtesy of NASA)*

the *Mariner* spacecraft showed erosional and depositional features that appear to have been caused by streams that once flowed on the Martian surface (Figure 1.13). Liquid water cannot exist in great quantities on the Martian surface now because the near-vacuum of the atmosphere would cause it to boil away. However, evidence that water once flowed on the surface of the planet suggests that temperatures and pressures may have been more Earth-like during the early history of the planet. What happened between then and now remains a mystery. We know that water vapor exists in the thin atmosphere and water ice apparently exists below the frozen carbon dioxide caps of the poles, but the water that once existed on the surface has long since vaporized into space or frozen beneath the regolith.

As we have seen, the largest mountain mass on Venus is a volcano, Monte Maxwell, and the largest single mountain mass on Earth is the island of Hawaii, a shield volcano. The largest single mountain mass on Mars is also a volcano, *Olympus Mons* (Figure 1.14). The base of the volcano would nearly cover the combined areas of Pennsylvania and New

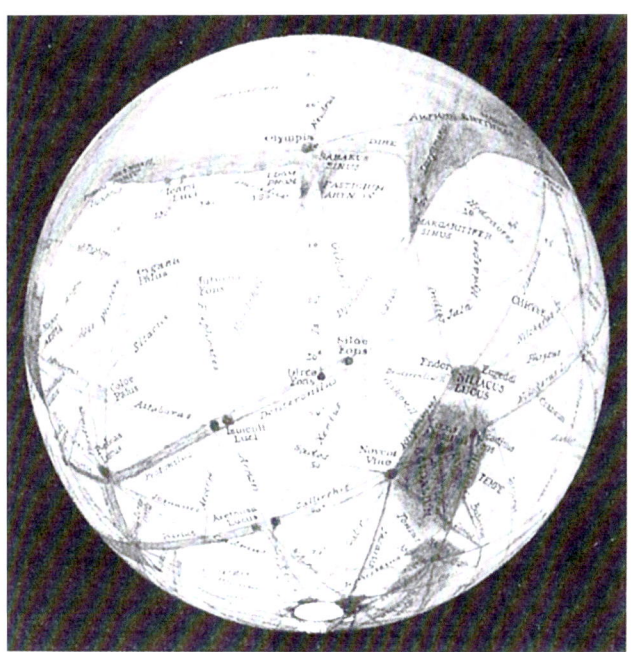

Figure 1.12 *From the very earliest observations of Mars, one of its most distinctive aspects were the linear features originally described by their Italian discoverers as* channels, *implying features produced by natural processes. The Italian word was erroneously translated into English as* canals, *which are features constructed to transport water; so began the stories of Martian spacepeople. (Lowell Observatory Photograph)*

Figure 1.13 *Channels, such as this example located just north of the Martian equator, are common on the surface of Mars and appear to have been carved by running water. The presence of running water on the surface of Mars indicates that the atmosphere must have been Earth-like sometime during its early history. With the present near-vacuum atmosphere, however, water could not exist. The obvious question is, what happened to change an Earth-like atmosphere to a vacuum? (Courtesy of NASA)*

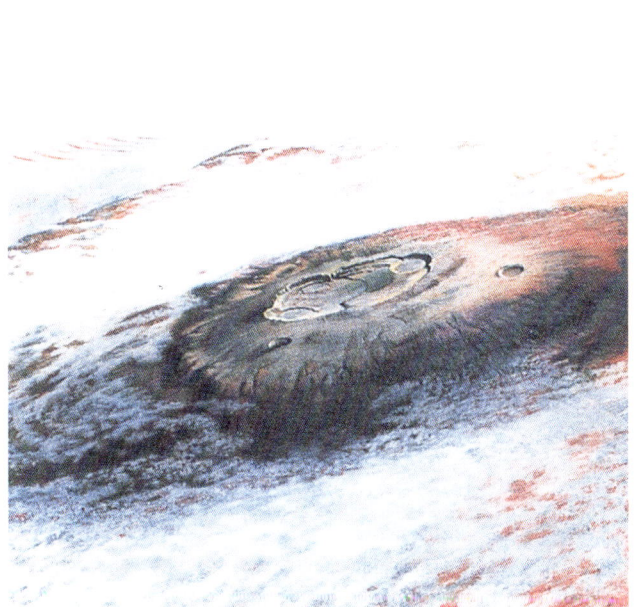

Figure 1.14 *The largest mountain mass on Mars, the volcano* Olympus Mons, *would cover the combined areas of Pennsylvania and New York and rise to an elevation of 80,000 feet (24,000 m) above its base. (Courtesy of NASA)*

York, and its summit rises 80,000 feet (24,000 m) above the surrounding surface. It is important to note that although volcanoes exist on Mars, no *active* volcanism has been observed on the planet. In addition to Olympus Mons, Mars also possesses an impressive valley, *Valles Marineris,* with a depth twice that of the Grand Canyon and a length that would span the United States from coast to coast.

The Jovian Planets

The **Jovian Planets** include *Jupiter, Saturn, Uranus,* and *Neptune.* These planets are huge, ranging from 40,000 to 86,000 miles (64,000–138,400 km) in diameter. Their densities are much lower than those of the terrestrial planets, averaging only 1.1gm/cm^3, and they consist primarily of gases and liquids. All have dense atmospheres, many satellites, rings made of small bits of rock or ice, high rotational velocities, and long orbital periods.

Jupiter

Jupiter is the giant of the planets with a mass greater than all the other planets combined. (Figure 1.15). Its diameter of 86,000 miles (138,400 km) is more than 10 times that of Earth. Structurally Jupiter is thought to have a small solid core, possibly made of rock, surrounded by a layer of metallic liquid hydrogen, overlain by a layer of liquid hydrogen, with an atmosphere of hydrogen and helium gas. The dominance of hydrogen and helium explains the low density of 1.32 gm/cm^3. Jupiter has been visited by several U.S. spacecraft, *Pioneer 10* in 1973, *Pioneer 11* in 1974, and, most recently, *Voyager 1* and *2* in 1979.

Saturn

Until the *Voyager* flights, the features that distinguished **Saturn** from the other Jovian planets were the rings that were discovered by Galileo when he first turned a telescope skyward (Figure 1.16). Now we know that all of the Jovian planets have rings, although the others are not as distinct as those of Saturn. Saturn is only slightly smaller than Jupiter with a diameter of about 75,000 miles (120,700 km), but its density of 0.7 gm/cm^3 is only about half that of Jupiter. Like Jupiter, Saturn is composed largely of hydrogen, but with its lower density, it probably does not contain as much liquid metallic hydrogen and has only a small rock core, if one is present at all.

Saturn was visited by the *Pioneer 11* spacecraft in 1979, by *Voyager 1* in 1980, and by *Voyager 2* in 1981 when the rings and its 23 moons were of much interest to astronomers. The data collected during the flyby indicate that Saturn's rings are likely made up of small particles of water ice or rock covered with water ice; the largest particles are 3 to 4 feet (1 m) in mean diameter.

Uranus

Until **Uranus** was visited by the *Voyager* spacecraft in 1986, little was known of the planet (Figure 1.17); it is too far away from Earth for its surface to be clearly visible. In fact, because of its indistinct image when viewed from Earth, Uranus's exact diameter was not known until 1877 when it passed in front of a star, allowing an accurate measurement to be made. We now know the diameter to be 32,000 miles (52,000 km). The planet is thought to consist of a rock core, which explains its relatively high density of 1.25 gm/cm^3, surrounded by a deep, hot layer of water and an atmosphere consisting of hydrogen and helium. Perhaps the most distinctive feature of Uranus is its

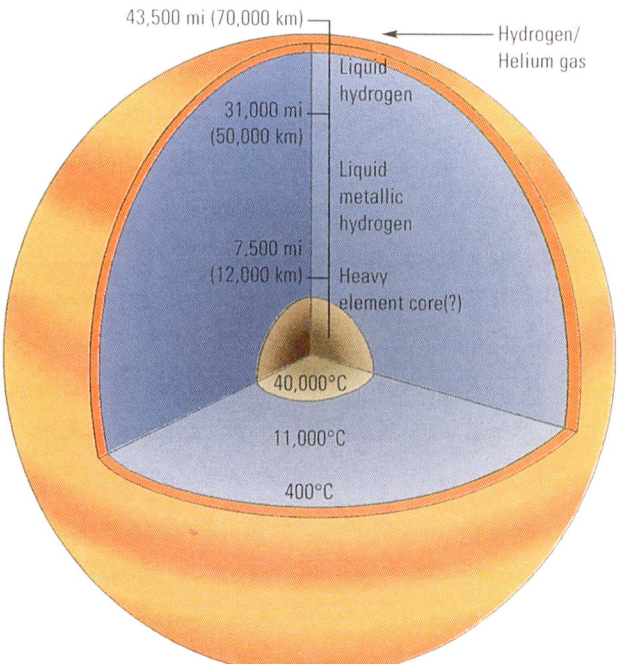

Figure 1.15 *Although* Jupiter *may have a small rock core, most of the planet consists of layers of liquid hydrogen with an atmosphere of hydrogen and helium, the same materials from which stars are made. The photo of the planet was taken by Voyager 1 on February 5, 1979, from a distance of 17.5 million miles (28.4 million km). Jupiter's innermost satellite,* Io, *can be seen against the disc of the planet. The bright spot to the right of the planet is the satellite* Europa *while another satellite,* Callisto, *is just barely visible at the bottom left of the photograph. (Photo courtesy of NASA)*

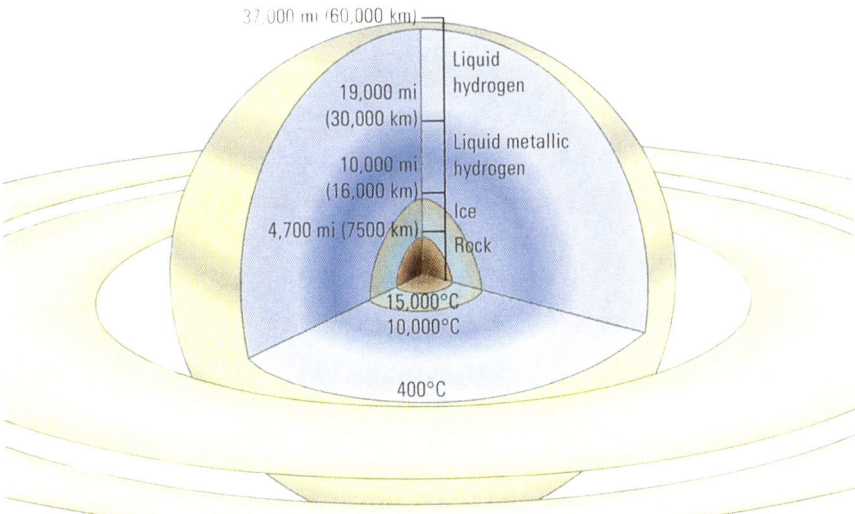

Figure 1.16 *The internal structure of* Saturn, *which is similar to that of Jupiter, consists of layers of liquid hydrogen possibly surrounding a very small rock core. The photograph of Saturn was taken by Voyager 2 on August 4, 1981, from a distance of 13 million miles (21 million km). Three of Saturn's icy moons,* Tethys, Dione, *and* Thea, *can be seen just beyond the edge of the planet; the dark spot on the disc of the planet just above the ring is the shadow of Tethys. The rings, once thought to be characteristic of Saturn, have been shown by the data sent back by the* Voyager *spacecraft to be common to all of the Jovian planets, although not so clearly visible as those surrounding Saturn. (Photo Courtesy of NASA)*

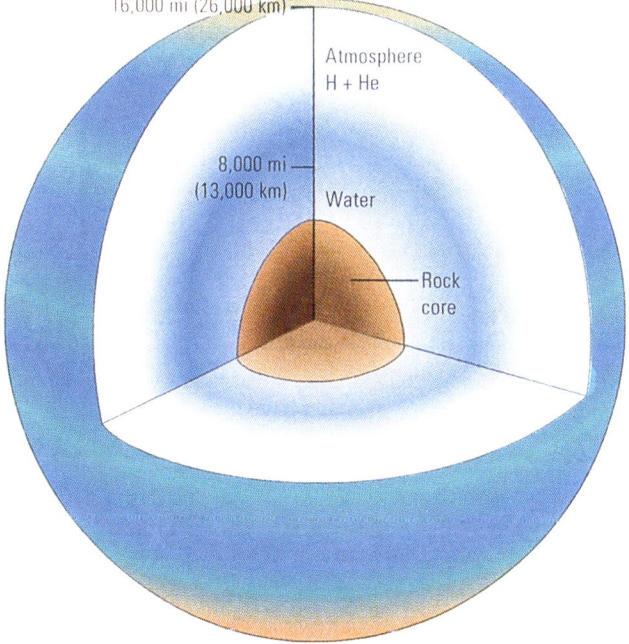

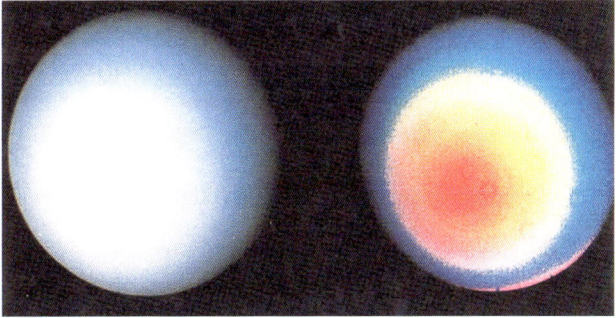

Figure 1.17 *The composition of Uranus is different from that of Jupiter and Saturn, and consists of water overlying a rock core with an atmosphere composed primarily of hydrogen and helium with some methane. The blue color, as seen in the photo taken by Voyager 2 on January 17, 1986, is the true color of the planet and is due to the absorption of red light by the methane within the atmosphere. The photo was taken from a distance of 5.7 million miles (9.1 million km) as the spacecraft approached the planet. The false-color image to the right of the true-color image is used to bring out details in the polar region of the planet and shows what has been interpreted as concentric banding within the atmosphere. (Photo courtesy of NASA)*

rotational axis, which is nearly parallel to the orbital plane or ecliptic; that is, the planet is tilted on its side as it orbits the Sun.

Neptune

Until *Voyager 2* arrived in August 1989, we knew only that **Neptune** was about the same size as Uranus with a little higher density (1.8 gm/cm^3) and probably the same composition (Figure 1.18). Voyager discovered six new satellites and showed that the planet has continuous rings.

Pluto, the Maverick Planet

Of all the planets, we know the least about **Pluto**. Although Pluto is usually described as the most distant planet in our Solar System, its extremely elliptical orbit causes it to spend time within the orbit of Neptune (Figure 1.19). Pluto moved inside the orbit of Neptune in 1978 and remained there until 1998. Thus, at the present time, Pluto is the most distant planet from the Sun.

Little is known about Pluto, and, since *Voyager 2* did not visit the planet, it will remain distant and mysterious. Although Pluto was not discovered until 1930, its existence had been hypothesized earlier, and its location calculated mathematically based upon eccentricities in the orbit of Neptune.

Pluto's highly elliptical orbit, its orbital path outside the ecliptic, its small size compared to the Jovian planets, and other characteristics strongly indicate that it is not one of the original planets. Some astronomers have suggested that Pluto was once a satellite of Neptune and that a near-collision with some celestial object removed it from orbit around Neptune and placed it into orbit around the Sun. However, the recent discovery that Pluto has a satellite of its own, a moon named Charon, may preclude this scenario inasmuch as it is difficult to explain how a satellite could acquire a satellite of its own. Others have suggested that both Pluto and Charon are asteroids. Paired asteroids in solar orbit are known to exist. The processes by which Pluto and Charon would have joined forces and entered solar orbit have not yet been explained, however. With no spacecraft visits to Pluto planned for the near future, its origin will continue to be debated, and it will remain little more than a distant faint point of light.

In 1993, NASA announced that, based on data sent back by the Voyager 1 and 2 spacecrafts, the boundary between the solar system and interstellar space, the **heliopause**, was located at a distance of between 8.3 billion and 11.1 billion miles from the

Figure 1.18 *This color-enhanced photo of* Neptune *was taken by* Voyager 2 *on August 14, 1989, from a distance of 9.2 million miles (14.8 million km). With a composition similar to that of Uranus, the slightly higher density of Neptune indicates that it may have a somewhat larger rock core. (Courtesy of NASA)*

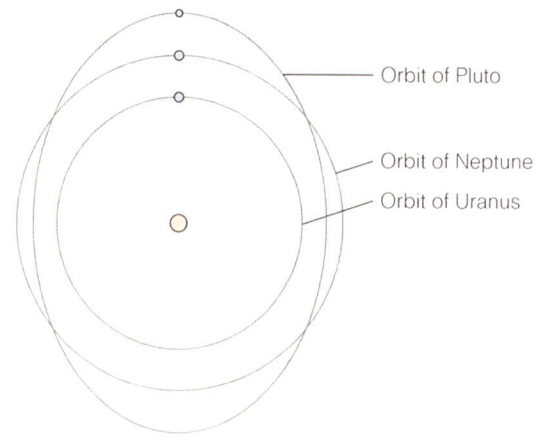

Figure 1.19 *The highly eccentric orbit of Pluto crosses the orbit of Neptune. As a result, Pluto and Neptune share the role of being the outermost planet of the Solar System.*

Sun. The boundary was established by detecting radio waves caused by the collision of solar and interstellar particles. These data indicate that the diameter of the solar system is more than double the presently accepted size of approximately 3.6 billion miles, the distance from the Sun to the orbit of

Pluto. Although Voyager 1 and 2 are now far beyond the orbit of Pluto, at their present speed of 279 million miles per year, it will take them another 20 years before they reach the newly-established heliopause and leave the solar system.

SPOT REVIEW

1. Prepare a chart that compares and contrasts the compositions and physical characteristics of the planets.
2. Indicate how the composition of the atmosphere affects the surface climate of each of the terrestrial planets.

ASTEROIDS

The **asteroids** are hundreds of thousands of rock fragments that orbit the Sun, mostly between the orbits of Mars and Jupiter (refer to Figure 1.7). Because some asteroids are too small to be observed even with the most powerful telescope, the actual number is unknown. The asteroids are all relatively small, the largest being about 700 miles (1,100 km) across. Because even the largest asteroid does not possess sufficient mass to be formed by gravity into a sphere, all are irregularly shaped. Also, because of their low mass, even the largest asteroid is too small to possess an atmosphere.

The origin of the asteroids has been a point of debate for years. At present two theories are prevalent. One theory postulates that the asteroids are the remains of a disrupted, terrestrial-type planet that was once located between the orbits of Mars and Jupiter. The latest and most favored theory holds, however, that the competing gravitational attractions of Jupiter and Mars would have prevented a planet from ever forming between Mars and Jupiter and that the asteroids are actually planetesimals.

METEORS, METEOROIDS, AND METEORITES

Nearly everyone has seen a "shooting star" or **meteor**, which is the streak of light that flashes across the night sky recording the penetration of Earth's atmosphere by a **meteoroid** (Figure 1.20). Although sightings of shooting stars had been recorded for millennia

Figure 1.20 *The most common of visitors from outer space are* meteoroids, *which announce their entry into our atmosphere by the streak of light called a* meteor. *(Courtesy NOAO, Kitt Peak Observatory)*

and rumors abounded that rocks fell from the sky, no accepted evidence existed that meteoroids fell to Earth to become **meteorites** until 1803 when numerous witnesses saw a large meteoroid explode over France and shower the countryside with thousands of bits of rock.

Most meteoroids are thought to originate from fragments generated by collisions among the asteroids, although some are derived from the debris sloughed off the surface of comets as they orbit the Sun. Most meteoroids do not survive the blazing entry into our atmosphere. As they plunge into the ever-thickening gas of the lower atmosphere at speeds of up to 100,000 miles per hour (161,000 kph), the heat generated due to friction becomes so intense that they vaporize.

Meteorites are classified into one of three basic types based upon their composition: (1) **iron meteorites** are 80% to 90% iron with the remainder being largely nickel, a composition similar to that thought to make up Earth's core; (2) **stony meteorites**, which may contain a little iron or nickel, are composed primarily of the same minerals present in Earth's mantle; and (3) **stony-iron meteorites**, as the name implies, are mixtures of the minerals in the other two. The stony meteorites are by far the most plentiful, making up 90% or more of all meteorites. The dominance of stony meteorites is based on the relative numbers of the various kinds of meteorites found in the Antarctic where they can be readily identified because they were imbedded in snow or ice rather than rock.

Although most meteoroids do not survive the fiery trip through our atmosphere, thousands of meteorites strike Earth's surface each year. Fortunately for all of us, most of the survivors are about the size of a speck of dust or perhaps, at most, the size of a grain of sand. As a result, their arrival goes without notice. Occasionally, however, a meteorite large enough to create a crater will impact Earth. One of the most famous meteorite impact craters is the Barringer Crater near Winslow, Arizona (Figure 1.21). The Barringer meteorite struck Earth about 50,000 years ago and blasted out a crater that measures nearly a mile across and is more than 500 feet (150 m) deep. From the size of the crater, scientists have estimated that a meteorite was about 130 feet (40 m) in diameter, weighed about 300,000 tons (295,000 metric tons), and was still traveling at a speed of about 25,000 miles per hour (40,600 kph) when it impacted. Many other meteorite craters can be found on Earth, but none is as well preserved as the Barringer Crater. Perhaps the largest meteoroid ever to impact Earth struck at the very close of the Cretaceous Period and is believed by many geologists to be responsible for the extinction of the dinosaurs.

COMETS

Perhaps the oddest members of the solar family are **comets**. Although very impressive when seen in space, comets are actually quite small, irregular shaped chunks of ice encapsulated within a tarry outer crust; they measure about 5 to 10 miles in diameter (8–16 km) and are composed primarily of water (80%), carbon dioxide, carbon monoxide, methane, and ammonia (Figure 1.22). Unlike asteroids and meteoroids, comets are relatively rare, and only a hundred or so have been named.

Comets are often confused with meteoroids although they are quite different and can be very easily distinguished. A meteoroid will be seen to move rapidly through the sky and disappear all within a matter of seconds whereas a comet will seem to be stationary in the sky at any moment and will only be seen to move from one horizon to the next over a period of days.

A characteristic feature of comets is the huge, extremely elliptical orbit. The orbit of Halley's comet, for example, extends from just within Earth's orbit at the comet's closest approach to the Sun to near the orbit of Pluto at its farthest point from the Sun, a trip requiring 75 years to complete.

The incandescent tail of the comet is the result of the impact of the solar wind on the comet's sunlit surface, but the tail develops only when the comet has approached the Sun to about the orbit of Jupiter. At that distance, the intensity of the solar wind is sufficient to vaporize and ionize the ice on the approaching face of the comet and drive the gas away from the comet in the direction away from the Sun. As the comet orbits the Sun, the tail swings around, always pointing away from the Sun (Figure 1.23).

Figure 1.21 *The* Barringer Crater *near Winslow, Arizona, commonly called* Meteor Crater, *was formed by the impact of a meteoroid about 50,000 years ago. At the time of impact, the meteoroid was estimated to weigh about 300,000 tons and be traveling at a speed of about 25,000 miles per hour (40,600 kph). Consider the effects of such an impact were it to occur in a relatively populated area today. (Courtesy of D.J. Roddy/USGS)*

Comets may die in a number of ways. Because a comet loses mass with each orbit of the Sun, each completed orbit brings the comet one step closer to annihilation. Halley's comet, for example, loses an estimated 25 to 30 tons of mass per second at its closest approach to the Sun. Even so, Halley's comet loses less than 1% of its mass with each orbit. Eventually, however, the mass of a comet will be completely vaporized and its material spread throughout space as gas and dust. Other comets are destroyed as they are drawn by gravity into the Sun, while others may be reduced to bits of rock as their ices are totally vaporized.

The origin of comets is surrounded by controversy. In 1950, a Dutch astronomer, Jan Oort, theorized that a cloud of comets, called the **Oort Cloud**, completely surrounds the Solar System and orbits at a distance far greater than the diameter of the Solar System, possibly a third of the way to the next nearest star. He postulated that individual comets are deflected from these distant orbits by the gravitational attraction of a passing star and are thrown into their characteristic elliptical orbits, some of which will carry a comet into the confines of our Solar System.

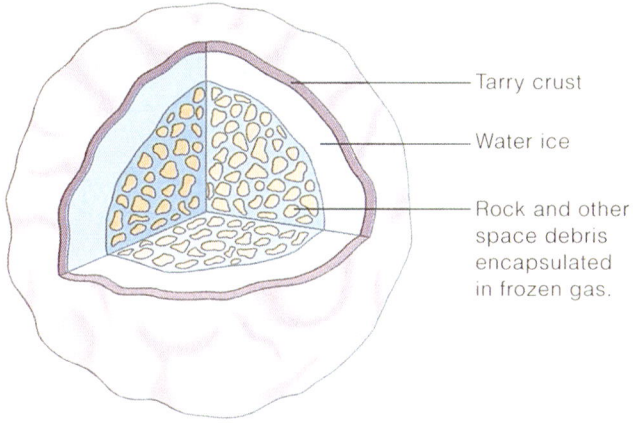

Figure 1.22 *Rare members of the Solar System, comets are thought to be primarily composed of frozen water ice intermixed with carbon dioxide, carbon monoxide, methane, and ammonia.*

SPOT REVIEW

1. How do meteoroids, asteroids, and comets differ?
2. What is the difference between a meteor, a meteoroid, and a meteorite?

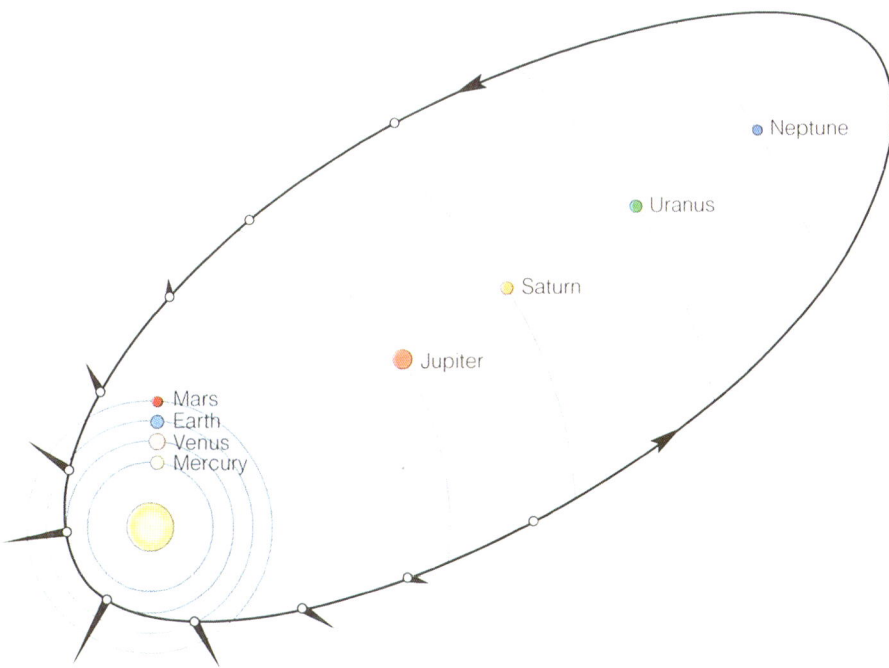

Figure 1.23 *The comets pass through the Solar System along huge, highly elliptical orbits that typically take them from the region of the Sun to the far reaches of the Solar System. The characteristic tails usually form only when the comet is closer to the Sun than the orbit of Jupiter and always point away from the Sun.*

CONCEPTS AND TERMS TO REMEMBER

Galaxies
  galaxy
  Milky Way Galaxy
  nebula
Distances in space
  light-year
Formation of the Universe
  Big Bang
Formation and death of stars
  stars
  white dwarf
  red giant
  supergiant
  supernova
Formation of the Solar System
  solar nebula
  solar wind
  planetesimals

protoplanet
Planets
  ecliptic
  terresterial planets
    Mercury
    Venus
      greenhouse effect
    Earth
      core
      mantle
        peridotite
        olivine
        pyroxene
      crust
        oceanic crust
        basalt
        continental crust
    Mars

Jovian Planets
  Jupiter
  Saturn
  Uranus
  Neptune
  Pluto
  heliopause
Smaller member of the Solar System
  asteroids
  meteors
  meteoroids
  meteorites
    iron meteorites
    stony meteorites
    stony-iron meteorites
  comets
    Oort Cloud

REVIEW QUESTIONS

1. The _____ were the first to attempt an explanation of the observed movements of the stars and planets.
   a. Mayans   c. Romans
   b. Greeks   d. Incas

2. Who was the first Greek astronomer to suggest that the Earth is spherical in shape and that all the heavenly bodies orbit the Earth in circular orbits?
   a. Galileo   c. Plato
   b. Pythagoras   d. Ptolemy

3. Who was the first astronomer to suggest that the heavenly bodies orbit the Sun?
   a. Plato   c. Copernicus
   b. Aristotle   d. Aristarchus

4. Until the work of Einstein, which astronomer's work most affected modern thinking about the Universe?
   a. Galileo   c. Newton
   b. Ptolemy   d. Aristotle

5. The telescope was first used to view the heavens by
   a. Aristotle.   c. Copernicus.
   b. Galileo.   d. Kepler.

6. Within the Universe, the stars are grouped together into clusters called
   a. novas.   c. nebulas.
   b. galaxies.   d. supernovas.

7. The major component of the stars is
   a. carbon.   c. iron.
   b. hydrogen.   d. oxygen.

8. Which planet is thought not to be one of the originally formed planets?
   a. Mars   c. Mercury
   b. Neptune   d. Pluto

9. Which planet has the highest surface temperature?
   a. Venus   c. Mercury
   b. Mars   d. Pluto

10. The largest planet is
    a. Jupiter.   c. Saturn.
    b. Pluto.   d. Neptune.
11. The asteroids orbit the Sun between the orbits of
    a. Mars and Jupiter.
    b. Mars and Earth.
    c. Venus and Mercury.
    d. Jupiter and Saturn.
12. Which spacecraft recently completed the "Grand Tour" of the Jovian planets?
    a. *Pioneer 12*
    b. *Voyager 2*
    c. *Viking 1*
    d. *Venera 6*
13. The ecliptic is
    a. the sphere enclosing the Earth from which the comets are derived.
    b. the angle that a planet's rotational axis makes with respect to the Sun.
    c. the solar equatorial plane within which or near which all the planets except for Pluto orbit the Sun.
    d. The time required for one revolution of a planet around the Sun.
14. According to theory, the Oort Cloud is the source of
    a. meteoroids.   c. asteroids.
    b. new stars.   d. comets.
15. What does it mean to say that in space, distance is time?
16. Why is the surface temperature of Venus hotter than that of Mercury even though Venus is farther from the Sun?
17. Explain the theory that holds that the asteroids are parts of a potential planet that never formed.
18. Compare and contrast comets and meteors.
19. How would a change in Earth's angle of inclination affect the lengths of daylight and darkness through the year? How would it affect the seasons?
20. Much reference is made to the "weightlessness" of space. Are astronauts orbiting Earth actually weightless?

THOUGHT PROBLEMS

1. Where would you weigh the most, at the equator or at either the North or South Pole? Why?
2. Once the *Voyager* spacecraft escaped from the gravitational pull of Earth, no engines were required to propel it to the outer planets. Where did *Voyager* get the energy to make the voyage?
3. Why does the same side of the Moon face Earth at all times?

FOR YOUR NOTEBOOK

Since most of us are not direct observers of the heavens, our information about celestial events is more likely to come from the news media. Weekly news magazines often have stories complete with photos about meteor showers, the comings and goings of the most recent comet, information gleaned from data sent back from the most recent U.S. and Russian space probes, and the latest discovery of some mysterious feature such as a black hole. Including such articles or summaries of the articles will help you develop a more thorough understanding of our companions in space.

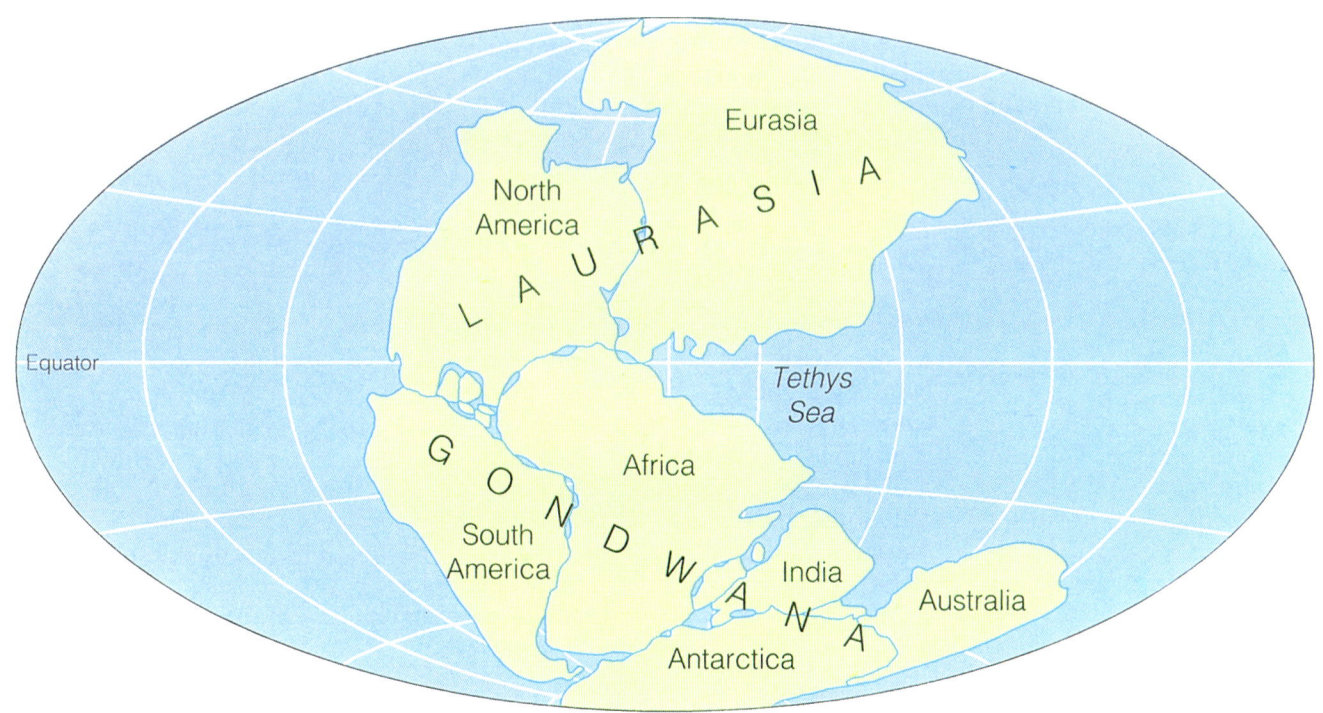

CHAPTER OUTLINE

INTRODUCTION

HISTORICAL DEVELOPMENT
   F. B. Taylor
   H. B. Baker
   Alfred Wegener

MODERN DEVELOPMENTS
   Rock Magnetism
   Topography of the Ocean Bottom
   Seismic Investigation of
     Earth's Interior

THE PRINCIPLE OF PLATE TECTONICS

CHAPTER 2

Plate Tectonics

INTRODUCTION For each science, one concept can be identified that, more than any other, has changed how that science visualizes its domain. For geology, *plate tectonics* is that concept. Plate tectonics has been called the *"unifying principle of geology."* Its impact on geology has been comparable to the effects of Newton's laws of motion and universal gravitation on physics, Darwin's concept of organic evolution on biology, and Copernicus's view of the Solar System on astronomy.

Before the concept of plate tectonics was proposed in the mid-1960s, we could not logically (or correctly) explain the distribution of earthquakes and volcanoes, the source of energy for the creation of mountains, or why most of Earth's major mountains were located along the margins of continents while a few, such as the Urals of Russia, were not. The origin of the deep-sea trenches was a mystery as were the origins of the most extensive mountain ranges of Earth, the oceanic ridges. Although many competing theories attempted to explain the origin of the ocean basins, none of them took into account the fact that ocean basins are born, grow in size, and then close, all within a period of a few hundred million years.

Many of Earth's surface features, as well as its surface and internal processes, remained mysteries until the formulation of the concept of plate tectonics. Looking back now with the clarity and wisdom of hindsight, we may wonder why the concept of plate tectonics was not proposed earlier. Certainly, some of the evidence that supports it had been known for decades. The answer perhaps is that the concept required the combination of modern techniques that looked more critically at Earth and enabled geologists to reevaluate old data and scientific insight that recognized the single thread that tied the various kinds of data together.

HISTORICAL DEVELOPMENT

Ever since cartographers produced maps with a sufficient degree of accuracy to portray the shapes of the continents realistically, individuals have noticed the correspondence in the shapes of coastlines on opposite sides of the Atlantic Ocean, in particular, the similarity between the eastern coastline of South America and the western coastline of Africa (Figure 2.1). As early as 1620, *Francis Bacon* commented on the similarity of the Atlantic coastlines of South America and Africa. Almost certainly, at least a few of those who made the same observation must have wondered whether the similarity was just coincidental or meant that the two continents had once been joined like two pieces of a gigantic jigsaw puzzle. In 1668, *François Placet* proposed that the continents of North and South America were once joined to Europe and Africa by islands or by a continent that foundered. Based on the similarity of present-day animals and plants, the *Compte de Buffon*, a French naturalist, suggested in 1750 that North America and Europe were once joined. In 1858, *Antonio Snider-Pellegrini* commented on the similarity of the fossil plants associated with coal-rich rocks on opposite sides of the Atlantic. In the late 1800s, *Edward Suess* pointed out the nearly identical geological records of the continents in the Southern Hemisphere. Suess proposed that all of the southern continents were once joined together in a supercontinent he called **Gondwana**.

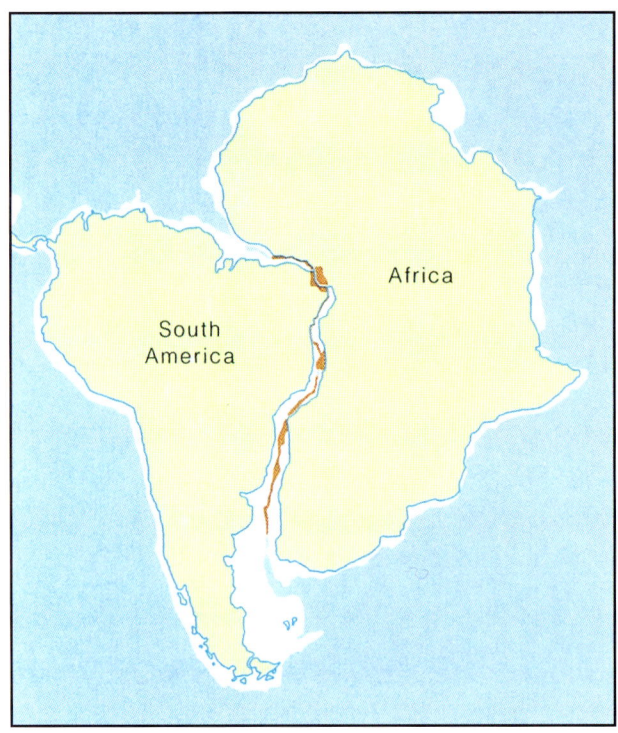

FIGURE 2.1 *The marked similarity between the Atlantic coastlines of South America and Africa led many to wonder whether the two continents had once been joined. (After A. G. Smith, "Continental Drift." In* Understanding the Earth, *edited by I. G. Gass. Courtesy of Artemis Press.)*

Although it was intriguing, few pursued Suess's suggestion that the continents had been joined. First of all, it went against all that had been taught about Earth. From the beginnings of the science of geology in the late eighteenth century, the size, shape, and location of the continents as we see them now were considered to have been established during the early days of Earth's creation when the brittle outer layer we call the crust was first formed. Secondly, if the continents were indeed once connected, some mechanism must have caused them to migrate to their present positions. What conceivable force could tear a continent apart and move it? As a result of these difficulties, the obvious similarity in the shapes of the Atlantic coastlines remained little more than an interesting and puzzling observation.

F. B. Taylor

It was inevitable that someone would take on the intellectual challenge of fitting and moving the continents. The first to suggest that the continents had moved over time was *F. B. Taylor* who, in 1908, argued that the location of young, folded mountains long the margins of certain continents was strong evidence that the continents had moved laterally. He theorized that the mountains had been created when the leading edge of the moving continent encountered the immobile rocks of the ocean basin in much the same way that the front end of a car crumples when it encounters an immovable wall. Taylor even suggested a possible source of energy to drive the movement, namely, strong lunar tides that existed during a theoretical period when the Moon's orbit was closer to Earth. His ideas were not widely accepted by the professional community.

H. B. Baker

The first geologist to actually suggest that the continents had once been joined together, had broken apart, and had moved with time was *H. B. Baker* in 1911. Impressed by the similarities in the shapes of the continental margins, Baker proposed the existence of a supercontinent made up of the joined masses of the present continents. In support, however, he offered little evidence other than the similarity in coastal shapes. He also failed to suggest a mechanism by which the proposed supercontinent fragmented and the resultant continents ultimately reached their present locations.

Alfred Wegener

Alfred Wegener, a German meteorologist, had also noted the similarities in the coastal outlines, especially the outlines of the opposing coastlines of South America and Africa. Picking up on Baker's idea, he proposed that all the continents were joined together in a supercontinent called **Pangaea** (Figure 2.2). Wegener, however, went further than any of his predecessors and presented evidence for the supercontinent. He showed not only that the continents fit together physically, but that mountain ranges, rock types, fossils, glacial deposits, directions of glacial striations, and other features from the eastern margin of South America matched those of the western margin of Africa (Figure 2.3). In short, he presented what would be accepted today as strong, if not irrefutable, evidence that the continents, now separated by an ocean, were once connected.

Wegener first published his hypothesis on continental drift in 1912. A few years later, while recuperating from wounds suffered during World War I, he wrote his famous work *The Origin of Continents and Oceans,* which was published in 1915. In this paper,

Plate Tectonics 51

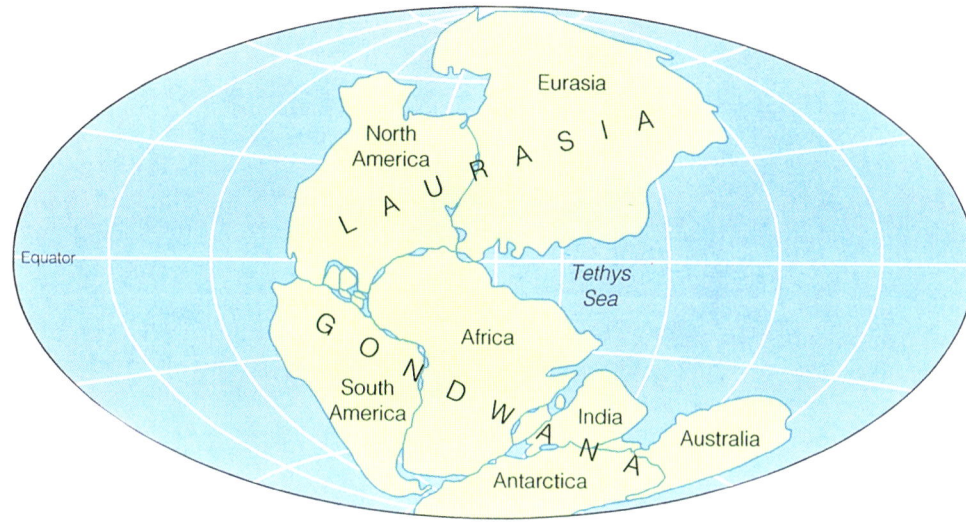

FIGURE 2.2 *Alfred Wegener proposed that all of the present continents were once joined into a supercontinent he called* Pangaea.

 Wegener noted that fossils of *Mesosaurus* were found in Argentina and Africa but nowhere else in the world.

 Fossil evidence tells us that *Cynognathus*, a Triassic reptile lived in Brazil and Africa.

 Remains of *Lystrosaurus* were found in Africa, Antarctica, and India.

 Fossil ferns, *Glossopteris* were found in all the southern land masses.

☐ Distribution of late Paleozoic glaciers
▨ Locations of major plateau basalts

FIGURE 2.3 *Various kinds of data exist to support the idea that the southern continents were once joined into a continent called* Gondwana, *some 200 million years ago.*

he postulated the existence of the supercontinent he called Pangaea, which based upon fossil evidence, began to break up about 200 million years ago. However, he could not identify a scientifically defensible mechanism to move the continents. Wegener's best attempt at explaining the movement was to suggest that the lower-density masses of granitic continental rocks somehow plowed through the denser basaltic rocks of the ocean basins. Unfortunately for Wegener, no deformation of the oceanic crust was ever found to support his proposal.

Like the proposals of his predecessors, Wegener's ideas found a mixed reception in the scientific community. Though widely accepted by South American and African geologists, they were ridiculed by most British and American geophysicists. Geologists were probably reluctant to accept Wegener's hypothesis for several reasons. Undoubtedly, professional jealousy and ego were involved. How could Wegener, a mere meteorologist, have the nerve to tell the geological community something so fundamental about Earth? Furthermore, if geologists were to consider the possibility of continental drift, let alone accept it as fact, they would be admitting that for the past hundred years they had been wrong in believing that Earth possessed a brittle, immobile crust and a never-changing face. Perhaps another reason why Wegener's thesis received limited acceptance was that most of his evidence was from the Southern Hemisphere while most geologists of the day lived and worked in the Northern Hemisphere. Nevertheless, the only scientifically sound argument against Wegener's proposal was his inability to find a reasonable mechanism for the movement. Although the majority of the geological community refused to accept Wegener's theory, it was not rejected by everyone. Like any radical new thesis, it found a few disciples willing to keep the idea alive.

Unfortunately, Wegener did not live to see his hypothesis vindicated. He died on a scientific expedition to the Greenland ice cap, presumably from a heart attack brought on by a strenuous march. From Wegener's death in 1930 until the late 1950s, the idea of continental drift remained in limbo. Except for geologists from the Southern Hemisphere and paleontologists (geologists who study fossils), relatively few geologists were willing to espouse the concept. Consequently, classroom instructors gave the topic little time or attention. All the while, the continents continued to drift.

SPOT REVIEW

1. What kinds of evidence did Wegener use to support his idea that the present continents formed as a result of the rifting of a much larger supercontinent, Pangaea?
2. Why were Wegener's ideas not universally accepted by the geologists of his day?

MODERN DEVELOPMENTS

In the 1950s, as bits of evidence appeared from various sources, continental drift was resurrected, and ultimately, Wegener and his predecessors were shown to have been basically correct. The evidence came from three totally different areas of investigation: (1) *rock magnetism,* (2) *ocean bottom topography,* and (3) *seismology.*

Rock Magnetism

In the 1950s, geologists found that solidifying basaltic lavas recorded the orientation of Earth's magnetic field when the lava cooled. Basalt is an igneous rock rich in magnesium-iron silicates. As the molten rock solidifies, some of the iron crystallizes in the mineral *magnetite* (Fe_3O_4), which, as its name indicates, is ferromagnetic (ferro = iron). As the tiny crystals of magnetite cool below a certain temperature called the **Curie point,** they become magnetized with their magnetic fields aligned with Earth's magnetic field. Not only do the magnetic fields of the individual crystals line up along the field of Earth's magnetic force, but the south pole of the magnetite crystals points toward Earth's north magnetic pole. As a result, the magnetite crystals "lock in" and preserve a record of Earth's magnetic field at the time of solidification. The orientation of the crystals' magnetic field acts both as a *compass* and a *dipmeter.* In a compass, the needle moves in a horizontal plane and determines the direction to the magnetic north pole (Figure 2.4). In a dipmeter, the needle moves in a vertical plane and determines the **magnetic inclination,** which is the vertical angle between the direction of the magnetic field and the horizontal (Figure 2.5). The magnetic inclination is used to calculate the latitude of the rock's location by the simple relationship:

$$\tan \lambda = 1/2 \tan \Theta$$
$$\text{where } \lambda = \text{the latitude}$$
$$\Theta = \text{the inclination}$$

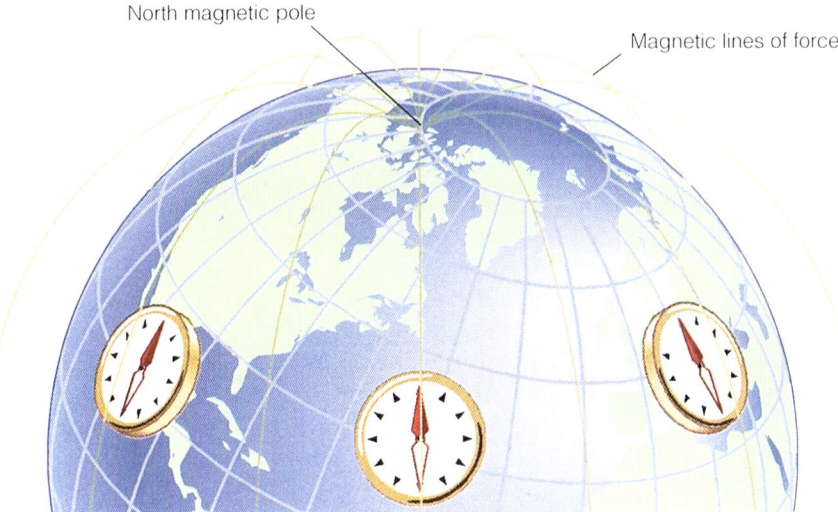

FIGURE 2.4 *The horizontally oriented needle of the compass aligns parallel to Earth's magnetic lines of force and thereby points toward the magnetic poles.*

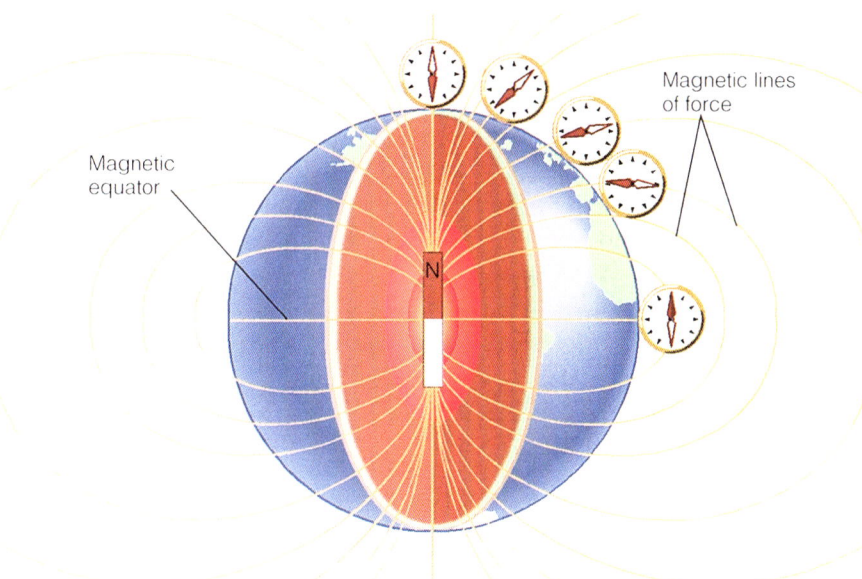

FIGURE 2.5 *The needle of the dipmeter also aligns parallel to Earth's magnetic lines of force, but in a vertical plane. Note that at the magnetic equator, the needle is parallel to Earth's surface (no dip angle). At locations closer to the magnetic poles, the angle between the needle and Earth's surface increases with increasing latitude until, over the magnetic pole, the needle is vertical (90° angle of dip).*

In this way, the magnetite crystals in the basalt record not only the orientation of the magnetic field but also the direction and distance to the magnetic poles. Both of these orientations can be determined in an oriented rock specimen with a sensitive instrument called a *magnetometer*. Because basaltic lavas are the most abundant kind of lava extruded onto Earth's surface, a large database of magnetic data can readily be amassed on any continent.

Magnetic Reversals

In establishing the magnetic orientation in successive lava flows (Figure 2.6), early paleomagnetists (workers who investigate rock magnetism) found that Earth's magnetic field had been **reversed** at times; that is, the north magnetic pole became the south magnetic pole and vice versa. The investigators found that the periods of reversed magnetism lasted a few hundred thousand years or more in duration and that the time interval between any two reversals was essentially random.

To this day, we do not understand the exact process by which the reversals take place. However, the fact that they *do* take place and are recorded simultaneously in the basaltic lavas around Earth was to play an important role in proving the existence of plate movement.

FIGURE 2.6 Stacked lava flows, *such as those that characterize the Columbia Plateau, record the orientation of Earth's magnetic field at the time of solidification and allow geologists to determine changes in the magnetic field over time. (Courtesy of P. Weis/USGS)*

Polar Wandering

Another important discovery made by researchers in rock magnetism was "**polar wandering.**" It is generally accepted that Earth's magnetic field is the result of movements within Earth's molten iron core and that these movements are generated by Earth's rotation. We also know that the geographic locations of the present north and south magnetic poles are not constant. Although the magnetic poles do migrate over geologic time, they have always been located near the rotational poles. As far as we know, the magnetic reversals are not related to the movements within the liquid portion of the core thought to be responsible for Earth's magnetic field.

As worldwide paleomagnetic data began to accumulate, researchers began to plot the locations of Earth's north magnetic pole over time for the individual continents. The maps showed that the north magnetic pole for each continent had followed a *different* pathway (Figure 2.7). Note that if the continents had remained fixed in location while the magnetic poles wandered, each continent would have shown the *same* polar track. On the other hand, if the locations of the magnetic poles had remained relatively fixed over time, the different polar pathways shown by the data from the individual continents could only be due to the movement of the continents. The conclusion was that most of the observed polar wandering was the result of continental movements. The "wandering" of the magnetic poles was only an *apparent* movement; most of the movement was actually due to changes in the relative positions of the continents. Interest in continental drift was suddenly revived.

SPOT REVIEW

1. How do basaltic rocks record the magnetic field that exists at the time of their formation?
2. How can a compass and a dipmeter be used to locate the position of the north magnetic pole?
3. How can magnetic data from different continents indicating different tracks of polar movement be used as evidence that the continents moved and not the magnetic poles?

Topography of the Ocean Bottom

Beginning with military investigations of the 1940s and continuing throughout the 1950s and 1960s, studies of the ocean bottom began to reveal unexpected features. Although limited **bathymetry** (Greek *bathus* = deep + metry) (topography of the ocean bottom) of the Atlantic Ocean was known by the late 1800s, until the 1950s, many geologists had taught, for no good reason, that the ocean bottom was a broad featureless plain extending from continent to continent. With the exception of a few volcanic peaks that mysteriously rose from the ocean floor, the ocean bottom was viewed as being flat as a billiard table. That picture was soon to change. As a result of echo-sounding studies originally begun by the U.S. Navy during World War II, an entirely different picture of the ocean bottom evolved. The data showed that the ocean bottom was hardly a flat, featureless surface (Figure 2.8).

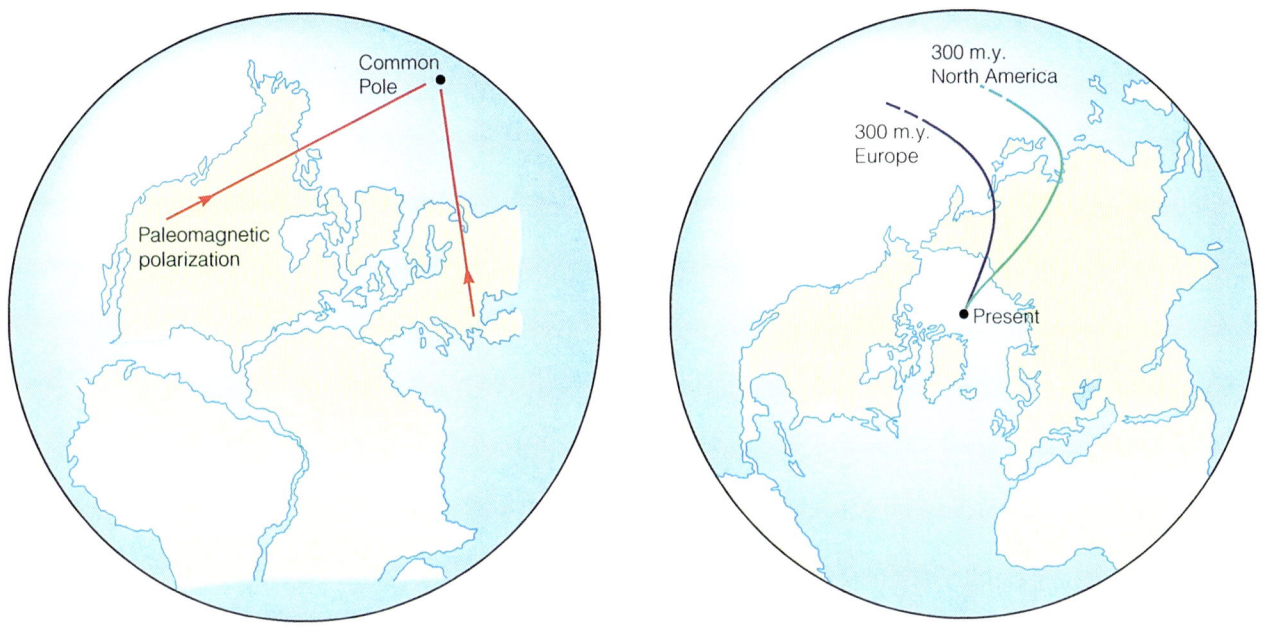

FIGURE 2.7 *Paleomagnetic data for individual continents indicate that the magnetic pole has had different locations over time. The patterns of apparent movement differ for individual continents. Possible explanations include the following: (1) the continents remained fixed in position as the magnetic poles moved, (2) the magnetic poles remained fixed in position as the continents moved, or (3) both the magnetic poles and the continents moved with time. Because the magnetic pole is believed to have always been located close to the geographic pole, the second option is the most likely.*

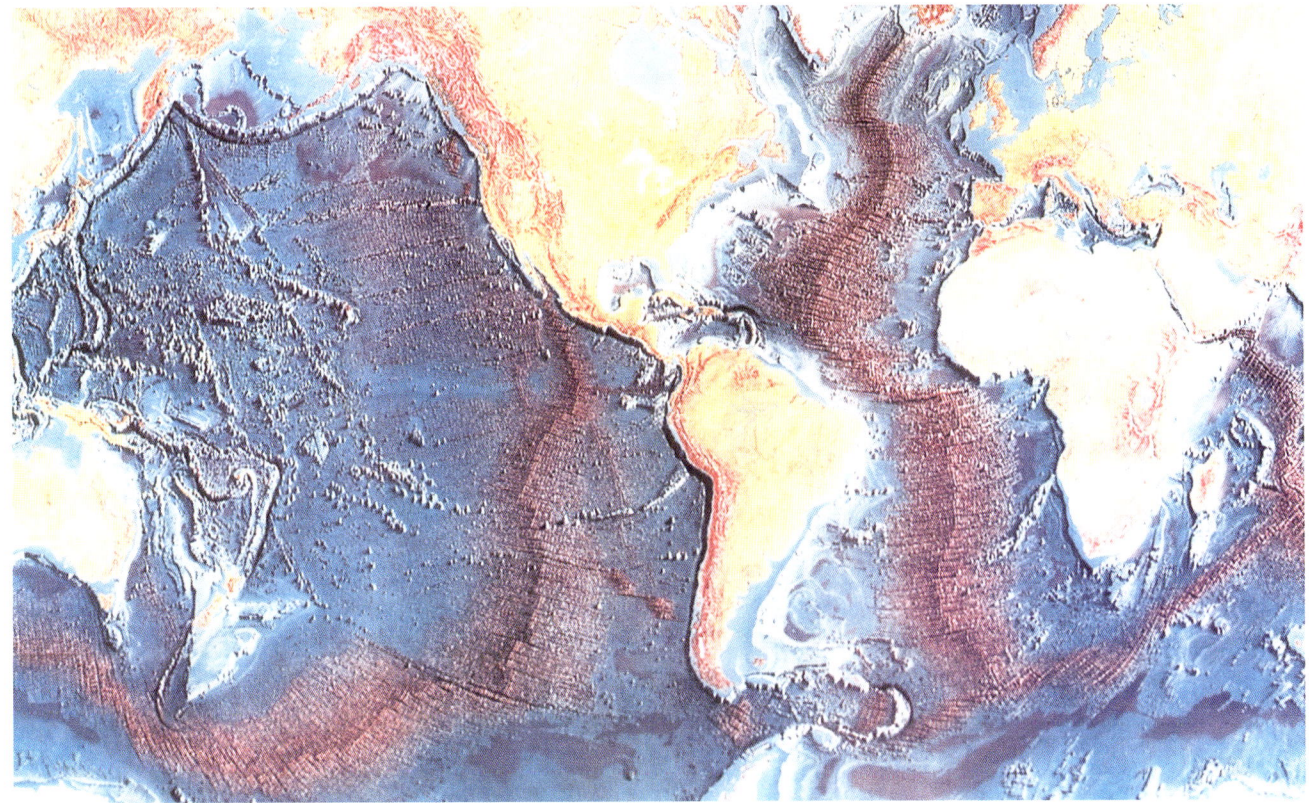

FIGURE 2.8 *Data largely amassed since the invention of echo sounding have revealed a varied landscape for the ocean bottom. (World Ocean Floor by Bruce C. Heezen and Marie Tharp, 1977 © Marie Tharp 1977. Reproduced by permission of Marie Tharp, 1 Washington Avenue, South Nyack, NY 10960.)*

Oceanic Ridges

The most impressive topographic features discovered by echo sounding were volcanic mountain ranges that ran the length of each ocean basin (Figure 2.9). The first **oceanic ridge** to be discovered was the one running down the middle of the Atlantic Ocean. Researchers soon learned that all oceans had oceanic ridges and, furthermore, that all the ridges were connected from one ocean to the next. As indicated in Chapter 1, the oceanic ridges are submarine; that is, with a few exceptions, notably the island of Iceland, they do not extend above the surface of the ocean.

Deep-Sea Trenches

Other major topographic features found on the ocean bottom are the **deep-sea trenches,** very deep, narrow troughs that trend parallel to some continental margins and chains of volcanic islands (Figure 2.10). Most of Earth's trenches are found around the margin of the Pacific Ocean. Invariably, the trenches are associated with a volcanic chain, either on the adjoining continental margin, such as the Andes Mountains along the western margin of South America or the Cascade Mountains of the Pacific Northwest, or just offshore of the continental margin, such as the Japanese Islands

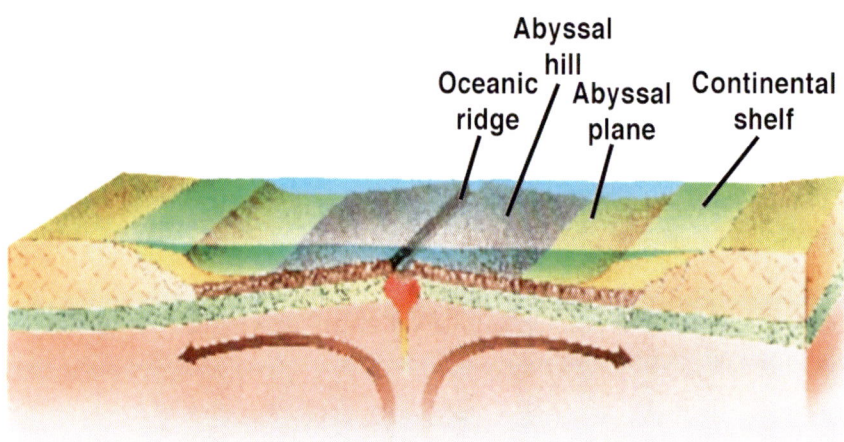

FIGURE 2.9 *An ocean basin begins with the formation of a continental rift zone and develops through rift valleys and linear oceans to the point where the landmasses finally separate completely to form the ocean basin.*

FIGURE 2.10 *A deep-sea trench forms at a convergent plate boundary where the accretionary wedge at the edge of the overriding plate encounters the downwarping oceanic plate.*

off the eastern margin of the Asian mainland. In the late 1950s and early 1960s, no one had an acceptable explanation for these trenches and their associated volcanism, at least not an explanation that met with the approval of the majority of the geological community.

Mantle Convection Cells

In 1960, *Harry H. Hess,* proposed a mechanism to explain both the oceanic ridges and the deep-sea trenches. Hess resurrected a hypothesis that had first been presented by *Arthur Holmes* in 1928. Holmes had proposed that convection cells in the mantle were the source of energy that drives the process of continental drift. According to Holmes, the convection cells in the mantle were driven by heat flow from Earth's interior much as currents are generated in a pot of boiling water (Figure 2.11). He postulated that as heat was conducted outward from Earth's center, convection cells caused a slow movement in the rocks within the mantle. He further postulated that if a convection cell became positioned beneath a continent, the lateral movement of the underlying mantle rocks would break the continent into fragments and drive them apart (Figure 2.12). Considering how little information Holmes had about the inner structure of Earth, his suggestion is amazingly close to the most popular mechanism that we now have.

Hess proposed that the oceanic ridges represented the rising portion of Holme's convection cell. He hypothesized that, as the mantle rock moved upward and outward beneath a continent, the continent would break apart under tensional forces and separate, just as Holmes had envisioned. As the two continental masses drifted apart, a rift zone would form and, in time, develop into a rift valley that would eventually open to the sea. The valley would then begin to flood, and as the continental masses continued to separate, a new ocean basin would be born (Figure 2.13). According to Hess's hypothesis, basaltic magmas welling up along fractures would continuously create new crust for the new, opening ocean basin. Once solidified, the newly formed oceanic crust moves laterally away from the oceanic ridge, and the continents continue to recede, a process called **sea-floor spreading.**

Hess further reasoned that with new oceanic rock continuously being created at the oceanic ridges, one of two events had to occur: (1) Earth had to expand to accommodate the volume of new oceanic crust, or (2) somewhere, an equal volume of old oceanic crust had to be consumed. He considered the idea of an expanding Earth, but dropped that possibility because there

FIGURE 2.11 *The convection cells within the mantle are commonly modeled by the movement of water in a heated pan. As the water is heated at the bottom of the pan, the water expands and its density decreases. The more buoyant hot water rises over the heat source and spreads out along the surface of the water where it cools. As the water cools and its density increases, the water sinks along the sides of the pan to complete the convection cell.*

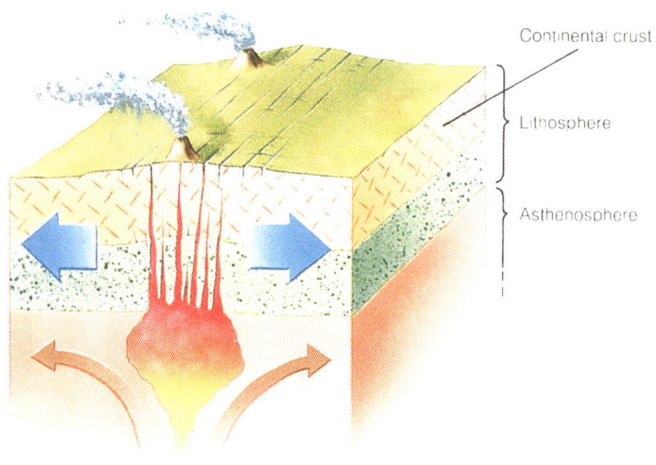

FIGURE 2.12 *Tensional forces that develop in the lithosphere above the rising portion of a mantle plume cause the continental lithosphere to fracture or rift. Molten rock forms at the top of the asthenosphere and moves upward along the fractures to produce the volcanism that characterizes a rift zone.*

was no evidence to indicate that Earth had been much smaller in the past and had expanded to the extent necessary to accommodate all the oceanic crust generated over geologic time. The most obvious candidate for a location where oceanic crust was likely to be consumed was the deep-sea trench. Hess proposed that the trenches are located over the

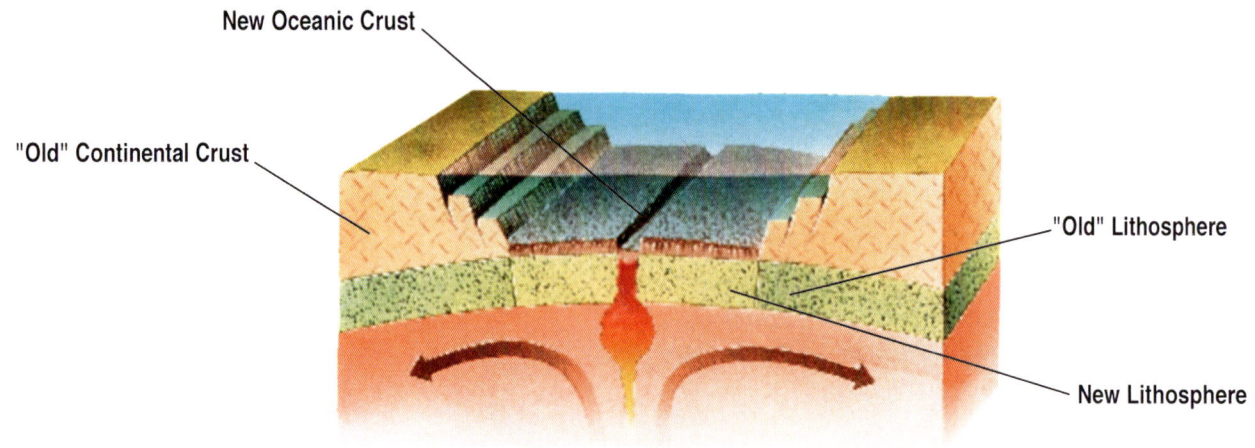

FIGURE 2.13 *Oceans are born as a flooded rift valley completely traverses the continent, finally separating the original landmass into two smaller landmasses.*

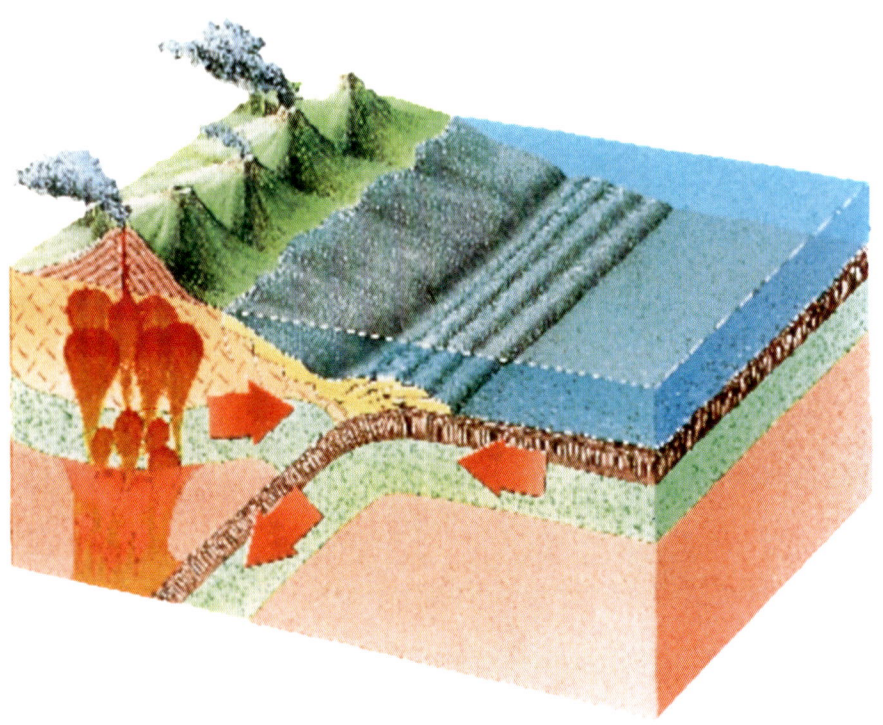

FIGURE 2.14 *A deep-sea trench forms when a zone of subduction develops at converging plates.*

downward-moving portion of a mantle convection cell (Figure 2.14). As the crust is compressed by the movement of the mantle rocks and breaks, the oceanic crust sinks into the mantle and is assimilated. Hess therefore hypothesized that a balance exists between the creation of new oceanic crust at the oceanic ridges and the destruction of old oceanic crust at the trenches. According to this idea, the continents are carried as passive passengers as the oceanic crust moves from the oceanic ridges to the trenches. Could this be a mechanism that Wegener had so desperately sought? We now know that it is.

SPOT REVIEW

Why must there be a balance between the volume of crustal rocks formed at the oceanic ridges and the volume of crustal rocks consumed at the trenches?

Magnetic Zonation of the Oceanic Crust Within a few years, other geologists published research that confirmed Hess's hypothesis. In 1963, *F. Vine* and *D. H. Mattews* published a paper describing their finding of symmetrical bands of varying width and magnetic intensity in the rocks of the ocean floor (Figure 2.15).

L. Morely published a similar, but independently developed, explanation of the magnetic striping on the ocean floor. Although submitted for publication earlier than Vine and Matthew's paper, Morely's paper had at first been rejected as too speculative. Some now suggest that the concept should be referred to as the *Vine-Matthews-Morely* hypothesis.

In an attempt to explain what they had found, Vine and Matthews utilized Hess's idea of a spreading center along the oceanic ridge. Consider the drawing in Figure 2.16. Basaltic magma wells up through fractures parallel to the axis of the oceanic ridge. We have already discussed how the magnetite grains record the existing magnetic field of the Earth as the molten rock solidifies. Once formed, the rocks split and move laterally in opposite directions to make way for new basaltic magma. In time, an everwidening band of basalt forms, recording both the orientation of the existing magnetic field and the direction to the north magnetic pole.

Consider what would happen if a magnetic reversal were to occur. The lavas welling up along the oceanic ridge would record the magnetic reversal by reversing the poles of the magnetite grains. In other words, the magnetic fields of the magnetite grains would flip end for end. As long as the magnetic reversal remained in effect, newly formed basaltic rocks would continue to record the magnetic field as a new band parallel to the previously generated band. Subsequent reversals would be recorded in similar fashion. The result would be a set of symmetrical bands with the *width* of each band reflecting the *duration* of the respective reversal episode and the *spreading rate* of the oceanic crust.

If magnetic measurements were made in a direction *perpendicular* to the axis of the oceanic ridge, they would show a variation in the **magnetic intensity** of the rocks. The magnetic intensity of a rock depends primarily on the strength of the original magnetic field and the magnetite content of the rock. The maximum magnetic intensity would be recorded for rocks where the popular orientation when the lava cooled was the *same* as the polar orientation at the time of measurement, that is, when the directions of the north magnetic pole were the same. The minimum magnetic intensity would occur when, because of magnetic reversals, polar orientations with the rocks were *opposite* to the polarity of the magnetic field at the time of the measurement (Figure 2.17). Note that the patterns of the magnetic bands on opposite sides of the oceanic ridge would be symmetrical. After the presentation of the Vine and Matthews paper, researchers found that the magnetic bands associated with the Atlantic Oceanic Ridge showed the predicted symmetrical arrangement, a discovery that was strong evidence in support of continental drift.

Soon researchers showed that the patterns of magnetic intensity in bands from different ocean basins were similar; the bands differed only in width, reflecting different rates of sea-floor spreading. Analysis of the bands showed that the Atlantic and Indian oceans are spreading at rates of 0.5 to 1.5 inches (1–4 cm) per year while the Pacific Ocean is spreading at about 4 inches (10 cm) per year.

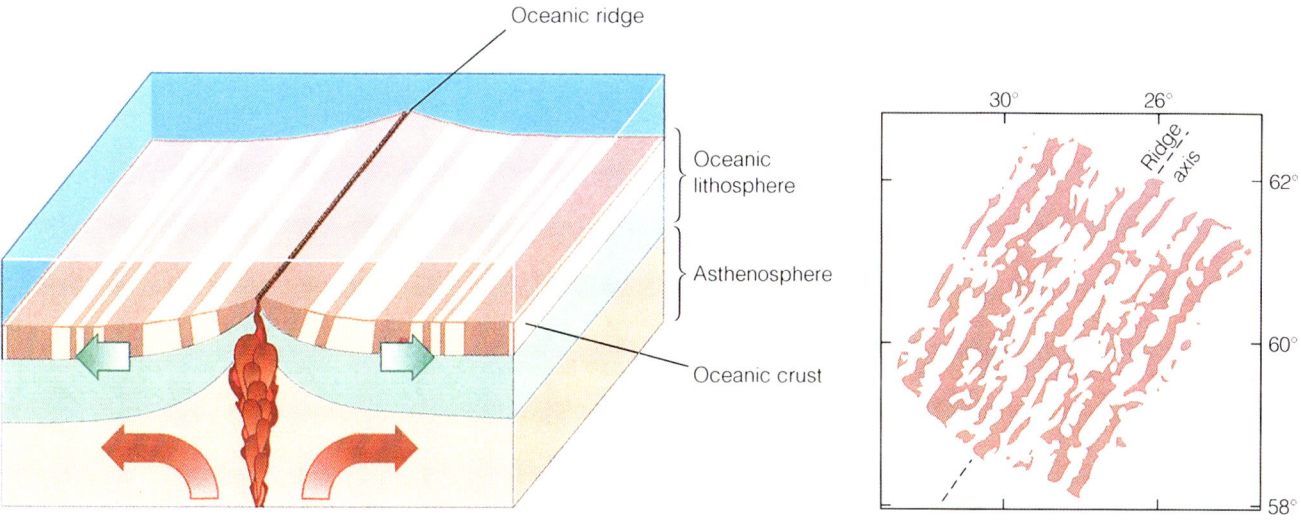

FIGURE 2.15   *As newly formed oceanic lithosphere moves away from an oceanic ridge in opposite directions, magnetic reversals are recorded in the oceanic basalts in the form of symmetrically disposed pairs of stripes of varying magnetic intensity oriented parallel to the trend of the ridge.*

SPOT REVIEW

Why do the rocks of the oceanic crust show parallel bands of reversed magnetism? Why do the widths of the bands differ, and why are they symmetrical on opposite sides of an oceanic ridge?

Seismic Investigation of Earth's Interior

With the magnetic data clearly indicating that the oceanic crust is moving at a measurable rate, geologists began to look for evidence to support the convection cell theory. In the 1960s, seismic (earthquake) data began to provide a new picture of Earth's internal structure that was somewhat more

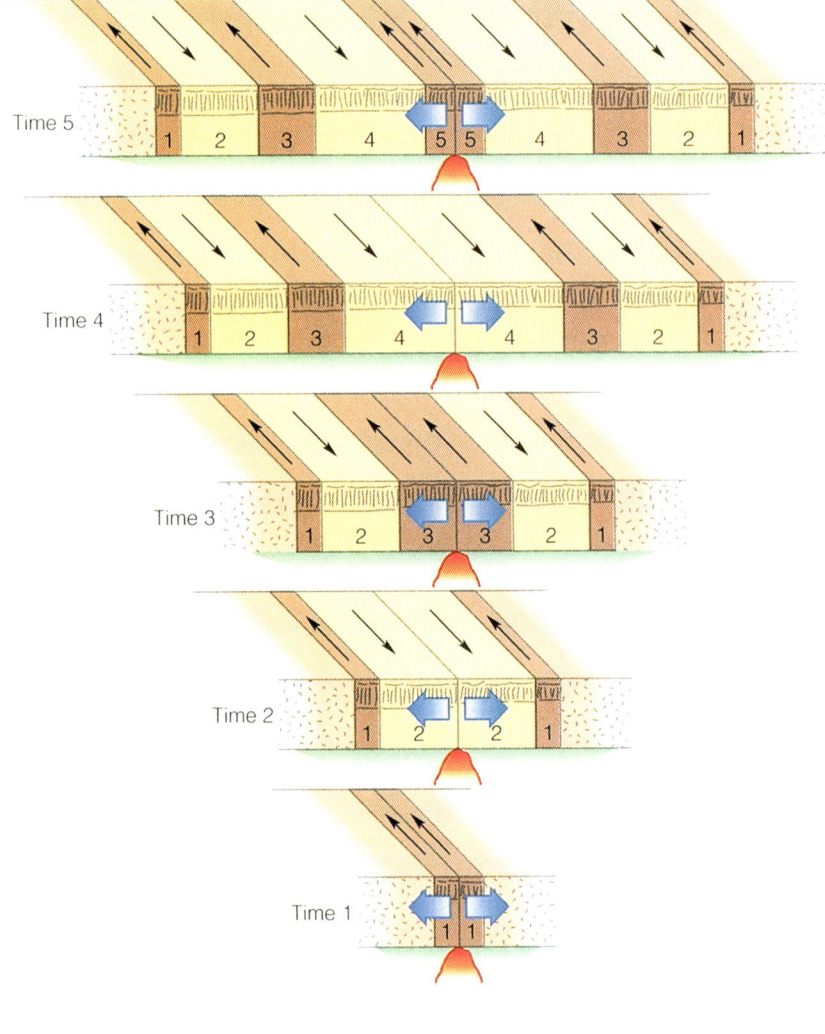

FIGURE 2.16 *The width of each pair of stripes depends on the length of time from one magnetic reversal to the next and on the rate of formation of new oceanic lithosphere.*

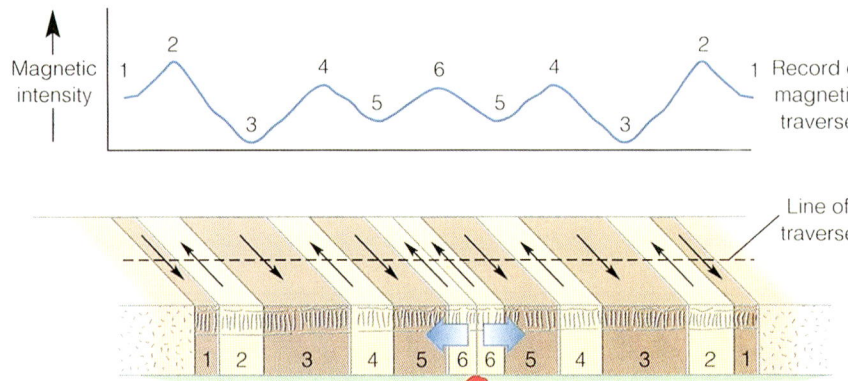

FIGURE 2.17 *A traverse perpendicular to the trend of the oceanic ridge shows a systematic variation in the magnetic intensity of the magnetic stripes.*

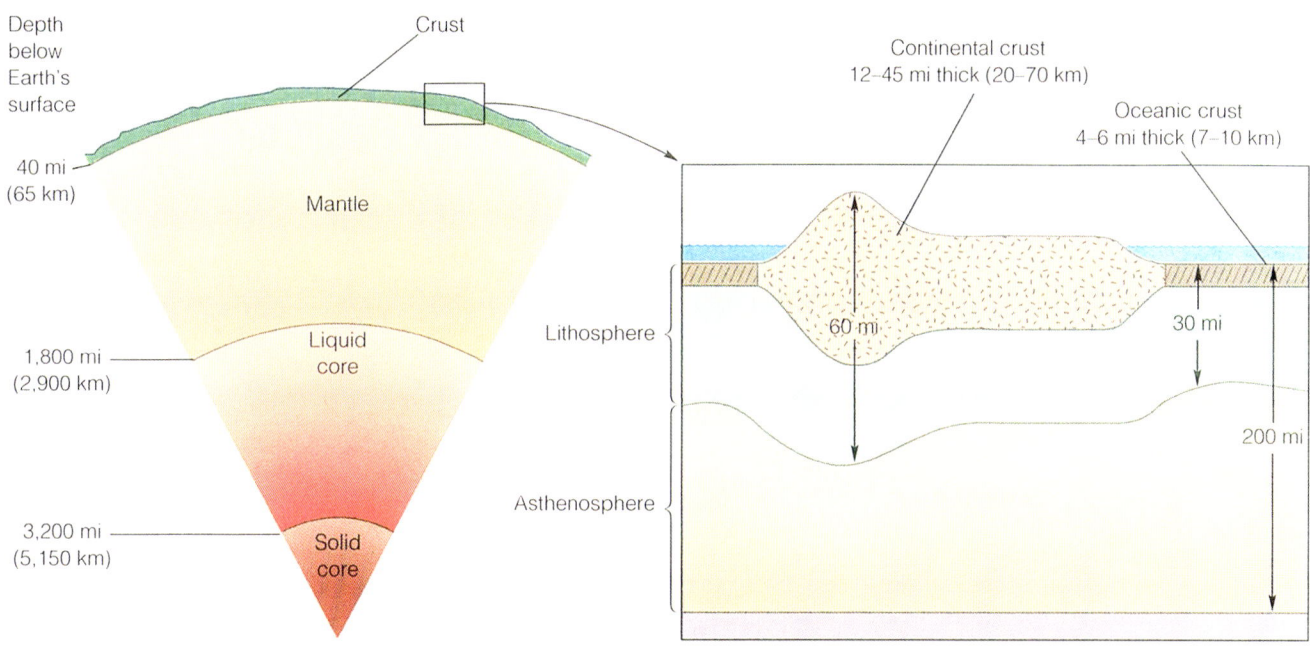

FIGURE 2.18 *Earthquake data have shown that the outer portion of Earth is divided into a brittle layer called the* lithosphere, *made up of the crust and the outermost part of the mantle, and an underlying layer called the* asthenosphere *within which the rocks show liquidlike properties.*

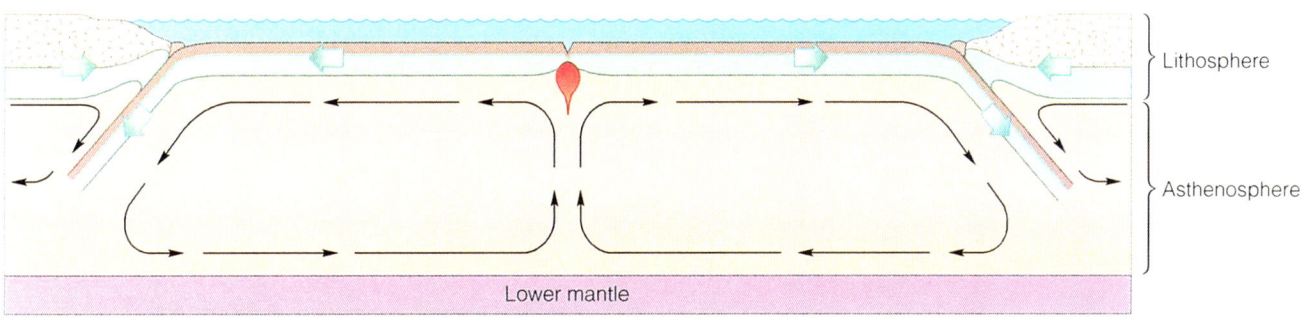

FIGURE 2.19 *Some geologists believe that the convection cells associated with plate tectonics are located within the asthenosphere.*

complex than the older core-mantle-crust model. Specifically, seismologists discovered that the combined oceanic and continental crust and the outermost portion of the mantle acted together as a single brittle layer they called the **lithosphere** (Figure 2.18). Note that the lithosphere varies in thickness. It is thickest where it contains the thick granitic continental crust.

Perhaps the most significant discovery made by the seismologists was a zone below the lithosphere through which earthquake waves move at significantly slower velocities. They interpreted the slowing of the earthquake waves as indicating that the layer was *plastic;* a plastic material is a solid that deforms by *flowing* like the ice at the bottom of a moving glacier. With this discovery, evidence had finally been found to support the idea first presented by Holmes in 1928 that rocks within the mantle could move as convection cells. The plastic layer within the mantle is called the **asthenosphere.**

Although most geologists agree that the driving force for the plate movements is mantle convection, there is no consensus on the convection mechanism. One model suggests that the convection cells are located within the asthenosphere (Figure 2.19). According to this model, the rising portion of a convection cell induces tensional forces in the overlying lithosphere. The tensional forces cause the lithosphere to break along fractures that eventually extend to the surface, producing **rift zones** and **rift valleys** that may eventually open to form ocean basins (Figure 2.20). Molten rock form the asthenosphere

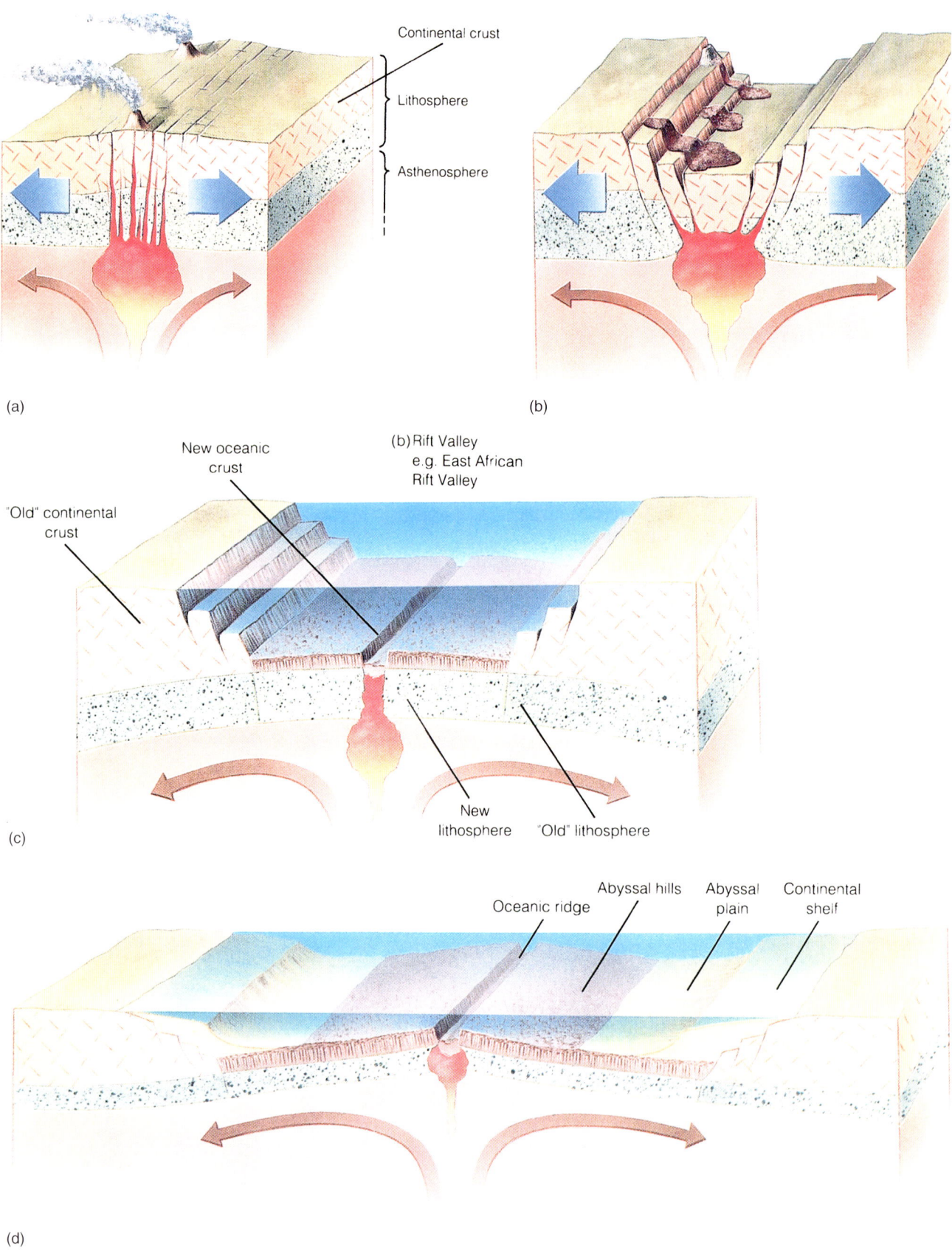

FIGURE 2.20 *Some geologists believe that the movement of the asthenospheric rocks* away *from the top of the* rising *portion of the mantle convection cell generating the tensional forces within the overlying lithosphere that cause it to break into plates, which then move away from each other.*

fills the fractures, solidifies, and adds to the lithosphere, which moves off in opposite directions from the zone of rifting. According to this model, the major force causing the lateral motion of the lithosphere away from the zone of rifting, called **mantle drag**, is generated by the movement of the asthenospheric rocks beneath the lithosphere. With time, the rocks of the asthenosphere cool, increase in density, and descend to form the downward-moving portion of the convection cell. At this point, forces generated by the downward-moving portion of the convection cell break the overlying lithosphere parallel to the continental margin. The portion of the lithosphere bearing the higher-density oceanic plate is driven downward beneath the edge of the lighter continent by a combination of compressional forces and descending mantle drag (Figure 2.21). The downward-moving portion of the convection cell drags the edge of the lithosphere into the asthenosphere where it is assimilated. The location of the downward-moving lithosphere is marked on the ocean floor by the presence of a deep-sea trench. The volume of lithosphere converted to asthenosphere at the descending portion of the convection cell compensates for the volume of asthenosphere converted to lithosphere at the rising portion of the convection cell. In this way, constant volumes of asthenospheric and lithospheric rocks are maintained. To summarize, summary, in this model, the lithosphere is passively carried or dragged along by the convective movement of the underlying asthenosphere with the major force being supplied by mantle drag.

Others disagree with this model, arguing that the degree of cohesion between the asthenosphere and the lithosphere is insufficient to drag the lithosphere laterally. These geologists maintain that the lithosphere is actually the uppermost, cooling portion of the convection cell and that the convection cells are not limited to the asthenosphere but involve a large portion of the mantle (Figure 2.22). According to this scenario, as the heated mantle rocks rise, they undergo changes in both composition and physical

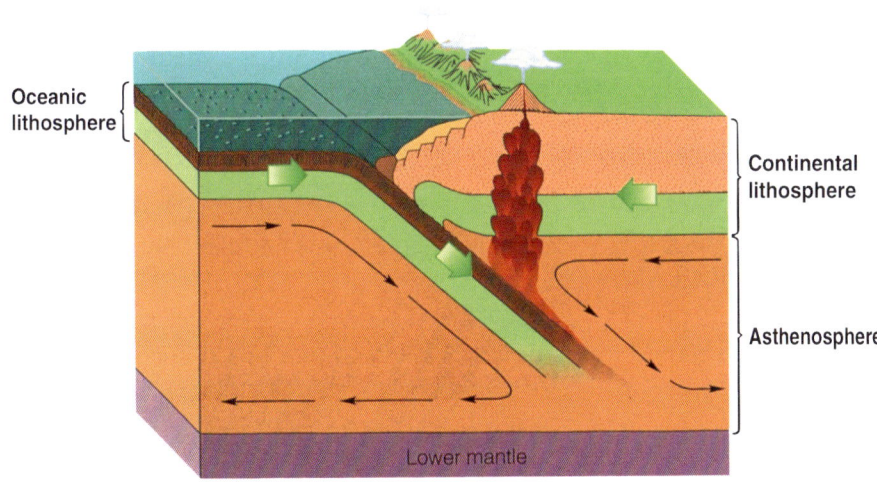

FIGURE 2.21 *The forces generated at convergent plate margins that result in the formation of a zone of subduction are associated with the* descending *portion of the mantle convection cell.*

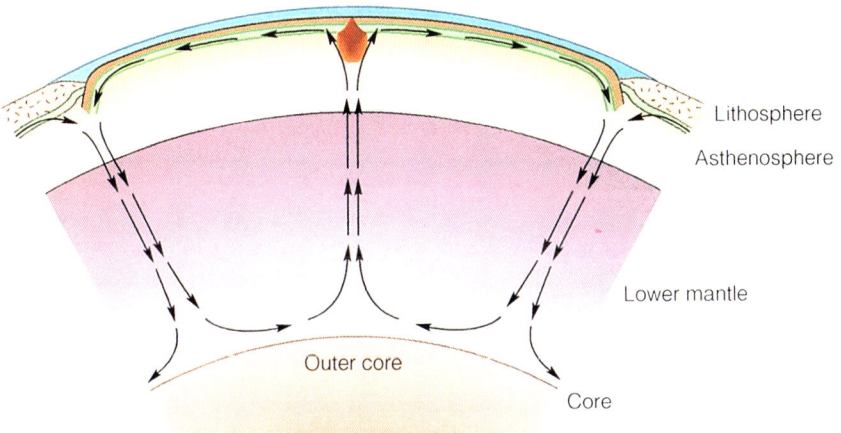

FIGURE 2.22 *Some geologists believe that the convection cells associated with plate tectonics are not restricted to the asthenosphere but rather involve the entire mantle. In this model, the lithosphere is the uppermost portion of the cell. Rather than simply responding to the movement of underlying rocks, the lateral movement of the lithosphere is the result of a combination of forces, which are produced as buoyant, molten rock is injected at the oceanic ridge and the cooled lithosphere sinks at the zone of subduction.*

properties as the conditions of heat and pressure change. Beneath rift zones, rift valleys, and oceanic spreading centers, the dense rocks of the asthenosphere melt to form the lower-density lithospheric rocks. As the new lithospheric rocks form and move laterally, tensional forces are generated that cause fractures to develop within the newly formed lithosphere. Molten rock moving up from below continuously fills the fractures and adds to the lithosphere. The elevated mass of the new lithospheric rocks formed at the rift zone, rift valley, or oceanic ridge exerts a gravitational force, called **ridge push,** on the older, adjacent lithosphere that causes the lithosphere to move laterally over the underlying plastic asthenosphere. In this model, mantle drag resulting from the movement of the underlying asthenospheric rocks may aid in the lateral movement of the lithosphere, but this force is not thought to be a major source of energy for plate movement.

The buoyancy of the hot, lower-density rocks rising beneath the oceanic ridge lifts the ridge higher than the adjacent cooler, denser oceanic lithosphere (Figure 2.23). Because of this difference in elevation, some geologists have suggested that both the lateral motion of the lithosphere and ridge push may result, at least in part, from *gravitational sliding* of the more rigid lithosphere down and away from the ridge over the more plastic asthenosphere.

Eventually, as the spreading oceanic lithosphere cools and increases in density, it breaks and begins to sink, generating a force called **slab pull** that draws the lithosphere downward into the mantle. Proponents of this model consider slab pull the most important of the three sources of energy involved in the movement of the lithosphere and regard mantle drag and ridge push as less significant. A major difference between this model and the model discussed previously is that here the lithosphere is part of the convection cell and both ridge push and slab pull *contribute* to the convective option, whereas in the first model the convection cell is restricted to the asthenosphere and mantle drag is the primary force.

The most recent data generated by three-dimensional seismic **tomography**, a seismic technique comparable to three-dimensional X-ray scans (CAT scans) of the human body, suggest that movements within the mantle are far more complex than any of the existing models portray (Figure 2.24). We have known about plate tectonics for fewer than 25 years, and it is apparent that much remains to be learned about Earth's interior.

SPOT REVIEW

1. How is convection within the mantle thought to be involved in the movement of the lithospheric plates?
2. What are the various mechanisms that have been proposed to explain how mantle convection results in the movement of the plates?

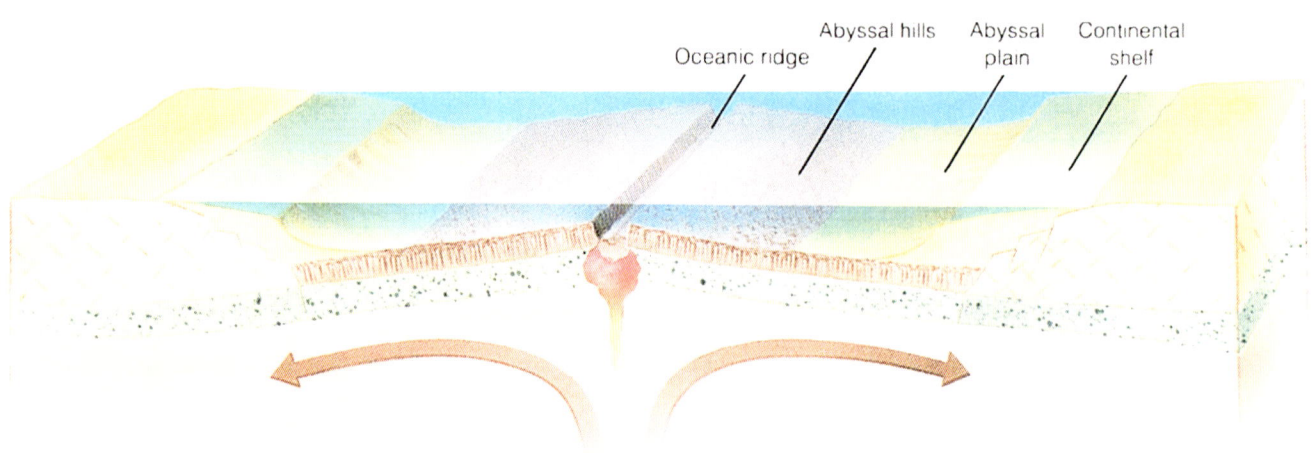

FIGURE 2.23 *The buoyancy of the molten rock that forms at the top of the asthenosphere results in an elevation of the ocean bottom along the trend of the oceanic ridge.*

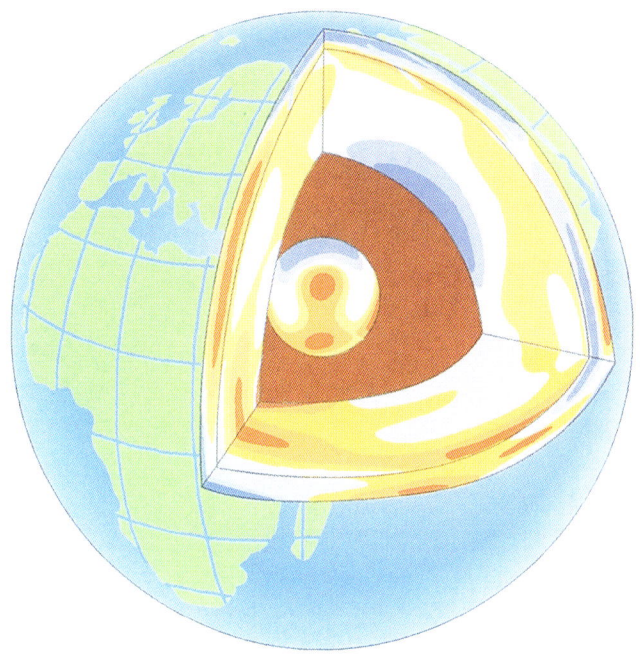

FIGURE 2.24 *The most recent information produced by a three-dimensional seismic technique called* tomography *reveals that Earth's interior is far more complex than the simplistic crust-mantle-core model would indicate. Such data will play an increasingly important role as our understanding of the driving forces behind plate tectonics evolves over coming decades. (Andrea Morelli et al. 1986, Geophys. Res. Lett. 13, 1545–1548).*

THE PRINCIPLE OF PLATE TECTONICS

A comprehensive principle to explain all that had been observed was the result of the combined efforts of four geologists, *D. P. McKenzie* and *R. L. Parker* of Cambridge University, *W. J. Morgan* of Princeton University, and *X. LePichon* of the Lamont Geological Observatory. Their theory, which became known as **plate tectonics**, views Earth's surface in a new and different way.

Earlier workers such as Wegener had regarded the continents and the ocean basins as the fundamental structural units of Earth's crust. They attempted to explain how the continents could first break up and then move about in this "sea" of oceanic basalt. McKenzie and his co-workers considered the possibility that the continental and oceanic crusts are not basic structural units, but instead are actually components of the lithosphere. They then proposed that the lithosphere is broken into about a dozen pieces, which they called **plates,** that fit together like a huge jigsaw puzzle (Figure 2.25). According to their theory, the plates rather than the continental and oceanic crusts are the basic structural units. They theorized further that the plates are moving relative to each other (Figure 2.26). As a result of this movement, certain

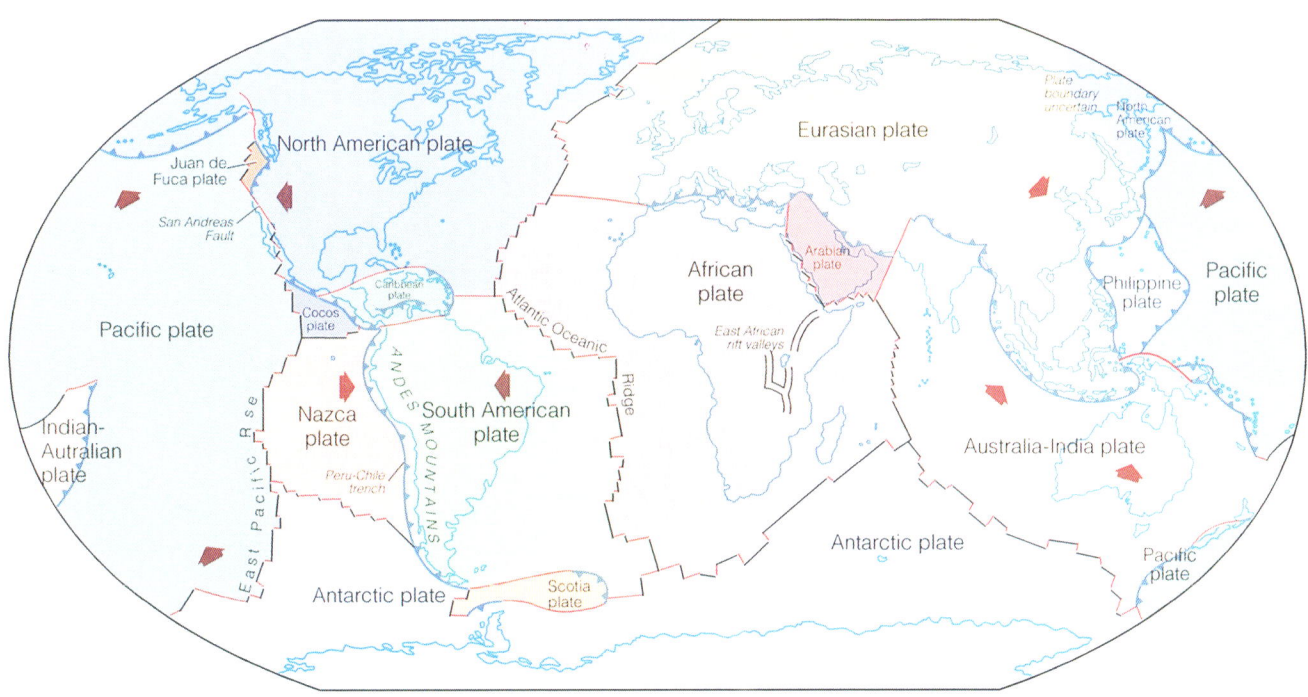

FIGURE 2.25 *The lithosphere is broken into about a dozen plates. The number of plates changes with time as large plates break apart and smaller plates collide and weld together. Sawtooth lines indicate the convergent boundary; the double color lines indicate divergent boundaries; and transform boundaries are shown with single color lines.*

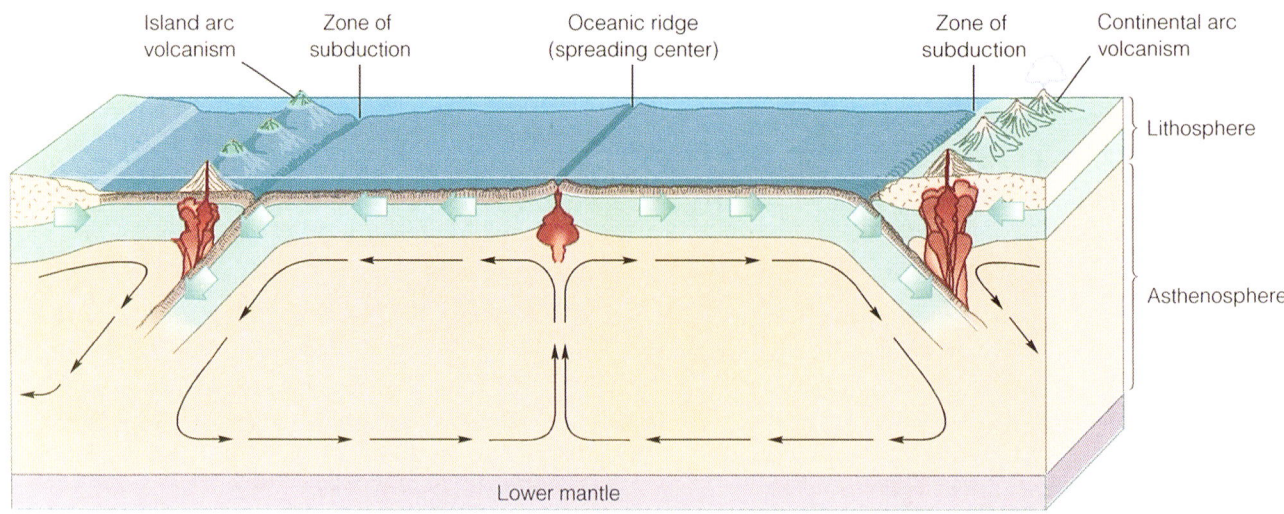

FIGURE 2.26 *We now know that oceans are born, open, and close, and that the continents move as continental crust is carried as the passive component of a lithospheric plate. Although most geologists agree that the driving force behind plate tectonics is mantle convection, the depths from which the convection cells rise is still a matter of debate.*

plates are moving away from each other. The margins of these plates, called **divergent margins** or **spreading centers,** are the sites of the rift zones, rift valleys, and the oceanic ridges where new oceanic lithosphere is being formed. At other margins, called **convergent margins** or **zones of subduction,** the plates are moving toward each other. Along these margins, the denser oceanic plate is driven or pulled downward beneath the lower-density continental plate or the edge of the opposing oceanic plate. These plate margins are the locations of the deep-sea trenches. Volcanic mountains located either on land, such as the Andes and the Cascades, or just seaward of the continental margins, such as the Japanese Islands and the Aleutians, mark the convergent boundaries.

In order to allow the plates to move along the surface of a sphere, other fractures called **transform faults** develop perpendicular to the linear trend of the oceanic ridges (Figure 2.27). The plates move laterally along the transform faults. No volcanism is associated with the transform boundaries, but later on we will see that the transform faults are the locale of shallow earthquakes. When we study earthquakes in Chapter 18, we will see that, with few exceptions, the locations of most of Earth's earthquakes coincide with the plate margins and that the most powerful earthquakes are concentrated along the zones of subduction.

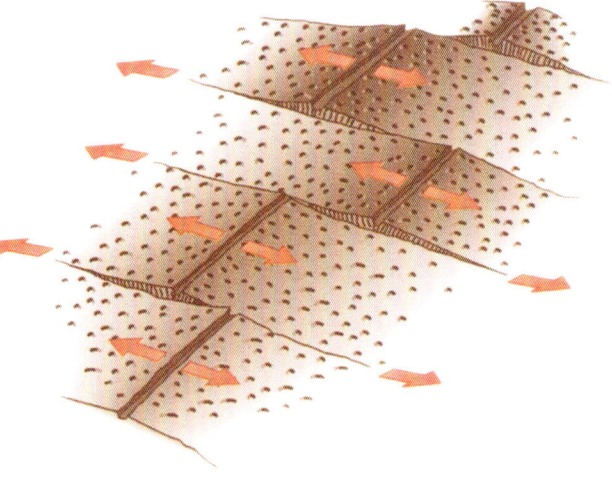

FIGURE 2.27 *Fractures called* transform faults, *which develop perpendicular to the trends of the oceanic ridges, allow the plates to move across the spherical surface of Earth.*

The introduction of plate tectonics in the late 1960s revolutionized the way we view Earth. Many geologic processes that had defied explanation now became understandable. For many years, the coincidence in the distribution of volcanoes and earthquakes had given rise to many a heated discussion over which was the cause and which was the effect.

Some had maintained that "quite obviously" the direct association of these phenomena indicated that either the onset of a volcanic eruption was initiated by an earthquake or that volcanic eruptions caused earthquakes. Although the rise of molten rock to Earth's surface does initiate earthquakes, such earthquakes are usually low in magnitude and are not the earthquakes that cause death and destruction. Plate tectonics finally explained why these two geologic phenomena are so intimately associated. Rather than a cause and effect relationship, earthquakes and volcanoes are related simply because they are associated with the same feature, namely, plate margins.

Now we also understand how the continents "*wander*." We know that they do not plow through the oceanic crust as Wegener had imagined. Instead, the continents are passive passengers, transported as part of a moving lithospheric plate.

CONCEPTS AND TERMS TO REMEMBER

Gondwana
Pangaea
Rock magnetism
  Curie point
  magnetic inclination
  magnetic reversals
  polar wandering
Topography of the ocean floor
  bathymetry
  oceanic ridges

deep-sea trenches
sea-floor spreading
Magnetism of the oceanic crust
  magnetic intensity
Seismological data
  lithosphere
  asthenosphere
  rift zone
  rift valley
  driving mechanisms
    mantle drag

ridge push
slab pull
tomography
Principle of plate tectonics
  plates
  divergent plate margins
  spreading centers
  convergent plate margins
  zones of subduction
  transform faults

REVIEW QUESTIONS

1. Ocean basins open at the rate of a few centimeters per
   a. year.
   b. century.
   c. 1,000 years.
   d. 1 million years.
2. Which of the following Earth features is an "ocean in the making"?
   a. the Grand Canyon
   b. the East African Rift Valley
   c. the Mediterranean Sea
   d. the Great Lakes
3. Which of the following landmasses is an exposed portion of a mid-oceanic ridge?
   a. Hawaii
   b. the Aleutian Islands
   c. Iceland
   d. Greenland
4. Oceanic trenches are associated with
   a. convergent plate margins.
   b. divergent plate margins.
   c. transform plate margins.
   d. both convergent and divergent plate margins.
5. The convection cells that drive the plates are thought to be located within the
   a. lithosphere.
   b. continental crust.
   c. oceanic crust.
   d. asthenosphere.
6. The most distinctive characteristic of the asthenosphere is its
   a. thickness.
   b. composition.
   c. lack of rigidity.
   d. content of molten rock.

REVIEW QUESTIONS continued

7. Why are basaltic rocks primarily involved in recording Earth's magnetic field?
8. How are the magnetic data recorded in the rocks of the ocean floor used to determine the rates of plate movement?
9. In what way is the asthenosphere involved in the movement of the plates?
10. What is the basic difference between the plate boundaries of the eastern and western Pacific Ocean basin?
11. How do you explain that while continental rocks more than 3 billion years old have been found, no oceanic rocks older than about 250 million years have been found?
12. Explain why basaltic rocks of the same age on different continents indicate different locations for the magnetic poles.
13. How can rocks of the asthenosphere constantly be converted into lithospheric rocks without depleting the asthenosphere?

THOUGHT PROBLEMS

1. Much of the data used to prove that the continents move relative to each other came from studies of the Earth's magnetic field, in particular, the phenomenon known as "polar wandering." A hypothetical polar wandering diagram is shown in Figure 2.28. The drawing illustrates the "apparent" wandering of the north magnetic pole based on magnetic data locked into the rocks of the continent at 10 different times (t_1 through t_{10}). How are magnetic data used to determine the direction and distance to the north magnetic pole at any time t? The drawing assumes a stationary continent and a moving pole. Prepare a second drawing to illustrate how the same magnetic data can be used to prove continental wandering. (*Hint:* Assume a stationary pole and a moving continent. Note that in the original drawing, the direction to the

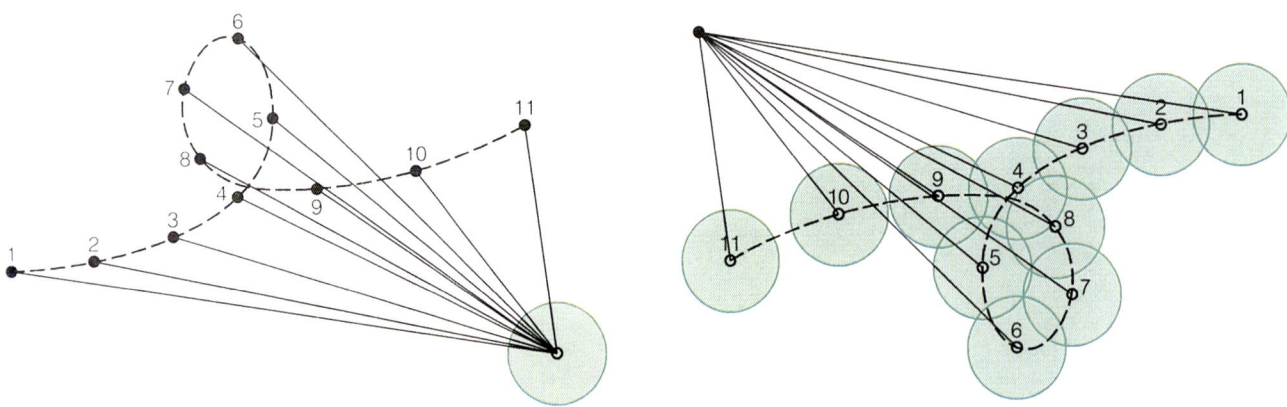

Figure 2.28 *Magnetic data interpreted on (a) fixed continent—moving magnetic pole, and (b) fixed magnetic pole, moving continent.*

Note: The lines between the continent and the magnetic pole are parallel and are of equal length but the relative positions of pole and continent for each time are opposite.

north magnetic pole and the distance between the continent and the pole at any time t were determined by the orientation of the Earth's magnetic field at that time.)

When you have completed your solution to the problem, compare the pattern of "polar wandering" in the original drawing to the pattern of "continental wandering" in your drawing. How do they compare?

2. Assuming that the rifting of Pangaea that produced the North Atlantic Ocean began approximately 200 million years ago, what is the approximate rate of opening of the North Atlantic Ocean in centimeters per year?

3. If the Earth's major mountain ranges are the result of the collision of two continents, such as the recent (45 million years ago) collision of India and Asia that produced the Himalayas, what two continents are most likely to be involved in the next collision? Following that collision, what two continents do you predict will be next to collide and where will the mountains rise?

FOR YOUR NOTEBOOK

The news media increasingly are using illustrations of plate movements to explain earthquakes and volcanic eruptions to their readers and viewers. A perusal of recent newspapers and weekly news magazines in your library will provide you with excellent examples of plate tectonics in action. Including the stories or summaries of the stories will help you understand more fully the profound effect that the movement of plates has on our everyday lives.

CHAPTER OUTLINE

INTRODUCTION

MINERALS DEFINED

CHEMICAL COMPOSITION OF MINERALS
- Compounds
- Chemical Bonding
- Crystal Structure

MINERAL IDENTIFICATION
- Physical Properties
- Instrumental Identification Techniques

MINERAL CLASSIFICATION
- Silicate Crystal Structures
- Nonsilicate Minerals

CHAPTER 3

Minerals

INTRODUCTION The rocks of Earth are assemblages of **minerals.** If we understand minerals, we can better understand Earth, how it was created, how both internal and external forces are constantly changing its face, how its history has been recorded throughout the eons of time, and how the history of eons yet to come will be preserved. Minerals in igneous rocks give us information about the temperature, composition, and perhaps the source of the original magma. The mineral assemblage of a metamorphic rock provides insight into the pressures and temperatures that existed when and where the rock formed. Sedimentary rocks contain clues as to the kinds of rocks that were weathered to produce sediments. The relief of the land reflects, at least in part, the relative resistance of the minerals in rocks to the relentless processes of weathering and erosion. All this information and more is engraved in the kinds and abundances of minerals that make up the rocks of Earth.

MINERALS DEFINED

A mineral is defined as *"a naturally occurring inorganic element or compound having an orderly internal structure of atoms and characteristic element composition, crystal form, and physical properties."*

The requirement that a mineral be naturally occurring precludes the use of mineral names to describe any number of synthetic materials, such as synthetic rubies or diamonds, which may be exact copies of their natural counterparts. The requirements that minerals be inorganic and have an orderly internal or solid structure eliminate a number of naturally occurring materials commonly referred to as "mineral" resources such as coal, oil, and water. There is no question that coal, oil, and water are *natural resources,* but by definition, they cannot be considered *mineral resources.* It should be pointed out that not all geologists agree that minerals should be restricted to inorganic compounds. Some geologists argue that natural occurring organic compounds such as calcium oxalate (CaC_2O_4), a common ingredient in plants, should be considered minerals.

The most important parts of the definition are the orderly internal structure and the characteristic elemental composition. A mineral's chemical composition and crystal structure are products of its environment of formation and also influence how the mineral will react at Earth's surface.

CHEMICAL COMPOSITION OF MINERALS

Chemists define an **atom** as the smallest subdivision of matter that retains the chemical characteristics of that particular material. The atom consists of three major subatomic components: (1) **protons,** which are positively charged particles with a mass of 1 AMU (atomic mass unit, equal to $1/12$ the mass of a carbon atom), (2) **electrons,** which are negatively charged particles with infinitesimally small size and mass, and (3) **neutrons,** which are an electrically neutral combination of a proton and an electron with a mass of 1 AMU (Figure 3.1).

The atom is often pictured as a miniature solar system with most of the mass concentrated in a tiny central nucleus composed of protons and neutrons. The electrons circle the nucleus at specific distances within layers called *shells* (Figure 3.2). The maximum number of shells within the naturally occurring elements is seven. The shells are designated with sequential letters of the alphabet starting with K for the innermost shell. The high orbital speeds of the negatively charged electrons keep them from being pulled into the positively charged nucleus.

The number of protons in the nucleus determines the **atomic number** of the atom, which in turn dictates the *identity* of the element. Of the first 92 elements (hydrogen to uranium), element 61

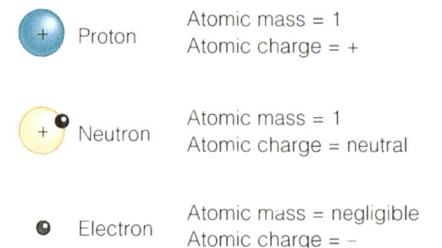

FIGURE 3.1 *The basic components of the atom are the* proton, *the* neutron, *and the* electron. *Note that a neutron is a combination of a proton and an electron.*

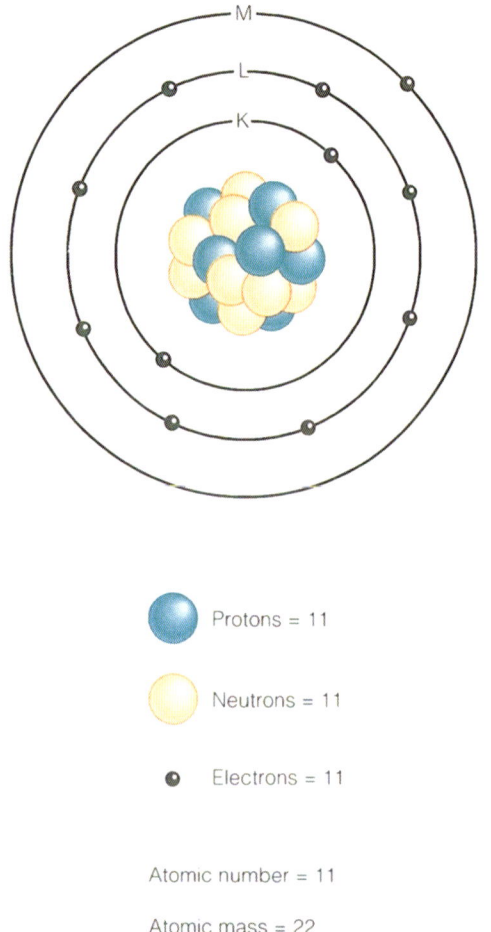

FIGURE 3.2 *The schematic of the sodium atom. Note that the outermost (M) shell has only one electron.*

The sum of the protons and neutrons in the nucleus is the **atomic mass** of the atom. The number of neutrons in atoms of the same element can vary, resulting in **isotopes** of the same element with differing atomic masses. For example, most atoms of hydrogen have only one proton in the nucleus and consequently have an atomic mass of 1, but the hydrogen isotopes deuterium and tritium have one and two neutrons, respectively, resulting in atomic masses of 2 and 3. Although the isotopes of a particular element differ in atomic mass, they exhibit most of the same chemical properties.

As protons are added to the nucleus, electrons are added to the electron shells, generally filling the shells from the innermost shell outward. The number of electrons allowed in any shell is limited. The two lightest elements, hydrogen and helium, possess only one orbital shell, which can hold two electrons. Other elements possess additional shells that may contain up to 32 electrons. However, the *outermost* shell of elements with higher atomic numbers can contain a maximum of eight electrons. With increasing atomic number, the atom adds electrons either by forming a new outer shell or by adding electrons to an existing inner shell (Figure 3.3).

The number of electrons in the outermost orbital shell is very important in determining the chemical reactivity of atoms. The **octet rule** states that when the outermost shell is filled with eight electrons, the atom becomes quite stable and does not readily participate in chemical reactions. The elements referred to as the noble or **inert gases** have an outermost shell filled with either two or eight electrons. All the other elements have fewer than the maximum allowable electron complement in the outermost shell are chemically reactive, although some elements such as gold and platinum react very slowly.

(promethium) does not occur naturally, and elements 43 (technetium), 85 (astatine), and 87 (francium) undergo very rapid radioactive decay and have only been identified in stellar spectra. Elements with atomic numbers greater than 92 do not occur naturally, but several have been made in atomic reactors. They are very unstable (radioactive) and disintegrate with time to atoms with atomic numbers less than 92.

SPOT REVIEW

1. Prepare a list of naturally occurring materials that would not fit the definition of a mineral and indicate what property or properties preclude each from being considered a mineral.
2. Explain the effect that each of the following would have on the atomic number and atomic mass of an atom:
   a. Gaining a proton in the nucleus.
   b. Losing a neutron from the nucleus.
   c. Losing an electron from the nucleus.
   d. Losing an orbiting electron.

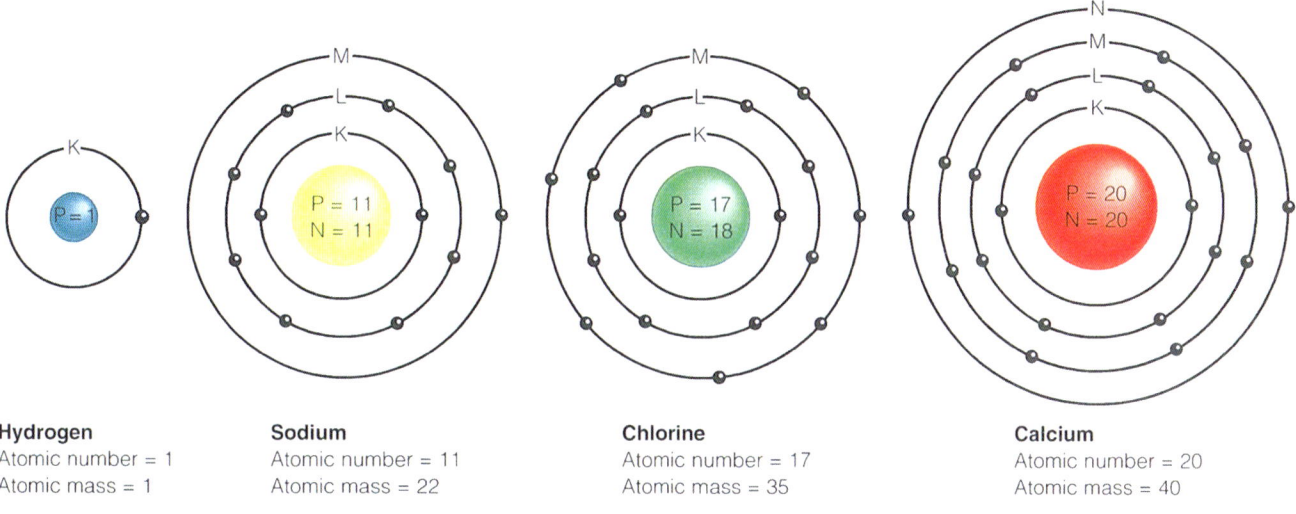

FIGURE 3.3 *The identity of an atom is determined by the number of protons in the nucleus; its mass is equal to the sum of the number of protons and neutrons.*

RADIOACTIVE DECAY OF ATOMS

The elements were created in the fiery cores of stars and once formed, most remain unchanged. However, some isotopes of certain elements are unstable and disintegrate by emitting *alpha* and *beta* particles from their nuclei. An alpha particle consists of two protons and two neutrons. Each emitted alpha particle decreased the atomic number of the element by two and the atomic mass by four. The beta particle is an electron released from a neutron. The emission of a beta particle therefore changes a neutron into a proton with a subsequent increase of one in the atomic number of the element and no change in the atomic mass (refer to Figure 21.15 in Chapter 21). Isotopes that undergo these nuclear decays are the **radioactive** isotopes.

In addition to losing protons, neutrons, and electrons from the atom, the nucleus of the potassium isotope ^{40}K (atomic number 19) can be changed into argon (atomic number 18) if one of its orbital electrons is captured by a proton in the nucleus, a process called **electron capture**. As Chapter 21 explains, this property of ^{40}K enables it to be used in radiometric dating.

Compounds

Atoms of elements react with each other to form **compounds**. A chemical reaction results in each atom filling its outermost shell. The formation of a compound is therefore the mechanism whereby individual atoms can become chemically stable. When some atoms react with each other and lose or gain outer electrons, they are converted into ions. Electron loss converts an atom to a positively charged ion called a **cation** while electron gain produces a negatively charged ion called an **anion**.

An atom will lose or gain electrons depending upon the number of electrons in its outermost shell. Atoms with 1, 2, or 3 electrons in the outermost shell tend to *lose* electrons while those with 4 or more electrons tend to *gain* electrons. For example, the sodium atom (atomic number 11) has 11 electrons: 2 in the innermost shell, 8 in the second shell, and 1 in the outermost shell (refer to Figure 3.2). Should the sodium atom lost the lone outermost electron, its outermost shell would be filled with 8 electrons. In losing the electron, the atom would become a cation with a single positive charge (Na^{1+}) because it would possess one more proton in the nucleus than electrons in its electron shells. The atom of chlorine (atomic number 17) has 17 electrons: 2 in the innermost shell, 8 in the second shell, and 7 in the outermost shell. Chlorine can accept one more electron to fill its outermost shell and in the process become an anion (Cl^{1-}). Ions with opposite charges strongly attract each other and, if other criteria that we will discuss shortly are fulfilled, can unite to form a compound (Figure 3.4). Sodium (Na^{1+}) and chlorine (Cl^{1-}), for example, combine to form sodium chloride, the mineral *halite* (NaCl) or common table salt. Once the ions combine, the compound will be electrically neutral; that is, the total positive charge will equal the total negative charge.

The element that account for 99% of Earth's crust are listed in Table 3.1. Of the 90 available natural elements, only 8 make up almost all of the rocks and minerals of Earth's crust. Note that of the elements in Table 3.2, only oxygen is prone to become an anion (O^{2-}). The remainder will form cations. Given the large number of elements that prefer to form cations, the number of elements that will ionize as anions seems insufficient to balance the cations and form the variety of minerals found in nature. The answer to this seemingly anomalous situation is that certain elements, although themselves prone to become cations, will combine with oxygen to form a multielement anion that chemists call a *functional group*. Silicone (Si), for example, readily joins with oxygen to form the silicate functional group, $(SiO_4)^{4-}$, a major component of a group of minerals called the *silicates,* which make up most of the rocks of the mantle and crust.

But even with the inclusion of the silicate functional group, there still appear to be only two important anions, limiting the kinds of minerals to oxides and silicates. If, however, we add *carbon* (C), *nitrogen* (N), *phosphorus* (P), *sulfur* (S), and *chlorine* (Cl) to the original list of eight, the situation improves

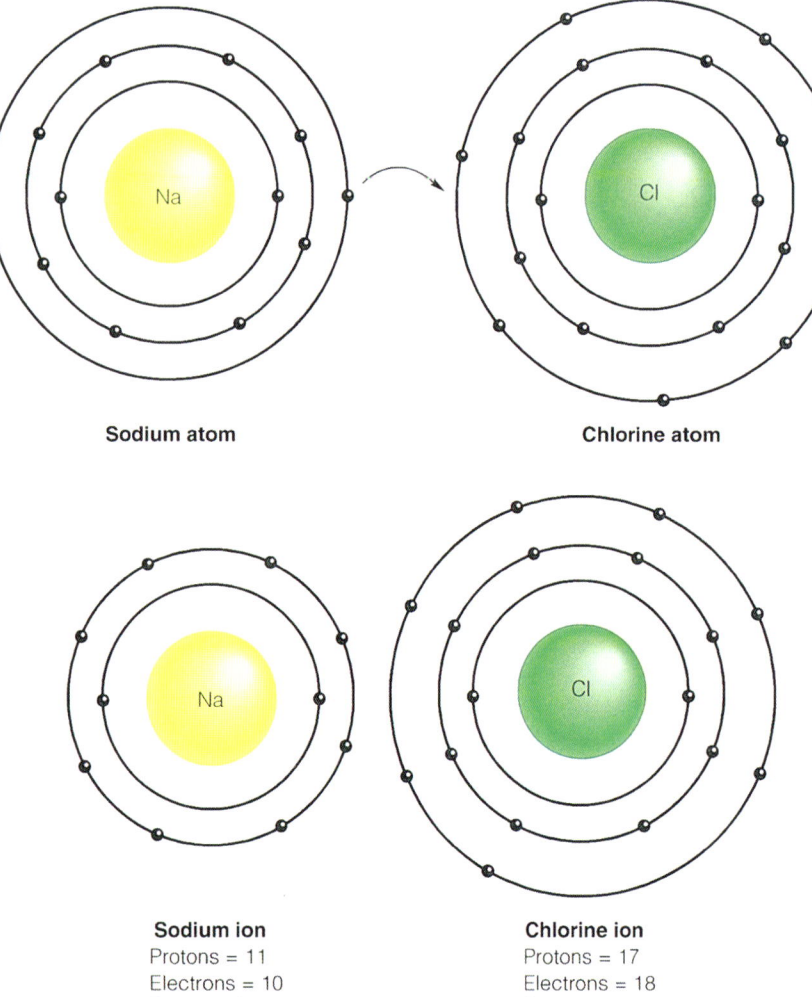

FIGURE 3.4 *The sodium and chloride ions in halite are held together by* ionic bonding *with the electron exchange fulfilling the octet rule for each atom. When the sodium atom donates the single electron from its M shell to chlorine, its L shell, containing 8 electrons, becomes the outermost shell. At the same time, the donated electron fills the outermost shell of chlorine with 8 electrons.*

significantly (Table 3.3). The first four elements—carbon, nitrogen, phosphorus, and sulfur—will also combine with oxygen to form multielement functional groups: the *carbonate* $(CO_3)^{2-}$, *nitrate* $(NO_3)^{1-}$, *phosphate* $(PO_4)^{3-}$, and *sulfate* $(SO_4)^{2-}$ functional groups. In addition to forming the sulfate functional group, sulfur can form two other anions, the *sulfide* (S^{2-}) and *disulfide* (S_2^{2-}) ions. Chlorine will readily accept electrons to become the *chloride* (Cl^{1-}) ion. With our expanded list of elements, not only is the division of positive and negative ions more equable, but we can account for nearly all the common minerals and, in most cases, the not-so-common minerals.

In order for a mineral to form, the necessary ions must first exist, usually in a solution. An important exception to this generalization is the formation of metamorphic minerals, many of which may form without the intervention of a solution. Once in solution, conditions, which we will discuss shortly, come into play that result in the union of the various ions to form a *crystal lattice*. Once the lattice grows to sufficient size, a crystal of the mineral will begin to precipitate from solution.

The role of the solvent is to mobilize the ions so that they may be recombined into compounds. In nature, water is a powerful solvent. Water may contain a wide variety of ions ranging in concentration from the salt-rich waters of a perennial desert lake to the pristine waters of a mountain stream containing very low concentrations of relatively few kinds of ions.

Many minerals consist of ions that were at one time contained in a water solution; calcite $(CaCO_3)$ and halite $(NaCl)$ are two examples. The ions that formed most of the *silicate* minerals, however, were not provided from water solution but rather crystalized as molten rock (magma or lava) cooled. Inasmuch as it is the medium that serves to mobilize the ions from which most of Earth's rocks are made, molten rock is as close to a "universal" solvent as could be achieved.

Once in solution, ions move at random through the solvent with a limited tendency to form permanent combinations. A number of conditions promote the union of cations and anions or functional groups to precipitate a compound from a solution.

A basic requirement for precipitation to occur is that the concentration of the ions be sufficient to ensure that they come in contact within the solution. Ionic concentration and subsequent ion contact can be increased either by adding more ions to the solution or by reducing the volume of the solvent.

TABLE 3.1
AVERAGE ABUNDANCE OF THE MAJOR ELEMENTS IN EARTH'S CRUST

Element	Symbol	Percentage (By Weight)
Oxygen	O	46.4
Silicon	Si	28.2
Aluminum	Al	8.2
Iron	Fe	5.6
Calcium	Ca	4.1
Sodium	Na	2.4
Magnesium	Mg	2.3
Potassium	K	2.1

TABLE 3.2
ION AFFINITIES OF THE MAJOR ELEMENTS IN EARTH'S CRUST

Element	Cation	Anion
Oxygen		O^{2-}
Silicon	Si^{4+}	
Aluminum	Al^{3+}	
Iron	Fe^{2+}, Fe^{3+}	
Calcium	Ca^{2+}	
Sodium	Na^{1+}	
Magnesium	Mg^{2+}	
Potassium	K^{1+}	

TABLE 3.3
EXPANDED LIST OF CRUSTAL ELEMENTS SHOWING THE CATION-ANION AFFINITIES

Element	Cation	Anion
Oxygen		O^{2-}
Silicon	Si^{4+}	$(SiO_4)^{4-}$
Aluminum	Al^{3+}	
Iron	Fe^{2+}, Fe^{3+}	
Calcium	Ca^{2+}	
Sodium	Na^{1+}	
Magnesium	Mg^{2+}	
Potassium	K^{1+}	
Carbon	C^{4+}	$(CO_3)^{2-}$
Nitrogen		$(NO_3)^{1-}$
Phosphorus		$(PO_4)^{3-}$
Sulfur	S^{4+}	$S^{2-}, S_2^{2-}, (SO_4)^{2-}$
Chlorine		Cl^{1-}

Precipitation also depends on the temperature of the solution. The higher the temperature, the faster the ions move and the less tendency they have to combine to make a compound. This is why most materials are more soluble in hot water than in cold. As the movement of the ions in solution slows, the chances of a union improve.

In summary, precipitation of compounds from solution is favored by (1) increasing the concentration of ions either by adding ions to the solution or by reducing the volume of the solvent and (2) by decreasing the temperature.

Another requirement for the union of ions of dissimilar charge is the establishment of **electric neutrality** within the compound. In order for a union to be made, the total positive and negative charges must balance. Electric neutrality must be maintained at all times.

Lastly, the ions must be able to combine and form a three-dimensional crystal lattice. Ion size is an important constraint in building crystal lattices. For example, a specific ion with all the proper electrical qualifications may be excluded from a particular lattice because the diameter of the ion is too large to fit into the available space. To illustrate with a simple example, imagine a table top covered with tennis balls where each tennis ball is in contact with all adjacent tennis balls (Figure 3.5). Can you replace certain tennis balls with baseballs? Yes, because the two balls are about the same size. Can you replace some tennis balls with ¼-inch ball bearings? You can, but when you do, there will be nothing to support the tennis balls adjoining the site occupied by the ball bearing, and they will begin to move from their original positions. Thus, substituting ball bearings for tennis balls will disrupt the pattern. In this case, the intended replacement is too small. How about replacing some tennis balls with basketballs? It can be done, but the basketball will make the layer excessively thick. As a rule of thumb, in order for an ion to occupy an available space within a lattice, the diameter of the ion and the diameter of the available space must not differ by more than 15%.

SPOT REVIEW

1. What determines whether an atom will tend to become a cation or an anion?
2. What conditions promote the combining of ions and the subsequent precipitation of a compound from solution?

FIGURE 3.5 *Assuming all other criteria are acceptable, ions with ionic radii within 15% of each other may readily interchange, just as baseballs and tennis balls may be substituted for each other. If the ionic radius is too small, subsequent lattice deformation may prevent substitution from occurring. Substitution of very large ions would result in unacceptable lattice distortions.*

Chemical Bonding

Once the concentration of ions needed for precipitation has been exceeded, compounds can form by the **bonding** of positive and negative ions into an orderly three-dimensional structure. Bonding results from the attempt of each atom to maintain the maximum number of electrons in its outermost orbital shell. There are four basic kinds of bonds: (1) *ionic*, (2) *covalent*, (3) *metallic*, and (4) *van der Waals*.

During **ionic bonding** (ionization), the outermost electrons of one atom or functional group are relinquished and given to another. The transfer of electrons results in each atom in the compound achieving the desired number of eight electrons in the outermost shell. Let's once again take the example of halite (NaCl), common table salt. As shown in Figure 3.4, the electron donated by the sodium atom fills the outermost shell of chlorine with eight electrons and at the same time leaves sodium with eight electrons in its outer shell. Both atoms now have a filled outermost shell. Once the two ions are present in sufficient concentration, they attract each other and create a crystal lattice, producing a compound with the required state of electric neutrality.

Compounds in which atoms are joined by ionic bonding are usually water soluble.

Covalent bonding forms the strongest bonds. In true covalent bonding, the outermost electrons of both atoms are *shared*. True covalent bonds can only be formed by two atoms of the same element. For example, carbon with four electrons in the outermost shell joins with other carbon atoms to form *diamond* and *graphite* (Figure 3.6) by the mutual sharing of their outermost electrons. The union produces an extremely strong bond by establishing the desired eight electrons in the outer shell of each atom. Because of its strength, covalent bonding produces some of the most chemically resistant compounds, including most minerals. Covalently bonded compounds are only slightly soluble in water, and more severe chemical conditions are required to return them to solution. Such conditions can be provided by a mass of molten rock, for example.

A continuum exists between ionic bonding where electrons are *completely exchanged* and covalent bonding where the electrons are *equally shared*. Whether a bond will have an ionic or covalent character depends on the degree to which the electrons are attracted to one of the two atoms. As the ability of the atoms to attract electrons becomes more equal, the covalent character of the bond increases.

Metallic bonding, as the name implies, is the bonding occurring in metals. In metallic bonding, the atoms are locked into place within a rigid lattice, but the outer electrons roam easily throughout the crystal structure. The characteristic properties of metals such as high electrical and thermal conductivity are due to this unique bonding style.

Van der Waals bonding differs from the other three types of bonding in that it does not involve electron transfer or sharing but rather a weak attraction between subunits *within* the main crystal structure. In the mineral graphite, for example, carbon atoms are joined by strong covalent bonding to form sheets. The sheets are then stacked to form the structure of graphite (Figure 3.6b). Within the sheets, the orientation of the electron fields around the carbon atoms causes positive and negative charges to develop, generating an electrostatic attraction between the sheets. This attraction is an example of van der Waals

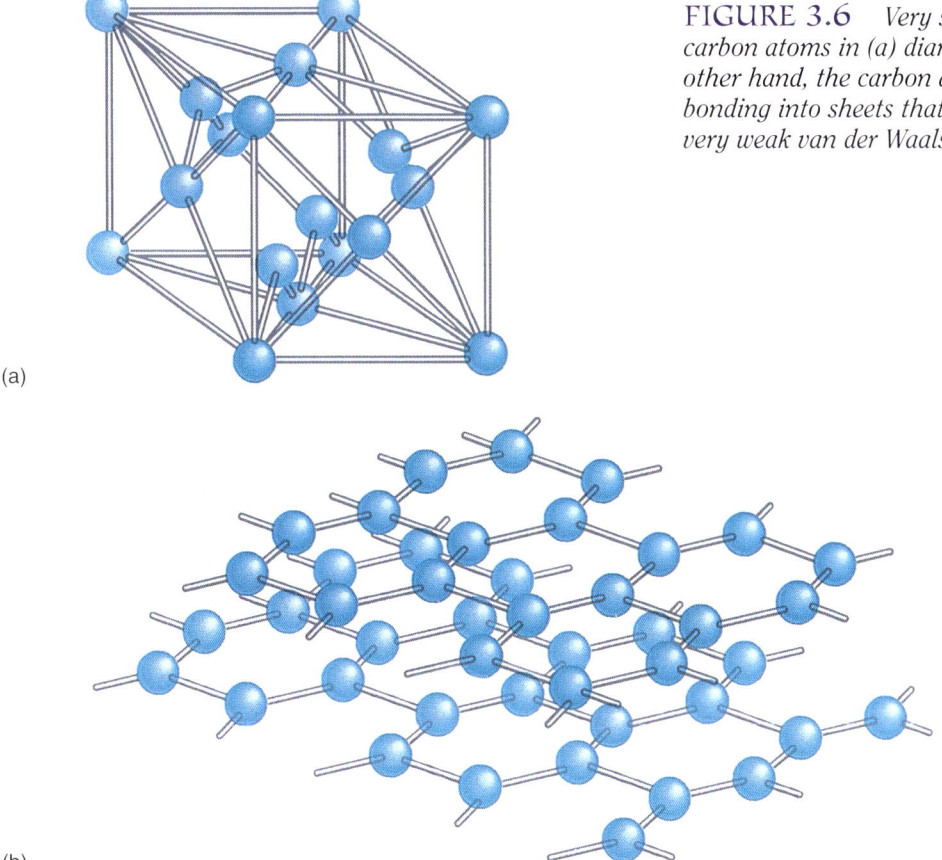

FIGURE 3.6 *Very strong covalent bonds join all the carbon atoms in (a) diamond. In (b) graphite, on the other hand, the carbon atoms are joined by covalent bonding into sheets that are in turn held together by very weak van der Waals bonds.*

bonding. Most van der Waals bonds are very weak. The bond in graphite is so weak it can easily be broken with the application of the smallest force generated between your thumb and forefinger. When you rub graphite or soft pencil lead, it feels greasy because sheets of carbon atoms are slipping over each other as you break the van der Waals bonding. Because of the weak van der Waals bonds, graphite can be used as a solid lubricant in door locks, electrical equipment, and similar instances when liquid lubricants are inappropriate.

Crystal Structure

Solids can form in two ways. The individual atoms can join together into an orderly three-dimensional array called a **crystal lattice** in which each atom, ion, or functional group is located at a specific site (Figure 3.7). Materials that form in such orderly arrangements are said to be **crystalline**, and if they form naturally, they can be called minerals. Atoms can also join together to form a solid where the individual atoms are arranged somewhat randomly with no specific order or assigned locations within the solid. The material is then said to be **amorphous** and is called a **mineraloid** or **glass**. *Opal* ($SiO_2 \cdot nH_2O$ where n is the number of water molecules) is an example of a mineraloid.

SPOT REVIEW

1. What is meant by the statement that a continuum exists between ionic and covalent bonding?
2. In what ways does the kind of bonding affect the chemical and physical characteristics of a compound?

MINERAL IDENTIFICATION

Minerals have orderly internal structures, characteristic composition, typical crystal forms, and physical properties. Any of these parameters may be used in identification. Sometimes, one can identify mineral grains with the unaided eye or a low-powered hand lens. At other times, mineral grains or crystals are too small to be observed and tested with a hand lens, and their identification requires other instruments such as powerful microscopes or X-ray analyzers.

When the mineral grains can be seen with the unaided eye or with low-power magnification, the

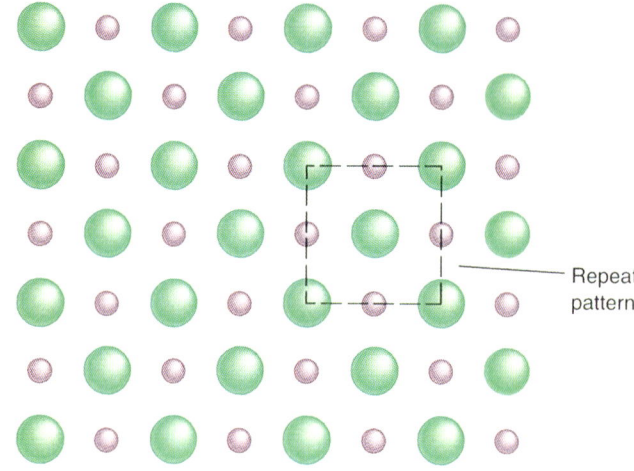

FIGURE 3.7 *Within every crystal lattice, the atoms or ions are arranged in a symmetrical, three-dimensional collection called a* unit cell, *here represented in two dimensions. The* crystal structure *is constructed by repeating the unit cell in the three crystallographic directions.*

identification technique employed is referred to as *visual identification*. With a low-powered hand lens and some experience, a geologist can usually identify most common mineral grains or crystals from an examination of one or more of its **physical properties**.

Physical Properties

Any property that can be determined by the senses is considered a physical property. A variety of physical properties ranging from color or taste are used in identifying minerals. Also included among these physical properties are such characteristics as hardness, specific gravity, and luster.

Color

Perhaps the most conspicuous physical property of minerals is *color*. Nevertheless, the first thing you learn as a beginning mineralogy student is not to rely on color for identification. The color of a mineral is the end product of the degree of absorption of white light as it is reflected back to the eye. Most minerals contain impurities of foreign atoms. Small concentrations of impurities can have significant effects upon the wavelengths of light absorbed and will therefore have a major effect on the color (Figure 3.8). Nevertheless, color is very useful for distinguishing between the dark-colored ferromagnesian (iron- and magnesium-rich) minerals such as augite and the light-colored nonferromagnesian minerals such as albite (Figure 3.9).

Minerals 81

  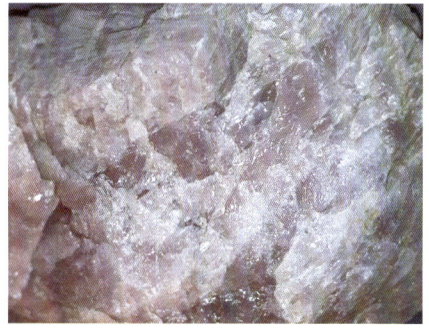

FIGURE 3.8 *Although colorless when pure, trace levels of various metals within the quartz lattice result in the mineral being found in a wide variety of colors. (a: VU/© Doug Sokell; b: © Tom McHugh/Photo Researchers; c: VU/© John D. Cunningham)*

(a)

(b)

(c)

(d)

FIGURE 3.9 *Ferromagnesian (a and b) and non-ferromagnesian minerals (c and d) can be easily distinguished by their colors. The former are usually dark in color while the latter are generally light in color. (a and b VU/© Albert Copley; c and d © E. R. Degginger)*

FIGURE 3.10 *As certain minerals are drawn across a piece of unglazed tile, small (powder-sized) particles of the mineral break off and are left behind as a* streak. *The color of the streak is diagnostic of the mineral. (© Runk/Schoenberger, Grant Heilman Photography)*

Streak The color of the powdered mineral (streak) is relatively diagnostic of some minerals, particularly metallic minerals such as *hematite* (Fe_2O_3). Experience has shown that the color of the powder is fairly constant. It is called streak because the most common way of obtaining a powder is by dragging the mineral across the surface of a plate of unglazed ceramic tile, thereby producing a line or streak (Figure 3.10).

Hardness An often-used physical property is **hardness.** Hardness, the resistance of a mineral to scratching, is a result of the strength of the bonds within the crystal structure. A geologist named *Fredrick Mohs* studied the relative hardness of minerals nearly two centuries ago (1822). When his study was completed, he selected 10 common or well-known minerals to represent the entire range of hardness. For the softest mineral, he selected *talc,* and for the top of his scale, he used the hardest mineral known, *diamond.* The minerals in **Mohs's Scale of Hardness** are listed in Table 3.4. The hardness of an unknown mineral is determined by finding the pair of minerals in Mohs's list, the harder of which will scratch the surface of the unknown mineral, and the softer of which will be scratched *by* the unknown. A geologist can carry a set of hardness mineral standards or use a few common items to estimate hardness. A *fingernail* has a hardness of 2.5, a *copper penny* a hardness of 3, and a piece of *glass* a hardness of 5.5.

TABLE 3.4

MOHS'S SCALE OF HARDNESS

Hardness	Mineral	Composition
10	Diamond	C
9	Corundum	Al_2O_3
8	Topaz	$Al_2SiO_4(F,OH)_2$
7	Quartz	SiO_2
6	Orthoclase	$KAlSi_3O_8$
5	Apatite	$Ca_5(PO_4,CO_3)_3(F,OH,Cl)$
4	Fluorite	CaF_2
3	Calcite	$CaCO_3$
2	Gypsum	$CaSO_4 \cdot 2H_2O$
1	Talc	$Mg_3Si_4O_{10}(OH)_2$

Before we leave the topic of hardness, perhaps some comments on a few of the minerals in Mohs's scale are in order. *Talc* is used to make baby powder and talcum powder while *gypsum* is widely used in the construction industry as plasterboard and dry wall. *Calcite* is the mineral that makes up the shells of animals such as clams, oysters, and snails, while *apatite* is used in the construction of bones and teeth, including those of humans. *Orthoclase,* one of the potassium feldspars, with a hardness of 6, is the abrasive used in most household cleansers. Because the abrasive has a hardness of 6, there is no chance of scratching such things as tile tubs and washbasins, which have a hardness of about 7. In fact, the logo of one cleanser includes the statement "hasn't scratched yet" encircling a baby chick just emerging from its shell. Orthoclase is also used as the abrasive agent in many toothpastes. Remember, the enamel of your teeth is made of apatite, which has a hardness of 5.

Quartz is the hardest of all the common minerals and has long been used as an abrasive. Antique grinding stones and wheels were usually carved from blocks of sandstone, which is made up primarily of quartz grains. Sandblasting, the method commonly used to clean the surfaces of stone or metal, often utilizes loose quartz sand grains blown at the surface by compressed air. At one time, quartz sand similar to the sand that makes up your favorite beach was glued to paper to make "sandpaper." Today, *corundum*, which is harder, has nearly replaced quartz as the major abrasive in stones, wheels, emery cloth, and sandpaper. Another common use of quartz is as the timing device in quartz watches.

It may come as a surprise that most *diamonds* are not used as gemstones, but as abrasives. The edges of saw blades used to cut rocks are embedded with diamonds. Drilling bits used to drill through extremely hard rock are encrusted with diamonds, and you can purchase diamond "paste" (powdered diamonds in an oil matrix) for fine polishing.

Form

The **form** of a mineral is the geometric shape of its crystal as defined by the angle between faces. The form is the outer expression of the crystal structure. If a crystal were able to grow without interference, it would have a distinctive crystal form. Many minerals have diagnostic forms that may be used in their identification (Figure 3.11). However, most mineral grains grow in restricted space, and the diagnostic crystal forms do not develop.

Mineral Cleavage

Mineral cleavage is a very diagnostic physical property of some minerals. Cleavage describes the tendency of certain minerals to break along specific planes within the crystal structure. These planes or surfaces develop where the bonding across some crystallographic planes is weaker than others. If stressed, the chemical bonds between these planes will break preferentially. An excellent example of cleavage is the ability of the *micas* to be peeled apart into sheets, giving rise to the term "books" of mica (Figure 3.12). Mica is made up of sheets in which atoms and ions are held together by strong bonding. The bonds that hold the sheets together are much weaker. In fact, they are so weak that they can easily be broken by prying with a fingernail. Other minerals with a single dominant cleavage plane due to weak van der Waals bonding are graphite and talc.

Many minerals have multiple cleavage planes. The mineral *halite* (NaCl), for example, has three mutually perpendicular cleavage planes, which explains why crushed salt grains have a cubic form (Figure 3.13). *Calcite* also has three sets of cleavage planes, but because they are not mutually perpendicular, the crystal form is rhombic. Other common minerals with multiple cleavage planes include *feldspar, augite,* and *hornblende,* each with two planes, and *fluorite* with four.

Fracture

Not all minerals show cleavage. Minerals such as quartz show no cleavage because the interatomic bonds are all of equal strength. If you shatter a piece of quartz, it will break or **fracture** along random, usually curved, surfaces like the breaking of glass rather than along planes of cleavage (Figure 3.14). Such curved fracture is called *conchoidal fracture.* Minerals such as the feldspars

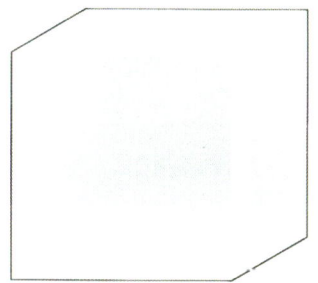 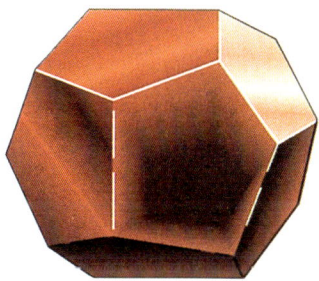 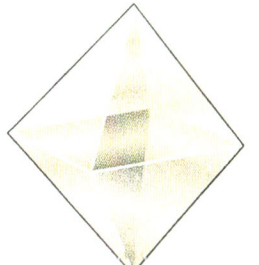 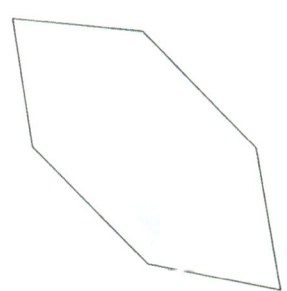

FIGURE 3.11 *The outer appearance of a crystal, called the crystal* form, *is the external expression of the systematic internal arrangement of atoms or ions within the crystal lattice.*

FIGURE 3.12 *The excellent cleavage in one direction exhibited by mica allows the mineral to be pulled apart into sheets. Because of the vertical stacking of the sheets, mica is referred to as being found in* books. *(© Runk/Schoenberger, Grant Heilman Photography)*

FIGURE 3.13 *The cubic arrangement of the atoms within halite results in its characteristic* cubic form. *(M. Claye/Jacana, Photo Researchers)*

FIGURE 3.14 *Because the bonds in quartz are equally strong in all directions, it does not cleave (break in any preferred direction); instead it breaks at random, a process called* fracture. *(VU/© A. J. Copley)*

that have two planes of cleavage will show fracture in noncleavage directions.

Specific Gravity The **specific gravity** is a useful physical property that can be estimated by "hefting" the mineral. Specific gravity is the ratio of the mass of a solid to the mass of an equal volume of pure water at 4°C. Specific gravity is determined by the kinds of atoms in the crystal structure and the closeness with which they are packed. Most rocks and minerals have specific gravities from 2.5 to 3. Specific gravity increases with increasing atomic number. For example, the specific gravity of *quartz* (SiO_2), is 2.65, *pyrite* (FeS_2) 5.02, *chalcocite* (Cu_2S) 5.8, *galena* (PbS) 7.6, and *gold* (Au) 19.3. The *specific gravity* and **density** of a substance are numerically equal. The only difference is that because the density of water at standard conditions of temperature and pressure is 1.0 gm/cm^3, specific gravity is dimensionless whereas density is mass per volume and has the units gm/cm^3.

Luster The **luster** of a mineral is the appearance of its surface under reflected light. Luster can be described as *metallic* or *nonmetallic*. Nonmetallic luster is further subdivided into descriptions that are self-explanatory such as *pearly, silky, glassy* or *vitreous,* and *earthy* and into others that are not so obvious such as *adamantine* (sparkles like a diamond). Metallic luster, such as that shown by galena, refers to the metallic appearance of the mineral.

Acid Reaction Certain carbonate minerals, especially calcite, can be readily identified by the fizz that results when a few drops of dilute acid are applied to the surface. The generation of the gas is explained by the following reaction:

$$CaCO_3 + 2HCl \rightarrow$$
$$CO_2 + H_2O + Ca^{2+} + 2Cl^{1-}$$

calcite + acid →
  gas + water + in solution + in solution

Few geologists would go into the field without a small bottle of dilute hydrochloric acid. Acid must always be used with care, however.

Taste *Taste* can be used (also with care) to identify some minerals. Examples include the minerals *halite* (NaCl) and *sylvite* (KCl). Halite is common table salt while sylvite is sold as a substitute for halite. Halite is identified by its familiar salty taste while sylvite, although salty, has a distinctive bitter taste.

SPOT REVIEW

1. List the mineral characteristics that are most responsible for each physical property.
2. How are crystal form, density, cleavage, and fracture related?

Instrumental Identification Techniques

Identification of minerals in very fine powders or multicomponent samples such as soil or rock requires analytical methods such as transmitted light microscopy and X-ray powder diffraction.

Light Microscopy Under the **light microscope**, each mineral has a diagnostic appearance when observed with plane-polarized light (Figure 3.15). In some cases, identification can be aided by the use of chemical dyes that selectively stain certain minerals.

Microscopy also provides data about the size, shape, and associated grains in the rock. These data are essential in classifying the rock and determining how it formed.

X-Ray Diffraction X-ray diffraction, invented in the early 1900s, has become a standard technique for identifying solid, crystalline materials. As X rays pass into solid, crystalline materials, they are turned (*diffracted*) from their original paths at certain angles depending upon the crystal structure of the material. The intensity of the diffracted beam will be influenced by the kind of atoms encountered by the X-ray beam. Although originally recorded on film, X-ray diffraction data are now usually recorded by a diffractometer and presented as a *diffractogram* on a strip chart or computer screen (Figure 3.16). The specific positions and intensities of the peaks on the diffractogram form a *diffraction pattern*. The mineral is identified by comparing its pattern to a standard pattern in much the same way as individuals are identified by their fingerprints.

MINERAL CLASSIFICATION

The more than 3,200 known rock-forming minerals are classified into two groups: (1) **silicates** and (2) **nonsilicates** (Table 3.5), a classification that once again emphasizes the importance of the silicate functional group. Depending on the *iron* (Fe) and *magnesium* (Mg) contents, the silicate minerals may be further subdivided by composition into two mineral subgroups: (1) the *ferromagnesian* and (2) the *nonferromagnesian* minerals. Subdivisions of the silicate subgroups are based primarily on structure.

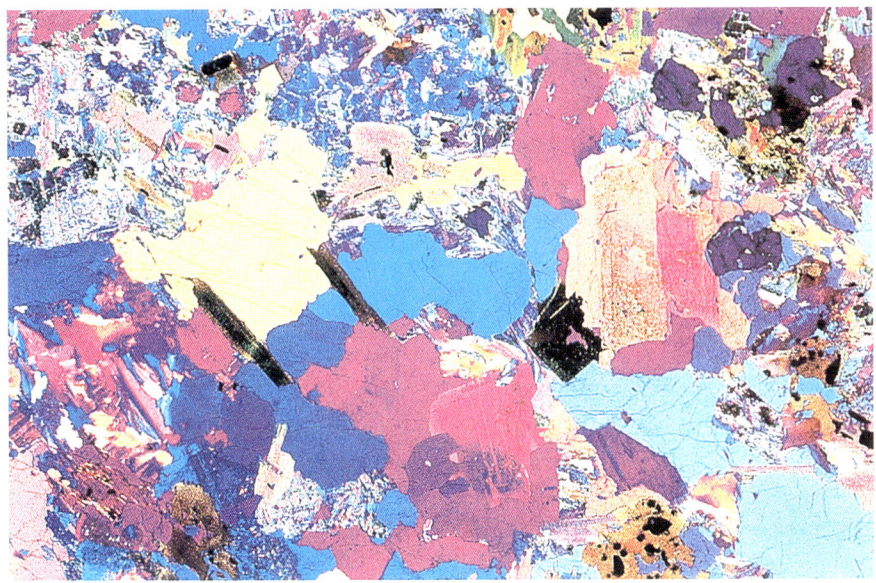

FIGURE 3.15 *When cut into an extremely thin slab called a* thin section *and viewed under polarized light, individual minerals exhibit diagnostic colors that can be used for their identification.* (© Alfred Pasieka/Science Photo Library)

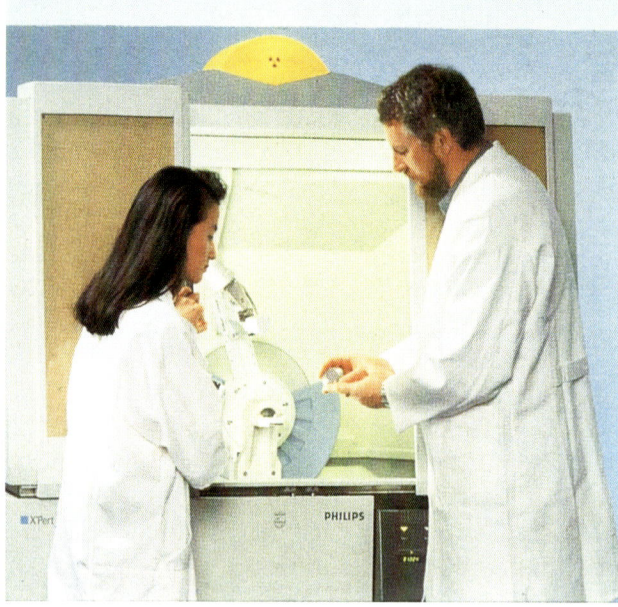

FIGURE 3.16 *Mineral grains that are too small to be identified under a microscope or the aggregate composition of minerals in a rock can be identified using a process called X-ray diffraction. Modern instruments have the necessary information in computer files to allow the rapid computer identification of the individual minerals in a multicomponent mixture. (Courtesy of the Philips Company)*

The nonsilicate minerals are subdivided into groups based on the anion or functional group. A common mineral representative is included in Table 3.5 for each anion or functional group.

Silicate Crystal Structures

Silicate minerals are the primary building blocks of rocks. The basic building block of all the silicate minerals is the **silicon-oxygen tetrahedron** (Figure 3.17), which consists of four oxygen atoms arranged at the corners of a tetrahedron around a central silicon atom. One of the negative charges of each of the four oxygen atoms is balanced by the 4 positive charges of the silicon, resulting in a net charge 4– charge for the silicate functional group.

The silicon-oxygen (silica) tetrahedra can be joined into five major structural arrangements: (1) *isolated tetrahedra*, (2) *single chains*, (3) *double chains*, (4) *sheets*, and (5) *framework structures*. Each structural type is illustrated in Figure 3.18.

TABLE 3.5
CLASSIFICATION OF MINERALS

Common Silicate Minerals

Ferromagnesium Minerals

Olivine group
  Fayalite Fe_2SiO_4
  Forsterite Mg_2SiO_4
Pyroxene group
  Augite $(Ca,Na)(Mg,Fe^{3+}, Fe^{2+}, Al)_2(Si, Al)_2O_6$
Amphibole group
  Hornblende
    $(Ca,Na)_{2-3}(Mg,Fe^{2+},Fe^{3+},Al)_5(Al,Si)_8O_{22}(OH)_2$
Mica group
  Biotite $K(Mg,Fe^{2+})_3(Al,Fe^{3+})Si_3O_{10}(OH)_2$

Nonferromagnesian Minerals

Feldspar group
  Plagioclase feldspar subgroup
    Anorthite $CaAl_2Si_2O_8$
    Albite $NaAlSi_3O_8$
  Orthoclase $KAlSi_3O_8$
Mica group
  Muscovite $KAl_2(Al,Si_3)O_{10}(OH)_2$
Quartz SiO_2

Common Nonsilicate Minerals

Oxides	Nitrates
Hematite Fe_2O_3	Niter KNO_3
Corundum Al_2O_3	Soda niter $NaNO_3$
Carbonates	Phosphates
Calcite $CaCO_3$	Fluorapatite $Ca_5(PO_4)_3F$
Dolomite $(Ca,Mg)CO_3$	Chlorapatite $Ca_5(PO_4)_3Cl$
Sulfates	Halides
Gypsum $CaSO_4 \cdot 2H_2O$	Halite $NaCl$
Anhydrite $CaSO_4$	Sylvite KCl
Sulfides	Fluorite CaF_2
Galena PbS	Native Elements
Sphalerite ZnS	Graphite C
Pyrrhotite FeS	Sulfur S
Disulfides	Gold Au
Pyrite FeS_2	Silver Ag
Chalcopyrite $CuFeS_2$	Copper Cu
	Platinum Pt

Isolated Tetrahedral Structure A crystal structure can be envisioned as a three-dimensional arrangement of anions or functional groups, *glued* together by cations. In the **isolated tetrahedral structure**, the silica tetrahedra are *isolated* from each other and held (glued) together by the cations. The silicon/oxygen ratio in these minerals is 1:4. An example is the **olivine group**, $(Fe,Mg)_2SiO_4$. The olivine group is an example of a **solid-solution series,** which is a group of minerals with a common crystal structure in which two positive ions of nearly the same size can substitute interchangeably. A solid-solution consists of two end-member minerals in which all the potential cation sites are occupied by one ion or the other. Between the two end-member minerals there is a series of minerals within which the cation sites are occupied by both ions, but in different proportions. Other minerals exist in the diverse group that contains both iron and magnesium, shown by the general formula $(Fe,Mg)_2SiO_4$. The olivines, for example, range in composition from the iron-rich end-member *fayalite*, Fe_2SiO_4, to the magnesium-rich end-member *forsterite*, Mg_2SiO_4. Other minerals with independent tetrahedral structures are *garnet* and *topaz*.

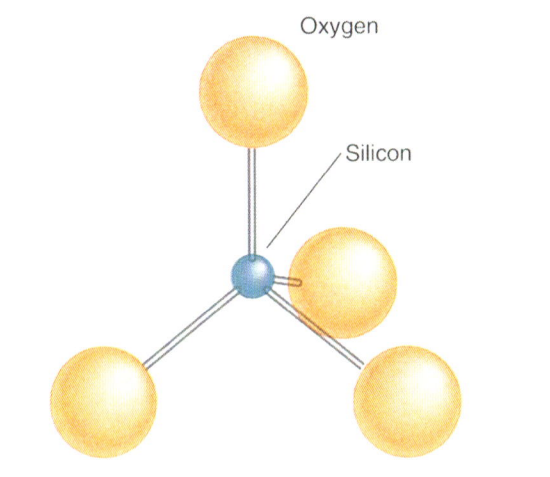

FIGURE 3.17 *The basic building block of all silicate minerals is the* silicon-oxygen tetrahedron.

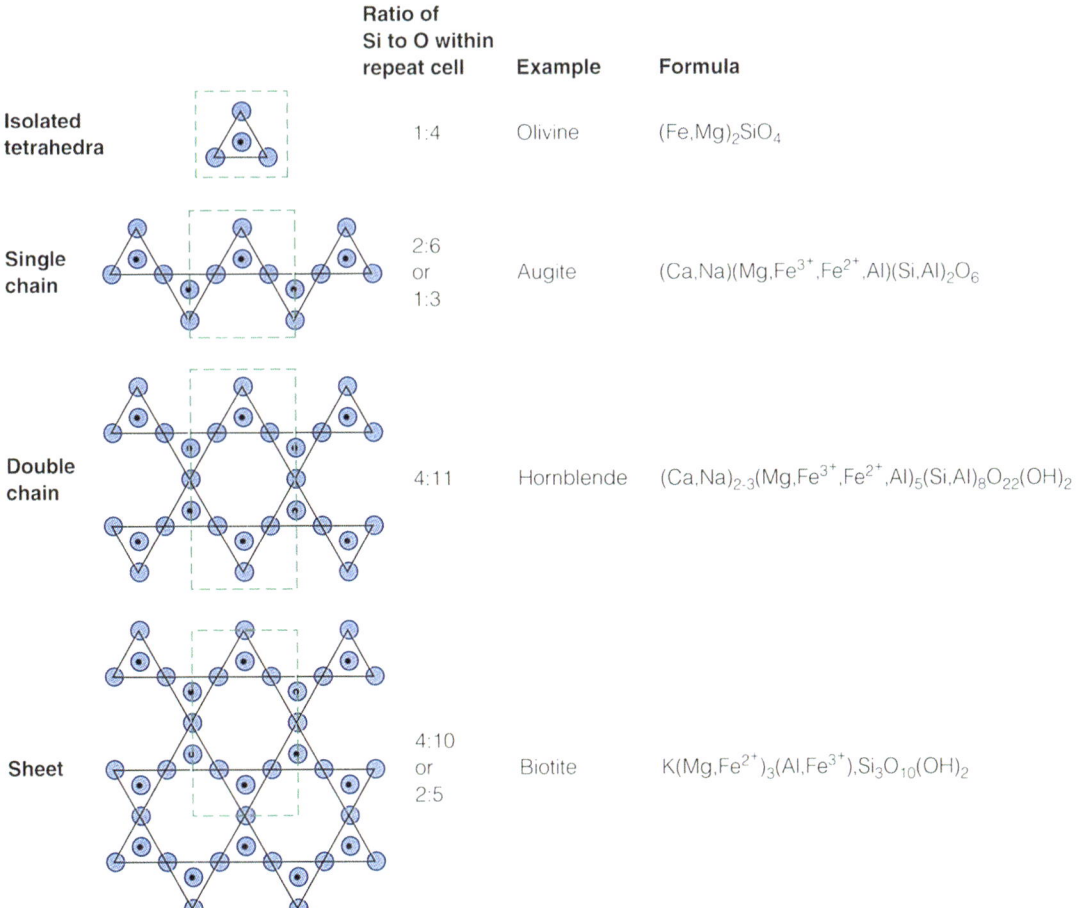

FIGURE 3.18 *Five different arrangements of silicon-oxygen tetrahedra, held together by cations, make up most of the important rock-forming silicate minerals.*

Single-Chain Structure   In the **single-chain structures**, the silicon-oxygen tetrahedra are joined into a chain in which two corners of each tetrahedron are shared, producing a unit with a silicon/oxygen ratio of 1:3. The single-chain structure is represented by the **pyroxene group** minerals. *Augite*, $(Ca,Na)(Mg,Fe^{3+},Fe^{2+},Al)_2(Si,Al)_2O_6$, is the most common representative. The single-chain minerals can be envisioned as stacked cordwood or logs with the long dimensions of the chains parallel to each other. Again, the cations bond the chains, producing a single mineral structure. The weaker bonding of adjacent chains by the positive ions as compared to the strong bonding within the chains results in the characteristic near–90° cleavage of the pyroxene group (Figure 3.19). Note also that as the crystal structure becomes more complex, the compositional makeup of the minerals becomes more diverse.

Double-Chain Structure   **Double-chain structures** are made by joining two single chains, once again by a sharing of oxygen atoms. This time, alternate third corners of tetrahedra are linked. The double-chain minerals are represented by the **amphibole group** of silicate minerals, of which *hornblende*, $(NaCa)_{2-3}(Mg,Fe^{2+},Fe^{3+},Al)_5(Si,Al)_8O_{22}(OH)_2$, is the most common representative. With the increasing compositional complexity as sodium, calcium, and potassium ions (Na^{1+}, Ca^{2+}, and K^{1+}) begin to enter the structure, aluminum (Al^{3+}) begins to substitute for silicone (Si^{4+}) in the silicon-oxygen tetrahedra to maintain the charge balance. The (silicon + aluminum)/oxygen ratio of the structure becomes 4:11.

The double-chain units are like planks with the crystal structure built up by overlapping the planks in successive layers, again *glued* together with cations. Note, once again, that compositional complexity increases with structural complexity. As in the single-chain minerals, the weak bonding of adjacent double chains by the positive ions results in the characteristic cleavage of the amphibole minerals (Figure 3.20).

Sheet Structure   In the **sheet structures**, the silican tetrahedra are joined into sheets with a symmetrical hexagonal (six-sided) pattern (Figure 3.21). Three oxygens of each tetrahedron are shared, leaving the one unshared oxygen pointing away from the sheet. The (silicon + aluminum)/oxygen ratio of the structure is now 2:5. Most sheet minerals are constructed by joining two sheets of tetrahedra together with cations bridging the negatively charged oxygen surfaces. The bond between the joined

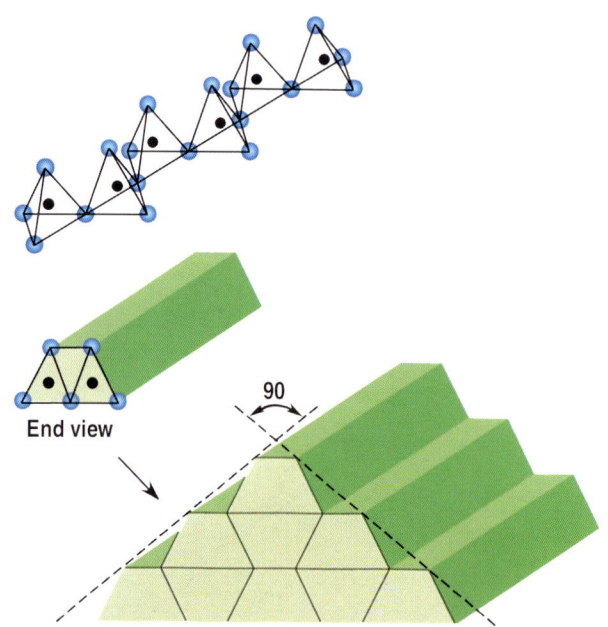

FIGURE 3.19   *The pyroxene minerals consist of parallel stacked, single chains held together by cations including iron and magnesium. The combination of the cross-sectional shape of the individual chains and the stacking arrangement results in two well developed cleavage planes that intersect at a 90° angle.*

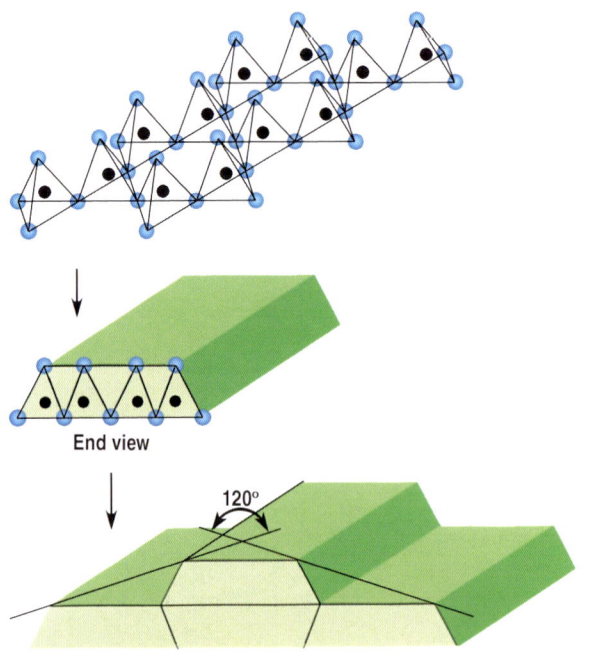

FIGURE 3.20   *The crystal structure of the amphibole minerals is similar to that of the pyroxenes except that the structural unit is the double chain rather than the single chain. The amphiboles also have two well developed cleave directions, but, because of the width of the double chain structure, the angle between the cleavage planes is 120°.*

tetrahedral sheets is very strong. The joined sheets are stacked like sandwiches to create the mineral structure. A representative mineral group is the **mica group**. In the ferromagnesian mica, *biotite,* $K(Mg,Fe^{2+})_3(Al,Fe^{3+})Si_3O_{10}(OH)_2$, the two tetrahedral sheets are bonded together with iron and/or magnesium whereas in the nonferromagnesian mica, *muscovite,* $KAl_2(AlSi_3)O_{10}(OH)_2$, aluminum links the two sheets (Figure 3.22). In both cases, the silicate sandwiches are bonded relatively weakly to each other by the potassium ion. The characteristic cleavage of mica is due to the weakness of the potassium bonds

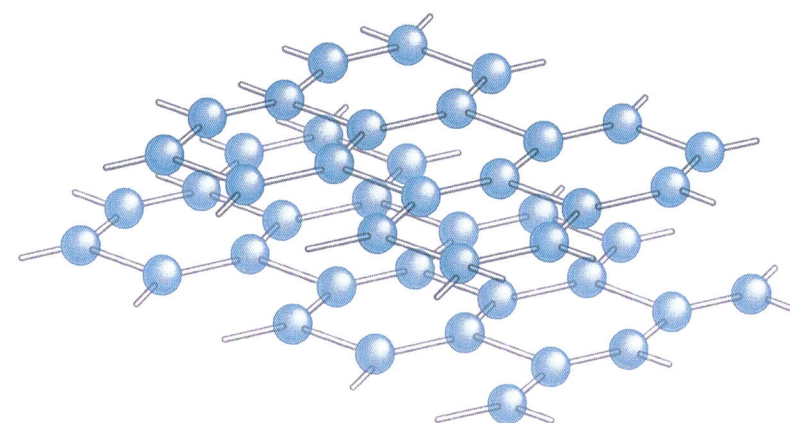

FIGURE 3.21 *The silicon-oxygen tetrahedra can also be joined in a two-dimensional hexagonal pattern, forming sheets. Pairs of sheets are then held together like a sandwich by a variety of ions including iron, magnesium, and aluminum. The sandwiches are then stacked and held together by ions such as potassium, calcium, and sodium to form the micas. The characteristic cleavage of the micas is due to the weakness of the bonds that join the sandwiches. The kinds of cations that join the sheets into sandwiches and hold the sandwiches together determine the identity of the individual mica mineral.*

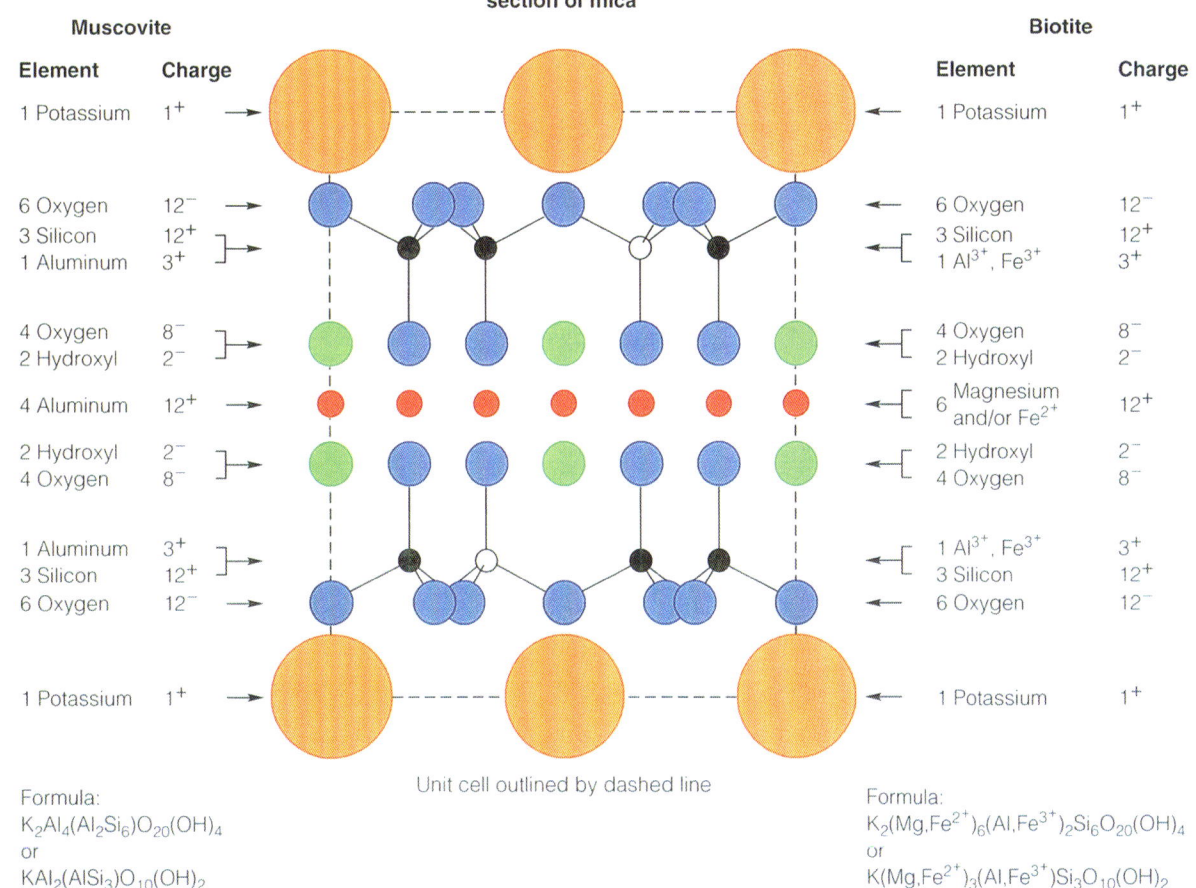

FIGURE 3.22 *The difference between biotite and muscovite is largely determined by the kinds of cations contained between the sheets. Biotite, the ferromagnesian mica, contains iron and/or magnesium while muscovite, the nonferromagnesian mica, contains aluminum. Note, however, that in both cases, the total cation charge is equal to the total anion charge; the lattice must be electrically neutral.*

between adjacent structural sheets. Note once again the substitution of aluminum (Al^{3+}) silicon for (Si^{4+}) in the tetrahedral structure.

Clay minerals are another important example of the sheet silicate structure. These minerals form by the chemical alteration of most of the silicate minerals, including the micas, and are a major constituent of many soils.

Framework Structure The **framework structure** is a three-dimensional arrangement of tetrahedra in which each oxygen is shared by adjoining tetrahedra giving a silicon/oxygen ration of 1:2 (Figure 3.23). The *feldspars*, the major rock-forming minerals, and *quartz* have this type of structure.

The **feldspar group** is subdivided into two subgroups, the **plagioclase** feldspars and the **alkali** feldspars. The plagioclase group is a solid-solution series in which composition varies from a *calcium-rich* (Ca^{2+}) end-member, *anorthite*, $CaAl_2Si_2O_8$, to a *sodium-rich* (Na^{1+}) end-member, *alabite*, $NaAlSi_3O_8$. The alkali feldspars include *orthoclase, microcline,* and *sanidine*. Orthoclase and microcline are identical in composition ($KAlSi_3O_8$), and sanidine differs in that sodium may substitute for potassium, (K,Na)$AlSi_3O_8$. All three minerals differ in crystal structure.

The two plagioclase end-members give us an opportunity to observe, with reasonably simple mineral formulas, an excellent example of the maintenance of electric neutrality in crystals. Note that the *sum* of aluminum and silicon atoms in anorthite and albite is the same although the *number* of individual aluminum and silicon atoms is different: two each in anorthite but one aluminum atom and three silicon atoms in albite. The substitution of one Al^{3+} for one Si^{4+} in the tetrahedral structure leaves a net

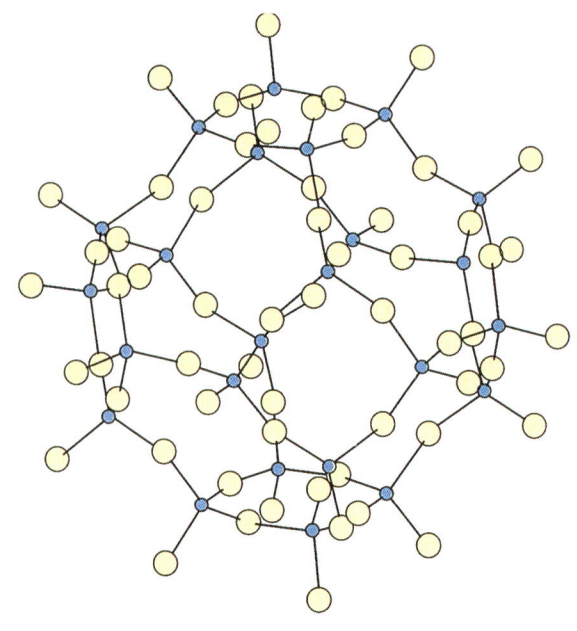

FIGURE 3.23 *Some silicate minerals, such as quartz, have a three-dimensional lattice where each silicon-oxygen tetrahedron is bonded to four other tetrahedra forming a very strong structure that is extremely resistant to chemical attack.*

imbalance of 1+, which is made up by a sodium ion (Na^{1+}). In similar fashion, the substitution of two Al^{3+} ions requires a calcium ion (Ca^{2+}) to maintain the charge balance (Table 3.6).

Orthoclase feldspar, $KAlSi_3O_8$, illustrates the importance of size in determining whether one atom can substitute for another in a crystal lattice. Although orthoclase is also a feldspar, the potassium ion (K^{1+}) is excluded from the plagioclase group, except at high temperatures, because its diameter is too large (Figure 3.24).

TABLE 3.6
CHARGE BALANCE IN THE PLAGIOCLASE FELDSPARS

Anorthite ($CaAl_2Si_2O_8$)			Albite ($NaAlSi_3O_8$)		
Element	Number	Charge	Element	Number	Charge
Ca^{2+}	1	+2	Na^{1+}	1	+1
Al^{3+}	2	+6	Al^{3+}	1	+3
Si^{4+}	2	+8	Si^{4+}	3	+12
O^{2-}	8	−16	O^{2-}	8	−16
	Total charge	+16 −16		Total charge	+16 −16
	Net charge	0		Net charge	0

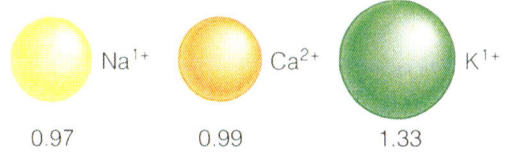

FIGURE 3.24 *The variety of silicate minerals is determined, at least in part, by substitutions that may go on within the lattices. For example, the similar diameters of the ions of sodium and calcium allow them to substitute freely within the feldspar lattice, giving rise to the suite of minerals known as the plagioclase feldspars. Note, however, that the potassium ion is excluded from the plagioclase feldspars because of its large diameter and as a result is a member of another group of feldspars called the K-spars.*

Quartz (SiO_2), is the second most abundant silicate mineral in Earth's crust after feldspar. In the crystal structure of quartz, each oxygen atom in the silicon-oxygen tetrahedra contributes a negative charge to two Si^{4+} atoms; consequently, no other positive ions are needed to attain electric neutrality. The structural formula of quartz, SiO_2, is therefore the same as the silicon/oxygen ratio of the network structure, 1:2.

The compositional subdivision of the silicates into ferromagnesian and nonferromagnesian minerals also separates the minerals in terms of a number of basic physical properties. Because the iron in crystal lattices is highly absorptive of light, ferromagnesian sililcates are all dark colored—black, dark green, and brown. The nonferromagnesian silicates, lacking iron, are light in color—colorless, white, or pink. Because of their iron content, the ferromagnesian minerals have a higher specific gravity than the nonferromagnesians. By utilizing these basic differences, mineral representatives of the two groups can usually be easily distinguished.

SPOT REVIEW

1. Many of the silicate minerals are members of a solid-solution series. What is a solid-solution series? Give examples of minerals that represent such a series.
2. Using the drawings in the text, demonstrate how the silicon/oxygen ratio for each of the silicate structures is determined.

Nonsilicate Minerals

Although thus far we have emphasized the silicate minerals, we cannot end a discussion of minerals without placing the nonsilicate minerals in proper perspective. Although the nonsilicate minerals make up less than 10% of Earth's crust, they are nevertheless of great importance. *Hematite,* Fe_2O_3, the representative of the oxide group in Table 3.5, is the major ore of iron. Where would modern civilization be without iron? *Calcite,* $CaCO_3$, is the major mineral of limestone. Limestone is the raw material used for the production of Portland cement, an essential material for the construction industry. *Soda niter,* $NaNO_3$, is an industrial mineral that serves as the raw material for a wide range of products including fertilizer. *Apatite,* $Ca_5(PO_4)_3(OH,F,Cl)$, is used in the production of phosphoric acid, which is a basic chemical ingredient in the manufacture of fertilizers. *Gypsum,* $CaSO_4 \cdot 2H_2O$, is used to make wallboard for home and building construction. The various crystalline and amorphous forms of Al_2O_3 found in *bauxite* are the source of aluminum, which is used for everything from aircraft bodies to beverage cans and aluminum foil. The sulfide minerals are major sources of a range of metals including copper, zinc, and lead. A number of elements, known as *native elements,* appear in nature uncombined in a solid or liquid state. Native elements can be either nonmetallic such as *carbon* and *sulfur,* semimetallic such as *arsenic* and *bismuth,* or metallic such as *gold, copper, mercury,* and *silver.*

CONCEPTS AND TERMS TO REMEMBER

Minerals	atomic number	octet rule
Atomic structure	atomic mass	inert gas
proton	isotope	Combining of Atoms
electron	radioactive	compound
neutron	electron capture	cation

CONCEPTS AND TERMS TO REMEMBER *continued*

anion
electric neutrality
Bonding
  ionic bonding
  covalent bonding
  metallic bonding
  van der Walls bonding
Crystal structure
  lattice
  crystalline
  amorphous
  mineraloid (glass)
Physical properties
  streak
  hardness

Mohs's Scale of Hardness
form
mineral cleavage
fracture
specific gravity
density
luster
Instrumental analysis
  light microscopy
  X-ray diffraction
Classification
  silicates
  nonsilicates
Silicate crystal structure
  silicon-oxygen tetrahedron

isolated tetrahedral structure
single-chain structure
double-chain structure
sheet structure
framework structure
Solid-solution series
Mineral groups
  olivine group
  pyroxene group
  amphibole group
  mica group
  feldspar group
    plagioclase feldspar subgroup
    alkali feldspar subgroup

REVIEW QUESTIONS

1. The tendency for a mineral to break along a plane of weakness is called
   a. cleavage.   c. hardness.
   b. fracture.   d. mstreak.
2. The loss of an electron from an atom will produce a(an)
   a. cation.   c. anion.
   b. isotope.   d. neutron.
3. The hardness of quartz on Mohs's Scale of Hardness is
   a. 4.   c. 7.
   b. 5.   d. 9.
4. An atom has 13 protons and 14 neutrons in its nucleus. Which of the following is true?
   a. The atomic weight is 13.
   b. The atomic number is 13.
   c. The atomic number is 27.
   d. The nucleus is surrounded by 27 electrons.
5. The most abundant element in Earth's crust is
   a. silicon.   c. oxygen.
   b. iron.   d. aluminum.
6. The Ca and K feldspars are _____, respectively.
   a. anorthite and orthoclase
   b. anorthite and albite
   c. albite and orthoclase
   d. orthoclase and anorthite
7. The silicate mineral group that is characterized by the single-chain structure is
   a. pyroxene.   c. plagioclase feldspar.
   b. mica.   d. amphibole.
8. The physical property known as "fracture" is shown by
   a. all minerals.
   b. minerals that do not show cleavage.
   c. silicate minerals with sheet-type structures.
   d. minerals with hardness in excess of 7.
9. The physical property that would most readily distinguish a pyroxene from a feldspar is
   a. density.   c. form.
   b. color.   d. hardness.

10. The most abundant group of minerals in Earth's crust are
    a. oxides.   c. ferromagnesians.
    b. feldspars.   d. carbonates.
11. Why does the proper identification of any object usually require information about both its composition and its structure? Give some examples where using only compositional or only structural information might result in an erroneous identification.
12. Several naturally occurring substances that are commonly referred to as "mineral resources" are by definition not minerals and therefore should be referred to as natural resources, not mineral resources. Give some examples and explain why each technically would not qualify as a mineral resource.
13. Why must atoms be converted into ions before they can be joined together into compounds?
14. Discuss the relative importance of composition and crystal structure in determining each of the physical properties used to identify minerals.
15. Of the feldspar minerals, anorthite and albite represent the end-members of a series of minerals referred to as the plagioclase feldspars. In what way are the plagioclase feldspars similar? In what way are they different? Why is orthoclase feldspar excluded from the plagioclase series?

THOUGHT PROBLEMS

1. Assume you have the following items: (a) a beaker graduated in tenths of milliliters, (b) a balance capable of determining mass to a tenth of a gram, and (c) a mineral specimen. How would you determine the density of the mineral?
2. What determines whether a particular atom will be prone to form a cation or an anion?
3. Using diagrams of the basic structures, explain why halite has three mutually perpendicular sets of cleavage planes, the pyroxene and amphibole minerals have two sets, and the micas have only one set.

FOR YOUR NOTEBOOK

Most of us come in contact with minerals each day. We have already commented on the use of feldspars in many household cleansers. It might be interesting to consider your home room by room to determine which, if any, minerals can be found. Don't forget certain health additives such as dolomite tablets. Many of us wear jewelry made of gold or silver, much of which is adorned with various kinds of gemstones. Although the names of many of the more common gemstones are familiar, their compositions and modes of origin are not. Using a list of birthstones as a starting point, collect information about the composition, mode of occurrence, and common sources of the gems. In addition to the information that may be available in a mineralogy textbook, a visit to a local jeweler or a rock and mineral shop will provide some more practical information such as their relative values and how they are most likely to be found as settings.

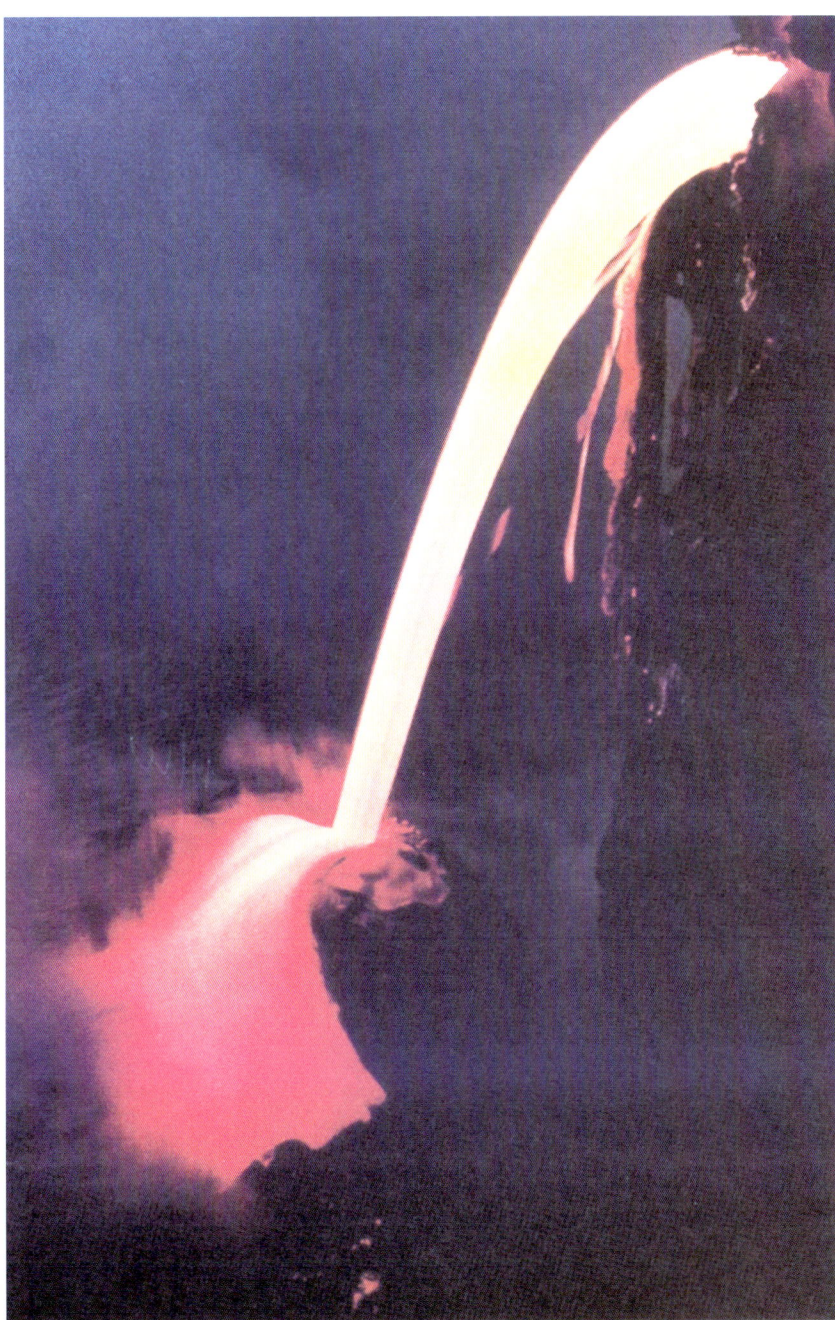

CHAPTER OUTLINE

INTRODUCTION

DISTRIBUTION OF VOLCANOES
- Relationship to Plate Boundaries
- Intraplate Volcanism

EJECTED MATERIALS
- Gases
- Liquids
- Solids (Tephra)
- Ash Flows
- Mudflows or Lahars

KINDS OF VOLCANOES
- Cinder Cones
- Spatter Cones
- Shield Volcanoes
- Stratovolcanoes or Composite Volcanoes

ERUPTIVE INTENSITY
- Classification of Eruptive Intensity

FISSURE ERUPTIONS
- Oceanic Ridges
- Flood Basalts
- Felsic Eruptions

CRATERS AND CALDERAS

OTHER VOLCANIC FEATURES
- Hot Springs
- Geysers
- Fumaroles

PREDICTION OF VOLCANIC ERUPTIONS

EXAMPLES OF OTHER NOTEWORTHY VOLCANOES
- Mount Vesuvius
- Santorini
- Krakatoa
- Mount St. Helens
- The Hawaiian Islands
- Sakurajima, Japan

CHAPTER 4

Volcanism

INTRODUCTION   During April 11 and 12, 1815, Earth experienced the largest single volcanic eruption in recorded time. Tambora, a volcano on the island of Sumbawa in Indonesia, erupted, blasting an estimated 36 cubic miles (150 km^3) of material into the air. The ash content of the atmosphere was so dense that for three days total darkness settled over an area of more than 900 square miles (2,500 km^2). The blast and the resultant seawave that inundated the surrounding islands killed an estimated 50,000 people with thousands more dying during the famine that followed. The winds of the upper atmosphere carried fine volcanic ash around the world, producing spectacular sunsets for several years. Unfortunately, the ash also filtered out the radiant heat of the Sun, and atmospheric temperatures dropped worldwide. The year 1816 became known as "the year without a summer." In Maine, frosts were recorded every month of 1816, and a snowstorm swept New England in June. Worse yet, around the world, at least 80,000 people starved as growing seasons were shortened and crops failed. Some blame the Tambora eruption for a famine in Ireland as well as for an outbreak of cholera in Asia that lasted until 1833 and claimed millions of lives. Can such a cataclysmic event happen again? The answer is yes. Where and when remain unresolved.

Before the advent of plate tectonics, the topic of volcanism was the center of a geologic debate. Geologists had long been aware that, for the most part, the distribution of active volcanoes coincided with zones of major earthquakes. The association sparked vigorous discussions. Some geologists argued that as rocks break under stress, they release energy in the form of earthquakes in much the same fashion as energy is released with the snap of a breaking stick. The earthquakes cause fractures to open in the lithosphere that allow molten rock to make its way to Earth's surface. In other words, earthquakes cause volcanic eruptions.

The other camp argued that volcanic eruptions cause earthquakes. As the molten rock forces its way to the surface, rocks are forced out of the way, stressed, and broken, giving rise to the release of energy responsible for the earthquake. It is true that the rise of magma does cause rocks to rupture, but the resultant earthquakes are all low magnitude. Major earthquakes are not caused by the ascent of magma.

We now know that the association of volcanism and major earthquakes is not a cause and effect relationship. Instead, both phenomena are associated with the same geologic setting, that is, hot spots and the edges of the lithospheric plates. Earthquakes occur where rocks are broken and moved as a result of being pulled apart or thrust together. Magma, generated beneath the plate edges, rises along the fractures that form as a result of these movements and emerges at the surface as a volcanic eruption. As the following discussion will explain, the composition of magma, the intensity of volcanic eruptions, and the magnitude of earthquakes associated with plate margins vary considerably, depending upon whether the margin is divergent or convergent.

Volcanoes are described as *active, dormant,* or *extinct.* Although the distinction between active and dormant is not precise, the term **active** is usually used to describe an erupting volcano. A **dormant** volcano, on the other hand, is one that is not now erupting but has erupted in historic time and is considered likely to erupt in the future. An **extinct** volcano is one that is not now erupting and is not likely to erupt in the future. At present, there are from 40 to 50 active and about 600 dormant volcanoes on Earth (Figure 4.1).

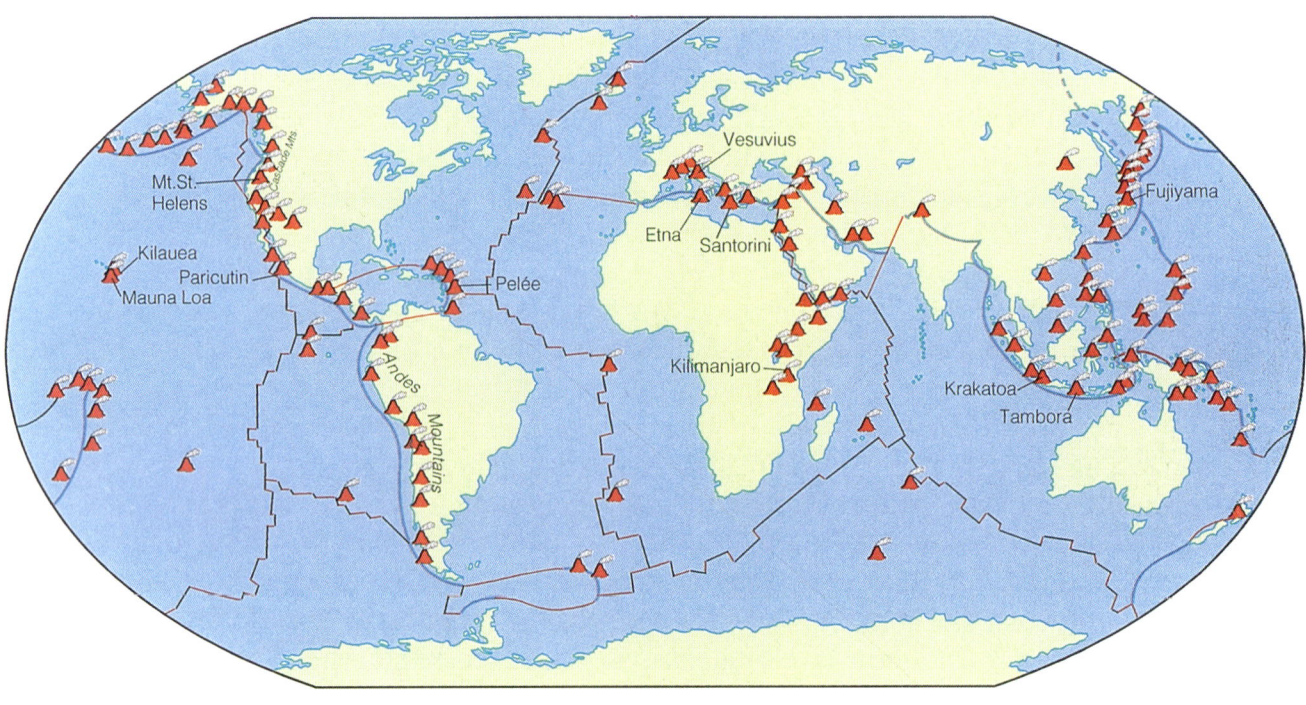

FIGURE 4.1a *Most of Earth's active and dormant volcanoes are associated with convergent plate margins with a few located within plate boundaries. The greatest concentration of volcanoes is found around the Pacific Ocean basin, giving rise to the name "Ring of Fire" for the Pacific rim.*

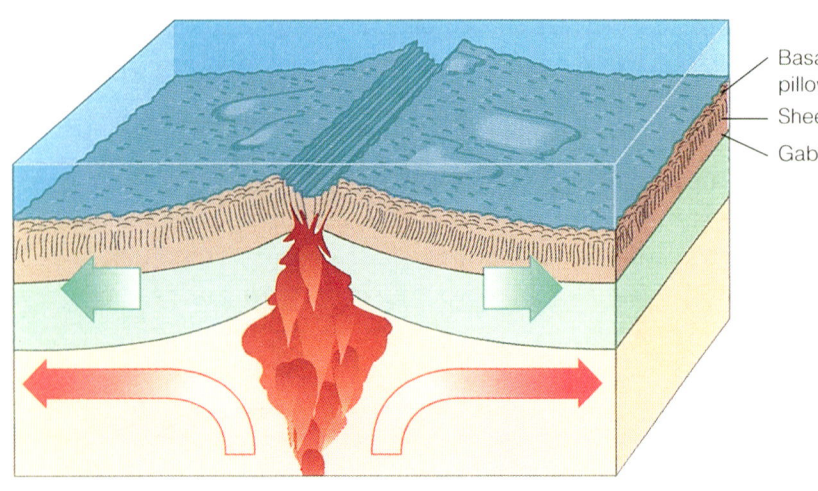

FIGURE 4.1b *Most of Earth's volcanic activity is associated with the oceanic ridges where molten rock constantly wells to the surface to create new oceanic lithosphere.*

DISTRIBUTION OF VOLCANOES

When the locations of Earth's active volcanoes are plotted on a map, a distinctive pattern emerges. Volcanoes are found along the margins of the lithospheric plates with some intraplate volcanism, especially beneath the Pacific plate.

Relationship to Plate Boundaries

Most volcanic activity is associated with **divergent margins** where basaltic magmas well up along the oceanic ridges to continuously form new oceanic crust. Because it is largely submarine, divergent boundary volcanism generally goes unnoticed. The island of Iceland where a small portion of the North Atlantic Oceanic Ridge builds above sea level provides a rare opportunity to observe oceanic ridge volcanism (Figure 4.2). Several examples of divergent boundary volcanism are located within continents, the best known being the East African Rift Valley (Figure 4.3). A possible example of an incipient divergent boundary is the Rio Grande Rift of the southwestern United States (Figure 4.4).

Most of the observable and familiar examples of active volcanoes are associated with **convergent margins**. In instances where two oceanic plates converge, chains of volcanic islands called **island arcs** form on the overriding plate (Figure 4.5). The Aleutian Islands are an example.

Where an oceanic lithospheric plate and a continental lithospheric plate converge at a plate boundary, the volcanism builds a chain of volcanoes called a **continental arc** on the edge of the continental plate (Figure 4.6). Examples include the Cascade Mountains of the northwestern United States and the Andes Mountains along the western margin of South America.

FIGURE 4.2   *The rift valley associated with the North Atlantic Oceanic Ridge can be seen on Iceland. (© Barry Griffiths/Photo Researchers)*

FIGURE 4.3   *The best-known example of a rift valley is the* East African Rift Valley. *At the present time, seawater is beginning to encroach into the northern end of the valley. Soon, the rift valley will begin to flood and may be converted into Earth's newest* linear ocean.

FIGURE 4.4 *The* Rio Grande Rift Zone *of the southwestern United States is possibly the first stage in the breakup of the North American continent. (Data made available by the New Mexico Bureau of Mines/Mineral Resources)*

Most of Earth's 600 or so active or dormant volcanoes are associated with the subduction zones that encircle the Pacific Ocean basin, a band known poetically as the **"Ring of Fire."** In contrast to the relatively quiet eruption of basaltic magmas along rift zones or oceanic ridges, subduction zone volcanism is characterized by the explosive eruption of more silica-rich magmas.

Intraplate Volcanism

A third source of volcanism is associated with mantle plumes generated by **hot spots** located deep within the mantle (Figure 4.7). Mantle plumes deliver heat to the lithosphere in much the same way as a Bunsen burner heats a beaker. The location of the hot spot and the mantle plume remains relatively fixed over geologic time while the overlying lithosphere moves laterally with the movement of the plate. Many of Earth's hot spots are located beneath the Pacific plate. As the mantle rocks rise within the plume and the pressure drops, the rocks at the top of the asthenosphere melt. The molten rock then moves upward and accumulates in a magma chamber within the lithosphere. From the magma chamber, the molten rock erupts to the surface where it gives rise to a large number of submarine volcanoes called *seamounts*.

The movement of the Pacific oceanic plate over three separate hot spots has created three parallel strings of volcanic islands and seamounts in the Pacific Ocean: (1) the Hawaiian Islands–Emperor Seamount chain, (2) the Tuamotu Archipelago–Line Island chain, and (3) the Austral, Gilbert, and Marshall Islands (Figure 4.8). Like divergent margin volcanism, hot spot volcanism associated with oceanic plates usually involves the relatively quiescent eruption of basaltic magmas. Hot spot eruptions involving continental crustal plates are generally quite different. For example, the hot spot located beneath Yellowstone Park resulted in the highly explosive eruption of viscous rhyolitic magma (Figure 4.9). The Yellowstone volcanism will be discussed later in the chapter.

SPOT REVIEW

1. Describe the geologic settings under which most of Earth's volcanoes form.
2. How do the geologic settings responsible for island arc and continental arc volcanism differ? In what ways are they similar?

EJECTED MATERIALS

The materials emitted from volcanoes include gases, liquids, and solids.

Volcanism   101

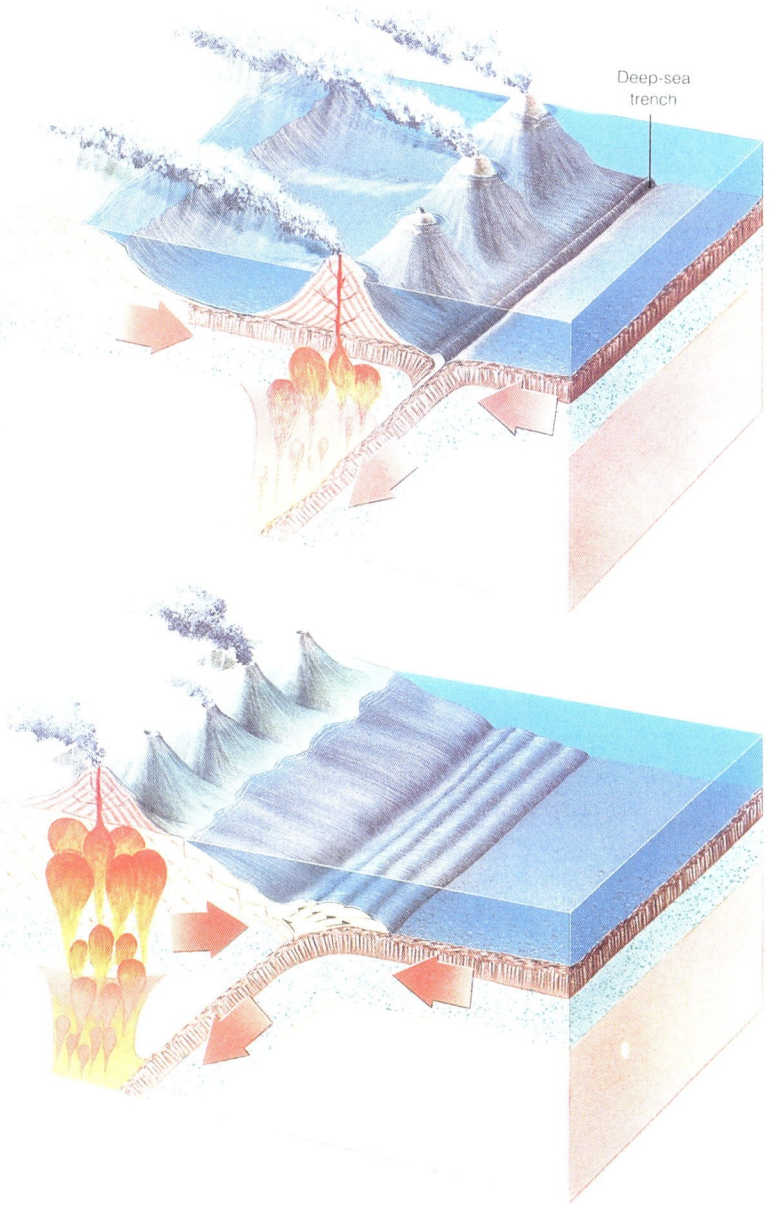

FIGURE 4.5   Island arcs, *such as the* Aleutian Islands, *form as chains of volcanoes build along the edge of the overriding plate at an ocean-ocean convergence.*

FIGURE 4.6   Continental arcs, *such as the* Cascade Mountains *of Oregon and Washington and the* Andes Mountains *of South America, form along the edge of the overriding plate at an ocean-continent convergence.*

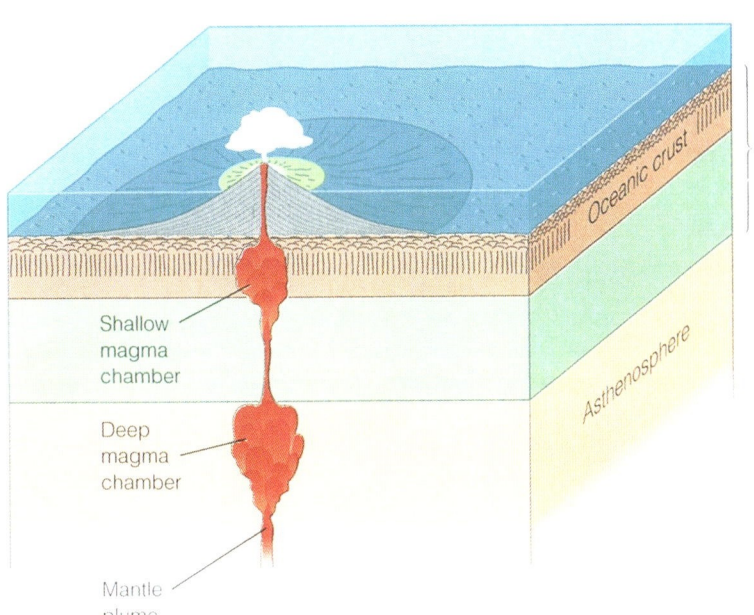

FIGURE 4.7   *Volcanism within the lithospheric plates is located over* mantle plumes *that produce hot spots in the overlying lithosphere. For reasons that are not fully understood, most hot spots are located under the Pacific Ocean lithospheric plate.*

FIGURE 4.8   *Three chains of volcanic islands that formed over stationary mantle plumes exist in the Pacific Ocean. Hawaii is a part of one of these chains.*

FIGURE 4.9   *The rocks exposed in* Yellowstone Park, *Wyoming, are pyroclastic rocks that were created by the violent eruption of highly viscous rhyolitic magma. (Courtesy of W. B. Hamilton/USGS)*

Gases

Obtaining an accurate analysis of the gases emitted by an erupting volcano is difficult due to the obvious problems involved in direct sampling and the rapidity with which the gases mix with the atmosphere (Figure 4.10). Data have shown that the evolved gases not only differ from one volcano to the next, but that the composition of gas evolved from a particular volcano will change during an eruption.

Water vapor is the most abundant gas emitted from an erupting volcano. Much of the billowing white cloud observed above an erupting volcano is condensed water vapor or steam. Some of the water is derived from the magma; however, most of the water is derived from groundwater, the water contained in the rocks beneath the surface of the Earth. As the magma rises toward the surface and encounters groundwater, the water is heated to boiling and emerges as water vapor. If the volume of water is large, a phreatic (steam) explosion can occur.

During the primeval days of Earth, most of the water emitted from volcanic eruptions was derived from the magma. The consensus among geologists is that nearly all of the water now contained within Earth's hydrosphere has derived from the volcanic eruptions that dominated the surface of Earth during

FIGURE 4.10 *An erupting volcano, such as Mount Pinatubo in the Philippine Islands, spews out a mixture of hot gases, magma, and rock fragments. (Courtesy of T. J. Casadevall/USGS)*

its early evolution. Scientists once thought that the primeval Earth was completely molten at some time in its early history. Although this idea is no longer accepted, many geologists are convinced that "magma oceans" may have existed during the very early history of Earth. During these early stages, Earth's solid surface was very hot and dominated by widespread volcanism, which expelled enormous volumes of water from Earth's interior. With the surface too hot to allow the rainwater to accumulate, the water collected in a thick cloud cover. Eventually, Earth cooled to the temperature at which rainwater was not vaporized. The water then accumulated and accelerated the cooling of the surface rocks; torrential rains washed the high areas and flowed into the low areas. In time, the low areas filled with water to form the oceans, and the high areas became land surfaces with lakes and streams. As Earth's hydrosphere formed, the basic geologic processes of erosion were initiated.

In addition to water, volcanic eruptions release other gases in significant volumes. Of these, the most common are *sulfur dioxide* and *carbon dioxide*. Sulfur dioxide, the acrid smell of "fire and brimstone," has an unpleasant odor that is characteristic of areas of active volcanism. Carbon dioxide is the atmospheric component that is an important contributor to the "greenhouse effect," in which heat reflected from the Earth's surface is trapped and retained by the atmosphere. Throughout geologic time, the two major natural sources of carbon dioxide have been volcanic eruptions and the decay of organic matter. The rate at which carbon dioxide was added to the atmosphere eventually became balanced with the rate at which it was removed by dissolution in water, the formation of the shells of many marine organisms, incorporation into the body tissues of plants and animals, and storage in limestone and in the fossil fuels, oil, gas, and coal. With this balance, Earth achieved a mean atmospheric temperature of about 58° F (14.5°C) that has been maintained over most of geologic time. The current increase in the rate at which carbon dioxide is being introduced into the atmosphere due to the burning of fossil fuels has the potential to significantly unbalance the carbon dioxide cycle, enhance the greenhouse effect, and subsequently increase Earth's atmospheric temperature.

Several gases emitted during eruptions, including chlorine and sulfur dioxide, react with water to produce strong acids. If inhaled, the fumes can cause acute discomfort. For this reason, individuals who suffer from respiratory problems are usually advised to avoid areas of active volcanism.

Liquids

Most lavas are basaltic in composition. Because of their relatively low viscosity, basaltic magmas rise to the surface through fractures with relative ease (Figure 4.11). From our discussions of plate tectonics, you will remember that basaltic lavas are associated with oceanic ridges, oceanic hot spots, fissure eruptions, and continental hot spots.

FIGURE 4.11 *Typical of Hawaiian eruptions, low viscosity basaltic magma is extruded quietly onto the surface and flows downslope as a river of molten rock. (Courtesy of USGS/HVO)*

Although lavas erupting in subduction zones vary widely in composition, the great majority are andesitic. Basaltic lavas are associated with some subduction zone volcanism, for example, in the Cascades, and late-stage subduction zone eruptions may evolve highly viscous rhyolitic lavas that form thick, semisolid plugs within the volcanoes' craters (Figure 4.12).

Depending on whether the lava solidifies exposed to the atmosphere (subaerial) or under water (subaqueous), basaltic lavas take on one of three forms: (1) *pahoehoe*, (2) *aa*, or (3) *pillow lava*. Subaerial solidification of hot, highly fluid lavas produces a flow with a smooth surface often with the appearance of twisted ropes (Figure 4.13). Although this type of lava is sometimes called "ropy," volcanologists refer to it by the Hawaiian name, **pahoehoe**.

As the basaltic lavas cool subaerially and lose dissolved gas, they become more viscous and flow more slowly (Figure 4.14). The surface of the flow solidifies, but because the interior of the flow is still molten and moving, the solidified layer breaks into angular blocks that are carried along on the surface of the molten flow. Eventually, when the entire mass is solidified, the lava is characterized by a rough, angular surface (Figure 4.15). Because of its appearance, this type of lava is sometimes referred to as "blocky." Volcanologists, however, almost universally use the Hawaiian name, **aa**, to describe this type of lava.

FIGURE 4.12 *Following the typically violent eruption of volcanoes associated with zones of subduction, such as Mount St. Helens, highly viscous rhyolitic magmas often move into the crater and solidify to form a plug. (Courtesy of USGS/CVO-F)*

FIGURE 4.13 *Low-viscosity basaltic magmas typically solidify to form a smooth or "ropy" surface that volcanologists refer to as* pahoehoe *lava.*

FIGURE 4.14 *As basaltic magma cools and loses its gas content, it becomes more viscous and moves more slowly. (Courtesy of USGS/HVO)*

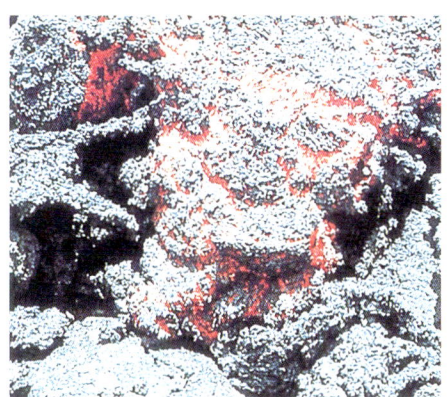

FIGURE 4.15 *When the more viscous basaltic magma solidifies, it forms a lava with a sharp, angular surface called* aa. *(a Courtesy of USGS/HVO, b Courtesy of Joe Donovan)*

Pillow lava solidifies under water. It derives its name from the flow's resemblance to a pile of pillows (Figure 4.16). Pillow lavas form at oceanic hot spots and at oceanic ridges where the lavas spread laterally to make up much of the oceanic crust. Because the oceanic crust represents 71% of Earth's surface, basaltic pillow lavas can be considered to be the most common rock type on or near Earth's surface. In addition to the lavas that are extruded under water, pillow lavas may also form from lavas that originate on land and subsequently flow into the water (Figure 4.17).

Solids (Tephra)

The solid fragments ejected from an erupting volcano are collectively called **tephra**. Such an eruption and the rocks formed by it are considered to be *pyroclastic*. The prefix *pyro-* comes from the Greek word meaning "fire," and *clastic* refers to broken bits and pieces of minerals and rocks. Pyroclastic material is derived from a number of sources including the old volcano walls, shattered fragments of solidified magma, and pieces of lava that were thrown into the air and solidified in flight or on landing. Pyroclastic materials are of two kinds: (1) *tephra falls* and (2) *ash flows*.

Tephra-fall materials are classified based on shape and size into five types: (1) *blocks*, (2) *bombs*, (3) *cinders*, (4) *ash*, and (5) *dust*. **Blocks** and **bombs** are particles with mean diameters in excess of 2 1/2 inches (64 mm), about the size of a baseball (Figure 4.18). Blocks are angular fragments that are generated when portions of the crater walls or volcanic cone are blown apart by explosions. Bombs are blobs of molten rock that are thrown out of the crater, take on a streamlined shape as they pass through the air, and solidify only partly before returning to Earth's surface. **Cinders** are particles that range in diameter from 2 1/2 inches (64 mm) down to about 1/16 inch (2 mm), about the size of a pinhead. Cinders form

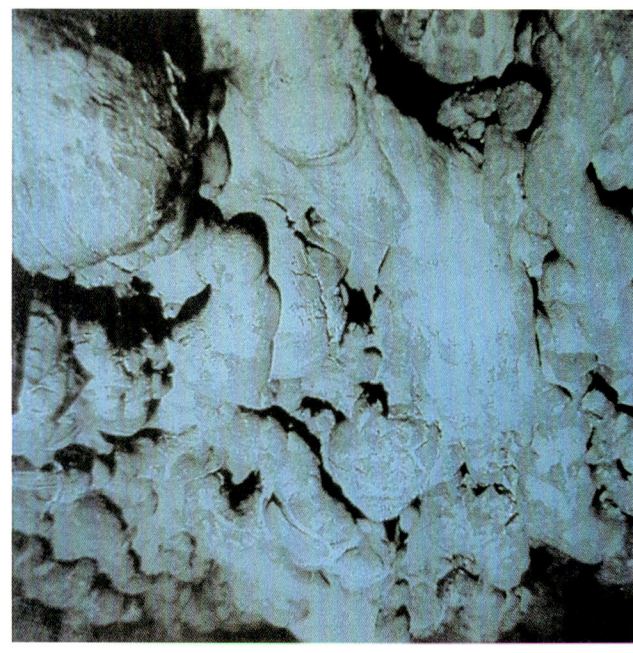

FIGURE 4.16 *When lava cools underwater, it forms* pillow lava, *so-called because of its pillow-shape. Since it typically makes up the* oceanic crust, pillow lava must be the most extensively exposed rock type on Earth. (Courtesy of Angus/Woods Hole Oceanographic Institution)

FIGURE 4.17 *Pillow lava also forms where molten lavas pour into the ocean, as is now occurring along the southern coast of Hawaii. (Courtesy of USGS/HVO)*

FIGURE 4.18 *When relatively large blobs of molten lava are thrown into the air, they may solidify after taking on an aerodynamic shape and land as* bombs. *(Courtesy of USGS/HVO)*

primarily molten spatter than cools either in midair or upon striking the ground (Figure 4.19). **Ash** and **dust** are particles with diameters smaller than 1/16 inch (2 mm), with dust being the finer.

Most tephra-fall materials larger than ash accumulate in the immediate vicinity of the volcano. The fine-sized materials, however, especially dust, can be carried long distances by the prevailing winds (Figure 4.20). Dust-sized materials from major volcanic eruptions may be carried around Earth in the upper atmosphere where they are responsible for many brilliant sunsets before finally settling back to the surface.

Volcanically derived dust in the upper atmosphere has also been responsible for significant reductions in mean atmospheric temperatures. One of the

FIGURE 4.19 Cinders *are commonly generated as basaltic magmas fountain, break into small pieces, and solidify in midair. Should the eruption continue for a period of time, the cinders will accumulate around the vent to form a* cinder cone. *(Courtesy of USGS/HVO)*

FIGURE 4.20 *Ash- and dust-sized materials thrown into the atmosphere by an erupting volcano can be carried by the prevailing wind for long distances. (Courtesy of USGS/MSH, July 22, 1980)*

first scientists to suggest that volcanic eruptions could affect weather was Benjamin Franklin who, in 1784, suggested that the extreme winter of 1783–1784 was the result of eruptions in Japan and Iceland. We have already commented on the climatic effects of the 1815 eruption of Tambora.

The rocks that form from the accumulated tephra-fall materials are called **tuff** in the case of the fine-grained ash and dust (figure 4.21) and **volcanic breccia** when made up of the larger particles.

Ash Flows

If the concentration of pyroclastic materials ejected during the eruption is very high, the eruption column may collapse, the ash will mix with superheated gases, and the combined mass, called an **ash flow**, will move down the mountainside like an avalanche (Figure 4.22). Because they ride on a cushion of hot air, ash flows move with little frictional resistance and, depending upon the angle of the slope, can attain speeds of over one hundred miles per hour. In some cases, an exceptionally hot cloud of gas called a **nuée ardente** (French for "glowing cloud"), often heated to glowing, may override the ash flow. Needless to say, such a mass of fast-moving, hot materials can be enormously destructive.

The solidified debris from ash flows or nuées ardentes are called **ash-flow tuffs**. If the fragments are partially fused by compaction after movement stops, the deposit is called a **welded tuff**. The term **ignimbrite** is used to describe a deposit consisting of a variety of tuffs.

Mudflows or Lahars

Even in equatorial regions, volcanoes such as Mount Kilimanjaro commonly build to such elevations that the summits are covered with snow and capped with ice. During an eruption, some snow and ice may melt, producing large volumes of water. In addition, volcanic eruptions are commonly accompanied by torrential rains triggered by the atmospheric disturbances above the erupting volcano. These waters mix with the loose tephra-fall materials and materials already accumulated on the slopes of the volcano and create a flow called a **lahar** that moves downslope, following an existing stream valley. Lahars usually move more slowly than ash flows, but they have still been clocked at 50 to 60 miles per hour (80–95 KPH). Lahars accounted for much of the destruction resulting from the 1980 eruption of Mount St. Helens (Figure 4.23).

FIGURE 4.21 *When the ash- and dust-sized materials blown into the atmosphere return to Earth, they commonly fall over a wide area and form extensive layers of a pyroclastic rock called* tuff. *(Courtesy of W. B. Hamilton/USGS)*

FIGURE 4.22 *The Pinnacles in Crater Lake National Park, Oregon are carved from tuff that formed from an ash flow. (© Raymond G. Barnes/TSW)*

FIGURE 4.23 *Most of the damage caused by the eruption of Mount St. Helens was the result of* mudflows *or* lahars. *(Courtesy of MSH-Banks/USGS)*

SPOT REVIEW

1. Explain how three different kinds of lava can form from molten rock of the same composition.
2. In what ways have the gases emitted from volcanoes affected the atmosphere since the formation of Earth?
3. What do ash flows, nuées ardentes, and lahars have in common?

KINDS OF VOLCANOES

Volcanologists define a volcano as the conical structure that builds around a vent or the vent itself. As in the case of intrusive igneous rock bodies, the definition of a volcano does not include a size requirement, another example of geologists being more concerned with the process by which a feature forms than with how large or small it may be. We will discuss four kinds of volcanoes: (1) *cinder cones*, (2) *spatter cones*, (3) *shield volcanoes* and (4) *strato-volcanoes* or *composite volcanoes*.

Cinder Cones

As the name implies, a **cinder cone** is an accumulation of largely cinder-sized pyroclastic material around a vent. The cinders commonly form by the midair solidification of molten rock spatter expelled by a lava fountain during the eruption (Figure 4.24). Most of the materials within the cone are unconsolidated, although some welding of semisolid particles may take place as the materials accumulate. Most cinder cones are relatively small, usually not growing to elevations of more than about 1,000 feet (300 m) above the surrounding land surface. The slope angle of cinder cones, usually about 40°, is the *angle of repose*, or the highest angle at which unconsolidated, relatively large, irregularly shaped particles can reside without sliding downslope. The summit of a cinder cone is commonly occupied by a crater that may be enlarged following the eruption by the collapse of the inner walls.

Spatter Cones

Vents erupting basaltic magma under low gas pressure typically produce large blobs of molten rock or spatter fragments that fall to the outer edge of the vent where they congeal and solidify (Figure 4.25).

FIGURE 4.24 *Short-lived eruptions of basaltic lavas commonly result in the formation of* cinder cones *around the vent.*

FIGURE 4.25 *When insufficient gas exists to create a fountain of lava, blobs of lava may lap out of the vent and solidify as a* spatter cone. *(Courtesy of USGS/HVO)*

Repeated expulsion of spatter builds a conical structure called a **spatter cone**. Like cinder cones, spatter cones do not grow to great height, usually not more than a few tens of feet (meters).

Shield Volcanoes

Shield volcanoes form as massive volumes of basaltic lavas are extruded from central vents and flank eruptions; they are usually associated with oceanic hot spots (Figure 4.26). Because of the low viscosity of the basaltic lavas, the molten rock readily flows long distances from the vent and solidifies. Repeated eruptions build the conical structure of the volcano. The volcanic cone is broad at the base with average slopes on the submarine portion of about 17°. In some cases, the cone may build above the surface of the ocean and become a volcanic island. Because the lavas cool more slowly in the atmosphere than under water and are therefore more fluid, the subaerial (exposed) portion of the cone typically has slopes of from 5° to 10°. The structures are called "shields" because they resemble the defensive armor carried by warriors. Shield volcanoes that build over oceanic hot spots remain submarine are called *seamounts*.

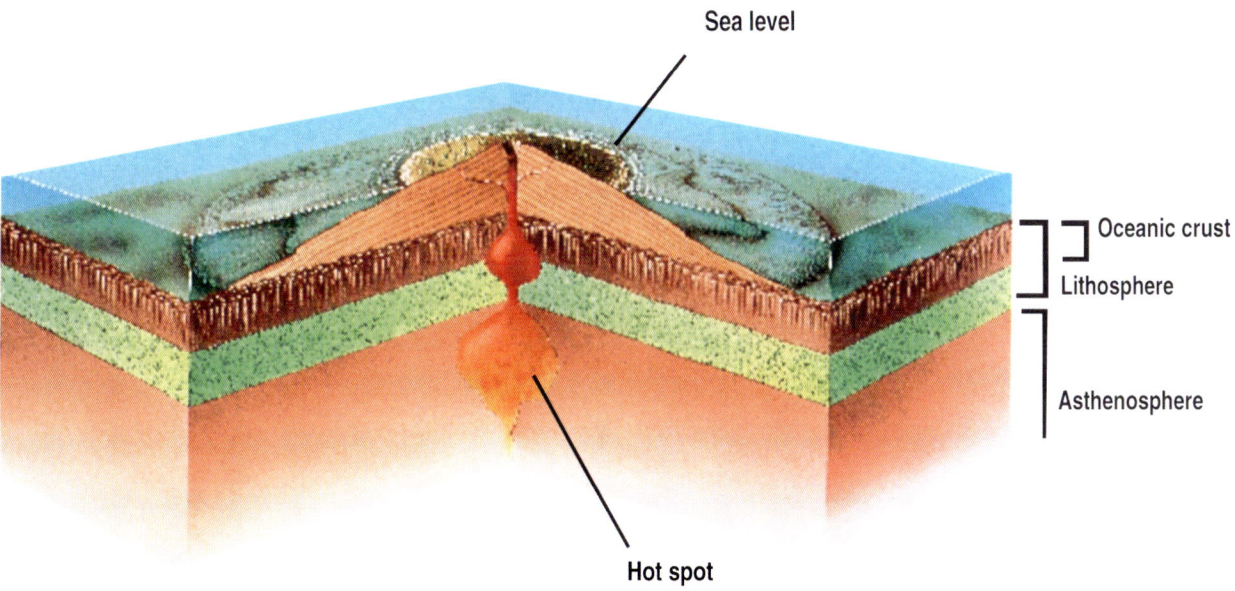

FIGURE 4.26 Shield volcanoes *are constructed of repeated flows of* basaltic lavas, *usually from vents associated with* hot spots. *Most of Earth's shield volcanoes are in the Pacific Ocean basin. The best-known examples of shield volcanoes are the* Hawaiian Islands. *(Photo courtesy of USGS/HVO)*

The best-known examples of shield volcanoes that have grown to become islands, the Hawaiian Islands, will be discussed in detail later in this chapter.

Stratovolcanoes or Composite Volcanoes

Stratovolcanoes or composite volcanoes are commonly associated with zones of subduction. The lavas involved in the construction of the cone are primarily andesitic. Because of the higher viscosity and inherently higher gas content of subduction zone magmas, andesitic magmas tend to contain a greater volume of dissolved gases when they erupt to the surface. As a result, the eruptions are more violent than those associated with shield volcanoes.

The explosive eruptions of subduction zone volcanoes generate vast quantities of pyroclastic material. The coarse pyroclastic materials fall in the immediate vicinity of the vent to create a steep-sided cone that is coated and sealed by the lavas of subsequent eruptions. Because the cone can accumulate with a relatively high angle of repose, the combined pyroclastic-lava layering develops a cone with a relatively steep slope. As successive eruptions repeat the pyroclastic-lava layering, the cone builds rapidly to relatively great heights compared to the diameter of the base (Figure 4.27). The "composite" or "strato" designations refer to the interlayering of the pyroclastic rocks and lava. Lateral penetrations of magma through fractures in the cone produce radiating dikes and result in flank eruptions that add to the complexity of the cone structure. Many of the best-known volcanoes are stratovolcanoes or composite structures, including the famous Fujiyama near Tokyo, Japan.

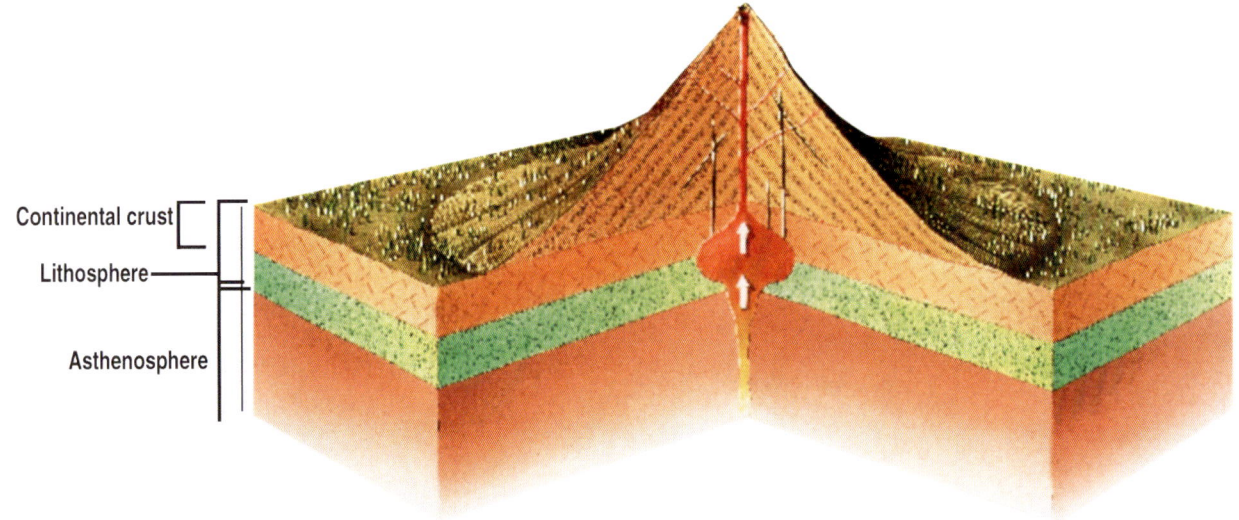

FIGURE 4.27 *Constructed of alternating layers of andesitic lavas and pyroclastic materials,* stratovolcanoes *or* composite volcanoes *form the island and continental arcs associated with zones of subduction. Because of the higher viscosity and gas content of andesitic magmas, eruptions of stratovolcanoes are usually violent. Perhaps the best-known example of a stratovolcano is Fujiyama. (Photo courtesy of Japan National Tourist Organization)*

ERUPTIVE INTENSITY

The differences in the eruptive intensities of volcanoes are graphically portrayed in Figure 4.28, which compares the 1980 eruption of Mount St. Helens with a typical eruption of Kilauea volcano on the island of Hawaii. The question that arises is, how can two volcanic eruptions differ so much in style? Why, for example, does one volcano erupt with the potential to destroy everything within miles while the other's eruption can be watched with relative safety from a few hundred feet? The answer lies in the fundamental difference between basaltic and andesitic-to-rhyolitic magmas and the solubility of gases in solutions.

The eruptive style of a volcano is primarily determined by two magma parameters: (1) **gas content** and (2) **viscosity**. The eruption of a volcano has been likened to the uncorking of a bottle of warm champagne. Indeed, the chemistry and physics involved are quite similar. Simply stated, explosions are the result of the violent expulsion of gas from the magma as it reaches the surface. Magmas that contain little or no dissolved gas when they reach Earth's surface will erupt quietly. If the magma has high viscosity because of low temperature and high silica content, the potential for violent eruptions increases with increasing gas content.

In a solvent, the solubility of any gas increases with increasing pressure. Dissolved gases can escape from solution when the pressure decreases. Consider, for example, your favorite carbonated beverage. The effervescence is due to the release of carbon dioxide the manufacturer forced into solution during the bottling or canning process. As long as the container is closed, the gas remains in solution, but when the container is opened and the pressure is relieved, the gas comes out of solution and bubbles toward the surface. In addition, the warmer the solution, the faster the gas will escape; this explains why keeping the beverage cold prolongs the effervescence and why the drink becomes "flat" when the container is opened and allowed to warm.

Magma viscosity controls not only the rate at which the gas is released from the magma but also the ease with which the gas bubbles can exit the magma mass. As low-viscosity basaltic magmas begin their ascent, gas begins to escape at greater depth, at a higher rate, and more easily than from more viscous andesitic or rhyolitic magmas. By the time basaltic magmas reach the surface, most of the dissolved gas has already been released and vented. As the magmas emerge from vents or fractures, enough gas may still remain to drive the molten rock a few thousand feet into the air, creating an impressive lava fountain, but the gas content is not sufficient to cause truly violent explosions. The result is a relatively quiet eruption. Volcanism associated with oceanic ridges, oceanic hot spots, fissure eruptions, and most continental hot spots involves basaltic lavas and is therefore rather quiescent.

FIGURE 4.28 *The difference in eruptive styles is best illustrated by comparing the eruption of a stratovolcano such as* Mount St. Helens *with that of a shield volcano such as* Kilauea. *(a Courtesy of MSH-Swanson/USGS; b Courtesy of USGS/HVO)*

The eruptive intensity of volcanism associated with subduction zones is quite different. Most of the magmas associated with subduction zone volcanism are andesitic to rhyolitic. Because of their higher viscosities, the release of the gas from solution as these magmas rise toward the surface is retarded. Large volumes of gas may be maintained in solution until the magma breaks through to the surface and the pressures are totally relieved. At this time, the gas is violently released, triggering an explosion such as the 1980 eruption of Mount St. Helens.

The explosions associated with eruptions of rhyolitic magma are often even more violent than those associated with andesitic eruptions and produce huge volumes of pyroclastic materials. Examples include the eruptions that formed most of the rocks in Yellowstone National Park.

Classification of Eruptive Intensity

Volcanologists use five classifications to describe individual volcanic eruptions (1) *Hawaiian*, (2) *Strombolian*, (3) *Vulcanian*, (4) *Pelean*, and (5) *Plinian*.

Hawaiian Type Eruptions   The **Hawaiian** type of volcanic activity has been described as "the quiet evolution of lava" and is typical of basaltic eruptions (Figure 4.29). As the name suggests, this eruptive style is typical of the volcanoes on Hawaii. The few explosive eruptions reported on the Hawaiian Islands have been ascribed to groundwater entering the magma conduit. The ensuing explosions were steam or phreatic explosions and were not the result of either the magma composition or the original gas content of the magma.

FIGURE 4.29   *A typical eruption on Hawaii usually involves little more than the quiet evolution of basaltic magmas. The few violent eruptions that have occurred on Hawaii have been attributed to the rising magma having come into contact with groundwater, resulting in a steam explosion. (Courtesy of USGS/HVO)*

Strombolian-Type Eruptions   The second type of volcanic activity is characterized by frequent but mild explosions that discharge incandescent pyroclastic materials and commonly build cinder cones around the eruption vents. The name, **Strombolian**, is derived from the volcano Stromboli, an island just north of Sicily. Stromboli is one of several volcanoes associated with the zone of subduction between Africa and Europe. Historically, Stromboli has erupted with a small explosion about every 15 minutes. Since gas is released in each of these frequent small eruptions, the gas content of the magma never builds to the point where a violent explosion can occur. The frequent eruptions constantly stir the magma in the crater, and a cloud of ash, dust, and steam is maintained above the volcano. The cloud can be seen for miles during the day; at night it glows due to the reflection from the red-hot magma below. As a result, Stromboli has been known for centuries as the "Lighthouse of the Mediterranean."

Vulcanian-Type Eruptions   The third type of volcanic activity is characterized by "infrequent but severe explosions." **Vulcanian** eruptions explosively eject fragments of incandescent lava, which cool and solidify in midair into bombs, along with large volumes of ash. Vulcanian activity is more typical of the eruptions associated with the zones of subduction. The name is derived from the island of Volcano located about 35 miles (56 km) southwest of Stromboli. The distinctly different eruptive styles of the two volcanoes demonstrate that eruptive style is one of the unpredictable characteristics of volcanoes.

Volcanism   115

FIGURE 4.30 *The city of St. Pierre on the island of Martinique was destroyed along with 29,000 inhabitants during the 1902 eruption of Mont Pelée. Only two individuals survived the eruption, which allowed geologists to observe a* nuée ardente *for the first time. (Courtesy of I. C. Russell/USGS)*

Pelean-Type Eruptions An explosive type of volcanic activity, the **Pelean** eruptions is named after Mont Pelée on the island of Martinique in the West Indies (Figure 4.30). Mont Pelée is one of the active volcanoes associated with the Caribbean subduction zone. The 1902 eruption of Mont Pelée allowed volcanologists to observe for the first time the most violent of explosive phenomena, the nuée ardente.

Mont Pelée had been active for centuries before the 1902 eruption, but had exhibited only minimal signs of activity such as the release of gas and steam. The relative peace of the island was broken in April 1902, when the volcano began to erupt with increasing intensity. The eruptions continued into May, culminating on May 8.

Many of the residents of the island had moved into the port of St. Pierre to vote in a local election. Overnight, the population of the town had grown from a few thousand to an estimated 29,000 people. Unfortunately, St. Pierre was located at the base of Mont Pelée. Frightened by the growing violence of the eruptions, many attempted to leave the island.

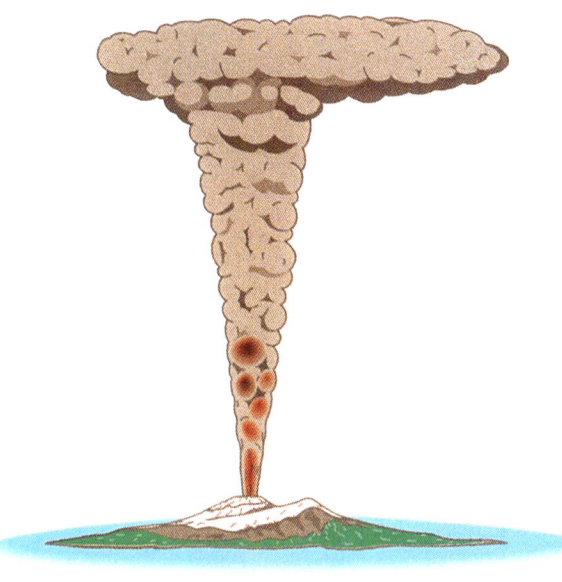

FIGURE 4.31 *The* Plinian-type *eruption is characterized by a column of pyroclastic material driven to great heights by a high-velocity jet of superheated gas.*

The events of the day were reported by a ship's captain who was able to set sail and escape the climactic eruption, but nevertheless lost more than half of his crew. He reported the volcano was erupting violently, sounding like hundreds of cannons firing simultaneously. Finally, with one gigantic explosion, the mountain was enveloped in a dense, glowing cloud.

As we now interpret what must have happened, from the safety of hindsight, the magma chamber literally disemboweled, erupting an enormous mixture of superheated gases, molten rock, and pyroclastic material that formed a super ash flow. This nuée ardente, or "glowing cloud" descended on the town of St. Pierre, destroyed everything in its path, and continued to the harbor where it capsized and set fire to the anchored ships. Within moments, many ships were destroyed, the town was gone, and the entire population was dead, except for two. The first, Leon Compere-Leandre, has been described as one of those individuals who would survive any disaster. Fortunately for him, he was in the basement of his home at the time of the disaster. The other survivor, Auguste Ciparis, was the town reprobate. Having been convicted of murder, Ciparis was incarcerated in the dungeon beneath the town hall, waiting to be hanged. Ironically, the eruption of Mont Pelée occurred on the day set for his execution. He was found by search parties after the ruins of St. Pierre had cooled. Apparently feeling that he had suffered enough, the authorities suspended his sentence. Stories differ as to his fate, but the most popular tale is that he joined Barnum and Bailey's Circus and toured as the "Man Who Defied the Wrath of Pelee, the Goddess of Fire."

Plinian-Type Eruptions   Plinian eruptions may be even more violent than Pelean. In the typical Plinian eruption, a highly turbulent, high-velocity stream of intermixed fragmented magma and superheated gas is released from a vent, creating eruption columns of great height (Figure 4.31). The name derives from Pliny the Younger who described the eruption of Mount Vesuvius in A.D. 79. Modern examples of Plinian eruptions include the eruption of Tambora in 1815 and Krakatoa in 1883. We will discuss the eruptions of Mount Vesuvius and Krakatoa later in the chapter.

FISSURE ERUPTIONS

Fissure eruptions occur along fractures in Earth's surface. Most fissure eruptions involve highly fluid basaltic lava, but a few produce felsic lavas.

Oceanic Ridges

The best examples of basaltic fissure eruptions are those along the oceanic ridges, where fractures are constantly being formed along the margins of the divergent plates (Figure 4.32). As the plates drift apart, fissures open and become the conduits of the basaltic magmas being generated at the top of the asthenosphere. The lavas well out from the fractures along the crest of the oceanic ridge and flow laterally down the shallow ridge slopes where they solidify to create new oceanic crust. Beneath the ridge, magma solidifies within the fractures to create gabbroic sheet dikes (Figure 4.33). New fissures then open along the ridge to perpetuate the process as the oceanic plates continue to move away from each other.

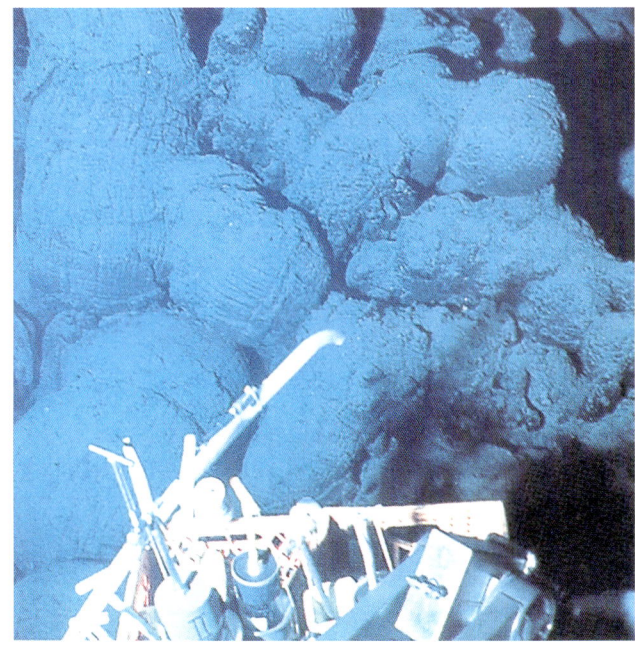

FIGURE 4.32   Pillow lava *is formed on the ocean bottom as basaltic magmas are extruded from fractures along the* oceanic ridges. *(Courtesy of W. B. Bryan/Woods Hole Oceanographic Institute)*

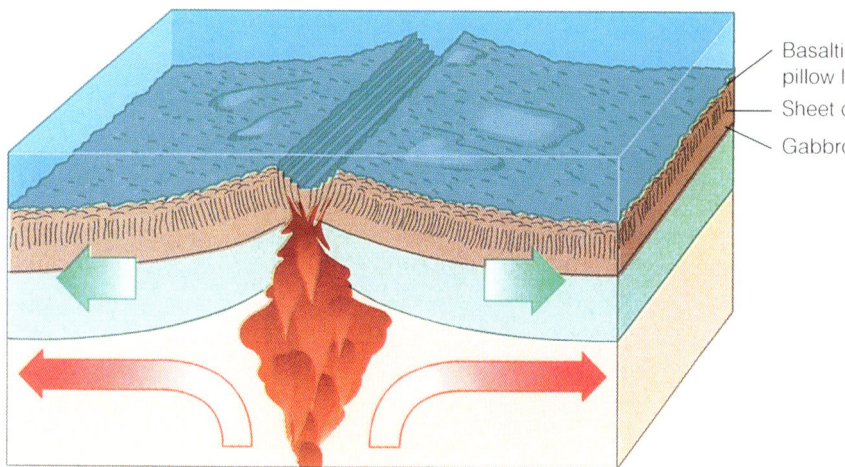

FIGURE 4.33 *Oceanic crust consists of three distinct layers. The uppermost layer is made up of basaltic pillow lava and flows; the intermediate layer is made of sheet dikes that form as basaltic magma solidifies in the fractures that are constantly being formed as the plates diverge; and the lowest layer is made up of coarser-grained gabbro.*

FIGURE 4.34 *Among the best examples of flood basalts are those that make up the Columbia Plateau of Washington, Oregon, and Idaho. (Courtesy of P. Weis/USGS)*

Historically, the only active fissure eruption ever observed was a fissure 20 miles (32 km) long that opened on the island of Iceland in 1783. Before the eruption ended, an estimated 3 cubic miles (12.5 km^3) of basaltic lava had poured out onto the island. Because Iceland is, in part, an exposed portion of the Atlantic Oceanic Ridge, the Icelandic eruption is thought to be comparable to those occurring along the oceanic ridge system except that the ridge eruptions are submarine and produce pillow lavas.

Flood Basalts

No continental fissure eruption has been recorded in historic time, although there are several geologic examples. The **flood basalts** of the Columbia Plateau of Washington, Oregon, and Idaho are an excellent example (Figure 4.34). Beginning about 20 million years ago, enormous volumes of basaltic lavas flowed from fractures that had opened throughout the area. These eruptions may have been associated with the movement of the North American plate over the Yellowstone hot spot. Over a period of several million years, repeated eruptions resulted in hundreds of layers of basaltic lavas. The accumulated flows buried the existing landscape under thousands of feet of lavas; individual lava flows ranged in thickness from less than 10 feet (3m) to more than 300 feet (90 m). It is estimated that over a period of 10 million years, a total of about 25,000 cubic miles (100,000 km^3) of lava spewed from the fissures, covering an area of about 75,000 square miles (200,000 km^2). The low viscosity of the lava allowed some of the flows to travel up to 100 miles (160 km) from the vent.

Felsic Eruptions

Although most fissure eruptions involve basaltic lavas, some produce more felsic lavas. An excellent example is the area around Yellowstone National Park in northwestern Wyoming. Beginning about 2 million years ago, and then repeating 1 million and 600,000 years ago, the area was the site of enormous eruptions. Volcanologists theorize that basaltic magmas formed over a hot spot at depths of perhaps 60 miles (100 km) and were injected into the granitic continental crust. There rhyolitic magmas were generated as heat from the rising basaltic magmas partially melted the continental rocks. Magma chambers containing the highly viscous, gas-charged rhyolitic magmas broke violently through to the surface. Hundreds of cubic miles of pyroclastic materials were erupted, generating ash flows and ash falls that covered hundreds of square miles with deposits hundreds of feet thick. The remains of these eruptive materials can be seen in the area today as extensive tuff and breccia deposits. Although the hot spot is still volcanically active, present activity is restricted to hot springs, fumaroles, and geysers.

CRATERS AND CALDERAS

Craters and **calderas** are circular to elliptical depressions characteristic of volcanic areas. Some workers distinguish them by size, with craters being the smaller. Most craters have diameters of less than a half to three quarters of a mile, while calderas may have diameters of several tens of miles.

Although craters and calderas do differ in size, most geologists would argue that the important distinction between them is not size but rather the mode of origin. Craters are primarily *constructional* structures, being sculpted in the summit area of the volcano by the products of eruption as they emerge from the vent. Calderas, on the other hand, are *destructional* structures formed by the collapse of the crater walls after an eruption has voided a significant portion of the underlying magma chamber. Although most craters are explosive structures, they may be enlarged by a collapse of the inner crater walls following the eruption. In this sense, the origin of some craters includes a certain element of collapse.

Calderas are the result of exceptionally massive volcanic eruptions that significantly reduce the volume of magma within the underlying magma chamber. With the loss of volume, the overlying rocks no longer have sufficient support, and the roof of the magma chamber collapses into the void (Figure 4.35). Calderas can be very large. The disappearance of the island of Krakatoa following the 1883 eruption is attributed to the collapse of the cone into the voided magma chamber, forming a caldera on the ocean floor that extended 1,000 feet (300 m) below sea level.

Calderas are common summit features in the Hawaiian Islands. Unlike most calderas, which form by collapse after a major explosive eruption, the summit calderas of the Hawaiian volcanoes form when magma moves laterally from the summit magma chamber into adjacent, radiating rift zones to feed new eruption sites (Figure 4.36). Collapse is initiated when supporting magma is removed from under the rocks of the crater floor. Thus, like other calderas, the

FIGURE 4.35 *Once thought to be the result of enormous eruptions that blew away the tops of volcanoes, we now know that* calderas *are* collapse structures *and can form anywhere large volumes of magma are erupted from a magma chamber.*

FIGURE 4.36 *The* Kilauea caldera *did not form after a major eruption of the volcano. Instead the caldera is the result of large volumes of magma being withdrawn from the magma chamber underlying the crater floor and transferred to the rift zone that has become the site of the recent eruptions. (Courtesy of USGS/HVO)*

FIGURE 4.37 *Much of what we know about volcanism is the result of work carried out by the scientists of the* Hawaiian Volcano Observatory *for nearly a century. (Courtesy of J. D. Griggs/USGS)*

Hawaiian calderas form when the volume of magma is reduced, leaving the rocks overlying the magma chamber without support. In the Hawaiian case, however, the collapse is not preceded by a massive eruption from the crater of the volcano itself, but rather by a flank eruption remote from the summit.

Kilauea Caldera is the best known of the Hawaiian calderas, largely as a result of the much-visited Hawaiian Volcano Observatory perched on its rim (Figure 4.37). Most of Kilauea Caldera has formed since the 1790 eruption although detailed mapping by volcanologists from the observatory shows that an earlier collapse followed a massive eruption about 1,500 years ago. Within the caldera, Halemaumau Crater has been the site of almost continuous activity for more than a century, building the lava surface that now fills the bottom of the caldera.

In the continental United States, the best-known caldera is Crater Lake, Oregon (Figure 4.38). The name illustrates the difficulty of distinguishing between craters and calderas. Originally, the depression occupied by the lake was thought to be a massive explosive structure formed when a huge eruption literally blew the top off a volcano named Mount Mazama. Detailed mapping in the vicinity of the volcano, however, showed that a very small fraction of the total volume of rock fragments accumulated around the mountain had been derived from the original cone. Volcanologists then recognized that the depression was not primarily the direct result of a gigantic explosion; rather, it had formed when the summit collapsed following a massive eruption. Wizard Island, located within the lake, is a cinder cone formed long after the collapse of the structure.

Calderas are not always associated with the collapse of the summits of volcanoes. Calderas may form anywhere magmas come close enough to Earth's surface to create surface fractures around the margins of the chamber (Figure 4.39). The fractures then become the conduits of massive ash-flow eruptions

FIGURE 4.38 *One of the best-known, and misnamed, calderas in the United States is* Crater Lake, Oregon. *Our understanding of how calderas form came when scientists realized that the amount of pryoclastic material surrounding the present mountain could not account for the volume of the cone that had apparently been lost during the eruption of ancient Mount Mazama. (Courtesy of USGS/CVO-F)*

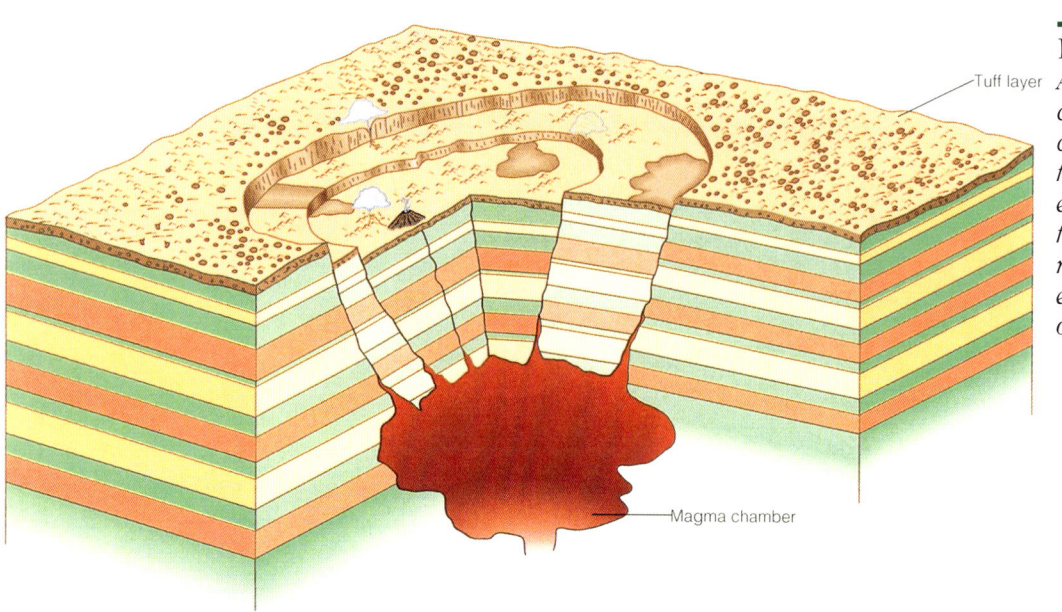

FIGURE 4.39 *Although most calderas form by the collapse of a volcano following a massive eruption, calderas can form anywhere a magma chamber erupts a large portion of its contents.*

that cover the surrounding area with tuff and breccia without constructing any conical structures that would be identified as volcanoes. As before, with the removal of the chamber contents, the surface rocks collapse into the voided chamber. Each of the three eruptions in the area of Yellowstone Park described previously resulted in the formation of a caldera, one of which is the site of Yellowstone Lake.

SPOT REVIEW

1. Compare the geologic settings under which shield volcanoes and stratovolcanoes form.
2. What factors control the intensity of volcanic eruptions?
3. Eruptions of felsic magmas are not as common as eruptions of basaltic magmas. Using the Yellowstone eruption as an example of a felsic eruption, explain how felsic eruptions are thought to occur.
4. Compare the modes of formation of craters and calderas.

OTHER VOLCANIC FEATURES

A number of other geologic features are also of volcanic origin. For the most part, these features are the result of the interaction between the groundwater and the heat derived from the magma, although some of the water involved may be of direct magmatic origin.

Hot Springs

As the name indicates, **hot springs** are emanations of heated waters from vents. Hot springs are very common in volcanic areas. Water temperatures in some springs are barely above body temperature. Other springs, such as those in Lassen National Park in northern California and in Yellowstone National Park in northwestern Wyoming, produce boiling water (Figure 4.40).

Geysers

Some hot springs boil so violently that at regular or irregular intervals, the conduit leading to the vent is explosively cleared by a burst of steam. The hot spring then becomes a **geyser**. Following the eruption of steam and hot water, the conduit requires a period of time to refill with groundwater before the sequence of events can be repeated.

Geysers are not common volcanic features, being found only in New Zealand, Iceland, and Yellowstone Park. The best-known Yellowstone geyser is "Old Faithful," which erupts "faithfully" for expectant tourists about every 70 minutes. During each eruption, Old Faithful (Figure 4.41) blows about 12,000 gallons (45,000 liters) of water to a height of about 150 feet (45 m). Old Faithful is only one of about 200 geysers in Yellowstone Park. Most, however, are smaller and neither as dependable nor as spectacular as Old Faithful. Yellowstone Park is quite unusual in that it contains about 3,000 thermal vents of one kind or another.

FIGURE 4.40 *Travertine terraces formed by the evaporation of mineral-laden waters surrounds the vents of* Mammoth Hot Springs *in Yellowstone Park, Wyoming. (Courtesy of Joe Donovan)*

Fumaroles

Fumaroles are pipelike vents that emit heated gases, usually water vapor (Figure 4.42). Fumaroles are common features found in all volcanic areas and may persist long after the active emission of molten rock has subsided. The gases emitted are very hot, not uncommonly reaching temperatures of 1,100°F to 1,300 F (600°C to 700°C). Fumaroles that emit hot sulfurous gases are called *solfateras*.

Geothermal Power

In many areas where fumaroles are common, superheated steam at sufficiently high temperatures and pressures is tapped by drilling wells and diverted to drive turbines, which in turn generate electricity (the standard steam-powered turbine requires steam at 950°F (500°C) and 150 atmospheres pressure). Natural steam is extensively used in Iceland, not only for power generation, but also to provide hot water and to heat homes and commercial buildings. The use of natural steam in this way is referred to as **geothermal power** and will be discussed in more detail in Chapter 21. In the United States, the largest commercial natural steam driven power plant is located near the town of Geysers, California. The town name is a misnomer inasmuch as the vents in the area are fumaroles, not geysers.

FIGURE 4.41 *Although* Old Faithful *is neither the only geyser in Yellowstone Park nor the largest, it is the most faithful in that it erupts on schedule once every 70 minutes. (Courtesy of J. R. Stacy/USGS)*

FIGURE 4.42 *Although* fumaroles *can emit any superheated gas, most fumaroles, such as these in the* Norris Basin *of Yellowstone Park, erupt superheated water vapor. (Courtesy of Joe Donovan)*

PREDICTION OF VOLCANIC ERUPTIONS

Volcanoes do not erupt without warning. A number of events precede an eruption, most notably increased **seismic activity**. As magma rises and gas pressure builds below the mountain, rocks are broken and earthquakes are produced, increasing in both frequency and intensity as the magma continues to rise. Networks of seismic sensors allow seismologists to construct three-dimensional computer drawings of the earthquake sources that help to outline the rising magma body. In addition to the seismic activity, the rising magma and increasing gas pressure cause the surface to **bulge** and **tilt**. These movements can be monitored with delicate instruments called **tiltmeters**. All of these events were recorded in detail before each of the Hawaiian eruptions and before the eruption of Mount St. Helens, but except for announcing that an eruption was "imminent," scientists were unable to pinpoint either the time or the severity of the final eruption.

At the present time, scientists are monitoring volcanic areas that, based on experience, seem ripe for an eruption. In the United States, for example, Mammoth Lakes, California, is considered by some to be the most likely candidate for the next major volcanic eruption (Figure 4.43). Mammoth Lakes is located in the Long Valley Caldera, which was formed by a massive eruption 730,000 years ago. The Long Valley eruption generated an estimated 146 cubic miles (600 km³) of tephra, covering an area of

(a)

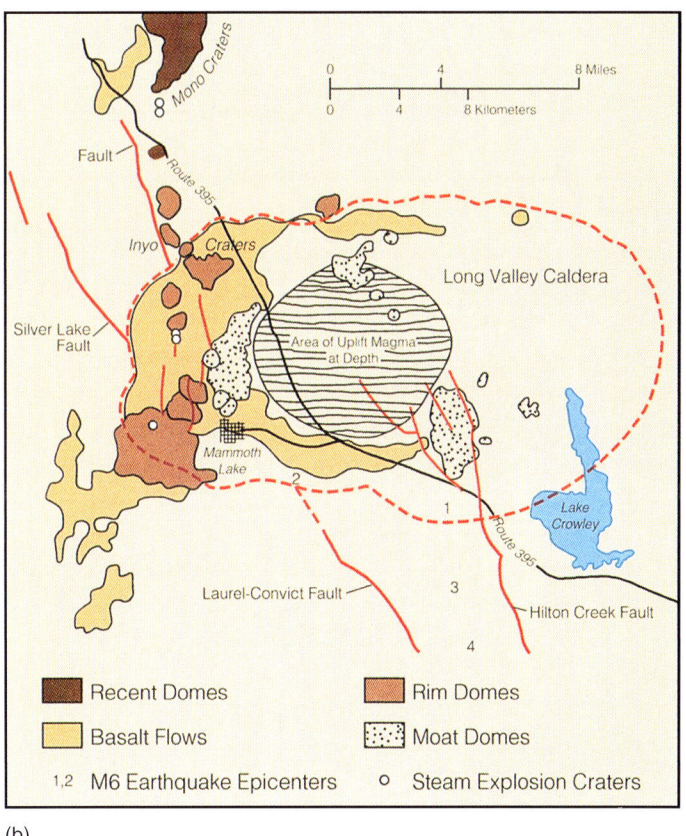

(b)

FIGURE 4.43 Mammoth Lakes, *California, may be the site of the next supereruption in North America. Located near the California-Nevada border, the geology of the area shows that it has been volcanically active for quite some time. The area of recent uplift indicates quite clearly that magma is rising beneath the site with significant new intrusions within the past decade. If the distribution of ash from the last major eruption at Mammoth Lakes is any indication, an eruption of comparable magnitude will be destructive beyond comprehension. (Parts* b *and* c *from R. A. Bailey, 1983. Mammoth Lakes Earthquakes and Ground Uplift: Precursor to Possible Volcanic Activity? In* U.S. Geological Survey Yearbook, Fiscal Year 1982; *Part* d *from C. D. Miller, D. R. Mullineaux, D. R. Crandell, and R. A. Bailey, 1982. Potential Hazards from Future Volcanic Eruptions in the Long Valley-Mono Lake Area, East-Central California and Southwest Nevada—A Preliminary Assessment. U.S. Geological Survey Circular 877.)* **(c)** *and* **(d)** continued on next page

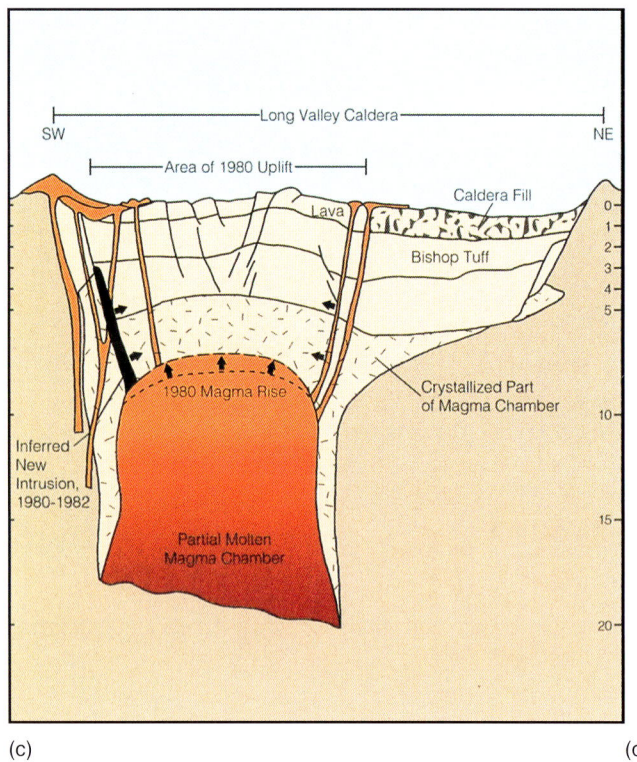

(c) (d)

FIGURE 4.43 *Continued*

570 square miles (1,500 km²). Some of the fine ash has been detected in soils as far away as Kansas and Nebraska. In the central portion of the caldera, the tuff is up to 4,920 feet (1,500 m) thick. Tephra falls up to 40 inches thick (1 m) accumulated 120 miles (75 km) away.

Since 1975, geologists have recorded increased seismic activity and have measured a bulging in the area of nearly a foot (25–39 cm). Since 1980, more than 3,000 earthquakes have been recorded in the area. Although most were low magnitude, at least four strong earthquakes occurred in 1980, and in 1986, a magnitude 6 earthquake occurred just east of the caldera. All of the data clearly indicate that magma is forming 3 to 4 miles (5–6 km) below the caldera and is rising beneath the area.

By 1982, the activity had increased to the point where the U.S. Geological Survey deemed it necessary to issue a formal warning of a potential eruption. Many local residents, fearing the loss of tourists, objected to the warning. No eruption has yet occurred.

In another part of the world, the fishing community of Pozzuoli near Naples, Italy, and Mt. Vesuvius, has been experiencing both seismic activity and uplift for more than 10 years (Figure 4.44). During the last 10 years, the area has experienced more than 4,000 earthquakes, and the land has been uplifted nearly 10 feet (3 m). Many of the homes and buildings have been extensively damaged. The population has been repeatedly warned of a potential eruption, and most of the inhabitants have moved to safer ground. Although the residents abandon the town at night for fear of being caught by an eruption while asleep, during the day it is "business as usual" as fishermen go about their daily work. All the while, the land quakes and rises.

Both Mammoth Lakes and Pozzuoli point out some of the problems that prediction still faces. Until scientists can pinpoint with reasonable accuracy *when* and *where* the eruption will occur and *how severe* it will be, they can only warn of an impending eruption, and the inhabitants can only hope that once the eruption begins, they will have time to escape.

FIGURE 4.44 *The increasing frequency of seismic activity combined with changes in the relative elevations of the land surrounding Mount Vesuvius indicate that the mountain may be building another major eruption.*

EXAMPLES OF OTHER NOTEWORTHY VOLCANOES

We cannot leave the topic of volcanoes without discussing a few of the most famous eruptions.

Mount Vesuvius

One of the best-known volcanoes, Mount Vesuvius, is the only active volcano on the European mainland. Located just outside Naples, Italy, Mount Vesuvius has had a long, eventful, and closely watched history. Before Mount Vesuvius came into being, another volcano called *Mount Somma* stood on the exact same spot. Mount Somma was created about 100,000 B.C. After an eruption ended its activity about 17,000 years ago, Mount Vesuvius grew from the remnants. The remains of Mount Somma can still be seen around the northern flank of Mount Vesuvius.

Since the beginning of historic time, the Romans had considered Mount Vesuvius to be extinct. With vineyards covering its slopes and crater floor, the mountain had never shown even the slightest indication that it was active. In 150 B.C., some minor internal rumblings and newly opened vents showed that Mount Vesuvius had not been extinct but simply dormant. Over the next two centuries, an occasional tremor or minor vent eruption let everyone know that the mountain was still alive. Beginning in 63 B.C. and continuing for 142 years, the vicinity was subjected to repeated and relatively intense earthquakes that were signaling the rise of molten rock below the mountain.

On August 24, in A.D. 79, Mount Vesuvius erupted in a violent Plinian-type eruption. At the time of the eruption, Naples did not yet exist, but two small coastal towns were located at the base of the mountain. *Herculaneum*, a name few would recognize today, lay to the west of the mountain, while to the south was the famous town of *Pompeii*.

The eruption on the morning of August 24 blew away the western flank of Mount Vesuvius and created a combined ash and mud flow that overwhelmed the town of Herculaneum within a matter of minutes. The inhabitants, although aware that an eruption was going on, had little warning of the impending avalanche and thus had virtually no time to escape before the town was completely buried.

The demise of Pompeii was quite different. Its fate was recorded by a young Roman historian, *Pliny the Younger,* who described the town being pelted with hot falling ash. He recalled frightened citizens desperately rushing to the harbor in the hope that they would be safe in the water only to find that the hot ash falls had heated the water in the harbor nearly to boiling. Pliny's uncle, *Pliny the Elder,* an admiral in the Roman navy, sailed his ship into the harbor. In an attempt to calm the people and demonstrate that they were in no real danger, he ignored all warnings and went ashore where he died of a heart attack.

As the ash fall slowly inundated the town, noxious gases asphyxiated the inhabitants as they tried to escape. Within a few hours, both Herculaneum and Pompeii were buried. Although most of the inhabitants of the two towns escaped, about 2,000 individuals were trapped and buried. Both towns lay beneath the ash until they were discovered in 1748. Since that time, both Herculaneum and Pompeii have been largely uncovered and restored, adding enormously to our understanding of the life of the times (Figure 4.45).

The remains of many individuals were preserved in the form of cavities or molds as the ash packed tightly around their bodies. After they were found, the molds were filled with plaster of paris to make a cast or statue of the original body (Figure 4.46). In many cases, the degree of preservation was so good that details of facial and body features could be distinguished as could impressions of hairstyles, jewelry, and clothing.

Since A.D. 79, Mount Vesuvius has experienced two major eruptions, one in 1631 and the other in 1906. Between these two dates, the mountain was more or less continuously active, although the individual eruptions were rather mild. The 1906 eruption, a Plinian type, lasted 18 days and resulted in a crater nearly 2,000 feet (600 m) wide and 2,000 feet deep. The crater was primarily the result of an enormous gas explosion that erupted continuously for an entire day and blew ash to an altitude of 8 miles (13 k). Another eruption of Mount Vesuvius occurred on March 12, 1944, just as the Allied troops liberated Naples during World War II. The mountain is still active today. The people of Naples and the surrounding areas live under the potential wrath of Mount Vesuvius.

FIGURE 4.45 *The towns of* Pompeii *and* Herculaneum *lay buried for nearly 1,700 years following the eruption of Mount Vesuvius in* A.D. *79. (Courtesy of Italian Government Travel Office)*

FIGURE 4.46 *Much of what we know about life during the Roman era has been obtained from the remains of individuals and artifacts buried during the eruption of Mount Vesuvius in* A.D. *79. (© Leonard von Matt/Photo Researchers)*

Santorini

North of Crete in the Aegean Sea is a group of islands called Santoria (Figure 4.47). The largest of the islands, Thera, surrounds a picturesque bay adjoined by steep cliffs exposing layers of volcanic ash. A more detailed study shows that the islands are the remnants of a volcano historians call *Santorini*. Until 1500 B.C., Santorini was the home of a colony

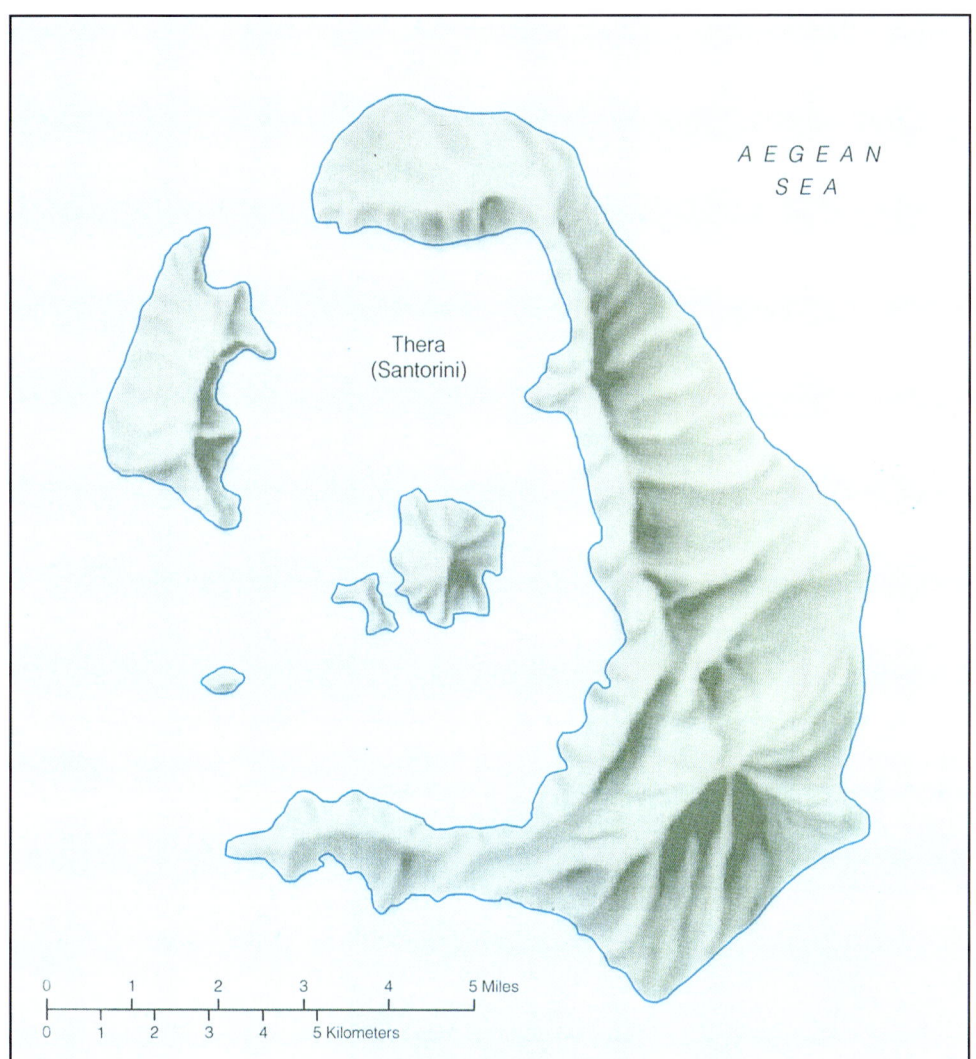
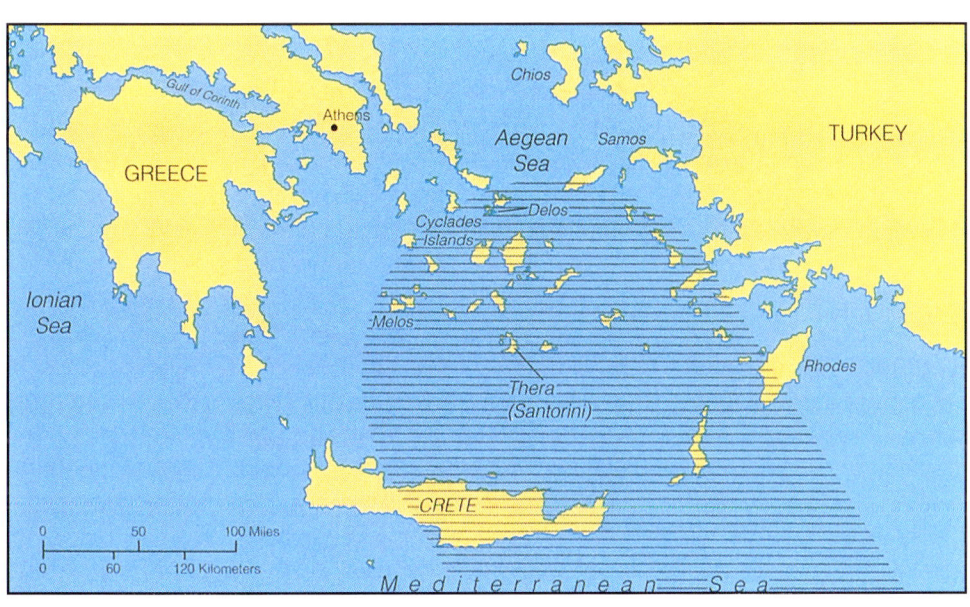

FIGURE 4.47 *The formation of a caldera and the subsequent collapse of the sea bottom following the eruption of Santorini is probably the basis for the tale of the* Lost Continent of Atlantis. *The shaded area on the upper map shows the zone of significant ash fall from the eruption.*

founded by one of the most advanced civilizations of the time, the *Minoans*. Based on the island of Crete and deriving their name from the legendary King Minos, the Minoans were among the first to study the night sky. The names of the constellations, first recorded in Homer's *Iliad* and *Odyssey* about 900–800 B.C., are thought to be based on legends originating with the Minoans. About 1500 B.C., the Minoan civilization on Thera came to an abrupt end with the eruption of Santorini. Earthquakes and eruptions preceding the main event convinced the Minoans to abandon their capitol of Akrotira and seek safety elsewhere. But such was not to be. An eruption rivaling that of Mount Somma destroyed most of the island, buried what remained in ash, and generated a seawave that devastated the surrounding islands. Although the eruption damaged the Minoan centers on Crete, the main Minoan civilization survived for another half century until it was conquered by the Mycenaeans from mainland Greece.

The eruption created a caldera covering 30 square miles (80 km^2), and most of what was left of Santorini sank to the sea bottom. Some historians have suggested that the disappearance of the flourishing island was the origin of the legend of the lost continent of Atlantis. The islands of Santoria now stand as a monument to the cataclysmic event and to the civilization that perished.

Krakatoa

One of the most violent volcanic eruptions ever witnessed was the eruption of Krakatoa on August 27, 1883. Krakatoa was a volcanic island located between the islands of Java and Sumatra, all part of the Java Trench subduction zone that exists between the Australia-India plate and the Asian mainland (Figure 4.48). The eruption was seen from the British ship *Charles Bal* as it sailed about 10 miles (16 km) to the south. The crew reported seeing lightning flashing around the summit of the mountain and being pelted with tremendous volumes of hot ash. The gale force winds that arose allowed the ship to set sail and escape what would have been certain destruction. The next day, the entire island was destroyed in a massive, culminating explosion.

Volcanologists theorize that during one of the mountain's final convulsions, a fracture opened on the sea floor that allowed seawater to gain access to the magma chamber, creating a steam explosion that threw an estimated 5 cubic miles (20 km^3) of rock into the air. For hundreds of miles around, the ash in the atmosphere was so dense that day became night as the Sun's rays were blotted out. Ash fell over an area estimated at 300,000 square miles (780,000 km^2). Fine dust encircled Earth and caused brilliant sunsets for several years. The worldwide decline in atmospheric temperatures shortened growing seasons in more northerly areas of Europe and Asia. The explosion generated a giant seawave that overwhelmed surrounding islands and killed nearly 40,000 people, while smaller seawaves were measured at seashores halfway around the world. The sound of the blast was heard in Australia 1,500 miles (2,400 km) away. After about 50 years, a new cone began to build from the remains and now stands 3,200 feet (980 m) above sea level; it is called Anak Krakatoa (Child of Krakatoa).

Mount St. Helens

Few volcanoes have been more studied than Mount St. Helens in the Cascade Mountains of the Pacific Northwest. Mount St. Helens is located about 45 miles north-northeast of Portland, Oregon, in southwestern Washington (Figure 4.49). The mountain is the youngest of the 15 major stratovolcanoes or composite volcanoes that make up the Cascade Mountains. The volcanic history of the mountain goes back nearly 40,000 years.

Studies of eruptive deposits indicate that Mount St. Helens has been active throughout its 40,000-year history. Following a 123-year period of dormancy since 1857, a few low-magnitude earthquakes on March 20, 1980, signaled its return to life. Within a week, fissures had opened in the summit area, and gases and ash began to be ejected. The summit area began to bulge outward like an inflating balloon as magma made its way upward beneath the mountain. By April, the bulge was expanding horizontally on the northern flank at a rate of about 6 feet (2 m) per day. Although the mountain was obviously still in a state of eruption, the signs of activity began to taper off. On the morning of the eruption, May 18, 1980, the mountain was quiet.

The main eruption was triggered by two closely spaced earthquakes that produced a landslide in the vicinity of the bulge. The removal of rock by the landslide took the roof off the magma chamber, released the pressure on the magma below, and resulted in the violent eruption of gases and magma. The event was observed by a young geologist, David Johnston, in an observation station nearly 5 miles (8 km) away. Johnston was to become the first known victim of the

Volcanism 129

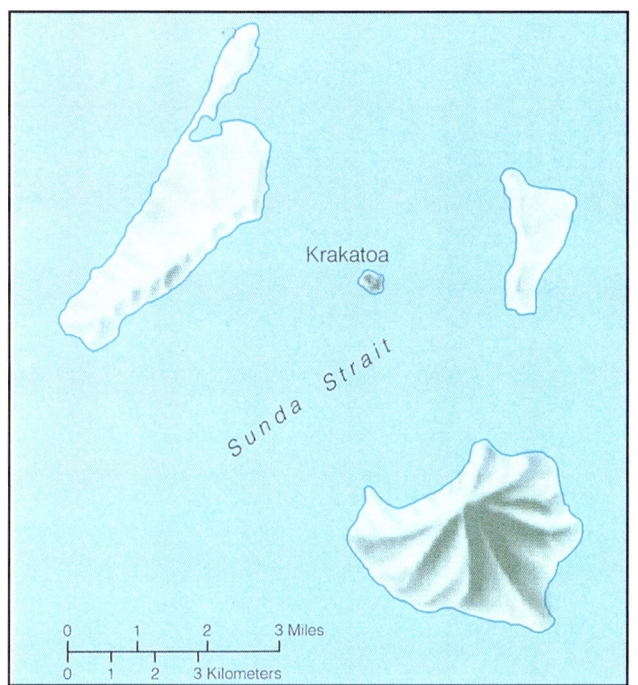

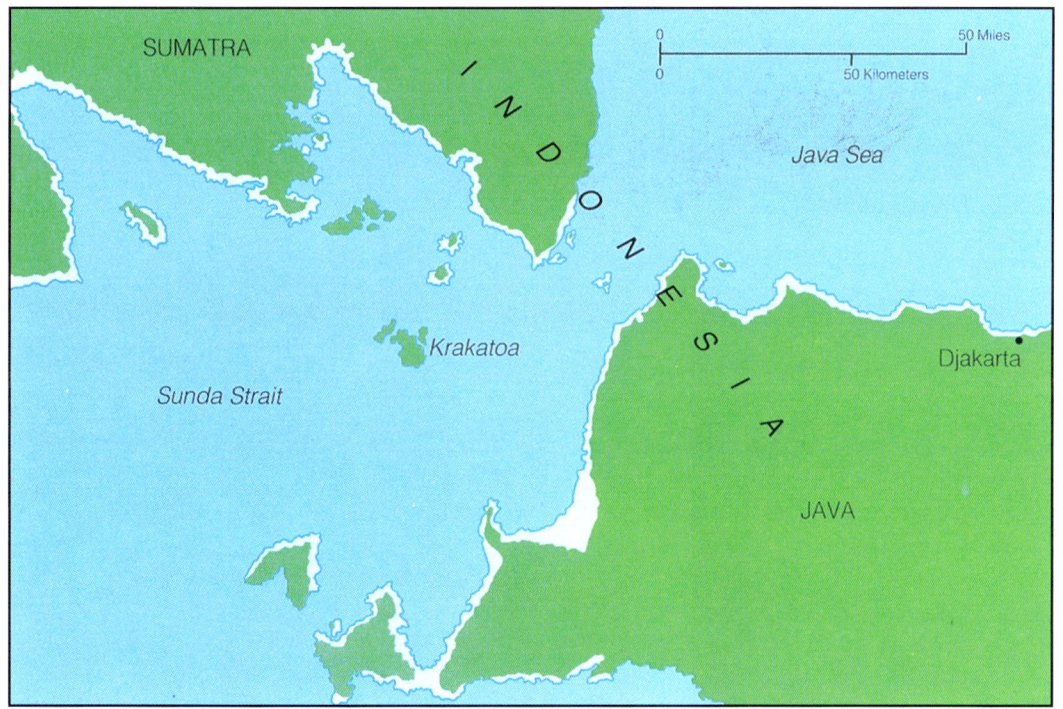

FIGURE 4.48 *All that now remains of the island of* Krakatoa *following one of the most violent volcanic eruptions ever witnessed is a tiny cone called* Anak Krakatoa.

eruption. Within moments of radioing his last communication to the base station, "Vancouver, Vancouver, this is it," he was overwhelmed by the blast that roared out of the breached flank of the mountain and devastated an area nearly 20 miles (32 km) wide extending out 12 miles (19 km) from the mountain (Figure 4.50).

The plume from the eruption rose to an elevation of 15 miles (25 km). The destruction was truly impressive. Out to a distance of 6 miles (10 km), the forest was covered with a blanket of searing hot ash. At a distance of 10 miles (16 km), trees were leveled, stripped of branches, and aligned like so many matchsticks (Figure 4.51). Water from the melting snowcap generated lahars that filled Spirit Lake and moved down the valley of the Toutle River, depositing debris 200 feet (61 m) deep on the valley floor.

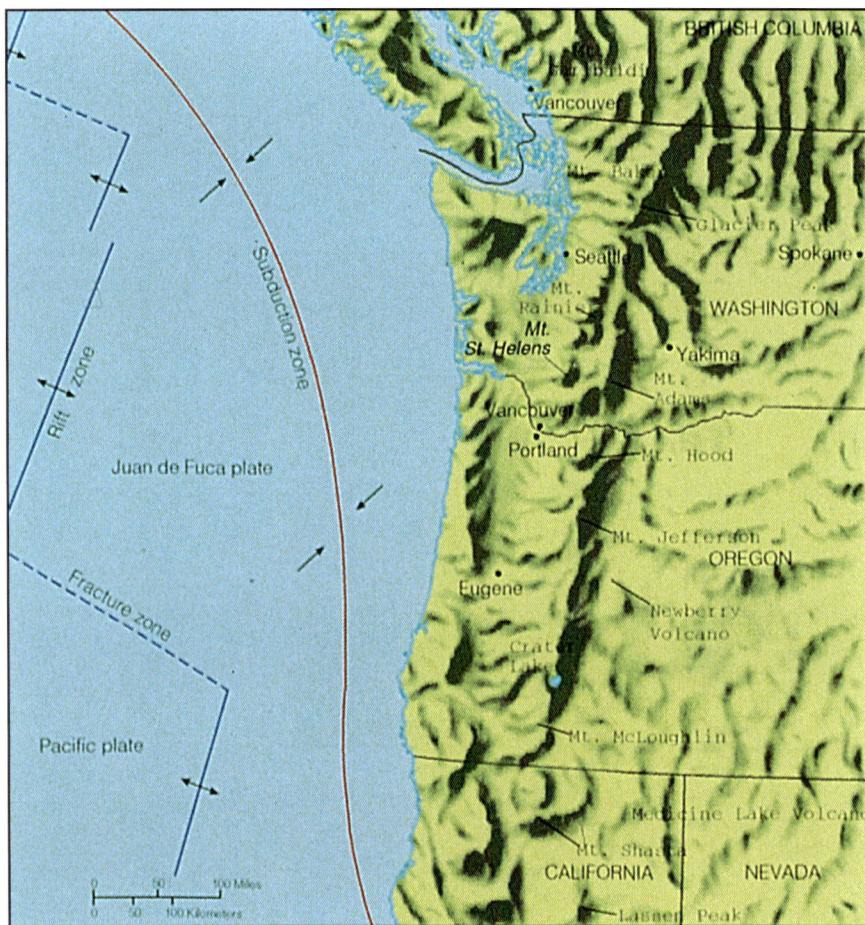

FIGURE 4.49 Mount St. Helens *is one of 15 volcanoes that make up the continental arc associated with the subducting Juan de Fuca oceanic plate.*

FIGURE 4.50 *The major eruption of Mount St. Helens was triggered by a landslide that removed rock from the roof of the magma chamber.*

FIGURE 4.51 *The damage resulting from the eruption of Mount St. Helens made many people, who possibly had become complacent, more aware of the potential for destruction of volcanic eruptions. It should be kept in mind, however, that the eruption of Mount St. Helens was only a minor event when compared to the eruption of another member of the Cascades that formed Crater Lake.*

Within an hour, ash began to fall in Yakima, Washington, about 80 miles (129 km) to the east. The ash became so thick that residents had difficulty breathing and were told to stay indoors. Auto engines stalled as ash clogged the air filters. Ash accumulated throughout eastern Washington and parts of Idaho and Montana. Within a few days, the dust cloud had crossed the continental United States.

As destructive as the eruptions was, the total amount of pyroclastic material produced was only about 20% of that generated by the Krakatoa eruption and only 7% of that produced by the Crater Lake eruption. Imagine what would happen today if an eruption of the magnitude that created Crater Lake or the Long Valley Caldera were to occur anywhere within the Cascade Mountains. Not only *could* such an eruption occur, it is almost certain that it *will* occur. The Cascade Mountains are still a young mountain range with a long and potentially violent future. Indications are that the present eruptive episode of Mount St. Helens is waning. Declining seismic activity indicates there has been no magma movement since 1986. As long as the North American plate continues its relentless westward movement, however, the Cascades will remain an active volcanic mountain range.

The Hawaiian Islands

The state of Hawaii has the distinction of being the only state constructed almost in its entirety from the products of volcanism. The Hawaiian Islands are a series of shield volcanoes that stretch west-northwestward 1,146 miles (1,844 km) across the Pacific Ocean from the island of Hawaii to Midway Island. The Hawaiian chain of volcanoes then continues for another 491 miles (790 km) as a series of submerged volcanic peaks or seamounts. At that point, the chain takes an abrupt turn northward and continues another 1,296 miles (2,085 km) as the *Emperor Seamounts* (Figure 4.52). Some geologists believe that the redirection occurred as India and Asia collided, changing the direction of movement of the Pacific plate. At its northern end, the Emperor Seamount chain reaches the Aleutian Island subduction zone.

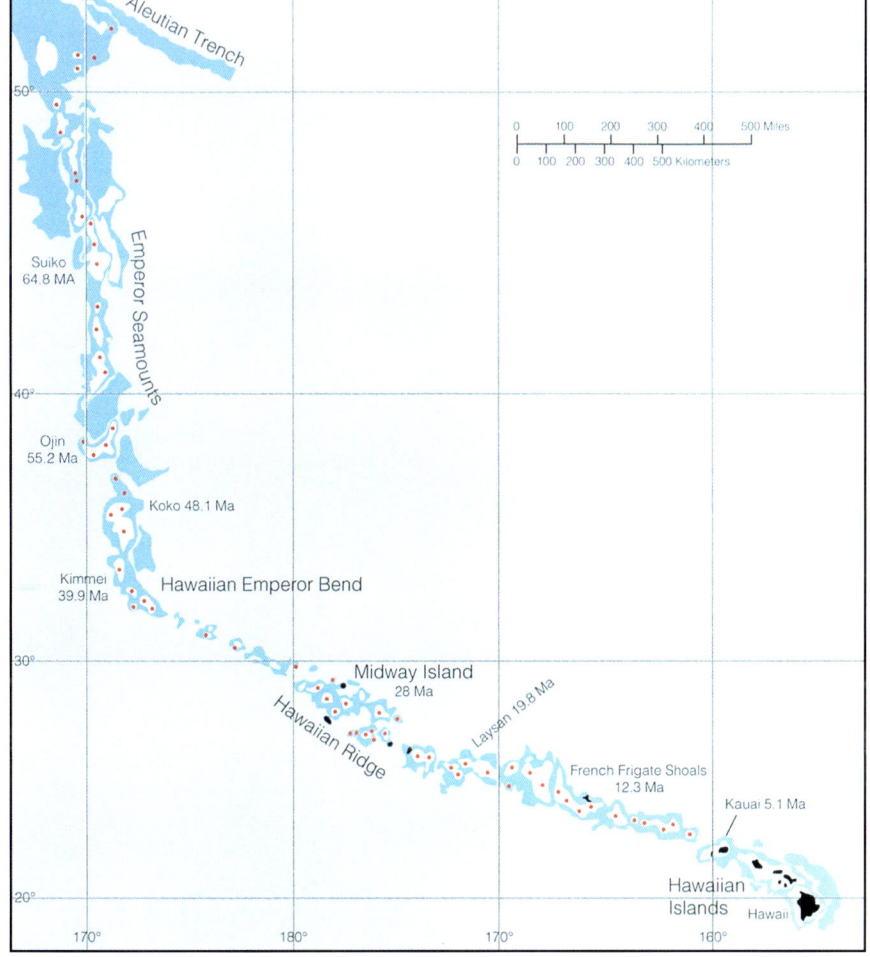

FIGURE 4.52 *Over the past 80 million years, a string of shield volcanoes has been produced by the Hawaiian hot spot that extends nearly 2,500 miles from Hawaii to the deep-sea trench off the southern coast of Aleutian Islands.*

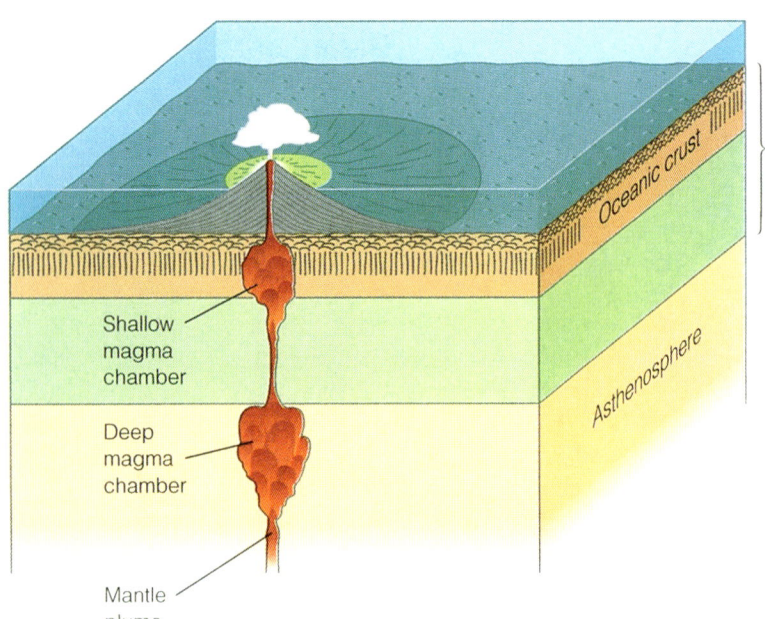

FIGURE 4.53 *Volcanologists believe that two magma chambers are associated with hot spot volcanism, one at the top of the asthenosphere, which serves as a source of the magma, and a second within the lithosphere, which stores the magma below the cone.*

The volcanic peaks of the Hawaiian Islands, the Hawaiian seamounts, and the Emperor Seamounts formed over a mantle plume located at the southeastern end of the chain (155° west longitude and 20° north latitude). Radiometric dating of the rocks from the most northerly seamount of the Emperor chain indicates that the plume has been active for at least 80 million years.

According to volcanologists, a basaltic magma body exists at the top of the asthenosphere and feeds a second magma chamber within the overlying lithosphere (Figure 4.53). The volcanoes are constructed as the magma rises from the second magma chamber and breaks through to the ocean floor. During the submarine phase of growth, the basaltic lavas construct a shield volcano with slopes of about 17°. Eventually, the cone breaks out above sea level and continues to build the summit of the volcano as a volcanic island. Because the lavas cool more slowly when extruded subaerially, the slope of the exposed cone becomes less steep with an angle of about 5° or 10° (Figure 4.54).

While the island undergoes construction, the Pacific plate moves northwestward at about 3 inches (8 cm) per year. As the lithosphere moves laterally relative to the feeder conduit in the asthenosphere, another volcanic cone is initiated adjacent to the first and begins to build upward while the original volcano remains active. Both volcanic cones are fed from the feeder conduit leading back to the asthenospheric magma chamber. The result of this process is that most of the Hawaiian Islands are constructed of multiple volcanoes. The island of Hawaii, for example, consists of five volcanoes (Figure 4.55).

Eventually, the early formed volcanoes move so far from the main feeder conduit that they are no longer provided with a supply of magma and become extinct. Of the five volcanoes on the island of Hawaii, the two oldest, *Kohala* and *Mauna Kea*, are now considered to be extinct. The next oldest, *Hualalai*, although still active, is thought to be nearing extinction. The two active volcanoes on the island are *Mauna Loa* and *Kilauea*. At the time of this writing, Kilauea is in active eruption. A new volcano, *Loihi*, is now building offshore to the south of Hawaii. It is estimated that Loihi will become part of the island in about 50,000 years.

After the construction of an island, which takes about 500,000 to 1 million years, the volcanic activity apparently shuts down. Why and how this happens is one of the most controversial topics in the science of volcanism today. Although the upward flow of magma ceases and the construction of the volcanoes stops, the oceanic plate is still moving. During the period of volcanic inactivity, the first-built island moves completely off the location of the mantle plume. The magma then begins to be generated once again and is delivered upward. Upon breaking through to the ocean floor, the erupting lavas initiate the construction of a new shield volcano that will eventually rise above sea level to become the second island of the chain. The chain of volcanic islands lengthens as successive islands are created.

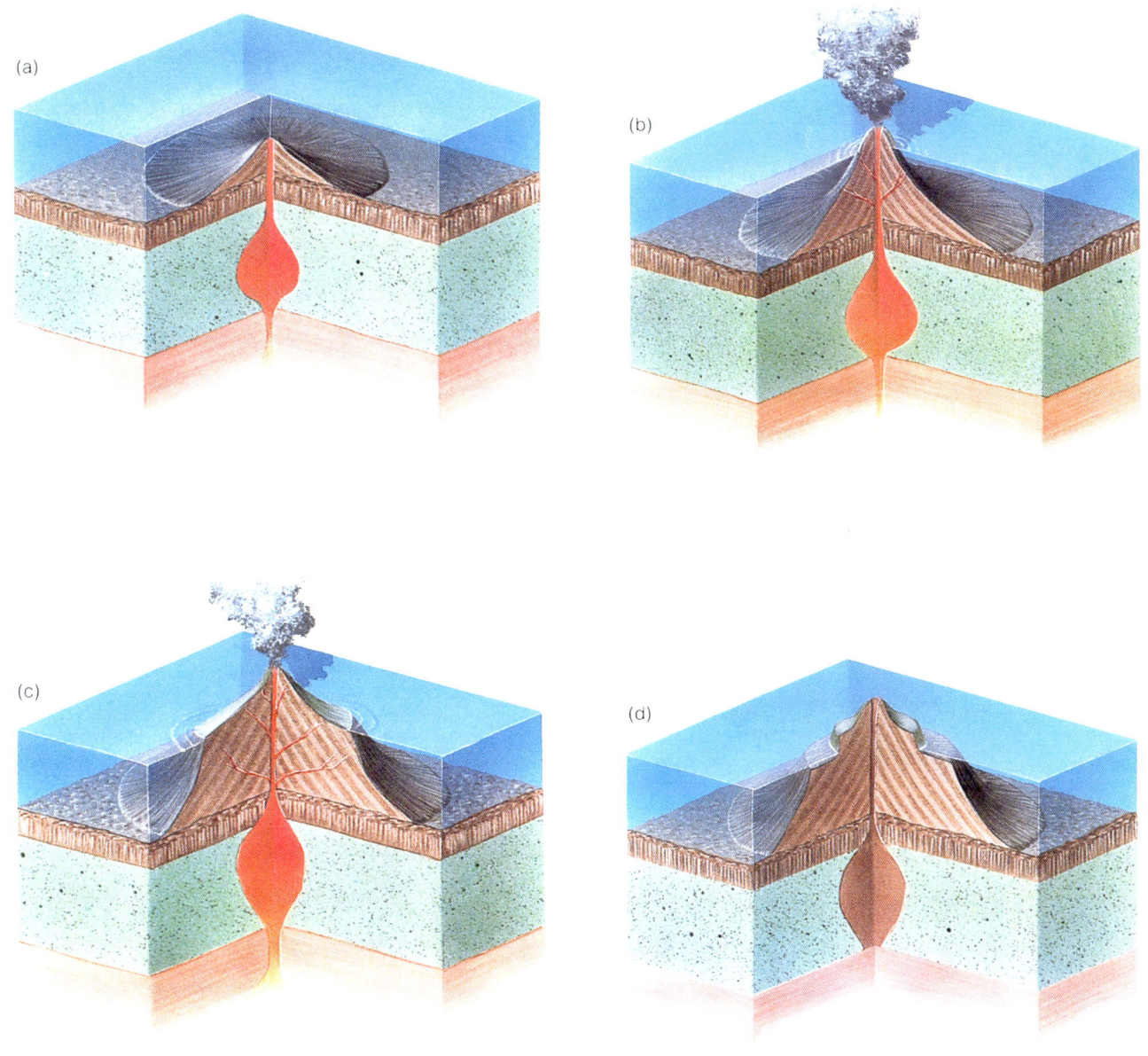

FIGURE 4.54 *The formation of the string of seamounts and volcanic islands associated with a mantle plume is believed to be due to cyclic heating and cooling of the mantle plume combined with the lateral movement of the oceanic lithosphere.*

As the lithosphere moves laterally, the individual islands begin to sink as the ocean floor is depressed under their weight. The subsidence, combined with erosion and slumping, results in the islands becoming progressively smaller. Eventually, each island will disappear beneath the sea (Figure 4.56).

Geologists agree that the Hawaiian Islands formed as the oceanic plate moved continuously over a mantle plume. Except for Hawaii, none of the islands currently has any direct connection to the magma chamber at the top of the asthenosphere, explaining why only Hawaii, the youngest island, has active volcanism. Small eruptions occurring on the other Hawaiian Islands have been attributed to the expulsion of magma masses that were entrapped within the rocks of the island when it moved off the mantle plume.

The scenario also explains why the islands are aligned and why they become older as one travels northwestward. The oldest rocks on the island of

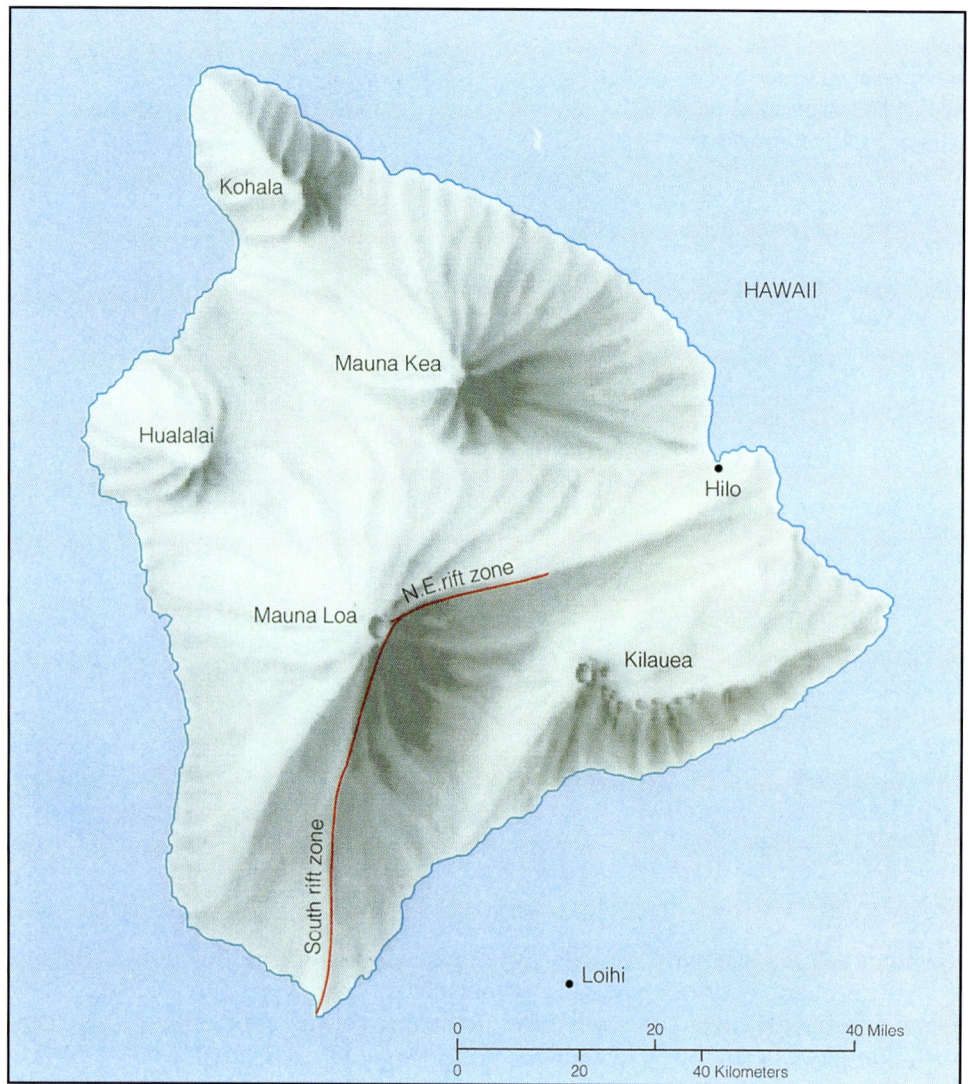

FIGURE 4.55 *The island of Hawaii is made of five volcanoes, only two of which are active. Just offshore, a new volcano named Loihi is building from the ocean floor and will become part of the island in about 50,000 years.*

Hawaii are just under 1 million years old while the oldest rocks of Kauai, the most northwestward island usually visited by tourists, are nearly 6 million years old. The most distant island in the chain before they disappear beneath the sea is about 30 million years old.

The marked change in the orientation of the trend of the Hawaiian Islands and seamounts and that of the Emperor Seamounts indicates the Pacific plate initially moved in a northward direction. After the last of the Emperor Seamounts was constructed as an island, the direction of plate movement changed to the present west northwesterly direction.

The volcano Kilauea is without doubt the most studied volcano on Earth, largely because of the work and dedication of the personnel of the Hawaiian Volcano Observatory who have been monitoring the activities of the mountain for more than 75 years. The present episode of eruption from the *Pu'u O'o* vent began in 1983 (Figure 4.57). Since then, the volcano has produced more than 27.5 billion cubic feet (850 million m³) of lava. Flows from the summit and from rift eruptions fed from the summit magma chamber have reached the sea 7 miles (11 km) away and have added more than 200 acres of new land to the island. At the same time, the flows have covered about 25 square miles (65 km²) of the existing island, causing property damage estimated at $10 million or more.

Throughout it all, however, no one has been killed, primarily because the Hawaiian eruptions are normally devoid of explosions of the intensity that destroyed Mount St. Helens. Typically, Hawaiian eruptions are characterized by lava fountains that tower from hundreds to nearly 2,000 feet (600 m) into the air and rivers of molten rock that flow down the mountain slopes,

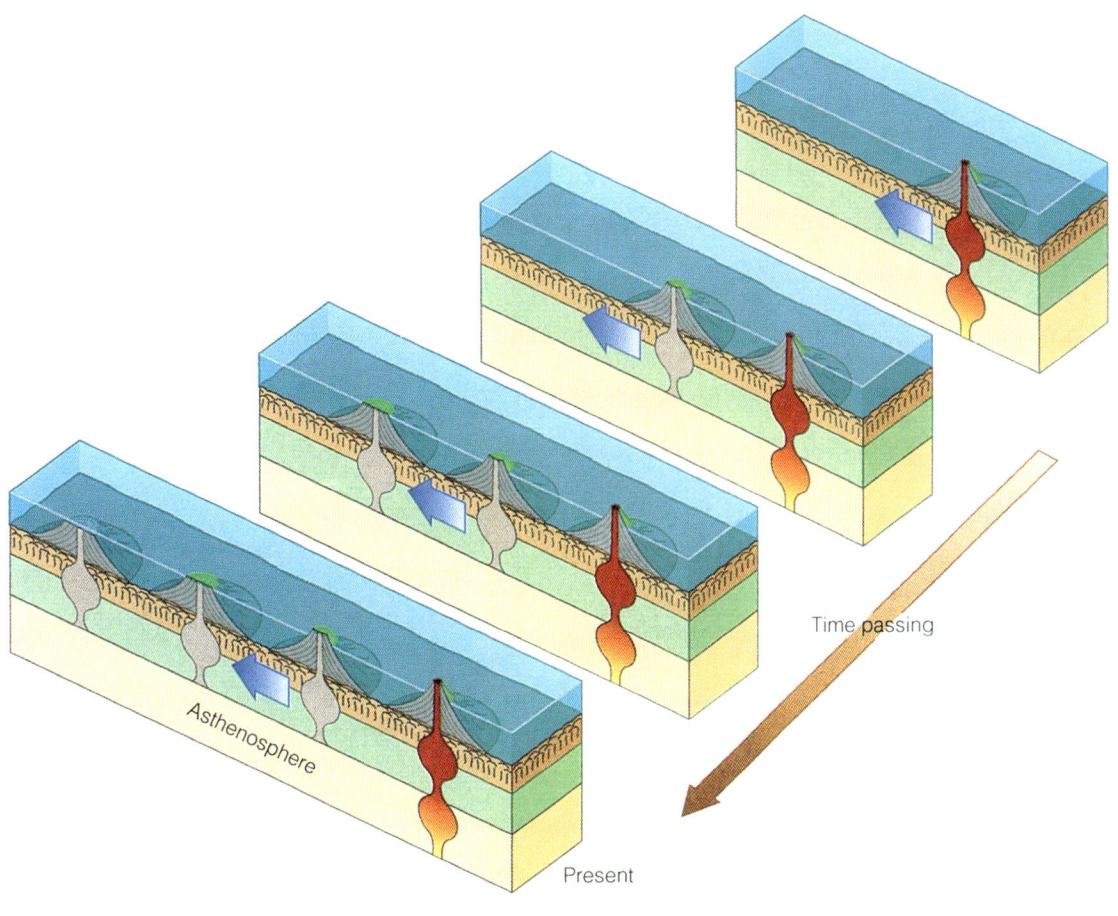

FIGURE 4.56 *With time, the islands become smaller due to the combined effects of the downwarping of the oceanic crust under the weight of the island, weathering and erosion of the exposed portion of the island, and slumping of the subaqueous flanks of the cone.*

FIGURE 4.57 *The present episode of activity on Hawaii began in 1983 with the eruption of the* Pu'u O'o *vent along the southeast fracture zone. (Courtesy of USGS/HVO)*

cascade over cliffs, and sometimes pour through clouds of billowing steam into the sea (Figure 4.58). The relatively quiescent nature of Kilauean eruptions allows geologists, and tourists, to stand within feet of a flowing tongue of lava, to venture out onto the surface of a newly solidified lava flow while it is still too hot to touch, and to watch the slow movement of the oncoming front of an aa flow while peering at the molten rock inside the solidified blocks. Few human experiences can be more awe inspiring than these.

Sakurajima, Japan

One last volcano deserves mention, if only because it has the highest frequency of eruption. The island of Sakurajima, Japan, is dominated by a volcano that has been erupting ash nearly 200 times per year (Figure 4.59). The 9,000 inhabitants of the island have adjusted to this ever-present threat to their lives. They live in homes built to exclude dust. Special channels line their streets so that the ash can be washed away. Parents send their children off to school with the reminder to "wear your hard hat." Five miles across the bay, Kagoshima, a city of half a million people, also thrives under the constant outpourings of Sakurajuma volcano. Here too, citizens are prepared to shovel away daily ash falls.

FIGURE 4.58 *New land is being created along the south shore of Hawaii as the lava reaches the sea and builds the edge of the land seaward. (Courtesy of USGS/HVO)*

FIGURE 4.59 *With few days going by without an eruption, the people of Sakurajima, Japan, have learned to live with the constant presence of the volcano. (Courtesy of Japan National Tourist Organization)*

SPOT REVIEW

1. How do hot springs, fumaroles, and geysers differ?
2. What kinds of information do geologists use to predict an impending volcanic eruption?
3. Using examples from noteworthy volcanic eruptions, describe how eruptions can affect human lives far beyond the immediate site of the eruption.

CONCEPTS AND TERMS TO REMEMBER

Active, dormant, and extinct volcanoes
Distribution of volcanoes
  divergent margins
  convergent margins
  island arc
  continental arc
  Ring of Fire
  hot spots
Ejected materials
  gases
  liquids
    magma
    lava
      pahoehoe
      aa
      pillow lava
  solids
    tephra
    tephra falls
      blocks
      bombs
      cinders
      ash
      dust
      tuff
      volcanic breccia
    ash flows
      nuée ardente
      ash-flow tuff
      welded tuff
      ignimbrite
      mudflow or lahar
Kinds of volcanoes
  cinder cone
  spatter cone
  shield volcano
  stratovolcano
  composite volcano
Eruptive intensity
  gas content
  viscosity of magma
Eruptive styles
  Hawaiian
  Strombolian
  Vulcanian
  Pelean
  Plinian
Fissure eruptions
  oceanic ridges
  sheet dikes
  flood basalts
  felsic eruptions
Craters and calderas
  crater
  caldera
Other volcanic features
  hot springs
  geyser
  fumarole
    geothermal power
Prediction of volcanic eruptions
  seismic activity
  bulging and tilting
    tiltmeter

REVIEW QUESTIONS

1. Most of the Earth's active volcanoes are associated with
   a. hot spots beneath oceanic plates.
   b. zones of subduction.
   c. transform faults.
   d. oceanic spreading centers.

2. The type of lava associated with stratovolcanoes or composite volcanoes is
   a. andesitic.   c. basaltic.
   b. rhyolitic.   d. felsic.

3. Stratovolcanoes or composite volcanoes are associated with
   a. mid-oceanic ridges.
   b. oceanic hot spots.
   c. continental hot spots.
   d. zones of subduction.

4. The eruptive feature known as a nuée ardente is a super
   a. ash fall.   c. fissure eruption.
   b. mudflow.   d. ash flow.

5. The most violent volcanic eruptive style is the
   a. Strombolian.   c. Hawaiian.
   b. Plinian.   d. Vulcanian.

6. The type of lava that forms by the underwater solidification of molten lava is called
   a. pahoehoe.   c. pillow.
   b. aa.   d. ropy.

7. The most abundant lava composition is
   a. rhyolitic.   c. basaltic.
   b. andesitic. d. felsic.

8. Subduction zone volcanism is more violent than hot spot volcanism because the magmas
   a. come from greater depths.
   b. are hotter.
   c. are more viscous.
   d. are less viscous.

REVIEW QUESTIONS continued

9. Calderas differ from craters in that calderas
   a. are smaller.
   b. form by collapse.
   c. form during the eruptive phase.
   d. are always associated with volcanic cones.
10. The 1980 eruption of Mount St. Helens was an example of a ____ style eruption.
    a. Strombolian   c. Pelean
    b. Hawaiian   d. Vulcanian
11. Why are island arc or continental arc volcanic eruptions more violent than those associated with most hot spots?
12. What is the basic difference between a shield volcano and a stratovolcano or composite volcano? Why are stratovolcanoes associated with zones of subduction while shield volcanoes are not?
13. Why are chains of volcanic islands such as the Hawaiian Islands aligned, and why is only one island volcanically active?
14. Name the different kinds of lava and describe the characteristics of each.
15. Explain how felsic and mafic magmas differ. How will the differences affect the style of an associated eruption?
16. What are the difficulties involved in predicting the future eruption of a volcano?
17. How does an increase in silicon content result in an increase in the viscosity of a magma?

THOUGHT PROBLEMS

1. Why are the Cascade Mountains limited in linear extent from northern California to southern British Columbia, Canada?
2. Even after plate movements have served the connection between an oceanic shield volcano and the underlying hot spot, volcanic activity can continue for a short period of time. Explain how this can happen.
3. What steps can the population of a town like Mammoth Lakes, California, take to protect themselves from a possible volcanic eruption?

FOR YOUR NOTEBOOK

Most of us do not live in areas of active volcanism or in the shadows of dormant volcanoes. Many of us, however, do live in areas where rocks recording ancient episodes of volcanic activity can be found. The Palisades along the west bank of the Hudson River, for example, is a sill. Perhaps the best source of information concerning the presence of volcanic rocks and the volcanic history of your area is the state geological survey or an office of the U.S. Geological Survey. They will be able to provide information about the locations of any volcanic rocks that may be in your area.

Hardly a year goes by without a volcanic eruption occurring somewhere in the world. Newspapers and weekly news magazines will provide much information that you can cut out and place in your notebook. Information about past eruptions can be acquired by a visit to the cataloged newspapers in your local library.

The activity on the island of Hawaii has certainly been well documented as has the 1980 eruption of Mount St. Helen. A number of videotapes of the Hawaiian eruptions are available from the Hawaiian Volcano Observatory, Hilo, Hawaii. The library at your college or university is also a possible source of video materials.

An excellent book written for a general audience is *Volcanoes,* by Robert and Barbara Decker, published by W. H. Freeman. Should you want to research the subject of volcanism further, the book includes a rather extensive bibliography.

CHAPTER OUTLINE

INTRODUCTION

MELTING AND CRYSTALLIZATION OF SOLIDS
  The Melting of Single Minerals of Constant Composition
  The Melting of Mixtures of Minerals and Minerals of Variable Composition
  Order of Mineral Crystallization

ROCK TEXTURE
  The Rate of Cooling

CLASSIFICATION OF IGNEOUS ROCKS

THE ORIGIN OF MAGMAS AND PLATE TECTONICS

IGNEOUS ROCK BODIES
  Intrusive Rock Bodies

CHAPTER 5

Igneous Rocks

INTRODUCTION Igneous rocks form from the solidification of molten rock that originated within Earth. Molten rock below the surface is called **magma.** Once formed, magma rises because it expands and is of lower density and therefore more buoyant than the surrounding rock. Some magmas solidify within Earth's crust and form **intrusive** igneous rocks. In other instances, the magma may move upward along cracks, eventually reaching the surface where the molten rock is extruded as **lava.** The solidification of lava forms **extrusive** igneous rocks. Other extrusive igneous rocks form by the accumulation of bits and pieces of lava or solid rock fragments blasted into the air during volcanic eruptions.

The major constituents of igneous rocks are silicate minerals. At this point, a review of the silicate minerals in Table 5.1 will help prepare you for the discussion that follows.

MELTING AND CRYSTALLIZATION OF SOLIDS

Every solid substance has a temperature above which it will melt and below which the melt will crystallize or solidify. This temperature is called either the *melting temperature* or the *crystallization (freezing) temperature,* depending upon the direction of the transformation. The melting-crystallization temperature is a physical property of every material and is dependent upon its composition and crystal structure. Because the minerals listed in Table 5.1 all differ in composition and/or crystal structure, it is reasonable to assume that they will have different melting-crystallization temperatures. Minerals with variable compositions melt and crystallize over a range of temperatures.

At temperatures above the melting-crystallization temperature, the amount of thermal energy (heat) available is sufficient to overcome the energy of the bonds that hold the crystal lattice together. As the bonds between ions and atoms within the crystal lattice are broken, the material melts. As long as the temperature is maintained above the melting point, the amount of thermal energy within the melt is sufficient to keep the ions and atoms from forming permanent bonds.

TABLE 5.1
THE MAJOR ROCK-FORMING SILICATE MINERALS

Ferromagnesian	NonFerromagnesian
Olivine group	Feldspar group
Fayalite	Plagioclase subgroup
Forsterite	Anorthite
Pyroxene group	Albite
Augite	Orthoclase
Amphibole group	Mica group
Hornblende	Muscovite
Mica group	Quartz
Biotite	

As the temperature drops below the melting-crystallization point, the attraction between the ions is sufficiently great to overcome the effects of the thermal energy. A crystal lattice starts to form (nucleate) as bonds begin to be established between the ions and atoms. As the crystal lattice grows, the mineral begins to precipitate from solution. Below the melting-crystallization temperature, the amount of thermal energy is insufficient to disrupt the chemical bonds, and the crystal lattice remains intact.

For a material of given composition, the melting-crystallization temperature can be affected by two external parameters: (1) *pressure* and (2) the presence of *volatile substances,* in particular, water. Volatile substances are those components that can be readily converted to gas. An increase in pressure increases the melting-crystallization temperature according to a basic precept in chemistry called LeChatelier's rule. Applied to this situation, this rule states that when a liquid-solid transformation is involved, increased pressure will favor the denser of the two phases (Figure 5.1). Therefore, an increase in pressure usually favors the solid phase. The melting temperature will rise because more thermal energy is needed to disrupt the chemical bonds. The common and important exception is water. Because water is denser than ice, increased pressure causes the melting point to decrease. As a result, in the presence of water, the melting temperature of ice decreases with increasing pressure.

One aspect of this relationship will be very important later in our discussions of magma formation. If a solid is already at high pressure and sufficiently high temperature, a rapid reduction in pressure without an associated decrease in temperature may cause the solid to melt (refer to Figure 5.1).

Volatiles, especially water, assist in breaking the chemical bonds, thereby reducing the amount of thermal energy needed for melting. In other words, the introduction of water lowers the melting temperature of a solid. Therefore, if a rock is dry and at a temperature just below the melting point, the addition of water may reduce the melting point and cause the rock to melt (Figure 5.2).

The Melting of Single Minerals of Constant Composition

A single mineral of a constant composition will begin to melt when the temperature reaches the melting point. Once melting begins, as long as heat is added, the temperature of the system will be maintained at the melting point of the mineral until it is entirely melted. When melting is complete, the temperature of the melt may then rise if heat continues to be added.

The Melting of Mixtures of Minerals and Minerals of Variable Composition

A mixture of minerals melts according to a quite different scenario. The melting of a mixture of minerals cannot be considered as simply the sequential melting of the individual minerals as their individual melting points are reached. First of all, with the exception perhaps of quartz, most of the minerals making up igneous rocks have variable compositions and therefore melt over a range of temperatures, rather than having specific melting points. In addition, because many of the minerals are members of solid solutions, melting points will vary depending on the specific composition of the mineral. Even combinations of minerals that are not members of a solid-solution series will melt over a range of temperatures depending on their relative concentrations. As melting progresses, the remaining solid and the highly reactive molten liquid become intermixed and continuously react with each other. As a result, as the rock melts, the compositions of both the melt and the remaining solid portion constantly change.

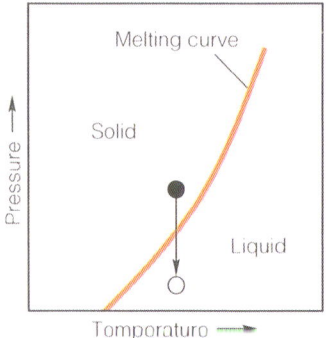

FIGURE 5.1 *When a solid is both under pressure and at high temperature, dropping the pressure may cause the solid to melt.*

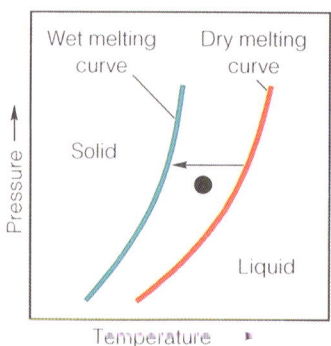

FIGURE 5.2 *When a solid is under both high temperature and pressure, the introduction of water may cause the solid to melt by providing an additional agent of bond-breaking.*

MAGMATIC FRACTIONATION

Magmas of various compositions may be generated from the crystallization of a *single* kind of magma, a process called **magmatic fractionation**. During the cooling of a magma, early formed crystals may separate from the magma; being of higher density, they sink to the bottom of the magma chamber, a process called **crystal settling**. The magma is thus depleted of the elements making up the new-formed minerals. If the rocks enclosing the magma chamber are then subjected to compressive forces, the magma may be squeezed from the mush of early formed crystals and injected into the surrounding rocks, a process called **filter pressing**. The process can then be repeated, producing different magma compositions each time.

As heat is applied and the temperature of the mineral rises, the material melts sequentially, beginning with the components with the lowest melting points; the result is a **partial melt**. An important point to note is that the composition of the melt at any point during the melting process will be preferentially enriched by the early melting components and will therefore be *different* from the composition of either the original rock or the remaining solid. Partial melting may, for example, enrich a melt in potassium and sodium while iron, magnesium, and calcium remain in the solids. Only if the rock completely melts and the melt does not separate from the remaining solid during the melting process will the elemental composition of the melt be the same as that of the original solids.

Order of Mineral Crystallization

Igneous rocks form by the crystallization of silicate minerals from magma or lava. Crystallization is the opposite of melting. Just as there is an order of melting, there is also an order of crystallization or a sequence in which the major silicate minerals appear as a molten rock mass cools. Like melting, the actual process of crystallization is complex and for the same reasons. As the individual minerals form, they may react with the remaining melt and change in both composition and structure.

The order of crystallization of the minerals that comprise an igneous rock can be investigated using several different approaches. One possibility is to observe the rock under a microscope by preparing a thin section (Figure 5.3). By identifying the individual minerals and observing their shapes, sizes, and grain boundary relationships, the order of crystallization can often be determined. Because the early formed minerals grow within the melt with no obstructions, they are able to grow relatively large and have well-developed crystal faces. As the later formed minerals begin to crystallize, their growth is partially impeded by existing crystals. As a result, their shapes are not as perfect. The last minerals to form crystallize in whatever openings are available between the existing crystals and are therefore limited severely in both size and shape.

FIGURE 5.3 *When rocks are cut very thin and observed with a polarizing microscope, not only can the shapes of the individual mineral grains be seen, but they appear in diagnostic colors by which they can be identified.* (© Alfred Pasieka/Science Photo Library)

A second method that has been used to establish the order of crystallization is to collect samples of cooling magma. Much of this work has been done by petrologists (geologists interested in the origin, occurrence, structure, and history of rocks) of the U.S. Geological Survey at the Hawaiian Volcano Observatory, a research facility on the island of Hawaii. Following an eruption, scientists venture out onto the solidified surface of the lava lakes and drill through the frozen crust of lava into the liquid lava below, an extremely dangerous adventure to say the least (Figure 5.4). (Over the years, some scientists have been seriously burned when the lava crust gave way beneath their feet like the ice breaking on a pond.) Once a hole has been opened in the lava crust, probes are extended into the lava to different depths, the temperature is recorded, and a sample of lava is retrieved. The lava is quickly cooled, converting the molten portion of the lava into a glass that entraps the mineral crystals that have already formed. Thin sections are then prepared and described. The percentage of glass relative to mineral grains allows the scientists to determine the extent to which crystallization has progressed at any particular temperature. The kinds and abundances of the individual minerals encased within the glassy matrix are then identified. By describing a series of samples taken from various depths within the cooling lava representing different temperatures and stages of solidification, the sequence of mineral crystallization can be determined. The results of such a study are summarized in Figure 5.5.

A third method of establishing the order of crystallization of the silicate minerals utilizes synthetic "magmas" prepared by melting powdered samples of various kinds of igneous rock. Portions of the melts are then slowly cooled to various temperatures; at that point, they are rapidly cooled and analyzed according to the same procedure used for the Hawaiian magma samples.

As was the case with the Hawaiian studies, the amount of glass indicates the degree of solidification that the "magma" has experienced as it cools to a particular temperature, and the minerals identification provides the order of mineral crystallization. Whereas the Hawaiian studies are limited in both magma composition and potential rock types, the preparation of synthetic magmas from different kinds of igneous rocks allows scientists to study a wider range of magma compositions and rocks. An example of the results from this kind of study is shown in Figure 5.6.

The champion of the experimental investigation of silicate mineral crystallization was *N. L. Bowen*. In 1928, based upon experimental data and the microscopic study of mineral associations in synthetic melts, Bowen established a sequence of mineral crystallization. He demonstrated that the major silicate minerals crystallize from a cooling molten mass in a specific order (Table 5.2). Studying basaltic magmas, Bowen showed that olivine forms first with the remainder of the ferromagnesian minerals crystallizing sequentially from augite to biotite. Following the crystallization of olivine, and simultaneous with the crystallization of the remaining ferromagnesian minerals, the plagioclase feldspars crystallize progressively from the calcium-rich anorthite at higher temperatures to the sodium-dominated albite at lower temperatures.

FIGURE 5.4 *Scientists at the Hawaiian Volcano Observatory on Hawaii study the crystallization of magma by taking samples of the molten lava for analysis. (Courtesy of R. T. Holcomb/USGS)*

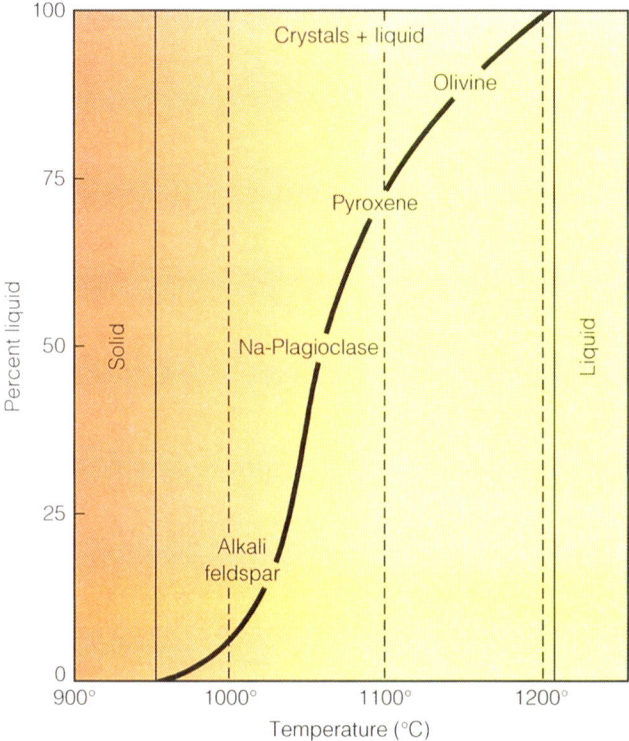

FIGURE 5.5 *Studies of the molten lava taken by the scientists at the Hawaiian Volcano Observatory have allowed them to determine the order in which the minerals form as the lava cools. (Adapted from U.S. Geological Survey Professional Paper 1004.)*

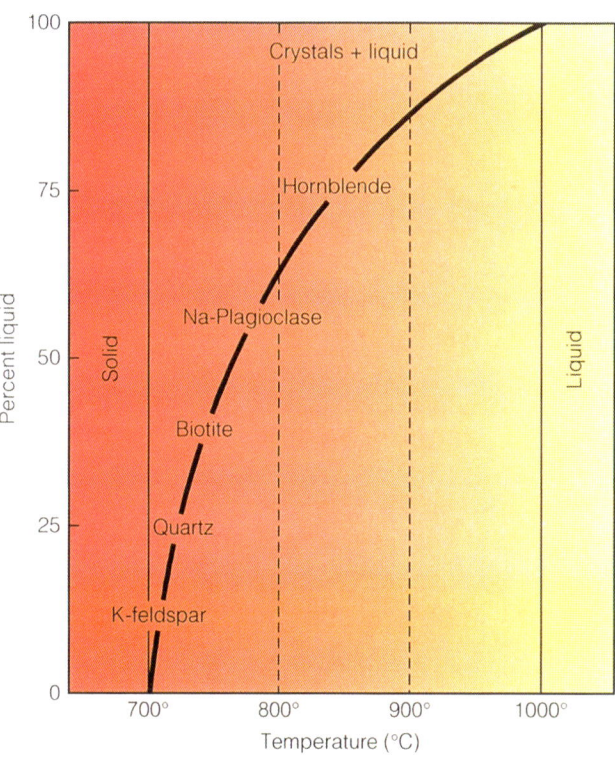

FIGURE 5.6 *Studies of synthetic magmas formed by melting various kinds of igneous rocks have verified the order in which minerals crystallize from cooling magmas. (Adapted from A. J. Piwinskii,* Journal of Geology *Vol. 76, University of Chicago © University of Chicago)*

TABLE 5.2
BOWEN'S CRYSTALLIZATION SERIES

High Temperature	Olivine	First to crystallize
	Augite	Anorthite
	Hornblende	
	Biotite	Albite
	Orthoclase	
	Muscovite	
Low Temperature	Quartz	Last to crystallize

The combined results of all these studies indicate that the major minerals that constitute igneous rocks crystallize in a specific order with minerals such as olivine crystallizing early at temperatures of 2,400°F to 2,200°F (1,300°C to 1,200°C) and minerals such as quartz and orthoclase feldspar forming last at temperatures of 1,300°F and 1,100°F (700°C to 600°C).

It is important to understand that the process of mineral crystallization is very complex. In his 1928 book, Bowen described his own work in a statement that is applicable to all experimental attempts to establish the order of crystallization of the silicate minerals: "An attempt is made…to arrange the minerals of the…rocks as reaction series. The matter is really too complex to be presented in such simple form. Nevertheless the simplicity, while somewhat misleading, may prove of service in presenting the subject in concrete form."

SPOT REVIEW

1. How can magmas of different compositions be formed during the cooling of a single magma?
2. How do pressure and the presence of water affect the melting point of rocks?
3. In what order do the major rock-forming silicate minerals form from a cooling magma?

ROCK TEXTURE

The **texture** of any rock refers to the size, shape, arrangement, and mutual interrelationships among the individual mineral grains.

There are five igneous rock textures: (1) *phaneritic*, (2) *aphanitic*, (3) *glassy*, (4) *porphyritic*, and (5) *pyroclastic* (Figure 5.7). In a rock with **phaneritic** texture, the individual grains are sufficiently large to be seen with the naked eye, usually larger than about 0.2 inches (5 mm) in diameter. **Aphanitic** texture indicates that the grains are too small to be resolved without the use of a hand lens or microscope, and **glassy** means that the rock has the amorphous (noncrystalline) structure of glass. **Porphyritic** texture describes a rock with phaneritic-sized crystals surrounded by either a finer phaneritic or an aphanitic matrix, and **pyroclastic** texture describes igneous rocks composed of rock fragments generated by volcanic eruptions.

The Rate of Cooling

The texture of igneous rocks is determined chiefly by the *rate of cooling* of the molten rock. Slow cooling favors the formation of fewer nucleation sites and the growth of larger crystals. Most intrusive magmas cool slowly because they are surrounded by rocks that conduct heat very poorly. As a result, most intrusive igneous rocks will be phaneritic (Figure 5.8). Lavas, poured out onto Earth's surface and exposed to either the atmosphere or water, and sometimes to ice, cool rapidly and are invariably either aphanitic due to the formation of many nucleation sites or glassy if solidification occurs without the formation of crystals (Figure 5.9).

Because the ions in a melt exhibit a certain degree of bonding, a melt can be readily converted to a glass if the cooling rate is exceptionally fast (Figure 5.10). Superfast cooling often occurs when lava is extruded

FIGURE 5.7 *Depending upon the rate at which molten rock cools, the texture of the rock that forms may be (a)* coarse-grained, *(b)* fine-grained, *(c)* porphyritic *(a mixture of coarse and fine grains) or (d)* glassy. *The texture of rocks that form from the fragments of rock produced by a volcanic eruption is called pyroclastic.* (a *N. K. Huber/USGS;* b *VU/© Albert J. Copley;* c *VU/© Albert J. Copley;* d *VU/© Doug Sokell*)

Igneous Rocks 147

(a)

(b)

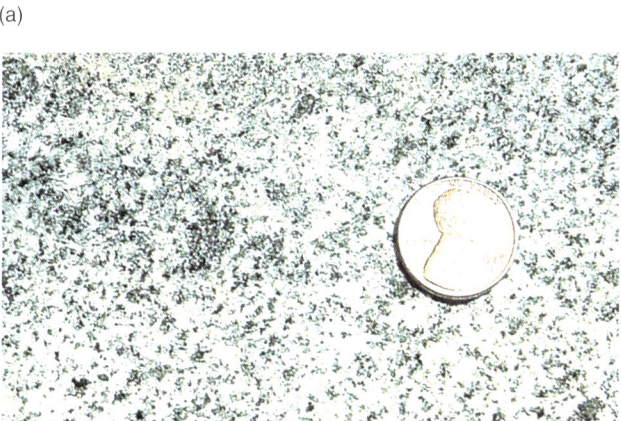
(c)

FIGURE 5.8 *Because granite forms by the slow cooling of magma, it is coarse-grained (phaneritic). The actual sizes of the crystals will be determined by the cooling rate. (Courtesy of N. K. Huber/USGS)*

FIGURE 5.9 *Because basalt forms by the rapid cooling of lava, it is fine-grained (aphanitic). (© E. R. Degginger)*

FIGURE 5.10 *When lava cools very fast, such as when it is extruded under glacial ice or into very cold water, the crystal lattices required to form mineral grains do not have time to form, and the lava solidifies as an amorphous glass showing glassy texture. (VU/© Doug Sokell)*

underwater or beneath a glacier. In addition, it is not uncommon for the surface of basaltic lava flows to be glassy.

Porphyritic texture is a combination of two textures, usually phaneritic and aphanitic. The rock, referred to as a **porphyry**, exhibits larger crystals called **phenocrysts** surrounded by a finer-grained **groundmass** (Figure 5.11). Such a rock may form as a result of an interrupted cooling sequence. The phenocrysts form as the magma begins to cool slowly within Earth's crust or mantle. At some time before solidification is complete, the mixture of crystals and remaining melt is then injected into cooler rocks, or perhaps extruded onto Earth's surface, where solidification is completed by more rapid cooling, forming a finer-grained groundmass.

SPOT REVIEW

1. How does the cooling rate affect the grain size of igneous rocks?
2. What is the significance of porphyritic texture?

CLASSIFICATION OF IGNEOUS ROCKS

Igneous rocks are named and classified on the basis of (1) mineral *composition* and (2) *texture*. The names of the major igneous rocks are summarized and illustrated graphically in Figure 5.12. The rocks are arranged horizontally in rows based on texture (cooling rate) and vertically in columns based on mineral composition.

Proceeding from right to left, the mineral fields in the lower portion of the diagram depict the order of crystallization of the major silicate minerals. The

FIGURE 5.11 Porphyritic texture *is a mixture of coarse-grained crystals surrounded by fine-grained crystals. The large crystals formed when the magma was cooling slowly. At some point in time, the slow cooling was interrupted and the remaining molten rock was cooled rapidly, possibly by being extruded and cooled by air or water or by being injected into narrow fractures where it was cooled rapidly by the surrounding, cold rocks. (Courtesy of N. K. Huber/USGS)*

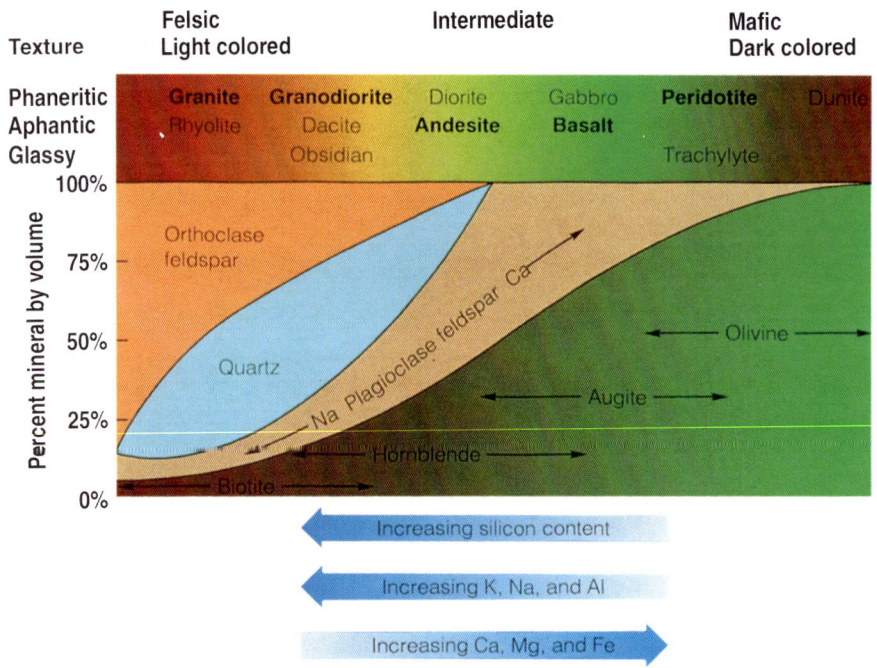

FIGURE 5.12 *The igneous rock chart summarizes the composition and cooling history of the major types of igneous rocks.*

rocks on the right-hand side of the chart contain minerals that crystallize at relatively high temperatures while rocks on the left are made from minerals that crystallize at lower temperatures. The diagram also illustrates from right to left the progressive change in the composition of the plagioclase feldspars and the ferromagnesian minerals, reflecting the decreased temperatures of crystallization.

The terms **mafic** and **felsic**, which appear at the top of the chart, refer to the compositions of igneous rocks, the silicate minerals, or the melts from which they form. Mafic rocks are rich in calcium plagioclase feldspars and in the ferromagnesian minerals olivine and pyroxene and have no minerals such as orthoclase feldspar and quarts. Rocks such as basalt or gabbro are mafic igneous rocks. On the other hand, igneous rocks enriched in minerals such as orthoclase feldspars, quartz, and sodium plagioclase but with few ferromagnesian minerals are felsic. Rocks, such as peridotite (especially the variety called **dunite**), that contain no felsic minerals and are composed almost entirely of olivine are said to be **ultramafic**.

These terms are also used to describe the compositions of magmas and lavas. A mafic magma is a molten mass that will crystallize to produce mafic minerals and rocks. Similarly, a felsic magma or lava will produce felsic minerals and rocks when it solidifies. These two terms are quite useful as a quick way to describe the composition of igneous materials, be they rocks, minerals, or molten materials.

The chart also graphically portrays the compositions of the various igneous rocks. To determine the mineralogy of any rock in the upper part of the chart, drop a line down from the rock type into the portion of the chart containing the minerals and observe the relative proportions of the various mineral fields cut by the line. A gabbro or basalt, for example, is made up of about 50% ferromagnesian minerals and 50% plagioclase feldspar, with no quartz or orthoclase feldspar. Thus, as the chart shows, one would expect to find calcium-rich plagioclase feldspar in basalt along with the ferromagnesian minerals augite and olivine. In contrast, a granite is composed primarily of orthoclase feldspar and quartz, with lesser amounts of sodium plagioclase and ferromagnesian minerals; the latter would most likely be biotite with possibly a small amount of hornblende.

Rock combinations such as granite-rhyolite or gabbro-basalt that appear above and below each other are identical in composition and differ *only* in texture. Glasses can be of any composition. Felsic glasses are called **obsidian** while mafic glasses are called **tachylyte**.

Variations in elemental components are indicated at the bottom of the chart. One of the most important components is the *silicon* (Si) content, which increases in the direction of the more felsic rocks and magmas. The silicon content is important for its effect on the *viscosity* of the magma. Viscosity, a physical property of liquids, is a measure of the liquid's resistance to flow. Low-viscosity liquids such as water flow readily while more viscous liquids such as syrup flow more slowly. The viscosity of magma increases with increasing silicon content. As silicon becomes more available within the magma, more strong silicon-oxygen bonds are formed, causing the viscosity of the melt to increase.

Because of their low silicon content, mafic magmas are generally less viscous than felsic magmas. This explains, at least in part, why basalt is the most common extrusive rock type. The viscosity of basaltic magma is low enough to allow the molten mass to move to the surface along narrow fractures with minimal resistance to flow. More felsic magmas, in contrast, are commonly too viscous to flow through these fractures and may solidify within Earth before reaching the surface.

Another compositional parameter that affects viscosity is water content. Because water inhibits the formation of silicon-oxygen bonds, an increase in the abundance of water dissolved in the magma causes the viscosity of the magma to decrease. For this reason, it is possible for some *hydrous* (water-rich) felsic magmas to have viscosities as low as basaltic magmas.

Several rock names are in boldface type on the chart because of their singular importance. **Granite** and **granodiorite** are referred to collectively as *granitic* rocks, the most common intrusive igneous rock in the continental crust. **Andesite** is the dominant extrusive rock type associated with subduction zone volcanism. **Basalt**, the most abundant extrusive igneous rock, makes up much of the oceanic crust and is associated with rift zones and the volcanism associated with most oceanic and continental hot spots. Finally, **peridotite** makes up Earth's upper mantle and is the source for most of the magmas that intrude Earth's crust at rift zones, rift valleys, and oceanic ridges.

Komatiite, a rare fine-grained rock named after the Komati River in the Transvaal of South Africa, is the extrusive equivalent of peridotite or dunite.

Because of the low silicon content, one would expect the magmas that form komatiite to be even less viscous than basaltic magmas and therefore to rise to Earth's surface with greater ease and produce more voluminous lavas than basalt. Such is not the case, however, presumably because the mineral constituents of komatiite crystallize at such high temperatures that the magma solidifies before reaching the surface. When the komatiites now exposed in South Africa were emplaced 2 billion years ago, crustal and mantle rocks were hotter than they are now, allowing the magmas to rise to Earth's surface without solidifying.

SPOT REVIEW

1. What factors affect the viscosity of magma?
2. Why are mafic lavas more common than felsic lavas?

THE ORIGIN OF MAGMAS AND PLATE TECTONICS

Any discussion of the origin of magmas must be prefaced with the statement that theories about their formation are speculative. Because the origin of magmas cannot be directly observed, their exact modes of formation will always be somewhat uncertain.

Since no one has ever observed the natural formation of a magma and likely no one ever will, our ideas about the origin of magmas are based upon (1) laboratory experiments, (2) observed distributions of the various kinds of igneous rocks at or near Earth's surface, (3) our understanding of plate tectonics, and (4) a large dose of sound scientific reasoning. We saw earlier that magmas are associated with four crustal areas: (1) *continental rifts,* which herald the initial breakup of continental crust; (2) *oceanic ridges,* where new oceanic crust is constantly being formed; (3) isolated *hot spots* beneath oceanic and continental crust; and (4) *zones of subduction,* where oceanic crust is being subducted and mountains are being created.

From our understanding of Earth's crust, magma appears to have three basic compositions: (1) *basaltic* magma, which provides the igneous rocks extruded at rift zones, rift valleys, oceanic ridges, and most hot spots; (2) *andesitic magma,* which is associated with subduction zone volcanic arcs; and (3) *granitic magma,* which accounts for most of the massive intrusive igneous rock bodies that are associated with the convergence of plates.

Most basaltic magmas are associated with zones of continental rifting and oceanic ridges, both of which are positioned above the rising portions of mantle convection cells (Figure 5.13). The specific combination of temperature and pressure within the asthenosphere places the rocks just within the solid-liquid transition curve of Figure 5.1. As they move closer to Earth's surface, the pressure progressively decreases. By the time the rocks in the convection cell arrive at the boundary between the asthenosphere and the lithosphere, the decrease in pressure is sufficient for melting to be initiated. The partial melting of the peridotite produces a basaltic magma that, because of its buoyancy, rises through the lithosphere to the surface. The question is, how can the melting of peridotite generate a basaltic magma?

Referring once again to the rock chart in Figure 5.12, you will see that peridotite is more mafic than gabbro-basalt. As noted above, during the partial melting of multicomponent silicate rocks, the order of mineral melting is the reverse of the order of crystallization. As a result, the composition of the melt is always more *felsic* than the composition of the original rock. Referring to Figure 5.12, note that the next rock composition that is more felsic than peridotite (just to the left on the chart) is gabbro-basalt.

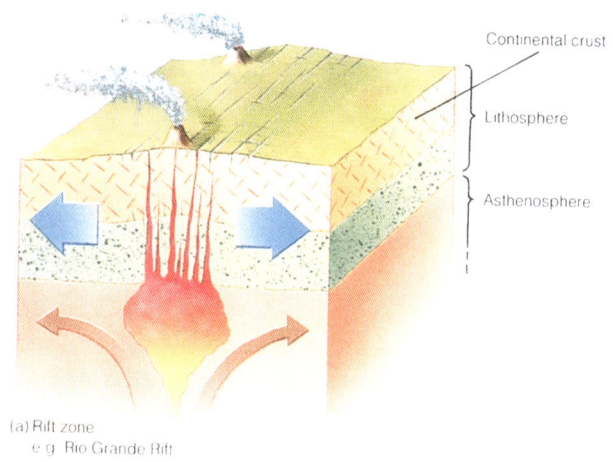

FIGURE 5.13 *At rift zones, basaltic magma, formed at the top of the asthenosphere, makes its way to the surface along fractures formed in the lithosphere. Some of the magma solidifies along the way to produce new lithospheric rocks. Other masses of the magma are extruded on the surface to produce the volcanic features associated with rift zones and rift valleys.*

Therefore, the melt formed by the partial melting of peridotite is gabbro-basaltic. In this case, because most of the molten rock makes its way to the surface to become lava and solidifies as an aphanitic rock, we refer to the magma as being basaltic. In summary, the basaltic magmas associated with rift zones, rift valleys, and oceanic ridges form from the partial melting of peridotite induced by a decrease in pressure.

Hot spots positioned below the lithospheric plate are the other major source of basaltic magmas. The hot spots are located at the tops of heat-driven plumes that, most evidence suggests, rise from deep within the mantle. Once again, the magmas form by the partial melting of peridotite as a result of a reduction in pressure. Once formed, the magmas rise due to their lower density and pierce the lithosphere at point sources to form seamounts and the spectacular shield volcanoes characteristic of oceanic volcanic islands such as the Hawaiian Islands (Figure 5.14).

The most complex magma-forming scenario involves the zones of subduction where we must explain the simultaneous generation of three kinds of magma (Figure 5.15): (1) the basaltic magmas that reach the surface along fractures that develop in the overriding plate behind the arc of volcanic mountains in what geologists call the back-arc basin; (2) the andesitic magmas that feed most of the arc volcanics associated with zones of subduction such as the Andes Mountains, the Aleutian Islands, and the

FIGURE 5.14 *A well-known locality where one can see igneous rocks being made is Hawaii where basaltic magmas are being extruded and cooled to form the basaltic rocks that make up the Hawaiian Islands (Courtesy of USGS/HVO)*

FIGURE 5.15 *Three compositions of magma are being formed within subduction zones: basaltic, andesitic, and granitic. The andesitic magmas are extruded and form the chains of volcanic mountains associated with converging plate margins. Most of the granitic magmas cool underground and add to the continental crust. Occasionally, basaltic magmas may reach the surface along fissure eruptions.*

Japanese Islands; and (3) the massive volumes of granitic magmas that account for the huge masses of igneous rocks that intrude into and add to the continental crust.

As the lithospheric plate bearing the oceanic crust subducts beneath an oncoming lithospheric plate bearing either an oceanic or continental plate, it is driven or pulled downward into the mantle. The major source of heat available within the zone of subduction is heat contained within Earth (refer to Figure 5.15).

Water is introduced into the mantle as rocks and sediments of the oceanic plate are subducted. As the hydrous minerals contained within the crustal rocks and oceanic sediments are heated, water is released and penetrates into the overlying mantle. With the combined effect of the heat provided at depth and the lowering of the rock melting points caused by the presence of water, massive quantities of basaltic magma are produced by partial melting of the mantle peridotites. Some of these basaltic magmas may rise to the surface and emerge along cracks in the crust to produce floods of basalt such as those that accumulated to form the Columbia Plateau of eastern Washington and Oregon (Figure 5.16) and the basaltic volcanics found throughout Nevada and adjoining states.

The silica-rich sediments commonly found on the deep-ocean bottom are derived from two major sources: (1) wind-transported dust from the continents, consisting primarily of clay minerals, quartz, and feldspar, and (2) the siliceous (silica-rich) shells of

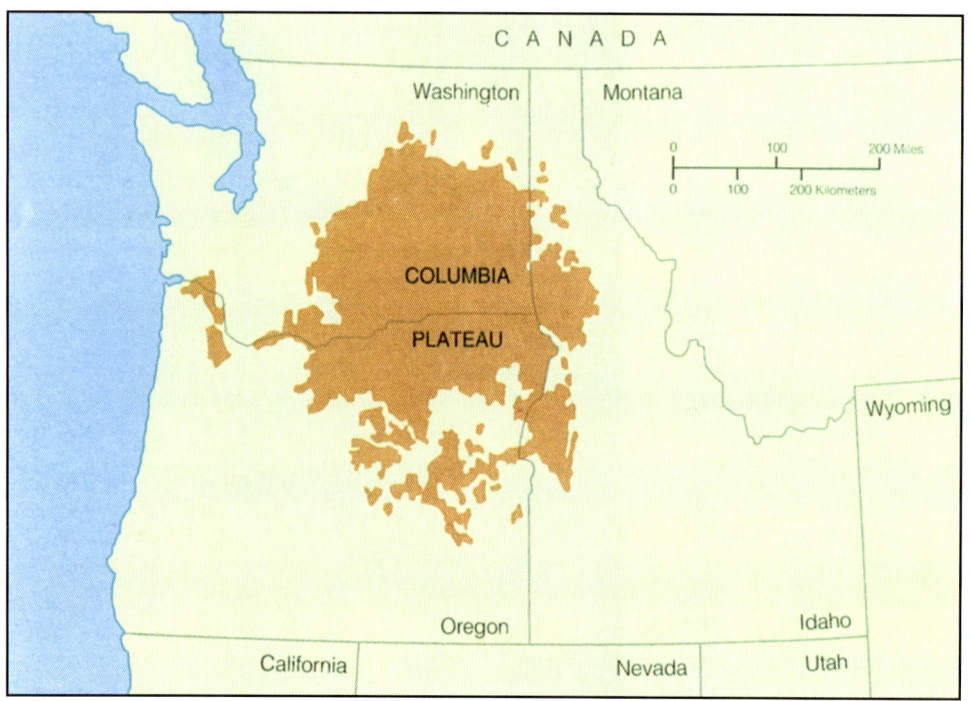

FIGURE 5.16 *Massive fissure eruptions of basaltic magma resulted in the formation of the layered basalt flows of the Columbia Plateau. (Courtesy of P. Weis/USGS)*

marine microorganisms such as diatoms and radiolarians (Figure 5.17). Some of these siliceous ocean bottom sediments are drawn down into the zone of subduction, melt, and become incorporated into the basaltic magmas, contributing to the increasingly felsic nature of the magma. Andesitic magmas may also form from the partial melting of subducted basaltic oceanic crust. Although andesitic magmas are more viscous than basaltic magmas due to their higher silica content, they can still rise along fractures to the surface. Because of the higher viscosity of the andesitic magmas, large volumes of gases remain dissolved in the magmas until they emerge onto the surface. Once the magmas are exposed to the atmosphere, the gases escape rapidly, producing the violent explosions characteristic of subduction zone volcanism. The large amounts of fragmental materials generated during the eruptions are responsible for the steep-sided volcanoes of the Japanese Islands, the Aleutians, the Cascades, and the Andes. We will discuss the various kinds of volcanoes in the next chapter.

The magmas that are the most difficult to explain are the huge masses of highly felsic granitic magmas that intrude into the overriding plate (refer to Figure 5.15). Partial melting of oceanic crustal basalt cannot generate granitic magmas. Even the intermixing of the silica available from the subducted oceanic sediments will not alone produce granitic magmas because the elements necessary to create the other required mineral constituents are not present. Two scenarios have been suggested for the production of the granitic magmas:

1. Intense compressional deformation within the subduction zone generates metamorphic rocks from a combination of lithospheric plate rocks and sediments. Because the oceanic sediments are enriched in silica, the overall elemental composition of the metamorphic rock-oceanic sediments assemblage is dominantly felsic. The melting of this mixture may be the source of large volumes of granitic magma.

2. The source of the granitic magmas may be the granitic continental crust itself. As the basaltic magmas generated by the partial melting of the mantle peridotite rise toward the surface and encounter these more felsic (and therefore lower melting point) rocks, the basaltic magma pools below the crust because of the crust's lower density. As the basaltic magma accumulates, heat is transferred from the magmas to the granitic rocks and causes them to melt. As the heat is removed from the basaltic magmas and they begin to solidify, their upward movement is arrested within the crust. This may explain why basaltic lavas are not widely represented at the surface of the continents even though massive volumes of basaltic magmas are thought to be generated within zones of subduction.

FIGURE 5.17 *The shells of microscopic animals such as radiolarians are major contributors to the siliceous sediments that accumulate on the ocean floor. (Courtesy of Deep Sea Drilling Project/Scripps Institution of Oceanography)*

Once generated, the low-density granitic magmas move upward and intrude into the overlying crust. Because of their high silica content, granitic magmas are usually highly viscous and tend to solidify within the crust, generating the huge masses of granitic rock that add to the growing mass of the continental crust. Upon occasion, however, these felsic magmas do break through to the surface and emerge as rhyolitic lavas. Because of the exceptionally high gas content of the magmas, the eruptions are typically highly explosive and produce enormous volumes of fragmental material. An example is the eruption that produced the rhyolitic rocks in the area of Yellowstone Park 600,000 years ago. The felsic magmas that erupted to produce these rocks are thought to have been generated by the melting of granitic crustal rocks with heat supplied by basaltic magmas rising from a subcontinental hot spot. The presence of the lower-density granitic magma prevented the denser basaltic magmas from moving upward. Following the solidification of the granitic magmas, it is believed that the basaltic magmas, having been continuously produced at the hot spot, rose to the surface along fractures in the solidified granities to emerge as the posteruption basaltic lava flows found in the area.

SPOT REVIEW

1. Why are the magmas associated with divergent plate boundaries and most hot spots dominantly mafic in composition while those associated with convergent plate boundaries range in composition from basaltic to granitic?
2. What are the two scenarios commonly presented for the formation of granitic magmas?

IGNEOUS ROCK BODIES

To fully understand how igneous rock masses are emplaced, we need to concern ourselves with the form of the igneous rock bodies. Three criteria are used to categorize igneous rock bodies: (1) where they are formed relative to Earth's surface, (2) their shape, and (3) their orientation relative to layered features in the surrounding rock.

We have already introduced the terms *intrusive* and *extrusive,* which categorize igneous rock bodies according to whether they are on or below Earth's surface. Magmas solidify to produce intrusive rock bodies while lavas and the fragmental materials produced by volcanic eruptions solidity to form extrusive rock bodies.

Intrusive Rock Bodies

Intrusive rock bodies are classified by shape as either **tabular** or **massive**. Because most rock bodies of any kind are usually observed in a two-dimensional exposure such as a cliff, canyon wall, or road cut, establishing the shape involves measuring the (1) maximum and (2) minimum dimensions. The dinstinction between tabular and massive shape is arbitrarily based upon the relative magnitude of these dimensions of the rock body (Figure 5.18). If the maximum dimension is larger than 10 times the minimum dimension, the body is tabular, and if it is less, the body is massive. Compare the difference in shape between a table top (tabular) and a football (massive).

In many cases, the host rock into which molten rock intrudes and solidifies will have some *planar feature* such as the layers that characterize sedimentary rocks, layered lava flows, pyroclastic deposits and the foliation planes of some metamorphic rocks. An igneous rock body is said to be **discordant** if its surface *cuts across* the layers, but if the body *parallels* the layers, it is **concordant** (Figure 5.19).

In summary, igneous rock bodies are classified as (1) *intrusive* or *extrusive* according to where they formed, (2) *tabular* or *massive* based on their shape, and (3) *discordant* or *concordant* based on their relationship to the surrounding rock.

Note that igneous rock bodies are *not* distinguished by *size*. Geologists are usually less concerned about the size of any an object or feature than they are about its origin. In only one instance, the designation of the *stock* and the *batholith* (to be discussed), is size a consideration in classifying an igneous rock body.

Tabular Intrusive Igneous Bodies

The two kinds of tabular intrusive rock bodies are illustrated in Figure 5.18a. A **dike** is defined as a "tabular discordant, intrusive igneous body" whereas a **sill** is a "tabular, concordant, intrusive igneous body." Again, it should be pointed out that there is no size restriction to these rock bodies. Except for the aphanitic portion of tabular rock bodies where they make contact with the host rock, most dikes and sills are phaneritic.

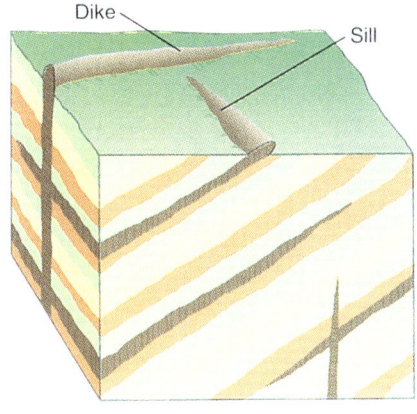
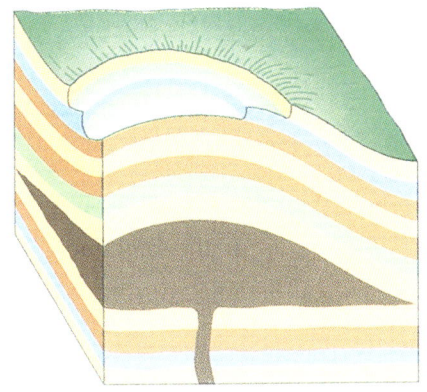

(a) Tabular igneous bodies   (b) Massive igneous body

FIGURE 5.18 Intrusive igneous bodies *are distinguished by shape as either* tabular *or* massive.

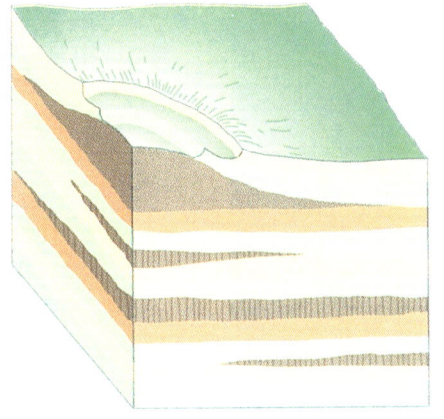
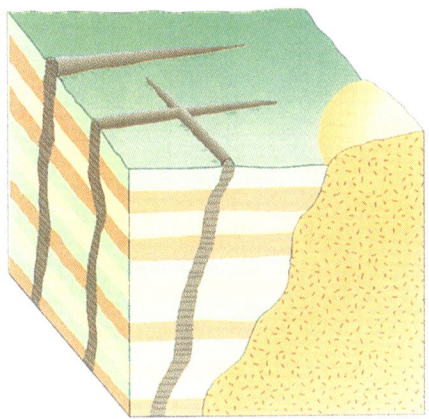

(a) Concordant contacts   (b) Discordant contacts

FIGURE 5.19 Intrusive igneous bodies *are referred to as either* concordant *or* discordant *depending upon whether the boundary of an intruded igneous body* parallels *or* cuts across *layering in the host rock.*

Massive Intrusive Igneous Bodies A **laccolith** develops when molten rock, fed by a feeder dike, is injected between layered rocks at a rate faster than the molten mass can move laterally between the layers of the intruded rocks. As the mass becomes thicker over the feeder site, the intrusion arches the overlying rocks and, at some point, changes in shape from concordant *tabular* to concordant *massive* (Figure 5.18b). Once the small dimension becomes more than one-tenth of the large dimension, the original sill turns into a laccolith. Once again, there is no size requirements. Laccoliths can be of any size. An example of exposed laccolithic intrusions that resulted from the removal of the overlying rock by erosion are the Henry Mountains of southern Utah (Figure 5.20). In a similar structure called a **lopolith**, the central portion of the body is sunken due to the sagging of the underlying rock (Figure 5.21).

The only two massive igneous bodies that are distinguished by size are the **stock** and the **batholith**. A stock is defined as a massive intrusive igneous body with a surface area of exposure of less than 40 square miles (100 km^2), whereas a batholith is a massive intrusive igneous body with a surface exposure greater than 40 square miles. In fact, the term *batholith* is synonymous with large size, often being defined as "a huge body of intrusive igneous rock." The concordant-discordant relationships of these massive bodies give an indication of the mode of intrusion. In some instances, the host rocks are deformed and arched over the top of the rock body, indicating a *forceful entry* of the molten mass. In other instances, the layering in the host rock is truncated at the body margin with little or no indication of deformation (Figure 5.22). In this case, the emplacement of the molten mass is interpreted as

FIGURE 5.20 *The core of the Henry Mountains in Utah is a* laccolith.

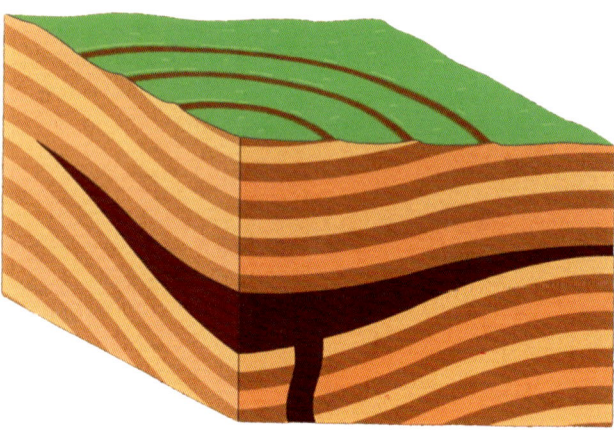

FIGURE 5.21 *A* lopolith *is a massive, concordant, igneous body whose center is downwarped due to the sagging of the underlying host rocks.*

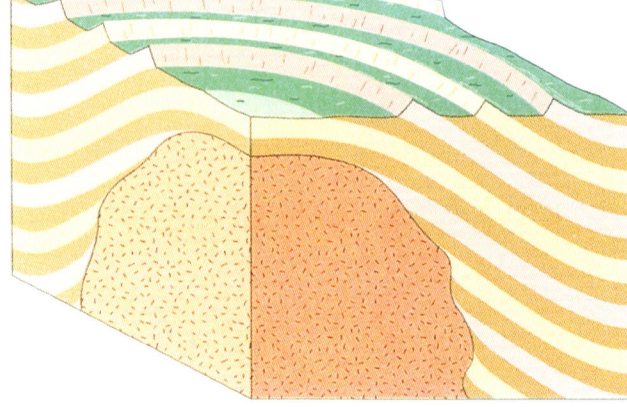

(a) Introduction by forceful entry

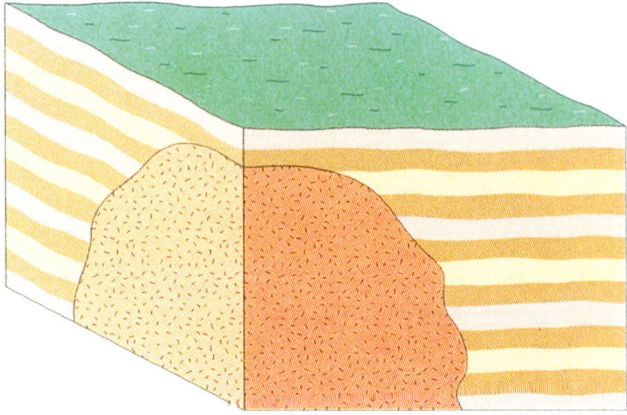

(b) Introduction by stoping

FIGURE 5.22 *Most plutons are forcefully injected into the host rocks as shown by the deformed layers of the host rocks. Some plutons, however, may be introduced by the* assimilation *or* stoping *of the host rock by the rising magma.*

having involved **stoping**, a process where the host rock is detached and engulfed within the rising magma. These pieces of host rock either sink to the bottom of the magma chamber, become entrapped in the magma near the boundary of the igneous body as an **xenolith**, or are assimilated within the magma (Figure 5.23).

Perhaps the most spectacular, and familiar, intrusive igneous rock body is the **volcanic neck**, which is the remains of an extinct volcano. In time, every volcano becomes extinct as the heat source maintaining the molten rock in the source magma chamber is exhausted. At this point, the constructive phase of the volcano ends. Erosion will then remove the outer

FIGURE 5.23 *Not uncommonly, intruding magma will engulf pieces of the host rock, which then become* inclusions *within the igneous rock called* xenoliths. *(VU/© Albert J. Copley)*

FIGURE 5.24 El Capitan *in Arizona is an example of a volcanic neck.*

conical structure of the volcano and expose the igneous rock in the core of the cone as a vertical spire. A well-known example of a volcanic neck is El Capitan, Arizona (Figure 5.24).

Extrusive rocks were discussed in the previous chapter.

SPOT REVIEW

1. How can a sill be distinguished from a buried lava flow?
2. How can one determine whether an intrusive rock body has been emplaced forcefully or by stoping?

CONCEPTS AND TERMS TO REMEMBER

Kinds of molten rock
  magma
    intrusive
  lava
    extrusive
Melting of solids
  partial melting
  magmatic fractionation
  crystal settling
  filter pressing
Cooling of molten rock
  texture
    phaneritic
    aphanitic
    glassy
    porphyritic

porphyry
phenocrysts
groundmass
pyroclastic
Classification of igneous rocks
  mafic
  felsic
  ultramafic
Kinds of igneous rock
  dunite
  obsidian
  tachylyte
  granite
  granodiorite
  andesite
  basalt

peridotite
komatiite
Kinds of rock bodies
  tabular
    discordant
      dike
    concordant
      sill
  massive
    laccolith
    lopolith
    stock
    batholith
      stoping
      xenolith
    volcanic neck

REVIEW QUESTIONS

1. The most common igneous rock found in intrusive igneous rock bodies is
   a. gabbro. c. basalt.
   b. granodiorite. d. peridotite.
2. According to Bowen's crystallization series, the first mineral to form from a crystallizing mafic magma would be
   a. quartz. c. hornblende.
   b. olivine. d. albite.
3. The igneous rock that makes up the oceanic crust is
   a. rhyolite. c. granite
   b. basalt. d. peridotite.
4. At which of the following localities is magma produced primarily by the pressure drop mechanism?
   a. zone of subduction (convergent boundary)
   b. hot spot
   c. mid-oceanic spreading center (divergent boundary)
   d. transform boundary
5. Which of the following statements will *always* be true?
   a. Intrusive igneous rocks are always coarse grained.
   b. Extrusive igneous rocks are never coarse grained.
   c. Intrusive igneous rocks can be glassy.
   d. Obsidian forms by the crystalllizataion of either magma or lava.
6. Which of the following pairs of igneous rocks have the same composition and differ only in grain size?
   a. gabbro-basalt c. diorite-peridotite
   b. granite-andesite d. gabbro-rhyolite
7. Which of the following is defined as a "tabular, concordant, intrusive igneous body"?
   a. sill c. laccolith
   b. stock d. dike
8. The difference between magma and lava is that magma
   a. is hotter.
   b. exists within the Earth while lava exists on the Earth's surface.
   c. is more viscous.
   d. has a higher silica content.

9. If you were shown an igneous rock that was phaneritic and dark in color, which of the following could it be?
   a. granite   c. granodiorite
   b. basalt   d. gabbro
10. The composition of most lavas is
    a. andesitic.   c. felsic.
    b. rhyolitic.   d. basaltic.
11. If the asthenosphere is composed of the rock peridotite, how can the melting of the upper portion of the asthenosphere be a source of the basaltic magmas that rise to the surface at mid-oceanic ridges and hot spots?
12. If less viscous, silica-poor mafic magmas rise more easily to Earth's surface, why are komatiites not extensively exposed at Earth's surface?
13. Why does the rapid rate of crystallization of magma favor the formation of small crystallites while slow rates of crystallization promote the formation of large crystals?
14. Explain why the inner portion of a crystal of plagioclase feldspar is commonly a calcium-rich variety while the outer portion of the crystal is a more sodium-rich variety.
15. Why is the mineral quartz not found in mafic igneous rocks even though mafic magmas contain as much as 50% silica by weight?
16. Can you give a possible explanation why the crystal structures in the ferromagnesian minerals become more complex with decreasing crystallization temperatures?

THOUGHT PROBLEMS

1. What evidence would you look for to distinguish between a basaltic still injected between layers of sedimentary rocks and an ancient, buried basaltic lava flow?
2. Why does the silica content critically control the viscosity of molten rock?
3. How can igneous rocks of different compositions form from a single magma?

FOR YOUR NOTEBOOK

If you live or go to school in an area where igneous rocks are exposed, a field trip to observe and describe some of the local occurrences would certainly be in order. A geologic map that you can obtain from your state geological survey or from your instructor will outline the distribution of the various kinds of rocks as well as the most likely place to find them well exposed. You may want to collect some specimens for a rock collection, and you will want to indicate the effect that different rock types have on the topography with a series of sketches or photos.

For the great majority of us who do not live in areas where igneous rocks can be observed in outcrop, a field trip through your town or city will undoubtedly produce some examples of igneous rocks used for construction or trim. (As you make your survey, not the use of other types of stone so that you may include them in your notebook under the appropriate chapter.) You may want to seek the help of your instructor in identifying some of the rock types, and he or she may suggest some examples you will not want to overlook. In addition to noting the kind of rock, indicate how the rock was used; that is, whether it was used as a major construction stone or as a decorative trim. A conversation with a local architect would be a source of important information concerning the use of different kinds of rocks in your area.

CHAPTER OUTLINE

INTRODUCTION
WEATHERING DEFINED
MECHANICAL WEATHERING
    Frost Action
    Exfoliation
    Spheroidal Weathering
    Effect of Temperature
    Plant Wedging
CHEMICAL WEATHERING
    Oxidation
    Dissolution
    Carbonation/Hydrolysis

RATES OF CHEMICAL WEATHERING
    Temperature
    Availability of Water
    Particle Size
    Composition
    Metastable Materials
REGOLITH
ENVIRONMENTAL CONCERNS
    Trace Elements
    Water Quality
    The Effect of Mining on Water Quality

CHAPTER 6

Weathering

INTRODUCTION To most people, rocks are permanent. Euphemisms such as "rock of ages" and "solid as the Rock of Gibraltar" reflect the common impression that rocks are extremely durable. Certainly, in terms of human lifetimes, some rocks do indeed seem to last forever. Rock of Ages, for example, is the trade name for a commercial brand of granite monuments. Granite was the rock used by the Egyptians to construct the cores of the pyramids, which many view as the epitome of antiquity (Figure 6.1). Today granite is still frequently used in the construction of buildings and monuments, especially those designed to last a long time (Figure 6.2). Granite is indeed a fairly durable rock for reasons that you will learn in this chapter.

The longevity of granite is, however, simply a question of perspective. Using a human lifetime as a measuring stick, granite does seem to be the "rock of ages," but in terms of geologic time, even granite soon succumbs. The process that dooms granite and all rocks is **weathering**. Weathering is the first of a series of processes collectively called **erosion**. Erosion itself is defined as any process by which Earth's surface is loosened, dissolved, and worn away, and simultaneously moved from one place to another, by natural agents. No matter how durable a rock may appear to be, no rock will survive these processes for very long.

Weathering is of critical importance to life on Earth. A major end product of the process of weathering is *soil*. Without soil, nothing but the lowest forms of plant life could exist on land. Without soil, the continents of Earth would be as lifeless as the Moon.

FIGURE 6.1 *Much of the longevity of structures such as the pyramids is due to the limited availability of water in their desert location. (Courtesy of E. D. McKee/USGS)*

FIGURE 6.2 *Because of its resistance to the various processes of weathering, architects commonly choose granite for monuments and edifices designed to survive for long periods of time. (Courtesy of Washington DC Convention and Visitors Association)*

WEATHERING DEFINED

Several definitions of weathering can be found, each of which concentrates on a specific aspect of the overall process. For example, weathering has been defined as any reaction between a rock surface and the agents of the atmosphere. This definition reflects the fact that the process can go on anywhere the atmosphere can penetrate, be it a hairline crack in a rock, a cave, or a mine.

Another definition specifies that rocks will undergo weathering wherever they are exposed to circulating surface water. In addition to emphasizing the role of water in the weathering process, this definition indicates that weathering can go on at depth and indirectly implies that the "agents of the atmosphere" responsible for the breakdown of rocks are transported dissolved in water.

From the standpoint of describing its mechanisms, weathering is best defined as any process whereby rocks either *disintegrate* or *decompose*. **Disintegration** is any process that breaks rocks into smaller pieces without changing the composition. **Decomposition** includes any process that results in either a partial or complete change in the mineral and elemental composition of the original rock.

Disintegration and decomposition are the basis for the two modes of weathering, **mechanical weathering** and **chemical weathering**, respectively. Although these two processes can and do occur independently, as one might expect, they usually go on simultaneously. Depending largely on the climate, however, one process usually dominates over the other.

MECHANICAL WEATHERING

Mechanical weathering is accompanied by a variety of physical processes, all of which break the rock materials into smaller pieces that retain the chemical composition of the parent rock.

Frost Action

During the spring, throughout the more temperate, humid parts of the world, roadways below road cuts and the bases of cliffs are commonly littered with boulders and rock fragments, the end products of one of the most common mechanical weathering processes, **frost action** (Figure 6.3). Frost action is the result of a special property of water. Water is one of the few materials that expands as it converts from the liquid to the solid state. A volume of water will expand about 9% as it freezes. This may not sound like much, but water freezing within a completely filled, enclosed container could theoretically generate a pressure of about 1,500 pounds per square inch (680 km/cm^2). If the container happens to be the water pipes in your home, the expansion of the ice will cause them to burst long before this pressure is attained. Expanding ice will also break the strongest rock. In fact, very few solid materials can long withstand the internal stresses generated by freezing water. When water penetrates even the narrowest crack or crevice within a rock and freezes, the crack is enlarged. When the temperature rises and the ice melts, more water enters the expanded crack and penetrates deeper into the rock. The next freeze opens the crack even further. Each successive **freeze-thaw cycle** continues to enlarge the crack, eventually separating the block of rock from the main rock body. An end product of frost action is the loose rock debris called *talus* that accumulates along the base of a cliff or a road cut.

FIGURE 6.3 Mechanical weathering *due to* frost action *is responsible for much of the talus that collects at the base of cliffs and road cuts in temperate, humid areas. (VU/© Steve McCutcheon)*

FIGURE 6.4 *Above the tree line where freezing conditions can occur every month of the year, the surface layers of the exposed rock are broken by the process of frost action. (Courtesy of Joe Donovan)*

Frost action is responsible for other examples of mechanical rock disruption. Most of the rock litter that frequently covers the ground in high mountainous areas is produced by frost action (Figure 6.4). This same process is a major contributor to the formation of the chuckholes or potholes that make driving such an exciting experience in areas where freeze-thaw is common.

It is important to note that the process of frost action requires numerous freeze-thaw cycles. Many cycles are needed to enable water to enter the enlarged cracks, refreeze, and ultimately cause the physical disruption of the rock. Consequently, regions subjected to prolonged freezing conditions such as the polar regions, do not exhibit extensive frost action.

Exfoliation

Exfoliation is a common process of weathering in which concentric layers are removed from the rock mass. The layers, ranging in thickness from fractions of an inch (a few millimeters) to several feet or meters, are formed and removed by pressures exerted on the outer layers of rock by a number of processes operating below the rock surface (Figure 6.5). An example of such a process involves the growth of water-soluble crystals. Water moving through layers of rock carries minerals in solution (we will study this underground movement of water when we examine groundwater in Chapter 14). At the face of a cliff or road cut where rock layers are exposed, the water may seep out and flow down the rock face (Figure 6.6). Often during times of low flow or on especially warm

FIGURE 6.5 *A number of processes result in the surface of the rock being removed layer by layer, a process called* exfoliation.

days, the water may evaporate and deposit the minerals, producing the white or reddish stains commonly observed on cliff faces (Figure 6.7). Consider, however, a special situation where the rock exposure faces toward the Sun and is heated by its radiant

FIGURE 6.6 *The seepage of groundwater along cliff faces and road cuts is a common sight. (VU/© Bill Kamin)*

FIGURE 6.7 *The white to reddish stains commonly seen on rocks exposed in cliffs and road cuts are the result of minerals deposited as seeping groundwater evaporates on the surface of the rock. (Courtesy of W. B. Hamilton/USGS)*

energy. Should the flow of water through the rock layers be sufficiently low, heat generated by the Sun's rays can cause the water to evaporate down to a depth of a few millimeters below the rock surface. As the water evaporates, the minerals crystallize in a layer and constantly grow in mass as they are "fed" by fresh solutions from behind (Figure 6.8). In time, the pressures of the growing crystal mass will cause the rock to split parallel to the rock surface and be removed, or **spall**, in sheets or flakes. As fresh rock surfaces are exposed, evaporation of water begins anew just beneath the surface and the process continues. Slowly, the cliff face retreats.

Oftentimes, local variations in rock composition result in "passageways" within the rock through which water is conducted to the surface more readily. At sites where these passageways reach the rock surface, the process of mineral growth is accelerated and causes the rock surface to retreat at a faster rate than the adjoining rock. The result is a weathering phenomenon called **honeycomb weathering** (Figure 6.9).

The effect of crystal growth can also be seen operating on the sunny side of stone walls or buildings. Water leaking from gutters may dissolve some of the mortar between the stone blocks. Driven by gravity, the water seeps out toward the face of the blocks and evaporates just beneath the surface, causing the rock surfaces to exfoliate in thin layers. If you observe the blocks closely, you can see the rock flakes being dislodged and just below, a white powder, most likely *gypsum* ($CaSO_4 \cdot 2H_2O$).

A related phenomenon can often be seen attacking the rocks exposed along the coast. Wind blowing onshore carries sea spray containing dissolved salts. As this moisture penetrates into the rocks along more porous zones or through cracks and evaporates, the pressure of growing salt crystals can loosen grains from the rocks. If you live near the ocean, you may try to find rocks that show evidence of physical deterioration due to the growth of sea salts.

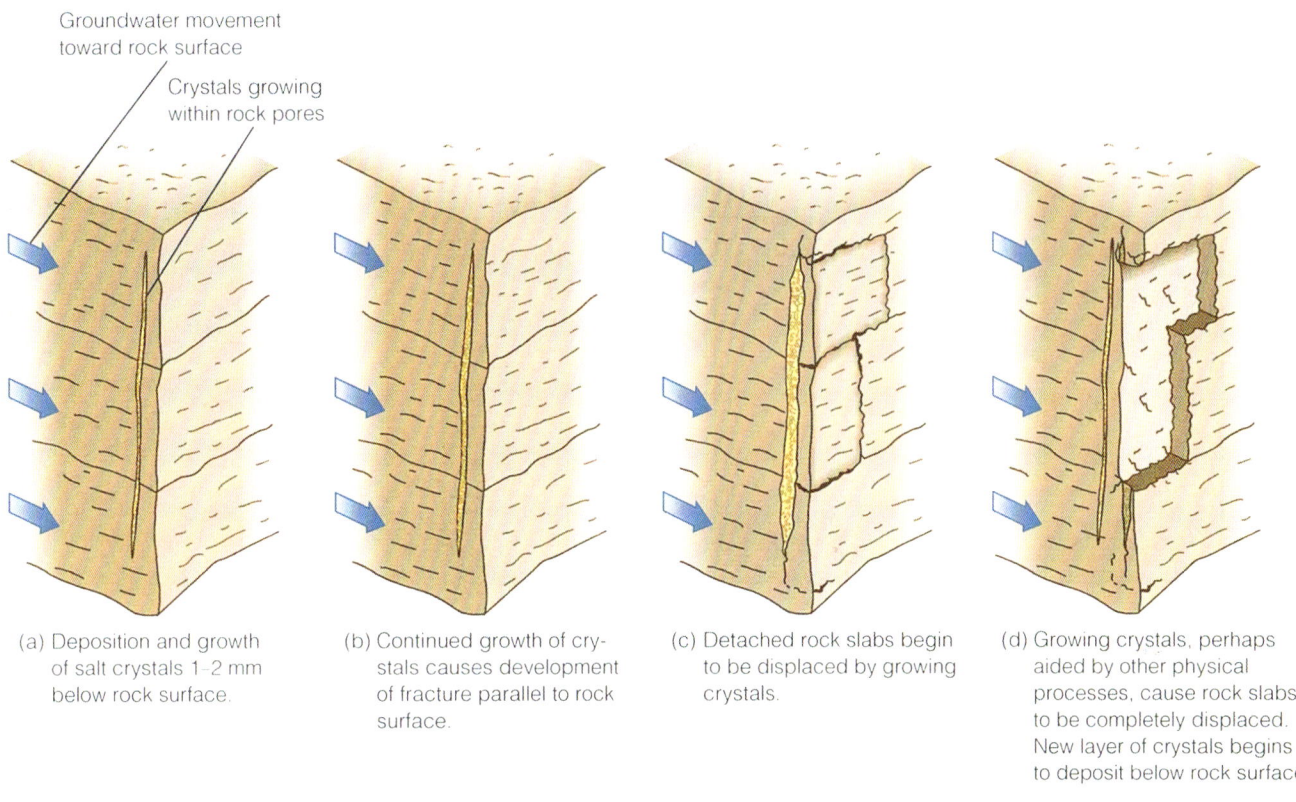

FIGURE 6.8 *Crystals that form and grow as groundwater evaporates just beneath the surface of sunlit outcrops are responsible for the spalling of the surface layers of rock.*

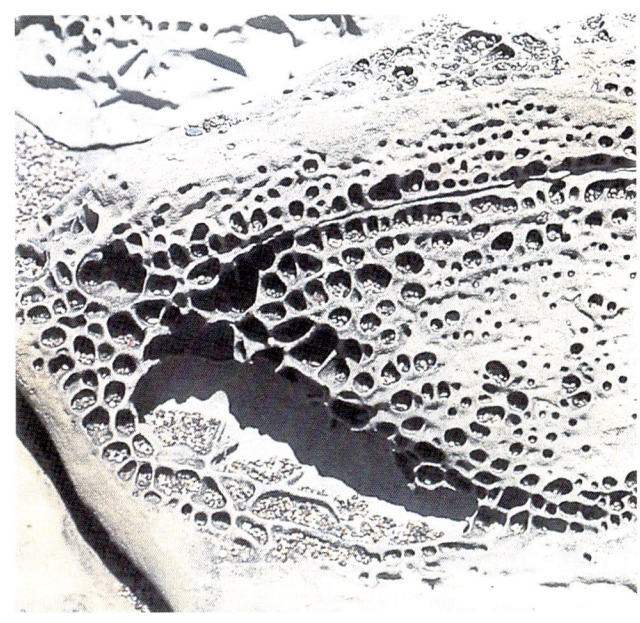

FIGURE 6.9 Honeycomb weathering *is a special type of weathering resulting from the subsurface growth of crystals in outcrops, usually sandstone, where variations in porosity control the amount of groundwater that is conducted to the surface.*

Sheeting Another kind of exfoliation that is especially well developed at the surface of exposed intrusive igneous rock bodies is called **sheeting**. As erosion removes the overlying rock and the stresses that accumulated within the igneous rock body during its initial emplacement and solidification within Earth's crust are released, cracks develop parallel to the rock surface. The rock surface may then fall prey to processes such as frost action, which removes the rock in layers or sheets. A well-known site that shows the process of sheeting is seen on a mountain in Yosemite National Park (Figure 6.10).

Spheroidal Weathering

The end result of weathering is the rounding of rocks. When first exposed by processes such as frost action, rock fragments of all sizes are invariably blocky with sharp edges and corners. Weathering processes preferentially attack the edges and corners, smoothing and rounding them (Figure 6.11). The same effect can often be seen where exposed rock masses are broken by intersecting sets of fractures called joints. As water

FIGURE 6.10 Sheeting *is a type of exfoliation that is commonly associated with exposed granitic plutons. (Courtesy of N. K. Huber/USGS)*

FIGURE 6.11 *Regardless of the size of the exposure, the combined efforts of mechanical and chemical weathering result in the* rounding *of rocks as sharp edges and corners are preferentially attacked.*

penetrates the rock and weathering attacks the rock adjacent to the fracture, the rock at the intersection of the fractures is particularly vulnerable. With time, as the weathered portions of the rock are removed grain by grain (Figure 6.12), or by exfoliation (Figure 6.13), the shape of the rock mass tends to become increasingly spherical. Because of this tendency toward sphericity, the process is called **spheroidal weathering**.

Effect of Temperature

Daily temperature cycles in hot desert areas are commonly presented as being especially effective at promoting the grain-by-grain disintegration of rock. Every solid compound expands and contracts at a specific rate with changes in temperature. Theoretically, as a result of this expansion-contraction process, the minerals within the rock may break apart. However, in experiments subjecting a variety of rocks to ranges of temperature equal to those experienced on Earth, no disintegration was observed. Under normal ranges of daily temperature fluctuation, the differences in the thermal responses of minerals are insignificant. Rock disintegration due to daily temperature fluctuations therefore may be questioned as a mechanism of physical weathering. If a rock is subjected to an extreme temperature change, however, such as being heated by a forest fire or a bolt of lightning and then quickly cooled by a sudden downpour, the rapid contraction may cause it to break.

FIGURE 6.12 *Granite is especially prone to grain-by-grain disintegration. The more chemically reactive minerals, such as the plagioclase feldspars and ferromagnesians, are removed by chemical weathering and produce openings for the entry of water and subsequent disintegration by frost wedging. (VU/© Valorie Hodgson)*

FIGURE 6.13 Spheroidal weathering *is the common result of the tendency of all weathering processes to eliminate sharp edges and corners from exposed rocks. (VU/© Paul Bierman)*

Plant Wedging

We should not forget to include the role of plants in our discussion of mechanical weathering. As plants grow and send their roots down in search of nutrients and water, they may enter cracks in rocks. As the plant roots extend and grow in diameter, they force the rock apart, a process called **plant wedging** (Figure 6.14). Plant roots are extremely efficient agents of physical weathering. Because their roots exude acids, they are also major contributors to certain processes of chemical weathering.

Before leaving the topic of mechanical weathering, take particular note of how water and changes in temperature are involved, either directly or indirectly, in nearly all mechanical weathering processes.

SPOT REVIEW

1. How do mechanical weathering and chemical weathering differ?
2. Why is the freezing of water such an effective agent of mechanical weathering? What are some common examples of this process?
3. Why are daily changes in temperature not thought to have any significant direct effect in mechanical weathering?

FIGURE 6.14 Plant roots *growing within cracks are extremely effective agents of mechanical weathering. Over time the roots result in the splitting of rocks by a process called* plant wedging. *(VU/© Nada Pecnik)*

CHEMICAL WEATHERING

Water and temperature also play major roles in all processes of chemical weathering. For example, very few chemical reactions go on without the intervention of water. If you have had a course in chemistry, think of all the laboratory experiments you have been asked to conduct. Very often, water was one of the reactants called for in the experimental menu. Similarly, few chemical experiments are conducted at room temperature; most are conducted at elevated temperatures. The importance of water and temperature will be borne out in the following discussions of the individual chemical weathering process.

Oxidation

A chemist might define the process of oxidation in several different ways. For our purposes, **oxidation** is any reaction combining an element and oxygen. An important point, however, is that, except when combustible materials are rapidly oxidized by fire, oxygen must be dissolved in water to be an effective oxidizing agent. A shiny iron nail kept in a closed container filled with oxygen will not rust. If the nail is thrown outside, however, and comes in contact with the moisture in the ground, rust will be visible within hours.

Of all the common rock-forming elements, iron (Fe) most readily reacts with oxygen. For this reason, oxidation is especially important in the chemical decomposition of the ferromagnesian minerals and mafic rocks. The process can be illustrated by the reaction of oxygenated water with the iron olivine, fayalite:

$$2Fe_2SiO_4 + O_2 + 4H_2O \rightarrow 2Fe_2O_3 + 2H_4SiO_4$$

fayalite + dissolved → hematite + silica in
          oxygen   solution

The silica is shown as being removed in solution as silicic acid although chemists are not sure precisely how silica exists in solution.

A common product of oxidation is the material known as *limonite,* which is a variable mixture of iron oxides, usually *goethite* (FeOOH or $HFeO_2$). Goethite is unstable and, in time, dehydrates to form hematite by the following reaction:

$$2FeOOH \text{ or } 2HFeO_2 \rightarrow Fe_2O_3 + H_2O$$

goethite → hematite + water

Once formed, the oxides of iron are extremely stable and accumulate with other products of weathering. Their bright colors are conspicuous, ranging from the red of hematite to the yellows and browns of the more hydrated forms. The iron oxides produced by chemical weathering are largely responsible for the colors of most sedimentary rocks.

Another reaction of environmental significance is the oxidation of the sulfide minerals, which always results in the formation of acid. The process can be illustrated with the reaction of oxygenated water with pyrite (FeS_2):

$$2FeS_2 + 7\tfrac{1}{2}O_2 + 4H_2O \rightarrow Fe_2O_3 + 4SO_4^{2-} + 8H^{1+}$$

pyrite + dissolved → hematite + sulfuric acid
        oxygen

The acid generated by the oxidation of the sulfide minerals is the source of the acid mine drainage that plagues the mining industry from the sulfide metals–mining regions of the western United States to the coal-mining areas of the east.

Dissolution

We usually do not think of rocks dissolving in the common sense of the word, but some rocks will dissolve almost instantly upon contact with water. For example, the salt with which we season our food may have existed as *rock salt* composed of the mineral *halite* (NaCl). Rock salt quickly dissolves in water. Combined with the fact that rock salt is not an abundant rock, its high solubility in water explains why we rarely see it exposed at Earth's surface except in desert areas where it may occur in and around seasonal lakes, salt licks, and salt domes (Figure 6.15).

Water dissolves more materials than any other known solvent. But before we discuss the process of dissolution further, we must consider another important characteristic of water that helps to explain its power as a solvent.

The formula for water is H_2O, which might seem to imply that a glass of water contains nothing except hydrogen and oxygen, that is, *pure* water. Pure water does not exist. Even the distilled water used in the chemistry laboratory is not pure water. As long as water is exposed to the atmosphere, it will contain all the atmospheric gases in solution, the "agents of the atmosphere" mentioned in one of the definitions of weathering. We have already seen that oxygen dissolved in (reacted with) water is an effective oxidizing agent. The other important gas that dissolves in water is carbon dioxide, which reacts to form **carbonic acid**:

$$H_2O + CO_2 \rightarrow H_2CO_3$$

water + carbon dioxide → carbonic acid

Carbonic acid is a very weak acid—so weak, in fact, that you can bathe in it or drink it. But regardless of how dilute or weak it may be, carbonic acid is nevertheless an acid and dissociates to provide hydrogen ions and bicarbonate ions according to the following equation:

$$H_2CO_3 \rightarrow H^{1+} + HCO_3^{1-}$$

carbonic acid → hydrogen ion + bicarbonate ion

The reaction between a mineral and dissociated carbonic acid is called **carbonation**. An important example of rock dissolution by carbonation is the chemical weathering of limestone. Limestone is largely made up of the carbonate mineral, calcite ($CaCO_3$). All carbonate minerals are acid soluble. When limestone is subjected to a dilute solution of carbonic acid, it dissolves according to the following equation, which summarizes several intermediate reactions:

$$CaCO_3 + H^{1+} + HCO_3^{1-} \rightarrow Ca^{2+} + 2HCO_3^{1-}$$

calcite + dissociated → in solution
carbonic acid

Although calcite dissolves at a much slower rate than rock salt, the rate is sufficiently fast for limestone,

FIGURE 6.15 *Water-soluble salts can only occur in abundance in arid regions such as Death Valley, California.*

FIGURE 6.16 *Exposed limestone surfaces, be they exposed in outcrops or as the surface of a statue or tombstone, weather very rapidly by dissolution. (a © John Buitenkant/Photo Researchers, b © Dan Guravich/Photo Researchers)*

FIGURE 6.17 *Limestone caves and caverns are among the most spectacular products of limestone dissolution. (Courtesy of W. B. Hamilton/USGS)*

newly exposed in a quarry or as a tombstone, to begin showing evidence of water dissolution within 10 or 20 years (Figure 6.16). Another impressive example of the dissolution of calcite is the formation of caves and caverns (Figure 6.17).

A very spectacular, but unfortunate example of acid dissolution is the deterioration of antique statuary where limestone and marble (a metamorphic rock also composed of calcite) are being attacked by rain acidified with strong acids such as sulfuric and nitric acid (Figure 6.18). These acids are created by the reaction between rainwater and the sulfurous and nitrous oxide gases produced by the burning of fossil fuels by industry and motor vehicles. Priceless artwork that would have survived for thousands of years is now being destroyed within a few decades by the as yet uncontrolled atmospheric pollution of modern civilization.

Although it is important to note that human activity has substantially increased the gases in the atmosphere and that dissolve in water to form "acid rain," it is equally important to appreciate that for billions of years the rainwater falling to Earth's surface has contained carbonic acid.

FIGURE 6.18 *Statues made of limestone or marble, both composed of calcite, experience rapid dissolution when exposed to the strong acids contained in acid rain. (© Adam Hart-David/Science Photo Library)*

Carbonation/Hydrolysis

As we saw previously, *carbonation* refers to chemical processes involving the reaction between a mineral and dissociated carbonic acid. **Hydrolysis** is a decomposition reaction involving water. Of all the chemical weathering processes, the combination of carbonation and hydrolysis is perhaps the most important because it is the chemical process by which the major rock-forming minerals, the silicates, decompose. The combined process of carbonation/hydrolysis may be represented by the following chemical equation:

$$2KAlSi_3O_8 + 2H^{1+} + 2HCO_3^{1-} + HOH \rightarrow$$

potassium + dissociated + water
feldspar   carbonic acid

$$Al_2Si_2O_5(OH)_4 + 2K^{1+} + 2HCO_3^{1-} + 4H^{1+} + SiO_4^{4-}$$

clay   +   metal ions and silica in
minerals   solution

In this equation, potassium feldspar is the representative silicate mineral, but it could be replaced with almost any common silicate mineral except olivine and quartz to illustrate the process. Olivine, $(Fe,Mg)_2SiO_4$, which decomposes by oxidation and carbonation, will not weather to produce aluminum silicate (the clay minerals) because it does not contain aluminum. The other ferromagnesian minerals may contain aluminum and therefore have the potential to produce aluminum silicates by carbonation/hydrolysis as well as iron oxides by oxidation. Quartz is chemically stable because it does not react with dissolved carbon dioxide.

In the carbonation portion of the process, the bicarbonate ion reacts with whatever major metal ion is contained within the mineral to form a soluble bicarbonate of that metal ion. In this case, the metal ion is K^{1+}. If the mineral had been one of the plagioclase feldspars, the metal ions would have been calcium or sodium ions (Ca^{2+} or Na^{1+}). Once the metal ion, the "glue" with which the crystal structure is being held together, is removed, the structure begins to disintegrate.

Hydrolysis then takes over as water reacts with the remaining silicate framework to form an important group of silicate minerals, the *clay minerals*. Generally defined as hydrous (water-containing) aluminum silicates, the clay minerals are represented by the formula for the clay mineral *kaolinite*. Later, we will see that the clay minerals are major constituents of soil. As was shown in the equations describing the oxidation of olivine, any remaining silica is usually removed in solution, probably as silicic acid, H_4SiO_4. In summary, we can say that the carbonation/hydrolysis of almost any common silicate mineral except olivine and quartz will reduce the original mineral to clay minerals and a variety of soluble materials that are carried off in solution.

SPOT REVIEW

1. Why are the ferromagnesian minerals so susceptible to the chemical weathering process of oxidation?
2. Describe the chemical weathering process by which most silicate minerals decompose.

RATES OF CHEMICAL WEATHERING

Up to this point we have not considered the rate at which the chemical reactions may proceed. Common experience tells us that some reactions are slow while others are fast. Gold, for example, reacts with the atmosphere so slowly that it will retain its luster for centuries. Silver, on the other hand, will tarnish so quickly that silver table settings usually need to be polished between uses. Rocks are much the same, in that they decompose at widely differing rates depending upon their own characteristics and those of the environment to which they are exposed. We will now consider the main controls of chemical weathering.

Temperature

Certain instructions included in the menu of a chemistry experiment are designed to affect the rate of the chemical reaction. In many experiments, for example, you are told to assemble the ingredients in a beaker or a test tube, which you then heat over a Bunsen burner. The reason for heating the ingredients is simply to increase reaction rates and shorten the experiment. In most cases, the reaction would have occurred without the application of heat, but the *rate* of the reaction would have been so slow that it might have taken years to complete.

The rates of most chemical reactions increase with an increase in temperature. With one notable exception, which we will discuss later, this is also true for the chemical weathering of rocks and minerals. Due to this dependence of chemical reactions on temperature, chemical weathering proceeds at the fastest rates in the tropics, decreases through the temperate areas, and occurs at the slowest rates in the polar regions.

Availability of Water

Our discussions of both mechanical and chemical weathering have emphasized the importance of water. Because of this dependence on the availability of water, areas of high rainfall show the highest rates of chemical weathering while areas of low water availability have correspondingly low weathering rates. Once again, the highest rates of chemical weathering are found in the tropics where yearly rainfalls are measured in hundreds of inches or centimeters. Combined with the high tropical temperatures, it is no wonder that rock outcrops in the tropics are conspicuous by their absence.

Conversely, chemical weathering is relatively slow in desert areas because of the scarcity of water. This explains why the remains of iron tools of the early pioneers are found throughout the deserts of the southwestern United States; although rusted, they are reasonably intact after lying on the desert floor for a hundred years or more. An equally impressive, though unfortunate, example of the ability of the desert to preserve metals is the story of the "Lucky Lady," a World War II bomber that ran out of fuel and crash-landed in the Sahara. Twenty years later, a geologic mapping crew found the aircraft in almost new condition with virtually no indication of weathering; it had been protected from oxidation by the lack of water.

Because temperatures are low and the available water is confined in ice, rocks exposed in the polar regions show the least effect of either mechanical or chemical weathering of any place on Earth. For this reason, geologists who are interested in observing rocks with minimal alternations due to weathering have gone to the ice-free regions of the Antarctic.

Particle Size

Another common instruction in chemistry experiments is to reduce all solid materials to a powder by grinding. For any mass of material, as the particle size decreases, the surface area increases (Figure 6.19). The rates of all chemical reactions involving solids are directly proportional to the total surface area of the solid components. For this reason, all processes of mechanical weathering serve to increase the rates of chemical weathering by reducing the particle size of rocks.

The effect of particle size on the rate of chemical weathering also means that, all other things being equal, fine-grained rocks should weather faster than coarse-grained rocks. Actual field observations, however, usually indicate that this is not always the case. Granite, for example, is often observed to disintegrate and decompose to a loose assemblage of sand grains faster than its fine-grained equivalent, rhyolite. The reason for this seemingly anomalous situation is that the weathering of the coarser-grained granite removes relatively large crystals of feldspar and ferromagnesian minerals, thereby creating larger holes and passageways than will be formed by the weathering of the finer-grained rhyolite. As a result, water penetrates deeper into granite and enhances both chemical and physical weathering processes.

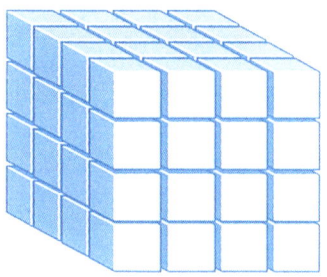

(a) Number of cubes = 1

(b) Number of cubes = 8
Increase in surface area = 2×

(c) Number of cubes = 64
Increase in surface area = 4×

FIGURE 6.19 *As the particle size of solids is reduced by either mechanical grinding or mechanical weathering, the material experiences an increase in the total surface area and, thereby, in the rate of chemical reactivity.*

Composition

If one parameter had to be chosen as the prime controller of chemical reaction rates, it would be composition. Because of fundamental differences in composition, when exposed to the atmosphere, a shiny new iron nail will rust quickly while silver will develop a coating of tarnish much more slowly, and gold will remain bright and shiny for centuries. The same is true of rocks and minerals. To understand the effect of mineral composition on chemical reaction rate, let's consider the minerals that make up the great volume of the rocks of Earth's crust, the silicates.

In the discussion of igneous rocks in Chapter 5, the major rock-forming silicate minerals were divided into two groups, ferromagnesian and nonferromagnesian, and listed in order of their crystallization temperatures. Let us now consider these same minerals in terms of their relative rates of decomposition when affected by a combination of oxidation and carbonation/hydrolysis.

If one were to experimentally determine the relative reaction rates at which the major silicate minerals succumb to oxidation and carbonation/hydrolysis by reacting them with air-saturated water and observing the rate at which soluble ions form, the ranking of the minerals based upon the observed chemical reactivity would be generally the same as the order of crystallization from a melt (Table 6.1). The minerals that crystallize *first* at the *highest* temperatures will show the *highest rates* of chemical weathering. Those minerals that crytallize *last* at the *lowest* temperatures will show the *slowest rates* of chemical weathering. Most of the major silicate minerals in igneous rocks, except for quartz, decompose by a combination of oxidation and carbonation/hydrolysis to form iron oxides and/or clay minerals, indicating that clay minerals and the oxides of iron are stable minerals at atmospheric temperature and pressure.

TABLE 6.1

GOLDICH WEATHERABILITY SERIES

High Temperature	Olivine		First to crystalize
	Augite	Anorthite	
	Hornblende		
	Biotite	Albite	
	Orthoclase		
	Muscovite		
Low Temperature	Quartz		Last to crystalize

Of the common rock-forming silicate minerals, quartz is the most resistant to chemical weathering. Probably the main reason quartz exhibits such low rates of chemical reactivity is that its crystal lattice does not contain metal ions such as sodium, potassium, calcium, magnesium, and iron that tend to react readily with either dissolved oxygen or the bicarbonate ion produced by the dissolution of carbonic acid. Another reason for the chemical stability of quartz is that it can crystallize at surface temperatures. Although most of the silica precipitated from solution at atmospheric temperatures consists of amorphous opaline materials ($SiO_2 \cdot nH_2O$ where n is the number of water molecules), quartz crystallizes in modern marshes and swamps from silica placed into solution by the carbonation/hydrolysis of other silicate minerals and from the decomposition of plants. Crystals grow and become embedded in the edge of "saw grass" where they may cut the skin of some unwary intruder. Crystals also precipitate within empty stem cells of certain aquatic plants to produce what are commonly called "scouring rushes." In fact, the name *scouring rush* comes from the early settlers who mashed rush stalks and used them as homemade "Brillo pads."

In summary, the major solid products of weathering include (1) the *clay minerals,* (2) residual *quartz,* (3) *oxides of iron* from the chemical weathering of the silicate minerals, and (4) *rock fragments* remaining from mechanical weathering.

Metastable Materials

Before leaving the topic of chemical weathering, a comment is in order concerning minerals such as graphite, diamond, and gold that form at high temperatures and pressures, yet can exist indefinitely at Earth's surface. These minerals survive the attack of weathering because, like quartz, they do not react with either oxygenated or carbonated water at any significant rate; as a result, they remain inert over geologic periods of time. A chemist would say that these materials have extremely high **activation energies**, meaning that unless they exist in an environment of very high energy, they will be reluctant to react with any other element. An example of an environment with levels of energy sufficiently high to result in such reactions is the interior of a mass of molten rock. The amount of energy available at Earth's surface is nowhere near sufficient.

SPOT REVIEW

1. Why is water essential for nearly all chemical weathering processes?
2. Although chemical reaction rates increase as particle size decreases, why are some fine-grained rocks more resistant to chemical weathering than their coarse-grained equivalents?
3. Why are certain silicate minerals relatively resistant to the chemical weathering process of carbonation/hydrolysis?
4. Why is quartz, one of the most common silicate minerals, resistant to all processes of chemical weathering?

REGOLITH

Geology is the study of rocks, yet students often comment that when they look about, they usually don't see any rocks. They are quite correct. In most places, the rocks are covered with the products of weathering, commonly referred to as "dirt," a material more scientifically termed **regolith** (*rego* = blanket; *lith* = stone). Regolith is the loose material atop the bedrock; it combines the products of both mechanical and chemical weathering and consists primarily of rock fragments, clay minerals, and quartz. In addition, certain other minerals such as the iron oxides (Fe_2O_3) and oxyhydroxides (FeOOH), calcite ($CaCO_3$), and gypsum ($CaSO_4 \cdot 2H_2O$) may precipitate from groundwater, coat individual mineral grains and rock fragments, and fill spaces between the solid materials. These precipitated minerals may not only determine the color of the materials, but may also significantly influence their chemical characteristics.

The thickness of regolith varies. In areas where weathering rates are extremely slow or where the products of weathering are being rapidly removed, there may be little or no regolith and bedrock will be exposed (Figure 6.20). Because all weathered material, once formed, begins a downhill journey that eventually ends in the sea, regolith in general will be thinnest over hill tops and thickest over valley floors (Figure 6.21). The processes by which the regolith is moved will be considered later in this text.

In most areas, if the regolith is allowed to accumulate undisturbed by natural events such as landslides or human activities such as plowing, a **soil** will develop (Figure 6.22). Soil has been defined as "that part of the regolith that supports plant life" or "that part of the regolith down to the deepest penetration of plant roots." Soils will be discussed in the next chapter.

FIGURE 6.20 *Because of the slow rates of weathering in arid regions, desert regolith is characteristically thin and often nonexistent. Except for the small patch of regolith that is protected by a few plants, most of the surface in the photo is exposed bedrock. VU/© Doug Sokell)*

FIGURE 6.21 *Because gravity causes weathered material to move downslope, the regolith generally thickens from the hilltops to the valley floor.*

ENVIRONMENTAL CONCERNS

When we think of elements entering the environment, especially metals, pollution arising from human activities come to mind. No doubt, many human activities are responsible for the contamination of the environment. However, what we fail to consider is that elements are constantly being released into the environment as chemical weathering decomposes the mineral components of rocks.

With the exception of iron, the elements released by weathering in the greatest abundance *are light elements,* that is, elements with atomic numbers of 20 or less. These elements are among the elements that make up more than 90% of most living tissues; the list includes sodium, magnesium, potassium, calcium, carbon, hydrogen, nitrogen, oxygen, phosphorous, sulfur, and chlorine. The only heavy element associated with life that is released in abundance by chemical weathering is iron, a major ingredient in blood hemoglobin.

Trace Elements

Unless concentrated by mineralization, most elements with atomic numbers higher than 20, the *heavy elements,* are contained in rocks as *trace elements* with concentrations of less (often, much less) than 1 weight percent of the original rock. Once released into the environment by chemical weathering, however, trace elements can become concentrated by various geologic and biologic processes. For example, it is not uncommon for the concentration of an element in a sedimentary rock to be significantly greater than its concentration in the rock from which the sediment was derived.

FIGURE 6.22 *The biochemical activity of plants results in the conversion of regolith to soil. (Courtesy of John Sencindiver)*

Once released into the groundwater, trace elements may become concentrated on the clay minerals by the process of cation adsorption. We will discuss cation absorption in the next chapter. As the soil is subsequently leached by acidic rainwater, these elements may be remobilized, become part of the groundwater, and pass up through the food chain, where they experience further concentration in plant and animal tissues. Eventually, these elements at their elevated levels may be ingested by humans as they eat the plants and animals.

Fluorine   Many trace elements are beneficial to our well-being. A well-known example, flourine, increases the strength of bones and tooth enamel by increasing the crystallinity of apatite, a major constituent of both.

Most of us acquire our daily requirement of flourine by consuming water supplied by municipal water treatment plants where flourine is maintained at concentrations of about 1 part per million (10,000 ppm=1%). Over the past several decades, the daily consumption of flourine-treated drinking water has been responsible for significant reductions in the incidence of tooth decay in the United States and for a reduction in the number of bone fractures suffered by the elderly.

Selenium   Some elements that are beneficial to life at trace abundances become toxic when ingested by animals at higher levels. An example is the metal selenium. The products of volcanic eruptions are the main source of selenium. Widely dispersed by the prevailing winds, volcanic ash and dust may become incorporated into soils where they undergo chemical weathering.

Once selenium is released from the volcanic materials, its mobility in the environment depends on the pH of the soil. In acid soils, selenium is highly insoluble and is therefore not available to either plants or animals. In alkaline soils such as those found throughout the Great Plains, however, selenium oxidizes to a highly soluble form and is taken up by the plants that serve as food for grazing animals. At concentrations above 4ppm, selenium becomes toxic to most animals.

Selenium is also concentrated in organic-rich rocks. This explains why elevated concentrations of selenium are commonly found in the acid mine drainage associated with the weathering of coal and coal-related rocks, especially when they are exposed to the atmosphere by mining.

A much-publicized example of pollution by high levels of selenium occurred in the San Joaquin Valley of California. The San Joaquin Valley is one of the most productive agricultural areas in the world. With an annual rainfall of less than 10 inches (25 cm), however, its productivity is dependent upon extensive irrigation.

One of the problems encountered in irrigated areas, especially where the downward percolation of the water is inhibited, is that the water levels may rise to the root zone of the plants and literally drown them. Such a problem exists in the western side of the valley where clay layers at shallow depths within the soil interrupt the percolation of the water. The solution to the problem is to install drains to carry the excess water off.

Another problem that occurs in the areas of low rainfall is that soluble salts accumulate in the soil. To keep the concentration of salts from rising to toxic levels, the soils are periodically flushed with large volumes of fresh water. The salt-laden water is then removed by the drainage system.

Between 1968 and 1975, a concrete canal was constructed to carry off the agricultural water from the valley. Originally, the canal was to empty into San Francisco Bay, but lack of funding prevented the canal from being completed, and the agricultural waters drained into Kesterson Reservoir. In the first years after the reservoir was built, when most of the water entering it was fresh, the Kesterson Wildlife Refuge purchased the water to establish a protected area for waterfowl. By 1981, however, the water entering the reservoir was the salt-laden agricultural drainage from the valley.

Among the elements that the irrigation water leached from the agricultural soils was selenium derived from the weathering of the rocks in the Coastal Range that forms the western margin of the valley. High levels of selenium were detected in fish from the reservoir in 1982, and by 1983, dead and deformed waterfowl chicks were being reported in the wildlife refuge. Suspecting selenium poisoning, scientists analyzed the waters in both the refuge and the drainage canal. Concentrations of selenium as high as 4,000 parts per billion (ppb) were reported in the drainage waters and up to 400 ppb in the ponds of the refuge. The Environmental Protection Agency limit for selenium in drinking water is 10 ppb. The Kesterson drainage water was declared a hazardous waste in 1985. The problem still remains with conflicting views as to how it should be solved. Even if a plan of action is eventually implemented that eliminates the selenium contamination of the agricultural waters draining into Kesterson Reservoir, the fact remains that selenium is still being leached from the rocks of the Coastal Range by natural weathering processes and may become a problem in other areas.

A problem faced by those who deal with the potential impact of specific trace elements such as selenium on human health is that insufficient data exist to determine the concentration at which a particular element stops being beneficial and becomes toxic. For example, while the concentration at which selenium becomes toxic to cattle has been well documented, the comparable concentration for humans has not been established, and the overall effects of selenium on human health are still not well understood.

Iodine   In many cases, health problems develop in areas where there is a *deficiency* of a particular trace element. It is known, for example, that the soil and water in areas of the United States affected by the most recent Pleistocene glaciation are deficient in iodine. Deficiencies of iodine are directly related to the incidence of goiter, the swelling of the thyroid gland located at the base of the neck. It has also been shown that babies born to mothers suffering from iodine deficiency are more prone to exhibit stunted growth and mental retardation. Another medical problem that has been linked to the iodine deficiency of the same area is breast cancer, the leading cause of death from all types of cancer among women between 35 and 55 years of age.

Although there is little agreement as to why the soils in these areas are deficient in iodine, some geochemists believe that the deficiency is due to the fact that insufficient time has lapsed since the retreat of the ice and deposition of the moraines for adequate iodine to accumulate in the newly developed soils.

Zinc   Another example of health problems related to elemental deficiency involves zinc. Zinc is considered an essential element for both plants and animals. Zinc deficiencies in plants cause a variety of plant diseases that result in a range of problems from low crop yields to total crop failure. In animals, zinc deficiency is linked to skin disorders, the slow healing of wounds, and disorders of bones and joints that have the potential to affect growth, especially among the young.

Water Quality

Unless strongly influenced by pollution generated by human activities, the chemical character of surface and ground water in any region is primarily due to the chemical weathering of the exposed rocks. For example, in regions immediately underlain by limestones, water is generally *hard* due to the high concentrations of calcium and magnesium while waters in areas deficient in limestones are generally *soft* because of the lack of these two elements. The original definition of water hardness is "the ability of the water to precipitate soap." If either calcium or magnesium is present in the water when soap dissolves, an insoluble precipitate forms and produces the familiar ring that collects around the washbasin or bathtub. Water softeners remove calcium and magnesium from the water by having them displace either sodium or potassium ions that are held on the cation exchange positions of resins specifically designed to attract and exchange the alkali and alkaline earth elements.

Water Quality and Heart Disease   Studies have shown a higher number of deaths resulting from heart disease in areas characterized by soft water. The real significance of this correlation is debated, however. Some argue that the increased heart disease is not due to the drinking of soft water per se, but rather to heavy metals that the acidic soft water leaches out of the supply pipes.

Another example of a health problem associated with water quality has been documented in Ohio where a correlation has been shown between increased incidence of deaths due to heart attacks and elevated levels of sulfate ion (SO_4^{2-}) in the drinking water. The affected area is the coal-producing counties of southeastern Ohio where sulfate ion is generated by the weathering of the pyrite contained in the coal and coal-associated rocks. In comparison, the incidence of deaths due to heart attacks and the sulfate content of drinking water in the glaciated portions of northern Ohio are both lower.

As was the case with the correlation between heart disease and soft drinking water, some argue that the increase in the number of deaths due to heart attacks is not caused by the higher concentration of sulfate ion in the drinking water, but rather is the result of another component, or components, that are present or absent in the sulfate-rich water.

Other studies have shown lower incidences of death due to heart disease in areas where there are trace levels of certain elements. For example, fewer cases of heart disease are found in areas where the drinking water contains trace levels of manganese, chromium, vanadium, and copper. Presumably, deficiencies in these metals would result in an increase in the heart disease that promotes heart attacks. As these few examples clearly demonstrate, the relationship between heart disease and water quality is poorly understood at best.

Water Quality and Stroke A study in Japan reported a correlation between water chemistry and health. The study related the ratio of sulfate and bicarbonate ions (SO_4^{2-}/HCO_3^{1-}) in the water to the incidence of stroke. Stroke is a loss of body functions resulting from either a rupture or a blockage of blood vessels in the brain. In northern Japan, where the ratio was greater than 0.6, the death rate from stroke was more than 120 per 100,000 of population. In southern Japan, where the ratio was less than 0.6, the death rate was below 120 per 100,000 with many areas experiencing fewer than 80 deaths per 100,000 population. The increased sulfate content of the waters in northern Japan was attributed to the chemical weathering of abundant sulfur-rich volcanic rocks compared to southern Japan where more of the exposed rocks are sulfur-poor sedimentary rocks.

The Effect of Mining on Water Quality

Human activities commonly result in an increase in the concentration of metals in surface and ground waters. Notwithstanding both air and water pollution associated with a wide variety of industrial operations such as processing and manufacturing plants, the generation of large volumes of rock fragments by mining operations and the disposal of ore processing refuse contribute directly to increased production of metal ions by increasing the surface area of the rocks exposed to the processes of chemical weathering. This is especially true in areas where metallic sulfide minerals, such as those of lead, zinc, copper, and iron, are mined. The chemical weathering of sulfide minerals, especially pyrite, FeS_2, produces sulfuric acid that both increases the dissolution rate of metals and provides a chemical environment in which most metals remain in solution.

CONCEPTS AND TERMS TO REMEMBER

Weathering
Erosion
Definition of weathering
   disintegration
   decomposition
   mechanical weathering
   chemical weathering
Processes of mechanical weathering
   frost action

freeze-thaw cycle
exfoliation
   spalling
   honeycomb weathering
   sheeting
   spheroidal weathering
   plant wedging
Processes of chemical weathering
   oxidation

carbonation
   carbonic acid
   hydrolysis
Metastable materials
   activation energy
Products of weathering
   regolith
   soil

REVIEW QUESTIONS

1. The major agent of physical/mechanical weathering is
   a. plant roots.
   b. daily temperature changes.
   c. freezing water.
   d. growing salt crystals.
2. Oxidation primarily involves the element
   a. silicon
   b. iron.
   c. calcium.
   d. sodium.
3. Water is such an effective agent of chemical weathering because it
   a. is present everywhere on Earth.
   b. contains dissolved oxygen and carbon dioxide.
   c. is itself a strong acid.
   d. can penetrate into small fractures.

4. Which of the following igneous rocks would you expect to have the fastest rate of chemical weathering?
   a. granite
   b. rhyolite
   c. basalt
   d. granodiorite
5. Which of the following minerals would you expect to weather at the fastest rate?
   a. olivine
   b. albite
   c. quartz
   d. orthoclase
6. The major mineral that forms by the chemical weathering of most silicate minerals is
   a. quartz.
   b. regolith.
   c. clay.
   d. feldspar.
7. The term "regolith" refers to
   a. the accumulated products of chemical weathering.
   b. the unweathered rock that is actively undergoing weathering.
   c. the total accumulated products of weathering.
   d. the layers of rock that result from the process of spalling.
8. In which of the following areas would you expect the process of frost wedging to be most effective?
   a. the polar regions
   b. regions of temperate, humid climate
   c. deserts
   d. subtropical areas
9. What are the special properties of water that make its presence so vital to both physical and chemical weathering processes?
10. If one of the major controllers of chemical weathering processes is temperature, why are chemical weathering processes not nearly as important as physical weathering processes in deserts?
11. What makes granite deserve the accolade "rock of ages"?
12. Why is the order in which the silicate minerals crystallize from molten rock identical to the order of the rates at which they chemically weather?
13. Why is the mineral quartz excluded from the process of carbonation/hydrolysis that chemically decomposes the great majority of the silicate minerals?
14. Why is it significant that the clay minerals are the end product of the chemical weathering of most of the silicate minerals?

THOUGHT PROBLEMS

1. In many areas of northwestern North America and throughout much of Europe, the rates of chemical weathering are being accelerated by the presence of acid rain. What are the strong acids in acid rain and what are their sources? Short of totally eliminating the sources for these acids, what can be done to reduce the impact of acid rain? Why is the statuary throughout Europe so susceptible to the ravages of acid rain and what can be done to protect it from further damage?

2. Until the evolution of green plants, the composition of Earth's primeval atmosphere was quite different from that of the modern atmosphere. Consider what possible effects these compositional differences would have on both the kinds and rates of weathering.

   The composition of the present atmosphere is being affected by pollution. Assuming that we continue to modify the composition of the atmosphere, discuss possible effects the changes might have on the processes of weathering in the future.

FOR YOUR NOTEBOOK

Another field trip is in order to seek examples of the chemical and physical attack on rocks. Because the processes of weathering attack not only natural rock outcrops but also rocks exposed at the surfaces of buildings and other structures, a trip through your town should reveal numerous examples of both chemical and physical weathering. The same processes that attack rocks are effective at reducing construction materials such as concrete to rubble. All-too-familiar examples are the potholes and the slow destruction of concrete driveways. You might want to consider what steps can be taken to minimize these kinds of problems.

CHAPTER OUTLINE

INTRODUCTION

SOIL DEFINED

FACTORS CONTROLLING SOIL FORMATION
- Biological Factors
- Physical Factors
- Climate
- Time

THE ROLE OF THE CLAY MINERALS
- Cation Adsorption
- Cation Exchange Capacity

SOIL HORIZONS
- The O Horizon
- The A Horizon
- The E Horizon
- The B Horizon
- The C Horizon

HORIZON DEVELOPMENT
- The Influence of Rainfall
- The Effect of Water Volume and Chemistry
- Tropical Regions

TYPES OF SOIL

ENVIRONMENTAL CONCERNS

CHAPTER 7

Soils

INTRODUCTION Few natural materials are more important to the existence of every terrestrial species than soil. Soils support the growth of the green plants that make up the basic components of the land-based food chain. Without these species, the remaining members of the terrestrial food chain, including humans, could not exist. The study of soils, a science called **pedology** (Greek: *pedon* = earth, soil), deals with the composition, formation, and distribution of soils and with their use and management. Soils often provide geologists with a means of determining the relative age of rocks, sedimentary deposits, and special landforms as well as providing information about past climate.

SOIL DEFINED

Soil scientists define **soil** as *"that part of the regolith that contains living matter and supports, or is capable of supporting, plant life out of doors."* You will remember that regolith is the layer of weathered material that accumulates above bedrock. A geologist might define soil as *"the outermost layer of the terrestrial crust composed of organic material, gases, water, and regolith."* In this sense, soil is the overlap of Earth's biosphere, atmosphere, hydrosphere, and lithosphere; the combination is called the *ecosphere* (Figure 7.1).

The base of the soil is usually the depth to which the roots of perennial plants penetrate the regolith (Figure 7.2). An important point made by the definition is that soil is a *part* of the regolith. While regolith may exist without either the development of soil or the presence of plants, both regolith and plants must be present for soil to develop.

In warm, humid areas where the rates of chemical weathering are high, the regolith is generally thick, may be totally converted to soil, and supports a continuous plant cover. In drier climates, either warm or cold, both plant cover and soil development may be minimal (Figure 7.3).

FACTORS CONTROLLING SOIL FORMATION

Soil is the end product of a complex interaction between the minerals contained within the regolith and a variety of chemical, physical, and biochemical processes. Its development is affected by factors such as the parent material, the topographic relief of the area, the orientation of the slopes with respect to the Sun, and the climate, especially temperature and the availability of water.

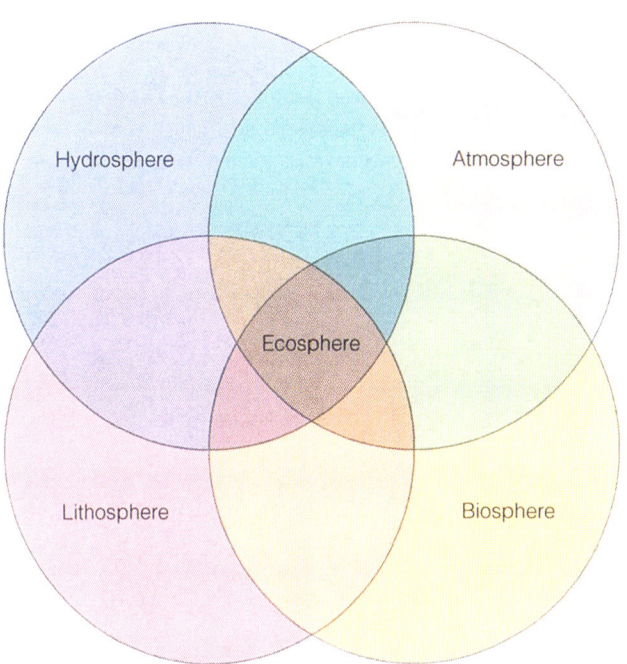

FIGURE 7.1 *The* ecosphere *is created by the interaction of the* atmosphere, *the gases that surround Earth; the* hydrosphere, *the water that exists on and within Earth; the* lithosphere, *the rocks of which Earth is made; and the* biosphere, *those regions of Earth that support life.*

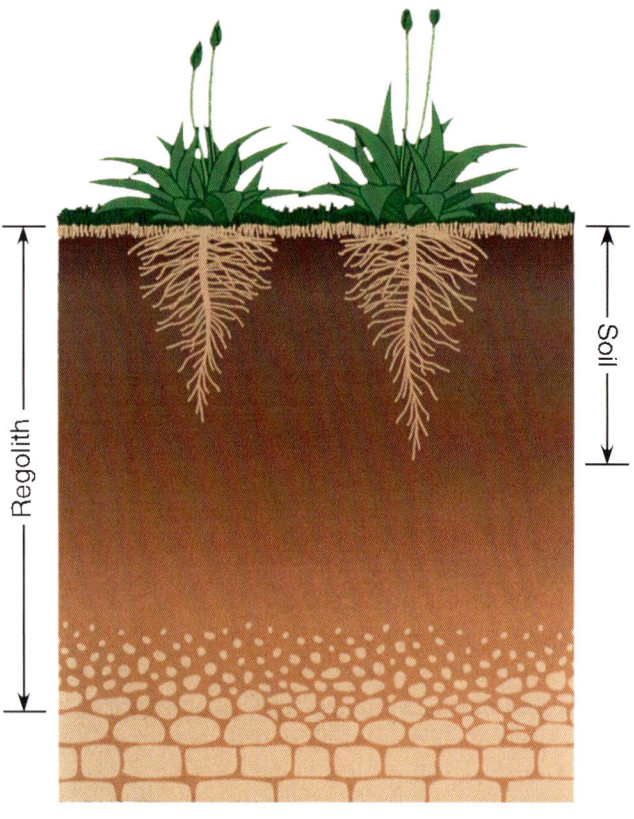

FIGURE 7.2 *Any accumulation of weathered material above bedrock is* regolith; soil *is that part of the regolith that supports plant life. When plant roots extend to bedrock, the entire regolith becomes soil.*

Biological Factors

The biological processes are especially important in the formation of soil. Although the primary role of the soil is to support plant life, certain organisms, in particular, microorganisms such as bacteria, play an essential role in both the initiation and the continued development of the soil. As microscopic organisms decompose the organic debris accumulated at the surface of the soil, essential plant nutrients are released to be used by the existing plant population. Various chemicals, including organic acids, are generated by the biochemical activity and attack the particles of regolith as agents of chemical weathering. As larger plants become established and send their roots into the regolith, acids exuded from the root surfaces promote chemical weathering of the regolith; at the same time, the roots penetrate into fractures within the larger fragments of regolith where they serve as an agent of mechanical weathering, further preparing the regolith for soil formation. Thus, plants, once established, play a key role in the development of soil.

In addition to plants, certain animals play an important part in soil formation. Foremost among these are earthworms, which ingest the clay-rich component of soil in search of nutrients and chemically alter the material as it passes through their digestive track. The end product is an organic-rich material that possesses all the characteristics of a

FIGURE 7.3 *Because the rates of weathering are limited by the lack of water, the regolith of arid areas is usually thinner than that found in humid regions. (VU/© Dick Poe)*

good fertilizer. In addition to modifying the chemical makeup of the soil, the activity of earthworms tends to loosen and stir the soil and in general makes it more porous and conductive for the penetration of plant roots, gases, and nutrient-bearing water.

To complete our discussion of organisms, we should mention the contribution of burrowing animals that overturn the soil within the area of their activity. Moles, voles, groundhogs, prairie does, and a host of other burrowers all contribute to the formation of this most important product.

Physical Factors

In addition to these biological factors, several physical factors affect soil formation including topography and climate. Topography affects the formation of soil in two ways: (1) by controlling the rate of downhill movement of the regolith and (2) by affecting the temperature and moisture content of the regolith.

On steep slopes, the downhill movement of materials under the force of gravity, a process called *mass wasting,* and the movement of a thin surface layer of water called *slope wash* combine to remove the smaller particles, leaving behind a layer largely composed of rock fragments (Figure 7.4). This coarse material requires additional weathering in order to provide both the finer particles required for soil development and the nutrients needed for initial plant growth. As we discussed in Chapter 6, the thickness of regolith generally increases with decreasing angle of slope. The thickest regolith is usually found over valley floors where the fine-grained materials removed from the slopes by mass wasting combine with those deposited on the valley floors by floods. Because these deposits commonly contain relatively high concentrations of clay minerals and because ample supplies of water are available, floodplains have been extensively used for crop production. In some of the more poorly drained areas within the floodplain, however, certain soil development processes may be inhibited, and as a result, agricultural soils may not be well developed.

The orientation of the slopes is extremely important in soil formation. Because slopes that face the Sun are warmer, the rates of all chemical processes tend to be higher. These include chemical weathering and the processes involved in the formation of soil, in particular, microbial activity that promotes the decomposition of organic matter and subsequent soil development. On the other hand, sunlit slopes are also drier, a condition that tends to retard chemical reactions including those that form soils. The effect of slope orientation on soil temperature and moisture is also apparent in the kinds of abundances of plants that populate the slopes (Figure 7.5). North- and south-facing slopes commonly have different plant populations whereas the orientation of east- and west-facing slopes has an intermediate impact upon soil development.

FIGURE 7.4 *Because the finer particles are most easily removed by mass wasting and erosion, the regolith that accumulates on steep slopes commonly consists of large rock fragments. (VU/© Richard Thom)*

FIGURE 7.5 *Because less water is available due to accelerated rates of evaporation, the vegetative cover of slopes that face the Sun commonly have different populations than those that face away from the Sun. (VU/© Brooking Tatum)*

Climate

Of all the parameters affecting the development of soil, *climate* is the most important. Climate controls both the temperature and the availability of water, which, in turn, control not only the rates of weathering and erosion and the kinds of plants that will dominate, but also the rates of the various chemical soil-forming reactions. In general, the thickest soils form in the moist tropics with soil thickness decreasing toward the polar regions as both temperature and the availability of water decline.

Time

Time is a soil-forming parameter that is often overlooked. By now, you have likely been impressed with the importance of time in all geologic processes and appreciate that a slow process acting over a long period can perform incredible feats. In general, the formation of soil is a slow process usually requiring hundreds of years. In regions where the temperatures are high and water is abundantly available, less time is required for soil to form on newly generated regolith than in areas where temperatures are low and water availability is limited. For this reason, soils develop most rapidly in the tropics, decreasing in rate toward the poles. Note, however, that in many areas, such as the midwestern United States, soils that have taken as long as 10 million years to develop are now being removed by erosion faster than they are being formed.

SPOT REVIEW

1. What is the difference between regolith and soil?
2. In what ways do organisms, topography, and climate affect the formation of soil?

THE ROLE OF CLAY MINERALS

Because of the singular importance of the **clay minerals** in supplying plant nutrients, let us pause in our discussion of soils to consider the chemistry of the clay minerals in some detail. The clay minerals are silicates with sheet structures similar to those of the micas. The clay minerals are the only minerals whose crystal lattices do not achieve the status of internal *electric neutrality* that is required for most other crystalline compounds. Most clay minerals have a deficiency of positive ions within the crystal lattice. As a result, the individual clay mineral particles are negatively charged. A small percentage of clay minerals develop with an internal surplus of positive ions and are therefore positively charged. Because the negatively charged clay minerals are by far the more abundant, we will discuss them in detail. Once the chemistry of the negatively charged clay particles is understood, the comparable mechanism involving the positively charged clays should be evident.

Cation Adsorption

Even though the clay minerals crystallize without fulfilling the usual requirement of electric neutrality, they are not exempt from the rule that crystal lattices must be electrically neutral. If electric neutrality cannot be attained internally, it must be attained *externally*. Exposed to the dissolved components of soil water and groundwater, the *negatively charged* clay particles attract *positively charged* cations, which are *adsorbed on* (adhere to) the surfaces of the clay particles (Figure 7.6). The number and concentration of cations adsorbed to the surfaces of the clay particles depend upon the magnitude of the negative charge within the particles and the charge of the particular cations. The process by which cations are attracted and adhere to the surface of the clay minerals is called **cation adsorption**.

Cation Exchange Capacity

Cations will remain adsorbed to the surface of a clay particle until other cations appear in the groundwater with a higher concentration or a higher positive charge than those occupying the clay mineral adsorption sites. The adsorbed cations will then be *displaced* and *replaced* by the second cations (Figure 7.7). The process by which a cation on an adsorption site is replaced by another cation from solution is called **cation exchange**. Depending on the magnitude of the internal charge deficiency, individual clay minerals have a greater or lesser capacity to engage in the process of cation exchange. Through a laboratory procedure, soil scientists can determine the **cation exchange capacity (CEC)**, which measures the ability of a particular clay mineral to exchange one cation for another. The cation exchange capacity of the clay minerals is important because it provides a mechanism by which plant nutrients can not only be stored within the soil but also can be made available to the plants upon demand.

SPOT REVIEW

1. Explain what is meant by "cation exchange." What characteristics of the clay minerals favor its formation?
2. How is cation exchange involved in providing plant nutrients?
3. What role could cation exchange play in chemical weathering?

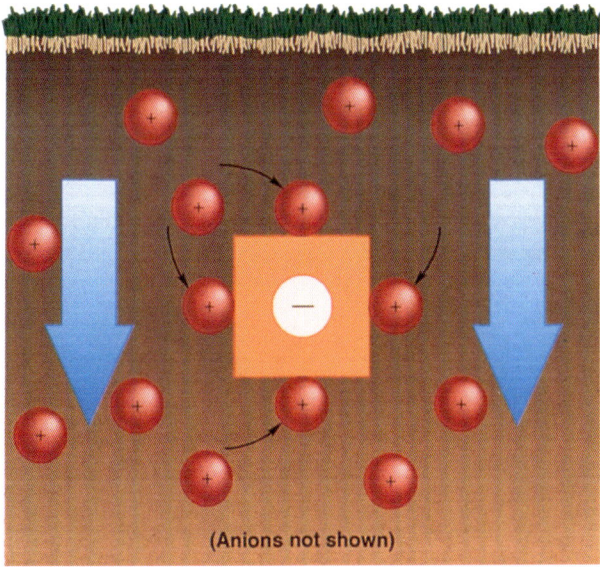

FIGURE 7.6 *During the process of* cation adsorption, *cations are removed from the groundwater and adsorbed to the surface of the clay particles to make up for the cation deficiencies that exist within the crystal lattices of most clay minerals.*

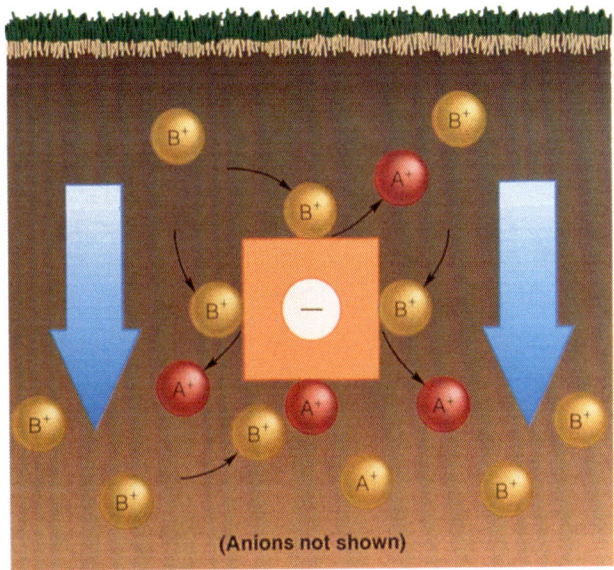

FIGURE 7.7 *Depending on their relative concentrations, cations carried by the groundwater may displace and replace those previously adsorbed to the surface of the clay minerals, a process called* cation exchange.

THE INTERACTION BETWEEN SOIL AND PLANTS

To illustrate how plants and clay minerals interact, let us consider a situation that is the topic of numerous advertisements for soil additives. Soils formed on non-carbonate rocks in temperate areas that receive more than 20 inches (50 cm) or rainfall per year are said to be "*acid.*" In the United States, this includes most of the country east of the Mississippi River and a small portion of the extreme Pacific Northwest (Figure 7.B1). But, what actually is an acid soil?

With more than 20 inches (50 cm) of rainfall per year, the dominant direction of water movement through the soil is downward. From our previous discussion of chemical weathering, you will remember that rainwater is really a dilute solution of carbonic acid. Therefore, the dominant cation being carried through the soil is the hydrogen ion. As the acidified rainwater moves downward through the soil, the hydrogen ions displace whatever cations were being held on the adsorption sites of the clay particles. Through this process, the clay particles become *hydrogenated*.

Once the cation exchange sites are occupied by hydrogen ions, each clay particle could be presented by the chemical formula H^{1+} (clay)$^{1-}$. If this clay particle were placed into a solution containing other ions, the clay particle would *dissociate* and give up the hydrogen ion into solution by *exchanging* it for a different ion. To a chemist, one of the basic definitions of an acid is "any compound that will provide a hydrogen ion into solution." Because the hydrogen-clay compound may contribute a hydrogen ion into solution as a result of cation exchange, it is by definition an acid.

Advertisements for various kinds of agricultural products stress that acid soils are not necessarily the best for growing the kinds of plants we often like to see in our gardens or lawns. Grass, in particular, will not grow well in an acid soil. Consequently, advertisements for lawn products tell us that before grass will grow with any real success, the acid soil must be *neutralized,* and

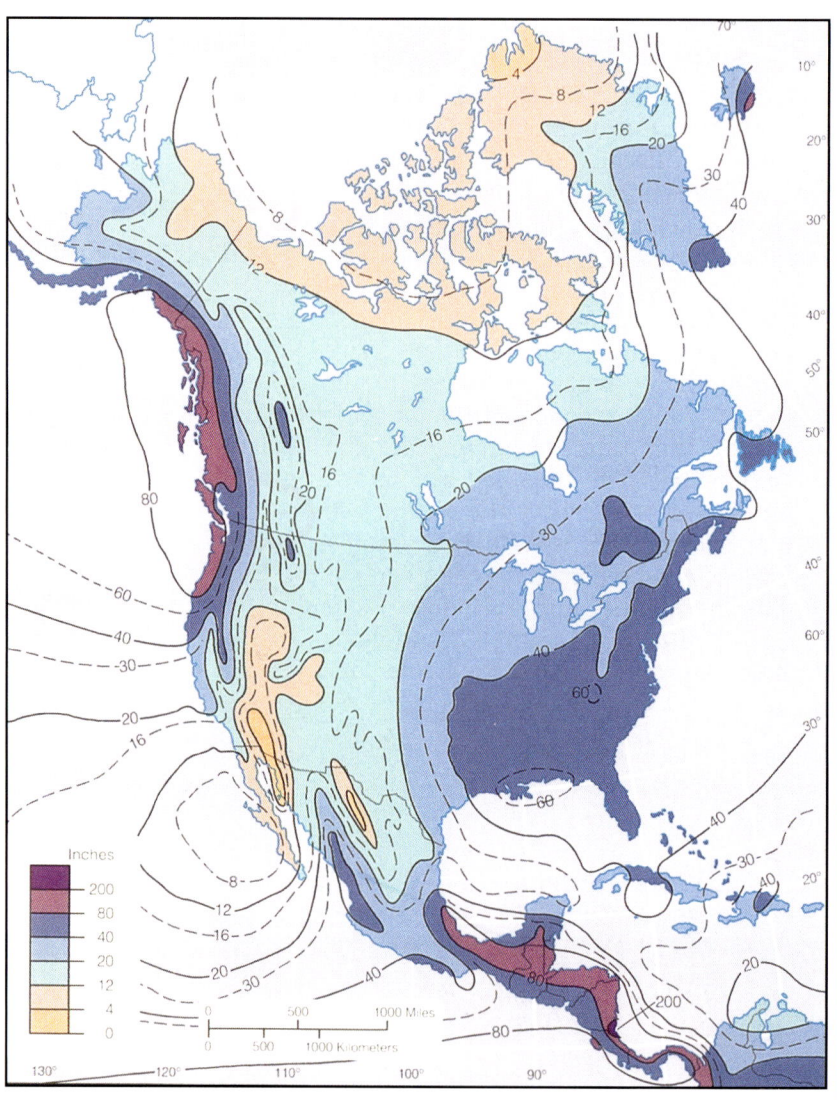

FIGURE 7.B1   *The distribution of rainfall throughout North America is the result of basic patterns of worldwide air movements.*

FIGURE 7.B2 *Rainwater reacts with carbon dioxide (CO_2) in the atmosphere and in soil humus, to produce carbonic acid. As the rainwater moves down through the soil, hydrogen ions produced by the dissociation of carbonic acid are adsorbed to the surface of individual clay mineral particles and neutralize the internal negative charge. Because the hydrogen ions can be released into solution, the clay minerals are sources of hydrogen ions. Thus the soil is acid.*

The surface of the lawn is treated with agricultural lime (powdered $CaCO_3$), which reacts with rainwater to produce Ca^{2+} and HCO_3^{1-} ions. Because of their greater concentration and charge, the calcium ions replace the hydrogen ions on the clay mineral particles. The displaced hydrogen ions combine with the hydroxyl ions (not shown) to produce water. Since the clay particles are no longer sources of hydrogen, they are no longer acid. The soil has been neutralized.

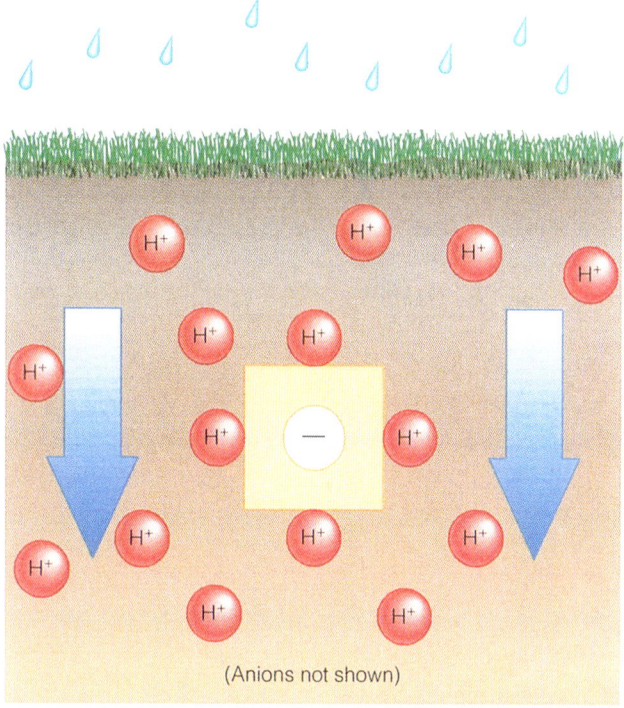

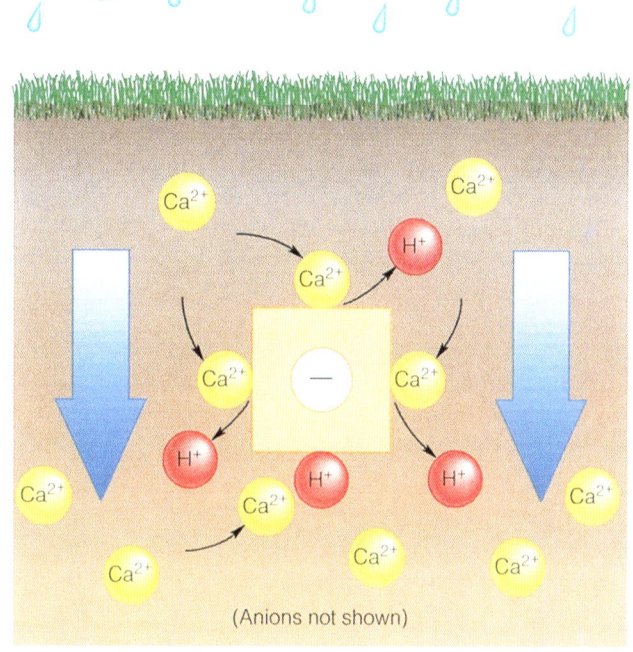

they suggest that this task can be accomplished by utilizing the *lime* that the manufacturer is trying to sell.

Agricultural lime is calcium carbonate ($CaCO_3$) and is manufactured by pulverizing limestone. The agricultural lime that the advertiser suggests that you spread on the ground dissolves with the next rain or in water provided by a lawn sprinkler to produce a solution concentrated in Ca^{2+} ions. As the solution moves down through the soil, the H^{1+} ions originally present on the surfaces of the *acid* clay particles are displaced and replaced by the Ca^{2+} ions (Figure 7.B2). Once the Ca^{2+} ions are adsorbed, the new clay

(continues on next page)

THE INTERACTION BETWEEN SOIL AND PLANTS *continued*

compound can be represented by the formula Ca^{2+} (clay)$^{2-}$. Note that once the H^{1+} ions are replaced by the Ca^{2+} ions, the clay particles are no longer sources of H^{1+} ions and are therefore no longer acid. They have been neutralized.

Not only have the clay particles been neutralized, but the H^{1+} ions displaced from the clay particles combine with the $(OH)^{-1}$ ions formed by the dissolution of the lime to form water. At this point, the entire system has been neutralized.

The final question is, how does the neutralized soil provide the nutrients necessary for the growth of grass? Once again, the mechanism is an example of cation exchange. Calcium ions are important nutrients for grasses of all kinds. Once grass seeds germinate, the roots penetrate the soil seeking the Ca^{2+} ions adsorbed on the surfaces of the clay particles. In order to obtain the Ca^{2+} ions, the plant roots secrete carbonic acid. Hydrogen ions generated as the acid dissociates are exchanged for the Ca^{2+} ions located on the surfaces of the clay particles. The Ca^{2+} ions are then taken up by the plant for its continued growth. Note, however, that as the Ca^{2+} ions are progressively replaced by H^{1+} ions, the clay particles once again become *hydrogenated*. In other words, as the grass acquires the Ca^{2+} ions needed for growth, the soil becomes progressively more acid. Eventually, the grass will begin to suffer from a lack of nutrients until the situation is alleviated by applying more agricultural lime to the surface.

In general, the successful growth of any plant, other than those that grow naturally, requires that the chemistry of the soil be modified by the addition of the necessary nutrient ions. The negatively charged clay minerals are the recipients of the added positive ions such as Ca^{2+} and K^{1+} while the positively charged clay minerals are the storehouses for the negative functional groups such as phosphates and nitrates. Because the average soil contains significantly fewer positively charged clay minerals than the negatively charged varieties and thus has a low anion exchange capacity, negative ions must be added to the soil more frequently. As the plants consume the added nutrients, they must be periodically replaced with further treatments.

SOIL HORIZONS

If a soil is allowed to develop undisturbed, that is, without being overturned by a natural process such as a landslide or by human activities such as deep plowing, it will develop distinct layers called **horizons** (Figure 7.8). Horizons usually develop parallel to the surface of the soil and are characterized by distinctive compositions and textures.

The O Horizon

The uppermost horizon is called the **O horizon** (the O stands for *organic*). The O horizon consists of plant debris that accumulates on the surface, including the leaves that are shed each fall, dead grass, broken twigs, limbs, and fallen trees. Once accumulated, these materials are immediately attacked and decomposed by bacteria and fungi to gain nutrients for their own growth and energy needs. As part of the decomposition process, the microorganisms begin to convert the plant material, especially cellulose and lignin, into a variety of organic compounds that combine to make up a dark brown to black substance called *humus*, which accumulates and colors the upper portion of the soil. Humus is an important source of nutrients for many plants.

The A Horizon

Below the O horizon is the **A horizon**, which is a mixture of humus and chemically decomposed regolith, mainly clay minerals intermixed with quartz. The A horizon is usually black to dark brown in color due to its humic content. The combination of the O and A horizons is often referred to as *topsoil*.

The E Horizon

Below the A horizon is the **E horizon**, commonly referred to as the *zone of leaching*. Acids generated by the bacterial decomposition of the organic debris in the overlying O and A horizons move down

FIGURE 7.8 *When allowed sufficient time to develop, layers called* horizons *develop within the soil. Within any soil, the thickness, compositions, and colors of the individual horizons depend in great part on the climate of the region, in particular, the amount of rainfall. (*a *VU/© Doug Sokell;* b *VU/© Albert J. Copley)*

through the E horizon layers and remove most materials except quartz. Consisting almost entirely of quartz, the E horizon is easy to recognize by its coarse texture and white to gray color. Because of its low clay mineral content, the E horizon has little nutrient-holding capacity.

The B Horizon

Below the O, A, and E horizons is the heart of the soil, the **B horizon**. The B horizon is commonly referred to as the *zone of deposition* because many of the materials leached from the overlying horizons accumulate here. In most soils, the B horizon contains the greatest concentration of the clay minerals, making it the major storehouse of plant nutrients. Other components of the B horizon are relatively small concentrations of iron oxides and oxyhydroxides, which are responsible for the brown to red to yellow colors of most soils.

Depending upon the type of soil, the B horizon may contain water-soluble minerals, most commonly, calcite. The presence or absence of calcite is important, not only in determining the soil type, but also in establishing the chemical characteristics of the soil.

The C Horizon

The bottommost soil horizon is the **C horizon**, which is usually referred to as the *parent material*. The C horizon is the part of the regolith where the soil-forming process is initiated.

HORIZON DEVELOPMENT

The thickness and composition of the different soil horizons vary substantially from region to region. Here we examine not only the factors that contribute to these differences but also the types of soil horizons characteristic of regions ranging from the humid tropics to arid deserts.

The Influence of Rainfall

One of the major factors controlling the development of the individual soil horizons and the type of soil that subsequently develops is the *amount* of rainfall. Water moves both upward and downward within the regolith. The relative volumes of water moving in the two directions depend upon the amount of rainfall. In moist tropical areas where it rains nearly every day and the rainfall is measured in hundreds of inches or centimeters per year, the movement of water is totally downward. In humid, temperate areas receiving more than 20 inches (50 cm) of rainfall per year, the dominant movement is downward as rainwater makes its way to the water table. During the low-rainfall periods of the year, however, upward movement of groundwater by *capillary action* becomes increasingly dominant. Capillary action is generated when the attraction of a liquid to a solid surface is greater than the internal cohesion within the liquid itself. The force causes the liquid to rise along vertical surfaces. Drawn up by capillary action, the groundwater evaporates in the upper layers of the soil. In arid regions where the rainfall is less than 10 inches (25 cm) per year, the upward movement of groundwater followed by evaporation dominates over the downward movement of rainwater.

The Effect of Water Volume and Chemistry

The marked differences in zonation and composition that evolve within a soil profile depend upon both the relative volumes and the compositional differences of the rainwater and groundwater moving within the regolith. Rainwater, as you will recall, is a dilute solution of carbonic acid. In many areas, the natural acidity of the rain has been significantly increased by the introduction of strong acids formed from the reaction of rainwater and the oxides of sulfur and nitrogen introduced into the atmosphere by industry, power-generating stations, and motor vehicles. These acidic solutions, the so-called *acid rain,* leach most soluble salts from the soil profiles and may render the soil exceptionally **acid**. While acid rain may be causing some soils to turn more acid, it is important to point out that not all soils are becoming acid because of acid rain. Neutral or calcite-rich soils may be only marginally affected by acid rain, if at all, while some soils turn acid naturally regardless of the rain quality.

In contrast to the acidic downward-moving rainwater, the upward-moving groundwater contains the soluble products of chemical weathering and is mostly neutral to alkaline. As groundwater evaporates in the soil, salts of the **alkali** (Na^{1+} and K^{1+}) and **alkaline earth** (Ca^{2+} and Mg^{2+}) elements are deposited and become part of the soil composition. Within any region, the volumes of these two sources of water moving within the soil establish a *balance* that, combined with their respective chemical components, profoundly affects the character of the resultant soil.

In the subtropical and humid, temperate areas where the annual rainfall exceeds 20 inches (50 cm), the downward movement of rainwater dominates over the upward movement of groundwater. The rainfall is sufficient to support a continuous plant cover, consisting of a wide range of plant types including trees. With abundant plant cover, both the O horizon and the humic A horizon are well developed. The downward-moving rainwater also produces a well-developed E horizon.

The B horizon is also well developed with abundant clay minerals to serve as potential storehouses of plant nutrients. Because of the dominant downward movement of rainwater, however, the clay particles are hydrogenated and the soil is acid. Consequently, these soils have to be neutralized by the application of lime before they can be used for most agricultural purposes, except for growing acid-tolerant tuber plants such as potatoes or carrots.

In semiarid regions that receive 10 to 20 inches (25 to 50 cm) of rainfall per year, the amount of rainwater moving downward through the soil during the rainy season is approximately offset by the amount of groundwater moving upward during the dry season. The upward-moving groundwater brings various ions, in particular, calcium ions that displace cations held in the exchange positions of the clay particles. As a result, these soils are self-neutralized and normally do not require any kind of pretreatment for agricultural use. These are some of the most fertile soils on Earth.

With such limited rainfall, large plants such as trees cannot be sustained. However, the amount of rainfall and the available nutrients, particularly the availability of the calcium ions, are perfect for grasses. These soils are the great **grassland soils**. The Great Plains of North America stretching from Oklahoma into south-central Canada, the pampas of South America, the plains of Ukraine, the veldt of Africa, and the great grassland of Australian interior all are based on these soils. The original grasses that dominated these soils were native grasses. With the coming of settlers, the natural grasses were plowed under and replaced with other grasses such as wheat, oats, and rye. With the exception of rice, most of the world's grains are produced by these soils, and they are therefore often referred to as the *breadbasket soils*.

Another characteristic of these soils is their black color. Because the dense grasses die from one season to the next, plant parts both above and below the surface decompose and generate humus, which is distributed throughout the entire soil profile. To fly over the Great Plains during plowing season and see the endless stretches of coal-black soils awaiting the next growing season is always impressive (Figure 7.9). Under the old classification of soils, these soils were called *chernozems*, a Russian word meaning "black soil."

In arid regions where the rainfall is less than 10 inches (25 cm) per year, the development of soil is severely restricted if, in fact, a soil develops at all. Without seasonal accumulations of plant debris from a continuous plant cover, humus will be nearly nonexistent. Without the downward movement of rainwater, horizonation does not develop. In many arid regions, soil development is minimal, and the bedrock is either exposed or overlain by unaltered regolith.

FIGURE 7.9 *Some of the most productive agricultural soils are the* mollisols. *They are commonly known by the Russian word* chernozem, *which refers to the black color that develops in response to their high organic content. (VU/© Dana Richter)*

In areas that have a covering of regolith, the extremely low humidity of the atmosphere initiates the dominant upward movement of groundwater. Subsequently, the water evaporates, resulting in the deposition of large volumes of salts within the regolith. In some areas, the accumulation of salts is so extensive that it forms a rock-hard layer up to several inches thick called a **hardpan**. Consisting mostly of the salts of the alkali and alkaline earth elements, these deposits render the regolith totally inhospitable for the growth of most plants. Only the desert plants are, by design, able to survive under these normally toxic conditions. Cellular adaptions within their roots allow these plants to filter out the excessive concentrations of dissolved ions.

In some desert areas where an ancient soil profile may exist from a more humid time, the concentration of toxic salts within the paleo-soil can be reduced by extensive irrigation that reverses the relative dominance of the upward movement of groundwater within the soil profile by providing artificial "rainwater." As the irrigation waters dissolve the salts and carry them off, the soil becomes agriculturally productive. In many desert areas around the world including the southwestern United States and the Middle East, irrigation has converted deserts into highly productive agricultural regions.

High levels of desert irrigation are not without potential problems, however. As the irrigation waters dissolve and flush the toxic salts from the soil, the salts may enter both the groundwater and streams, rendering them too saline for use. In other areas such as the Imperial Valley of California, soils have become saline due to the intense evaporation of water containing trace quantities of salt added to the soil by irrigation.

Tropical Regions

In tropical areas where the rainfall is very high, only downward movement of rainwater occurs. The super-dominant downward movement of rainwater leaches the soils and, combined with the high temperatures, develops a unique soil profile. Little O horizon exists due both to extreme chemical and biochemical attacks upon the plant debris and to the fact that tropical plants acquire their nutrients directly from the developing humus. Clay minerals, which are dominant ingredients in most soils, are nearly nonexistent in tropical soils. Although generally considered to be chemically stable at surface conditions, under the extreme weathering conditions that exist in the tropics, the clay minerals decompose into the hydrated oxides of aluminum (*alumina*) and silicon dioxide (*silica*). In some areas, the accumulations of hydrated alumina are so concentrated that they constitute a commercial aluminum ore called *bauxite*. Silica precipitates as quartz and adds to the general low fertility of the tropical soil. Oxygenated water precipitates iron, if present in the soil, as hematite, which is responsible for the characteristic red color of the tropical soils. When iron is not present, tropical soils are usually yellow. As is the case with alumina, the concentrations of hematite in some tropical areas may become sufficiently high to be considered a commercial iron ore.

Although the casual observer might think the lush tropical rain forests reflect high soil fertility, such is not the case. The fertility of tropical forest soils is actually very low. Most plant nutrients are leached from the soil by the high downward flows of acidic rainwaters. Tropical plants acquire their nutrients either directly from the decomposing humus or from mineral dust that falls on the surfaces of their leaves.

In many tropical areas, rain forests are being cut down in order to obtain agricultural and grazing lands. Unfortunately, the soil is generally not suited to growing the high-protein-yielding plants that are subsequently planted. In only a few growing seasons, the limited concentration of plant nutrients in the original soil in the form of minerals or humus is exhausted, necessitating the cutting down of more rain forest. It is estimated that rain forests are being destroyed at the rate of an acre per second. Unfortunately, these lush forests not only serve as a source of atmospheric oxygen but also play a role in maintaining the carbon dioxide content of the atmosphere. Carbon dioxide, you will remember, is one of the major components controlling the atmospheric temperature via the greenhouse effect. The reduction in the areas covered by rain forests will certainly add to the already increasing greenhouse effect.

SPOT REVIEW

1. Compare the various soil horizons.
2. Under what conditions do soils become "acid?"
3. What characteristics of semiarid soils make them especially good for the production of grains such as wheat and corn?
4. Why are tropical soils ill-suited for agriculture and the grazing of cattle?

TYPES OF SOIL

Although no scheme for soil classification is universally accepted, the system that has been widely used since 1975 is **soil taxonomy**. This system is based upon the physical, chemical, mineralogical, and morphological properties of the soil, which are in turn related to the processes under which the soil formed. The classification scheme utilizes a hierarchy beginning with orders subdivided into suborders, great groups, subgroups, families, and series. As in most hierarchal classification systems, the number of components increases exponentially with each subdivision. At the series level, for example, there are about 14,000 different kinds of soils. We will only consider the 11 basic orders. The suffix, -*sol,* in each of the orders is derived from the Latin word *solum* meaning soil. A simplified soil map of the United States showing the regional distribution of the basic soil orders is shown in Figure 7.10. The basic soil orders are summarized in Table 7.1.

ENVIRONMENTAL CONCERNS

Although most environmental problems are perpetrated directly or indirectly by humans, some problems are the result of natural processes. The most serious problem involving soils is a case in point.

As Table 7.1 indicated, the vertisols are soils that contain appreciable contents of expandable clay minerals. In regions such as the southwestern United States where the soils are subjected to seasonal wetting and drying, the cyclic expansion and contraction of the clay minerals result in the natural overturning of the soil. Samples of pure expandable clays can expand up to 10 or 15 times their original volume as water is absorbed into the crystal lattices. Although vertisols may contain less than 5% expandable clay minerals by volume, the pressures that develop within the soil are often sufficient to cause damage to foundations of all kinds. In the United States, an estimated $2 billion of damage results each year from soil expansion.

The problems of expandable soils can be controlled by a combination of soil testing and proper construction methods. Because the expansion and contraction of the soil are critically affected by soil moisture, drainage of the water immediately adjacent to the foundation is essential. In many cases, the foundation can be protected from the pressures generated by soil expansion by employing buffers such as crushed rock that both absorb the forces generated by expansion and prevent the soil from coming in direct contact with the surface of the foundation.

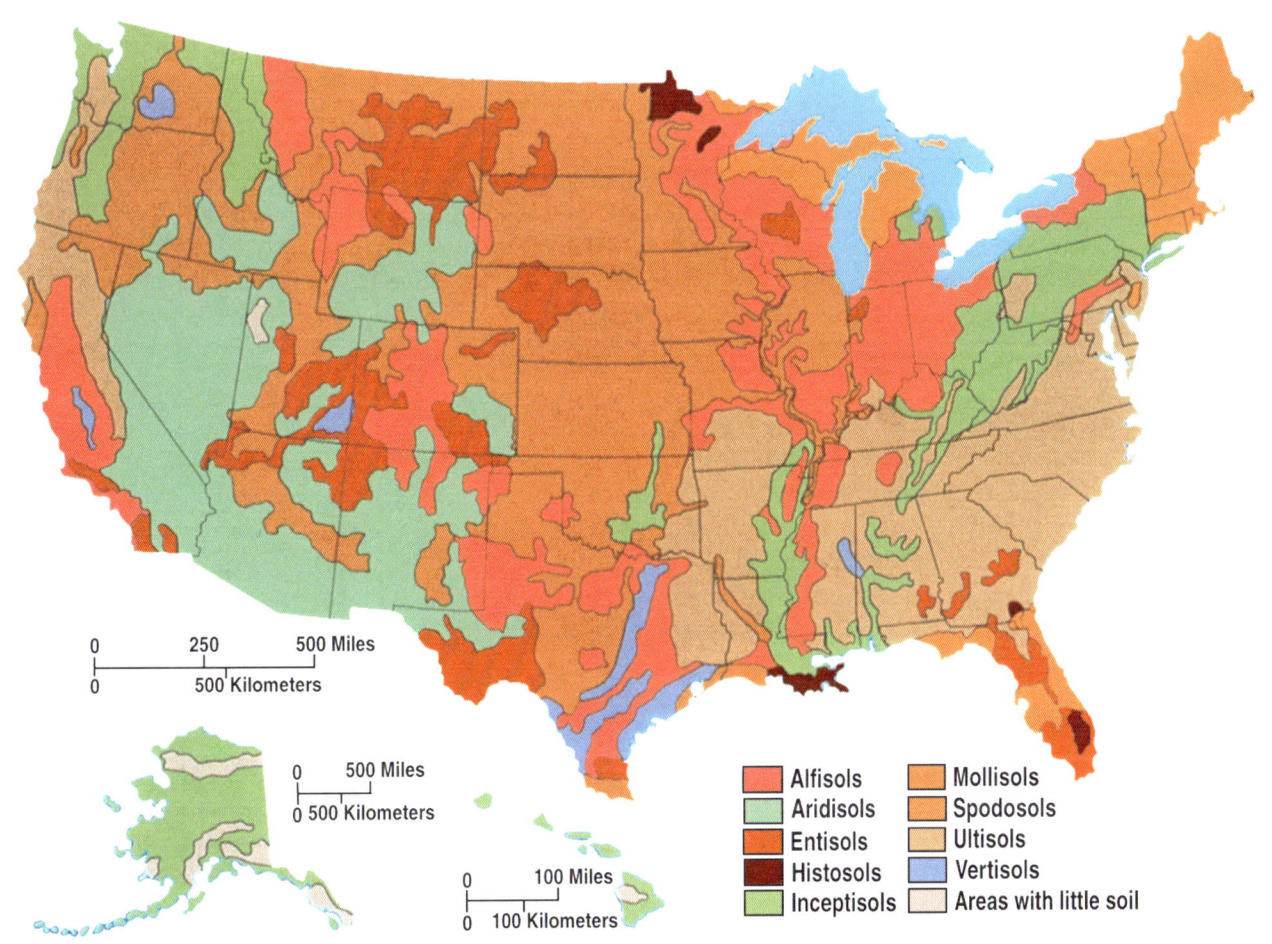

FIGURE 7.10   *The distribution of the basic soil orders throughout the United States reflects the regional variation in climate.*

TABLE 7.1
THE BASIC SOIL ORDERS

ENTISOLS	Entisols are regolith that is being subjected to the very beginnings of soil fermentation.
INCEPTISOLS	Inceptisols show the beginnings of horizonation and include materials accumulated on steep slopes, floodplains, and the surface of slowly weathering bedrock.
ARIDISOLS	Aridisols form in areas of low rainfall and are characteristically composed of rock fragments with little or no humus. High levels of salts may be present in some aridisols, rendering them highly alkaline.
MOLLISOLS	Mollisols generally develop in areas receiving 10–20 inches (25–50 cm) of rainfall per year. Vegetation is continuous and dominated by grass. Mollisols are usually neutral to slightly alkaline and are among the most fertile agricultural soils on earth.
SPODOSOLS	Spodosols develop under conifer forest cover in cool, humid areas and are characteristically acid. Spodosols can be converted to marginally fertile agricultural soils with proper lime treatment and the addition of nutrients.
ULTISOLS	Ultisols are similar to spodosols except that they develop under broadleaf forests in areas of higher mean temperatures. Ultisols may be converted to highly productive agricultural soils with appropriate lime neutralization and the addition of nutrients.
ALFISOLS	Alfisols are transitional soils between the mollisols of the grasslands and the spodosols and ultisols of the forested regions. Alfisols are widespread and are highly productive food producers.
OXISOLS	Oxisols develop in moist, tropical areas. They are the most highly weathered of all soils and commonly lack both clay minerals and humus. In general, they are quite infertile. In many cases, they consist of little more than quartz and the hydrated oxides of aluminum and iron.
VERTISOLS	Vertisols are unusual soils that develop in semiarid areas underlain by expandable clay minerals. The name refers to vertical cracks that develop in the soil during the dry season. As the soils wet and dry, they experience a natural mixing. When irrigated, vertisols are highly productive agricultural soils.
HISTOSOLS	Histosols develop in poorly drained areas such as swamps and marshes. They commonly contain peat, which when drained and dried can be used as a fuel or as a humic soil additive. When buried and subjected to millions of years of elevated temperatures, the peat may convert to coal.
ANDISOLS	Andisols develop on various kinds of volcanic materials such as lava or tuff. They are the newest soil order, previously being classified as inceptisols.

Although the problem of expandable soils stems from natural causes, other serious problems involving soils are attributable in varying degrees to human actions. The loss of soil by erosion, for example, is often the result of poor agricultural practices. The removal of a protective layer of vegetation combined with the loss of soil strength as a result of plowing and tilling accelerates the rates of erosion by both water and wind. The loss of soil from agricultural lands is aggravated by long periods of drought when the wind becomes increasingly effective. A case in point was the erosion that took place in the *Dust Bowl* of the central plains states during the early 1930s (Figure 7.11). The human impact was poignantly portrayed in John Steinbeck's *The Grapes of Wrath*, which describes the plight of a family of farmers driven from the land. During a single dust storm in 1935, an estimated 5 million tons of dust were in suspension over an area of about 30 square miles (78 km^2). Reports of nearly a foot of soil being removed within a single day were not uncommon. Although accelerated soil loss due to cultivation cannot be eliminated altogether, practices such as contour plowing and planting combined with the rotation of row and total coverage crops can alleviate the problem.

In many cases, overgrazing of pasture land has led to similar soil loss by erosion. The impact of too many animals, especially those that tend to close-crop the grass, during times of low rainfall can initiate the erosion of soil by destroying the protective layer of grass.

More recent sources of environmental damage are the estimated 12 million **off-the-road vehicles**

FIGURE 7.11 *Poor agricultural practices have accelerated the loss of soil, especially during times of drought. (Bettmann Archive)*

(ORVs) that are driven through forested areas, deserts, and dunes throughout the country (Figure 7.12). Aside from the direct damage caused by the passage of the vehicle, the disruption of the soil surface exposes the disturbed materials to increased erosion by water and wind. During periods of heavy rain, increased sediment loads clog streams and contribute to the infilling of ponds and lakes. As the popularity of ORVs increases, their encroachment into new areas poses a constant problem to those who manage public lands and dramatizes the clash between those dedicated to the preservation of wildlands and those who claim equal access.

Urbanization creates a host of environmental problems. Construction exposes soils to increased erosion. The near-continuous cover of pavement drastically changes the rates of runoff into streams and subsequent stream erosion. Growing populations increase the potential of soil pollution from improperly installed septic systems, leaking storage tanks, effluents from solid waste disposal sites, and the chemical treatments utilized by some homeowners to maintain perfect lawns. The adsorptive ability of the clay minerals renders the soil ready receptors and storage sites for a host of pollutants.

One of the most notorious cases of soil contamination is Love Canal. Located in Niagra Falls, New York, Love Canal was originally excavated in 1892 as a portion of a transportation system that was supposed to connect various industrial sites in the area. After the venture failed, the canal became the dumping ground for waste of all kinds. In particular, from the 1920s, the canal became the repository for a wide variety of chemical wastes, many of which were later determined to be carcinogenic. In 1953, the land was sold to the school board for the sum of one dollar, and homes were built on the old dump site. Unusually heavy rainfalls and snow melts in 1976 caused the toxic materials to surface. Vegetation in the area began to die. Rubber products such as the soles of shoes and bicycle tires began to disintegrate after coming in contact with the contaminated soil. Chemicals of unknown composition began to seep to the surface and collect in puddles.

FIGURE 7.12 *Off-the-road-vehicles (ORV's) disrupt fragile soils, eliminate the protective plant cover, and result in accelerated erosion, especially on steep slopes in semi-arid to arid regions. (a VU© Kevin and Betty Collins; b VU/© Frank M. Hanna)*

The federal government bought the homes along Love Canal and relocated the residents. Subsequently, $20 million was made available to more than a thousand current and former residents to compensate them for medical expenses, property damage, and the costs of relocation. At the present time, nearly $100 million has been spent to clean up the area, and the task is not yet complete. How much the ultimate cost will be is anyone's guess. The real question, however, is what is being done to ensure that there will be no more Love Canals.

CONCEPTS AND TERMS TO REMEMBER

Soil science
   pedology
   soil
Clay mineral chemistry
   clay particles and cations
   cation adsorption
   cation exchange
   cation exchange capacity (CEC)
Soil horizonation
   kinds of horizons
      O horizon

A horizon
E horizon
B horizon
C horizon
Soil formation
   rainwater versus groundwater
   acid soils
   alkali and alkaline earth elements
   alkaline soils
      grassland soils

superalkaline soils
hardpan
tropical soils
Soil taxonomy
Environmental problems
   soil erosion
   off-the-road vehicles (ORVs)
   urbanization

REVIEW QUESTIONS

1. The specific source of the cation exchange capacity of the clay mineral is
   a. the crystal structure.
   b. the particle size.
   c. deficiencies of cations within the crystal lattices.
   d. a surplus of cations within the crystal lattices.

2. Except for tropical soils, plant nutrients are stored within the
   a. O horizon.   c. B horizon.
   b. A horizon.   d. C horizon.

3. In general, soils that develop in regions receiving more than 20 inches (50 cm) of rainfall annually will
   a. be acid.
   b. be thin.
   c. be saturated with water.
   d. have low cation exchange capacities.

4. Commercial deposits of bauxite, the major source of aluminum, are found in
   a. mollisols.   c. entisols.
   b. oxisols.   d. aridisols.

5. The soils that support the world's grasslands are the
   a. vertisols.   c. inceptisols.
   b. mollisols.   d. aridisols.

6. In terms of soil horizons, laterites are characterized by an overdevelopment of the _____ horizon.
   a. E   c. B
   b. O   d. A

7. Chernozems are important because they
   a. are the most widespread of all the various kinds of soil.
   b. support the world's grasslands.
   c. support the tropical rain forests.
   d. are the beginnings of soil development.

8. Our major source of aluminum, bauxite, develops in _____ climates.
   a. cold, wet   c. hot, dry
   b. cold, dry   d. hot, wet

9. Most of the soils in the eastern United States are
   a. mollisols.   c. oxisols.
   b. ultisols.   d. histosols.

REVIEW QUESTIONS *continued*

10. Some desert soils are converted into productive agricultural soils by irrigation. What characteristics do many arid soils have that preclude their being used for agriculture without irrigation, and what changes are brought about by irrigation that allow these soils to become productive?
11. What is the cation exchange capacity of a soil? Explain what determines the magnitude of the CEC and how it is involved in the supply of plant nutrients.
12. Why is it common practice for homeowners in the eastern United States to spread lime on their lawns each year? Why don't homeowners in the Midwest do the same?
13. What are the major factors involved in the topographic control of soil formation?
14. How does climate affect the formation of soil?

THOUGHT PROBLEMS

1. Debate the statement that the introduction of clay minerals to tropical soils would allow them to be converted to permanently productive agricultural soils.
2. A conflict exists between the owners of ORVs who feel that they have a right to access to public lands and those who feel that ORVs should be banned because of the damage they cause. What steps could be taken to resolve the problem?

FOR YOUR NOTEBOOK

You will want to determine the kinds of soils in your area. Soil maps are available from the Soil Conservation Service (SCS) of the U.S. Department of Agriculture. If there is no local SCS office, copies may be available in your library or from the departments of soil science or agriculture of a university in your state.

Note the kinds of soils in your immediate area. Investigate the relationship between the local geology and topography and the distribution and types of soil.

If the soils in your area are used for agriculture, investigate whether specific soils are dedicated to certain crops and whether the soils require treatments before they are used. The general topic of chemical treatments of agricultural products and the potential that exists for soil and groundwater contamination is very controversial. You may want to find out whether there are any restrictions on the use of chemical additives in your area.

In light of our discussion of Love Canal, are there any local sources of soil contamination that may present health concerns? Not all sources of contamination are industrial. A growing problem in urban areas is the chemical treatment of lawns. In some areas, local governments have restricted and even prohibited the use of pesticides and herbicides. In other areas, treatments can only be applied if written releases have been acquired from neighbors.

CHAPTER OUTLINE

INTRODUCTION

SLOPE STABILITY

FACTORS AFFECTING COHESION AND FRICTION
- Water Saturation
- Ice Formation
- Oversteepening of Slopes

CLASSIFICATION OF MASS WASTING PROCESSES
- Creep
- Earthflow
- Landslides

ENVIRONMENTAL CONCERNS
- Causes of Slope Instability
- The Effects of Earth Moving
- Identification of Potentially Hazardous Sites
- Methods of Slope Stabilization
- Subsidence

CHAPTER 8

Mass Wasting

INTRODUCTION As soon as rocks are exposed at Earth's surface, they come under attack by the combined processes of erosion; they are loosened, dissolved, and worn away. Below sea level, these processes carve submarine canyons into the continental shelf and transport the debris to the abyssal ocean floor. Above sea level, the processes attempt to reduce the land surface to sea level. In all cases, the force responsible for this leveling or *gradation* is the relentless downward pull of gravity.

The first geologic process to operate is usually weathering, which converts bedrock into regolith. Once regolith forms, processes of erosion begin to move it downslope, ultimately to the ocean floor. Some of the more conspicuous agents of erosion are streams, glaciers, groundwater, shoreline processes, and the wind. However, the process that often initiates the downhill trip is **mass wasting**. In many cases, before any of the regolith is picked up and carried off by the other agents of erosion, the processes of mass wasting have largely completed their tasks.

Mass wasting is definitely the unsung hero of surface processes. Several of the most important processes operate so slowly that they go largely unnoticed. If it were not for the occasional landslide or rock fall, the entire process of mass wasting would probably go on without most people ever knowing it existed.

Several characteristics of mass wasting distinguish it from the other agents of erosion. Unlike materials moved by water, wind, or ice, the debris moved by mass wasting is not carried within, on, or under any other medium. The downslope movement is strictly the result of gravity.

Whereas the other agents of erosion may move materials for hundreds and perhaps thousands of miles, processes of mass wasting normally move materials only for short distances, usually no farther than from a hilltop to the valley floor. At that point, the task of transportation is transferred to an agent of erosion such as a stream or a glacier.

Because few, if any, natural land surfaces are perfectly horizontal, mass wasting in one form or another is constantly active over the entire exposed portion of the land's surface. In contrast, other forms of erosion are limited in both area and time: Streams are restricted to channels. Glaciers are restricted to those areas of Earth's surface where snow and ice can accumulate and persist in large volumes. The erosional effect of the wind is not only intermittent but is usually restricted to the desert.

Running water is the most important agent of erosion. As streams carve their channels downward and widen valley floors, slopes are formed, and the overall elevation of the land surface above sea level is reduced. Nevertheless, the reduction in topographic *relief*, the distance between hilltops and the valley floors, and the ultimate elimination of the divides that separate stream channels are primarily the result of the processes of mass wasting. Depending on the particular mass wasting processes in operation, as the valley walls retreat, the divides are removed either parallel to the original valley wall, as is common in arid regions (Figure 8.1), or by continuously reducing the slopes of the valley walls as in humid regions. (Figure 8.2).

Depending on the angle of slope, mass wasting moves rock materials at various rates ranging from the maximum velocity of a *free-fall* to imperceptible *creep*. In some arid regions where slopes remain steep or precipitous throughout the erosional cycle, mass wasting is dominated by rapid movements. In contrast, in humid regions where initially steep slope angles become more gentle as the erosional cycle progresses, the processes of mass wasting proceed at progressively slower rates.

In summary, the appearance of the land today is the result of the more or less *equal* and *concomitant* efforts of mass wasting and the other forms of erosion.

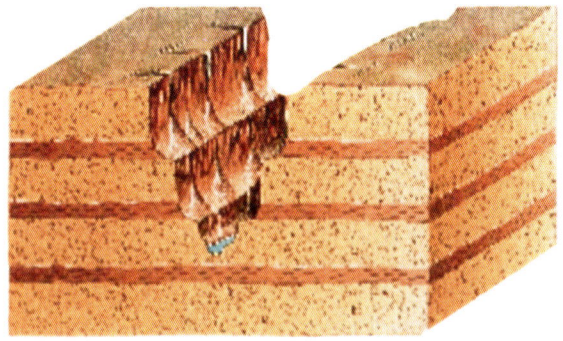
(a)

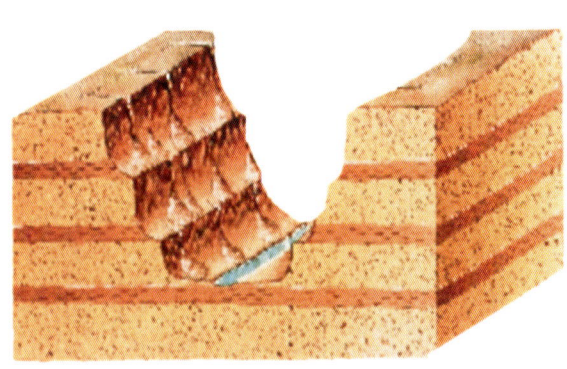
(b)

(c)

FIGURE 8.1 *In arid regions, as streams carve their channels deeper and widen the valley floor, the valley walls retreat parallel to the original vertical slopes. Under the influence of fast processes of mass wasting such as rock falls, rock edges remain sharp and angular, and divide elevations are little reduced. The overall appearance of the topography remains rugged as the valleys broaden and the divides are eliminated. Arid erosion is best illustrated by the mesas and buttes of Monument Valley. (Photo courtesy of E. D. McKee/ USGS)*

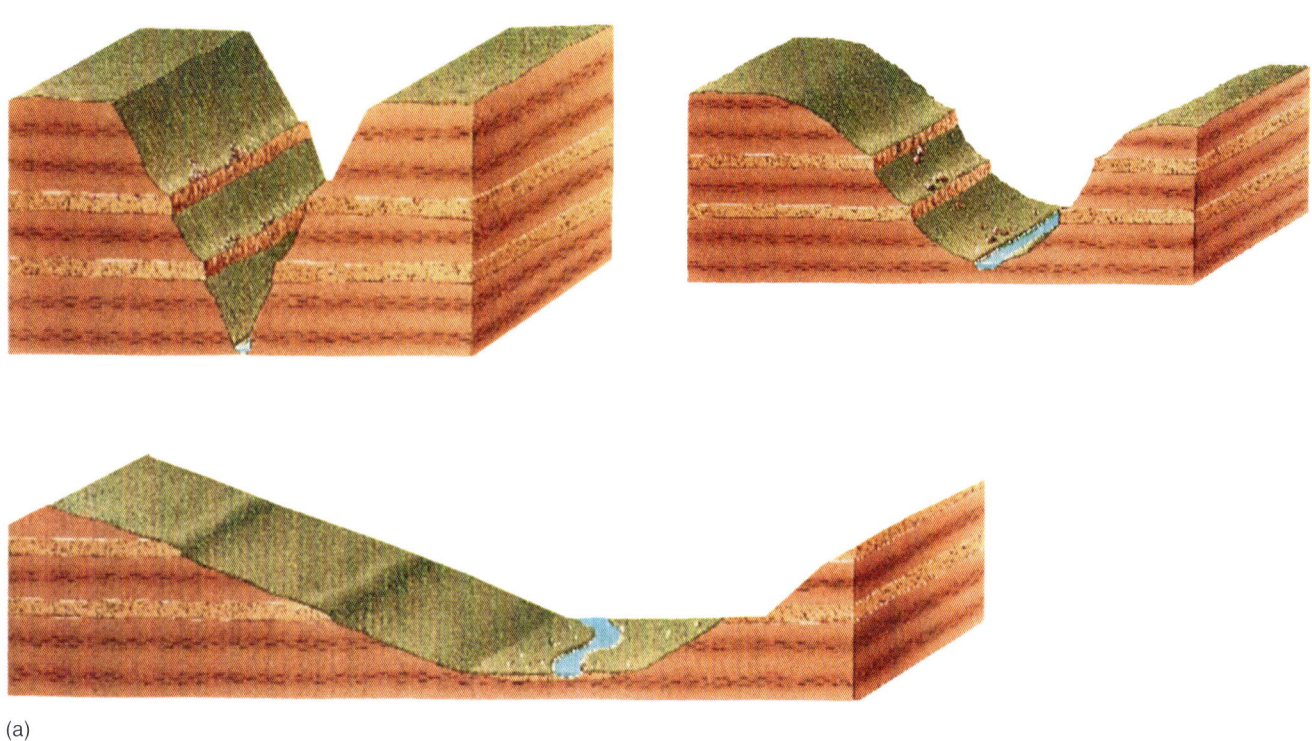

(a)

(b)

FIGURE 8.2 *In humid regions, the combined effects of stream erosion and slow processes of mass wasting, such as creep and slump, reduce the slopes of valley walls, round rock edges, and reduce the elevation of divides. The overall effect is a rounding and softening of the topography as typified by the topography of the Appalachians. (Photo VU/© William J. Weber)*

SLOPE STABILITY

Mass wasting involves the downslope movement of both regolith accumulated on slopes and, oftentimes, the underlying bedrock. The stability of regolith accumulated on slopes is a function of two forces: (1) a component of the force of **gravity** directed downslope parallel to the surface and (2) the force of **cohesion** and **friction**, both within the regolith and between the regolith and the bedrock surface.

The fundamental parameter that determines the stability of regolith accumulated on a slope is the **angle of repose**, which is the maximum angle at which loose material can accumulate before beginning to move downhill. The angle of repose of any material depends upon the cohesion and friction between its particles. An important factor in determining cohesion and friction is particle shape. In general, as the sphericity and smoothness of the particles decrease, the angle of repose increases. For most loose materials, the maximum angle of repose is about 40°. As a result, most slopes with angles in excess of 40° are bedrock surfaces free of regolith.

The dynamics of slope movement are illustrated in Figure 8.3. Figure 8.3a shows a particle resting on a horizontal surface. The vertical arrow represents the force of gravity operating on the particle, that is, its weight. The mass has no tendency to move because no force component is directed parallel to the surface. For any particle resting on a surface, the force of cohesion and friction is proportional to the force acting perpendicular to the surface. In this case, the force of cohesion and friction and the force of gravity are the same. For the particle in Figure 8.3a, the force of cohesion and friction is maximum.

Consider now Figures 8.3b and 8.3c, which show the particle resting on slopes of increasing angles. The force of gravity (the weight of the particle) in all three figures is vertical and identical in magnitude. In Figures 8.3b and 8.3c, the force of gravity is resolved into two components: (1) a force directed *downslope* and *parallel* to the surface and (2) a force of *cohesion* and *friction* directed *upslope* that is proportional to the force directed *perpendicular* to the surface. As long as the force of cohesion and friction directed up slope is greater than the force of gravity directed downslope, the particle will remain immobile.

Note, however, that as the angle of slope *increases,* the force directed downslope parallel to the surface increases, and the force directed perpendicular to the

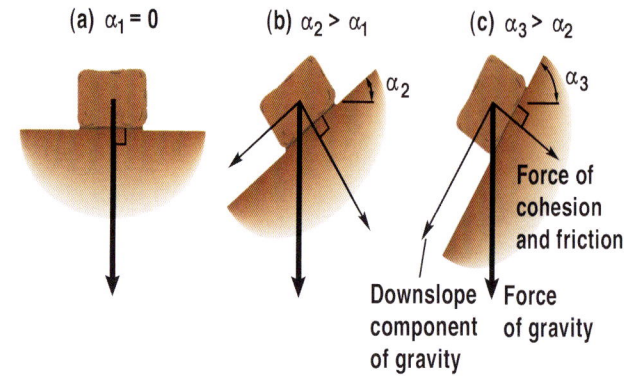

FIGURE 8.3 *The force that drives all processes of mass wasting is* gravity. *The force of gravity is resolved into two forces, one directed* parallel *to the slope and the other directed* perpendicular *to the slope. Whether a particle moves downslope depends upon which of these two force dominates.*

surface—and subsequently the force of cohesion and friction—decreases. Eventually, when the force directed downslope is greater than the force of cohesion and friction, the particle will begin to slide. The angle at which movement is initiated is the angle of repose.

FACTORS AFFECTING COHESION AND FRICTION

Materials accumulated at less than the angle of repose can be set in motion if the force of cohesion and friction can be reduced. Three mechanisms reduce cohesion and friction: (1) *water saturation,* (2) *ice formation,* and (3) *oversteepening* (undercutting) *of the slope.*

Water Saturation

The presence of water significantly affects cohesion and friction in unconsolidated materials. Small amounts of water *increase* cohesion and friction between the particles. Think of how much easier it is to make a sand castle at the beach from wet sand than from dry sand. Eventually, however, as the water content increases, both particle cohesion and internal friction within regolith will be reduced. As water fills the spaces between the rock and mineral particles, pressure forces the individual particles apart and reduces cohesion. In some cases, water may decrease friction by "lubricating" the points of contact between the particles. If clay minerals make up appreciable portion of the regolith, they tend to absorb water and further reduce particle cohesion and friction by providing a "greased" slide on which the particles may move. With the force of cohesion and friction reduced, and the downslope component of gravity increased by the added mass of water, the material may begin to slide downslope.

Mass movement induced by water saturation increases with heavier rainfall. In humid, tropical regions where high rates of chemical weathering have produced thick layers of regolith, water-induced mass movement is common. In humid, temperate regions, slope failures commonly occur after heavy rains have saturated the regolith and soil profiles.

Ice Formation

In humid, temperate regions, the formation of ice is responsible for several mass wasting mechanisms. *Frost heaving* is effective in moving the uppermost layers of relatively fine-grained regolith or soil. Moisture freezing between the particles of the surface layers produces ice crystals that grow perpendicular to the surface. As the crystals grow, the pressure overcomes particle cohesion and lifts small particles perhaps a half inch or so. With the thaw, the particle drops vertically to a location displaced downslope. Repeated *freeze thaw* cycles result in the surface layers of regolith moving downslope in a sawtooth pattern (Figure 8.4). Other processes may also result in the sawtooth movement of surface materials. The expansion and contraction of clay minerals in response to wetting-drying cycles will produce a movement similar to that induced by the freezing and thawing of ice crystals. In drier climates, similar surface movement may be induced as crystals in the surface layers precipitate and grow during droughts when groundwater evaporates and subsequently dissolve during rainfalls. In areas where the surface is populated by annual plants (those that live and grow for only one season), the surface materials may be dislodged as roots grow and decay.

In both arid and humid temperate regions, *frost action* is a major mechanism of physical weathering. The cyclic freezing of water and thawing of ice in fractures within vertical to near-vertical exposures of rock (Figure 8.5) eventually dislodge rocks from the surrounding rock mass, producing a precariously balanced "layer" of regolith (Figure 8.6). Mass wasting

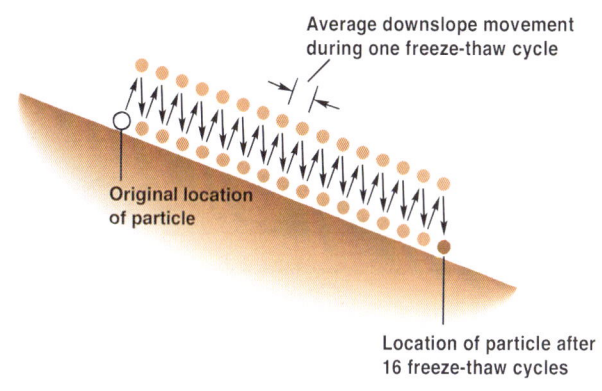

FIGURE 8.4 *In humid, temperate regions, the process of* frost heaving *is a very effective agent of mass wasting, inducing the slow downhill movement of the upper layers of regolith by a process called* creep.

FIGURE 8.5 *Cracks in rocks widen as water undergoes cyclic freezing and thawing, a process called* frost action. *Eventually, a portion of the original rock will be dislodged and break away to become a* rock fragment. *(VU/© Glenn M. Oliver)*

FIGURE 8.6 *Features such as* pedestal rocks *are often the end result of millions of years of weathering and erosion during which time* frost action *plays an important role. (© Peter Kresan)*

FIGURE 8.7 *As streams undergo a process called* meandering, *they move from one side of a valley to another. As the stream encounters and erodes the base of the slope, the slope becomes* oversteepened *and fails by slumping. The valley is widened as the debris is picked up by the stream and carried off.*

FIGURE 8.8 *As streams* undercut *their banks, especially during times of flood, the banks will fail. Eventually, the debris is carried off as part of the stream's bed load and suspended load. (VU/© Ted Whittenkraus)*

may then be initiated as less resistant underlying layers of rock are removed by erosion or by a horizontal force provided by a local shock such as an earthquake.

In colder regions, ice may form and persist within deeper layers of the regolith, even during the warmer months. The formation and expansion of the ice between the particles reduce particle cohesion. At the same time, the downslope movement of the rock mass is initiated as the weight of the overlying rock causes the ice between the particles to flow in a fashion similar to the movement of glaciers.

Oversteepening of Slopes

As a stream deepens and widens its valley, the slope of the lower valley walls immediately adjacent to the stream is increased beyond the angle of repose (Figure 8.7). As described above, the reduced particle cohesion and internal friction resulting from an increase in slope angle initiate the mass movement of the regolith and reestablish a slope angle less than the angle of repose.

Another example of oversteepening can be observed along the channels of streams that have begun to flow in the sinuous pattern called *meandering* (Figure 8.8). As the water flows around the bends in the stream channel, it undercuts the outer bank and removes the support for the overlying materials. With no support from below, the bank materials collapse under the full force of gravity. An identical situation can be observed along steep, high-energy coastlines where waves undercut the cliff (Figure 8.9).

FIGURE 8.9 *Cliffs along high-energy coastlines* retreat *as they are* undercut *by waves and fail by* slumping *and* rock falls. *(VU/© James R. McCullagh)*

In some cases, materials can accumulate on slopes steeper than the angle of repose. The most familiar example is snow, which frequently accumulates in overhanging slopes. When the slope materials are in such an unstable condition, an external source of energy such as an earthquake or the detonation of an explosive charge may be sufficient to initiate mass movement. According to reports, avalanches have been triggered by no more than the sound of a human voice in instances where snowbanks accumulated far beyond their angles of repose.

SPOT REVIEW

1. What factors determine the angle of repose of a particle?
2. What agents may cause the downslope movement of a particle resting on a slope with an angle less than its angle of repose?

CLASSIFICATION OF MASS WASTING PROCESSES

Mass wasting processes are classified according to (1) the *kinds of materials* they involve, (2) the *rate of movement*, and (3) the *amount of water*.

Creep

Creep refers to the slow (fewer than 1 to 1.5 inches or 3 to 4 mm per year) downslope movement of the upper layers of regolith. Of all the methods of mass wasting, creep is the most widespread and is responsible for the movement of the greatest mass of material.

Creep is caused by a number of factors. In humid, temperate regions, the combination of chemical and physical weathering results in a more or less continuous layer of regolith covering the bedrock surface. During the winter, when the regolith is subjected to repeated freeze-thaw cycles, frost heaving results in slow downslope movement of the uppermost layers (Figure 8.10). During the warmer months, soils that contain clay minerals, which are capable of expanding and contracting as they take on and release water, may experience the same kind of sawtooth downhill movement as frost heaving produces. Eventually, the materials are transported to the valley floor where they are picked up by a stream and carried away. In arid regions, creep may be initiated by the cyclic growth and dissolution of water-soluble crystals in the upper layers of regolith. Evidence for creep includes fence posts tilted downslope, overturned stone walls, trees with bent truck bases, leaning retaining walls and tombstone, and hummocky or undulating slopes (Figure 8.11).

Earthflow

During the spring and early summer, the upper layers of the regolith frequently become saturated with water, which reduces cohesion and friction and accelerates the slow downslope movement; the result is a mass movement process called **earthflow**. Unlike creep, which may affect an entire hillside, an earthflow involves only a limited area of the slope surface. In some cases where the movement of the regolith does not break the cover of vegetation, earthflow will result in a hummocky surface. In other cases, the regolith may break away at the upper edge of the earthflow (Figure 8.12). Rates of earthflow movement range from an imperceptible flow of a fraction of an inch per day to several feet or meters per day.

FIGURE 8.10 *As the uppermost layers of regolith slowly move downslope by* creep, *the vegetated surface often becomes wrinkled or hummocky.* (VU/© John D. Cunningham)

(a)

(b)

(c)

FIGURE 8.11 *Evidence of creep includes (a) the bent bases of trees, (b) tipped tombstones, and (c) the bending of rock layers, particularly shales.* (a Courtesy of W. B. Hamilton/USGS, b VU/© John D. Cunningham)

FIGURE 8.12   *Oftentimes after heavy rains, when the regolith has become saturated with water, the combination of increased mass and decreased cohesion and friction will result in a portion of the regolith breaking away and moving downslope as an* earthflow. *(VU/© L. Linkhart)*

FIGURE 8.13   *As the regolith between the surface and the underlying frozen layer becomes saturated with water, the material will literally liquify and begin to move on very shallow slopes by the process of* solifluction. *(VU/© Steve McCutcheon)*

In high latitudes and high elevations, the regolith at depth is permanently frozen. During the summer, the upper layers are warmed, and ice present between the rock and mineral particles melts. The permanently frozen layers below prevent the water from draining, and it collects in the thawed portion of the regolith. As the water accumulates in the upper layers of regolith, cohesion and friction are essentially eliminated, and the layer is converted to a viscous fluid mass that undergoes a type of earthflow called **solifluction** and flows slowly downhill at speeds of 0.25 to 2 inches (0.5 to 5cm per year, often on slopes of no more than 2° or 3° (Figure 8.13).

Solifluction is not restricted to the permafrost areas of high latitudes or higher elevations. In more temperate mid-latitudes, water in the surface layers of the regolith may freeze to considerable depth. With the spring warming, the upper layers of the regolith may thaw, becoming a viscous fluid mass, while the lower layers remain frozen. The water-saturated upper layers may begin to move over the frozen layers below. Eventually, when the entire regolith thaws and the water is able to drain, the materials will restabilize.

Landslides

One of the terms most commonly used to describe mass movements of rock and regolith is *landslide*. Geologically, a landslide refers to any of a wide variety of downhill movements of rock, regolith, and soil with no restrictions on the kind or amount of material involved or the rate of movement. Common usage, however, implies a relatively rapid movement of material within a confined area and would therefore include any of the following mass wasting processes.

Slump One of the most common examples of a fast mass wasting process is **slump.** Slump may affect the regolith, the bedrock, or both. Typically, when regolith becomes saturated with water and coherence and friction are diminished, the material within a limited area is converted to a viscous, almost liquid mass. Initially, the regolith may begin to move as an earthflow. A slump forms as continued movement causes a volume of the material to break away from the adjacent regolith and slide along curved (arcuate) slip surfaces as a coherent mass. As a result the affected mass rotates downward and out, forming curved scarps at the upper margin and bulging and flowage at the base (Figure 8.14). After moving a short distance, the material regains coherence and comes to rest. Slumping can also affect bedrock slopes. The coherence and interfriction within masses of rock can be reduced by freeze-thaw cycles and the pressure of water within fractures to the point where the rock mass fails along the same type of arcuate surface (Figure 8.15).

Slump can also be observed along stream banks or sea cliffs where regolith or rock masses are being undercut by running water or by the action of the waves. In both cases, the removal of material from below by undercutting results in an oversteepening of the slope. As the material at the base of the slump is removed by the stream or the waves, further slumping will occur. In the case of streams, the process results in the formation of the steep banks seen on the outside banks of stream meanders (refer to Figure 8.8). In the case of shorelines, the process produces steep cliffs (refer to Figure 8.9).

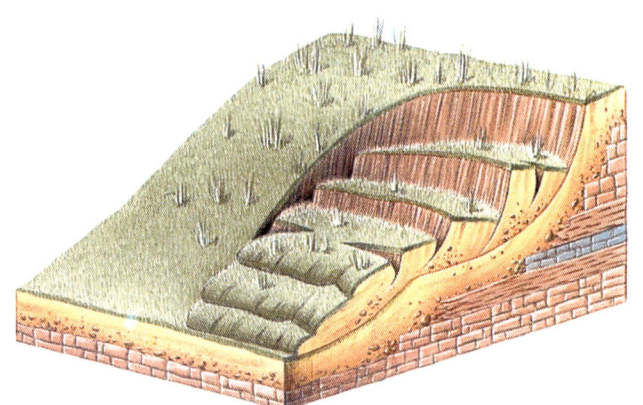

FIGURE 8.14 *In a* slump, *a portion of the regolith, sometimes including bedrock, breaks away along* curved or *arcuate fractures, rotates and moves downhill until enough water escapes to allow cohesion and friction to become reestablished within the mass. Typically, the regolith is exposed in a scarp along the uppermost portion of the slump and bulges out at the toe of the slump. (Photo courtesy of P. Carrara/USGS)*

FIGURE 8.15 *Slope failure along arcuate fractures, similar to those described in slumping but involving large masses of bedrock, results in major slope movements of material generally referred to as landslides. (Courtesy of J. T. McGill/USGS)*

Rock Falls   Another rapid mass wasting process is the free-fall or tumbling of rock dislodged from the face of a cliff, a process referred to as a **rock fall** (Figure 8.16). Rock falls may involve individual blocks of rock of any dimension or large masses of rock. Typically, rocks exposed at the surface of steep cliffs or road cuts are subjected to various processes of weathering that slowly break the layers along natural fractures. In both arid and humid temperate climates, frost action dislodges blocks of rock from cliff faces. Various processes of erosion may then remove the last remaining support from below, or the rock mass may be subjected to a local shock such as an earthquake that initiates the downward movement. With no support, the rock mass free-falls to the valley or slope below.

Vertical to near-vertical slopes of exposed bedrock are commonly subjected to rock falls. Upon impact, the larger rocks shatter into smaller fragments and collect as an accumulation of rubble called **talus**. Examples of talus can be seen along many road cuts (Figure 8.17). Accumulations of talus are especially apparent in arid regions where physical weathering processes dominate and slopes are not concealed by vegetation.

As the rock fragments in talus continue to be attacked by weathering and further reduced in particle size, other mass wasting processes move the surface materials downslope to the stream below. In cold, mountainous regions, especially on slopes that face away from the Sun, the frost action of ice forming in the lower layers of talus may result in a slowly moving mass called a *rock glacier* (Figure 8.18).

FIGURE 8.16   Rock falls *occur when masses of rock are displaced from near-vertical cliffs, commonly as the result of many cycles of* frost action. *(VU/© Albert J. Copley)*

FIGURE 8.17 Talus *is the rock debris that accumulates at the base of cliffs and road cuts. (Courtesy of N. K. Huber/USGS)*

FIGURE 8.18 Frost action *operating in the* talus *can result in the downsplper movement of masses of angular rock fragments called a* rock glacier. *(Courtesy of P. Carrara/USGS)*

Rock Slides When rocks contain surfaces of weakness such as bedding, fractures called joints, or foliations that are oriented parallel or nearly parallel to the slope of the hill, the scene is set for the mass wasting process called a **rock slide**, also referred to as a **rock avalanche** or **debris avalanche** (Figure 8.19). Of all the mass wasting processes, rock slides can be the most catastrophic and, potentially, the most destructive.

Rock slides have been clocked at speeds of more than 100 miles (160 km) per hour. Some geologists believe that the high rates of speed are due to layers of air trapped beneath the moving mass that significantly reduce friction.

Rock slides can be initiated by a number of processes. Typically, weathering, aided by a movement of groundwater, weakens the rock mass along the contact between two rock layers. As weathering proceeds, the two layers begin to separate. If one or both of the rock layers adjoining the weathering zone is a carbonate rock, the rate of separation is enhanced by the rock's solubility. In many cases, dissolution of carbonate rock layers leaves behind a residue of clay between the two rock masses that reduces friction. The same effect may be generated by a layer of water-saturated shale located beneath the rock mass. Eventually, the detached rock mass sits precariously on the underlying rock, awaiting some source of energy to initiate

FIGURE 8.19 Rock slides *or* rock avalanches *involve large masses of bedrock that fail when planes of weakness such as bedding, jointing, or foliation are oriented parallel to the slope. (Courtesy of P. Carrara/USGS)*

movement. If the slope of the rock layers is greater than the angle of repose, the force of gravity alone may be sufficient to initiate movement. Commonly, however, movement is triggered by heavy rainfalls that result in both a buildup of pressure and a decrease in cohesion and friction between the rock layers. In other cases, the energy may be provided by a local shock such as an earthquake. Once movement starts, the rock mass slides down the interlayer surface.

A spectacular rock slide dammed the Gros Ventre River in Wyoming on June 23, 1925 (Figure 8.20). The river valley was carved into sedimentary rocks with layers oriented parallel to the surface of the southern valley wall. Water, entering the layers along the crest of the ridge to the south, began to flow through and dissolve away layers of limestone located below layers of sandstone and shale. The process took many thousands of years, but eventually, cohesion between the limestone layers was decreased to the point where they began to separate along planes parallel to the layering. Lubricated by water provided by heavy rainfalls and snow melt, and triggered by the energy of a local earthquake, the overlying rock mass broke away from the weathered limestone below and slid down into the valley where the debris blocked the flow of the Gros Ventre River and created a lake 3 miles (5 km) long and 230 feet (70 m) deep. Later, uncontrolled overflow from the lake breached the debris dam and resulted in a flood that drowned several inhabitants of Kelly, Wyoming, a small town downstream.

Another massive rock slide dammed the valley of the Madison River in Montana on August 17, 1959. In this case, a steeply dipping layer of carbonate rock (dolostone) provided support for an underlying mass of weak metamorphic rocks (Figure 8.21). Dissolution of the carbonate rock eventually reduced the support for the mass of metamorphic rock to the point where the entire rock mass was in precarious balance. All that was needed was a source of energy to initiate the slide. The energy came in the form of a major earthquake that occurred about 20 miles (30 km) to the east. Movement of the surface caused a mass of rock estimated at more than 40 million cubic yards (30 million m^3) to break away from the south wall of the canyon, roar down into the canyon floor, and climb 300 feet (100 m) up the opposite canyon wall before coming to a stop. The debris dammed the Madison River and produced a lake 5 miles (8 km) long and nearly 100 feet (30 m) deep, now called Earthquake Lake. Unfortunately, the slide overran a public camping area and 28 people lost their lives. The experience of the failed debris dam at the Gros Ventre site caused the U.S. Army Corps of Engineers to construct a spillway across the debris dam to control the outflow from the lake.

Debris Slides *Debris slides, debris flows,* and *mudflows* represent a more or less continuous transition reflecting a progressive decrease in particle size and a corresponding increase in the amount of water involved. A **debris slide** is a slow- to

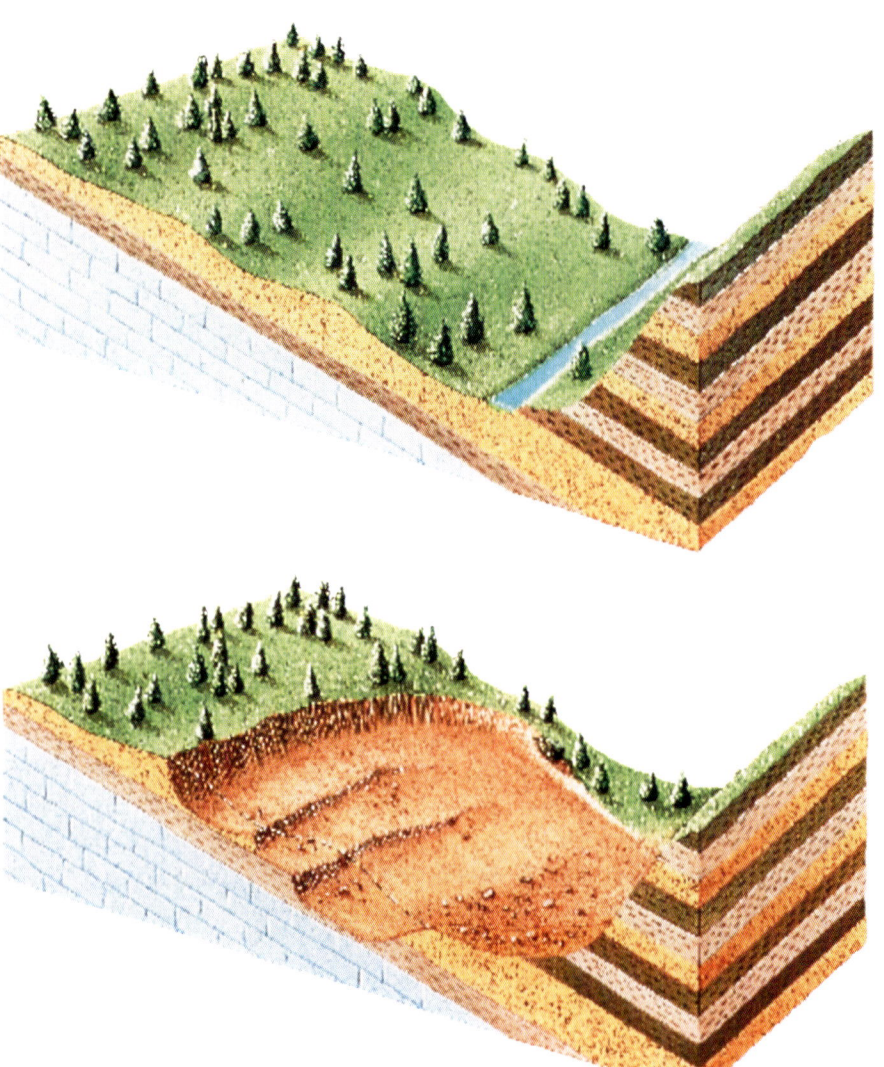

FIGURE 8.20 *In 1925, a massive slide occurred along the Gros Ventre River in Wyoming. Groundwater percolating downslope along the bedding of sedimentary rocks eventually caused the rocks to fail at the top of a layer of limestone, perhaps "greased" by an overlying, wetted, clay-rich shale.*

rapid-moving mixture of relatively dry rock, regolith, and soil (Figure 8.22). Debris slides generally form along valley walls. As various processes of weathering and erosion act on hillsides and continuously generate rock debris, the slope eventually becomes oversteepened to the point where the angle of repose is exceeded and the rock masses become physically unstable. From time to time, the accumulated debris detaches from the underlying rock, succumbs to the forces of gravity, and moves downslope. As in most mass wasting processes, the debris usually travels only a short distance downhill.

Debris Flows A **debris flow** is a moving mass of rock fragments, soil, and mud in which more than half of the particles are larger than sand size. Typically, both debris flows and mudflows originate when torrential rains soak slopes where the regolith has accumulated to near the angle of repose. Depending upon the mass and viscosity of the flow and the angle of slope, debris flows can travel as slow as a few feet per year or as fast as 200 miles (320 km) per hour.

From the standpoint of the loss of human life, one of the most disastrous natural events in recent years was the debris flow that killed an estimated 50,000 people in Peru on May 31, 1970. The debris flow was set in motion when a major earthquake caused a huge rock fall from the upper slopes of Mount Huascarán, the highest peak in Peru. Rocks, freefalling for more than 3,000 feet (1,000 m), impacted and shattered rocks on the lower slopes, creating a *rock avalanche.* Heat generated as the debris cascaded down the steep slopes melted sufficient snow and ice to transform the moving mass into a debris flow. Buoyed up by air trapped and compressed beneath the rock mass, the

Mass Wasting   219

FIGURE 8.21 *In 1959, the shock of a nearby earthquake caused a massive landslide along a segment of the* Madison River *in southwestern Montana. The earthquake initially caused the failure of a sequence of carbonate rocks whose bedding planes dipped into the valley. The carbonate rocks had been supporting highly weathered metamorphic rocks. With their support removed, masses of metamorphic rocks slipped into the valley, overran a public camping ground, buried 28 people, dammed the Madison River, and created Earthquake Lake. (Photo courtesy of J. R. Stacy/USGS)*

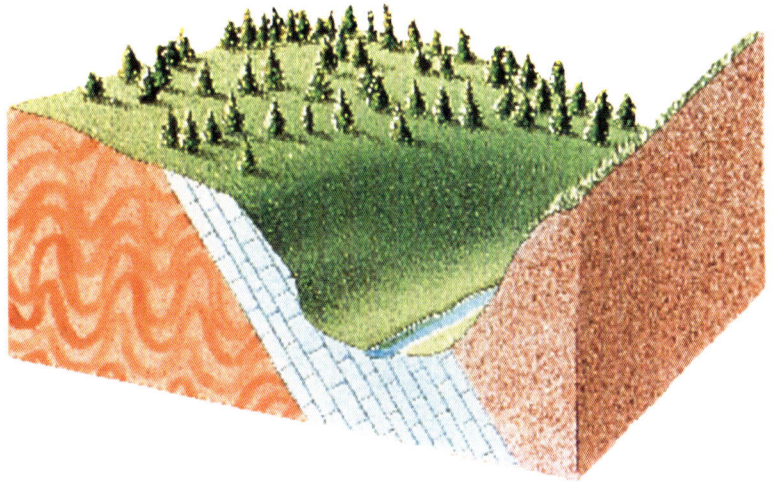

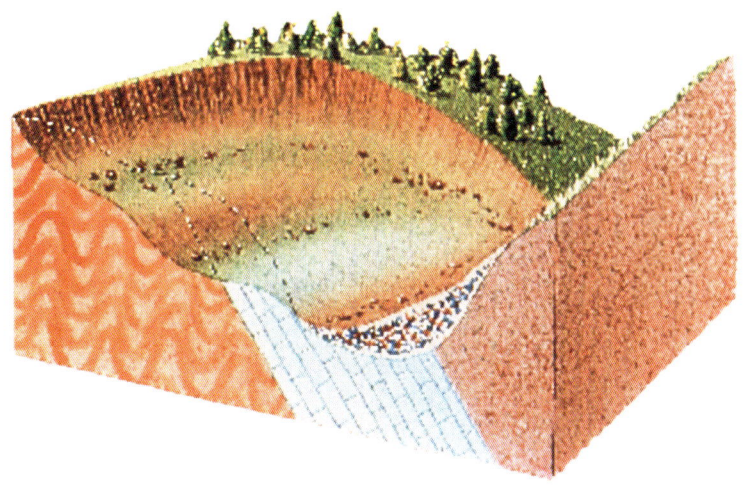

debris flow sped down the mountainside at speeds estimated at 200 miles (320 km) per hour, overrunning villages and towns in its path. The town of Yungay, along with its population of about 20,000 people, was buried within a matter of seconds. Today, all that remains of Yungay is the statue of Christ in the local cemetery, where the only survivors had fled to safety.

FIGURE 8.22 *A moving mixture of soil, regolith and fractured bedrock forms a* debris slide. *(Courtesy of P. Carrara, USGS)*

Mudflows Mudflows differ from debris flows in that they consist mainly of particles smaller than sand size and usually contain more water. Flowing like a highly viscous stream, mudflows usually follow existing stream channels. Because of their high viscosity, they are able to transport large objects such as boulders, cars, and houses. Mudflows are especially prevalent in arid and semiarid regions where slopes are largely unprotected by vegetation and on oversteepened slopes that have recently been deglaciated. Relatively fine-grained, unconsolidated materials covering the slopes are set into motion when heavy rainfall reduces cohesion and friction within the materials. Mudflow deposits are commonly found at the mouths of mountain canyons where they exit and spread out onto valley floors as part of a larger deposit called an alluvial fan (Figure 8.23).

The slopes of volcanoes are especially prone to the development of mudflows. During eruptions, the loose pyroclastic materials that typically cover the slopes of volcanoes may become saturated with water from torrential rains or with meltwater generated by the melting of ice and snow; the result is a mudflow called a **lahar**. Much of the damage from the 1980 eruption of Mount St. Helens resulted from lahars that moved down the Toutle River at speeds of more than 20 miles (30 km) per hour destroying two hundred homes that lay in their path (Figure 8.24). The lahars carried an estimated 65 million cubic yards (50 million m^3) of debris into the Columbia River. It might be pointed out that the largest recorded mudflows in the United States are both associated with other volcanoes in the Cascades, namely, Mount Rainier and Mount Shasta.

SPOT REVIEW

1. What evidence indicates that a hillside is undergoing creep?
2. What evidence may indicate that a portion of a hillside is about to experience a slump?
3. Why is the angle of repose so low for materials undergoing solifluction?
4. What may explain the exceptionally high speeds of some rock slides and avalanches?

ENVIRONMENTAL CONCERNS

Each year in the United States, the various processes of mass wasting account for nearly a hundred lives and more than $2 billion of damage. Although all exposed land surfaces are subject to mass wasting, destructive mass movements are usually restricted to regions of relatively great relief. Due to its particular combination of climate, rock type, topography, vegetation, and intrusive human activities, the area of southern California is especially notorious for problems involving slope instability.

Causes of Slope Instability

Regardless of where they occur or the type of slope failure, all processes of mass wasting result when the downslope force of gravity exceeds the

FIGURE 8.23 Mudflows *often emerge from the mouths of mountain canyons. (Courtesy of E. E. Brabb/USGS)*

FIGURE 8.24 *Mudflows associated with volcanic eruptions are called* lahars. *Lahars can be extremely destructive as evidence by the lahar that swept down the Toutle River following the eruption of Mount St. Helens in 1980. (Courtesy of MSH-C. D. Miller/USGS)*

resistance of movement presented by the friction and cohesion of the slope materials. Any factor or combination of factors that increases the downslope force of gravity or reduces friction and cohesion within the slope materials, thereby decreasing the stability of the slope, will promote movement. As we have seen, many factors are potentially involved, but two are of prime importance in nearly all mass wasting processes: (1) the **amount of water** contained within the slope materials and (2) the **presence or absence of vegetation**.

The Amount of Water As water infiltrates into slope materials, it promotes mass movement by (1) increasing the weight of the slope materials, (2) decreasing internal friction and cohesion within the materials, and (3) dissolving cementing agents.

Most landslides are the result of the accumulation of water within the slope materials. During periods of heavy rainfall, water may infiltrate and collect in the unconsolidated slope materials faster than it can permeate downward into the underlying bedrock.

The mass of accumulating water adds weight to the slope materials and simultaneously reduces friction and cohesion as increasing hydrostatic pressure separates the individual particles. When the downslope force of gravity exceeds the resistance to movement presented by friction and cohesion, the surface materials will move.

The Presence or Absence of Vegetation
The effect of vegetation is complex and depends on both the extent and kinds of plant cover. In general, the root system of a more or less continuous plant cover *promotes* slope *stability* by increasing the friction and cohesion of the soil. At the same time, however, continuous plant cover *promotes* slope *failure* by reducing surface runoff and enhancing infiltration.

In many cases, whether a plant cover will be effective in stabilizing a slope depends on the kind of plant. For example, due to its attractive appearance and ability to cover a slope totally within a relatively short time, the ice plant is commonly used for slope cover in warm climates such as California (Figure 8.25). Unfortunately, during periods of high rainfall, the ice plant tends to take up and store large volumes of water in its leaves, thereby adding significant weight to the slope materials. The added weight of the plant mass combined with its shallow root system often results in a slumping of the plant cover, exposing the slope materials to erosion.

FIGURE 8.25 Ground covers *are widely used to stabilize slopes. A plant commonly used in California is the ice plant. Although it is an attractive ground cover, slopes covered with ice plant are often denuded following heavy rains as the plants, heavy and swollen with absorbed water, shear from their roots at ground level.* (VU/© D. Long)

A continuous plant cover does not always afford the best protection against mass wasting. As pointed out, a continuous cover of grass tends to increase the rate of water infiltration, which, in turn, promotes slumping. On the other hand, slopes that are only partially covered with trees and bushes experience more direct runoff. Although some surface materials may be lost due to erosion, the reduction in water infiltration combined with the deeper penetration of tree roots promotes additional cohesion and friction within the remaining slope materials. As a result, slopes partially covered with trees and bushes generally experience a lower frequency of slope failures and have steeper angles of repose than those with a continuous grass cover.

Regardless of the type or extent of plant cover, most slope stability problems involving plants result when the plant cover is either disturbed or removed. An excellent example is clear-cutting (Figure 8.26). The lumber industry says clear-cutting is the most economical method of harvesting trees, but environmentalists condemn it as a prime cause of excessive slope erosion, slope failure, loss of wildlife habitat, and stream siltation. Although there is little question that clear-cutting is an efficient way to clear trees from a slope, evidence indicates that the total removal of trees increases the infiltration and accumulation of water in the slope materials. As the trees are removed, water that would have been returned to the atmosphere through transpiration accumulates in the slope materials. The increased water content of the slope materials, the increased angles of slope resulting from the construction of haulage roads, and the eventual elimination of the stabilizing effect of the tree roots as they decay all contribute to increased slope instability.

The Effects of Earth Moving
No topographic feature is too formidable for modern earth-moving equipment. Road cuts towering hundreds of feet above the roadway are testimony to the ability of modern engineering to literally move mountains. The removal of large volumes of rock in order to carve out the flat surfaces needed for the construction of buildings, parking lots, and roadways leads to increased slope angles and subsequent slope instability (Figure 8.27). The weight of a structure such as a building of a roadbed may further decrease the stability of the underlying materials. In the case of a home site, the drainage from a septic system

FIGURE 8.26 *Although clear-cutting is the most efficient way to harvest trees, the widespread removal of protective vegetative cover exposes steep slopes to increased rates of erosion and mass wasting. (VU/© Frank Hanna)*

FIGURE 8.27 *Earth-moving operations designed to provide large flat surfaces suitable for construction often result in the formation of steep slopes that have the potential to become the sites of future instability. (Courtesy of J. T. McGill/USGS)*

combined with the water infiltrated from lawns and gardens may add to the weight of the slope materials and lessen cohesion and friction. Subsequent failure of the slope, usually by slump, may partially or totally destroy any overlying structure.

Identification of Potentially Hazardous Sites

Features that indicate slope movement are usually easy to recognize. The development of a hummocky surface signals the beginning of a slump. Cracks in soil and rock surfaces parallel to a hillside indicate downslope movement (Figure 8.28). Cracks in foundations and walls are sure signs of earth movements. More subtle signs of slope movement are doors and windows that have become difficult to open and close because of the deformation of the jambs. We have already discussed the misalignment of fences and hedge rows and the tilting of posts, trees, and utility poles as evidence of creep. Another danger signal that may indicate the slope materials are becoming saturated is the seepage of water from the base of the slope (Figure 8.29). It certainly would behoove anyone considering purchasing property to examine the site thoroughly, looking specifically for evidence of earth movements, and to investigate whether slope stability maps are available for the area.

FIGURE 8.28 *The formation of* tension cracks *in the upper portion of slopes is a sign of future slope problems. (Courtesy of J. R. Stacy/USGS)*

FIGURE 8.29 *The seepage of water from the toe of a slope may be a danger sign. If the water is not artificially drained, the slope materials may begin to move a they become saturated, increase in mass, and experience an internal decrease in cohesion and friction. (Courtesy of J. T. McGill/USGS)*

Methods of Slope Stabilization

Once one or more of these telltale signs of slope movements have been identified, the pertinent question becomes, what can be done to avert future damage or possible disaster? Remembering that water is a key ingredient in most processes of mass wasting, any action that minimizes the infiltration of water into the slope is in order. A commonly used method of controlling water infiltration is the construction of *diversion drains* (Figure 8.30). In especially troublesome situations, infiltration can be eliminated by completely covering the slope with an impermeable layer of plastic or asphalt (Figure 8.31). Slopes that show evidence of water saturation can often be dewatered by drilling horizontal holes into the hillside and installing perforated tubing that will allow the water to drain.

A common procedure utilized in road construction to decrease slopes that have been oversteepened by excavation is *benching* with the uppermost bench protected by diversion drains (Figure 8.32). In many cases, the orientation of a road cut is determined by the geology of the area. In areas where the rocks have been deformed into folds or contain well-developed fractures or foliation, good engineering practice is to orient the road cut such that the planes of weakness do not slope into the excavation; otherwise, the surface of the road cut would be a prime candidate for rock slides (Figure 8.33).

Mass Wasting 225

FIGURE 8.30 Drainage ditches *are commonly used to divert water away from areas where infiltration may cause stability problems.* (VU/© W. Banaszewski)

FIGURE 8.31 *Various kinds of coverings can be used to stabilize slopes. In cases where infiltration is to be prevented, impervious materials such as plastic, asphalt, or cement are used. In other cases where controlled infiltration is desired and the need is to prevent the displacement of surface materials until a vegetative cover can become established, cloth can be used.* (VU/© Steven McCutcheon)

FIGURE 8.32 *Benching is used in nearly all road cuts. Note that the overall slope angle of the road cut is considerably less than the individual slopes between the benches. In this way, individual slope failures will be limited in size, the materials will collect on the benches, and in time, the entire exposure will attain a stable angle of repose.* (VU/© John D. Cunningham)

FIGURE 8.33 *Road cuts where the rock layers dip toward the roadbed will eventually experience rock slides that not only become an added expense for road maintenance crews, but pose a hazard to vehicles. (Courtesy of W. B. Hamilton/USGS)*

Slopes that are prone to failure can be stabilized by employing a number of construction techniques that increase the resistance of the slope materials to movement. Perhaps the most common is the **retaining wall.** In many cases, however, a retaining wall will be unable to resist the downslope force of gravity unless it employs **deadmen,** which take advantage of the weight of the slope materials to anchor the wall (Figure 8.34). Retaining walls should also be constructed with spaced openings to allow the drainage of water that might otherwise collect behind the wall.

Another common method of stabilizing slopes and reducing the effects of surface erosion by running water is the use of rock-filled wire baskets called **gabions** (Figure 8.35). Under certain conditions, where the high cost of installation can be justified, pilings can be driven down into bedrock, or the surface materials can be anchored to the underlying bedrock with bolts (Figure 8.36). Regardless of what preventive action is taken, one must always keep in mind that attempts to *stop* natural processes are likely to fail. The objective of most attempts to stabilize slopes is to *reduce* the *rate* of mass movement to a point where a degree of stability will be maintained for a reasonable length of time.

Subsidence

Subsidence is the collapse of the surface into an underground void. Both natural processes and human activities can lead to subsidence. An excellent example of natural subsidence is the creation of **sinkholes** in areas underlain by limestone. Layers of limestone can dissolve underground to form *caves* and *caverns*. During times of normal rainfall, water fills many of the suberranean caves and supports the weight of the overlying rock. During prolonged periods of drought, however, the water drains from the caves, thereby removing support for the cave roof; the roof collapses, and the overlying ground surface subsides to form a sinkhole. In a highly publicized example, a sinkhole formed in May 8, 1981, in Winter Park, Florida, causing more than $2 million of damages (Figure 8.37).

Another example of subsidence due to the removal of fluids is associated with the production of oil. The oil contained within the tiny pores of a reservoir supports the overlying rock layers in much the same fashion that water supports the roof of a water-filled cave. As the supporting oil is removed, the reservoir tends to collapse. Depending on the thickness of the reservoir and its proximity to the surface, significant subsidence may occur. A well documented case in point is the area around Long Beach, California (Figure 8.38). Oil production began in the area in the late 1920s. By the early 1940s, subsidence had caused near-shore areas to flood. By 1975, certain areas had subsided as much as 30 feet (9 m). Beginning in the late 1950s, efforts were made to reverse the trend by pumping water back into the reservoir as the oil was removed. In some cases, water injection not only stopped, but actually reversed the subsidence. Comparable subsidence problems have been experienced throughout extensive areas of the Great Valley of California, where large volumes of groundwater have been removed to provide water for domestic and agricultural needs.

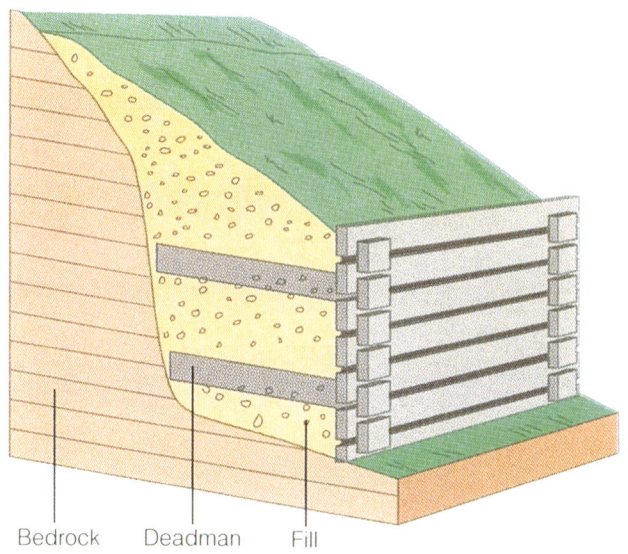

FIGURE 8.34 *Walls designed to stabilize slopes are constructed with components called* deadmen *that are fastened to the wall and extend into the fill behind the wall. The cohesion and friction between the fill material and the surface of the deadmen prevent them, and the wall to which they are attached, from moving. (b (VU/© John D. Cunningham)*

FIGURE 8.35 Gabions *are rock-filled wire baskets. Usually the size of a desk, they can be stacked or layered as necessary to provide additional protection to slopes. Typical applications are to prevent erosion along the cutbank of a meandering stream and to provide support and erosion protection for steep slopes. (a VU/© S. C. Reuman, b VU/© Kim Francisco, c VU/© W. Banazewski)*

FIGURE 8.36 *Another way to stabilize slopes is to drive steel pilings down and into bedrock. (Courtesy of J. T. McGill/USGS)*

FIGURE 8.37 Collapse sinkholes *such as this one in Winter Park, Florida are constant threats to landowners in areas underlain by limestones into which groundwater has carved subsurface caves and caverns. (© Phil Degginger)*

In regions of **underground mining**, subsidence due to the collapse of old workings can be a serious threat to surface structures. In many coal fields, subsidence has destroyed buildings, broken highway surfaces, and severed pipelines (Figure 8.39). Researchers in the field of mine subsidence are trying not only to predict the degree of surface subsidence from underground workings but also to devise methods of minimizing the resultant damage to surface structures.

In the case of deep coal mines, no hard-and-fast rules exist for predicting the impact of mine collapse on surface subsidence. The amount of subsidence depends on a number of interrelated factors including the type of mining operation, the thickness of the coal bed, the kind and thickness of the individual rock layers above the coal bed, and the depth to the top of the coal bed. In general, however, if the coal bed is deeper than 1,000 feet (300 m), surface subsidence will be negligible. Only when the coal bed is less than 1,000 feet deep will surface subsidence be significant. Serious damage from subsidence is likely to occur when the depth to the coal bed is less than 50 times the coal bed thickness.

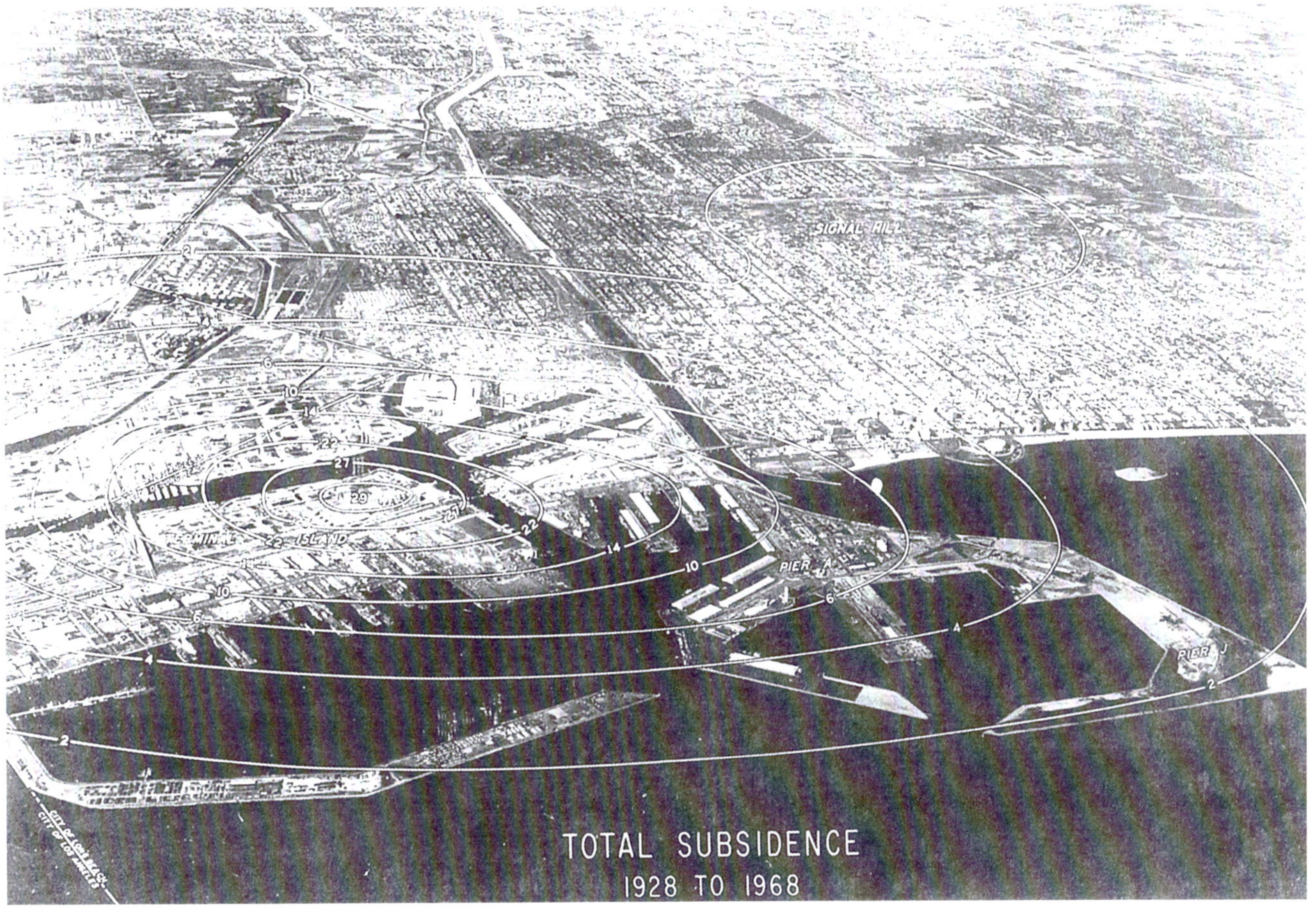

Mass Wasting 229

FIGURE 8.38 *The withdrawal of fluids from underground reservoirs, be they oil, gas, or water, may cause significant subsidence of the surface. While in place, these materials served as support for the overlying rocks. With their removal, the rocks are reduced in thickness as the empty pores and cavities collapse. The area around Long Beach, California experienced subsidence as a result of oil production.*

FIGURE 8.39 *The underground mining of coal leaves cavities that are eventually filled as the roof of the former mine collapses. The subsidence patterns that appear at the surface follow the old mine workings. Research is now underway to test the feasibility of preventing surface subsidence by filling abandoned underground workings with cement made from ash produced by coal-burning power plants. (Courtesy of C. R. Dunrud/USGS)*

SPOT REVIEW

1. How may the presence or absence of vegetative cover affect the stability of a slope?
2. What can be done to stabilize road cuts and other steep slopes?
3. What is the most important single factor that must be controlled in order to stabilize most processes of mass wasting?

In summary, every exposed land surface of Earth is undergoing mass wasting. Mass wasting is responsible for the retreat of slopes, the reduction of topographic relief, and the ultimate elimination of divides separating stream valleys. Although the work of streams and glaciers is spectacular and is responsible for moving weathered materials over long distances, we should not overestimate their significance. Even if no other processes existed, mass wasting would be sufficient to reduce the grandest of land surfaces to the level of the sea.

CONCEPTS AND TERMS TO REMEMBER

Mass wasting
Factors affecting slope stability
  gravity
  cohesion
  friction
  angle of repose
Kinds of mass wasting processes
  creep
  earthflow
    solifluction
  slump

rock falls
  talus
rock slides
  rock avalanche
  debris avalanche
  debris slides
  debris flows
  mudflows
  lahar
Environmental concerns
  amount of water

presence or absence of vegetation
earth moving
Slope stabilization
  retaining walls
  deadmen
  gabions
Surface subsidence
  sinkholes
  underground mining

REVIEW QUESTIONS

1. The mass wasting process that is particularly effective in permafrost regions is
   a. creep.
   b. solifluction.
   c. rock glacier.
   d. slump.
2. Frost heaving is a major agent involved in
   a. solifluction.
   b. creep
   c. rock glaciers.
   d. slump.
3. The energy for all processes of mass wasting comes from
   a. shocks such as earthquakes.
   b. gravity.
   c. solar heat.
   d. geothermal heat.
4. Which of the following particles would exhibit the highest angle of repose?
   a. small, spherical
   b. large, spherical
   c. small, irregular
   d. large, irregular
5. The most rapid form of mass wasting is
   a. rock fall.
   b. slump.
   c. debris slide.
   d. solifluction.
6. For most loose materials, the maximum angle of repose is about
   a. 10°.
   b. 20°.
   c. 40°.
   d. 60°.
7. Within the regolith, cohesion and friction
   a. decrease as the angle of slope increases.
   b. remain constant regardless of the angle of slope.

c. increase with the degree of water saturation.
   d. increase as the angle of slope increases.
8. Why is the angle of repose for irregularly shaped particles higher than for more spherical particles?
9. What is required for materials accumulated at less than the angle of repose to begin to move downslope?
10. What, if any, is the justification for the statement that mass wasting alone would eventually reduce the elevation of the land to sea level?
11. In how many different mass wasting processes is the intervention of ice essential?
12. Why can the process of solifluction operate on slopes where the angle of repose is so low that it can hardly be perceived?
13. Why are slopes in arid regions generally steeper than slopes in humid regions?

THOUGHT PROBLEMS

1. Assume that you are a homeowner. Over a period of a few months, you notice that certain windows are becoming harder to open and close, and you observe several cracks developing in your basement walls. There is little doubt that your home is the victim of earth movements. Assume several different scenarios for the origin of the problem, and perform the following for each scenario:
   a. Outline how you would go about determining the cause of the earth movement.
   b. Explain how you would evaluate the present severity of the problem and its potential to become more serious.
   c. Describe the steps you would take to arrest the movement and thereby save your home from more damage.
   d. Finally, determine the point at which you would decide that the problem cannot be solved and would abandon the structure.
2. Assume that you are a new resident of an area that has a history of landslides. You are in the market to purchase property on which you plan to construct a home. What would you do to evaluate a specific site's potential for future earth movements?
   Assume that you are considering purchasing an existing house rather than constructing a new one. What steps would you take to evaluate not only potential problems but also any effects the structure may already have experienced?

FOR YOUR NOTEBOOK

One or more field trips are certainly in order. In more urbanized areas, look for attempts to control slope stability including retaining walls and various ground covers. Make a survey describing the different procedures used as well as their apparent success (or lack thereof) at accomplishing their objectives.

In more rural areas, you will be able to study mass wasting processes that have been allowed to proceed uninhibited as well as some instances of control. Note the structure of road cuts and perhaps attempts at controlling the banks of streams, cliffs, and ravines. Look for evidence of creep, earthflow, and slumps. This is an excellent chapter to document with photos.

CHAPTER OUTLINE

INTRODUCTION
THE HYDROLOGIC CYCLE
DRAINAGE SYSTEMS
STREAM ORDER
STREAM PATTERNS
THE ENERGY OF A STREAM
    Gradient
    Channel Texture
    Water Volume
THE DYNAMICS OF WATER FLOW
THE PROCESSES OF EROSION AND TRANSPORTATION
    The Generation of Force
    The Effect of Particle Size and Shape
    Capacity and Load
THE DEVELOPMENT OF TOPOGRAPHIC RELIEF
    Baselevel
    Fluvial Landforms
    Meandering
THE DEVELOPMENT OF GRADE
    Climatic Influences
    Tectonic Influences
    Human Influences
ALLUVIAL DEPOSITS
    Channel Deposits
    Floodplain Deposits
    Deltas
    The Mississippi Delta
    Alluvial Fans
ENVIRONMENTAL CONCERNS
    Floods
    Water Pollution

CHAPTER 9

Streams

INTRODUCTION Approximately 30% of Earth's surface is land, created by mountain building, broad uplift of the continental crust, or changes in sea level. As soon as the land forms, forces of erosion begin to wear it away or dissolve it and return the material to the sea. We have already studied (1) *weathering,* which tears the rocks apart physically and decomposes them chemically in preparation for the journey, and (2) *mass wasting,* which begins the downhill movement of the rock debris. Once mass wasting has transported the material to the valley floor, other agents of erosion including water, ice, and wind carry the materials off. This chapter deals with the most important of those agents, namely, running water or **streams.**

The appearance of the land at any given moment represents an intermediate stage between the topography of the newly exposed land and the ultimate goal of erosion, a land surface at the level of the sea. When we think of lofty mountains towering thousands of feet above sea level, such as a task seems impossible. But no mountain is so grand nor any rock so durable that it can long withstand the continuous attack of erosion, in particular, the relentless attack of running water.

THE HYDROLOGIC CYCLE

Earth is unique among the terrestrial planets (the four inner Earth-like planets—Mercury, Venus, Earth, and Mars) in possessing a **hydrosphere** (Greek *hudor* = water + sphere). Included in the hydrosphere are (1) bodies of surface water such as oceans, lakes, ponds, and streams; (2) surface accumulations of snow and glacial ice; (3) all water, ice, or water vapor below Earth's surface; and (4) all forms of water in the atmosphere.

Most of the water (97%) is contained within the ocean basins, which cover about 70% of Earth's surface. The oceans are both the source and the ultimate repository of the water that is continuously recycled in the **hydrologic cycle** (Figure 9.1). In this cycle, which is driven by the energy of the Sun, water evaporates from the oceans and enters the atmosphere where it is carried by the prevailing winds in more or less fixed patterns over Earth's surface. As air masses in the lower atmosphere cool, the water vapor condenses and falls as precipitation. Most precipitation falls directly back into the ocean basins to complete the cycle. Some, however, falls over land as either rain or snow.

Water vapor that precipitates as snow may accumulate temporarily in snow fields or becomes glacial ice. Approximately 2% of Earth's water—about 80% of all the water outside the ocean basins—exists as glacial ice. Most glacial ice is found in the ice caps and ice sheets over the continent of Antarctica and the island of Greenland. In time, all snow and ice will melt, and the meltwater will return to the sea to complete the cycle.

Of the rain that falls on land, 70% returns to the atmosphere either through direct evaporation or by way of plants, which *transpire* water into the atmosphere through their leaves. The remaining water either runs off as *streams,* accumulates temporarily in *lakes* or *ponds,* or infiltrates into the ground to become *groundwater.*

Approximately 0.7% of the water outside the ocean basins is contained in lakes (0.02% of the total). Groundwater accounts for 20% of all the water outside the oceans (0.5% of the total volume). Of all the water outside the ocean basins, the smallest volume is contained in streams, which account for only 0.005% of the water outside the oceans and only 0.0001% of all the water on Earth. In fact, the atmosphere contains approximately as much water as is in all the

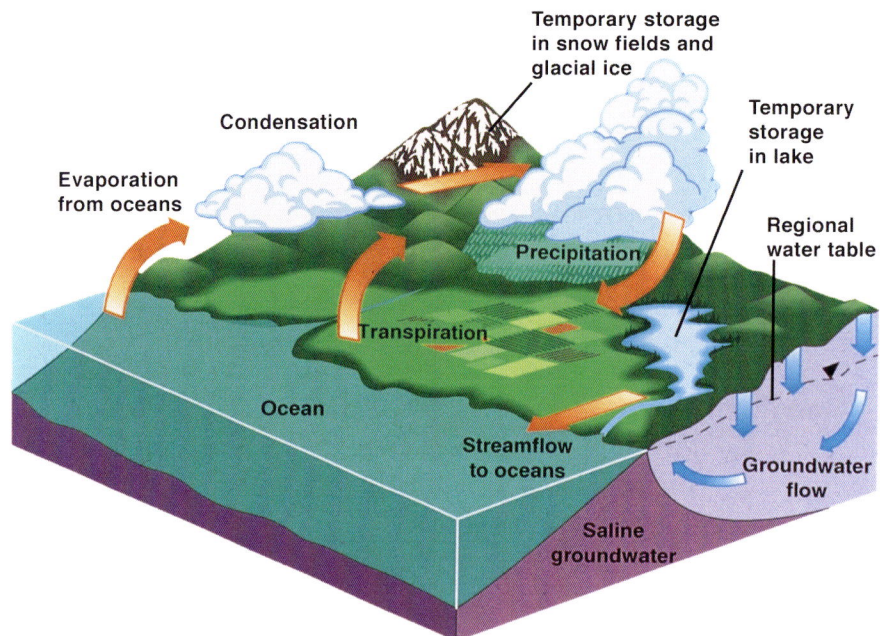

FIGURE 9.1 *The* hydrologic cycle *describes the movement of water on Earth. The source of all water is the* ocean. *Once deposited on land, the water may be stored temporarily in* glaciers *or in* impondments. *Ultimately, all water deposited on land returns to the ocean either overland by way of* streams *or underground as* groundwater.

streams together. Nevertheless, despite these seemingly unimpressive statistics, streams play a major role in the formation of nearly all Earth's landforms, a testament to their incredible power of erosion.

DRAINAGE SYSTEMS

Every stream occupies a position within a specific area of Earth's surface called a **drainage basin**. Smaller streams join larger streams as *tributaries,* and their combined drainage basins form a larger drainage area called a **drainage system** (Figure 9.2). In the most common case, the streams within a drainage system form a pattern that resembles the veins in a leaf where a central master vein splits into progressively smaller veins that reach to the most distant extremities of the leaf.

Drainage systems may be either (1) *external drainage* or (2) *internal drainage* systems. In an **external drainage system**, the water eventually flows to the ocean. Most streams belong to the external drainage systems. An **internal drainage system**, on the other hand, terminates in some inland basin. Internal drainage systems are more common in regions where the amount of rainfall is not sufficient to maintain stream flow for long distances. Internal streams, therefore, are usually found in deserts where they flow into and terminate in an interior basin.

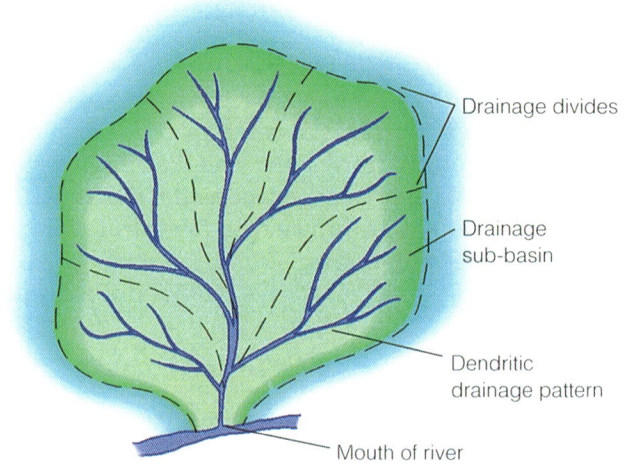

FIGURE 9.2 *Every stream is a member of a* drainage basin *that combines with other drainage basins to form a larger drainage area called a* drainage system.

Individual drainage basins and drainage systems are separated from adjacent drainage basins and systems by **drainage divides** or **interfluves** (Figure 9.3). The drainage divide connects the points of highest elevation between two drainage basins or systems. Whether it is the crest of a high mountain ridge or of a low hill barely perceptible to the human eye, the divide is the line where a drop of falling rain will run either to one side or the other.

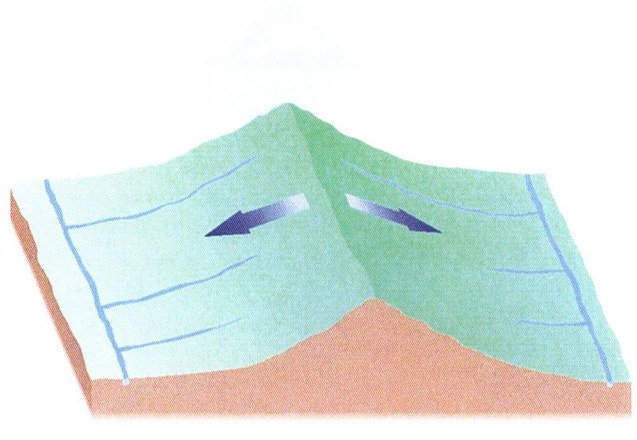

FIGURE 9.3 *Precipitation falling onto the surface is diverted into adjacent drainage basins by* drainage divides *or* interfluves.

Drainage divides exist at all scales from the watershed between two tiny adjacent rills flowing down a roadside bank to the **continental divides**; the latter separate the major drainage systems of a continent, which often flow into different oceans (Figure 9.4). North America contains three continental divides. The most prominent divide runs along the highest elevations of the Rocky Mountains and separates the streams that flow westward to the Pacific Ocean from those that flow southeastward to the Gulf of Mexico or northeastward to the Arctic Ocean. The second divide runs parallel to the Appalachian Mountains just west of the eastern edge of the Appalachian Plateau and separates the streams that flow westward to the Mississippi River and

FIGURE 9.4 *Within the area of any continent, there are a few major divides, called* continental divides, *that determine the direction with which water is shed from the continent to the ocean.*

subsequently to the Gulf of Mexico from those that flow eastward to the Atlantic Ocean. The third divide runs east-west just south of the Saint Lawrence Seaway and the Great Lakes, crossing the continent roughly along the border between Canada and the United States. This divide separates the waters that flow southward into the Gulf of Mexico via the Mississippi River from those that flow north and northeast to the Atlantic and Arctic oceans and Hudson Bay.

STREAM ORDER

Within a drainage system, each stream is identified by a **stream order** (Figure 9.5). A first-order stream (order = 1) is a stream that has no tributaries. First-order streams are the most headward streams in a drainage basin and would be represented by the most distal veins in a leaf pattern. Second-order streams have only first-order streams as tributaries. Third-order streams have second-order streams as tributaries and so forth.

Many of the basic stream parameters change systematically with increasing stream order. For example, within any drainage system, the streams that extend the drainage area of the basin in an upstream direction by *headward erosion* are first-order streams (Figure 9.6). In some cases, headward erosion by one stream may eventually intersect the channel of an adjacent stream and *capture* the more headward water flow of the intersected stream, a process called *stream piracy*.

Within a drainage system, the first-order streams have the steepest channel slopes, are the most turbulent, and carry the smallest volumes of water. As the stream order increases, the stream load increases, and channel slope and turbulence usually decrease. At the same time, the volume and velocity of the water increase as the water flows through progressively larger channels with smoother channel bottoms.

The appearance of the stream valleys also changes systematically as stream order increases. First-order streams typically occupy steep-sided valleys with relatively straight stream segments dominated by waterfalls and rapids (Figure 9.7). With increasing stream order, the valleys become wider and develop floodplains, the streams begin to flow along increasingly sinuous courses, and features such as waterfalls and rapids disappear (Figure 9.8). Other stream parameters also undergo systematic changes, as will be discussed later in the chapter.

The order assigned to the drainage system is the order of the main stream. For example, the Mississippi River system is tenth order while Earth's largest stream, the Amazon River system, is thirteenth order.

STREAM PATTERNS

When viewed from above, each stream system assumes a pattern largely determined by the character of the underlying rocks. The most common stream pattern, the **dendritic** (Greek *dendroeides* = tree) **stream pattern,** typically develops where the

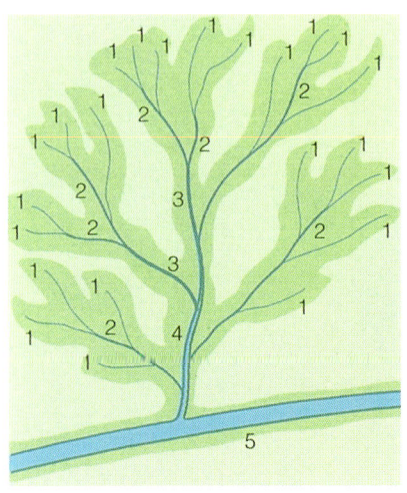
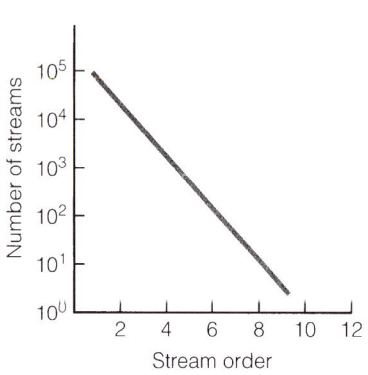

FIGURE 9.5 *Within a drainage system, the individual stream segments are assigned an* order. *Because stream order increases from the headwaters to the mouth, low-order streams are exponentially more numerous than streams of higher orders.*

FIGURE 9.6 *The area of a stream drainage basin is increased by the headward erosion of the first-order streams. (Courtesy of Joe Donovan)*

FIGURE 9.7 *First-order streams invariably occupy steep-sided valleys, flow for relatively long distances along straight segments, and typically are highly turbulent as the stream flows over rapids and waterfalls. (Courtesy of W. B. Hamilton/USGS)*

FIGURE 9.8 *Streams of higher orders flow in wider valleys with lower slopes, are less turbulent as rapids and waterfalls disappear, and begin to flow in a sinuous pattern called* meandering. *(Courtesy of W. B. Hamilton/USGS)*

underlying rocks lie flat and are fairly uniform in the composition and texture, such as horizontal sandstones and shales (Figure 9.9), although the pattern may also develop in some areas underlain by batholithic intrusions. It is also worth noting that dendritic patterns commonly exist in the lower-order streams of other patterns. For example, the first- and second-order streams of trellis patterns (discussed below) commonly are dendritic. When we likened a stream system to the pattern of veins in a leaf, we were referring to the dendritic pattern.

Where the underlying rocks possess a well-developed rectangular structure pattern, especially one that has been accentuated by weathering, the streams tend to follow the underlying structure and assume a **rectangular stream pattern** (Figure 9.10). Rectangular stream patterns are commonly found in areas underlain by horizontal layers of limestone where fracture patterns have been opened by solution and in areas underlain by granite and metamorphic rocks such as in the Canadian Shield and the Adirondack Mountains of New York.

When rocks have been folded into long, parallel ridges and valleys such as in the Appalachian Mountains, the characteristic pattern that develops is the **trellis stream pattern** (Figure 9.11). In the trellis pattern, low-order streams drain the slopes of the resistant rock ridges and flow into higher-order streams that flow parallel to the ridges within the valleys. These valley streams in turn become tributaries to major streams that cut across the ridges through **watergaps**, not uncommonly following major fracture zones and faults (Figure 9.12).

In areas of domal uplift such as the Black Hills of South Dakota and on conical structures such as volcanoes, the land surface is drained by a combination of the **radial** and **annular** (Latin *annularis* = ring) **stream patterns** (Figure 9.13). The radial streams drain off the slopes of the structures in a pattern resembling the spokes of a wheel and then become tributaries of the annular streams that flow around the structure like the rim of the wheel. The water then drains off and joins the major regional stream pattern.

In areas characterized by a uniform, regional slope and underlain by a relatively homogeneous lithology (a small-scale example being a newly exposed road cut), streams and their tributaries follow parallel to subparallel paths spaced more or less equally (Figure 9.14). The pattern is called **parallel drainage**.

 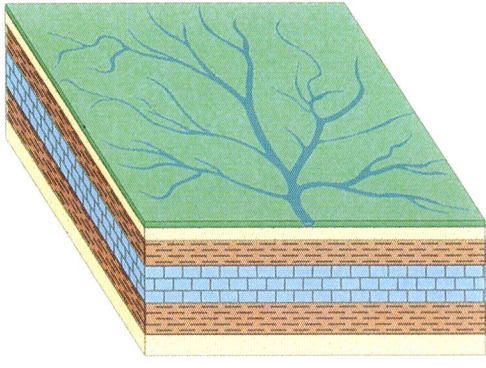

FIGURE 9.9 *Whenever streams flow across surfaces underlain by flat-lying, relatively homogeneous rocks, they develop a* dendritic *pattern that resembles the veins in a leaf.*

 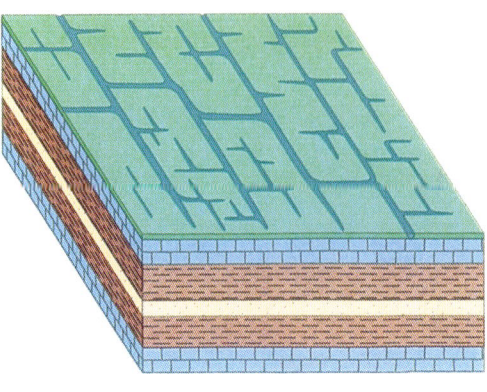

FIGURE 9.10 *When the rocks underlying the surface have a well-developed rectangular fracture pattern, such as the joints in horizontally bedded limestones, a* rectangular *stream pattern develops.*

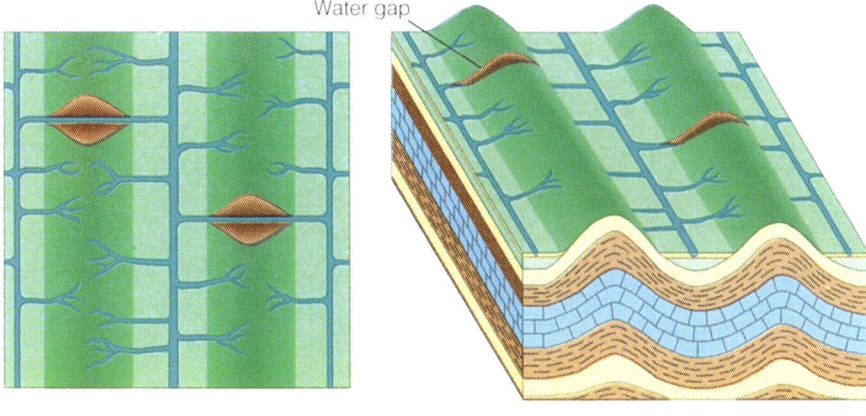

FIGURE 9.11 *When rocks are deformed into long parallel folds, a* trellis *stream pattern develops as higher-order streams flow through the valleys and are joined through watergaps that they have carved through the ridges. Greenland Gap is a watergap through Wills Mountain Anticline, the most westerly of the ridges of the Valley and Ridge Province of the Appalachian Mountains. (Photo courtesy of Peter Lessing). Note that the low-order streams draining the flanks of the ridges commonly have a dendritic pattern.*

FIGURE 9.12 *Streams cross topographic barriers by flowing through* watergaps *that they carved as the original surface was lowered by erosion. The Potomac River cuts through the Blue Ridge Mountains at the water gap at Harper's Ferry, West Virginia. (Courtesy of West Virginia Geological and Economic Survey, WVEGS)*

 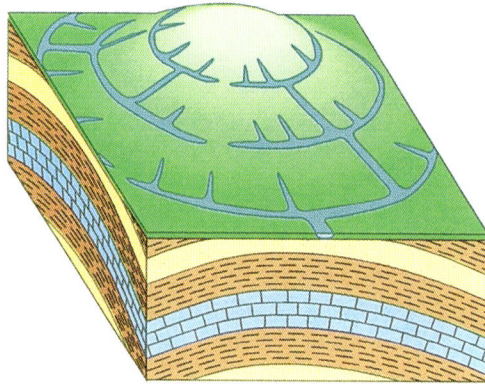

FIGURE 9.13 *A combination of* annular *and* radial *stream patterns develops on dome structures.*

 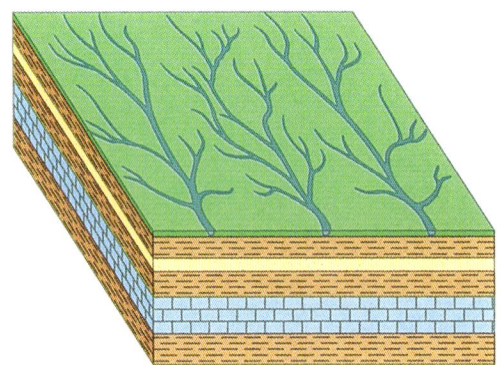

FIGURE 9.14 *When streams flow across relatively broad surfaces with uniform slopes underlain by homogeneous rocks, the individual streams commonly flow along parallel paths and form a* parallel stream pattern.

In areas where the underground dissolution of limestones has resulted in sinkholes and underground drainage, two characteristic drainage patterns are found. **Reversed radial** streams drain the interior of larger sinkholes (Figure 9.15), while the water of **disappearing streams** is diverted into subsurface channels (Figure 9.16).

Where streams heavily laden with sediment emerge from steep mountain valleys onto valley floors and experience a rapid change in gradient, a multichanneled stream called a **braided stream** commonly forms (Figure 9.17). As the ability of the stream to transport sediment rapidly decreases and the sediment load becomes too much for a single channel to carry, many individual channels develop and intertwine in an attempt to transport the load. Braided streams also are commonly associated with the drainage in front of melting glaciers where the volume of water is insufficient to transport the amount of sediment being introduced from the melting ice.

As this brief discussion indicates, geologists can acquire a great deal of information about the subsurface geology of an area by observing the stream drainage patterns.

SPOT REVIEW

1. Where does most of Earth's fresh water reside?
2. What is the difference between external and internal drainage?
3. In what ways do stream patterns reflect the subsurface geology of an area?

THE ENERGY OF A STREAM

The amount of energy available to a stream is proportional to the stream **discharge**, which is the product of two parameters: (1) the *cross-sectional area* of the wetted portion of the stream channel and (2) the *velocity* of the water.

With few exceptions, the cross-sectional area of the wetted portion of the stream channel increases downstream as tributaries add water to the main stream. The volume also changes seasonally with variations in rainfall and, in some areas, the melting of snow and ice.

 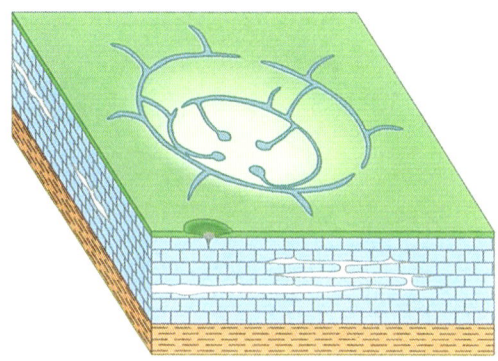

FIGURE 9.15 Areas underlain by limestone commonly develop surface features called sinkholes. The sinkholes are drained by reversed radial patterns. The water usually drains to the center of the sinkhole and disappears underground.

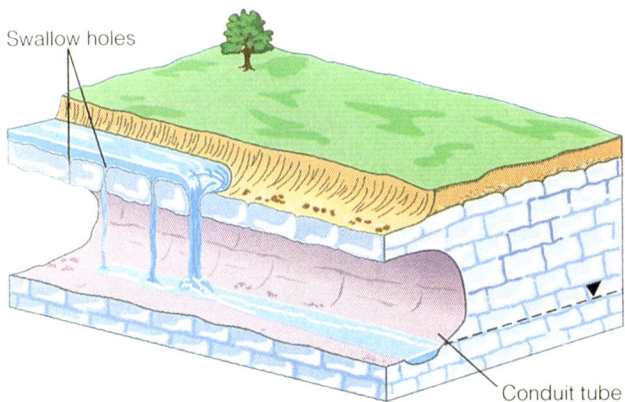

FIGURE 9.16 Disappearing streams are relatively common in areas underlain by limestone; the diverted water usually returns to the surface at some point downstream as the underground drainage terminates. The Gandy River disappears as it is temporarily diverted underground into a solution passageway in the Greenbrier Limestone at Sinks of Gandy, Pendleton County, West Virginia. (Photo courtesy of Allen B. Judy)

FIGURE 9.17 When exceptionally high loads are introduced into a stream, the main channel may break into a number of smaller channels in order to better transport the load. The result is a braided stream. (VU/© Ron Spomer)

The velocity of water generally increases downstream as the result of the combined influence of three factors: (1) the *gradient* of the stream channel; (2) the *width*, *depth*, and *roughness* of the channel; and (3) the *volume* of water.

Gradient

The **gradient** of a stream is a measure of the *slope* of the stream channel. Measured in "feet per mile" or "meters per kilometer," gradient is the drop in elevation over a horizontal distance (Figure 9.18). The gradient may change along the length of the stream, steepening over waterfalls and through rapids and decreasing in between. Averaged over the length of the channel, however, a typical stream gradient profile is a concave upward curve that steepens toward the *headwaters* and, ideally, approaches zero at the *mouth* (Figure 9.19). All other things being equal, water velocity will increase as the gradient steepens because of the increased downslope component of the force of gravity (see the discussion of slope stability on page 254 in Chapter 8). As we will see, however, "all other things" are not equal.

Channel Texture

The **texture** of the channel refers to the smoothness of the channel surface. The smoother the channel, the less resistance to flow and, therefore, the higher the velocity. When larger particles are introduced, coarsening the texture of the channel surface, the forward motion of the water is retarded as it is forced over and around the rock particles. In general, a channel becomes smoother downstream where the channel bottom is covered with smaller particles.

Water Volume

Velocity increases as the **water volume** rises. The increase in velocity is due to the increase in the downslope component of the force of gravity resulting from the larger mass of water. Except in streams where water is diverted to groundwater, the water volume increases downstream as tributaries enter the main stream. As a result, the velocity of the water increases downstream even though the gradient is decreasing.

Until researchers proved that velocity increases downstream, it was commonly thought that water velocity was highest in the high-gradient headwaters. For example, a steep, turbulent mountain stream certainly seems to be flowing faster than a lower-gradient

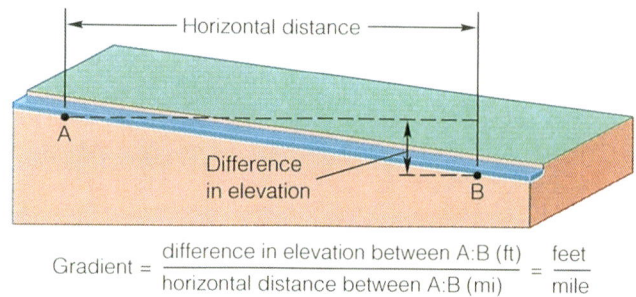

FIGURE 9.18 *One of the important stream parameters, the* gradient, *is the number of feet (or meters) the channel drops over a horizontal distance of one mile (or kilometer).*

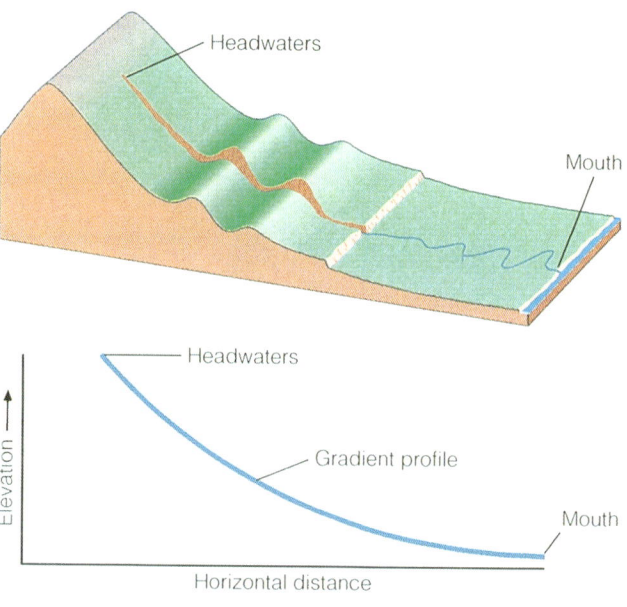

FIGURE 9.19 *Downstream from the headwaters, the gradient changes constantly, increasing over a waterfall or rapids and decreasing in between. Overall, however, the gradient decreases in a nonlinear fashion, reaching a minimum at the mouth of the stream.*

downstream segment. This seemingly anomalous situation may be a matter of perception. Undoubtedly, at certain places where the water free-falls over and around large rocks, a mountain stream attains high velocities because of the steep gradient. But consider the *average* forward motion of the entire mass of water. For each free-fall situation, there are places where the water encounters rocks and is thrown *upstream*, *vertically*, or *sideways* from the points of impact, all of which detract from the downstream movement. Thus, the overall forward motion of a turbulent mountain stream is not as fast as in segments downstream that have lower gradients but greater volume and smoother channels.

THE DYNAMICS OF WATER FLOW

The energy of a stream is used to accomplish two major tasks: (1) the *erosion* and *transportation* of the solid products of weathering, collectively referred to as *sediment,* and (2) the formation of a valley by the *downward* and *lateral carving* of the stream channel. Channel carving begins only after the primary tasks of sediment erosion and transportation have been accomplished. To understand how a stream performs the task of sediment erosion and transport, we must first consider the basic dynamics of water flow.

Within a stream, water may flow with two styles: (1) *laminar flow* and (2) *turbulent flow*. In **laminar flow,** energy is used to move individual water molecules with relatively constant velocities along parallel paths (Figure 9.20a). In **turbulent flow,** the energy is used to move the water molecules in all directions with constantly fluctuating velocities (Figure 9.20b.)

Generally speaking, the bulk of water in streams moves by turbulent flow. In most cases, irregularities within the stream channel caused by rock particles preclude laminar flow, especially when the particles are relatively large. Therefore, water flow in streams is generally turbulent with the degree of turbulence increasing toward the channel bottom and sides. The exception to this generality is a thin layer at the point of contact between the water and the channel surface. Called the **laminar** or **viscous sublayer,** the layer is an estimated 0.5 to 10 mm thick depending upon the water velocity and is considered to be the only place in a stream where laminar-like flow is attainable. The importance of this layer will be discussed later.

Except for the laminar sublayer, the water becomes more turbulent toward the channel surface and as the size of the sediment particles increases. For this reason, headwater segments of streams containing large rock particles are invariably more turbulent than downstream segments where the channel bottom is covered with finer particles such as sand and clay. The turbulence of all streams increases with higher velocity, explaining why streams are more turbulent during floods.

The amount of energy available to erode and transport sediment is determined by how much

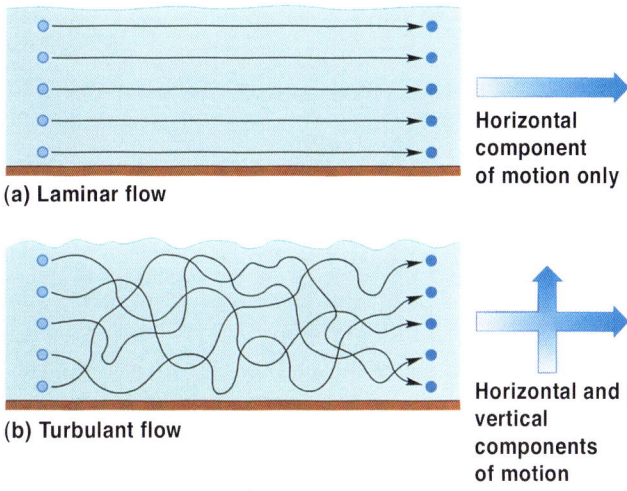

FIGURE 9.20 *Fluids may flow by either* laminar flow *or* turbulent flow. *Because of the water's low viscosity, the flow in streams is* turbulent *except for movement within the laminar layer.*

energy is consumed internally by various forces that resist flow. In water masses moving by laminar flow, the main consumer of energy is the internal friction between individual water molecules or layers of water, which increases proportionally as the velocity rises.

The greatest consumer of energy, however, is turbulence. Energy is consumed as the water constantly eddies, changing both velocity and direction. Resistance to flow due to turbulence and the subsequent consumption of energy increase proportionally to the square of the velocity. As a result, turbulence not only consumes more energy than laminar flow but does so at a much higher rate. Because most water in streams moves by turbulent flow, most of the original energy provided by stream discharge is consumed and is therefore unavailable for erosion and transportation. The small portion of the original energy that remains, however, performs an impressive task of landscape development.

THE PROCESSES OF EROSION AND TRANSPORTATION

The processes by which sediment is eroded and transported are very complex and at best poorly understood, largely because of the difficulty of obtaining meaningful measurements in either laboratory or actual field situations.

The mass of water can move particles along a stream by lifting them from the bottom and carrying them for various distances or by rolling or pushing them along the bottom. The force necessary to lift, roll or push the particles comes from the energy provided by the stream discharge.

The Generation of Force

Force is generated by a reduction in the velocity of a mass. Consider holding your hand in a moving stream of water or in the path of an oncoming baseball. As the water or the baseball encounters your hand and is slowed or redirected, a force is generated that pushes your hand in the direction in which the water or the baseball was moving.

Direct Lifting
As the water impacts a particle and is slowed or redirected, the reduction in velocity generates a force. The force that lifts a particle is generated when an *upward-moving* water mass is slowed. If the upward force is greater than the force of gravity, the particle will be lifted. Once the particle is lifted from the bottom, it is moved downstream by the horizontal force generated when the mass of water moving *downstream* is slowed. Because turbulence declines away from the channel surface, the amount of the force that can be developed by slowing the upward-moving water decreases away from the stream bottom. The decrease in the magnitude of the upward force, coupled with the ever-present force of gravity and the presence of downward-moving masses of water, will eventually return most particles to the bottom.

A particle can also be lifted into suspension when the water velocity increases across the top of the particle, thereby reducing the pressure. When the pressure on the top is lowered, the particle may be lifted by the same mechanism that is responsible for the lift developed under the wing of an airplane.

Traction
Particles too large to be lifted may be moved by a process called **traction** in which a particle is rolled, slid, dragged, pushed, or bounced along the bottom. The force for traction is generated by the reduction in the *horizontal* velocity of the flow. For a given stream or stream segment, the potential magnitude of the horizontal force is limited by the stream velocity. The largest particle that a stream can move by traction determines the *competence* of the stream. Competence always increases during floods because the increased discharge results in more energy being made available for both the horizontal and vertical velocity components of turbulence.

The Effect of Particle Size and Shape

The ability of a stream to erode, transport, and deposit sediment is summarized in Figure 9.21. The lower curve in the diagram represents the points for each particle size where the force acting to lift the particle and the force of gravity are equal. Below the curve, immobile particles will remain immobile, and moving particles will settle to the bottom and become immobile (*deposition*). The upper curve represents the conditions required to lift each of the various particle sizes (*erosion*). Note that in all cases, a greater upward force is needed to lift the particle than is required to overcome the force of gravity. The reason for the discrepancy depends upon the particle size. For sand-sized particles and larger, the additional energy is needed because the particles tend to shield each other on the stream bottom. Higher water velocities are necessary to increase the turbulence and make the upward forces between the larger particles more effective. Clay- and silt-sized particles have much small diameters and can reside completely within the laminar sublayer where turbulence is much lower and few vertical forces are developed to lift them. To lift these particles, both velocity and turbulence must be increased in order to reduce the thickness of the laminar sublayer.

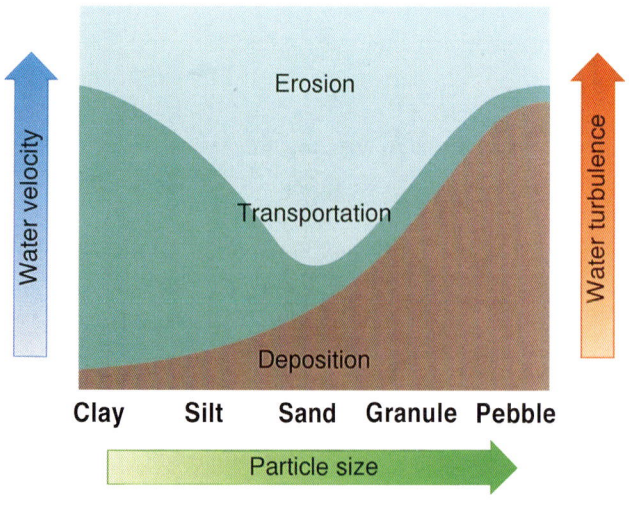

FIGURE 9.21 *The erosion of particles by streams is dependent on particle size and shape and the velocity and turbulence of the water. Sand is the easiest particle size to move; movement of sizes both larger and smaller requires higher velocities and turbulence. Once particles are transported, however, they are deposited in order of decreasing size as velocity and turbulence decrease.*

The three-dimensional shapes of the grains also influence the ease with which they will be lifted. For example, the platelike clay minerals tend to adhere to the stream bottom. Another factor inhibiting the erosion of silt- and clay-sized particles is the attraction they have for one another resulting from both chemical and electrical forces.

Observations indicate that the easiest particle size for streams to pick up and move is sand. The forces required to dislodge very small particles are comparable to those needed to lift much larger particles. Clay- and silt-sized particles are so small, however, that once lifted from the bottom, they will remain in suspension even though the velocity of the water drops to very low levels. This explains why streams stay muddy so long after a flood.

Once the velocity requirements represented by the upper curve in Figure 9.21 are exceeded and the individual particles are moving, the velocity can be dropped to the lower curve, and the particles will continue to be moved (*transportation*). Only when the conditions fall below the lower curve will deposition occur, beginning with the largest particles and continuing to the smallest.

Capacity and Load

The **capacity** of a stream is the total amount of material that the stream *can* carry and depends upon the discharge (total amount of available energy). The **load** of a stream is the amount of material that the stream *is* carrying. Although a stream's load may approach its capacity, the load is unlikely to ever equal the capacity except during deposition.

Because they depend upon discharge, both load and capacity change along the length of a stream and with time. Capacity and load generally increase downstream with the maximum for both being achieved during floods. As a flood ebbs and the discharge declines, some of the load will be deposited as the velocity decreases.

Given sufficient time, a balance will be established between capacity and load; at that time the stream is said to be at **grade**. When a stream has achieved grade, all stream parameters including gradient, channel size, channel shape, and discharge are in balance, providing the exact amount of energy needed to transport the load. In order for the stream to maintain constant load, erosion and deposition also come into balance with erosion at one point in the channel being offset by deposition nearby. We will return to the topic of grade later in the chapter when we describe the development of stream valleys.

Kinds of Load A stream acquires its load through three mechanisms: (1) *dissolution,* (2) *direct lifting,* and (3) *traction.* The load is then carried in three forms: (1) *dissolved load,* (2) *suspended load,* and (3) *bed load.*

Dissolved Load The **dissolved load** is acquired by the process of dissolution and contains primarily the soluble products of chemical weathering generated within the drainage basin. Most of the dissolved load is provided by groundwater, which is able to dissolve great volumes of material due to its long contact with the rocks. Streams that flow on carbonate bedrock such as limestone acquire additional dissolved load by dissolving the rock in the stream channel. The mass of material carried in solution is usually significantly less than that carried in either suspended or bed load, but it is still responsible for billions of tons of material being carried to the ocean each year.

The dissolved load determines the *quality* of the water. Whether the water quality is *good* or *bad* is very subjective and depends upon the water's intended use; specific demands or limitations are placed on the dissolved ingredients according to the ultimate use. Water intended for human consumption will have a different set of compositional constraints than water to be used for livestock or for irrigation. Unfortunately, many of the by-products of human activities entering streams as part of the dissolved load pollute this very important natural resource.

Suspended Load **Suspended load** is acquired by direct lifting promoted by turbulence. The suspended load can be readily seen clouding the water (Figure 9.22) and consists of particles that are carried for considerable distances within the moving mass of water before returning to the bottom. *Extremely* fine sizes may be carried constantly in suspension. Under normal conditions, only silt- and clay-sized particles are carried in suspension, although larger particles may be carried during floods.

Bed Load The **bed load** includes the particles that are being moved along the stream channel bottom by traction. The force that moves the particles is generated by the reduction in the horizontal velocity of the water. The particles carried in bed load range in size from those that are just slightly too large to be lifted into suspension (usually sand size) up to the largest particle the stream can move, the *competence* of the stream. The bed load moves by rolling, sliding, or bouncing along the bottom; the latter process is

FIGURE 9.22 *The difference between a crystal-clear stream and the murky waters of a stream at flood is the amount of* suspended load. *(Courtesy of W. B. Hamilton/USGS)*

called **saltation** (Latin *saltatio* = to leap). In most streams, the amount of material carried in bed load is only about 10% of that carried in suspension.

The sizes of the bed load particles decrease downstream due to the combined effects of solution and abrasion. Large particles such as cobbles and boulders will remain immobile on the stream bottom until floods move them or solution and abrasion reduce them to sufficiently small sizes that the stream can move. Particles stored in floodplain deposits continue to undergo weathering and reenter the stream channel smaller than when deposited. Once the particles are small enough for the stream to move, the mutual abrasion between the particles as they are moved along both reduces their size and changes their shape. Within relatively short distances, sharp edges will be abraded and the particles will begin to round. In fact, rounding is so characteristic of bed load materials that rounded particles in sediment or in rocks are usually interpreted as being the result of stream transport, the so-called *stream-worn pebbles* (Figure 9.23).

Once the particles approach sand size and especially when they attain silt and clay size, further reduction in size proceeds very slowly. Once taken into suspension, silt- and clay-sized particles are separated by water and, as a result, experience little abrasion that would physically reduce their dimensions. Further reduction in size occurs largely through chemical solution.

FIGURE 9.23 *Rock fragments will not be transported very far by a stream before they begin to* round. *Streams are the major agents for the production of rounded rocks, as indicated by the common reference to "stream-worn pebbles." (Courtesy of H. H. Barnes, Jr./USGS)*

SPOT REVIEW

1. What factors determine water velocity?
2. What is the difference between capacity and load?
3. What determines the amount of energy available to a stream?
4. What determines whether a particle will be moved by direct lifting or by traction?
5. Why is it difficult for a stream to erode the smallest particles?

THE DEVELOPMENT OF TOPOGRAPHIC RELIEF

Once the energy required for sediment erosion and transport has been provided, the remaining energy can be used for both the downcutting and lateral cutting of the stream channel. In rock, downcutting of the stream channel is primarily the result of the **abrasion** of the channel bedrock by the moving bed load and the lift the particles experience during increases in water velocity. The individual bed load particles scour and abrade the exposed bedrock and carve the channel deeper. In streams with carbonate bedrock, **dissolution** of the rock aids in downcutting.

Baselevel

The lower limit to which a stream can carve its channel is the **baselevel** or the elevation at its mouth (Figure 9.24). Baselevels can be of two kinds: (1) *temporary* and (2) *ultimate*.

Temporary Baselevel An example of a **temporary baselevel** is the level of water in a lake or pond into which a stream flows. The elevation at the top of a waterfall is also a temporary baselevel inasmuch as the lip of the waterfall is the "mouth" of the upstream segment of the stream. These baselevels are temporary because in time they will be eliminated. Lakes and ponds will eventually fill with sediment, and as the resistant rock that makes up the lip of a waterfall is undercut and collapses, the waterfall will work its way upstream, eventually disappearing as riffles in the stream bottom.

Ultimate Baselevel Each stream has only one **ultimate baselevel.** For most streams belonging to external stream systems, the ultimate baselevel is sea level. The few exceptions are streams that flow

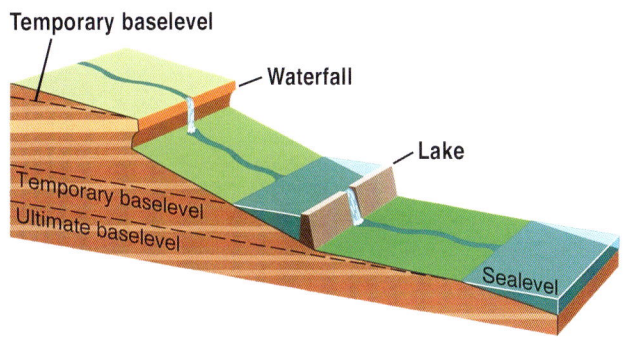

FIGURE 9.24 *Baselevel is the elevation to which a stream is actively carving its channel. A stream may have any number of* temporary baselevels *but only one* ultimate baselevel.

beyond the shoreline along submarine canyons and eventually deposit their loads in the deep-ocean basin. The ultimate baselevel of a stream belonging to an internal stream system, such as most desert streams, is the elevation of the lowest basin into which the stream flows. Desert streams can have an ultimate baselevel above sea level (for example, the streams that flow into the Great Salt Lake, Utah) or below sea level (for example, those that flow into Death Valley, California).

Fluvial Landforms

The end result of the combined efforts of mass wasting and downcutting by streams is the development of the **fluvial landforms** that dominate the surface of the land. Because streams have long been recognized as the main sculptors of the land, one might think that scientists would not only thoroughly understand but also agree upon the mechanisms by which the various stream-sculpted landforms evolve. This is not the case, however. Several theories about the origin of stream-derived landforms have been proposed, each with its advocates and critics. The theory that has most influenced modern thinking was set forth by *William Morris Davis* in the late 1800s.

Davis envisioned the landscape evolving in stages that he identified as *youth, maturity,* and *old age* (Figure 9.25). It is important to note that these terms describe the *valley* and do not in any way refer to the geologic age of any portion of the stream.

According to Davis, development of a landscape begins with rapid uplift of the ocean floor or continental landmass to form a new landmass. Although our modern ideas concerning the mechanisms by which various crustal forces can create landmasses

(a) Youthful landscape

- Rugged topography
- High gradient, turbulent streams in "V"-shaped valleys
- Little or no floodplain
- Waterfalls, rapids common

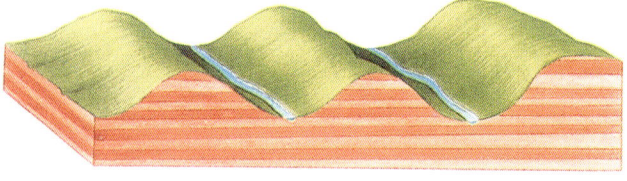

(b) Mature landscape

- Subdued topography
- Floodplains developed adjacent to low gradient meandering streams
- Waterfalls, rapids, turbulent streams not common

(c) Old-age landscape

- Nearly level topography
- Valleys many times wider than stream channels
- Extreme meandering oxbow lakes common

(a)

(b)

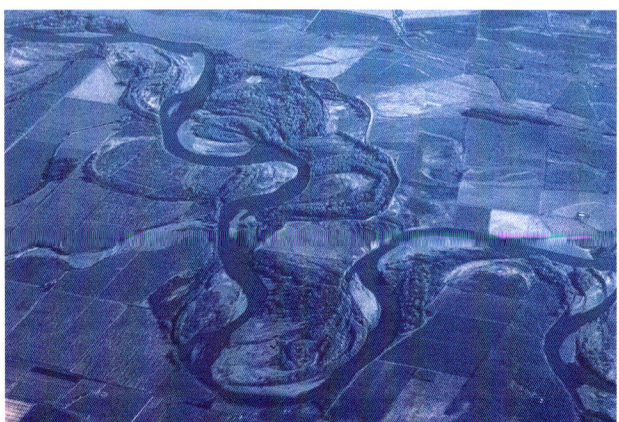

(c)

FIGURE 9.25 *The Davisian concept postulates a progressive change in the appearance of the land with time. Davis's choice of the terms (a) "youth," (b) "maturity," and (c) "old age" probably reflects an increased biological awareness at the time he formulated his ideas due to the recent introduction of Darwin's theory of organic evolution. (Photo a courtesy of Joe Donovan, b courtesy of W. B. Hamilton/USGS, and c VU/© Martin G. Miller)*

are more complex than those envisioned by Davis, the initial development of stream systems on newly emergent land surfaces and the subsequent erosional processes are not much different. Regardless of how the land originates, streams flow downslope following surface lows and create individual drainage basins that join to form drainage systems.

The Stage of Youth   On newly uplifted land, the distance between the surface of the land and the baselevel of the streams is at a maximum, whether the baselevel is temporary or ultimate. At this early stage of landform development, all of the energy available to the streams for erosion is used for *downcutting* by channel abrasion combined with erosion of the bed load by direct lifting and traction. The volume of rock carved out by the stream during this initial period of time is indicated by the two parallel dotted lines in Figure 9.26. Processes of chemical and mechanical weathering then attack the rock surface and generate a layer of regolith and soil. Various processes of mass wasting, including slow movements such as creep and more rapid movements such as landslides operating on the slopes, then move the materials downslope, widening the valley to the characteristic **V shape** of youth. Streams at this stage of development have steep gradients and are highly turbulent. Waterfalls and rapids are common features (Figure 9.27).

As time passes and the stream drainage system is extended by headward erosion by the first-order streams, the valley deepens and widens while maintaining the characteristic **V** shape, and the divide areas between adjacent stream valleys narrow (refer to Figure 9.25a). Eventually, the upper edges of adjacent valleys will meet. At this point, the *relief* of the area, defined as the vertical difference in elevation between the hilltops and the lowlands or valleys of a region, will be at a maximum. The landscape takes on the rugged topography associated with mountainous areas and represents Davis's stage of *youth* (refer to Figure 9.25a).

The Stage of Maturity
As the distance between the stream channel and the baselevel decreases, downcutting *and* lateral cutting begin to share the energy available for erosion. This twofold allocation of erosional energy signals the transition from youth to *maturity*. The basic features of the mature valley are illustrated in Figure 9.28.

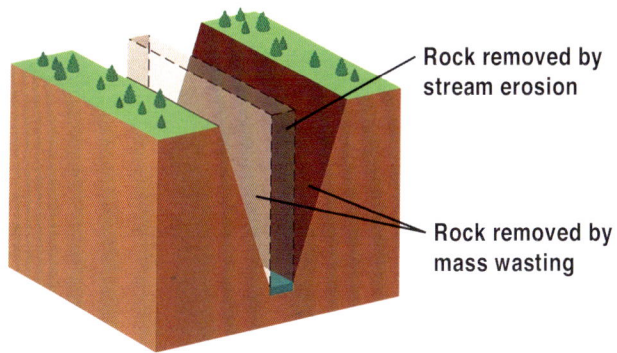

FIGURE 9.26   *Although headward erosion by first order streams is responsible for the initiation of valley carving, most of the rock removed to form* the V-shaped valley *of youth is the result of mass wasting.*

FIGURE 9.27   Rapids and waterfalls, *common scenes along youthful stream valleys, begin to disappear as the stream comes closer to its baselevel and enters the stage of maturity. (Photos courtesy of the WVEGS)*

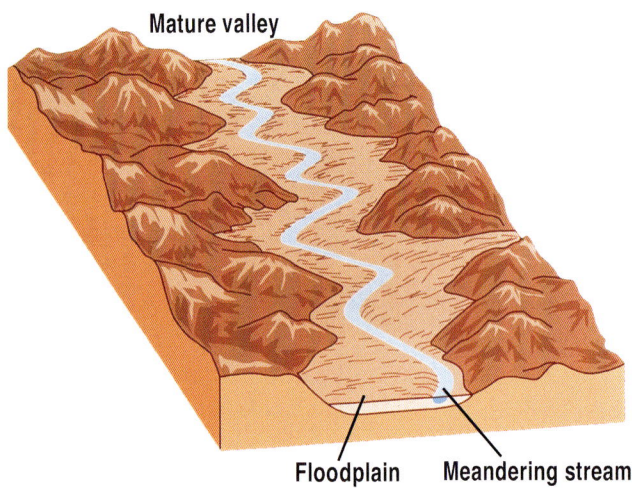

FIGURE 9.28 *By maturity, the stream valley has begun to widen and adjacent hills are reduced in elevation. Mass wasting decreases the slopes of the valley walls and rounds the hill tops, the stream becomes less turbulent as waterfalls and rapids disappear, and the stream begins to meander. All these changes occur in response to the decreasing distance between the stream channel and the baselevel.*

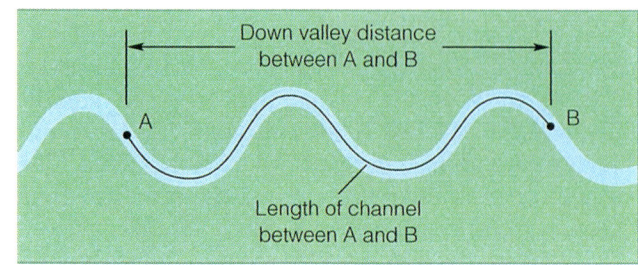

FIGURE 9.29 *The sinuosity of a meander pattern increases from maturity through old age as the stream gradient decreases.*

The lateral cutting initiates the development of a **valley flat**, a bedrock erosional surface flanking the stream channel. In most streams, the valley flat is covered with sediment and cannot be observed. When the stream periodically floods and temporarily requires a larger channel, the waters move out onto the valley flat. As the flood ebbs, the stream waters abandon the valley floor and return to the stream channel. Suspended load carried out onto the valley flat during the flood can no longer be transported and is deposited to form a **floodplain** that covers the valley flat. More will be said about the role of the floodplain later.

As the distance in elevation between the stream channel and the baselevel continues to decrease, progressively more of the energy available to erode is devoted to lateral cutting and less to downcutting. As the stream gradient decreases, the stream begins to follow a snakelike or sinuous pattern of flow called **meandering** (refer to Figure 9.28). The **sinuosity** of a meander pattern is measured by a ratio of the *actual distance* along the channel between two points of the horizontal distance between the two points (Figure 9.29). As the valley progressively develops from early maturity through old age, the sinuosity of the meander pattern increases. We will discuss meandering in more detail later in the chapter.

With greater maturity, the stream gradient and turbulence progressively decrease. Mutual abrasion of the particles within the bed load results in smaller particles in the bed load sediments downstream. Chemical weathering of the particles during both their temporary storage in the floodplain and their transport within the stream also reduces particle size. Waterfalls and rapids become fewer as the stream gradient decreases and the channel assumes an increasingly sinuous mode.

Throughout the erosional cycle, the continued headward erosion of the first-order streams expands the area of the stream drainage system and thereby increases both the stream discharge and load. During each flood, an additional veneer of sediment coats the valley flat. The sediment's residence on the valley flat is only temporary, however. With its next sweep across the valley, the stream will pick up the sediment and carry it further downstream.

The Stage of Old Age

As the stream channel approaches the baselevel, the erosive process enters *old age*. Lateral cutting of the channel is now far more dominant than downcutting as the meander belt wanders far and wide across the floodplain (Figure 9.30). The valley is now many times wider than the stream channel. Meandering and sinuosity of the stream take on extreme proportions. As sinuosity increases, adjacent meander loops enlarge and eventually are separated by only a narrow strip of land that is ultimately breached, usually during a flood. The stream then follows the newly formed, higher-gradient channel and abandons the cutoff meander, forming an **oxbow lake** (Figure 9.31). As happens with all lakes, the oxbow lake will fill with sediment provided by seasonal floods of the main

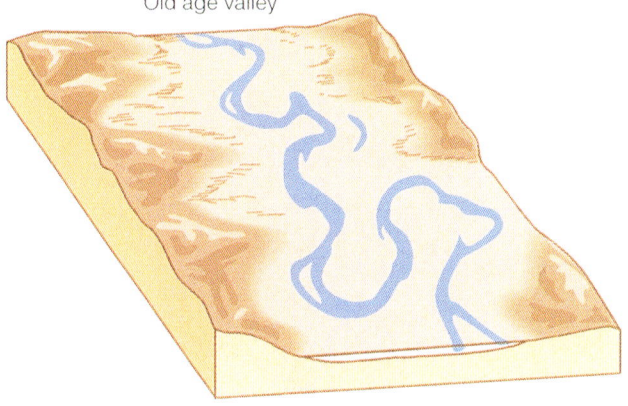

FIGURE 9.30 *As the stream channel nears the baselevel, the valley enters the stage of* old age. *The valley has become many times wider than the stream, adjacent hills are very low, the stream shows extreme meandering, and abandoned meanders from oxbow lakes and meander scars as the channel sweeps back and forth across the valley. (Photo by VU/© Martin G. Miller)*

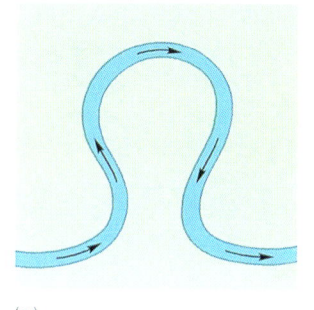 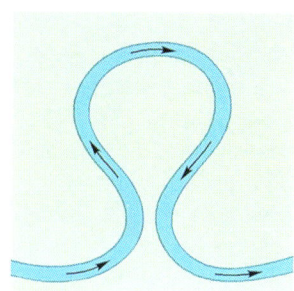

(a)   (b) Separation between stream across meander narrows.

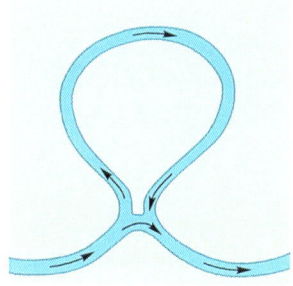 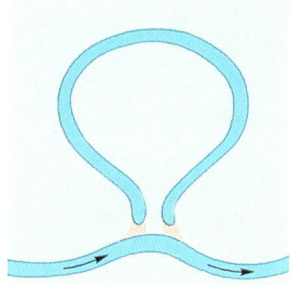

(c) Separation breached. Flow continues around meander.   (d) Sediments block stream from entering abandoned meander.

FIGURE 9.31 *An* oxbow lake *forms when a meander loop is cut off, usually during a flood, as the stream seeks a more direct route to its mouth along a steeper gradient. (Photo © Science VU/Visuals Unlimited)*

stream and, aided by encroaching vegetation, will eventually become a meander scar, recording the wanderings of the stream channel over the valley (Figure 9.32). Oxbow lakes and the scars of abandoned meanders are widely evident in old age valleys whereas waterfalls and rapids are nonexistent. Stream gradient and turbulence are quite low. The stream load is primarily fine silts and clay-sized materials carried in suspended loads. Hills bordering the valley are low. If the stream belongs to an external drainage system and is approaching the ultimate baselevel, the landscape becomes a relatively flat surface called a **peneplane** (Latin *paene* = almost + plane), which Davis envisioned as the end product of the erosional process. We will comment more on the peneplane later in this chapter.

Any stream system becomes increasingly mature downstream. First-order streams in the headwaters area invariably occupy youthful valleys with the valleys of higher-order streams growing progressively older downstream, even reaching old age.

FIGURE 9.32   *A stream's wandering during old age is recorded by the presence of* meander scars. *(VU/© Steve McCutcheon)*

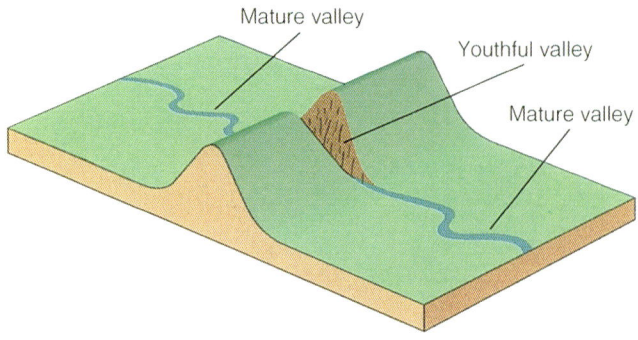

FIGURE 9.33   *Along a stream's course, its valley may change to a more youthful stage as the stream cuts across resistant rock formations and forms a* watergap.

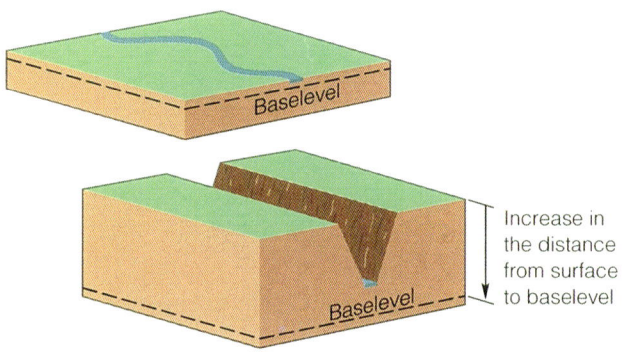

FIGURE 9.34   *As the distance between the stream channel and the baselevel increases, the stream undergoes* rejuvenation, *and the valley develops a more youthful appearance.*

In some situations, however, local reversals in age may occur. Consider, for example, a stream flowing through a mature valley that encounters a mass of particularly resistant bedrock (Figure 9.33). As the stream exerts more energy downcutting the resistant rock, the valley may narrow and the valley walls steepen, forming a more youthful valley segment. Once the stream emerges from the area of resistant rock, the valley will once again become mature.

Rejuvenation   One of Davis's most important ideas was that the erosional process is a *cycle*. This idea implies that during any stage of development, the stream can revert back to a more youthful stage and begin the erosional process anew, a process called rejuvenation. Rejuvenation is the result of an increase in the vertical distance between the stream channel and the baselevel, accomplished either by lowering the elevation of the baselevel or by reuplifting the land (Figure 9.34). As the stream reallocates more energy to downcutting, the valley develops more youthful characteristics, and the landscape enters another cycle of erosional development. An example of a rejuvenated stream with a youthful valley is the Colorado River and the Grand Canyon. The rapid uplift of an area containing a meandering stream can result in the formation of *incised meanders*. The incised meanders observed from Dead Horse Point in southeastern Utah is a well-known example (Figure 9.35).

In some cases, rejuvenation of a stream valley produces terraces that flank the stream. The upper surfaces of the terraces are the surfaces of the old floodplains. When the stream undergoes two or more episodes of *rapid* downcutting, the terraces on opposite sides of the valley are paired whereas more *gradual* downcutting produces unpaired terraces (Figure 9.36). Repeated downcutting can result in multiple terraces. Once terraces are formed, mass wasting and erosion can cause them to retreat toward the valley walls and eventually disappear.

FIGURE 9.35 *If the channel-baselevel distance increases relatively quickly, mature features such as meandering may be preserved as* incised meanders *as the stream rapidly erodes its channel and develops the V-shaped valley of youth. Dead Horse Point, Utah, is an excellent example.*

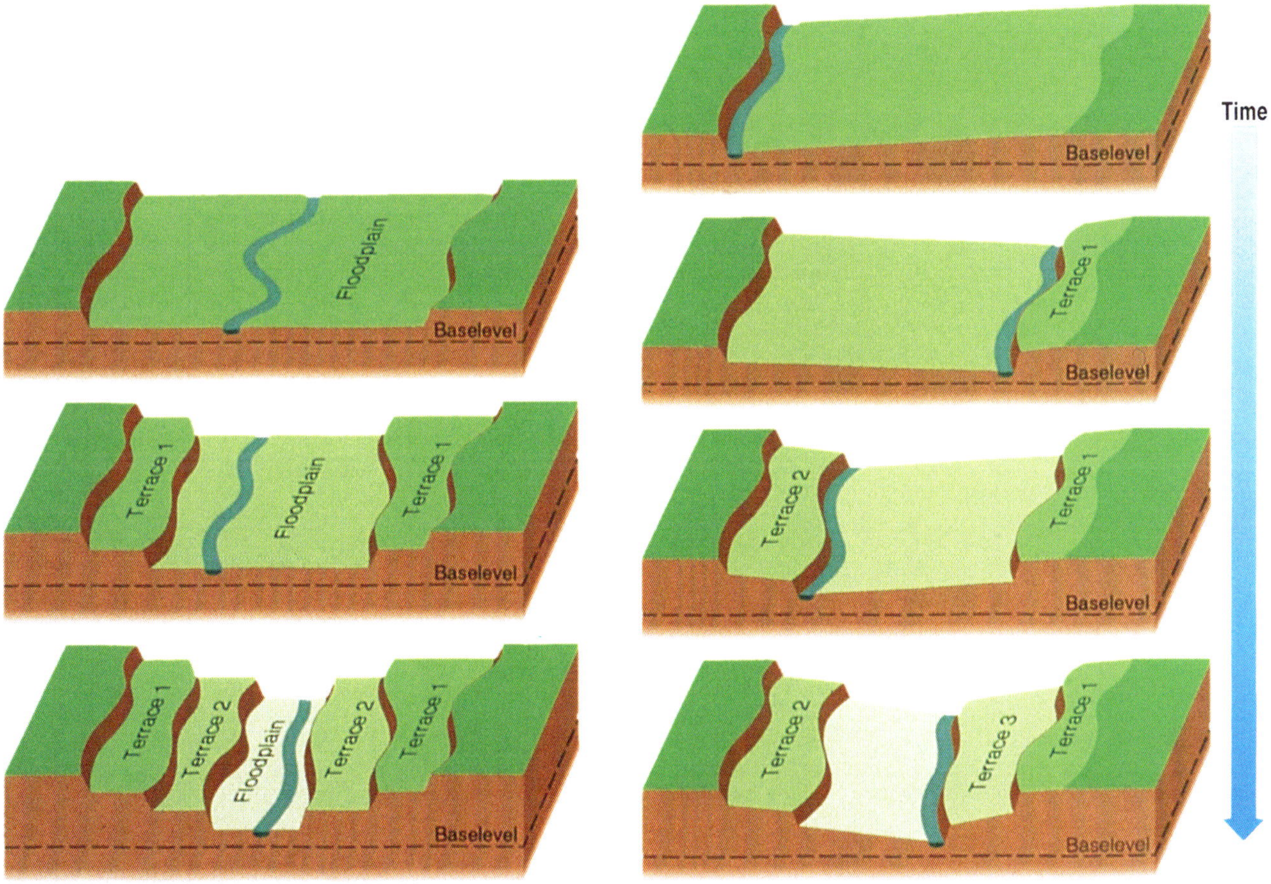

(a) Paired terraces formed by the step-wise lowering of the baselevel with periods of stability

(b) Unpaired terraces formed by continuous lowering of baselevel

FIGURE 9.36 *The way in which the baselevel drops may be recorded in the form of terraces flanking the stream. A more or less continuous, steady drop in baselevel is recorded as a set of* unpaired terraces *while* paired terraces *indicate that the baselevel dropped in stages separated by periods where the channel-baselevel distance remained relatively constant.*

Some Comments on the Davisian Concept

Before we continue our discussion of streams and landform development, several points about the Davisian concept of landform development should be made. First of all, it is a general concept, and as a result, there will always be specific cases that don't seem to fit. In desert areas, for example, the valley walls retreat parallel to the original vertical slope of the valley wall with little or no decrease in slope angle (Figure 9.37). Hilltops in arid landscapes undergo minimal rounding and lowering as the erosion progresses through maturity and old age. In many areas, the topography remains sharp and rugged rather than becoming rounded and subdued as it does in more humid areas. As we will see later in our discussions of desert erosion, this characteristic may be related to the fact that in humid areas the bedrock surface is attacked by chemical weathering and covered with regolith and soil whereas in desert areas mechanical weathering dominates. Without a continuous cover of

(a)

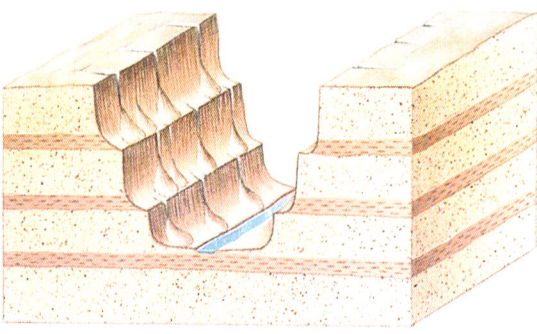
(b)

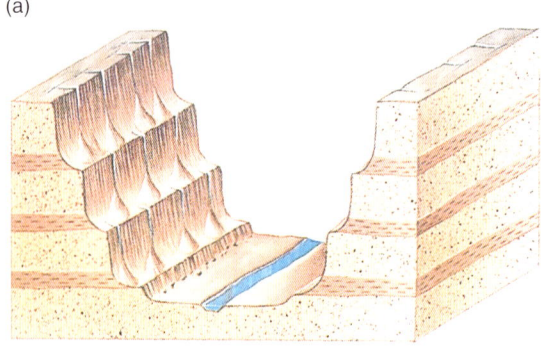
(c)

FIGURE 9.37 *The near-vertical retreat of valley walls in desert areas seems to contradict the Davisian concept in that neither the angle of the valley walls nor the elevation of the divides separating adjacent stream valleys is reduced and the rounding of rocks that is evident in the humid cycle does not occur. (Photo by VU/© Martin G. Miller)*

regolith, the desert bedrock is more exposed. This, in turn, affects the type of mass wasting that dominates. Rock falls and landslides, for example, are more common in arid areas while creep and slump are more typical of humid areas. Although the humid cycle of erosion clearly dominated Davis's thinking, these differences between arid and humid processes do not necessarily negate the validity of his ideas.

The theoretical ultimate end product of the humid cycle of erosion, the *peneplane* (originally *peneplain* when introduced by Davis), has also been the subject of some criticism. For example, some argue that there are no examples of peneplanes in North America while others are of the opinion that the Canadian Shield is close to a peneplane. Some geologists think that the extensive plains in the centers of the Gondwana continents may be peneplanes. If so, they indicate that peneplanes may require in excess of several hundred million years to develop. It is interesting to note that for generations after Davis first introduced his concept, many geologists who enthusiastically embraced the new theory "found" the remnants of peneplanes everywhere. Now we consider the peneplane to be an ideal situation—an end-member of the erosional process never attained and not attainable because uplift or subsidence of the land or changes in sea level usually occur before a peneplane can fully develop. Despite these criticisms and others, Davis's concepts have survived both critics *and* advocates for a century, which is perhaps the best testimony to the basic validity of his ideas.

Meandering

Meandering is the most commonly observed stream flow pattern, yet its exact cause or causes are unknown and remain a topic of considerable controversy. Although this debate is beyond the scope of this text, we can point out some basic relationships exhibited by meandering streams. The amplitude and wavelength of the meander pattern are a function of the stream width at bankfull conditions. Although the data vary somewhat from one investigation to the next, the maximum wavelength of a meander is about 6 to 12 times the bankfull stream width while the maximum amplitude ranges from 10 to 20 times the bankfull stream width.

The straight stream segment of a meandering stream is called the **reach,** while the outside and inside of the **bend** are called the **cutbank** and the **point,** respectively (Figure 9.38). Within the reach, as in any straight stream segment, the point of highest

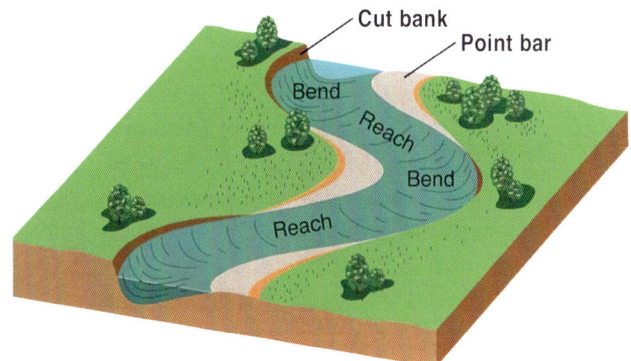

FIGURE 9.38 *The most commonly observed stream pattern,* meandering, *signals the onset of the stage of* maturity *and continues throughout* old age *where it becomes highly accentuated. (Photo by VU/© Glenn M. Oliver)*

velocity is at midchannel just below the surface where turbulence is minimal. As the water moves into the bend, centrifugal force shifts the point of highest velocity progressively from midchannel toward the cutbank, resulting in an asymmetric velocity profile across the stream. In addition, as the water moves from the reach to the cutbank, the maximum water velocity *increases* just as the speed of a skater on the outside of a whip increases. The maximum water velocity in a meandering stream is therefore at the cutbank, and the minimum velocity is located on the opposite bank at the point.

With a large number of velocity profiles drawn across the stream from the center of one reach downstream to the center of the next reach, we can plot a line of highest velocity. Beginning in the middle of the first reach, this line crosses the stream to the next downstream cutbank and recrosses the stream to the next downstream reach (Figure 9.39). Along this line of maximum water velocity, the velocity progressively increases from the middle of the first reach to the next downstream cutbank and then decreases as it moves to the center of the next downstream reach. The line that connects the points of maximum water velocity very closely approximates the *thalweg,* the line of maximum channel depth along the stream.

Consider now the processes of erosion, transportation, and deposition along a meandering stream. The point of maximum erosion will be at the point of highest velocity, namely, in the bend at the *cutbank.* As the water leaves the bend, moves downstream, and slows, some of the material eroded and transported from the cutbank is deposited in the reach and on the next downstream *point* where the overall stream capacity is at a minimum. The deposit that forms at the point, called a **point bar,** is the most visible deposit within the stream channel and has served as a campsite for many a hiker.

This simultaneous process of erosion and deposition within short segments of meandering streams is another example of a stream attempting to maintain the balance of grade. The *erosion* of sediment from the cutbank is offset by *deposition* in the next downstream reach and point.

As erosion of the floodplain deposits continues at the cutbanks with simultaneous deposition on the opposite point bar, the stream meanders back and forth across the valley, erodes the floodplain deposits, and transports and redeposits them further downstream. Eventually, the sediment temporarily stored on the valley flat will be carried to the mouth of the stream.

As the stream valley becomes more mature, the meandering of the stream becomes more *sinuous.* The sinuosity of the stream is measured by dividing the channel length by the valley length between any

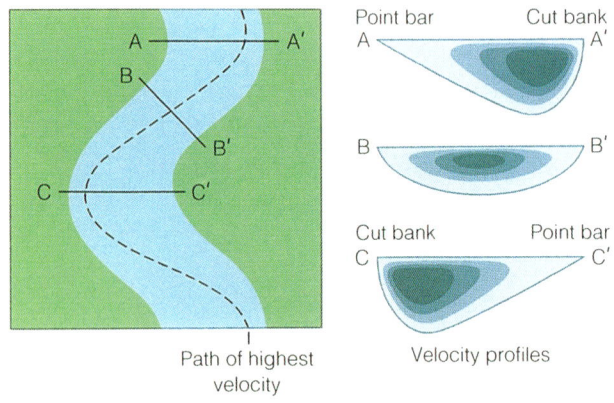

FIGURE 9.39 *An important aspect of meandering streams is the variation in the* water velocity, *and therefore the* erosion-deposition *characteristics, of the stream. Erosion goes on at the* cutbank *where the velocity reaches a maximum. The materials eroded from the cutbank are deposited in the next downstream* reach *and* point *as the velocity of the water decreases.*

two points along the stream (refer to Figure 9.29). In general, sinuosity increases downstream as the stream gradient decreases. The factors that control sinuosity or meander development are as elusive and arguable as the factors that initiate meandering itself. Although meanders may form in any stream, they seem to develop most readily when the channel is incised into floodplains composed of sand- to clay-size materials.

Meandering serves another important function in the development of the valley. The beginning of meandering and the development of increased sinuosity initiate the lateral movement of the stream that carves an ever-widening valley flat. Simultaneous with the widening of the valley flat by the meandering stream, mass wasting reduces the angle of the hill slopes, rounds the divides between the streams, and reduces the elevation of the hilltops. Once the stage of maturity is attained, the combined efforts of stream erosion and mass wasting continuously reduce the relief of the region.

THE DEVELOPMENT OF GRADE

Let us return to our discussion of *grade,* another concept introduced by Davis. Davis envisioned grade as a state of balance or equilibrium that develops along the profile of a stream between *all* the hydraulic compo-

NAVIGATING MEANDERING STREAMS

When people first began to navigate rivers, they soon recognized some of the characteristics of meandering streams, in particular, the variation in water depth along the stream. They found that the water was deep enough for navigation in the bend where the water velocity opposite the cutbank was sufficient to carve and maintain a deep channel, but that water depth often became a problem near the reach. Experience showed that the patterns of deposition within the reach shifted so that the deepest channel through the reach could rarely be either predicted or assumed. Because meandering streams are invariably laden with suspended load, the bottom could not be seen. The slight turbulence of the water surface might give a hint of water depth, but only someone with an intimate knowledge of the river could properly interpret the surface to determine where the deepest parts of the channel lay. Consequently, the reputations of riverboat captains could be made or broken by their ability to "read" the river and successfully navigate the reach.

To make the task easier, early river people built their boats with flat bottoms, which gave the boat a shallow *draft*, meaning that it would float high on the water. All riverboats and barges are therefore built with flat bottoms to allow passage in the shallowest water within a meandering stream.

A well-known meandering stream in the United States is the lower part of the Mississippi River. With its tributaries, the Missouri and the Ohio, the Mississippi River was an important route into the wilderness during early U.S. history and still serves as a major artery for commerce. In a boat with a shallow draft, the riverboat captain would enter the reach on the lower Mississippi, carefully watching the water's surface ahead for any indication of a shallowing bottom. Still not satisfied with what could be seen from the wheelhouse, the captain would station a crew member on the bow with a weighted line marked at 6-foot (1.8 m) intervals, that is, at every fathom. The crew member would throw the line overboard and count the number of marks that disappeared beneath the water. Since the draft of the boat was usually one fathom, a call of "mark three" indicated a water depth of 18 feet (5.5 m), which left 12 feet (3.5 m) of water under the keel. This exercise would be repeated as the boat made its way through the reach. All would be well until the water depth under the keel was only 6 feet, which brought the call "mark twain." At "mark one," of course, the boat would be aground. You will recognize *Mark Twain* as the pen name of *Samuel Clemens* who wrote of life on the Mississippi River. It is altogether fitting that his pen name represented a critical event on the Mississippi River, the success or failure of a passage through a reach.

---

nents of the stream including the channel characteristics of width, depth, gradient, sinuosity, and roughness; the flow parameters of discharge, capacity, and load; and the processes of erosion, transportation, and deposition. A balance or an equilibrium situation implies that a change in one parameter must necessarily initiate a change in one or more of the others.

When a stream comes to grade, all stream parameters will have adjusted to accomplish the major goal of the stream, which is to transport any sediment introduced into its channel. Channel width and depth increase downstream to accommodate increased discharge. The stream gradient that develops provides the necessary discharge, and thereby the energy, to transport the sediment introduced into the stream. The end result of this complex interaction is the concave upward longitudinal profile shown in Figure 9.19.

Grade is not achieved in youthful streams where erosion is the order of the day. Youthful headwater streams are more or less constantly out of balance as they carve away at the rock in their restricted narrow channels. A stream is also out of balance during a flood when it is subjected to very large volumes of water and abnormally high levels of suspended and bed load. Streams attempt to "compensate" for this imbalance during floods in several ways. For example, as we have already noted, the stream will overflow its bank to create a larger channel, and after the flood, the floodplain covering the valley flat will become a storage site for excess load until the sediment can be reeroded and carried downstream during periods of normal flow.

The dynamic nature of streams and the changes that are constantly taking place to maintain the dynamic equilibrium of grade are rarely recognized by the casual observer. For example, the drainage basin progressively expands by headward erosion, necessitating a larger stream channel to accommodate the expanding load; accordingly, the physical dimensions of the stream channel are enlarged by slow erosion of the sediment accumulated on the floodplain. If the stream undergoes rejuvenation, the increased energy made available by the steepened gradient may be greater than the energy needed to transport the sediment currently being introduced from the upstream drainage basin. In this case, the sediments from the floodplain can be easily eroded and added to the stream as additional bed and suspended load to utilize the increased energy. Terraces will record the episodes of downcutting.

In addition to short-term episodes of disequilibrium, a stream may experience events that cause it to go out of grade and require major changes in its parameters over longer periods of time to reestablish equilibrium. The major long-term events that cause streams to go out of grade stem from (1) *climatic,* (2) *tectonic,* and (3) *human* factors.

Climatic Influences

Climatic influences, in particular, changes in temperature and precipitation, affect the rates of weathering and erosion in the drainage basin and either increase or decrease the amount of sediment introduced into the stream. Either an increase in sediment load or a decrease in water volume will initiate deposition within the floodplain of the upstream segments of the valley with the deposition working progressively downstream. As the floodplain in the upstream valley builds vertically, the increase in the downstream gradient will provide the additional energy needed to transport the larger load. Conversely, a climatic change that results in a decrease in load will initiate downcutting of the valley flat and erosion of floodplain deposits, which reestablish a lower stream gradient that decreases the amount of available energy. In both cases, the readjustments required to accommodate the new load conditions imposed by the climatic change take place over a relatively long period of time.

Tectonic Influences

Tectonics largely affect the relationship between the stream channel and the ultimate baselevel. Rapid uplift of the land surface in the headwaters region will result in stream rejuvenation and a subsequent increase in downcutting that steepens the stream gradient. Subsidence of the land will cause the floodplain to build up beginning at the mouth of the stream and working progressively upstream, thereby reducing the stream gradient in the downstream segments. The stream meander patterns may also become more sinuous in an attempt to reduce the stream gradient. This last scenario could also result from a climatically induced rise in sea level. An example of such a readjustment was the deposition and development of the delta in the lower reaches of the Mississippi River Valley in response to the rise in sea level that occurred during the waning of the last episode of the Pleistocene Ice Age.

Human Influences

Human manipulation of streams and stream channels inevitably causes streams to go out of grade. An example is the effect of dam building. Dams create lakes that serve not only as temporary baselevels for the upstream stream segments but also as sediment traps for the upstream load. Because the sediment is trapped behind the dam, the stream below the dam becomes sediment starved. Energy once allocated for the transport of sediment is no longer needed and becomes available for downcutting. As the stream channel below the dam is downcut, the stream gradient decreases, reducing the total amount of energy available in the downstream segments to accommodate the decreased load.

Other examples of human impact can be seen throughout the mining areas of the western United States. While the mines were active, increased sediment load caused the streams to deposit. When the mining operations ceased, the old stream regime was reestablished, and the streams cut into the valley fill, leaving remnants of the old deposits as terraces.

SPOT REVIEW

1. What are the characteristics of youthful, mature, and old age stream valleys?
2. What is "rejuvenation" and how is it initiated?
3. What is meant when a stream is said to be "at grade"?
4. How do certain long-term factors cause a stream to go out of grade?
5. How do the processes that produce paired and unpaired stream terraces differ?

ALLUVIAL DEPOSITS

An **alluvial deposit** is any deposit laid down by a stream. Before we discuss the individual kinds of alluvial deposits, some general comments are in order. Our discussion of erosion, transportation, and deposition has stressed that streams are quite selective in the size of particle they will pick up, transport, and deposit. For example, we have seen that a stream will preferentially pick up and move sand-sized materials first. On the other hand, as a stream begins to lose capacity and competence, it will deposit its load beginning with the largest particles and ending with the smallest; that is, it will form a graded bed. The process by which streams separate particles by size is called *sorting*. As a result of sorting, the range of particle sizes contained within a particular stream deposit is usually limited. Deposits containing the least number of different sizes are said to show *good sorting* or to be *well sorted*.

Alluvial deposits can be of three types: (1) *channel*, (2) *floodplain*, and (3) *delta/alluvial fan*. The types are classified according to whether the materials accumulate (1) *within* the stream channel, (2) *along* the flanks of the stream and on the valley flat, or (3) *at the mouth* of the stream, respectively.

Channel Deposits

The coarsest material carried by a stream is moved slowly downstream as bed load while the fine silt- and clay-sized material is carried along more rapidly, suspended within the moving water mass. Materials larger than coarse silt and fine sand are usually transported totally within the channel. Even during times of flood, it is quite unusual for materials larger than silt and clay to leave the channel and be carried out onto the floodplain. Furthermore, except for the highly competent streams in the headwaters of the drainage basin, which may contain very large particles, the materials in the bed load of the average stream are pebble size and smaller with sand size dominating as the particles are continuously reduced in size as they make their way downstream.

In our discussion of erosion, transportation, and deposition within a meandering stream, we saw that erosion occurred at the cutbank with deposition at the downstream reach and point. Of these two depositional environments, the *point bar deposit* is the more dominant. As the meandering stream slowly sweeps back and forth across the valley, bed load materials, predominantly sand size, are accumulated in laterally transgressing point bar deposits. In outcrop, point bar deposits take the form of sandstone bodies characterized by flat tops, scalloped bottoms, and inclined layers called cross-bedding that represent the slope from the point into the deep channel nearest the cutbank (Figure 9.40). Note that the direction at which the cross-bedding dips indicates the direction of lateral movement of the meandering stream. The attitude of the cross-bedding in the channel sandstone shown in Figure 9.40 indicates that the stream was moving laterally from left to right.

Floodplain Deposits

As the stream overflows its banks during floods, the suspended load of silt and clay is swept out onto the floodplain. As the water slows immediately upon going overbank, the largest particles within the suspended materials are deposited along the stream banks to create a **levee** (Figure 9.41). In the segments farthest downstream, the levees build upward, increase the depth of the channel, and thus enable the stream to carry the high discharges of floods without going overbank. In some areas such as the lower reaches of the Mississippi River Valley, the levees have grown to the point where the adjacent floodplain is actually below the level of the water in the channel (Figure 9.42). Because waters cannot readily drain from these areas, they are commonly occupied by marshes and swamps.

As the flood ebbs, the suspended materials carried out onto the valley flat are deposited as mud on the surface of the floodplain. Because most of these materials are clay minerals, floodplain deposits eventually enter the geologic record as thinly bedded shales. The floodplains of streams in the stages of late maturity and old age often contain shallow lakes, marshes dominated by grasses, and swamps with

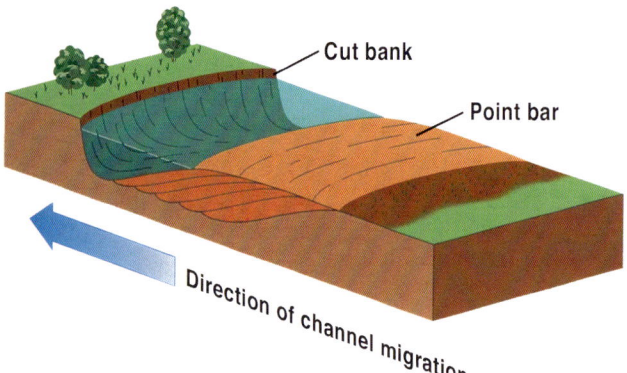

FIGURE 9.40 *One of the most commonly observed sandstone bodies associated with alluvial deposits forms from the sand that accumulates in point bars as the channel of a meandering stream migrates laterally across a valley.*

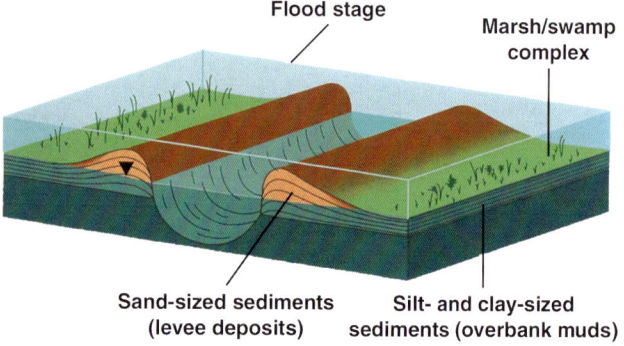

FIGURE 9.41 *As floodwaters go overbank and slow, the coarser materials being carried are deposited along the banks to produce a natural levee. Because they prevent the water from reentering the main channel following the flood, the land adjacent to the stream commonly consists of swamps and marshes. Artificial levees are built along areas such as the lower reaches of the Mississippi River; however, if the levees give way the flooding will devastate the area as it did here in Monroe County, Illinois in 1993). (Photo by AP/Wide World)*

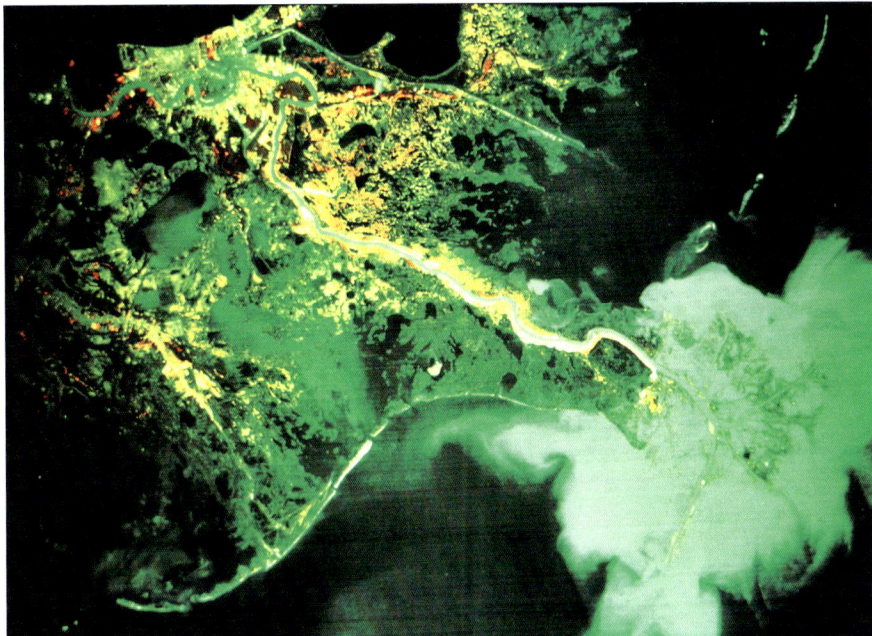

FIGURE 9.42 *Because of the levees that contain the water within the main channel, the areas adjacent to the lower reaches of the Mississippi River are lower than the river and are occupied by extensive wetlands. (© Science VU/Visuals Unlimited)*

FIGURE 9.43 *Wetlands such as swamps and marshes are common features at the mouths of old-age streams.*

their population of more woody plants. Marshes and swamps are most commonly found near the mouths of streams where the gradients become very low and the channels are flanked by levees (Figure 9.43). Freshwater carbonate minerals may begin to accumulate in the oxbow lakes adjacent to the streams and eventually produce limestones. In the marshes and especially in the swamps, plant debris may accumulate to form peat. If the peat is made largely from woody plants, it may, if buried, be converted to coal.

As this brief description indicates, floodplains contain a wide variety of different depositional environments. As a result, the rocks that form from floodplain deposits are often a complex mixture of freshwater limestones, organic-rich black shales, coals within an enclosing matrix of black shales, and siltstones. The sand-rich channel deposits may then cut into the floodplain materials as the stream sweeps from one side of the valley to the other. To the casual observer, the end product may appear to be only a

FIGURE 9.44 *Because several different depositional environments exist in close proximity to each other, the rocks that record* floodplain deposition *may appear to be a complex assortment of rock types. However, with a knowledge of the floodplain environments, the geologic history recorded by the rocks can be easily deciphered.*

bewildering collection of rocks (Figure 9.44). Nevertheless, with a basic understanding of how streams work and of the variety of depositional settings that exist within a floodplain, the assemblage of rocks that can be seen along many road cuts will begin to make sense.

Deltas

As the waters of a stream approach the mouth, the major task of the stream, the transport of sediment, is fulfilled. If the stream is a *tributary* to a larger stream and the currents in the larger stream are sufficiently strong at the point of entry, the load of the tributary is simply incorporated into the growing load of the main stream. If, however, the stream enters a larger body of water at a point where the currents are not strong enough to pick up all of the added load, the water mass slows, loses capacity and competence, and deposits the bed and suspended loads in a deposit called a **delta** (Figure 9.45). Deltas may therefore form along streams, at the point of entry into a lake or pond, or at the edge of a continent. The name *delta* refers to the roughly triangular shape some delta deposits assume in plan view; in the Greek alphabet, the letter delta is Δ. Herodotus, the Greek naturalist and historian, first used the term around 430 B.C. to describe the deposit at the mouth of the Nile River.

The characteristic triangular shape of the delta is due to the outward branching of the main stream into smaller *distributary streams* that distribute the stream discharge and load more efficiently as the stream gradient approaches zero (Figure 9.46). Without the distributaries, the lower reaches of the stream would go into a state of constant flood in the delta area as the water velocity slows.

Within deltas that form at the mouths of bed load-dominated streams, such as those emerging from glaciers, the bed load is deposited first with the coarsest particles cascading down the front of the growing deposit to form the **foreset beds** (Figure 9.47). The fine-grained portion of the bed load and the suspended load are carried progressively further into the water body and come to rest in the horizontal **bottomset beds** that form the outermost edge of the delta. As the delta builds outward, the distributaries move back and forth across the delta surface and deposit a surface layer of bed load sediment called the **topset beds**. The topset beds build the delta surface up above water level during floods as a basinward extension of the floodplain.

Most streams, however, are dominated by suspended loads. The deltas formed by such streams do not show the development of top-, fore-, and bottomset beds but rather construct a trilevel delta structure consisting of a **delta plain**, a **delta front**, and a **prodelta** (Figure 9.48).

The delta will continue to build out into the larger water body until currents within the body are strong enough to pick up and redistribute the

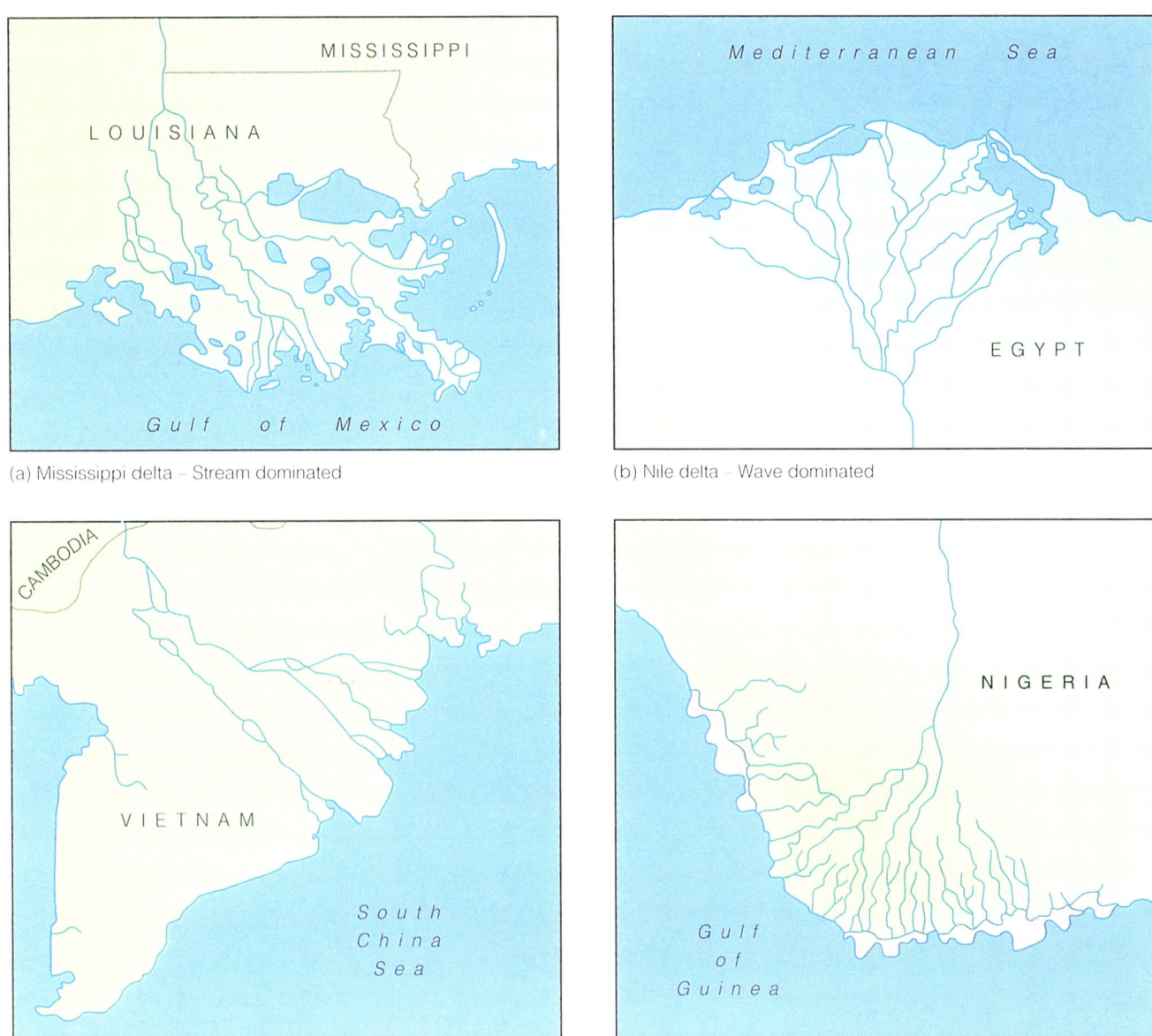

FIGURE 9.45 *The deposit that forms when a stream flows into a larger body of water is called a* delta. *The shape of the delta front depends on the relative influence of the constructive power of the stream and the erosive power of the waves and tides.*

sediments. In lakes, ponds, and lagoons where few if any currents exist, a delta can build until it eventually fills the water body. This is the process that eventually fills the lake with sediment and transforms the lake into land. Deltas that form where streams enter the ocean will build out from land until the combined effect of ocean currents, waves, tides, and longshore currents becomes sufficiently strong to pick up the deposited sediments and distribute them on the ocean bottom.

The Mississippi Delta

The Mississippi River, active since the Late Jurassic, has been delivering sediment to the Gulf of Mexico since the Cretaceous. Its drainage basin covers 1,291,898 square miles (3,344,570 km²), extending from the western continental divide in the Rocky Mountains to the eastern continental divide in the Appalachians and from the southern margin of the Canadian Shield near the Canadian-U.S. border to the

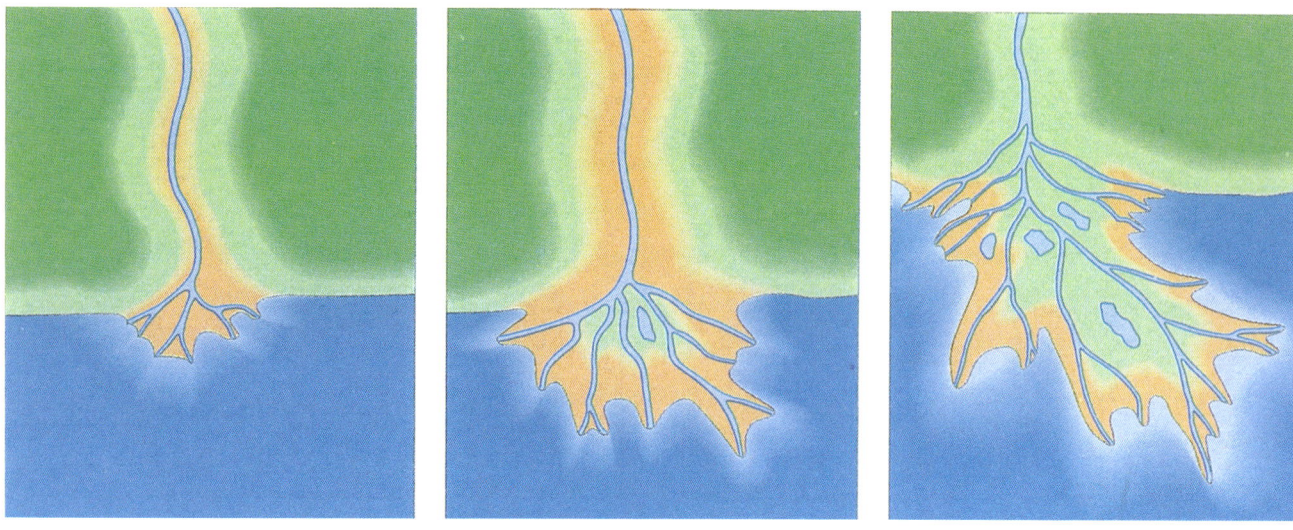

FIGURE 9.46 *The structure of the delta builds as the main stream* bifurcates *(branches) into smaller streams called* distributaries *as the gradient of the stream approaches zero. The delta will* prograde *(build out) into the larger body of water until the currents within the body become strong enough to redistribute the sediments as fast as they are deposited.*

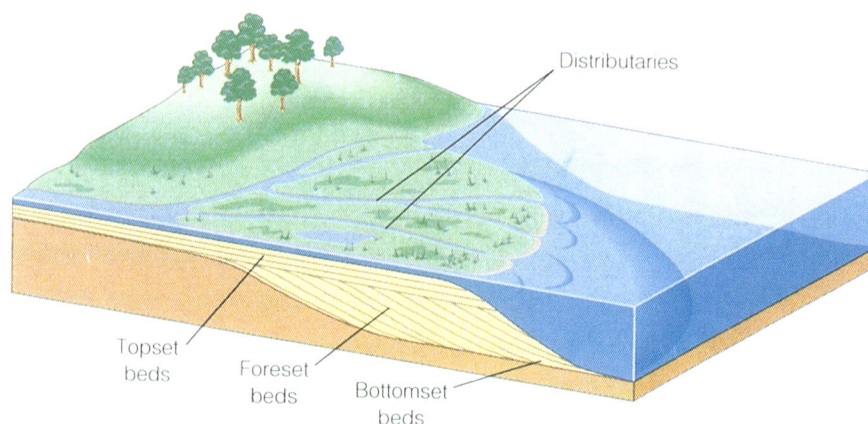

FIGURE 9.47 *Deltas constructed by bed load-dominated streams show the development of* topset, foreset, *and* bottomset *beds as the particles become immobile in order of decreasing size and come to rest on the bottom.*

Gulf of Mexico (Figure 9.49). Each year, the river discharges water into the Gulf at an average rate of 542,433 cubic feet (15,360 m³) per second along with 684,528,300 tons (6.21 × 10¹¹ kg) of sediment. The bed load of the river is fine sand while the suspended load is a mixture of silt (35%) and clay (65%). The sediments are deposited in a complex of alluvial, deltaic, and basin environments.

During the low stand of sea level about 18,000 to 20,000 years ago, the river carved its channel across the exposed surface of the continental shelf and constructed a delta that extended the edge of the shelf seaward. With the melting of the Pleistocene ice caps and the subsequent rise in sea level, deposition shifted back to the alluvial valley. About 9,000 years ago, the river began to construct the present deltaic complex, which now extends nearly to the edge of the continental shelf (Figure 9.50). The delta covers an area of 11,035 square miles (28,568 km²) of which 9,232 square miles (23,900 km²) are subaerial (exposed above sea level).

Constructional or Progradational Phase During the *constructional phase,* a delta rapidly *progrades* (builds out) and constructs a *delta lobe* around a major distributary channel and associated lesser distributaries. Except during times of flood, most of the stream's load is carried by the major distributary and deposited at the mouth of the stream.

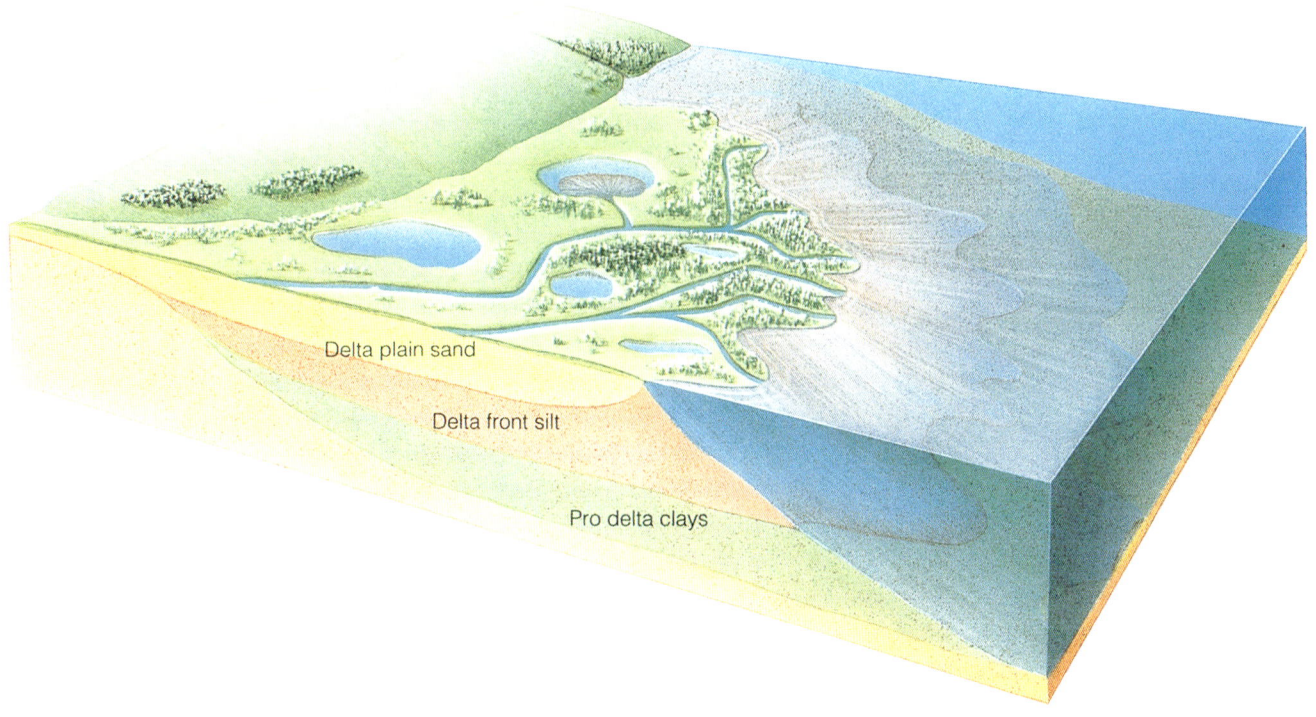

FIGURE 9.48 *The deltas constructed by streams whose loads are dominantly* suspended materials *also consist of three parts. The coarsest materials, usually fine sand, accumulate in* distributary mouth bars *that become the* delta plain. *The plume that characteristically forms around the delta mouth consists of silts that settle to form the* delta front *and clay that forms the* prodelta.

FIGURE 9.49 *The Mississippi River drains nearly 1.3 million square miles (3.3 million km²) of the North American continent.*

The depositional processes occurring at the mouth of the Mississippi River are typical of those going on at the mouths of most rivers that enter the ocean. Although the Mississippi is carrying about 2,000 parts per million of sediment load as it enters the Gulf of Mexico, the density of the water is not appreciably affected. As a result, the lower-density, fresh river water (1.000 gm/cm³) floods out over the top of the denser, saline waters (1.028 gm/cm³) of the Gulf (Figure 9.51). As the sediment plume moves out over the saline water, the particles begin to settle. The fine sands of the bed load accumulate nearest the mouth of the river as a *distributary mouth bar* (Figure 9.52) and eventually form the layer of sand that makes up the *delta plain*. Beyond the distributary mouth bar, the silt settles to form the *delta front* while farthest from the mouth, the clays accumulate to form the *prodelta* (refer to Figure 9.48). As the delta progrades, these three depositional units form the layered sequence that is commonly referred to in discussions of the principle of superposition and Walther's law.

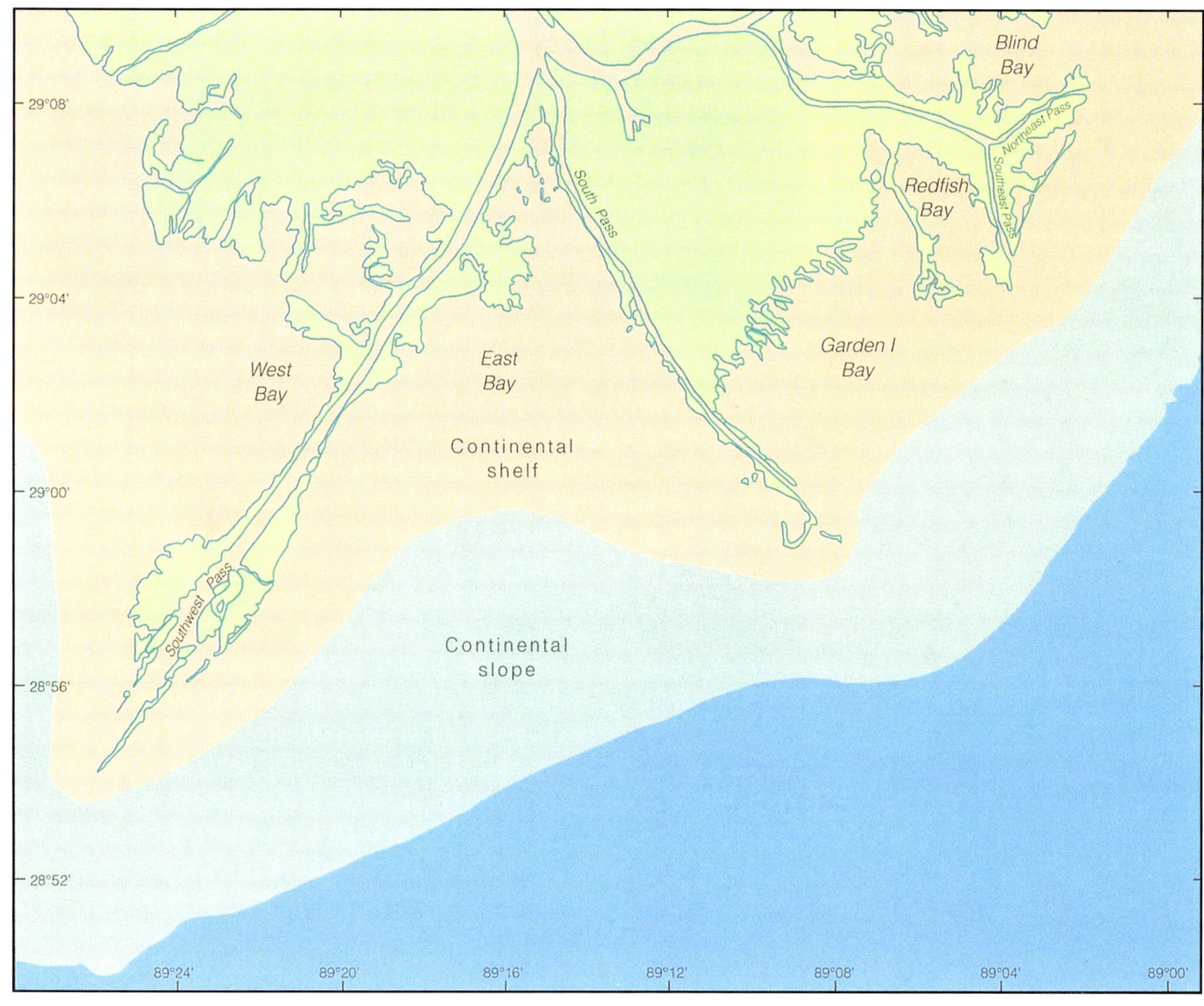

FIGURE 9.50 *The active Balize delta lobe extends to the edge of the continental shelf.*

Bay Filling During the progradational phase, seasonal floods carry sediment overbank into the innumerable *bays* that flank the distributaries. In addition, as the natural levees are periodically breached, a portion of the discharge is diverted into the interdistributary bays. The sediments carried through the breach are deposited as *a crevasse splay* or *subdelta*. In the active Balize delta lobe, the river has breached the natural levee in four places since 1939 and has built crevasse splays into the interdistributary bays (Figure 9.53). On the average, these subdeltas remain active for 10 to 15 years during which time 10 to 15 feet (3–5 m) of sediment accumulate over an area of about 4 to 6 square miles (10–15 km^2). In other areas of the Mississippi delta where breaching of the levees has diverted the stream into large bays, the process of *bay filling* continues for 100 to 200 years and distributes up to 60 feet (20 m) of sediment over an area of 100 to 150 square miles (250–400 km^2).

As the bays fill with sediment, the open water of the bay is converted to a wetland as the open water progressively gives way to grass-dominated *marshes* and tree-dominated *swamps,* which eventually build above sea level. Compaction of the underlying sediments results in *subsidence* of the surface. Although the rate of subsidence depends upon both the thickness and kinds of sediments, natural subsidence rarely exceeds 0.5 inches (10 mm) per year. As long as sediments are provided at a rate that offsets natural subsidence, the wetland is maintained.

Streams 269

FIGURE 9.51 As the Mississippi River enters the Gulf of Mexico at the end of the active Balize delta lobe, the lower-density river water with its suspended load spreads out as a plume over the higher-density saline waters of the Gulf. (Courtesy of NASA)

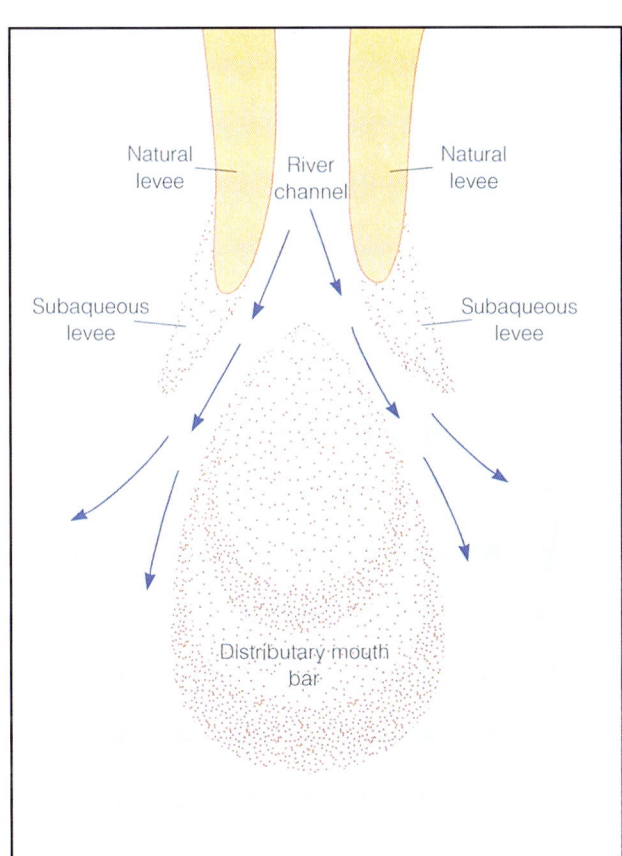

FIGURE 9.52 The bed load, consisting of fine sand, deposits at the mouth of the active distributary and forms a distributary mouth bar.

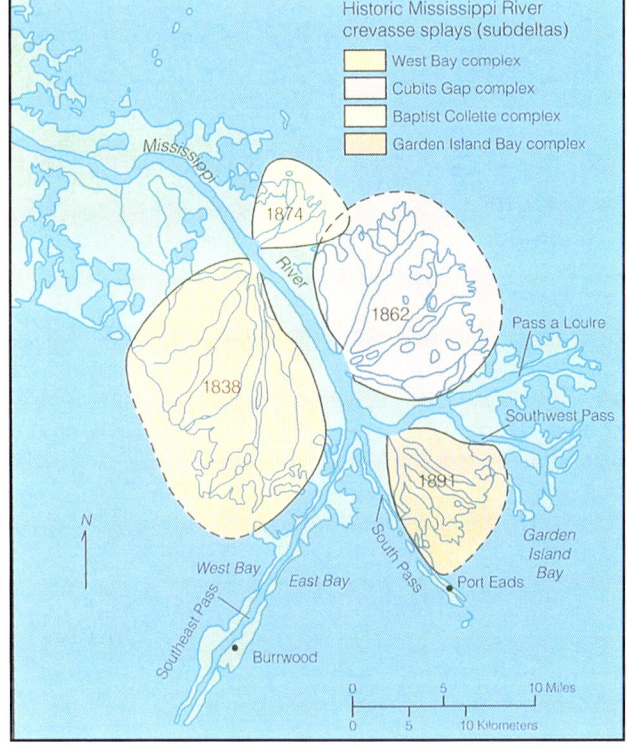

FIGURE 9.53 Interdistributary bays are filled as the river breaches the natural levee during times of flood and forms a crevasse splay or subdelta. Since 1838, four major crevasse splays have broken through the natural levee and have built subdeltas into the interdistributary bays. Modified after Coleman and Gagliano, 1964.

Channel Switching With time, the major discharge will switch to another distributary as the stream seeks a more direct and steeper pathway to the sea. In the case of the Mississippi River, the oldest delta lobe is the Maringouin/Sale (Figure 9.54). Since then, the river has switched distributaries several times with each new delta lobe prograding for 1,200 to 1,700 years before the river sought a new channel.

Each of the Mississippi delta lobes covers an area of about 12,000 square miles (30,000 km²) depositing a layer of sediment about 100 feet (35 m) thick. If extrapolated over the past 2.5 million years, the total thickness of sediments that have accumulated in the Gulf Pleistocene depocenter exceeds 11,000 feet (3,600 m).

The active Balize delta lobe began to form about 1,000 years ago. Beginning about 100 years ago, an increasing percentage of the river's discharge began to be diverted down the Atchafalaya distributary. If the river had not been controlled by a series of artificial levees constructed by the U.S. Army Corps of Engineers following the disastrous flood of 1927, experts believe that by now, most of the discharge of the Mississippi River would be going down the Atchafalaya, a new delta lobe would be prograding into the Gulf, and the Balize delta lobe would have been abandoned. One of the main reasons for artificially maintaining the river in its present course is to provide a constant supply of fresh water to the industries along the lower reaches of the river. If the river were allowed to abandon the Balize delta lobe for the Atchafalaya distributary, fresh water downstream from the entrance to the Atchafalaya distributary would be displaced by salt water encroching along the present channel.

Destructional or Abandonment Phase As channel switching begins to divert the flow to a new distributary, the former delta lobe enters the *destructive* or *abandonment phase*. As the amount of sediment provided to the interdistributary areas diminishes, subsidence begins to dominate over

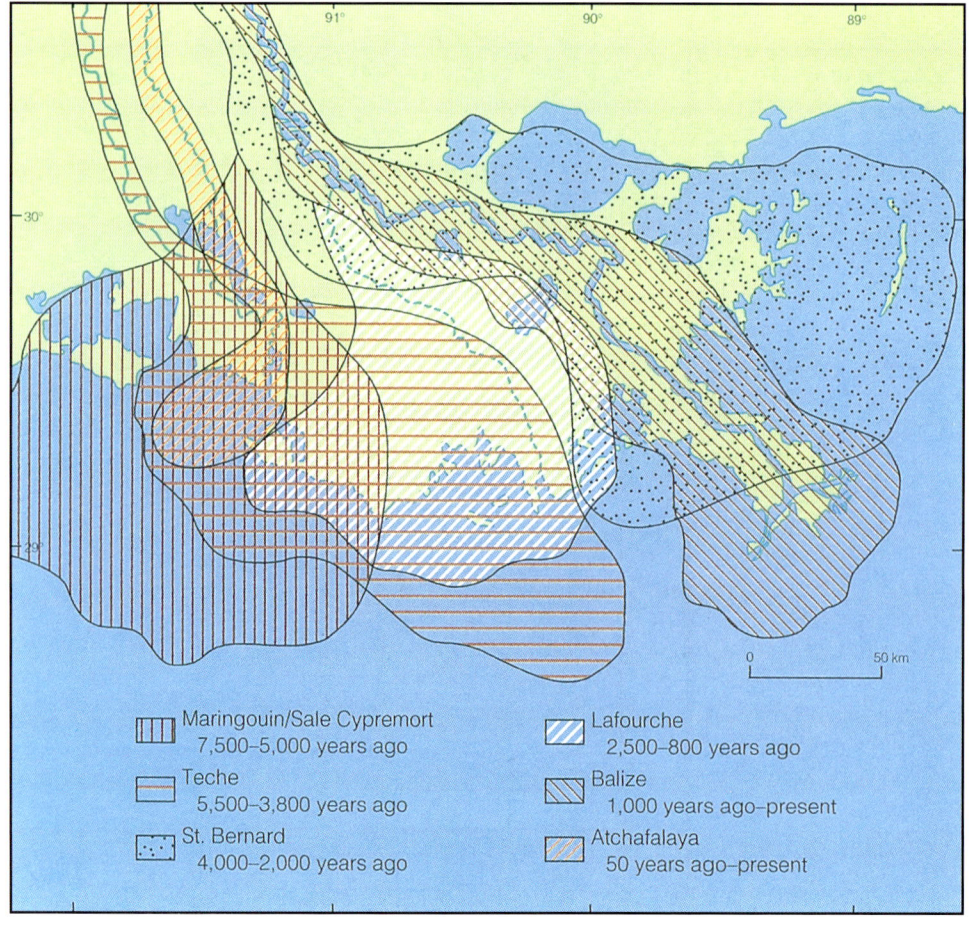

FIGURE 9.54 *The present Mississippi River delta complex consists of four abandoned delta lobes and two active lobes.*

deposition, and seawater begins to encroach into the interdistributary areas. In response to the salt water, marsh plants become stressed and die; in time, the swamp-marsh complex reverts back to open water. The edge of the delta, now open to the sea, begins to undergo wave erosion. The sands of the distributary mouth bar are redistributed into spits. With continued subsidence, seawater eventually surrounds and isolates the spits converting them into barrier islands (Figure 9.55). In time, the process of delta switching will once again return the river to the area, a new delta lobe will prograde across the deteriorated delta lobe, and the cycle will repeat. About 2,000 to 4,000 years are required for the entire cycle from progradation to the formation of barrier islands.

Loss of Land Natural processes other than compaction-induced subsidence have contributed to the loss of land within the Mississippi delta. Over the past 65 years, sea level in the Louisiana delta has been rising at the rate of a little more than 0.5 inches (1.5 cm) per year. Land has been lost to catastrophic storms that breach and destroy barrier islands and inundate the wetlands. Over longer periods of time, the natural downwarping of the continental margin and long-term climatic changes have also influenced land loss.

Unfortunately, however, over the past century, various human activities have accelerated the rate of land loss and, in some areas, are primarily responsible for the loss of land. South of Baton Rouge, the U.S. Army Corps of Engineers constructed a series of artificial levees and spillways to protect the adjacent land from the catastrophic effects of floods. Unfortunately, the levees prevent the diversion of water and sediments from the main channel into the interdistributary areas. Without the periodic influx of sediment to offset subsidence, the surface of the land subsides below sea level. To keep the land from being flooded, water must be removed by networks of drainage canals or pumps. A well-known example is the city of New Orleans. Nearly 45% of the city is below sea level, with some areas as much as 18 feet (6 m) below. To keep the city dry, water is removed via 87 miles (140 km) of canals, 57 miles (92 km) of pipelines, and 21 pumping stations capable of removing 22.5 million gallons (85.1 million l) of water per minute from streets.

The dredging of canals for navigation and for access to oil and gas production sites has accelerated the loss of land by allowing salt water to encroach and destroy the vegetation of marshes and swamps. As a result, ponds begin to form that eventually enlarge and eliminate the marsh-swamp complex. Another major cause of surface subsidence, the subsurface withdrawal of oil, gas, and sulfur, removes support for the overlying sediments. The collapse of the petroleum reservoirs and the cavities left by the removal of sulfur results in the subsidence of the surface.

The combined effect of natural and human activities has been a sharp increase in land loss during the present century. In 1918, land was being lost at a rate of about 6.7 square miles (17.3 km^2) per year. By 1980, the rate of land loss had increased to nearly 40 square miles (100 km^2) per year, a figure that represents 80% of the total U.S. loss of wetlands.

Coastal Erosion Once a delta lobe is abandoned, erosion of the distributary mouth bar results in the formation of a series of spits and, subsequently, the system of barrier islands (refer to Figure 9.55). As the name implies, the barrier islands protect the wetlands that lie beyond from the energy of the sea, especially from storms. Barrier islands along the Louisiana coast are being lost at an alarming rate. The highest rate of loss by any barrier island system is being experienced by the Isles Dernieres (Figure 9.56), which have lost 77% of their area over the past 135 years. At the

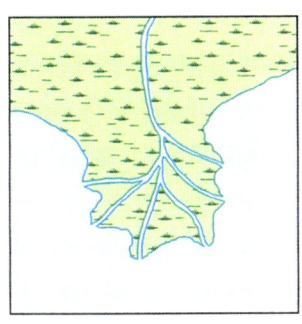
(a)

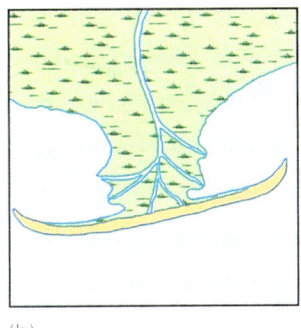
(b)

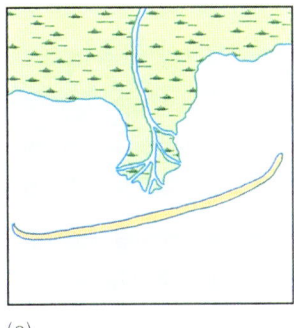
(c)

FIGURE 9.55 *As a delta lobe is abandoned, the deltaic deposits are eroded by wave action and storms. The sands of the distributary mouth bar are redistributed to form* spits*. With time, a combination of subsidence and sea level rise results in the spit deposits becoming isolated and converted into* barrier islands.

present time, the barrier is being eroded at a rate of 65 acres per year. At this rate, it is predicted that the barrier system will be totally gone early in the twenty-first century.

As barrier islands are destroyed, saltwater encroachment destroys the marsh vegetation and accelerates the loss of wetlands as ponds are extended into the delta plain. At the present time, the Louisiana Gulf coast has the highest coastal erosion rate in the United States losing an average of about 12 feet (4 m) each year. In comparison, the average erosion rate for the entire Gulf coast shoreline is 5.5 feet (1.8 m) per year; the average erosional rate for the Atlantic coast is only about 2.5 feet (0.8 m) yearly, while the Pacific coast experiences no net loss of land due to erosion.

Alluvial Fans

Alluvial fans are deposits that form at the mouth of a stream where it emerges from a steep-gradient, V-shaped mountain valley onto a valley floor (Figure 9.57). Although alluvial fans form in humid regions, they are best observed associated with desert streams. Because a desert stream usually disappears quickly by a combination of evaporation or infiltration, its load is dumped with less sorting than is characteristic of deltas. As the alluvial fans of adjacent streams build outward from the base of the mountain onto the valley floor, they commonly overlap to form a deposit along the base of the mountain range called **bajada** (Figure 9.58). Bajadas from the bases of parallel facing mountain ranges may then build across the

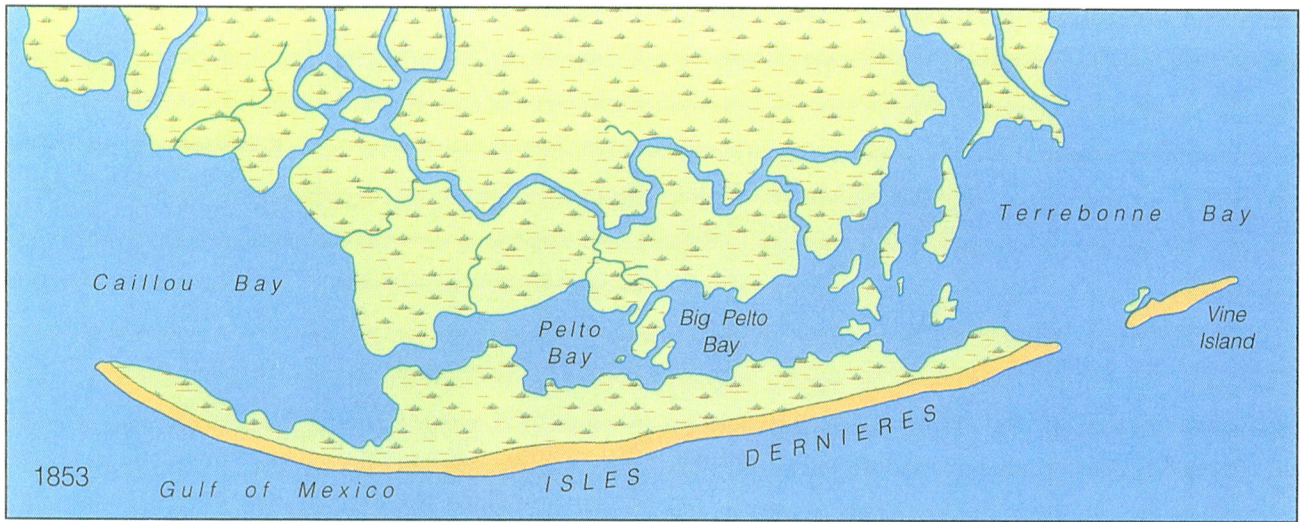

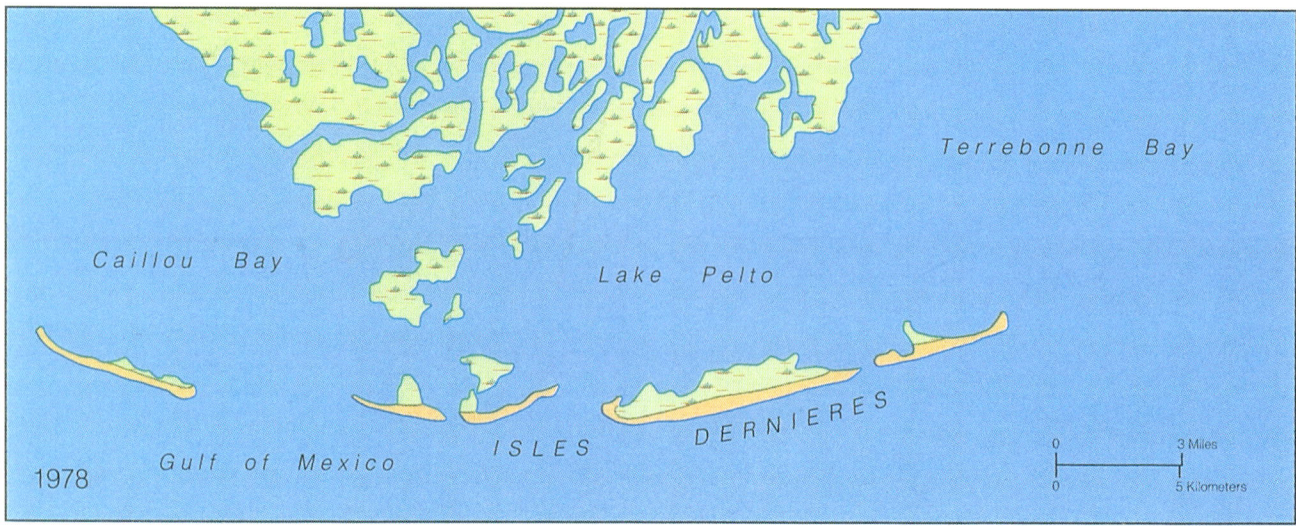

FIGURE 9.56 *The* Isles Dernieres, *a barrier island system off the abandoned Lafourche delta lobe, is undergoing extreme erosion. At the present rate of destruction, the islands are not expected to last more than 50 years. (Used with permission from S. Penland and R. Boyd, Shoreline Changes on the Louisiana barrier coast,* Oceans, *81:209–219, 1981)*

Streams 273

FIGURE 9.57 Alluvian fans *are the most commonly observed depositional form in desert areas. (Courtesy of Death Valley National Monument)*

FIGURE 9.58 Overlapping alluvial fans *along the base of a mountain range form a* bajada. *In time, bajadas growing from opposite sides of an intermontane valley eventually meet in mid-valley, completely covering the valley floor with sediment and forming a sediment-filled valley called a bolson. (Courtesy of NASA)*

intermontane valley and eventually fill the valley with debris; the result is a sediment-filled valley called a **bolson** (refer to Figure 9.57).

The retention of the products of weathering within the general area is one of the major differences between continental areas undergoing erosion by internal stream systems and areas dominated by external streams. By retaining the materials shed from surrounding highlands in adjacent basins, highland areas in arid regions are literally buried in their own debris. We will return to a discussion of desert erosion and deposition in Chapter 11.

SPOT REVIEW

1. What is meant by "sorting" and how is it accomplished by a stream?
2. What is the difference between a delta and an alluvian fan?
3. How would you expect the materials that accumulate in a stream channel to differ from those that accumulate on a floodplain?

ENVIRONMENTAL CONCERNS

Streams raise several environmental concerns, in particular, the problems associated with flooding and pollution.

Floods

The most serious environmental problem associated with streams is the loss of life and property due to **flooding**. In the United States, floods claim up to a hundred lives and cause $1 to $2 billion of damage each year. It is important, however, to distinguish between *upstream* and *downstream* floods.

Upstream floods are usually restricted to the more distal, smaller tributaries of a larger drainage basin. They are usually the result of short-duration, torrential rainfall, although in mountainous areas they may result from the rapid release of snow melt. Some of the most destructive upstream floods have resulted from a dam collapse (refer to Figure 16.17). Upstream floods can be destructive within the limits of the affected stream channel. However, because their destructive potential diminishes as the waters move downstream into larger stream channels, they cause little destruction in the downstream segments of the basin.

Downstream floods are more widespread and affect a much greater area of the drainage basin. Most downstream floods are caused by long-duration, widespread, seasonal rainfall. As the soil and regolith become saturated, additional rainfall runs off into the streams, which eventually overflow. While upstream flooding may occur almost without warning, as in the case of a failed dam, downstream flooding almost never occurs without adequate time to implement protective measures such as evacuation.

Frequency of Flooding Although a stream may run bankfull several times each year, the channel dimensions of most streams are such that the stream will overflow only once every 1.5 to 2 years. In other words, floods can be expected to occur within any stream basin once every year or two. Floods of increasingly greater discharge occur at progressively longer intervals. These are designated as *10-year, 25-year, 50-year,* and *100-year floods* (Figure 9.59). It must be remembered, however, that although a 50-year flood is statistically expected to occur only once every 50 years, two or more 50-year floods can occur within a much shorter time span, all of which points out the inherent unpredictability of most natural events.

Effect of Urbanization Almost *any* human intrusion into an area will increase the frequency of flooding. As land is plowed and tilled, the soil is exposed to more rapid rates of runoff and erosion.

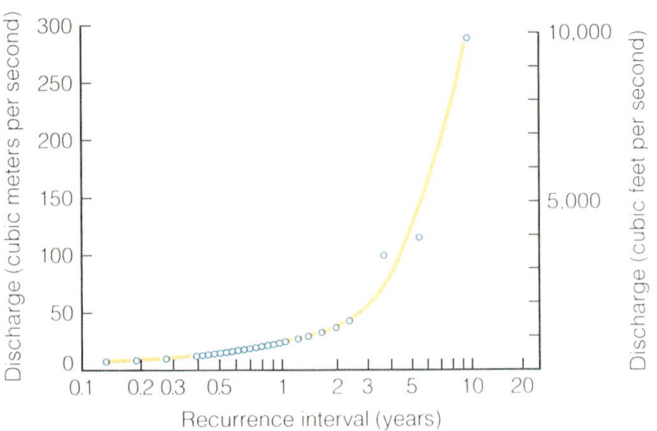

FIGURE 9.59 *Statistically, the time interval between floods of a given discharge increases exponentially with increasing discharge. As a result, one can expect years or tens of years to lapse between major floods while several small floods may occur in the same year. It must always be kept in mind, however, that a stream may experience two major floods within the same year.*

The planting of row crops alters the relative rates of infiltration and runoff and subsequently affects the rate at which water enters streams.

The greatest impact on stream discharge, however, comes when agricultural lands are urbanized and the land surface is covered with homes, driveways, roadways, shopping centers, and parking lots. With continued urbanization, structures appear that are impervious to water. Areas that were once covered with grass give way to commercialization. Apartment buildings replace single-family dwellings. More and larger shopping areas with ever-widening expanses of concrete and macadam are built, all of which prevent the infiltration of water. Increasingly, in urbanized areas, underground conduits replace streams as water is drained by storm sewers. In many areas, the storm sewers eventually discharge into the remaining streams. Because the streams have had insufficient time to prepare for the increased discharge by expanding their channel dimensions, even relatively small rainfalls will cause them to overflow. A nearly linear relationship exists between the ratio of flooding before and after urbanization and the degree of urbanization as measured by the percentage of the area covered by impervious surfaces. Urbanization primarily increases the number of relatively small floods and has little, if any, effect on the recurrence of 50- or 100-year floods.

Floodplains   *Floodplains* have always been desirable sites for human habitation. The presence of relatively flat, fertile soils attracts farmers. The access to transportation attracts merchants, and the ready source of water appeals to city builders. One might wonder why people choose to live where they will be inundated by major floods several times during their lives, but obviously those who live on floodplains believe the advantages outweigh the recognized disadvantages of flooding.

People's desire to live on floodplains has led to the construction of various devices that prevent or control flooding including dams, levees, and diversion channels. All of these structures serve only to provide the floodplain dweller with a false sense of security. Although they can control small floods, flood-control devices may be overwhelmed by 25- or 50-year floods. Any attempt to construct barriers designed to provide protection from a 100-year flood would be impractical.

Although attempts are still made to *control* the inevitable flood, the more modern approach to life on a floodplain is **floodplain management.** Accepting that people will not abandon the floodplain, planners have established rules for its use. The first step in floodplain management is to establish a *floodway* for a 100-year flood. This can usually be accomplished by using historical data aided by satellite photography. The outer margin of the 100-year floodway where the effects of the flood will be minimal is delineated as the *floodway fringe*. Within the floodway fringe, a permit is required for the construction of homes, apartments, and storage facilities, and the builder must show that the proposed structure is capable of resisting the impact of relatively shallow floodwaters. Within the inner floodway, where the full effect of the floodwaters will be concentrated, land utilization is restricted to uses such as farming, pastures, parks, and parking areas; structures are limited to facilities that are considered expendable, such as boat docks, or can be repaired at an acceptable cost. Most importantly, the area is restricted to uses where there is little chance for the loss of life.

Water Pollution

Pollutants can be categorized as either *undesirable* or *harmful*. An example of an undesirable pollutant is dissolved iron, which gives water an undesirable metallic taste and stains the glazed surface of washbasins and bathtubs but is not particularly harmful to the body. An example of a harmful pollutant is dissolved lead or mercury, which, even at very low concentrations, can cause severe neurological damage and death.

Water pollutants come from either *point sources* or *nonpoint sources*. Most urban water pollution comes from **point sources** such as the drainage from storm sewers and the effluents from industrial sites and sewage treatment plants (Figure 9.60). The advantage of point source emissions is that most may be isolated and treated before the waters are released into streams. All point source emissions are controlled by law with specific limits placed on the concentrations of individual dissolved materials. Upon occasion, these limits can be exceeded. For example, in many municipalities, the storm sewers drain into the sewage treatment plant so that contaminants such as oil and the lead from leaded gasoline can be removed. Unfortunately, during periods of high rainfall, the sewage treatment facility may be overwhelmed by the abnormally large discharge from the storm sewers. Unless the plant design allows the storm sewer drainage to temporarily bypass the treatment facility, untreated sewage may be discharged into the stream.

Nonpoint source emissions are much more difficult to control because their entry into surface and subsurface water supplies is more diffuse. Nonpoint sources include organic and chemical contaminants from livestock feedlots, chemical additives applied to crops, petroleum residues, salts and lead washed from streets and highways, and seepage from rural septic systems and solid waste disposal sites (Figure 9.61).

FIGURE 9.61   *The sources of water pollution that are most difficult to control or treat are those where the pollutant is diffusely introduced over a relatively large area and only becomes a problem when it becomes concentrated downstream. Feedlots are among the most serious non-point sources of pollutants. (Courtesy of the USGS)*

FIGURE 9.60   *Most urban and industrial sources of water pollution are from* point sources *and can therefore be controlled. The sewage emanating from our homes would be an example. The holding pond shown in view b is lined to prevent the water from entering the groundwater before it can be treated. (a VU/© Doug Sokell, b © Science VU/Visuals Unlimited)*

SPOT REVIEW

1. How has land use affected the natural frequency of flooding?
2. How does floodplain management differ from flood control?

CONCEPTS AND TERMS TO REMEMBER

Streams
Hydrologic cycle
  hydrosphere
Drainage basins
  drainage systems
    external
    internal
  drainage divides (interfluves)
    continental divide

stream orders
stream patterns
  dendritic
  rectangular
  trellis
    watergaps
  radial
  annular
  parallel

reversed radial
disappearing streams
braided
Stream parameters
  discharge
  gradient
  channel texture
  water volume

Water flow
  laminar flow
    laminar or viscous
    sublayer
  turbulent flow
Erosion
  force
    direct lifting
    traction
  capacity and load
    grade
Stream load
  dissolved
  suspended
  bed
    saltation
Topographic relief
  abrasion
  dissolution
Baselevel
  temporary
  ultimate

Fluvial landforms
  V-shaped valley
  valley flat
  floodplain
    meandering
    sinuosity
    oxbow lake
  peneplane
  rejuvenation
    terraces
      paired
      unpaired
  meandering streams
    reach
    bend
    cut bank
    point
    point bar
Development of grade
  climatic influences
  tectonics

Kinds of alluvial deposits
  channel
  floodplain
    levee
  delta
    foreset beds
    bottomset beds
    topset beds
    delta plain
    delta front
    prodelta
  alluvian fan
    bajada
    bolson
Flooding
  upstream
  downstream
  floodplain management
Water pollution
  point source
  nonpoint source

REVIEW QUESTIONS

1. Most of the water outside the ocean basins is contained in
   a. streams.
   b. lakes and ponds.
   c. glaciers.
   d. groundwater.
2. The product of width × depth × water velocity is the ____ of a stream.
   a. capacity
   b. discharge
   c. gradient
   d. load
3. Most of the load of a stream is carried in
   a. bed load.
   b. dissolved load.
   c. suspended load.
   d. invisible load.
4. Except for the laminar sublayer, the least turbulent part of a stream is found
   a. in the headwaters.
   b. midchannel.
   c. at the mouth.
   d. in the bend nearest the cutbank.
5. The particle size most easily eroded by a stream is
   a. silt.
   b. sand.
   c. clay.
   d. granule.
6. The competence of a stream is measured by
   a. the total volume of water moved during any period of time.
   b. the total amount of load being transported.
   c. the largest particle size being moved.
   d. the maximum water velocity.
7. Which of the following is not characteristic of a youthful stream valley?
   a. **V**-shaped valley profile
   b. waterfalls and rapids
   c. high turbulence
   d. levees

REVIEW QUESTIONS continued

8. Rejuvenation is caused by
   a. a rapid lowering of the baselevel.
   b. a rapid increase in gradient.
   c. increased rainfall.
   d. increased discharge.
9. Within a delta, the finest particle sizes accumulate in the
   a. bottomset beds. c. topset beds.
   b. foreset beds.   d. cross-beds.
10. Distributaries differ from tributaries in that
    a. distributaries are larger.
    b. distributaries have higher gradients.
    c. distributaries are concentrated in the headwaters.
    d. distributaries flow out from the main channel while tributaries flow into the main channel.
11. In what ways do basic stream parameters change with increasing stream order?
12. What stream parameters determine the amount of energy available to a stream?
13. How many changes in climate affect the state of grade within a stream system?
14. How are capacity, load, and competence related?
15. In what way does stream turbulence determine the amount of load carried by a stream?

THOUGHT PROBLEMS

1. Environmentalists opposed the construction of the Lake Powell Dam on the Colorado River, arguing that the dam would have profound, detrimental effects on the river. Were their concerns justified? What particular effects has the presence of the dam had on the downstream segments of the Colorado River? In general, what kinds of effects will the presence of an impoundment the size of the Lake Powell Dam have downstream from the dam? Your geology instructor should have information regarding the effect on sedimentation in the river while a faculty member in the Biology Department may have specific information about the effect on the plants and animals.
2. Consider the kinds of sediment deposited in the various alluvial environments. What kinds of sedimentary rocks do you think would characterize each environment? What sedimentary features would you expect to find in the rocks that would be diagnostic of each depositional environment? How would you go about determining the direction of stream flow?
3. How is it possible that the level of the water in the Mississippi River at New Orleans, Louisiana, is tens of feet (meters) above the elevation of the surrounding land? What kinds of problems do you think this situation creates for the residents of the city?

FOR YOUR NOTEBOOK

Start by determining the kind of stream pattern that is characteristic of your area. Compare the stream pattern to a geologic map, and attempt to determine what factors seem to control the local and regional stream patterns. Using topographic maps, outline the major divides, and identify and determine the area of each drainage system.

Plan a reasonably extensive field trip to look at some of the streams firsthand. If possible, walk segments of selected streams, and try to locate some of the features you have read about such as point bars and cutbanks. Thoroughly investigate the valley floor looking for evidence of channel migrations or terraces that record former floodplain elevations. Make note of the bed load sediments, in particular, the downstream changes in size and roundness.

In some areas, stream sediments are being mined for construction sand and gravel. If you have an opportunity to collect and study some of these materials, especially the larger gravels, look for particles of rock types that are not indigenous to your area. Should you find any, try to determine the most likely source.

If you happen to live in an area where the exposed sedimentary rocks are fluvial in origin, you might want to study selected road cuts and attempt to identify rocks that accumulated under different depositional environments. A good way to study outcrops exposed in road cuts is to make a photo montage that you can overlay with clear plastic to protect the photo surface while you mark and measure various features.

Investigate any methods of flood control that you have been utilized to protect people and property as well as the flood history of your area. You may want to consult cataloged newspapers in your local library to compare the descriptions of flood impact before and after the construction of local flood-control devices.

CHAPTER OUTLINE

INTRODUCTION

THE DISTRIBUTION OF GLACIERS

THE MOVEMENT OF GLACIAL ICE
  Ice Thickness
  Ice Temperature
  Slope of the Bedrock and Ice Surface

THE GLACIAL BUDGET

PROCESSES OF GLACIAL EROSION
  Erosion by Alpine Glaciers
  Erosion by Continental Glaciers

GLACIAL DEPOSITION
  Kinds of Glacial Deposits

PERIGLACIAL REGIONS
  Mass Wasting
  Weathering
  Streams
  The Wind

UNDERSTANDING CONTINENTAL GLACIATION
  The Great Ice Age
  The Great Lakes
  Glaciation in the Western United States during the Great Ice Age

THE CAUSES OF CONTINENTAL GLACIATION
  Celestial Causes
  Volcanic Activity
  Plate Tectonics

ENVIRONMENTAL CONCERNS
  Rising Sea Levels
  Goiter

CHAPTER 10

Glaciers

INTRODUCTION   Although running water is the major sculptor of Earth's land surface, few scenes can compare in sheer beauty to those produced by glacial erosion. The Matterhorn, the lofty slopes of Mount Everest, the Grand Tetons, and the rugged summits of the northern Rocky Mountains are all the work of glaciers (Figure 10.1).

But the mountains are not the only place to see the impressive results of glaciation. The broad expanses of the Bonneville Salt Flats, the Great Lakes, the beautiful Finger Lakes of upstate New York, Long Island, and much of the topography of New England, including historic Bunker Hill, are all the results of past glacial activity. Given the existence of such landforms, few would deny that glaciers have done their share in providing us with magnificent scenery.

Currently, glaciers are being studied to provide baseline information for air and water pollution. By analyzing air entrapped in glaciers and the trace element composition of the ice, scientists are able to establish the composition of the atmosphere before the impact of humans.

THE DISTRIBUTION OF GLACIERS

Glacial ice covers about 5 million square miles (15.6 million km^2) of Earth's surface today. About 90% of the total is found on the continent of **Antarctica** where the ice sheet reaches thicknesses of more than 15,000 feet (4,776 m) in places. The second largest ice mass on Earth, the **Greenland** ice sheet, covers about 80% of Greenland; it has an area of nearly 700,000 square miles (1.8 million km^2) and a maximum thickness of a little more than 10,000 feet (3,000 m) (Figure 10.2). The Antarctic and Greenland ice sheets are examples of **continental glaciers**. The remaining 250,000 square miles (648,000 km^2) of ice are scattered throughout the mountains of the world as **alpine** or **valley glaciers** and **piedmont glaciers**. Alpine glaciers form in high mountain areas and flow downslope following former stream valleys (Figure 10.3). Piedmont glaciers form when alpine glaciers emerge from mountain valleys and spread out beyond the base of the mountain (Figure 10.4).

Inasmuch as trips to the Antarctic or Greenland are not usually on the average vacation agenda, most of us are more likely to have a chance to observe an active alpine glacier than a continental glacier. Active alpine glaciers can be seen in many places from Montana to Alaska. In fact, some of the largest alpine glacier fields in the world can be seen just north of Juneau, Alaska, in the Saint Elias Mountains.

THE ORIGIN OF GLACIAL ICE

The formation of glacial ice requires the accumulation and preservation of massive volumes of snow. Glaciers are therefore limited to regions that receive adequate precipitation and are either at sufficiently high elevations or at latitudes where large quantities of snow can accumulate and be preserved. The elevation above which snow can accumulate and be preserved from year to year is called the *snow line*. In the polar regions, the snow line is found at sea level, approximately at the location of the Arctic and Antarctic Circles. Toward the lower latitudes, the elevation of the snow line rises although it is not necessarily highest at the equator. Because of the high levels of precipitation and cloud cover, the elevation of the snow line at the equator is generally lower than at the higher latitudes of the Tropics of Cancer and Capricorn where Earth is bathed in descending masses of very dry air.

FIGURE 10.1 *Some of the most spectacular mountain scenery throughout the world is the result of glaciation. (a courtesy of the Swiss National Tourist Office, b courtesy of Cecil W. Stoughton/National Park Service, c Courtesy of P. Carrara/USGS)*

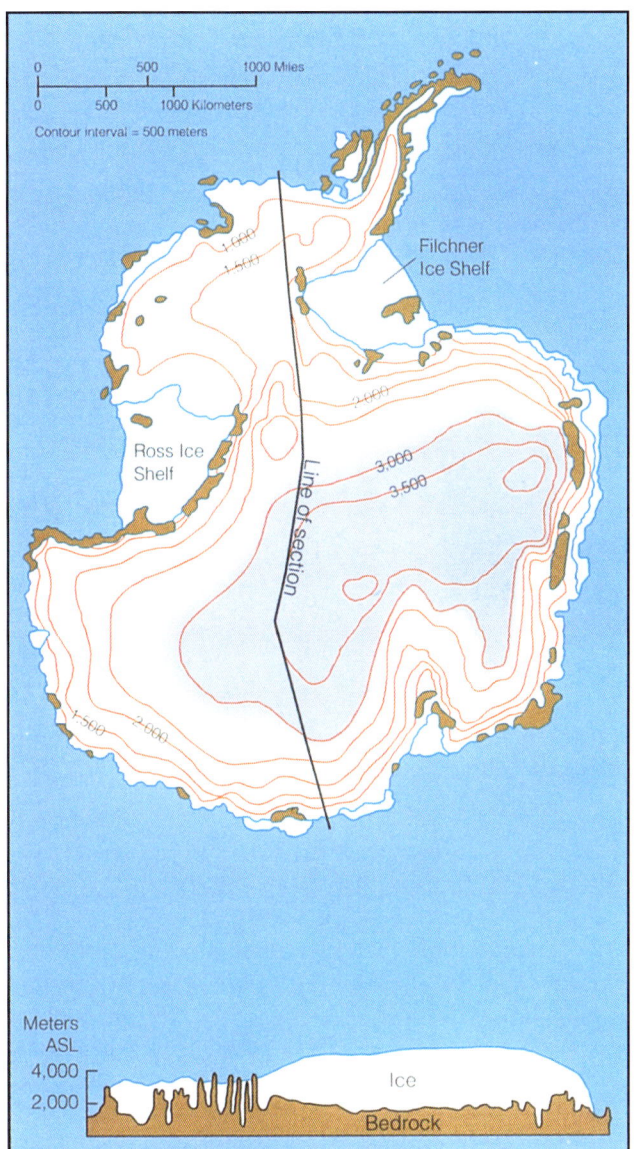
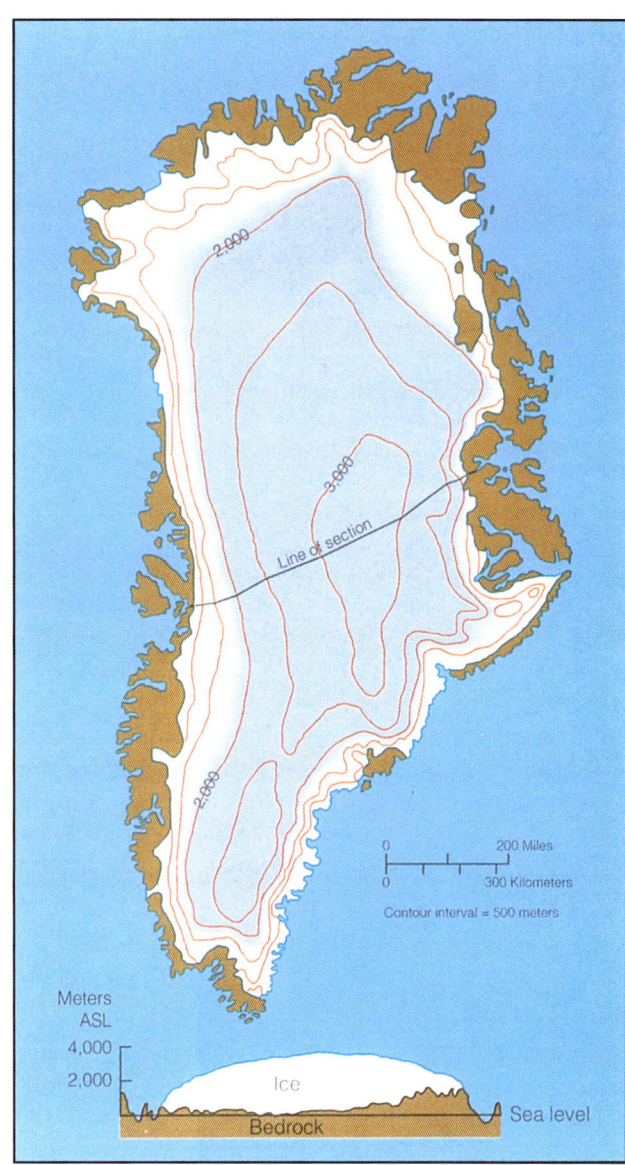

FIGURE 10.2 *About 90% of Earth's glacial ice is located over the landmasses of* Antarctica *and* Greenland.

FIGURE 10.3 Alpine *or* valley glaciers *form on mountain summits and flow downslope, following former stream valleys. (Courtesy of R. G. McGimsey/USGS)*

FIGURE 10.4 Piedmont glaciers *form when alpine glaciers flow beyond the mouth of a mountain valley and out onto the adjoining valley floor. (Courtesy of P. Carrara/USGS)*

The transformation of snow into glacial ice begins with the **melting** or **sublimation** (the conversion of a solid to gas without becoming liquid) of the individual snow flakes followed by crystallization of the water or water vapor into granular ice called **névé** or **firn** (Figure 10.5). Upon being buried and compacted, the grains of ice undergo **pressure melting**. Increased pressure affects the melting/freezing points of solids by favoring the denser phase, which, in this case, is water. As the ice melts and the meltwater moves into the pores between the grains and refreezes, most of the air is eliminated, and the grains of ice are cemented together to form glacial ice. Note that the process of formation qualifies glacial ice as a metamorphic rock.

The rate at which snow is transformed into glacial ice depends in large part on the climate and on the amount of snow accumulated. In temperate regions where the temperatures of the snow and firn may approach the melting point during the warmer seasons of the year, the availability of water allows the transformation to take place within a few years. In the polar regions, where the temperatures are constantly below freezing, the process may take hundreds of years.

THE MOVEMENT OF GLACIAL ICE

The movement of glacial ice is dependent on (1) the *thickness* of the ice, (2) the *temperature* of the ice,

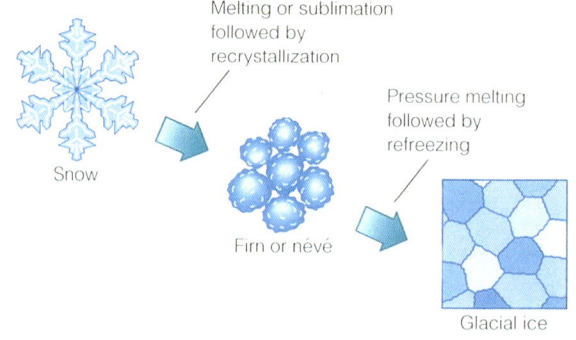

FIGURE 10.5 *The first step in the formation of glacial ice is the transformation of* snow *into granular ice called* névé *or* firn. *With burial and compaction, the firn undergoes* pressure melting; *the individual ice crystals are cemented together as the water freezes in the pores between the grains.*

and (3) the *slope* of both the bedrock and the surface of the ice mass.

Ice Thickness

At atmospheric pressures, ice is a brittle solid as the jab of an ice pick can easily prove. Once ice accumulates to a thickness of more than about 150 feet (46 m), however, the pressure generated by the weight of the overlying ice, combined with the directional forces induced by the slope of the underlying surface, causes the ice below 150 feet to respond as a plastic material and flow (Figure 10.6). The mechanism of plastic flow is not fully understood. If the temperature

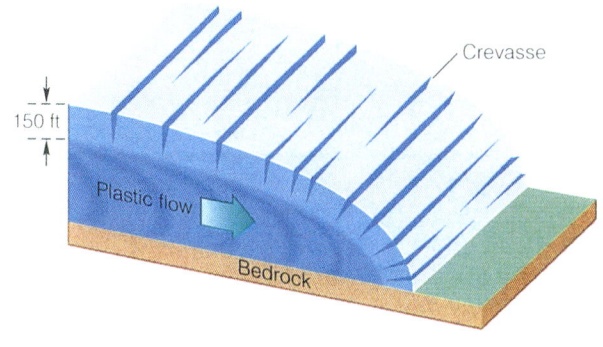

FIGURE 10.6 *Once the thickness of glacial ice exceeds 150 feet (46 m), the lower portion of the ice begins to undergo* plastic flow. *The brittle surface ice breaks into fractures called* crevasses *as it is carried along by the flowing ice below.*

of the ice is near the melting point, the process of **plastic flow** may involve pressure melting and recrystallization of the ice. On the other hand, where the ice temperature is far below the melting point, plastic flow may be the result of solid-state recrystallization or inter- and intra-granular movements. Again, an indication of the metamorphic character of glacial ice. While the layers of the ice below 150 feet move by plastic flow, the overlying, brittle ice breaks, forming fractures called **crevasses,** and is carried "piggyback" on the plastically deformed layer below.

Ice Temperature

The temperature of glacial ice increases with depth. Heat within the glacier may come from a number of sources including internal friction, geothermal heat conducted upward from the underlying bedrock, and, in some cases, the presence of underlying hot springs and near-surface magmas.

In most polar ice caps and ice sheets, the temperatures at the base of the ice are *far* below the melting/freezing point of ice, and as a result, the ice is frozen to the underlying bedrock. Such glaciers, called **dry-based** or **cold glaciers**, move primarily by plastic flow.

The pressure of burial, however, causes the melting/freezing point within a mass of glacial ice to decrease with depth at a rate of 1.2°F per 3,300 feet (0.7°C per 1,000m). If the temperature of the ice at the base of the glacier has been elevated to near the melting point, the ice can undergo *pressure melting* as it moves and encounters irregularities in the bedrock surface. In a process called **regelation**, the ice melts on the "upstream" side of obstructions, and the water flows around or over the obstruction and refreezes on the "downstream" side. The transfer of mass causes the ice at the base of the glacier to move by a process called **regelation slip.**

As a result of the combined processes of regelation and regelation slip, a film of water a millimeter or two thick may be generated at the base of the glacier. If the water layer becomes continuous under a large portion of the glacier, it may separate the ice from the underlying bedrock, resulting in the formation of a **wet-based glacier.** The layer of water beneath a wet-based glacier serves as a lubricant and promotes the movement of the glacier by **basal sliding.** It is estimated that at least half of the downslope movement of alpine or valley glaciers results from a combination of regelation slip and basal sliding. In glaciers resting on very steep slopes, basal sliding may account for *all* of the movement.

Wet-based glaciers are not restricted to temperate climates, however. At Byrd Station in the Antarctic, the bottom of a borehole drilled 7,100 feet (2,164 m) through the glacial ice to bedrock encountered a layer of water about 1 mm thick at the base of the glacier. The calculated melting point of the ice at the bottom of the borehole was 29.1°F (–1.6°C) while the surface temperature was –21.3°F (–28.8°C). Due to the presence of the wet base, the ice moved more than an inch per day. The rapid movement of the ice constantly caused the core drill to jam and frustrated attempts to retrieve a core from the underlying bedrock. The unexpected high temperature at the ice-bedrock contact at Byrd Station was explained by the fact that this portion of the Antarctic landmass is a volcanic archipelago.

At the margins of glaciers where the ice is too thin for either plastic flow or basal sliding, the ice may move as a result of **brittle fracture.** In the steep headwaters of alpine glaciers, for example, downslope movement of the ice may be the result of masses of ice rotating along arcuate (curved) fractures in much the same fashion as regolith moves during slumping. At the margins of glaciers and at the terminus, fractures develop and allow masses of ice to be driven up and over the underlying stagnated ice (Figure 10.7).

Slope of the Bedrock and Ice Surface

In addition to regelation slip and basal sliding, the rate of downhill movement of alpine glaciers is affected by the slope of the bedrock surface and of the ice mass. The movements of alpine glaciers have been measured at speeds up to 500 to 600 feet (150–180 m) per year, and some valley glaciers in Alaska have moved as much as 10 feet (3 m) per day.

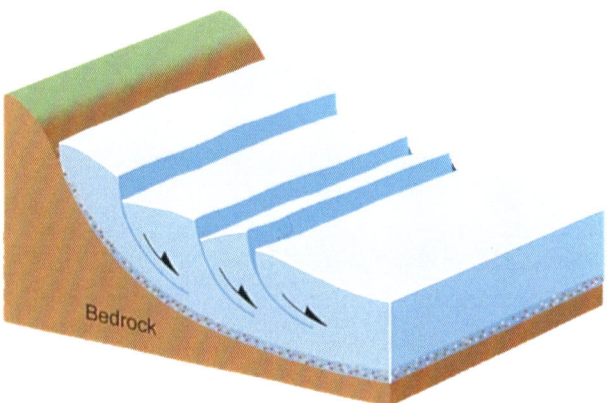

(a) Arcuate tension fractures developed in glacial cirques

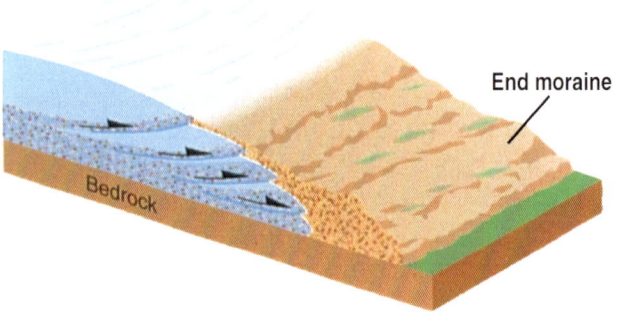

(b) Thrusting of ice along fractures at the toe of the glacier

FIGURE 10.7 *Toward both the headwaters and terminus of the glacier where the ice* thins to less than 150 feet (46 m), it reacts as a *brittle* material and forms fractures along which the ice moves. In the (a) headwaters, gravity drives the ice *downward* along curving fractures while in the (b) terminus, the pressure exerted by the moving ice *thrusts* layers of ice *forward and over the underlying ice.*

The highest recorded speeds of alpine glaciers have occurred during **surges** where short-term movements of as much as 1,000 feet (300 m) per day have been observed. Most glaciologists believe that surges are the result of pressures that develop in the upstream ice mass. The pressure buildup has been attributed to a number of factors including restricted downstream flow of the ice and the rapid accumulation of new ice in the source area. One study of a surge experienced by the Variegated Glacier in Alaska indicated that an abnormal buildup of meltwater, generated by regelation at the base of the glacier, resulted in accelerated basal slippage. Once thought to be rare, glacial surges are now known to be relatively common events. Surges in Glacier Bay National Monument in Alaska occur every 20 years.

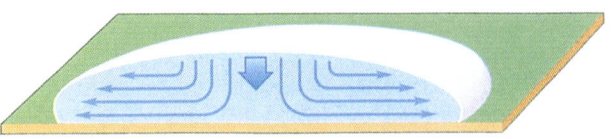

FIGURE 10.8 *In* continental glaciers *where the ice in the center of the ice mass may be thousands of feet (meters) thick, the weight of the overlying ice causes the ice mass to move downward and laterally by* plastic flow *in a fashion not unlike the spreading of pancake batter on the surface of a griddle.*

The rates of movement of glaciers resting on relatively gentle slopes, such as continental glaciers, are affected only by the slope of the ice surface, not by the slope of the underlying bedrock. As long as the thickness of the ice is greater than the relief of the underlying bedrock, the doming of the ice in the center of ice caps or ice sheets produces a surface slope toward the margins of the ice mass. The effect of gravity on the ice mass induces the ice to move laterally by plastic flow (Figure 10.8). As a result, continental ice sheets may ride up and over hills and obstructions as well as into and out of depressions in the bedrock surface.

Recorded speeds of continental glaciers are as varied as those for alpine glaciers. The ice sheet at the South Pole is moving toward the margin of Antarctica at a rate of about 30 feet (9 m) per year while portions of the Greenland ice sheet are moving through buried mountain passes at nearly 5 miles (8 km) per year.

THE GLACIAL BUDGET

The size of a glacier and whether it advances or retreats depend upon the relative rates of ice *accumulation* and *ablation* (wastage). For most glaciers, ice accumulation dominates during the winter season when the precipitation of snow is the highest. Although snow may fall over the entire surface of the glacier, the accumulated snow will be transformed to glacial ice only in the **zone of accumulation** (Figure 10.9). In alpine glaciers, the zone of accumulation is located in the source area while in a continental glacier, it is located in the interior of the ice sheet.

Beyond the zone of accumulation, in the **zone of ablation,** the glacier experiences a net loss in ice mass. In temperate regions, the primary cause of ablation is melting whereas in polar regions, where

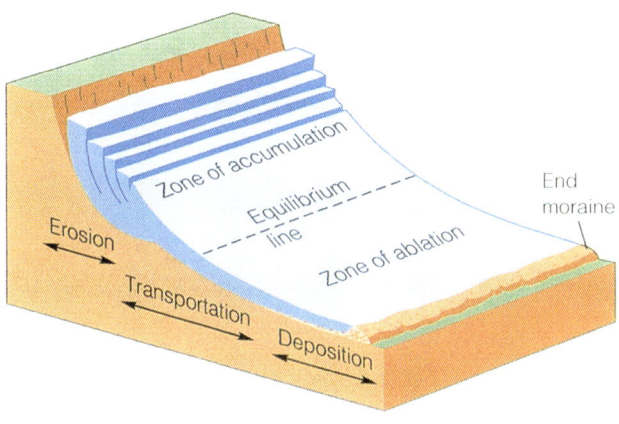

FIGURE 10.9 *Whether a glacier is* advancing *or* retreating *depends upon whether the mass of ice being formed* in the *zone of accumulation is greater or less than the amount of ice being* lost *in the* zone of ablation.

the temperatures are constantly below freezing, sublimation is the dominant ablation process. Glaciers that flow to the sea, into an inland body of water, or over a cliff also ablate by breaking off or **calving** at the terminus (Figure 10.10).

The zones of accumulation and ablation are separated by the **equilibrium line** or, in the case of some valley glaciers, the **firn line** (refer to Figure 10.9). From the glacier's point of origin in the zone of accumulation to the equilibrium line, the ice mass *increases* constantly whereas from the equilibrium line to the terminus of the glacier, the mass of ice constantly *decreases*. At the terminus, ablation equals the rate of ice advance.

When the rate of ice accumulation exceeds the rate of ablation, the glacier will advance. When the rates of accumulation and ablation are balanced, the glacier will come to a "stop." The word *stop* is not entirely accurate, however, because the ice mass is continuously moving from the source to the terminus even though the leading edge of the ice sheet may no longer be advancing.

When ablation dominates over accumulation, the glacier will retreat. During retreat, the loss of ice usually is not concentrated at the terminus. In fact, the ice sheet often is thinned, significantly reducing the volume of the ice mass, while the terminus undergoes minimal retreat. As the ice sheet thins and stagnates, blocks of ice may separate from the main glacier, become stranded, and be buried within the accumulating debris.

FIGURE 10.10 *Where a glacier flows into a body of water and loses support from below, it ablates by breaking off masses of ice, a process called* calving. *(VU/© Will Troyer)*

SPOT REVIEW

1. How is glacial ice distributed around the world?
2. How is snow converted into glacial ice?
3. Compare and contrast the mechanisms by which dry-based and wet-based glaciers move.
4. Describe the mechanism by which a glacier comes to a "stop."

PROCESSES OF GLACIAL EROSION

The processes of glacial and stream erosion are similar in many ways. For example, streams and glaciers both erode by the combined processes of direct lifting and bed load abrasion. The major difference between the erosional styles of glaciers and streams is due to their relative viscosities. Due to its higher viscosity, the competence of ice is much greater than that of water and allows a glacier to carry both larger objects and greater volumes of material than an equal volume of water. At the same time, the high viscosity of ice precludes the sorting of materials during the erosion, transportation, or depositional phases.

As a glacier advances, it strips the regolith from bedrock by direct lifting. Water generated at the bottom of wet-based glaciers penetrates into fractures within the bedrock, refreezes, and disrupts the rock. The loosened blocks of rock are rotated out of their original position and pulled into the ice mass by a process called **quarrying** or **plucking,** mixed with the regolith stripped from the bedrock surface, and carried off embedded in the bottom few feet of the ice as a glacial "bed load" (Figure 10.11).

As wet-based glaciers move, the bottom surface of the ice with its embedded debris moves over the bedrock surface like a gigantic piece of sandpaper and **abrades** the bedrock. Depending upon the size of the particles of embedded material and the thickness of the ice, the moving ice sheet may generate a variety of abrasion features in the bedrock from *smooth polish* to hairline *striations* or deep *grooves*. Striations and grooves have been used to map the direction of glacial flow (Figure 10.12). As the iceborne fragments abrade the bedrock, they are themselves abraded, producing flat surfaces called *facets* on pebble-sized and larger particles. Because the ice of dry-based glaciers is frozen to the underlying bedrock, they do not abrade the bedrock and erode almost exclusively by plucking.

Most of the load carried by alpine or valley glaciers is not directly generated by ice erosion. Instead it is made up of rock debris that falls onto the surface of the ice as avalanches and rock falls from the oversteepened valley walls and is transported on the surface of the ice. Because continental glaciers normally bury all but the highest bedrock surfaces, they carry little debris on the surface.

Erosion by Alpine Glaciers

Much spectacular mountain scenery is the product of erosion by alpine glaciers (Figure 10.13). Ice at the source of an alpine glacier erodes the underlying bedrock by quarrying and carves a bowl-shaped depression called a **cirque**. When adjacent alpine glaciers are operating simultaneously along both sides of a ridge, the adjacent cirques begin to overlap and form a knife-edged mountain ridge called an **arête**. Two glaciers eroding from opposite sides of a ridge may eventually carve through the ridge and form a high mountain pass called a **col**.

A spectacular topographic feature that results from multiple cirques carving around a peak is a "sharp mountain spire" or **horn**. Perhaps the best-known horn is the Matterhorn in the Swiss Alps

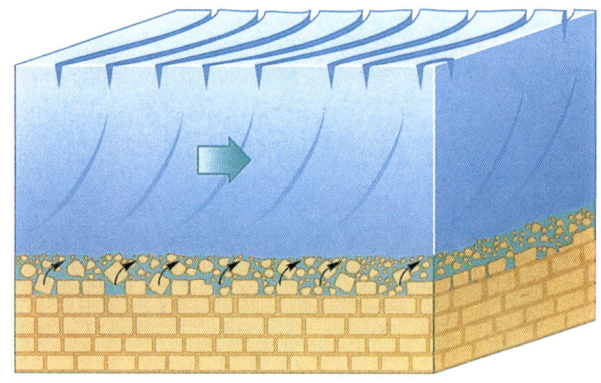

FIGURE 10.11 *As the underlying bedrock is disrupted by freezing water, rock fragments are pulled or* plucked *from the surface of the bedrock and carried off embedded in the bottom of the ice. These rock fragments then become the abrasive surface responsible for the eroding power of the moving ice.*

FIGURE 10.12 *As wet-based glaciers move, rock debris in the bottom of the ice* abrades *and* gouges *the rock surface, forming* striations *that allow geologists to determine the direction of movement of former glaciers. (Courtesy P. Carrara/USGS)*

(Figure 10.14), but equally impressive examples can be seen throughout the northern Rocky Mountains, in particular, the beautiful Grand Tetons just west of Jackson Hole, Wyoming (Figure 10.15).

Once alpine glaciers exit the cirque and move down the valley, the dominant process of erosion becomes abrasion. Rock debris in the underside of the glacier scours the valley floor and walls, straightens the valley and results in the formation of a **glacial trough** or **valley** with a diagnostic **U-shaped** profile (Figure 10.16). One of the best-known examples of a

FIGURE 10.13 *Some of Earth's most spectacular mountain scenery is the result of erosion by alpine glaciers (a). Results of alpine erosion include sharp mountain ridges called* arêtes *(b) and high mountain passes called* cols *(c). (a and c courtesy of W. B. Hamilton/USGS, b courtesy of R. G. McGimsey/USGS)*

U-shaped glacial trough in the United States is Yosemite Valley in the Sierra Nevada of central California. Tributaries to the main glacier carve smaller U-shaped valleys, which, after the retreat of the ice, form **hanging valleys.** These are often sites of spectacular waterfalls where tributary streams plunge hundreds of feet to the main valley.

After the retreat and final melting of alpine glaciers, basins carved by the glacier or depressions in the deposits laid down by the retreating glacier may fill with water to create mountain lakes. The most common of these are the **tarns,** which occupy abandoned cirques (Figure 10.17), while *paternoster lakes* fill depressions in the deposits laid down on the lower

FIGURE 10.14 *A delight to many rock climbers, horns are sharp mountain spires carved by the combined efforts of several glaciers operating around a mountain peak. The most well known of all horns is the* Matterhorn *in the Swiss Alps. (Courtesy of the Swiss National Tourist Office)*

FIGURE 10.15 *The* Grand Tetons *west of Jackson Hole, Wyoming are excellent examples of horns. (Courtesy of Cecil W. Stoughton/National Park Service)*

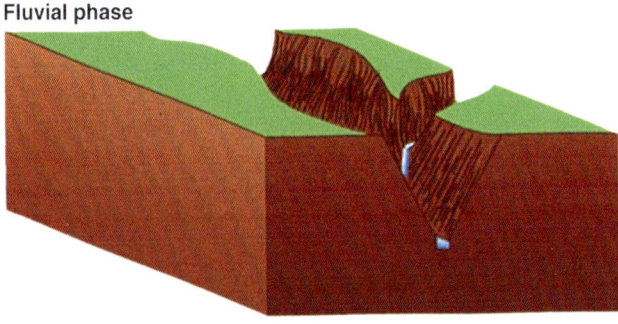
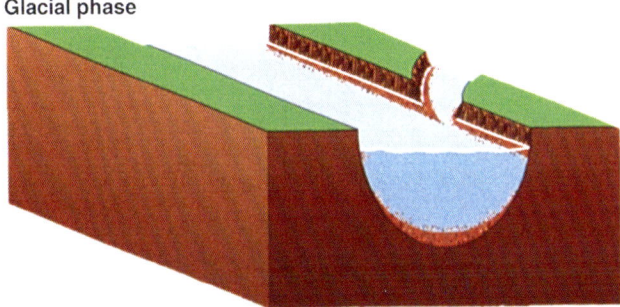
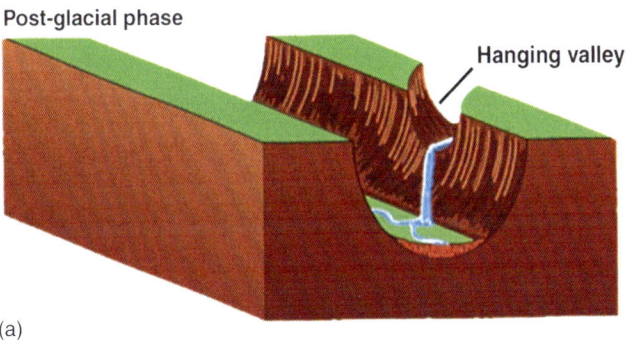

FIGURE 10.16 *One of the most characteristic topographic features formed by glaciation is the U-shaped valley carved by a main valley glacier with its tributary hanging valleys carved by smaller tributary glaciers. (Photo courtesy of N. K. Huber/USGS)*

FIGURE 10.17 *As glacial ice melts from a cirque, the basin commonly becomes filled with water to form a scenic mountain lake called a* tarn. *(Courtesy of W. B. Hamilton/USGS)*

end of the valley floor. Each lake drains into the next lower lake like a string of rosary beads, giving rise to the name paternoster (Latin for *Lord's Prayer*) (Figure 10.18).

In some areas, glacial troughs fill with water behind a rock barricade or a terminal moraine to form *trough lakes*. The Finger Lakes of central New York are examples of trough lakes (Figure 10.19).

Where glacial troughs enter the sea, the valley floods to form a *fjord*. In most fjords, the water becomes shallower toward the sea. In some cases, the shallowing is due to the presence of a terminal moraine. In most cases, however, it occurs because the ability of the glacier to scour the bedrock of the

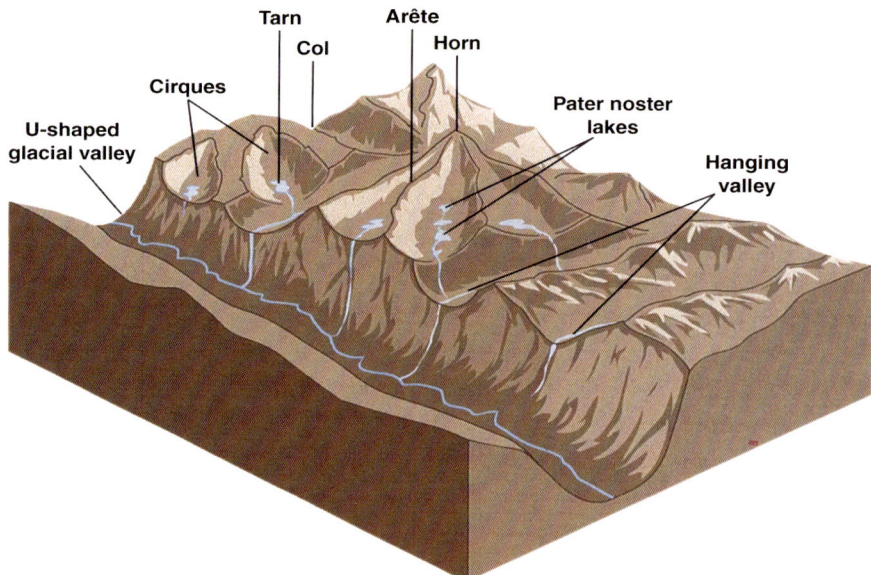

FIGURE 10.18 *Strings of lakes called* paternoster lakes *form down the glacial valley as water imponds behind morainal materials. The name comes from their resemblance to strings of prayer beads as streams join one lake to the next.*

FIGURE 10.19 *The best-known examples of* trough lakes *in the United States are the* Finger Lakes *of upstate New York. (VU/© John Meuser)*

trough floor decreases as the ice enters the sea and begins to float.

Erosion by Continental Glaciers

Erosion by continental glaciers is dominantly by abrasion and results in a much more **subdued topography** than that produced by alpine glaciation. Where alpine glaciers sharpen and accentuate ridges and carve valleys with vertical to near-vertical walls, continental glaciers overrun ridges and remove sharp spires and ridge lines by abrasion. A comparison of the topography of the Adirondack Mountains of upstate New York with that of the northern Rockies demonstrates the difference in the erosional styles of continental and alpine glaciation (Figure 10.20).

Topographic high points overrun by continental glaciers are both reduced in elevation and smoothed, while valleys are carved deeper because of the greater thickness of ice that overlies the valley floor (Figure 10.21). In other words, erosion by continental glaciers *increases* the relief of the bedrock surface because elevations of the valleys are deepened more

FIGURE 10.20 *The difference in topography resulting from erosion by* alpine *and* continental *glaciers is well illustrated by comparing the rugged peaks in the northern Rockies and the subdued summits of the Adirondacks of upstate New York. (a courtesy of P. Carrara/USGS, b VU/© W. A. Banaszewski)*

than the hilltops are reduced. As the glacier retreats, however, glacial debris may fill in the valleys to produce a gently rolling, subdued topographic surface. An example is the low, rolling countryside seen throughout Ohio and Indiana, which hides deeply scoured, gravel-filled valleys in the underlying bedrock surface.

Because the ice mass is so huge, continental glaciers have accomplished monumental feats of scouring, several examples of which are found in North America. Foremost among these are the basins now occupied by one of the largest accumulations of fresh water on Earth, the Great Lakes (Figure 10.22). With the possible exception of the basin of Lake Superior, which may have predated the coming of the glacier, the basins occupied by all the Great Lakes were gouged as the North American ice sheet moved southward. As the last ice age waned about 10,000 to 15,000 years ago, the basins began to fill with meltwater behind the edges of the retreating ice. We will examine the Great Lakes in more detail later. Another example of lakes whose basins are the result of the scouring of the last ice sheet as it advanced southward are the Finger Lakes of central New York (refer to Figure 10.20).

An erosional feature formed by a combination of abrasion and quarrying is a **roche moutonnée** (French for "woolly rock" or "rock sheep"). Commonly found in areas that have been subjected to continental glaciation, roches moutonnées are relatively small, elongate bedrock hills that align with the direction of ice flow (Figure 10.23). The more

296   Chapter 10

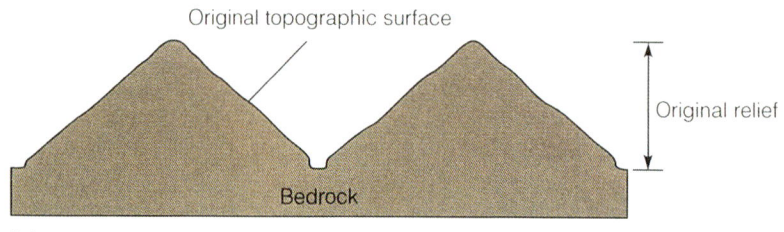

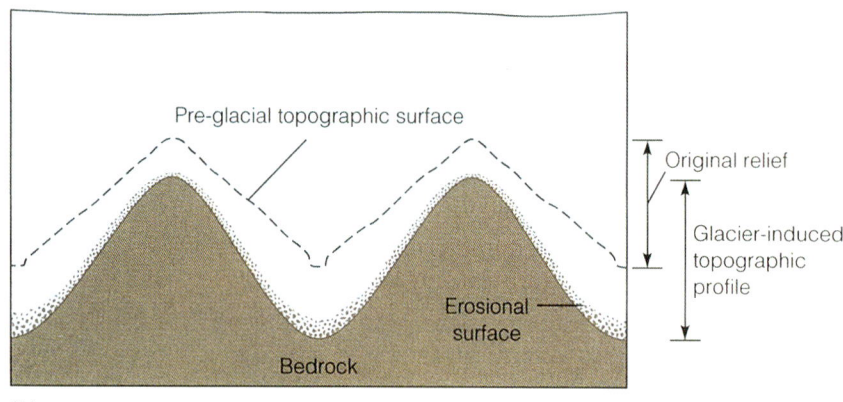

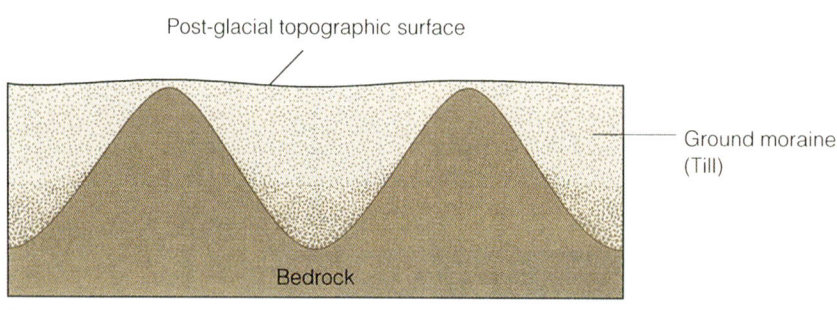

FIGURE 10.21 *Much of the subdued topography characteristic of regions that have undergone continental glaciation is due to the filling of deeply scoured valleys by ground moraine as the glacier retreats.*

FIGURE 10.22 *One of the largest accumulations of fresh water on Earth, the* Great Lakes, *illustrates the enormous erosive power of the advancing Pleistocene ice sheet.*

FIGURE 10.23 Roche moutonnées *are rock masses sculptured by an advancing continental glacier by a combination of abrasion and plucking. These examples, located near Black Bay, Lake Athabasca, Saskatchewan, Canada, are another illustration of the erosive power of the Pleistocene ice sheet. (Courtesy of Geological Survey of Canada, No. 28521)*

gently sloping end, smoothed by abrasion, points "upstream," and the blunt end, terminated by quarrying, points "downstream."

GLACIAL DEPOSITION

Because ice does not have the ability to discriminate between particle sizes during erosion, transportation, or deposition of material, most glacial deposits are poorly sorted. Two general terms are used to describe glacial deposits. The term **drift** dates back to the days when geologists thought glacial deposits were of flood origin. We will discuss the origin of the term later. Drift now refers to *any* material deposited either directly or indirectly from the ice. **Till** is a general term that refers to all unstratified (unlayered), unsorted materials derived from a glacier.

Kinds of Glacial Deposits

The most common glacial deposit is the **moraine**, which is any mound, ridge, or other distinct accumulation of unsorted, unstratified material (till) deposited *directly* from glacial ice. There are five types of moraines: (1) *terminal* or *end*, (2) *ground*, (3) *recessional*, (4) *lateral*, and (5) *medial* (Figure 10.24). Terminal, ground, and recessional moraines are associated with *both* alpine and continental glaciers, but lateral and medial moraines are associated only with alpine glaciers.

Terminal or End Moraines When a glacier reaches the point of maximum advance and the location of the terminus stabilizes for a period of time, the debris released from the melting ice accumulates in a **terminal** or **end moraine**, a ridge bordering the leading edge of the ice. A well-known example of a terminal moraine is the north shore of Long Island.

Ground Moraines During a glacier's retreat, the material released from the melting ice is spread onto the surface of the ground as a layer called **ground moraine** in much the same fashion as gravel is spread by a moving dump truck. The gently rolling topography of western Ohio and Indiana, which we mentioned as characteristic of areas that have been subjected to continental glaciation, is due in part to the infilling of ice-gouged valleys by ground moraine as the glacier retreated.

Recessional Moraines As a glacier retreats, it may temporarily stop and deposit materials at its leading edge. The materials have the same general appearance as a terminal or end moraine, but because they were deposited when the glacier was in retreat, the deposit is called a **recessional moraine.**

Lateral and Medial Moraines As alpine glaciers advance, rock debris falling off the oversteepened valley walls accumulates on the ice surface along the margins of the glacier (Figure 10.25). When the glacier melts, this material is deposited as a ridge along the sides of the valley as a **lateral moraine.** Not uncommonly, it is plastered high on the valley walls.

Where two tributary glaciers flow together to form a main glacier, the bands of rock debris collected on the converging margins form a single debris deposit down the middle of the main glacier (refer to Figure 10.26). When the ice melts, this material is deposited as a single ridge in the middle of the valley called a **medial moraine** (Figure 10.26).

(a)

(b)

(c)

(d)

(e)

FIGURE 10.24 *A variety of* moraines *are deposited by glaciers. All glaciers deposit a* terminal *or* end moraine *recording the point of maximum advance (a). As all glaciers retreat, they deposit* ground moraine *in much the same fashion as icing is spread on a cake (b).* Recessional moraines *record hesitations in the retreat of any glacier (c).* Lateral *(d) and* medial moraines *(e) are associated only with alpine glaciers and consist of materials that fell from the steep walls of the valley. (*a, c *and* d *courtesy of D. R. Crandell/USGS;* b *VU/© Glenn Oliver; and* e *courtesy of Geological Survey of Canada, No. KS197).*

Glaciers   299

FIGURE 10.25 *The material that falls from the valley walls and accumulates on the surface of an alpine glacier will eventually be deposited along the sides of the valley as a* lateral moraine *when the glacier retreats. (VU/© Glenn Oliver)*

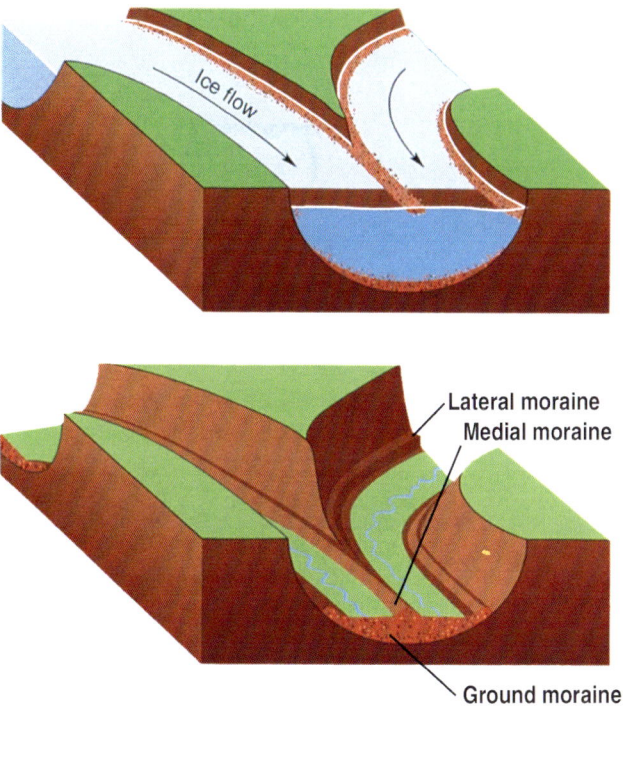

Lateral moraine
Medial moraine

Ground moraine

FIGURE 10.26 *The debris that accumulates along the margins of two intersecting tributary glaciers will eventually be deposited as a* medial moraine *when the ice retreats. (Photo courtesy of Joe Donovan)*

Kames A **kame** is a stream-deposited mount of isolated, stratified drift. As large as 150 to 160 feet high (45–50 m) and 1,200 to 1,300 feet long (360–400 m), kames can have various origins. Some are small deltas or sediment fans that build out from or against the edge of the ice mass. Others are stream sediments deposited in low places on the surface of stagnating ice. When the deposit accumulates between the edge of the ice and a valley wall, the deposit is called a **kame terrace**.

Kettles As glaciers retreat and the ice mass thins, blocks of ice often become detached and buried in the ground moraine and kame fields. Eventually, when the ice melts, a surface depression called a **kettle** is formed (Figure 10.27). Kettles range in diameter from a few feet to miles across. Some kettles may fill with water and form **kettle lakes** as the water table rises within the moraine.

Erratics The competence of ice allows it to transport rock fragments over long distances. When the ice melts, many of the larger fragments, ranging in size from pebbles to boulders as large as a house, are deposited at localities far from their source, not uncommonly in an area where the bedrock lithology is different from that of the particle (Figure 10.28). Such particles are called **erratics**. When many erratics

FIGURE 10.27 *As the glacier begins to recede, blocks of ice often become stranded in moraine. (a) A depression called a* kettle *forms in the surface of the moraine as the ice melts, which, if filled with water, produces a* kettle lake *(b) Thousands of kettle lakes can be found throughout the northern Great Plains. (a VU/© John D. Cunningham, b © Bob Firth)*

FIGURE 10.28 *Continental glaciers can carry rocks, often of great size, for long distances. Should such rocks be deposited in areas characterized by totally different rock lithologies, the rocks are called* erratics. *(Courtesy of N. K. Huber/USGS)*

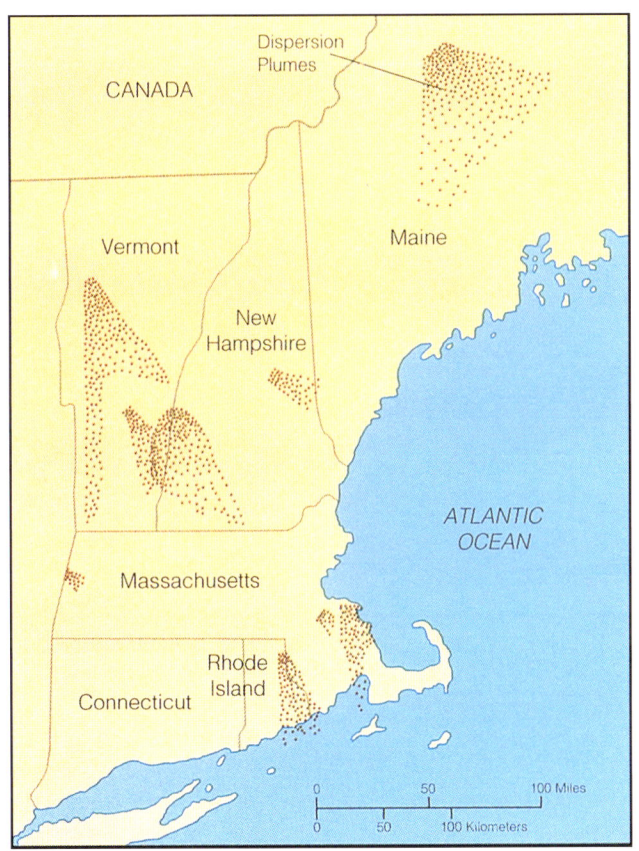
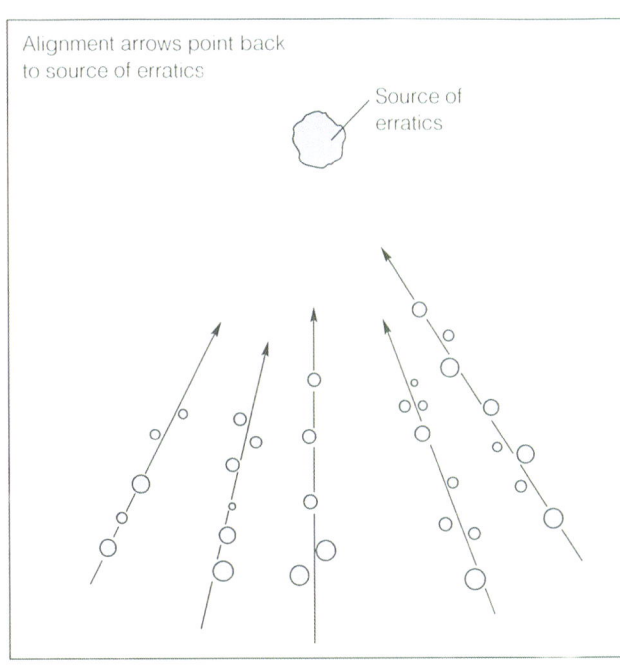

FIGURE 10.29 *As continental glaciers overrun rock outcrops, rock fragments can be carried off in fan-shaped patterns to become erratics. By plotting the locations of erratics on a map, geologists are often able to locate the outcrop from which the erratics were derived. (a after J. W. Goldthwait in R. F. Fling, "Glacial Map of North America," Geol. Soc. Am. Spec. Paper 60, 1945)*

of a particular rock type are distributed over a wide area, their spatial distribution can often be used to determine not only the direction of ice flow but also the distance to the source of the rock (Figure 10.29).

Drumlins Several deposits are more characteristic of continental glaciers than alpine glaciers. One such deposit, the **drumlin,** is a low, elongate hill usually found in groups or *fields*. All of the drumlins within a field are oriented with the long dimension parallel to the direction of ice flow and the blunt end facing the direction from which the ice approached (Figure 10.30; compare with the roche moutonnee in

FIGURE 10.30 *Drumlins form under the advancing glacier and align with the direction of ice movement. The blunt end faces the oncoming glacier. (VU/© Steve McCutcheon)*

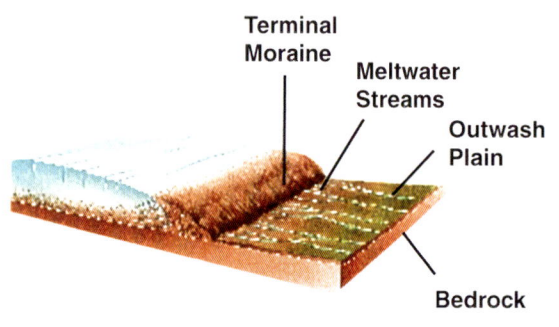

FIGURE 10.31 *As meltwater emerges from the terminus of a glacier, it carries materials that are eventually deposited as* well-sorted stream deposits. *The meltwater stream deposits that accumulate in front of a continental glacier are called an* outwash plain.

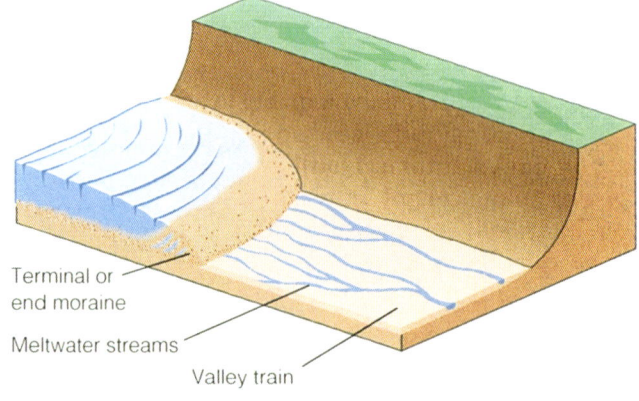

FIGURE 10.32 *The well-sorted meltwater stream deposit in front of an* alpine glacier *is called a* valley train.

Figure 10.23). In some cases, drumlins are composed entirely of till while others have rock cores covered by a thin layer of ground moraine. Drumlins commonly average a mile or less in length and 100 feet (30 m) or so in height. Their shape indicates that they formed under the ice sheet as the ice advanced. Perhaps the most famous drumlins in the United States is *Bunker Hill*. Another drumlin that was the site of an early Revolutionary War battle is Breed's Hill.

Outwash Plains and Valley Trains Not all glacially derived deposits are poorly sorted and unstratified. At the terminus of the glacier, materials picked up by emerging meltwater will be sorted, transported, and deposited as a typical well-sorted stream deposit called an **outwash plain** (Figure 10.31). In the case of alpine glaciers, the outwash deposited down-valley from the terminus of the glacier is called a **valley train** (Figure 10.32).

Eskers Another example of a moderately well sorted glacial deposit is the **esker,** which is more commonly associated with continental glaciation than with alpine glaciers (Figure 10.33). Eskers are the channel deposits of meltwater streams that originally may have flowed on, within, or below the surface of a stagnated mass of ice. With time, the stream eroded its way down through the ice and deposited the accumulated ice-transported debris onto the land's surface. Eskers are long and sinuous, ranging in height from a few feet to 100 feet (30 m) high and, in some cases, up to hundreds of miles long with frequent breaks. In many areas, eskers are important sources of commercial sand and gravel.

FIGURE 10.33 Eskers *are the* channel deposits *of streams that flowed on, through, or under the ice.* (Courtesy of the Geological Survey of Canada, No. GSC 201596)

SPOT REVIEW

1. What are the mechanisms by which glaciers erode bedrock?
2. Why is the appearance of the topography resulting from alpine glacial erosion so different from that created by continental glaciation?
3. Why are most glacial deposits poorly sorted?
4. Give examples of well-sorted glacial deposits and explain why they are well sorted.
5. Why are lateral and medial moraines not found in association with continental glaciers?

PERIGLACIAL REGIONS

The area beyond the terminus of the glacier that is affected by cold temperatures due to the proximity of the ice and by the climate responsible for the ice is referred to as **periglacial**. Periglacial regions are characterized by intense freezing and seasonal thaws. The subsoil is permanently frozen (permafrost), and the topsoil thaws during the short summer warming. Vegetative cover is usually discontinuous and dwarfed.

Mass Wasting

Mass wasting in periglacial regions is dominated by **solifluction** with rock falls, rock glaciers, and rock streams occurring in areas of higher relief. Solifluction is normally restricted to the top 6 feet (2 m) of the soil with movements taking place on slopes as low as 1°. During the summer thaw, mudflows and debris flows sometimes break out from the terminus of solifluction lobes.

In east-west–oriented valleys, preferential solifluction on the valley wall facing the Sun often results in the development of distinctly asymmetrical valley profiles, with the steeper slope facing away from the Sun. North-south–trending valleys are not affected.

Weathering

Physical or mechanical weathering dominates in periglacial regions with freeze-thaw cycles the driving force behind most processes. Frost heaving and frost wedging are major weathering processes even though the volume of water is limited. Frost heaving tends to loosen and lift soil particles and promote soil creep in the same fashion as in more temperate climates. Wedges, veins, and lenses of ice form in cracks and disrupt the frozen topsoil into polygonal slabs. The result is the formation of the **patterned ground** that is so commonly observed in periglacial areas (Figure 10.34).

Lens-shaped masses of ice grow within the sediment along the edges of alluvial plains and old lake beds and at the base of steep slopes. Eventually, the ice forces the sediment to arch up and form a dome-shaped structure called a **pingo** (Figure 10.35).

The combination of freeze-thaw cycles and plastic deformation of ice within masses of rock debris results in the formation and movement of **rock glaciers** (Figure 10.36). As the climate moderates and the ice melts, meltwater flushes the fine-grained

(a) (b) (c)

FIGURE 10.34 Frost action *within the upper layers of regolith in periglacial regions often causes the surface to break into* polygonal slabs, *a feature referred to as* patterned ground *(a). Other examples of pattered ground are the* rock stripes *(b) and* rock rings *(c). (*a VU/© Steve McCutcheon, b and c courtesy of Joe Donovan)

FIGURE 10.35 *Another feature formed by the subsurface formation of ice is the* pingo. *This example is located along the McKenzie River in the Northwest Territory, Canada.* (VU/© Steve McCutcheon)

FIGURE 10.36 *Frost action within rock debris can cause it to move downslope as a* rock glacier. (VU/© Steve McCutcheon)

FIGURE 10.37 *When the fine sediments are flushed from a rock glacier, the deposit remaining is called a* rock stream. *A well-known example is* Hickory Run *in Carbon County, Pennsylvania. (VU/© J. Serrao)*

materials from the rock glaciers to form **rock streams**. An excellent example of a rock stream is the Hickory Run boulder field in Carbon County, Pennsylvania (Figure 10.37).

Because periglacial areas experience subfreezing temperatures throughout most of the year, the rates of all chemical weathering processes are severely reduced. Field evidence clearly indicates that nearly all rock disintegration is there result of physical processes.

Streams

Except for the short summer thaw, stream discharge in periglacial areas is never high. Although large rivers and some mountain streams may continue to flow under an ice cover during the winter, most streams freeze solid. During the short period of thaw, stream discharge increases dramatically and the streams go into flood. Because a very large volume of sediment is made available by the various mass wasting processes, the streams quickly become choked with sediment and commonly form **braided** patterns as they move the sediment downstream. Because the period of stream activity is so short, stream erosion is limited, and the valleys consequently become filled with rock debris.

The Wind

In general, the wind is more effective at transporting sediment in periglacial areas than are streams. The arid to semiarid conditions combined with the freezing and drying of the surface sediments make them prone to wind erosion. **Dunes** are common features in the vicinity of streams, and very large areas are covered with **loess**. In fact, most of the world's grain-producing regions are in the areas covered with loess deposited during the Great Ice Age.

UNDERSTANDING CONTINENTAL GLACIATION

Although geologists of the 1800s were vaguely acquainted with alpine glaciers, they were not at all familiar with continental glaciers. Only since the middle of the nineteenth century have we come to understand continental glaciation.

One person who understood the processes and deposits of alpine glaciers was a Swiss civil engineer named *Ignatz Venetz-Sitten.* Having been born and raised in Switzerland, a country with many alpine glaciers and valleys filled with glacial debris of all kinds, Venetz was quite knowledgeable about glaciers and the characteristics of glacial deposits. At one point, Venetz took a job building canals in northern Europe where he became intrigued with a layer of material the locals called "drift." When questioned, his co-workers explained that the material was a flood deposit. Some claimed that the flood in question was the Great Flood of the Bible that occurred in Noah's time.

Venetz, however, was convinced that this material was ground moraine, not the aftermath of a gigantic flood. In 1821, he published a paper in which he presented his theory of the glacial origin of drift. Venetz visualized that at some time in the recent geologic past, the atmosphere had cooled and the glaciers in the Alps had flowed out across Europe. As Earth's atmospheric temperatures warmed, the glaciers retreated, depositing the moraines and outwash plain debris that constituted the materials known as drift. Needless to say, his ideas were not well received, especially by those who preferred the biblical origin.

In 1832, a German professor of forestry named *Bernhardi* presented a paper, agreeing with Venetz's basic premise. Bernhardi, however, suggested that the source of the ice was an ice cap over the northernmost portion of the Scandinavian Peninsula.

Louis Agassiz, a Swiss zoologist, geologist, and the leading naturalist of the day, undoubtedly read these radical ideas with a certain amount of skepticism. After all, no one had ever suggested an episode of glaciation of such dimensions. Fortunately, Agassiz did not reject the ideas forthwith. After looking more critically at the deposits of drift, Agassiz was soon convinced of the glacial origin of drift, and in 1837, he proposed that Europe had been engulfed by a huge ice sheet that had moved out of the Arctic and left drift behind in its wake. Agassiz went on to prove that the ice actually made four advances, separated by periods of glacial retreat. Figure 10.38 shows how Agassiz recognized that the ice had made multiple advances rather than a single advance.

Consider the sequence of events that would produce the deposits shown in Figure 10.38. The figure shows two terminal moraines, TM 1 and TM 2, and two ground moraines, GM 1 and GM 2, deposited by glaciers that advanced into the region during two ice ages, Ice Age 1 and Ice Age 2, separated in time by perhaps hundreds of thousands of years. The figure shows that glacier 1 advanced further than glacier 2. The point of maximum advance of the first glacier is recorded by the deposition of the older terminal moraine, TM 1. As the interglacial period developed and glacier 1 receded, it deposited the older ground moraine, GM 1. During the long interglacial period, a soil profile developed on both the terminal and ground moraines.

As the second ice age developed, the glacier returned, stripping the land of the older ground moraine up to the point of its furthermost advance. Note that had the second glacier advanced *further* than the first, *all* of the moraine deposited by the first glacier would have been eroded by the advancing ice, leaving no record of the first glacial advance. The drawing, however, shows the second glacier not advancing as far as the first. As the second glacier comes to a stop, its terminal moraine, TM 2, is deposited on top of the older ground moraine. Note, however, that the new terminal moraine is separated from the older ground moraine by the soil profile.

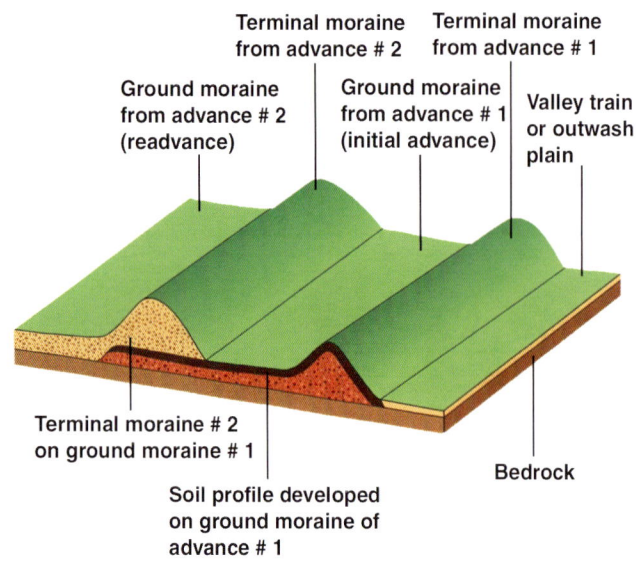

FIGURE 10.38   *The identification of a* terminal moraine *separated from an underlying* ground moraine *by a* soil *allowed Agassiz and other geologists to show that the Pleistocene Ice Age consisted of glacial advances separated by interglacial periods during which the glaciers retreated.*

The presence of the soil profile between the two moraine deposits is *all* important. Had the soil not been present, or had it not been *recognized,* the younger terminal moraine could have been misidentified as a recessional moraine of the older glacier, in which case there would have been no way to recognize the multiple advance. Thus, the discovery of the soil profile between the two deposits proved they were from two different ice ages. Based on data retrieved from deep-sea sediments, we now know that there were at least 20 distinct ice advances.

Agassiz came to the United States in the mid-1800s and discovered that while Europe was being subjected to the Great Ice Age, North America had undergone a comparable event, complete with the same multiple advances. It became apparent to Agassiz that this episode of continental glaciation was more global than he had previously thought, and that glaciation was not restricted to Europe but was dominant throughout the Northern Hemisphere. This was indeed the *Great Ice Age.*

In all fairness to those who investigated the interrelationships of glacial deposits before the time of Agassiz, we should note that many of the ideas attributed to him were derived from the work of his older Swiss colleagues.

WHAT IF THE ICE CAPS WERE TO MELT?

These past advances and retreats of the ice sheet suggest an important question: What would happen if most or even all of the 5 million square miles of continental ice now on Earth's surface were to melt? Scientists have estimated that if the ice now present as continental ice caps were to melt, sea level would rise at least 200 feet (60 m). The consequences can readily be seen by noting how many major cities in the United States alone would be submerged by the rising sea level. Consider that the population of greater New York City, equal to the entire population of Canada, would have to move inland to find living space in a constantly shrinking landmass. When this scenario is duplicated for all of the population centers of the world located at elevations less than 200 feet above sea level, the result is an explosion in population density without the addition of a single new life.

Not only would living space be significantly reduced, but coastal lands that now produce large quantities of food for an already overcrowded Earth would be lost. Because most of the land surface capable of crop production is already being used, the chances of replacing these areas with new lands dedicated to the production of food are slim, although some previously barren land might become arable as a result of climatic changes accompanying the melting of the ice caps.

The potential consequences of mass movements of populations, subsequent increases in population densities, and declining food supplies are not pleasant to contemplate. It is worth noting that the quantities of carbon dioxide we are introducing into the atmosphere through our consumption of fossil fuels may well be accelerating the melting of the glaciers and contributing to the projected atmospheric warming. At the present time, however, we cannot tell with certainty whether the warming has begun.

The Great Ice Age

Earth has experienced several episodes of widespread continental glaciation. The latest and best understood is the Pleistocene glaciation, historically referred to as the *Great Ice Age*.

The **Great Ice Age** began approximately 1.6 million years ago, signaling the beginning of an interval of geologic time known as the **Pleistocene Epoch**. During each of the four major episodes of maximum glacial development as recorded by terminal moraines, ice covered about 15 million square miles (40 million km^2) or about one-third of the total land surface of Earth.

Although known as the *great* ice age, the Pleistocene Ice Age was neither the longest nor the most extensive episode of widespread continental glaciation. The longest episode of continental glaciation, known from deposits in South Africa, South America, and India, occurred from 2.5 to 2.1 billion years ago. The most extensive episode occurred in the later Proterozoic. Its deposits have been found on every continent, indicating that it was truly global in extent.

The Great Lakes

The **Great Lakes** are one of the most impressive products of the Pleistocene Ice Age on the North American continent. Among the largest bodies of fresh water on Earth, the Great Lakes occupy basins gouged out by the advancing North American ice sheet. The historical development of the Great Lakes is summarized in Figure 10.39.

During preglacial times, the basins occupied by the lakes were stream valleys. Because of the exceptional depth of Lake Superior, some geologists think that a lake basin may have existed at the present site before the glacial advance. As the final ice sheet began to retreat, water filled the basins as they emerged from under the receding ice. Initially, the lakes drained to the south into the Mississippi River by way of the Illinois and Wabash rivers (refer to Figure 10.39). As the ice receded from the eastern lakes, a new outlet was temporarily established southward across New York by way of the Mohawk River. Eventually, about 8,000 years ago, as the ice receded beyond the Niagara Escarpment and exposed the Saint Lawrence Valley, the present drainage system

FIGURE 10.39 *The Great Lakes formed as the last advance of the Pleistocene ice sheet retreated and water filled the troughs that the glacier had gouged during its advance. (Adapted from Leverett and Taylor, USGS)*

was established over Niagara Falls and via the Saint Lawrence River to the Atlantic.

Niagara Falls formed where the Niagara River encountered the outcrop of the Lockport Dolostone along the northward-facing Niagara Escarpment (Figure 10.40). When first exposed from beneath the ice, the falls were approximately 7 miles (11 km) farther downstream (northeastward) toward Lake Ontario. Since then, the falls have been receding at the rate of about 3 to 4 feet (1 m) per year. Niagara Falls will continue to retreat upstream, diminishing in height until, in about 30,000 years, they will be nothing more than a riffle in the Niagara River north of its present exit from Lake Erie. The geology of Niagara Falls is illustrated in Figure 10.40.

Glaciers 309

FIGURE 10.39 (continued)

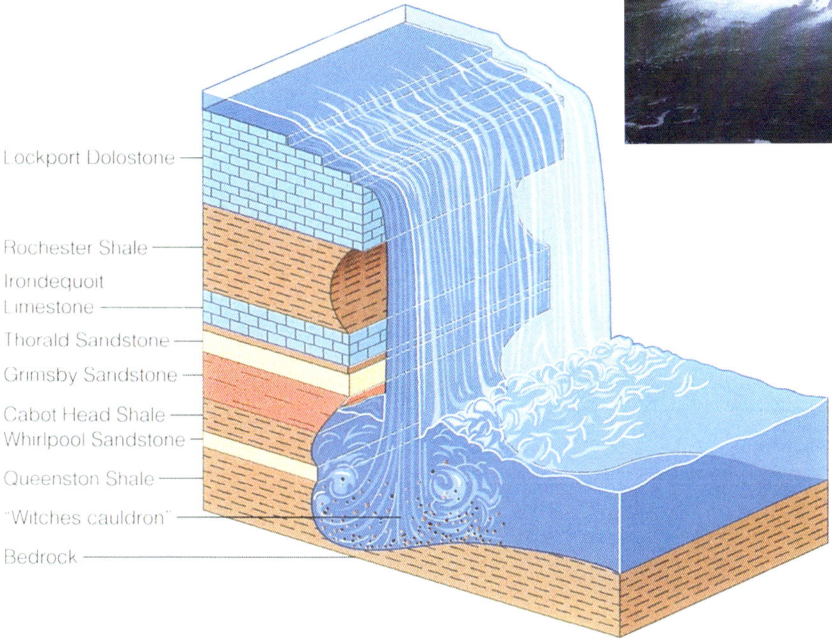

FIGURE 10.40 One of the best-known features to form as the last Pleistocene ice sheet retreated from the area of the Great Lakes is Niagara Falls. Retreating at the rate of 3 to 4 feet (1 m) per year, the falls will diminish in height and, in about 30,000 years, will be little more than a riffle in the Niagara River. The geology of Niagara Falls is shown in the block diagram. (Photo by VU/© William J. Weber)

Glaciation in the Western United States during the Great Ice Age

While much of the North American continent east of the central and northern Rocky Mountains was covered by the North American continental ice sheet, extensive, interconnected alpine ice caps developed throughout the central and northern Rocky Mountains (Figure 10.41). Erosion by these ice caps was responsible for much of the rugged, scenic beauty of these mountains.

Pluvial Lakes

During the period of maximum development of the Rocky Mountain ice caps, heavy rainfall to the south resulting from the southwestward displacement of cyclonic storms formed a large number of **pluvial lakes.** The lakes were especially numerous throughout Nevada in a region called the Basin and Range Province where parallel mountain ridges and adjacent basins formed catchments for the meltwater.

FIGURE 10.41 *At the climax of the Pleistocene Ice Age, more than half of North America was under ice. The Rocky Mountains were covered by interconnecting alpine glaciers while the landmass to the east was under continental glaciers.*

Most of the lakes existed only during the glacial episodes. During the interglacial periods, those lakes that did not completely dry up were significantly reduced in size. Portions of an especially large pluvial lake complex, Lake Bonneville, still survive today. Once Lake Bonneville covered more than 19,300 square miles (50,000 km^2) with depths as great as 300 feet (90 m), but all that remains are the Great Salt Lake, Utah Lake, and Sevier Lake (Figure 10.42). Some sense of the depth of the water in Lake Bonneville can be gotten by noting the elevations of successive lake shorelines that can be seen flanking the now-dry lake bed (Figure 10.43). Due to its perfectly flat surface, a portion of the old lake bottom, known as the Bonneville Salt Flats, is used for testing high-speed vehicles.

The Channeled Scablands A particularly interesting, and for many years baffling, geological area is the Channeled Scablands, which cover nearly 15,000 square miles (38,000 km^2) in the Columbia Plateau of east-central Washington (Figure 10.44). Characterized by braided stream channels that incise up to 100 feet (30 m) into the layered lava flows, huge ripples on the surfaces of sand and gravel deposits, and landforms unlike anything in the immediate area, the Scablands long defied reasonable explanation.

Most geologists now accept the explanation given by Harlan Bretz. According to Bretz, the entire landscape is the result of the breaching of an ice dam that originally developed on the Clark Fork River, a tributary of the Columbia River, as the North American ice sheet moved southward (Figure 10.45). During each glacial episode, glacial Lake Missoula formed behind the dam. In places, the lake was over 1,000 feet (300 m) deep. As the ice receded during the interglacial

FIGURE 10.42 *The largest of the pluvial lakes to form in the Basin and Range Province during the Pleistocene Ice Age was* Lake Bonneville. *The best-known remnant of this once extensive lake is* Great Salt Lake.

FIGURE 10.43 *Former shorelines of glacial Lake Bonneville are now recorded as terraces such as those seen from the margins of Great Salt Lake. (VU/© D. Newman).*

FIGURE 10.44 *One of the most fascinating features resulting from the last ice age are the* Channeled Scablands *of eastern Washington. The dimensions of the landforms such as giant ripples in gravel deposits baffled geologists for years. We now believe these and other surface features to be the result of huge floods that surged through the area as ice dams broke during the retreat of the ice, releasing water that had been imponded in glacial* Lake Missoula. *(Courtesy of P. Weis/USGS)*

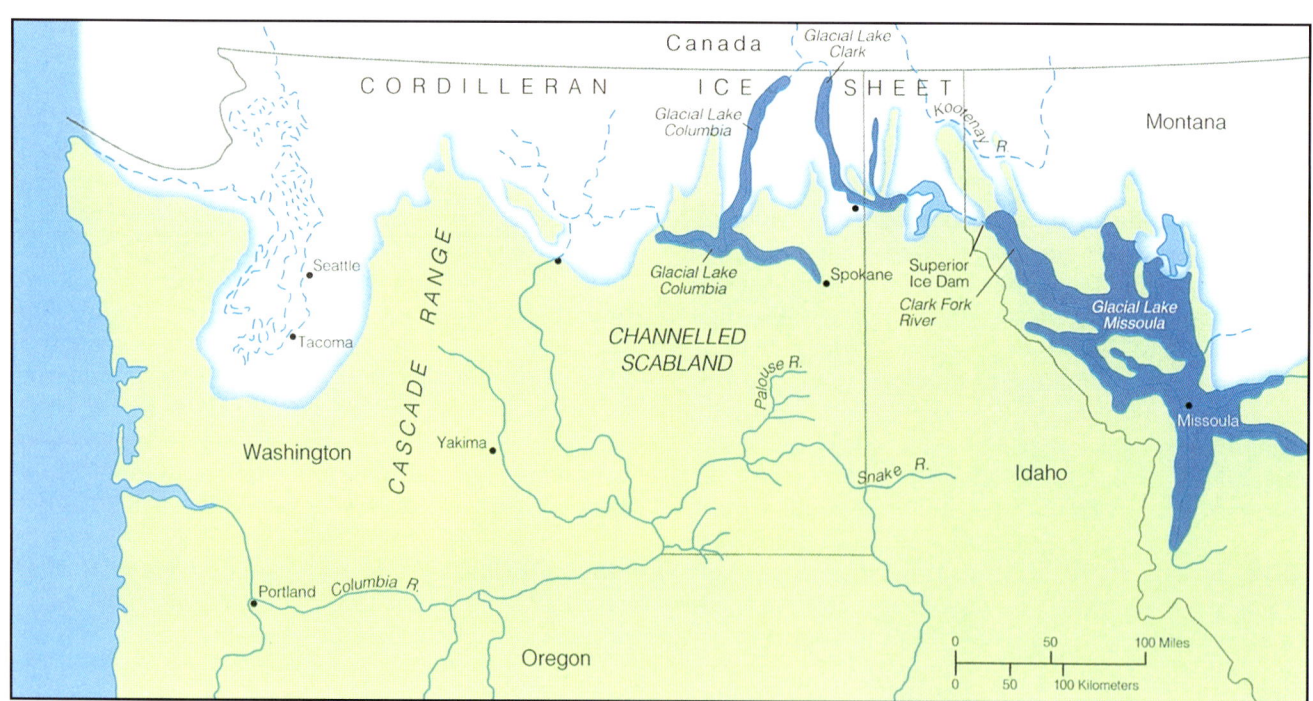

FIGURE 10.45 *Lake Missoula formed as ice dammed the* Clark Fork River. *Geologists now believe that Lake Missoula may have formed and drained as many as seven times as the Pleistocene ice advanced and retreated.*

periods, the dam broke, releasing the contents of Lake Missoula, about 480 cubic miles (2,000 km³), in a huge surge of water that flooded the area, perhaps within a matter of hours. The water released when the dam broke would have exceeded the combined volume of all the world's rivers. This scenario, repeated after each of the glacial episodes, was not only responsible for the topography of the Channeled Scablands, but also carved the channels of the Lower Snake and Columbia rivers.

SPOT REVIEW

1. What are the characteristics of periglacial regions?
2. What are the major agents of sediment transport in periglacial regions?
3. How did Louis Agassiz and other researchers determine that the Pleistocene ice sheets had undergone multiple advances separated by periods of retreat?
4. What potential problems will the human population of Earth face if a significant portion of the present ice cover melts?
5. What is unique about the deposits found in the Channeled Scablands? How are the Scablands thought to have formed?

THE CAUSES OF CONTINENTAL GLACIATION

The mechanism by which alpine glaciers form is well understood. All that is needed are mountains with a plentiful supply of moisture and the appropriate combination of elevation and latitude to provide the cold climate necessary for the year-long accumulation and preservation of snow. Alpine glaciers have probably always existed somewhere on Earth's surface throughout the billions of years of geologic time.

The occurrence of continental glaciers is not as easy to explain. One of the fundamental problems facing any theory that attempts to explain the occurrence of worldwide continental glaciation is the rarity of the event. Episodes of continental glaciation are known to have occurred six times in the nearly 5 billion years of Earth's history. Little record remains of the first two episodes, both of which occurred more than a half-billion years ago during the *Precambrian*.

The third episode occurred about 480 million years ago during the late *Ordovician Period* and affected the area of North Africa, while the fourth occurred in the Devonian. The fifth episode of continental glaciation, and the one for which we have a good record, occurred about 250 million years ago during the *Pennsylvanian* and *Permian Periods*. It affected the supercontinent of Gondwana before it broke up to form today's continents. Erosional features and deposits from this glacial episode were among the evidence Alfred Wegener used to support his argument for the existence of a supercontinent and continental drift. The last episode, the great *Pleistocene Ice Age,* began about 1.6 million years ago and extends to the present time. Some geologists are of the opinion that we are now in the waning of the Pleistocene Ice Age while others would say that we are in an interglacial period.

As we noted, the difficulty in explaining worldwide continental glaciation is that it has occurred only six times in 4 to 5 billion years. Several theories have been set forth, none of which is universally accepted. In fact, we may never be able to explain how or why continental glaciation comes about. Nevertheless, intellectual curiosity demands that we try.

Celestial Causes

One of the popular hypotheses about the onset of worldwide continental glaciation involves the relationship between Earth's orbit and the Sun. In the early 1900s, a Yugoslavian geophysicist, Milutin Milankovitch, determined that three cyclic changes take place as Earth orbits the Sun, each of which could cause climatic changes: (1) a 100,000-year cycle during which the distance from Earth to the Sun changes, (2) a 41,000-year cycle during which Earth's angle of inclination relative to its orbital plane varies from 22° to 24°, and (3) a 21,000-year cycle during which Earth's axis precesses (Earth wobbles on its axis). Milankovitch calculated that the combined effects of these three variations would result in a cyclic temperature variation of Earth's atmosphere with the coldest portion of the cycle occurring on a 40,000-year cycle. Cores brought back from deep-sea drilling projects seem to support Milankovitch's ideas of cyclic temperature fluctuations in Earth's atmosphere. The 100,000-year and 21,000-year cycles are well recorded in the ocean sediments, but any record of the 41,000-year cycle is difficult to detect.

It should be noted that Milankovitch was only attempting to explain the cyclicity of Pleistocene glaciation. Although there is little doubt that such cycles have been going on since the formation of the Solar System, no attempt has been made to use them to explain the six episodes of glaciation that have occurred over the last 5 billion years.

Volcanic Activity

Others have suggested that the temperature changes needed to bring on glaciation were due to increases in volcanic activity, which introduced large quantities of dust into the upper atmosphere. Certainly, volcanic activity can and has affected Earth's atmospheric temperature. You will recall that the eruptions of Tambora in Indonesia and Mayon in the Philippines in 1815 were sufficient to affect Earth's mean atmospheric temperature and bring on the "year without a summer" in 1816.

A more recent theory involving volcanism matches episodes of volcanic activity along the plate margins, changes in sea level, and the chemical weathering of rocks. The theory is based on three facts: (1) volcanic eruptions are major *producers* of atmospheric carbon dioxide; (2) chemical weathering of rocks is a major *consumer* of atmospheric carbon dioxide; and (3) the temperature of Earth's atmosphere is controlled, at least in part, by the *concentration* of atmospheric carbon dioxide via the greenhouse effect. According to the theory, the combination of increased volcanic activity, high sea level, and subsequent reduction in the area of land exposed to chemical weathering will result in increased atmospheric carbon dioxide and a subsequent rise in global warming. Conversely, reduced volcanic activity combined with low sea level and increased chemical weathering of an expanded landmass will result in a decrease in atmospheric carbon dioxide and subsequent global cooling. The problem with any attempt to correlate continental glaciation with worldwide volcanism, however, is that there is no evidence that Earth was any more (or less) volcanically active during the six known episodes of continental glaciation than when continental glaciers were absent.

Plate Tectonics

There is little doubt that plate tectonics is a major determinant of continental glaciation in that it controls the distribution, number, and elevation of the landmasses and the topography and interconnections of the ocean basins. As a result of plate tectonics, continents move into polar regions where they experience continental glaciation. Paleomagnetic data indicate that at the time of the glaciation of Gondwana 250 million years ago, a portion of the supercontinent was moving through the region of the South Pole. Glaciation in Pangaea came to an end when the continent rifted and portions moved out of the polar region.

From the time Pangaea broke up, and for nearly 200 million years thereafter, oceans occupied Earth's polar regions. As the water moved freely among the various ocean basins, and warm, equatorial waters mixed with the waters of the polar seas, the temperature of the oceanic waters became more equable. With time, however, the continents began to move toward the poles. Antarctica moved southward and became the first continent to move into a polar location. The Antarctic ice sheet began to develop perhaps 35 million years before any continental ice appeared on either North America or Europe.

As North America and Europe moved northward, the flow of southern waters into the north polar sea became increasingly restricted, and the water began to cool. Perhaps as early as 20 million years ago, the North American and European continents had moved far enough north that precipitation began to fall as snow. Continued northward movement of the continents further restricted the flow of southerly waters into the Arctic Ocean. Although the Arctic Ocean was cooling, movement of water from both the Atlantic and Pacific oceans was sufficient to keep the Arctic Ocean ice-free. Without a covering of sea ice, the Arctic Ocean was able to continue to be a source of moisture for the growing mass of snow and ice that accumulated over the northern portions of North America, Europe, and Siberia. The scene was set for the onset of the Great Ice Age.

As snow and ice accumulated over the northern continents and the ice sheets grew larger, the reflection of the Sun's radiant energy from the white, frozen surface caused the atmospheric temperatures in the polar region to fall so low that the Arctic sea eventually froze over. As the Arctic Ocean became ice-bound, it was eliminated as a source of moisture. Deprived of this major source of moisture, the volume of snow and ice over the continents began to decrease by ablation, initiating the beginning of the first interglacial period.

As the ice and snow were removed and the darker, heat-absorbing rock surfaces were exposed, solar energy began to warm Earth's surface, which in turn warmed the atmosphere. Eventually, the Arctic sea thawed and again became a source of moisture. The scene was now set for the second ice advance. This cycle was then repeated to produce the remaining ice advances and interglacial periods.

ENVIRONMENTAL CONCERNS

At first glance, glaciers may not seem to have much current environmental impact upon the lives of most of us who live far from the Greenland and Antarctic ice sheets. Further consideration, however, indicates that glaciation is indeed a matter of environmental concern.

Rising Sea Levels

Perhaps the most publicized environmental concern relative to glaciation is the potential effect of rising sea levels should global warming result in the melting of significant volumes of glacial ice. Although not all scientists are convinced that Earth is at present undergoing global warming, advocates most commonly express concerns about an enhanced greenhouse effect due to the volumes of carbon dioxide, CO_2, and methane, CH_4, being introduced into the atmosphere as a result of human activities, especially the burning of fossil fuels.

It should be pointed out that there are natural cycles of atmospheric temperature change. Within historic time, for example, a worldwide atmospheric warming trend between A.D. 800 and 1200 allowed the Vikings to settle areas such as Iceland, southern Greenland, and several sites in North America. By A.D. 1400, however, most of these settlements had been abandoned as Earth's atmosphere cooled and Arctic conditions returned, an episode commonly referred to as the "Little Ice Age."

Beginning in the mid-eighteenth century, Earth again experienced a warming trend, which has resulted in a mean annual atmospheric temperature increase of about 0.5°C since the beginning of the twentieth century. Although predictions about what the future holds are topics of vigorous debate in the scientific community, nearly all models predict continued warming throughout the next century with mean atmospheric temperatures increasing as much as 1.2°C. Should this take place, two events are most likely to occur: (1) worldwide changes in climate, and (2) a rise in sea level due to the combined effects of the expansion in the volume of oceanic waters and the melting of glacial ice.

Changing patterns of rainfall and temperature have the potential to affect agriculture worldwide. Although regions now too cool to provide acceptable growing seasons may be rendered more favorable for the production of crops, others that are now in production may be lost as the amount of rainfall decreases. It is not clear what the overall effect of such changes on world food production may be.

Warmer ocean waters will make more thermal energy available to storms born at sea. Hurricanes will not only become more numerous, but individual storms will become more powerful and destructive. With large populations living in coastal areas worldwide, the potential both for the loss of life and for property damage will increase significantly.

The most obvious impact of global warming, however, is the rise in sea level as the ice sheets melt. Estimates of the potential sea level rise in the next century range from 16 inches (40 cm) to nearly 7 feet (200 cm). Even the low estimate would significantly increase erosion along the coastlines. A sea-level rise of 7 feet would certainly inundate coastal property and structures. One need only consider the impact on the lives of those living in the megalopolis that extends from New York City to Washington, D. C., to appreciate the consequences of a rising sea level. Some areas may be protected, at least temporarily, by the construction of dikes and sea walls, but at enormous cost. Other areas deemed not so essential would necessarily have to be abandoned. The combined effect of the mass migration of people now living in those areas, the increased population densities in the remaining land areas, and the loss of productive croplands is not pleasant to contemplate.

Goiter

The increased incidence of goiter throughout the glaciated regions of the United States may also be an example of the environmental influence of glaciation. Goiter is the chronic enlargement of the thyroid gland, located at the base of the neck, due to a deficiency of iodine in the diet. Iodized salt has been widely used to control the disease.

Although a topic of some debate as to cause and effect, the patterns of Pleistocene glaciation, iodine deficiency in soil, and the incidence of goiter in the United States show distinct similarities. Iodine is released by the weathering of rocks and is concentrated in the upper layers of soil by biological processes involving the plants and the soil. Because of its high solubility, most iodine is removed to the oceans. As it is being removed, however, it becomes a component of surface and groundwater, and, thereby, drinking water.

According to a current hypothesis, the soil and its content of iodine were removed as the Pleistocene glaciers advanced. Since the glaciers retreated and soil developed on the ground moraine, the biologic processes responsible for the concentration of iodine have not had sufficient time to accumulate the normal content of iodine in the soil. As a result, the amount of iodine necessary to protect against goiter does not yet exist in the soils, surface waters, and ground water of the region.

SPOT REVIEW

Play the devil's advocate and compare the major hypotheses that have been proposed to explain the onset of worldwide glaciation.

CONCEPTS AND TERMS TO REMEMBER

Distribution of glacial ice
   continental glaciers
      Antarctica
      Greenland
   alpine or valley glaciers
      piedmont glaciers
The origin of glacial ice
   accumulation
   conversion of snow to glacial ice
      sublimation
      névé or firn
      pressure melting
Movement of glacial ice
   brittle fracture
      crevasses
   plastic flow
      dry-based or cold glaciers
      regelation
         regelation slip
      basal sliding
         wet-based glaciers
      surges
Zones of accumulation and ablation
   calving
   firn line/equilibrium line
Glacial erosion
   processes of erosion
      quarrying or plucking
      abrasion

alpine erosion
   cirque
   arête
   col
   horn
   glacial trough
   U-shaped valley
   hanging valley
   tarn
continental erosion
   subdued topography
   roche moutonnée
Glacial deposition
   drift
   till
      moraine
         terminal or end
         ground
         recessional
         lateral
         medial
      kame
         kame terrace
      kettle
         kettle lake
      erratic
      drumlin
      outwash plain
      valley train
      esker

Periglacial regions
   freeze-thaw cycles
   mass wasting
      solifluction
   weathering
      ice wedging
         patterned ground
         pingo
      rock glaciers
      rock streams
   stream erosion
      braided streams
   wind erosion
      dunes
      loess
The Great Ice Age
   Pleistocene Epoch
   multiple advances
   topographic features
      Great Lakes
      pluvial lakes
      Channeled Scablands
Causes of continental glaciation
   frequency
   causes
      celestial variations
      volcanic activity
      plate tectonics

REVIEW QUESTIONS

1. Most of the glacial ice on Earth today exists over
   a. Siberia.
   b. Arctic Canada.
   c. Greenland.
   d. Antarctica.
2. Material intermediate between snow and glacial ice is called
   a. kame.
   b. roche moutonnée.
   c. firn.
   d. tarn.
3. A bowl-shaped mountain depression is a (an)
   a. arête.
   b. cirque.
   c. col.
   d. horn.
4. Regelation is the process by which
   a. bedrock is scoured and eroded by alpine glaciers.
   b. rocks are disrupted and removed by the ice at the headwaters of alpine glaciers.
   c. snow is transformed to glacial ice.
   d. ice at the base of a wet-based glacier moves by undergoing pressure melting and refreezing.
5. Glacial valleys can be recognized by their
   a. near-vertical walls.
   b. steep slopes.
   c. cross-sectional shape.
   d. long, straight segments.
6. Which of the following types of moraine is limited to alpine glaciers?
   a. recessional
   b. ground
   c. end or terminal
   d. lateral
7. How has plate tectonics been used to explain episodes of worldwide continental glaciation?
8. Why is the topography resulting from alpine and continental glaciation so different?
9. In what ways do the mechanisms of motion and erosion of dry-based and wet-based glaciers differ?
10. How can a glacier be in retreat even though the ice mass is still moving toward the terminus?
11. How are cirques, arêtes, cols, and horns related?

THOUGHT PROBLEMS

1. Some geologists are of the opinion that we are out of the last ice age but could be about to enter a new ice age. Consider the scenario of events that could take place around the world should they be right.
2. Before the advent of plate tectonics, nearly all the theories explaining continental glaciation proposed that continental glaciation was the result of worldwide atmospheric cooling; the theories differed mainly in the method they proposed for the cooling. Using what you know about the hydrologic cycle, consider the difficulty of using such theories to explain the cyclicity of the ice ages and the maintenance of the high levels of snowfall required to develop and maintain continental ice sheets.

FOR YOUR NOTEBOOK

Although widespread glaciation has long since disappeared, assemble examples of glacial features that serve to remind us of its former dominance. If you live in the Northeast, adjacent to the Great Lakes, in the northern Great Plains, in the northern Rockies, or on the island of Hawaii, you should have no trouble finding examples of glacial erosion or deposition. Those of us who live in other areas may have to look for features that are indirectly the result of glaciation. The remnants of the pluvial lakes throughout Nevada and the loess deposits of the Mississippi Valley are examples. You may also look for features of local significance. For example, in my area, terraces along the major river system are the remnants of a lake that formed when the last ice sheet dammed the river nearly 100 miles to the north, and there are still isolated valleys in the upland areas where the descendants of tundra-type plants can be found growing.

CHAPTER OUTLINE

INTRODUCTION
RELATIVE HUMIDITY
KINDS OF DESERTS
    Rainshadow Deserts
    Subtropical Deserts
    Fog Deserts
    Isolation Deserts
EROSION IN THE DESERT
    The Effectiveness of Water
    The Arid Cycle of Erosion

EROSION BY THE WIND
    Deflation
    Abrasion
DEPOSITION FROM THE WIND
    Loess
    Dunes
ENVIRONMENTAL CONCERNS
    Overgrazing
    Overcultivation
    Fuel
    Salinization

CHAPTER 11

Deserts and the Wind

INTRODUCTION Even though deserts cover nearly a third of the land surface of Earth, they remain relatively unknown regions to the general population. Most travelers choose either not to visit deserts at all or to cross them in the shortest possible time. Most likely, they make the trip at night when, unfortunately, the fascinating landscape of the desert is lost in the glare of headlights.

As a result, several misconceptions have developed about what characterizes a desert. Deserts, for example, are commonly perceived as being shifting, simmering seas of sand while, in reality, sand is a relatively scarce commodity in many desert areas. In the deserts of the southwestern United States, for example, sand covers less than 10% of the surface. Most desert lands are either exposed bedrock or are covered with coarse rock fragments. Sand is conspicuous by its absence.

Another common misconception is that all deserts are hot. It is true that the highest surface temperature ever recorded, 136°F (58°C), was in the Sahara of Libya, in North Africa, while the highest temperature recorded in North America was 134°F (57°C) in Death Valley, California. It should be pointed out, however, that although a desert may attain a high daytime temperature, nighttime temperatures commonly drop below freezing.

High temperature is not a prerequisite for a desert. The interior of Antarctica is a desert, and its temperatures rarely go above freezing. In summary, there are both hot and cold deserts.

Another commonly held view is that deserts are basically devoid of life. Deserts are certainly not devoid of plant life. Although plant populations may be sparse, a visit to a desert when the plants and wildflowers are in bloom is a truly spectacular experience, albeit if only for a short time (Figure 11.1).

A desert visit also provides an opportunity to see plants specifically adapted to grow in this harsh environment. Most desert plants have extensive root systems that enable them to acquire an adequate supply of the scarce water. Usually, only a small portion of the plant extends above ground, and the leaves are relatively few and small; both characteristics are adaptations to minimize the loss of water (Figure 11.2). The reduction in the number of leaves reaches an extreme in cacti where the chlorophyll needed for photosynthesis is contained in the stems that make up the body of the plant (Figure 11.3). Cacti also exhibit another characteristic of many desert plants, namely, spines or thorns that discourage animals from eating the plants to obtain the water stored inside. These adaptations preserve water in an environment where water is hard to find and equally difficult to keep.

Animal life in the desert is not as scarce as it may seem. Because the intense heat of the daytime Sun can quickly cause dehydration and possibly death, most desert animals are nocturnal, spending the daylight hours hidden in burrows, sheltered beneath rocks,

FIGURE 11.1 *Few scenes are more beautiful than the desert in bloom. Here, for a short time, poppies surround the saguaro cacti in the* Sonoran Desert of southern New Mexico. *(VU/© Peter Dunwiddie)*

FIGURE 11.2 *Desert plants or* xerophytes, *have adapted to the scarce water supply by reducing the surface area they expose to the atmosphere. (Courtesy of W. B. Hamilton/USGS)*

FIGURE 11.3 *The extreme example of surface area reduction among desert plants is cactus, which has completely eliminated leaves and concentrated the food-producing chlorophyll in the surfaces of its stems. (Courtesy of W. B. Hamilton/USGS)*

or tucked away in rock crevices (Figure 11.4). The time to see desert animals is at night. Be warned, however, that roaming the desert at night can be dangerous. Many desert animals can inflict painful stings and potentially lethal bites.

The adaptations made by desert plants and animals are the result of a characteristic common to all deserts, namely, that they receive very limited precipitation, commonly less than 10 inches (25 cm) yearly.

Although most deserts are generally not particularly attractive to the average person, to a geologist a desert is a wondrous place where bedrock is often not obscured by regolith, soil, and plants and where rock vistas can be seen for miles in all directions (Figure 11.5).

FIGURE 11.4 *Desert animals protect themselves from the heat of the Sun by remaining in the* shade *or by* burrowing *underground. (VU/© John Gerlach)*

FIGURE 11.5 *Because of the extremely low humidity of the desert air, one can often see for a hundred miles or more from high vantage point.*

RELATIVE HUMIDITY

To understand how desert conditions develop, one first needs to understand the meaning of **relative humidity**. Most people have the intuitive feeling that relative humidity has something to do with moisture in the air, but few know its exact definition. Relative humidity is the ratio of the *actual* water vapor content of the air to the *capacity* of the air to hold water vapor at a *particular temperature*. A relative humidity of 50% means that the water vapor content of the air is half of the maximum.

The amount of water vapor that any mass of air *can* hold is directly proportional to the temperature, which means that the capacity of the air to hold water vapor *increases* with *increasing* temperature and *decreases* with *decreasing* temperature.

For a *given* water vapor content, the relative humidity of air changes *inversely* with the air temperature. In other words, an *increase* in temperature results in a *lower* relative humidity (the air becomes dryer) because the capacity of the air to hold water vapor increases. Conversely, a *decrease* in temperature results in a *higher* relative humidity (the air becomes *wetter*) as the capacity to hold water vapor decreases.

A relative humidity of 100% means that the air is saturated with water vapor. The temperature at which a cooling air mass becomes saturated is called the **dew point**. Condensation must occur when the temperature of the air drops below the dew point because the relative humidity cannot exceed 100%.

KINDS OF DESERTS

Deserts can form in four different locations: (1) in a topographic *rainshadow,* (2) in the *subtropics,* (3) in association with near-shore *cold ocean currents,* and (4) in *isolation* from major sources of water. The conditions under which the four kinds of deserts form differ primarily in the mechanism by which the moisture is removed from the prevailing winds.

Rainshadow Deserts

Just as the intensity of light is diminished in the (*light*) *shadow* of an object, the amount of rain is decreased in the **rainshadow** of a mountain range.

The deserts of the southwestern United States are excellent examples of rainshadow deserts. Consider the drawing in Figure 11.6. The prevailing westerly winds arrive at the coast with relative humidities near saturation. As the air moves inland, it rises over

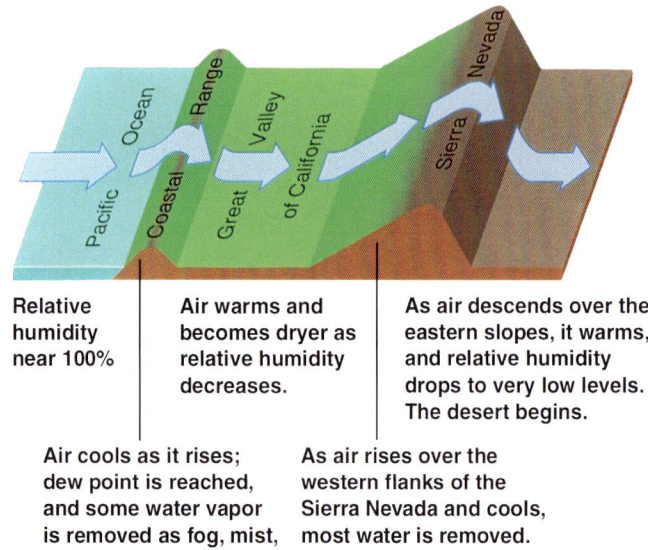

FIGURE 11.6 *One of the scenarios responsible for the formation of deserts is the* rainshadow *that exists on the* leeward side *of a topographic* obstruction *that crosses the path of the prevailing, moisture-laden wind.*

the coastal mountains, expanding and cooling in the process. Because the initial relative humidity is near saturation, the dew point is quickly reached and condensation occurs. Average elevations in the Coast Ranges are only about 5,000 feet (1,500 m), so the temperature drop is relatively small, and the condensed moisture is usually limited to the fog and mists that commonly enshroud the Pacific coast (Figure 11.7). The western slopes of the Olympic Mountains west of Seattle, Washington, are an exception; they receive up to 250 inches (625 cm) of precipitation per year, the highest in the continental United States (Figure 11.8). At the same time, the eastern slopes receive only 12 inches (30 cm) of precipitation per year and are the driest areas along the Pacific coast north of Los Angeles.

After the air masses cross the coastal mountains, they descend into the valleys to the east, either the Great Valley of California or the Willamette Valley in Oregon. As the air descends and warms, the relative humidity decreases. The warmed air then crosses the valley. Since temperatures are above the dew point, little or no precipitation occurs. In fact, the relative humidity is often so low that clouds cannot form, allowing the natives to brag that they receive 340 days of sunshine each year. Abundant sunshine combined with long growing seasons and extensive irrigation have resulted in the California and Willamette valleys being highly productive farmlands.

FIGURE 11.7 *A common sight along the Pacific Coastal Range is the fog that forms as the moisture-laden winds ascend the western slopes and become chilled to the dew point. (VU/© John Gerlach)*

FIGURE 11.8 *The highest rainfall in the continental United States occurs in the Olympic Mountains along the Pacific coast west of Seattle, Washington. (Courtesy of E. D. McKee/USGS)*

After crossing the valleys, the air masses rise over the Cascade Mountains in the north and the Sierra Nevada in the south. As the air ascends to elevations in excess of 12,000 feet (3,700 m), temperatures drop below the dew point, and much of the remaining moisture is removed as rain or snow. The western slopes of the Cascades and Sierra Nevada receive high rainfalls in the summer and heavy snowfalls in the winter.

As the air masses descend the *leeward* side of the mountains, they compress and warm, and the relative humidity drops. Upon reaching the valley floors, the relative humidity of the air is commonly so low that some of the driest localities in North America are found here (see Figure 11.6). Death Valley, an area whose name is synonymous with extreme aridity, is located east of the Sierra Nevada (Figure 11.9). A similar situation exists in the Pacific Northwest where the deserts of eastern Washington and Oregon are located in the rainshadow of the Cascade Mountains.

Subtropical Deserts

The largest deserts in both the Northern and Southern Hemispheres are **subtropical deserts**. To understand their formation, consider the movement of the air masses over the surface of Earth (Figure 11.10).

Near the equator, the air is warm, moisture-laden, and of low density. As the air rises and is chilled, most of the water vapor is released as rain. When the air attains elevations of 50,000 to 60,000

FIGURE 11.9 Death Valley, *along the eastern base of the Sierra Nevada, is one of the driest and hottest places on Earth. (Courtesy of E. N. Hinrichs/USGS)*

feet (15,000–°18,000 m), it flows to the north and south. The cool air then descends, producing two belts of high pressure located at approximately 25° north and south latitude. As the air descends, compresses, and warms, the relative humidity drops and the air becomes extremely dry. Where the air descends over an ocean, it quickly acquires moisture via evaporation. But when the air masses descend over land, available moisture evaporates from the land, producing some of the most extensive and driest deserts on Earth. The largest such desert region extends continuously from the northwestern coast of Africa to northwestern India (Figure 11.11). Included in this belt is the largest single desert on Earth, the *Sahara,* as well as the deserts of Arabia, Iran, Afghanistan, and Pakistan.

In the Western Hemisphere, the deserts of central Mexico are located under these descending, dry air masses. In the Southern Hemisphere, subtropical deserts cover nearly three quarters of the continent of Australia (see Figure 11.11) and also include the *Kalahari Desert* of southern Africa.

Fog Deserts

Deserts and cold ocean currents might seem to be rather odd partners. Nevertheless, deserts along the western coast of South America and the southwestern coast of Africa are the result of two cold, surface ocean

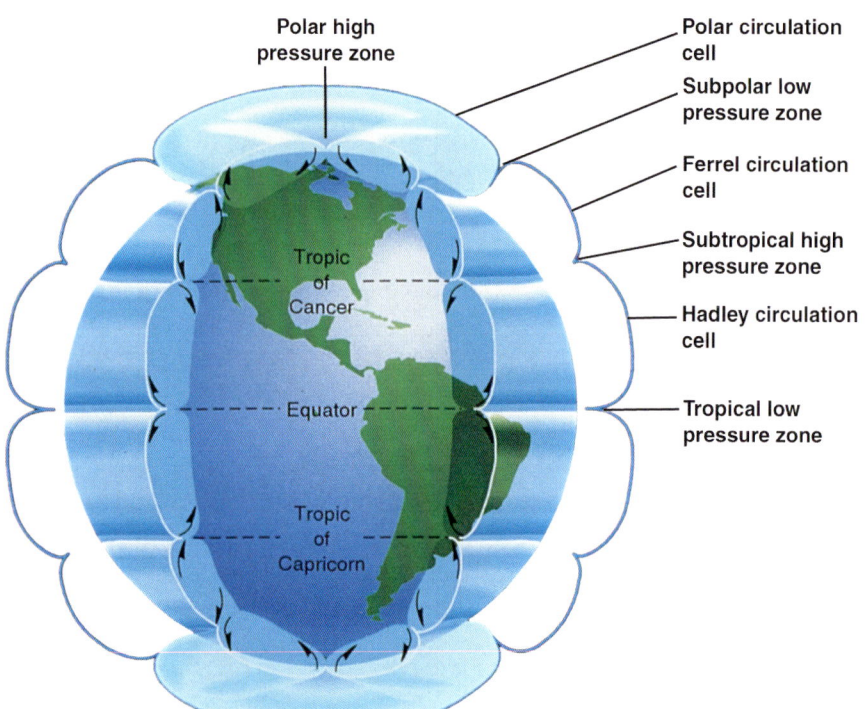

FIGURE 11.10 *Along the* Tropics of Cancer *and* Capricorn, *located approximately 25° north and south of the equator respectively, Earth is bathed in columns of descending, extremely dry air. When these air masses descend over land, any available moisture is removed from the surface, resulting in the formation of Earth's most extensive deserts. (For further discussion, see page 447.)*

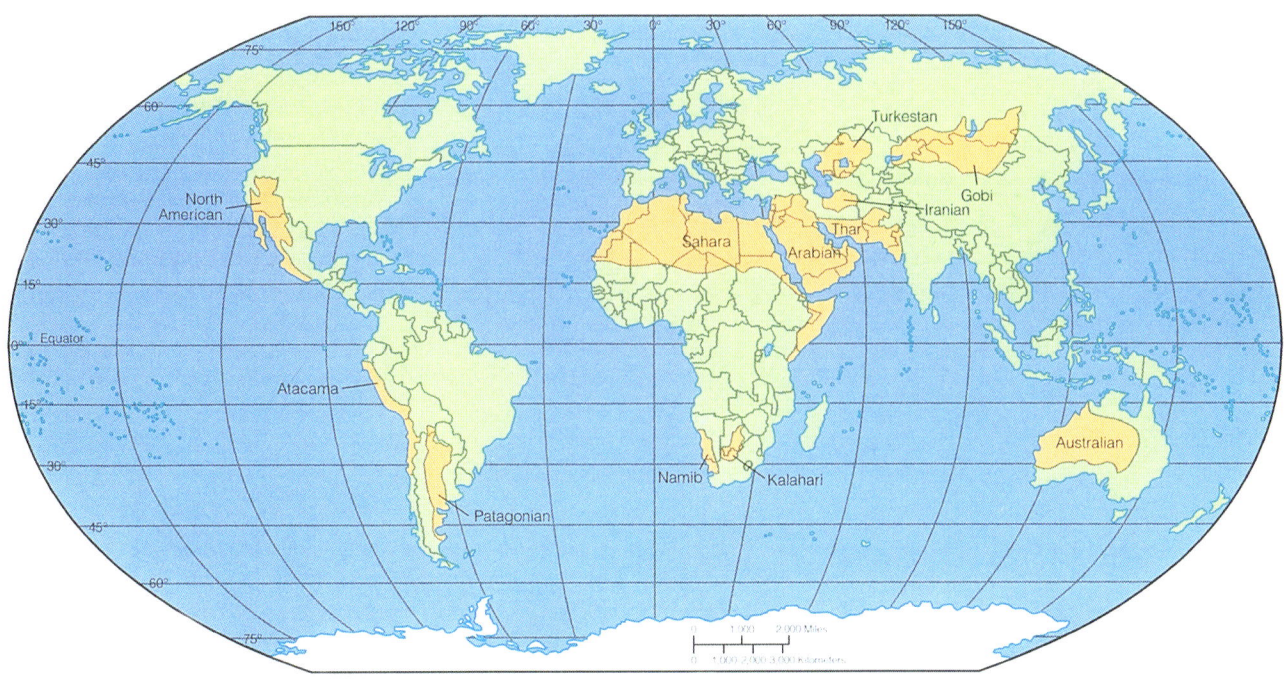

FIGURE 11.11 *With the exception of the Gobi Desert and the interior of Antarctica, most of Earth's large deserts are in areas where landmasses are located beneath the descending warm, dry air masses of the subtropical high pressure zones in both the Northern and Southern hemispheres.*

currents, the **Humboldt** and **Benguela** currents, respectively (Figure 11.12). Both currents consist of cold water and flow counterclockwise just offshore from their respective continents.

To illustrate how a cold offshore current can produce a desert, consider the Benguela Current (Figure 11.13). As the moisture-laden air masses from the south Atlantic move toward the African coast, they cross the cold surface of the Benguela Current and are chilled to the dew point. Since much of the moisture originally contained in the air is removed as thick fog and heavy rains, the associated deserts are called **fog deserts.** As the air masses move landward beyond the Benguela Current and rewarm, the relative humidity drops. Eventually, the warm, dry air reaches the southwestern coast of Africa where it is responsible for the formation of the coastal *Namib Desert* (Figure 11.14). In similar fashion, the Humboldt Current is at least partially responsible for the formation of the *Atacama Desert* of Chile. One of the driest places on Earth, the Atacama Desert is the result of the combined effect of the cold Humboldt Current and the rainshadow of the Andes (Figure 11.15).

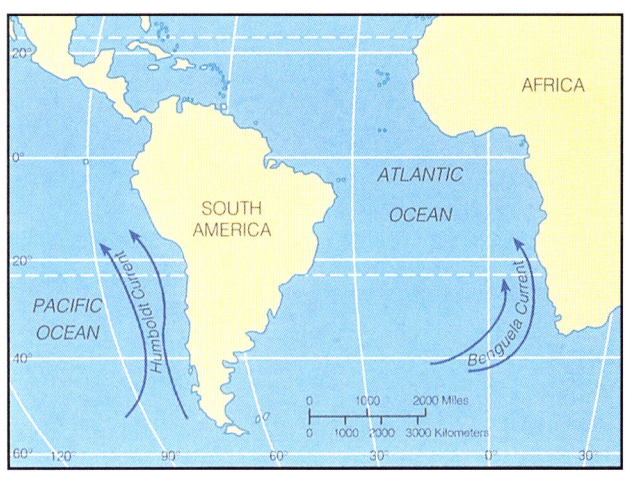

FIGURE 11.12 *The presence of the cold* Humboldt *and* Benguela currents *results in the formation of* fog deserts *along the west coasts of* South America *and* South Africa *respectively.*

Isolation Deserts

Only two desert areas fit the **isolation** scenario, the *Gobi Desert* of central Asia and the interior of

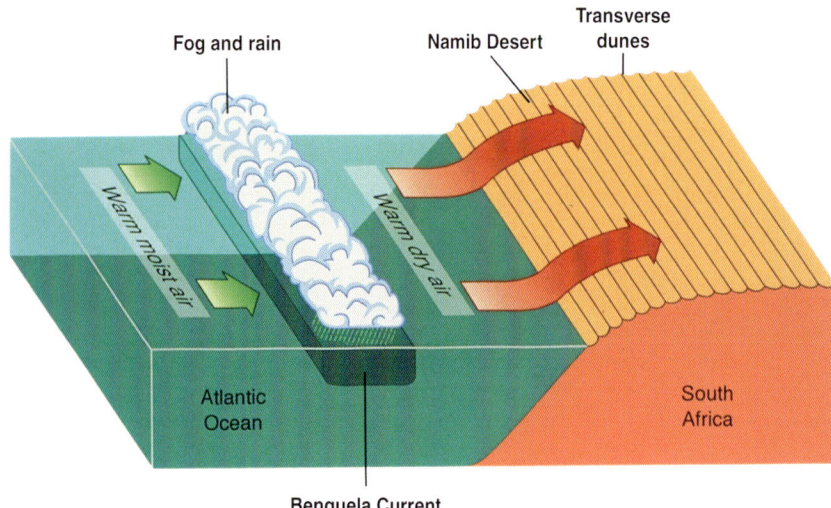

FIGURE 11.13 *As the prevailing, moisture-laden air masses cross the cold surface ocean currents, they are chilled to the dew point and lose much of their water content in the form of fog and rain. Once across the currents, the air masses warm and become drier as they approach the coast.*

FIGURE 11.14 *The* Namib Desert *along the west coast of* South Africa *is a* fog desert *associated with the cold* Benguela Current. *(Courtesy of NASA)*

FIGURE 11.15 *The driest desert on Earth, the Atacama Desert of Chile, is the result of the combined efforts of the cold Humboldt Current and the rainshadow of the Andes Mountains. (Courtesy of USGS/ Young Faults C1)*

Antarctica. The Gobi Desert exists because it is far removed from any source of water (see Figure 11.11). Westerly winds blowing off the Atlantic Ocean have long since lost all their moisture by the time they reach the Gobi. The arctic Ocean to the north cannot provide much moisture because it is frozen over for much of the year. When the surface waters of the Arctic Ocean do thaw during the short summer, the air is too cold to carry much moisture. Air masses moving off the Indian Ocean to the south are blocked by the Himalaya Mountains. What little moisture remains after the air masses have crossed the Himalayas is lost over the high Tibetan Plateau. Although the Pacific Ocean is located a relatively short distance to the east, the prevailing winds are unfortunately moving in the wrong direction. The Gobi Desert therefore exists as a result of being "isolated" from any source of moisture.

The interior of the Antarctic continent is often not included in general discussions of deserts because its moisture is limited by a somewhat different mechanism from that operating in other desert regions. In the Antarctic and other polar areas, the amount of moisture is severely limited simply because the moisture-transporting capacity of the frigid air masses is so low. The glacial ice covering the continent results in large part from the low rate of evaporation due to the cold climate.

SPOT REVIEW

1. How is relative humidity affected by changes in temperature?
2. What is the significance of the dew point?
3. Why is the relative humidity low on the leeward side of mountain ranges
4. What circumstances are responsible for the formation of deserts such as the Sahara and the interior of Australia?
5. How may the presence of ocean currents affect the amount of moisture available to coastal landmasses?

EROSION IN THE DESERT

Arid and humid landscapes are notably different in appearance, which suggests that different agents of erosion must be at work, operating within different cycles of erosion. As we will see, however, the reality is somewhat more complex, and the erosional processes in the two areas exhibit some similarities as well as differences.

The Effectiveness of Water

Another common misconception about deserts is that the major agent of erosion is the wind. Indeed, the sharp differences between desert landscapes and those of humid areas, where streams are known to be responsible for the various landforms, might suggest that some agent of erosion *other* than streams must be responsible for the desert landscape. Because the wind is so prevalent in deserts and our image of dust storms is so clear, the wind naturally comes to mind as the most likely sculptor of the land. Although the wind is certainly an agent of erosion in deserts, its importance is often overrated. The major agent of erosion in the desert is running water. A point we made earlier is worth repeating: streams are the major agent of erosion anywhere that water exists. Although, admittedly, the amount of water in deserts is limited, most desert topography is the result of the combined efforts of mass wasting and erosion and deposition by streams.

It is true that desert landscapes look different from those of humid areas. For one thing, physical evidence for the geologic work of running water is easier to observe in deserts than in humid areas where the surface is covered with regolith, soil, and plants. In humid areas, the only evidence for the past presence of running water may be the rills and rivulets that develop on the exposed regolith or soil of a road cut, embankment, or denuded construction site. The situation is quite different in the desert where little or no plant cover obscures the surface. Evidence that the surface has been subjected to running water can be seen almost everywhere. Rills and rivulets are incised (cut) into nearly every shale bank, hillside, and slope (Figure 11.16). Fan-shaped deposits of debris, obviously transported and

FIGURE 11.16 *Evidence of running water is not difficult to find in the desert. Most slopes show the result of water erosion from the tiny grooves cut by the rills that follow a brief shower to canyons incised by a stream in flood. (Courtesy of H. E. Malde/USGS)*

FIGURE 11.17 *The alluvial fans that build out from the mouths of valleys where they open onto the desert floor are the most conspicuous evidence of running water in the desert. (VU/© Martin G. Miller)*

deposited by streams, build out from the mouths of canyons (Figure 11.17). Mud-cracked surfaces can be seen wherever water once occupied a shallow pond or puddle (Figure 11.18). In just a few minutes, even a casual observer will find ample evidence that the desert surface has been affected by running water.

The low rainfall totals in deserts are offset by the much *greater* erosion potential of each raindrop. In a humid area, much of the kinetic energy of raindrops is dissipated by vegetation. When the rain reaches the surface, the combination of regolith, soil, and plants further reduces the erosive potential of the water.

Such is not the case in the desert. Because of the near-absence of vegetation, not only are the bedrock and regolith exposed to the full energy of the falling rain, but the runoff of the surface water is rapid. As a result, the full kinetic energy of each raindrop is expended on the surface. The runoff quickly drains through rills and rivulets into stream channels, which rapidly fill. Within a short time, a dry streambed can become a raging torrent, transporting large volumes of rock debris (Figure 11.19). Because of the high discharges and turbulence, an enormous amount of material can be eroded and moved during a desert flash flood.

Some of the differences between the landscapes of humid and arid regions occur because the streams belong to different types of drainage systems. Whereas external drainage systems dominate in humid areas, most desert streams belong **to internal drainage systems**. Rather than ultimately flowing to

FIGURE 11.18 *Layers of mud commonly shrink and crack upon drying and form mud cracks, which may in turn be buried by subsequent sedimentations and preserved during lithification. (Courtesy of Robert Behling)*

the sea, streams belonging to interior drainage systems drain into an inland basin where they may (1) terminate by evaporation and infiltration, (2) flow into a permanent lake, or (3) flow into a seasonal or **playa lake** (Figure 11.20). Upon occasion, streams do flow through deserts to the sea but they are quite rare. Two notable examples are the Nile River, which successfully crosses the Sahara to the Mediterranean Sea without a single tributary, and the Colorado River, which flows through the dryer areas of the southwestern United States to the Gulf of California.

Deserts and the Wind 331

FIGURE 11.19 *Most desert stream channels, or arroyos, are dry most of the time. However, shortly after a flash rainstorm in the mountains, it is not uncommon for a dry arroyo to fill with a raging torrent of water within a matter of minutes only to return, just as quickly, to a dry streambed as the flood waters ebb. For this reason, a good piece of advise is "never camp in an arroyo." (© VU/John D. Cunningham)*

Since interior drainage streams do not transport their loads to the ocean, the sediment they carry accumulates in intermontane basins adjacent to the highlands. With time, the basins progressively fill with debris as the adjacent highlands are lowered by weathering and erosion.

The Arid Cycle of Erosion

Desert landscapes and those of humid areas also seem to undergo different cycles of erosion. As the humid cycle of erosion progresses from *youth* through *maturity* to *old age,* the landscape becomes increasingly subdued. Hilltops become rounded, the angles of slope decrease, and rocks lose their sharp edges and angularity as they are continuously attacked by both mechanical and chemical weathering. The landscape is rugged only in the youthful stage when streams incise deep V-shaped valleys into the land surface.

In the arid cycle of erosion, the appearance of the land throughout the cycle depends primarily on whether the area is drained by interior or through-flowing streams. In some areas with interior drainage, the topography becomes progressively more subdued as the highlands are attacked by weathering and erosion, and their remains are buried in the debris (Figure 11.21). In the rare areas drained by through-flowing streams, the topography often maintains a sharp, angular, rugged appearance throughout the entire erosive cycle. As the topography progresses from youth toward old age, rock outcrops at the top edges of valleys retain their angularity; slopes retreat parallel to the original, steep valley walls; and individual rock fragments remain as sharp edged as they were when they were first torn from the main rock mass by processes of mechanical weathering. Examples are the canyonlands of southern Utah including the Grand Canyon and Monument Valley (Figure 11.22).

FIGURE 11.20 *Few desert lakes would hold water year-round. Most are* playa lakes *that contain water only during times of seasonal rainfall. Throughout the remaining time, playa lake beds are covered with salts deposited as the water evaporated.*

FIGURE 11.21 *Erosion by the streams of an internal drainage system results in the remnants of the mountains literally being buried in their own debris as the sediments fill the intermontane valleys. (Courtesy of NASA)*

These observations may at first seem to contradict the Davisian concepts of erosion and landscape development presented in Chapter 9. Yet, although Davis undoubtedly had humid areas in mind when he formulated his concepts, his ideas are still applicable to arid erosion. The different appearance of the landscape in humid and arid regions is perhaps due more to differences in the relative *effectiveness* of the individual agents of erosion than to differences in the *mechanism* of erosion.

One of the reasons the topography of humid areas becomes more subdued as erosion progresses is that chemical weathering plays a dominant role in the process. Chemical weathering quickly attacks and rounds the edges of rocks and rock fragments, smoothing the overall angularity of the topography. Chemical weathering also reduces rocks to a mixture of fine-grained clay minerals and sand-sized quartz grains that accumulate as regolith. Continued attack of chemical weathering upon the regolith, combined with the activities of microorganisms, results in the development of a clay-rich soil. Eventually, the soil supports a continuous cover of plants whose roots penetrate into the rocks below, furthering the chemical and physical breakdown of rocks. As the cover of relatively fine-grained regolith and soil accumulates, slow processes of mass wasting such as creep move

FIGURE 11.22 *In desert areas drained by through-flowing streams, cliffs remain vertical as the valley floors widen and the debris is carried off, resulting in the type of topography seen in the Grand Canyon. (Courtesy of E. D. McKee/USGS)*

the debris downhill to the streams, which carry the materials away. The combined actions of chemical weathering, slow mass wasting, and erosion by streams progressively change the appearance of the landscape in accordance with the Davisian concept of landscape development.

The scenario in desert areas differs only in the relative importance of the various agents of weathering and erosion. Because water is scarce, the effectiveness of chemical weathering is diminished. Processes such as frost wedging operate within vertical fractures where they slowly pry rock slabs away from the main rock mass, generating rock fragments with angular shapes. Because mineral-bearing solutions evaporate rapidly, crystal growth is an important physical weathering process as minerals are deposited and grow just beneath the rock surface. As minerals grow in cracks, they force rocks apart just as water does when it freezes.

Because the slopes are steep, mass wasting removes the regolith and deposits it at the base of the slopes as talus. Without the protection of regolith, the rock surface is subjected to continuous attack. Slow processes of mass wasting such as creep, which operate primarily on fine-grained regolith and soils in humid climates, are of minor importance in the desert. In contrast, the major mass wasting processes are *fast* processes such as rock falls and rock slides. Relatively large masses of rock, loosened from steep slopes by weathering, finally succumb to the effects of gravity and crash to the base of the slope, where they shatter on impact into the angular rock fragments that accumulate as talus (Figure 11.23). With the effects of chemical weathering reduced, the edges of rock fragments within the talus do not round with time, but rather retain their original sharp, angular outlines even when their size is reduced by continued mechanical weathering. Only after the size of particles is considerably reduced can slow mass wasting processes such as creep take an active part in erosion and move the smaller particles to the toe of the talus where they will be picked up by a stream and carried off.

In desert areas with through-flowing streams, gravity-induced rock falls are the main mode of slope retreat. Valley walls tend to retreat with little change in slope angle, maintaining vertical to near-vertical slopes throughout the erosional cycle (Figure 11.24). Because horizontal surfaces such as the tops of mesas and buttes experience minimal weathering and erosion, their elevations are reduced at extremely slow rates. The overall result is erosion dominated by (1)

FIGURE 11.23 *The* talus *that accumulates at slope bases in desert areas is largely composed of angular rock fragments generated by frost action and moved by fast processes of mass wasting such as rock falls.*

the lateral retreat of vertical to near-vertical cliffs, (2) minimal reduction of elevated surfaces, and (3) the retention of sharp, angular edges on hard rock surfaces throughout the cycle.

One of the best examples of the combined effects of weathering, mass wasting, and stream erosion in desert areas is *Monument Valley* (Figure 11.25). Located at the southern edge of the Colorado Plateau along the Utah-Arizona border, Monument Valley is dominated by essentially horizontal sedimentary rocks. Thick sandstone beds incised by streams have produced steep-sided valleys. Monument Valley is dominated by vertical cliffs of sandstone. As mechanical weathering attacks the rock cliffs along vertical fractures, the rock is removed from the cliff face by rock falls and accumulates as talus. As the cliffs retreat parallel to their original slopes, impressive rock masses such as **mesas** (Figure 11.26), **buttes** (Figure 11.27), and **needle rocks** are left behind (Figure 11.28).

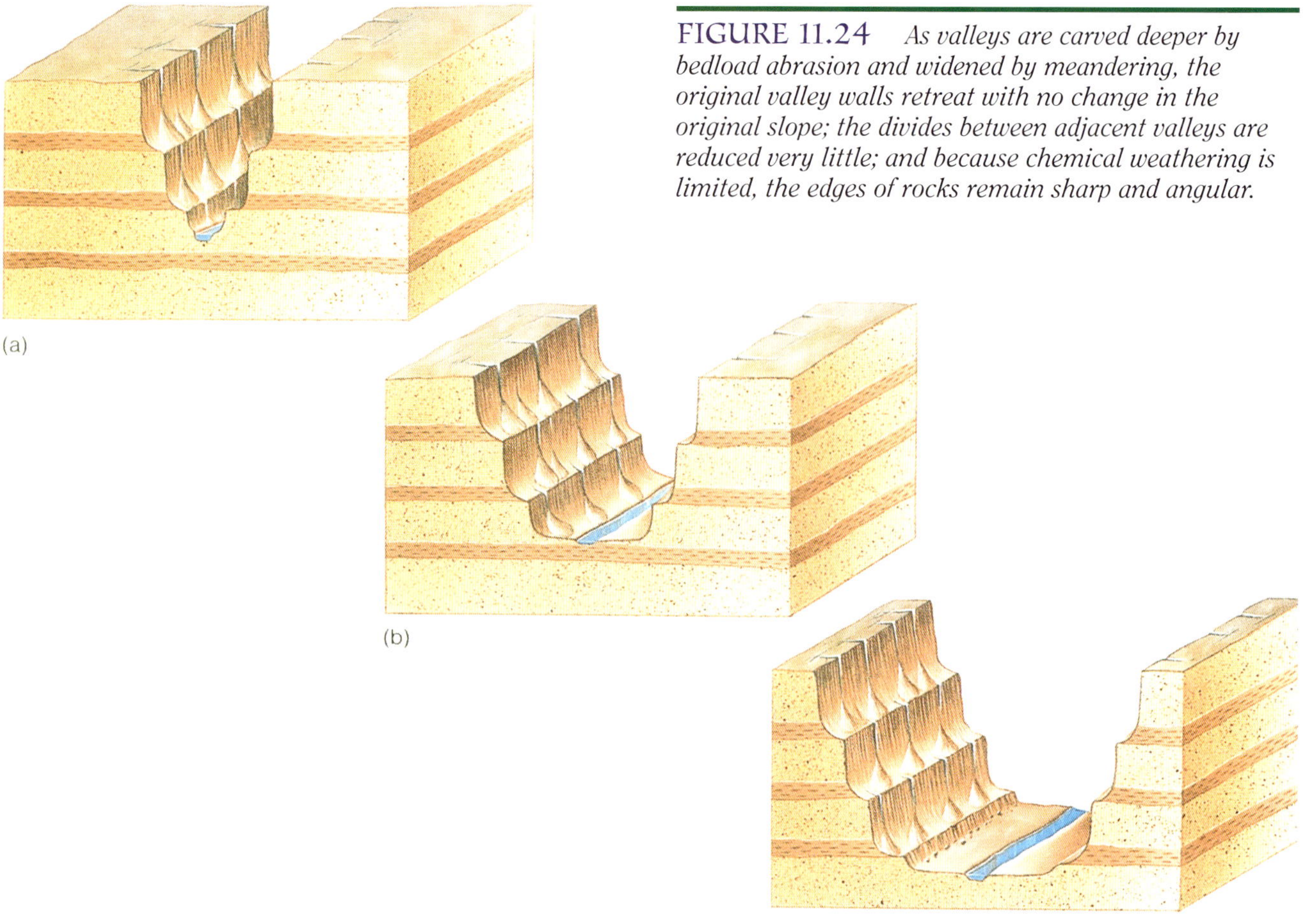

FIGURE 11.24 *As valleys are carved deeper by bedload abrasion and widened by meandering, the original valley walls retreat with no change in the original slope; the divides between adjacent valleys are reduced very little; and because chemical weathering is limited, the edges of rocks remain sharp and angular.*

FIGURE 11.25 *No place illustrates the incredible eroding power of desert streams better than Monument Valley. One needs only to realize that the solid rock seen in the mesas, buttes, and needle rocks once existed as a continuous layer throughout the area. (Courtesy of E. D. McKee/USGS)*

FIGURE 11.26 *The largest masses of rock that remain as the valleys of through-flowing streams widen are called* mesas. *(Courtesy of W. B. Hamilton/USGS)*

FIGURE 11.27 *As the valleys continue to widen, the mesas are reduced in size and eventually become* buttes. *There is no agreement as to when a mesa becomes small enough to qualify as a butte; apparently, it is in the eye of the beholder. (Courtesy of E. D. McKee/USGS)*

The casual observer may find it hard to imagine that the mesas, buttes, and needle rocks of Monument Valley are the remnants of a continuous layer of rock that once extended throughout the area. Although the name refers to these sentinels of rock, the true monuments are the *valleys,* which are testaments to the processes that created the landscape: namely, the relentless attack of slow weathering, rapid mass wasting, and, above all, the streams, which during their short and infrequent bursts of life removed all the material that once filled the space that now exists between the buttes, mesas, and needle rocks.

Basin and Range Topography

Another major type of desert landscape in the southwestern United States is called **Basin and Range topography.** Basin and Range topography is dominated by long parallel ridges and valleys that formed by vertical movement along parallel fractures called faults. A detailed discussion of the process by which the fractures formed will be found in Chapter 15. The best examples of this topography are found in Nevada (Figure 11.29).

As is characteristic of the arid cycle of erosion, the combined processes of weathering, mass wasting, and stream erosion have resulted in the parallel retreat of the original slopes. High-gradient, youthful mountain streams incise the slopes and upland areas of the ridges. As the streams emerge from the mountains, they lose water by evaporation and infiltration, and the stream load is rapidly deposited in an **alluvial**

FIGURE 11.28 *In time, the rock masses are reduced to slender fingers of rock called* needles *or* needle rocks. *In fact, one well known needle in Monument Valley is named "ET's Finger." The fate of needles is eventually to be toppled, much like a tree. (Courtesy of S. W. Lohman/USGS)*

fan (see Figure 11.17). Because the sediments are deposited quickly, they are not as well sorted as those that accumulate in deltas. The coarser materials are deposited higher up on the fan, and the finer materials are carried farther out into the valley. With time, the individual alluvial fans increase in size and impinge on adjacent fans, forming a **bajada** (Figure 11.30). Bajadas built out from opposite sides of the valley meet to form a **bolson.**

As the mountain slopes retreat and the debris is transported to the bolson, an erosional rock surface called a **pediment** develops upslope from the alluvial fans. A pediment is usually covered with a thin, discontinuous layer of sand or gravel. During the old age stage of the arid erosional cycle, the erosional rock surface and the alluvial fan form a **pediment slope,** the arid cycle equivalent of a peneplane. As the rocks of the mountain range slowly succumb to weathering and erosion, pediment slopes from opposite sides of the mountain range eventually meet. At this point, all that remains of the original mountain range are a few small hills; the mountain range is literally buried in its own debris (refer to Figure 11.30).

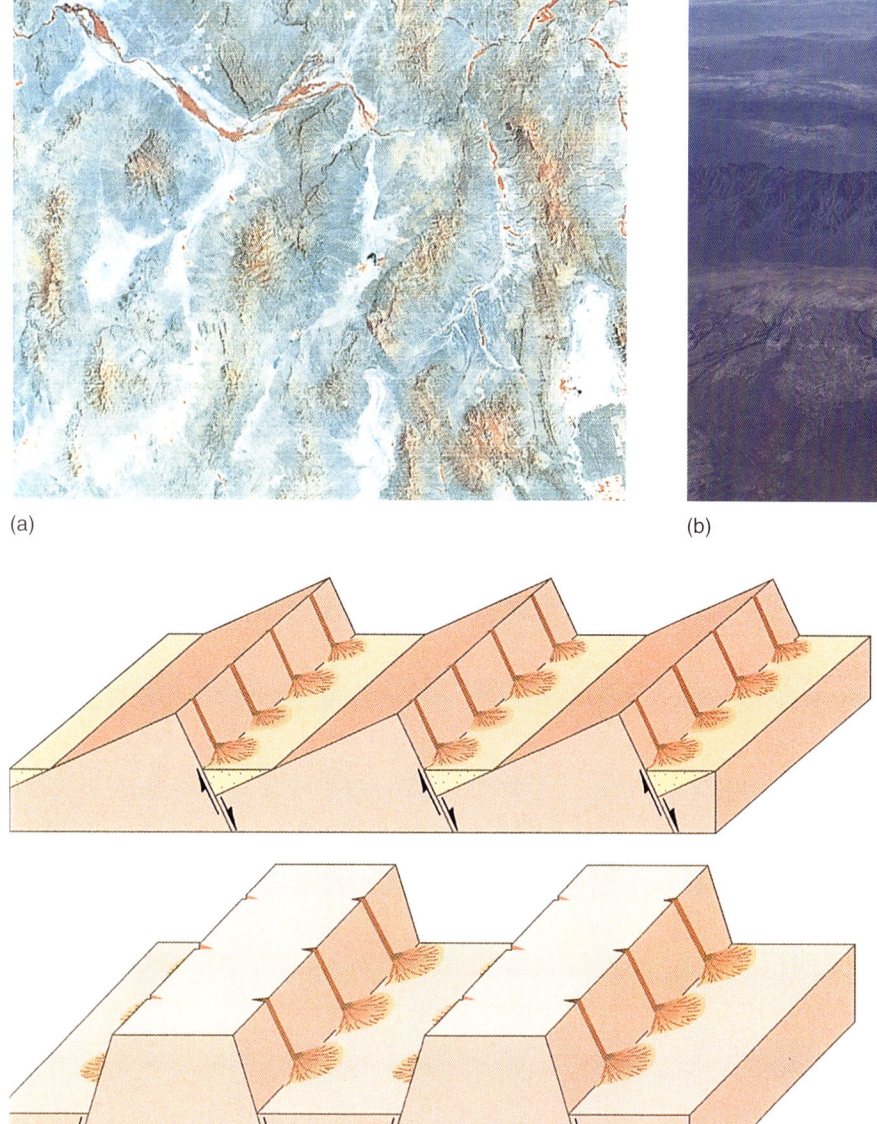

FIGURE 11.29 *The* Basin *and* Range Province, *centered over Nevada, is dominated by parallel, north-south trending mountain ranges and intermontane desert valleys. Two different styles of faulting are responsible for the formation of the mountain ranges and intermontane valleys. This terrain is among the most difficult in the United States to cross even today; early pioneers en route to the Pacific Coast were required to flank the Basin and Range either to the south or the north. (a Courtesy of NASA;* b *VU/© Martin G. Miller)*

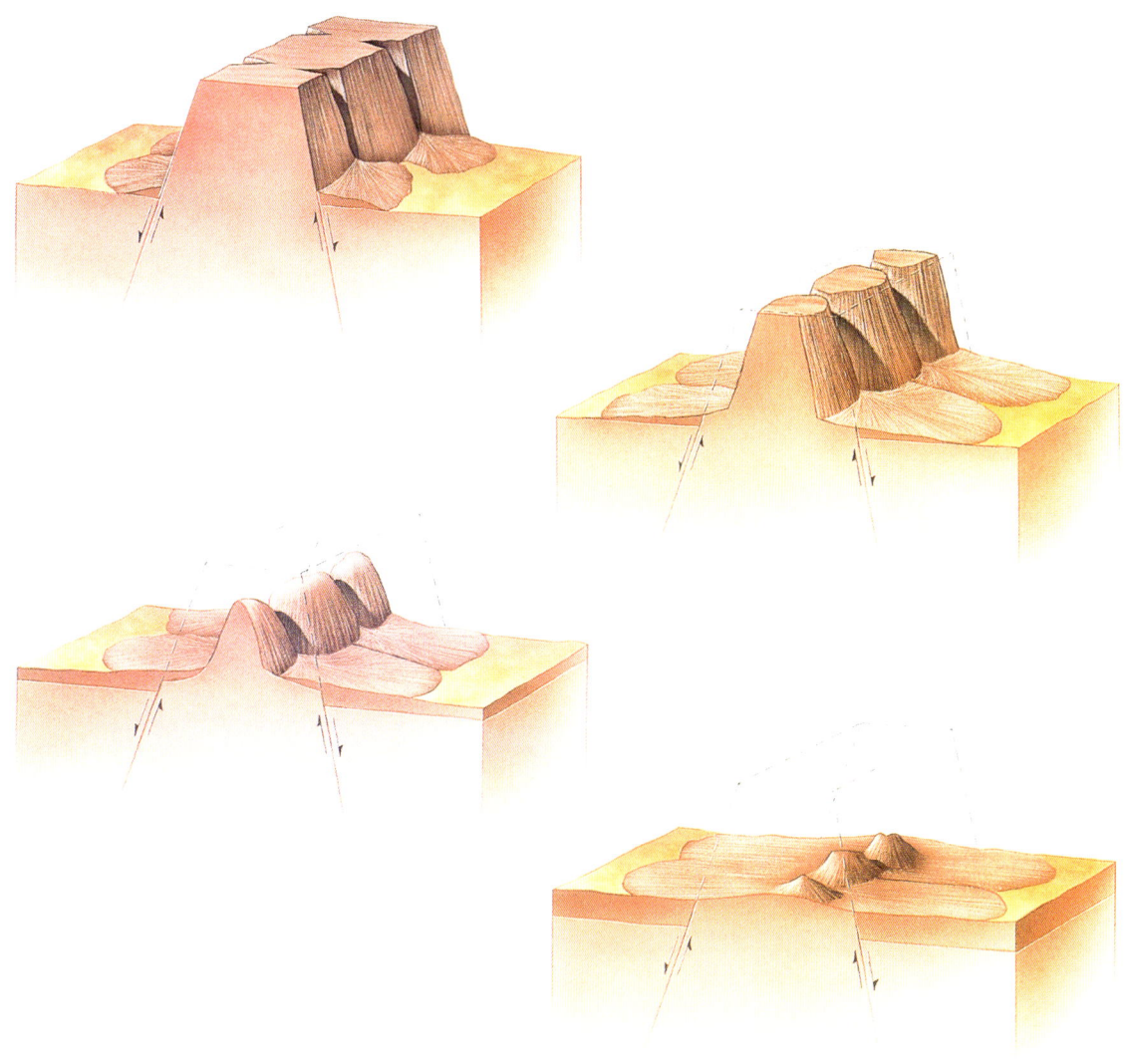

FIGURE 11.30   *As streams of an* internal stream system *wear mountains away, the debris is deposited into the adjacent valleys as* alluvial fans *that eventually coalesce to for* bajadas, *which in turn build from opposite sides of the valley to form a sediment-filled valley called a* bolson. *In time, the mountain range is reduced to low hills literally buried in its own debris.*

SPOT REVIEW

1. In what ways do the operations of streams in internal and external stream systems differ?
2. What evidence can you present that demonstrates that streams are the most effective agents of erosion in deserts?
3. Explain the following statement: in many desert areas, mountains eventually become buried in their own debris.
4. How are alluvial fans, bajadas, bolsons, pediments, and pediment slopes related?

EROSION BY THE WIND

Compared to streams, the wind occupies a distant second place as an agent of desert erosion except in the driest deserts. The wind accomplishes the task of erosion by mechanisms that are similar in many ways to those used by running water. Any differences are due to the lower viscosity and density of air.

The major difference between the operation of water and wind is in their *competence*. Because of its lower density, the wind is generally limited to lifting and moving particles of sand size and smaller. The discussion in Chapter 9 of how water erodes particles by direct lifting is directly applicable to the wind.

As is the case with water, sand is the easiest particle size for the wind to move. This explains, at least in part, why sand is a scarce commodity in the desert. Also like water, the wind has difficulty picking up and moving clay-sized particles and for the same reason. Unless the surface of a clay-sized deposit is disturbed by some other moving object such as a moving sand grain, a running animal, or a rolling piece of tumbleweed, the wind cannot get under clay- and silt-sized particles and dislodge them from the surface.

As they are transported by the wind, sand-sized particles remain near the ground surface and move in a fashion similar to the bed load of a stream, either rolling along the surface or moving in a bouncing mode called **saltation** (Latin *saltatio* = to leap) (Figure 11.31). Even under the influence of high-velocity winds, it is unusual for sand-sized particles to be lifted higher than two or three feet (60 to 90 cm) from the surface.

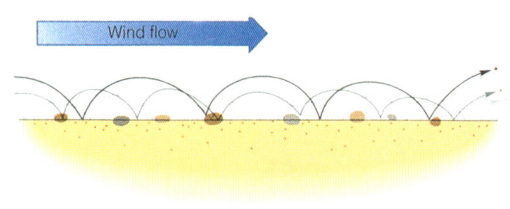

FIGURE 11.31 *The process by which the wind moves sand-sized particles is called* saltation. *The sand does not usually rise more than a fraction of an inch above the surface; even in very strong winds, sand particles rarely rise more than a foot or two.*

Once taken into suspension, clay- and silt-sized materials can be lifted to considerable heights and carried for long distances. Most of the material in what is commonly called a "sand storm" is, in fact, not sand but dust composed of silt- and clay-sized materials (Figure 11.32).

Deflation

Wind erosion is called **deflation**. Because the competence of the wind is limited to sand-sized particles, larger particles are left behind and form a deposit called **desert pavement** (Figure 11.33). In most deserts where the bedrock is not exposed, desert pavement rather than sand is the most common material covering the rock surface.

In areas covered with sand, deflation produces shallow, closed depressions called **blowouts** (Figure 11.34). Ultimately, the depth of a blowout is limited when it intersects bedrock or the water table and erosion stops. Water stabilizes the sand by producing cohesion between the grains, and in some cases, vegetated oases may develop.

Abrasion

In the process called **abrasion**, tiny rock fragments are removed from the exposed surface of bedrock by the impact of windblown sand grains. In other words, abrasion is a natural sandblasting process. Rocks shaped by the action of windblown sand are called **ventifacts**, and the flat sand-cut surfaces are called **facets** (Figure 11.35). Multiple facets on individual

FIGURE 11.32 *Although commonly referred to as a "sand storm," most of the material that fills the air is silt- and clay-sized material. (VU/© Martin G. Miller)*

Deserts and the Wind 339

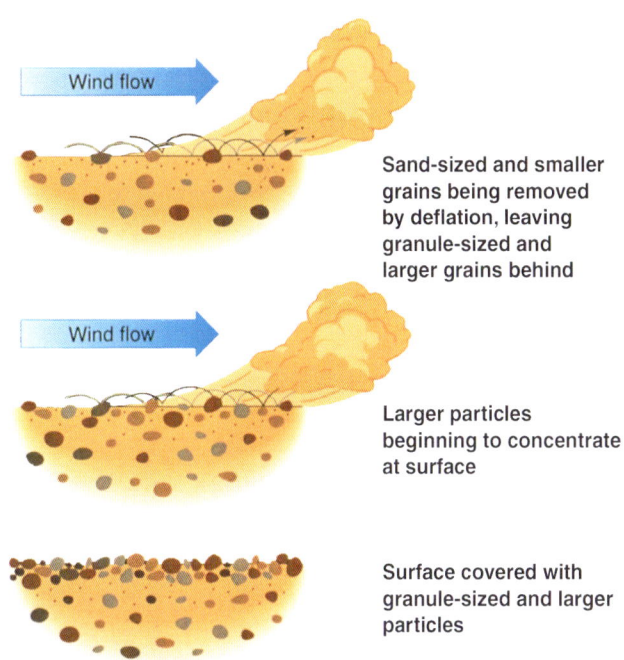

FIGURE 11.33 Desert pavement, *the most common material found covering the desert surface, consists of a layer of granule-sized and larger particles. Desert pavement forms as the process of* deflation *preferentially removes sand-, silt-, and clay-sized particles.*

FIGURE 11.34 *Where abundant sand exists, the process of deflation may form depressions of various sizes called* blowouts. *(Courtesy of E. D. McKee/USGS)*

pebbles may record different wind directions, or they may form when pebbles are carried or moved by a flash flood, exposing a new side to the wind.

A topographic feature often attributed to wind abrasion is the **pedestal rock** where a softer rock layer has been preferentially removed from beneath a more competent rock (Figure 11.36). Although abrasion

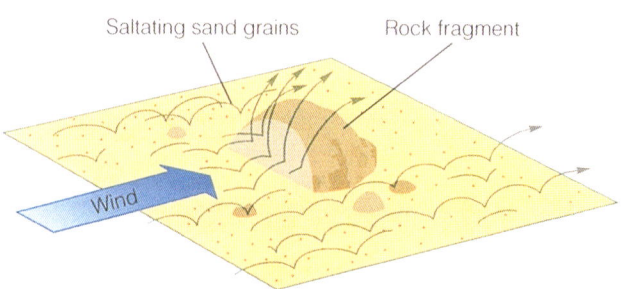

FIGURE 11.35 *Any object formed by the abrasion of windblown sand is called a* ventifact. *The most common ventifacts are rock fragments that possess one or more* facets *carved by sand abrasion. (Photo by VU/© Martin G. Miller)*

may play a minor role in the evolution of pedestal rocks, physical weathering, rainfall, and mass wasting in combination are usually primarily responsible.

DEPOSITION FROM THE WIND

Two kinds of sedimentary deposits originate from wind action: (1) silt- and clay-sized materials called *loess* and (2) accumulations of sand called *dunes*.

Loess

Although **loess** is a common depositional feature in many deserts, several regions that contain vast deserts, such as North Africa and Australia, have almost no loess deposits. The absence of loess is presumably due to the depressed rates of chemical weathering, which have produced only minimal amounts of clay minerals.

The clay and silt within loess are not necessarily of desert origin. In the continental United States, for example, the extensive loess deposits along the Mississippi River Valley formed from silt and clay originally deposited in outwash plains and valley trains of the North American ice sheet (Figure 11.37).

FIGURE 11.36 *Although* pedestal rocks *are popularly attributed to the erosive power of the wind, in reality they are largely due to the combined efforts of frost action, mass wasting, and water erosion. (Courtesy of W. B. Hamilton/USGS)*

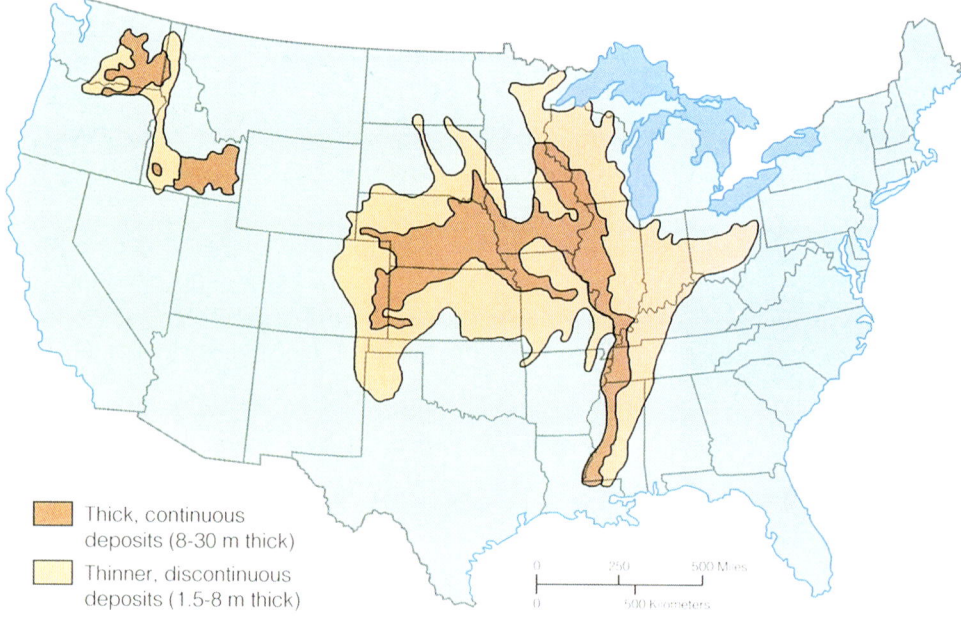

FIGURE 11.37 Loess *is a deposit of wind-transported silt- and clay-sized materials. Although the sediments in many loess deposits are of desert origin, some are not. The extensive loess deposits found along the Mississippi River valley, for example, are formed from silt and clay derived from the outwash plain of the Pleistocene glaciers.*

The most extensive loess deposits on Earth are found in northern China where the loess is several hundred feet (meters) thick in some places. The Chinese loess deposits consist of silt and clay blown in from the Gobi Desert. The fine dust constantly blowing in from the Gobi also accounts for the haze commonly seen throughout northern China. In fact, visitors to the area often suffer from respiratory distress brought on by the high concentrations of atmospheric dust.

In northern China, loess-covered hills are extensively terraced for agricultural use (Figure 11.38), and every small town has a brick plant that utilizes the loess as its basic raw material. Throughout the region, caves dug back into the loess cliffs serve as homes for hundreds of thousands of people (Figure 11.39). Streams are commonly yellow due to the high suspended loads that result from the erosion of loess. A noteworthy example is the Yellow River.

Dunes

The most familiar wind-derived deposit is the **dune**. Dunes can form anywhere there is an adequate supply of sand and a strong prevailing wind, such as at

FIGURE 11.38 *Loess produces excellent agricultural soils. The soils of many of the world's grain producing areas, including the Great Plains of the United States and Canada, are formed from loess. In northern China, loess-covered hills are terraced to take advantage of every possible surface capable of producing food for their enormous population.*

FIGURE 11.39 *Hundreds of thousands of people in northern China live in homes carved into loess cliffs.*

the beach where the wind picks up sand and moves it inland. A dune begins to form when the moving sand encounters a **windshadow**. A windshadow is a zone of decreased wind velocity on the leeward side of an obstruction (Figure 11.40). If you live in an area where the winters are cold and blustery, you have probably experienced a windshadow. Whenever you seek shelter behind a large object such as a tree or the corner of a building, you are in the windshadow of the object and feel much warmer because the decreased wind velocity significantly reduces the windchill factor.

As the wind enters a windshadow, its velocity drops and it deposits sand. The addition of sand to the windward side of the dune raises the effective windshadow, which in turn allows more grains to be deposited. Thus, once the process of dune formation is initiated, it is self-sustaining. As more sand is added, the dune grows and, eventually, moves or *marches* in the direction of wind flow.

Obviously, there is a limit of how large dunes can grow and how far inland they will move. Certainly, one of the limiting factors is the amount of available sand. In addition, the wind velocity will diminish as

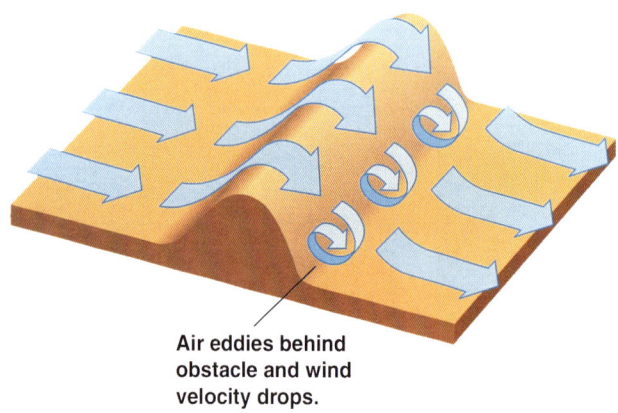

Air eddies behind obstacle and wind velocity drops.

FIGURE 11.40 Dunes *originate as sand accumulates in the* windshadow *that exists on the* leeward *side of obstructions.*

the dunes move inland and the topography impedes the flow of air. Another major factor that limits dune migration is vegetation, primarily grasses (Figure 11.41). The grasses prevent the sand grains from moving any further, so the dune can only grow vertically.

A precarious balance always exists between sand supply, wind velocity, and the effectiveness of grass stabilization. Should the sand supply or the wind velocity increase, the vegetation may be overwhelmed, and the dunes may begin to move again. Usually, when dunes resume their windward movement, the most common cause is that the grass cover has been destroyed, most likely by some human activity. Signs at the beach constantly warn people to "stay off the grass" for precisely this reason.

Sand dunes take on four basic styles: (1) *barchan,* (2) *parabolic,* (3) *transverse,* and (4) *longitudinal.*

Barchan Dunes Barchan dunes form in desert areas where sand is scarce and there is little if any vegetation. The dunes have a characteristic crescent shape with the *horns* pointing in the direction of wind flow (Figure 11.42). This type of dune is very common inasmuch as sand is scarce in most deserts. As is the case with most dunes, barchan dunes commonly form in "fields" consisting of many dunes moving in the direction of wind flow.

Parabolic Dunes Parabolic dunes are similar in shape to barchan dunes except that the horns point in the direction opposite to the wind flow (Figure 11.43). Another difference is that parabolic dunes commonly form in areas that have abundant supplies of sand, some vegetative cover, and strong winds such as those often found along coastlines. Parabolic dunes often develop downwind from a blowout (Figure 11.44).

Transverse Dunes Transverse dunes form in areas of plentiful sand and sparse vegetation. These dunes develop with the long dimension of the dune perpendicular or "transverse" to the wind direction and characteristically have asymmetrical cross sections with the steeper slope being the leeward side of the dune (Figure 11.45).

Longitudinal Dunes Longitudinal dunes, sometimes referred to as **seif dunes**, form in areas that have abundant sand and strong directional winds that blow in more or less the same direction throughout the year. Longitudinal dunes are oriented with the long dimension of the dune parallel to the direction of wind flow and characteristically have symmetrical cross sections (Figure 11.46).

FIGURE 11.41 *Vegetation is a major agent of dune stabilization. The hills to the right are dunes that have been stabilized by vegetation. As long as the vegetative cover is maintained, the dunes will remain immobile. (VU/© Victor H. Hutchison)*

Deserts and the Wind 343

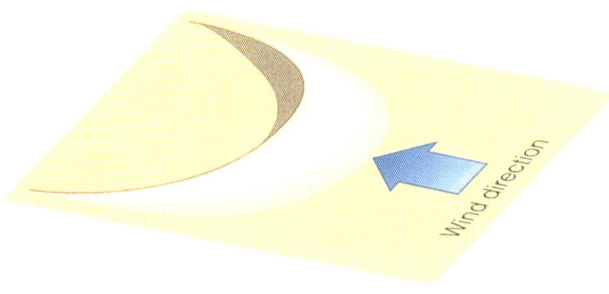

FIGURE 11.42 *Crescent-shaped* barchan dunes *form where both sand and vegetation are scarce. Barchan dunes can often be seen moving across a bedrock surface with their* horns *pointing in the direction of wind flow.*

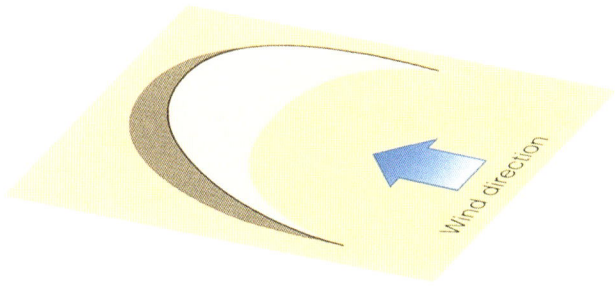

FIGURE 11.43 Parabolic dunes *are also crescent-shaped, but with the horns pointing in the direction from which the wind is blowing.*

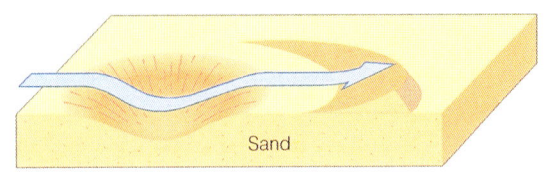

FIGURE 11.44 *Parabolic dunes are commonly found in association with* blowouts *where the dune forms along the downwind margin of the depression.*

FIGURE 11.45 *Most of the dunes observed along the beach, with their long dimensions oriented perpendicular to the direction of wind flow, are examples of* transverse dunes. *(Courtesy of NASA)*

FIGURE 11.46 Longitudinal *or* seif *dunes form in areas that have abundant sand and strong directional winds and orient parallel to the prevailing wind direction. (Courtesy of E. D. McKee/USGS)*

ENVIRONMENTAL CONCERNS

At the present time, arid land or deserts cover about 35% of the land surface of Earth. When the semiarid land is added, the figure increases to about 43%. Nearly 15% of the world's population (about 630 million people) live in arid and semiarid lands. These people depend upon these lands for sustenance even though their agricultural potential is low. Unfortunately, many of these people are citizens of developing countries where populations are increasing at the rate of 2.5 to 3.5% yearly. Eventually, the increasing populations demand more food and fuel than the land can provide.

One of the most serious environmental problems facing the world today is **desertification**, the expansion of deserts into areas where they did not previously exist (Figure 11.47). At the present time, deserts are increasing at the rate of 25,000 square miles (60,000 km^2) each year. The populations that are displaced by the expanding deserts move into adjacent regions. As a result, the population density of the adjacent region increases at the same time that the ability of the region to provide food is decreasing. In other words, a bad situation gets progressively worse. The result of this unfortunate sequence of events is the suffering and death now being experienced by the peoples of countries such as Somalia and Ethiopia.

Although natural processes can result in the expansion of deserts, there is little doubt that the major causes of desertification are human activities, in particular (1) overgrazing of semiarid grasslands, (2) overcultivation of semiarid lands, (3) the removal of woody plants from semiarid regions for fuel, and (4) the salinization of arid and semiarid lands resulting from irrigation.

Overgrazing

Throughout history, semiarid grasslands have been used for grazing livestock. For hundreds of years, the people living in such areas, being fully aware of the fragile nature of the land, limited the populations of their herds to suit the availability of vegetation. They also assumed a nomadic way of life, constantly moving with their herds to allow the vegetation of an area to recover before being subjected to another season of grazing.

In many areas of Africa, the livestock population is increasing at about the same rate as the human

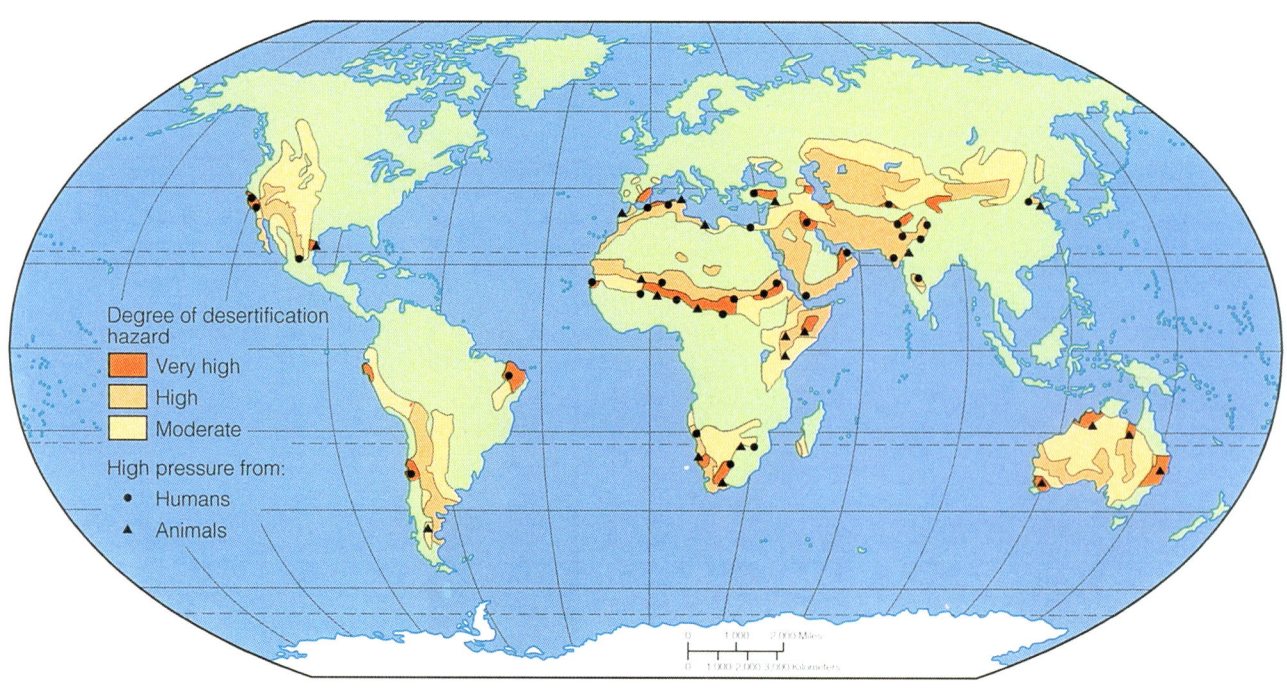

FIGURE 11.47   *Deserts around the world are undergoing expansion, a process called* desertification. *Although in some cases the expansion is the result of natural causes, most present-day desertification is the result of human influences, primarily* poor agricultural practices *and* over grazing *by cattle.*

population. Unfortunately, in some of these areas, the livestock are raised not as a source of food, but as a measure of the owner's wealth. Overgrazing by the growing herds is denuding the land of its natural grass cover. Eventually, not only are the plants cropped to the point where they can no longer survive, but the root systems are trampled by the animals' hoofs. With the grass cover destroyed, the land is exposed to erosion by water and wind, promoting the further advance of the desert and forcing the herds of livestock to migrate. The process will continue until additional grassland is no longer available.

Overcultivation

As human populations and the demand for food increase, more land must be converted to agricultural use. Much of the land converted to agriculture is located in semiarid regions where there is little doubt that the denuding of land for agriculture and the subsequent erosion of the soil are a major cause of desertification.

As the natural vegetation is removed, the soil is subjected to increased erosion by water during the rainy season and by wind during the dry periods. In areas where rainfall is low, crops often fail and leave the land with no vegetative protection at all. With the loss of vegetative cover, the soil is removed and the land is swiftly converted into a desert. Unfortunately, the soil removed by erosion during a single year may have required hundreds of thousands of years to produce. In this respect, soil is a nonrenewable resource; once removed by erosion, it will not be replaced within the foreseeable future.

Normally, areas such as central Africa come to mind when we think of agricultural land being lost due to erosion of soil and desertification. It is important to realize, however, that the same processes are at work in the United States. It has been estimated, for example, that over the past two centuries, poor agricultural practices have resulted in the loss of a third to a half of the soil that once covered the Great Plains, the western Midwest, the Southeast, and parts of the Southwest. Until the 1870s when the settlers began to arrive in large numbers, the land was protected by natural grasses. When the land was plowed and planted with annual crops, the soil was exposed to the processes of erosion for part of each year. Droughts that occurred during 1890, in 1910, and between 1926 and 1934 resulted in the removal of large volumes of topsoil from the Great Plains in an area referred to as the Dust Bowl. As a result of this disaster, the Soil Conservation Service (SCS) was established in 1935 to promote good conservation practices throughout the country. Conservation techniques were initiated such as *contour plowing,* where furrows are oriented perpendicular rather than parallel to the slope of the land; *strip cropping,* where two or more crops are planted on the same acreage; *minimum tillage,* where crops are planted with minimum plowing and often without any plowing at all; and *mulching,* which protects the land during periods of crop growth. These techniques are now common practice.

Fuel

In many developing countries, wood is the major fuel used for heating and cooking. Increased populations, especially in areas such as the Sahel, which borders the southern Sahara, have denuded the land of woody plants in their quest for fuel; these plants are an important agent of soil stabilization. Their loss not only results in decreased protection of the soil, but it also decreases the amount of moisture that would be transferred from the land to the atmosphere by evapotranspiration. Both processes contribute directly to the continued desertification of the land.

Salinization

We are often impressed by the conversion of arid and semiarid lands to agriculture by irrigation—the conversion of deserts into oases. What we often forget, however, is that irrigation can be a double-edged sword that may not only destroy the agricultural potential of arid and semiarid land but may actually promote desertification of semiarid land.

Of all the major causes of desertification, salinization resulting from irrigation is the most ancient. For example, the Sumerian civilization that prospered in the Tigris-Euphrates Valley, the "Cradle of Civilization," for nearly two thousand years was based on an agricultural economy supported by irrigation. Eventually, the concentration of salt in the soil became so toxic to plants that crops could no longer be produced. The area was abandoned and the deserts took over; the deserts are still there today.

Unlike the Sumerians, we now understand the process of salinization. Because of the high degree of evaporation, the groundwater in arid and semiarid regions commonly contains elevated concentrations of dissolved salts, especially NaCl or common table salt. Except for desert plant that are able to select and control their uptake of alkali and alkaline earth ions

(Na^{1+}, K^{1+}, Ca^{2+}, and Mg^{2+}), elevated levels of dissolved salt are toxic to most plants. In most desert semiarid regions, the water table is quite deep, often hundreds of feet. Irrigation causes the groundwater table to rise. As the groundwater table approaches the surface, capillary action draws the salt-laden waters into the soil where it evaporates and deposits the salts. Unless precautions are taken, the concentration of salts within the soil will eventually render it toxic to any crop. Irrigation-induced salinization of the soil is a major contributor to the desertification of many areas of the world including the Middle East, southeastern Asia, North Africa, and large parts of the western United States.

A number of techniques can be used to minimize the accumulation of salts within irrigated soil. *Drains* can be installed to removed excess irrigation water and thereby minimize the elevation of the groundwater table. A commonly used method of water application, called *drip irrigation,* significantly reduces the amount of irrigation water by slowly applying the water directly to the plant roots. More recently, scientists are using genetic engineering to develop salt-resistant crop plants that allow more saline water to be used for irrigation.

SPOT REVIEW

1. What is deflation and what kinds of desert features does it produce?
2. Why is wind a more effective agent of erosion in arid regions than in humid areas?
3. How are dunes eventually stabilized and what changes may result in their remobilization?
4. Describe the different conditions responsible for the formation of the four kinds of dunes.

CONCEPTS AND TERMS TO REMEMBER

Relative humidity
  effect of temperature
    dew point
Kinds of deserts
  rainshadow deserts
  subtropical deserts
  fog deserts
    Humboldt Current
    Benguela Current
  isolation deserts
Erosion in deserts
  erosion potential of raindrops
  internal drainage systems
    playa lake
  cycle of erosion
    arid versus humid

mesas
buttes
needle rocks
Basin and Range topography
  depositional features
    alluvial fan
    bajada
    bolson
  erosional features
    pediment
  erosional/depositional feature
    pediment slope
Erosion by the wind
  saltation
  deflation
    desert pavement

blowout
abrasion
ventifacts
facets
pedestal rock
Deposition from the wind
  loess
  dunes
    windshadow
    stabilization
    barchan dunes
    parabolic dunes
    transverse dunes
    longitudinal (seif) dunes
Desertification

REVIEW QUESTIONS

1. Which of the following statements is true?
   a. Relative humidity increases as the ability of the air to hold moisture increases.
   b. Relative humidity decreases as the temperature of the atmosphere increases.
   c. Relative humidity will only increase if the moisture is added to the air.
   d. As the temperature of the air changes, the relative humidity will remain constant until the dew point is reached.
2. Which of the following is an example of a rainshadow desert?
   a. the deserts of the southwestern United States
   b. the Sahara of North Africa
   c. the Namib Desert of southwestern Africa
   d. the deserts of western Australia
3. Most of the world's deserts are the result of being
   a. isolated from sources of moisture.
   b. located in the rainshadow of mountains.
   c. located near cold offshore ocean currents.
   d. located in regions bathed in descending columns of very dry air.
4. What feature that develops during the old age stage of the arid cycle of erosion is equivalent to the peneplane of the humid cycle?
   a. alluvial fan   c. bolson
   b. pediment slope   d. bajada
5. Which of the following desert features is the result of deflation?
   a. bolson   c. ventifact
   b. desert pavement   d. talus
6. A crescent-shaped dune with the horns pointing in the direction of wind flow is called a ____ dune.
   a. longitudinal   c. transverse
   b. parabolic   d. barchan
7. How is relative humidity related to the precipitation or lack of precipitation of atmospheric moisture? How is it related to the origin of deserts?
8. In the regions of the Tropics of Cancer and Capricorn and the equator, surface winds are often calm for long periods of time. Hence, sailors during the days of sailing ships referred to these regions as the "doldrums." Why does this phenomenon occur?
9. In what ways do the processes of weathering and erosion differ in arid and humid regions? In what ways are they similar?
10. Compare and contrast erosion by the wind and erosion by water explaining how they differ and how they are similar.
11. Why is sand scarce in most deserts?

THOUGHT PROBLEMS

1. What changes have taken place during the last few decades that could explain the southward migration of desert conditions in Africa? Can anything be done to arrest the encroachment?
2. Although the most obvious effects of the wind can be seen in arid regions, its effects can be observed in other settings as well. Discuss examples where the wind has been effective in shaping the land in nondesert areas.

FOR YOUR NOTEBOOK

Because of the demographics of the United States, few of us live in or near deserts. As a result, you will probably not be able to add first-hand information about deserts to your notebook. It might be interesting, however, to examine a few topics in more detail. One important example is the transformation of deserts into oases. With the increase in world population and the growing need for food, areas that normally are too dry for crops have been put into production through irrigation. These efforts have not been without problems, however, such as the increased salinity of irrigation waters. You might want to investigate some of the areas in the southwestern United States or the Middle East where desert areas have been converted into productive agricultural lands.

Later in the text, we will discuss various alternative energy sources to the fossil fuels. One alternative is solar power. Because of the demand for intense, direct sunlight, large-scale solar power stations will necessarily be built in desert areas. The U.S. Department of Energy recently completed a pilot-scale solar power plant in the Mojave Desert. This might be a logical time for you to investigate the various methods of tapping the energy of the Sun and to consider both the potential and the problems of utilizing solar power.

CHAPTER OUTLINE

INTRODUCTION
EXPLORING THE OCEAN BOTTOM
    Sonar
    Submersibles
    Deep-Sea Drilling
THE ORIGIN OF THE OCEAN BASINS
    Cosmic Origins
    Plate Tectonics
    Supercontinent Cycles
LANDFORMS ON THE OCEAN BOTTOM
    Oceanic Ridges
    Abyssal Hills
    Abyssal Plains
    Deep-Sea Trenches
    Shield Volcanoes

CONTINENTAL MARGINS
    Continental Trailing Edges
    Continental Leading Edges
MOVEMENT OF OCEAN WATER
    Waves
    Tides
    Longshore Currents
    Ocean Currents
COASTAL PROCESSES
    Emergent or High-Energy Coastlines
    Submergent or Low-Energy Coastlines
REEFS
    Kinds of Reefs
MODIFICATION OF SHORELINE PROCESSES
ENVIRONMENTAL CONCERNS

CHAPTER 12

Oceans and Shorelines

INTRODUCTION Earth is unique among the terrestrial planets in possessing a **hydrosphere** (Greek *hudor* = water + sphere), the layer of water that exists on and just below its surface. Most of Earth's water (97%) is contained in the ocean basins, a volume of approximately 280 million cubic miles (1.35 billion km^3).

Until 1872 when the H.M.S. *Challenger,* a British warship converted for research, made its historic voyage, relatively little was known about the oceans. The voyage, funded by the British government, was mandated to chart the depth of the ocean, measure the various oceanic currents, amass data on the composition of the ocean's water and bottom sediments, and collect information on oceanic life. At the time of the voyage, except for a few soundings, almost nothing was known about the ocean bottom. Most scientists had considered the bast expanses of the deep-ocean basins to be nothing more than flat, featureless surfaces extending from one continent to another, interrupted only by a few isolated volcanic islands. The discovery of a submarine mountain range by the H.M.S. *Challenger* was the first indication that the topography of the ocean bottom was not so simple. Although the data amassed by the *Challenger* expedition significantly expanded our understanding of the ocean, knowledge of the ocean was still very limited.

EXPLORING THE OCEAN BOTTOM

For centuries, seafarers determined the depth of water by dropping a **weighted line** overboard and measuring the length of line required for the weight to hit bottom. This technique works reasonably well in shallow water, but it is fraught with difficulties where the water depth exceeds 10,000 feet (3,000 m). Imagine managing a spool wrapped with 10,000 feet of hemp rope, determining exactly when the line hits the bottom, and keeping the line vertical while measurements are made.

The first dedicated effort to determine the topography of the ocean bottom began in the mid-1800s. In preparation for the laying of the first trans-Atlantic telegraph cable, the U.S. Navy requested ships sailing the Atlantic-crossing routes to take deep-ocean soundings and report the data. Still, with weighted lines the only method of determining the depth of the ocean bottom, the amount of information acquired was minuscule. Other methods of exploring the ocean bottom would not become available until the twentieth century.

Sonar

Invented at the U.S. Naval Experiment Station at Annapolis, Maryland, just after World War I, a depth-finding method called **sonar** (*so*und *na*vigation *r*anging) ushered in the modern era of deep-ocean studies. Sonar utilizes a sound wave reflected from the bottom and recorded as it returns to the ship (Figure 12.1). In shallow water, the sound wave may be generated by a hammer striking an anvil within the hull of the ship. To determine the depth of very deep water, the sound wave is usually generated by the detonation of a quantity of high explosives or by a sudden release of pressurized air from a device towed behind the ship. The returning sound wave is recorded by an array of waterproof microphones, called hydrophones, trailing the ship. The water depth is calculated by estimating the speed of sound in water and measuring the time required for the sound wave to make the trip from the source to the recording phone. Depending upon the amount of energy used to generate the sound wave, it may penetrate the ocean bottom. In addition to locating the water-sediment interface, sonar can then be used to

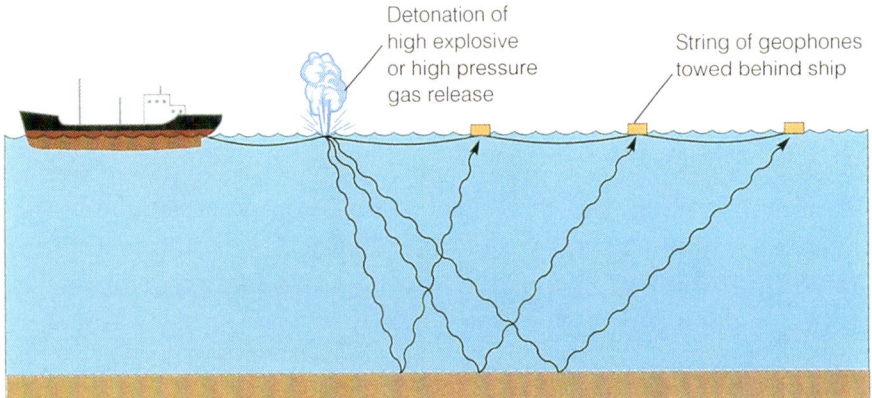

FIGURE 12.1 The invention of sonar *allowed scientists, for the first time, to "see" and map the topography of the ocean bottom.*

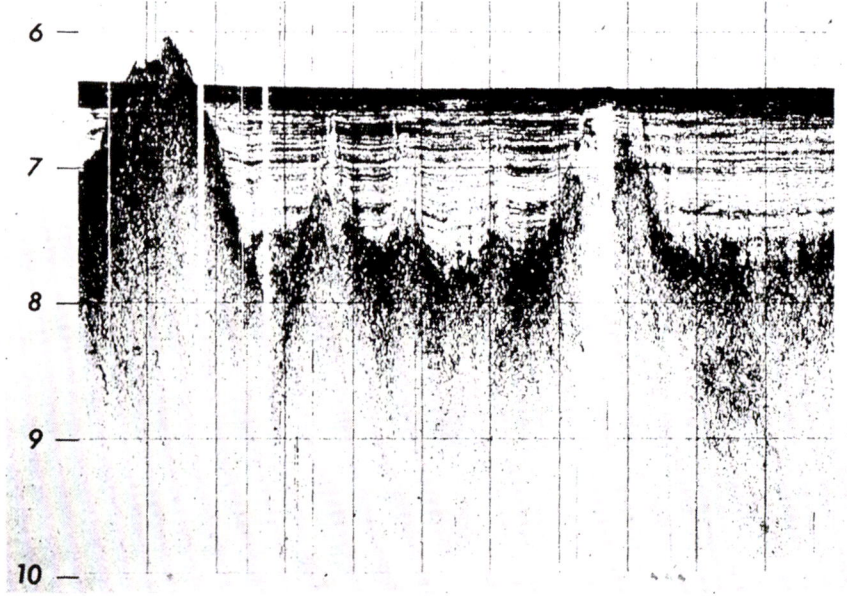

FIGURE 12.2 *If the sonar energy source is sufficiently strong, the sound waves may penetrate into the sediments on the ocean bottom and provide a picture of structures that exist below the sediment-water interface. (From Vogt et al. in Hart:* The Earth's Crust and Upper Mantle, *p. 574, 1969 © The American Geophysical Union.)*

record both layering within the bottom sediments and structures within the sediments and underlying rock (Figure 12.2).

Sonar technology revolutionized our picture of the ocean bottom. Maps of the ocean bottom based upon echo soundings began to appear in 1923, and the first contour map of a portion of the ocean bottom was made in 1958 when the nuclear submarine U.S.S. *Nautilus sailed* beneath the Arctic Ocean pack ice.

Sonar mapping of the ocean bottom showed a landscape as varied as any seen on land. Its features included mountain ranges longer than any on land, trenches that plunged tens of thousands of feet deeper than the adjoining ocean floor, and a great number of submarine peaks, including a strange flat-topped variety that defied explanation until the theory of plate tectonics. In addition to these heretofore unsuspected landforms, the ocean bottom also displayed the flat, featureless plains previously thought to exist at abyssal depths. In fact, some of the ocean bottom is a surface flatter and far more extensive than any plain or plateau on land.

Submersibles

Other techniques that have been used to study the ocean bottom include robot **submersibles** equipped with television cameras, such as *Argo* and *Jason*, and manned submersibles such as *Alvin* (Figure 12.3). Perhaps the most publicized use of *Alvin* was the 1985 discovery and investigation of the remains of the steamship *Titanic,* which struck an iceberg on its maiden voyage across the Atlantic in 1912 and sank in 12,5000 feet (3,800 m) of water. In 1989, *Alvin* discovered the German battleship *Bismarck,* which sank in 1941 in 15,600 feet (4,755 m) of water.

Submersibles not only allow first-hand viewing of bottom features, but some can collect samples and objects from the ocean floor. With such techniques at our disposal, we are able to "see," map, and sample the deep-ocean bottom as never before. The information we have obtained has revolutionized our understanding of the physical and biological world that exists at abyssal depths.

Deep-Sea Drilling

In 1968, a worldwide **deep-sea drilling** program was initiated with the launching of the *Glomar Challenger,* a ship equipped with a drilling rig designed to take sediment and rock cores from the deep-ocean bottom (Figure 12.4). Since then, newer research vessels have been built. Hundreds of cores have been retrieved for scientific study from all the oceans. It was data from these scientific efforts that ultimately provided confirmation for the concept of plate tectonics.

Thus, from the spectacular beginning with the voyage of the H.M.S. *Challenger,* a new science of *oceanography* has evolved. Nevertheless, despite all the knowledge amassed during the past century, the ocean and the ocean bottom still remain in large part unvisited and unknown. The ocean is indeed Earth's last great frontier.

THE ORIGIN OF THE OCEAN BASINS

Over the years, many hypotheses have been presented to explain the origin of the ocean basins. Here we examine one of the older hypotheses, which viewed the ocean basins as the product of cosmic forces, and the current hypothesis, which is based on plate tectonics.

Cosmic Origins

According to one of the older hypotheses, the ocean basins originated when vast amounts of rock were ripped from Earth's surface by the near-miss of some cosmic passerby such as an asteroid or meteoroid. The hypothesis went on to propose that the rock torn from Earth's surface was thrown into space to become the Moon. In the absence of any more reasonable explanation for these vast basins, especially the huge Pacific Ocean basin, this scenario actually seemed at the time to be within the realm of possibility, especially to those who were not advocates of continental drift. Not until the introduction of the concept of plate tectonics in the mid-1960s did we come to understand the mechanism by which the ocean basins form. It is interesting to note that the

FIGURE 12.3 Submersibles *such as* Alvin *allow scientists to observe and collect samples from the abyssal ocean floor. (Courtesy of W. R. Normark/USGS)*

cosmic hypotheses has recently been revived, but as an explanation for the origin of the Moon, not the ocean basins.

Plate Tectonics

Earlier in Chapter 2, we saw that the rifting and subsequent movement of continents were responsible for the formation of ocean basins. You will recall how Alfred Wegener showed that geologic structures, stratigraphy, fossils, and glacial deposits on the adjoining continental margins of South America and Africa were similar. He offered those findings as proof that the Atlantic Ocean had been born by the rifting of a supercontinent and had grown to its present size since the end of the Early Cretaceous Epoch as the Americans drifted away from Europe and Africa. The greatest impediment to the acceptance of Wegener's idea was his inability to provide a mechanism to move the continents. Wegener could not have conceived of the currently accepted mechanism by which

FIGURE 12.4 *Ships equipped with* drilling rigs, *such as the* Glomar Challenger, *allow scientists to collect cores of the sediments and rocks of the deep ocean floor. (Courtesy of Scripps Institution of Oceanography)*

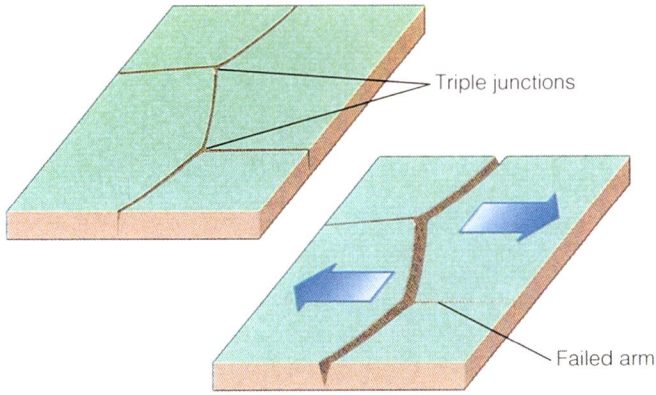

FIGURE 12.5 *The formation of three radiating fracture zones, called a* triple junction, *signals the initial splitting of the lithosphere above a mantle plume. With time, two of the three fracture zones may propagate and join fractures propagating from two adjoining triple junctions to form a* divergent plate boundary *and eventually open to from an ocean basin, while the third fracture zone, called a* failed arm, *may not progress beyond a rift zone or rift valley.*

the continents move, primarily because the data did not exist during his lifetime. Wegener was to die with his great vision yet to be proved.

It might be well to review what we know about the creation of the ocean basins. From data accumulated over the past two decades, we now know that the outermost component of Earth consists of a brittle layer called the lithosphere, which includes the combined oceanic and continental crusts and the outermost portion of the mantle. Separated into plates, the lithosphere moves on the underlying asthenosphere, driven by heat produced by the breakdown of radioactive isotopes within the mantle and residual heat generated during the formation of the planet.

Oceans are created by the breakup of continents, although there is no universal agreement as to the mechanism by which rifting begins. Some geologists have suggested that when continents have remained relatively stationary for long periods of geologic time, heat accumulates beneath the continental lithosphere and initiates the development of mantle plumes and hot spots. As the lithosphere begins to dome above the mantle plume, a three-pronged fracture called a **triple junction** develops in the continental crust, not unlike the fractures that develop in the surface of a muffin or cupcake as it rises during baking. According to the theory, two of the fractures propagate laterally and join similar fractures generated at adjacent hot spots to create a **rift zone**, which opens and causes the continent to break apart and form an ocean basin. The third fracture, called a **failed arm,** aborts and does not progress beyond the development of a rift zone or **rift valley** (Figure 12.5). Figure 12.6 shows an example of a triple junction where the two propagating fractures opened to form the Gulf of Aden and the Red Sea, resulting in the separation of Africa and the Arabian Peninsula, while the failed arm became the East African Rift Valley.

Another theory proposes that the continents rift along zones of weakness induced in the continental lithosphere by past tectonism and/or as the plate moves over a mantle plume. The movement not only forms a chain of volcanoes at the surface, but also thins the lithosphere, making the plate conductive to rifting by subsequent mantle convection.

At the present time, it is not possible to determine which, if either, of these theories is correct. We can only surmise that whatever scenario is responsible for the initial rifting, the breakup of continents is the result of tensional forces developing within the continental lithosphere (Figure 12.7).

The first sign of continental breakup is the development of rift zones, which subsequently evolve into rift valleys. Examples previously discussed include the Rio Grande Rift of the southwestern United States and the East African Rift Valley. If the plate margins continue to diverge and the ends of the continental rift valley eventually reach the sea, the valley will begin to flood.

Oceans and Shorelines 355

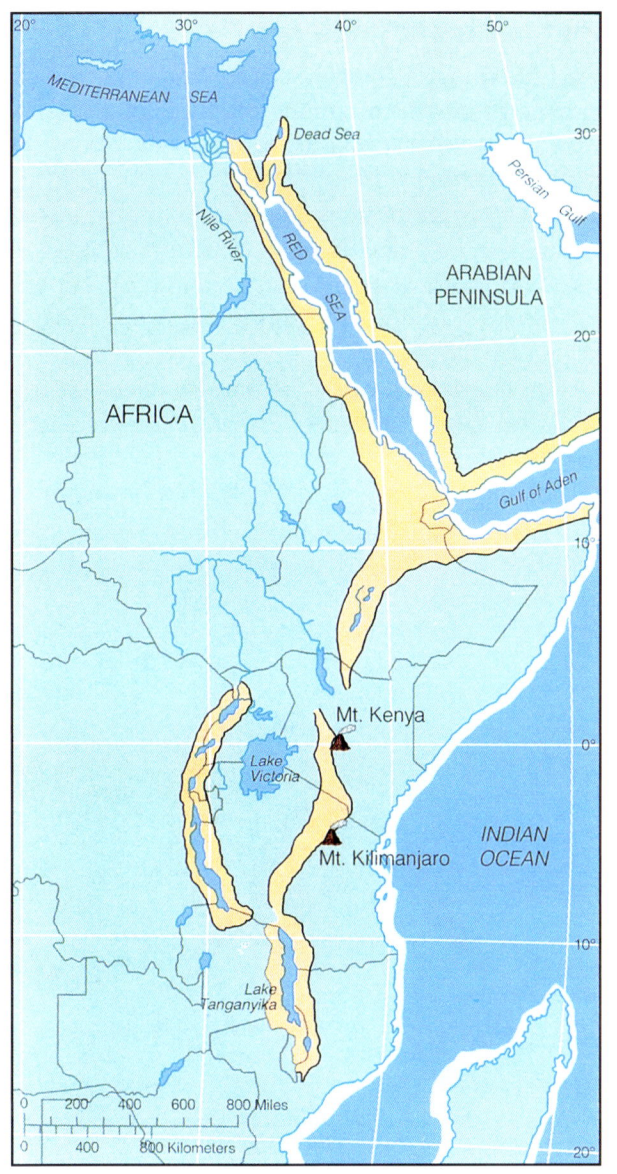

FIGURE 12.6 *The* Red Sea, *the* Gulf of Aden *and the* East African Rift Valley *are the three fracture zones of a* triple junction. *Of the three, the East African Rift Valley appears to be the* failed arm *while the Gulf of Aden and the Red Sea are linear oceans formed by diverging plate margins and will eventually become an ocean basin when Africa becomes completely detached from the northern continental mass. (Photo courtesy of NASA)*

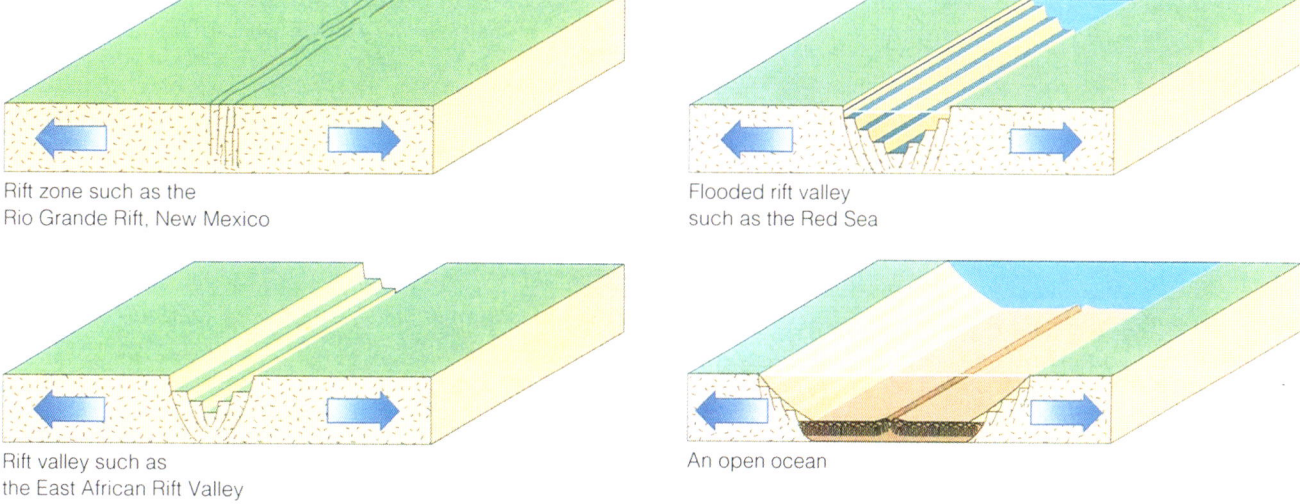

Rift zone such as the
Rio Grande Rift, New Mexico

Flooded rift valley
such as the Red Sea

Rift valley such as
the East African Rift Valley

An open ocean

FIGURE 12.7 *Although geologists do not universally agree on the mechanism by which rifting is initiated, they do agree that the divergence of plates is the result of tensional forces within the lithosphere.*

Two pieces of evidence indicate that the East African Rift Valley is active. First, it is the site of active volcanism. One of Earth's best-known volcanoes, Mount Kilimanjaro, is located within the valley. Secondly, within recent years, seawater has begun making its way into the northern end of the valley through the fracture system that connects the valley with the Red Sea, creating inland saltwater lakes that are growing larger.

Divergent Plate Margins

As the divergent plate edges continue to move apart, the rift is flooded and becomes a long, narrow arm of the ocean. Accordingly, the combination of the Gulf of Aden and the Red Sea may be the beginning of Earth's next major ocean basin. At the present time, the Atlantic and Indian oceans are opening by sea-floor spreading at rates that range from 1 to 2 inches (2–5 cm) per year (about the rate at which your fingernails grow), and spreading along the East Pacific Rise continues at a rate of as much as 6 inches (15 cm) per year.

Convergent Plate Margins

Eventually, an ocean will stop opening and as a zone of subduction forms at or near the edge of a continent, a new plate boundary is created. If, as the plates converge, oceanic lithosphere is consumed by subduction at a higher rate than it is being produced at the oceanic ridge, the ocean will begin to close.

As the ocean closes, new oceanic lithosphere continues to be produced at the oceanic ridge, and the sea floor continues to spread. For example, the rate of sea-floor spreading within the Pacific Ocean basin is higher than for any other ocean. The reason why the Pacific ocean basin is closing is that old rocks of the Pacific plate are being *consumed* by subduction *faster* than new rocks are being *created* at the oceanic ridge.

Oceans close as continents collide. Before the rifting of Wegener's supercontinent about 200 million years ago, the *Tethys Sea* existed between the southern continent and *Gondwana* and the northern *Laurasia* (see Figure 1.11). A portion of the Tethys Sea began to close as the supercontinent began to break up, and the newly formed continent of Africa moved northward toward Europe. The closure of this portion of the Tethys Sea is not yet complete. Its presence is still represented by the Mediterranean, Black, and Caspian seas. On the other hand, the initial collision of India and Asia 45 million years ago completely closed the portion of the Tethys Sea that once separated those landmasses.

Supercontinent Cycles

Three geologists, R. D. Nance, T. R. Worsley, and J. B. Moody combined tectonic, oceanographic, and geochemical data to propose a hypotheses contending that supercontinents formed six times during Earth's history. According to their hypothesis, the rifting of continents and the opening and closing of oceans are parts of cycles that were initiated about 2.6 billion, 2.1 billion, 1.7 billion, 1.1 billion, 650 million, and 250 million years ago.

A cycle begins as heat, accumulated beneath the continental lithosphere over a period of 80 million years, initiates the rifting of the supercontinent (Figure 12.8). After about 40 million years, new Atlantic-type ocean basins form and widen as older Pacific-type oceanic lithosphere is consumed. About 120 million years later as the continents attain their widest points of dispersion, the new ocean basins stop

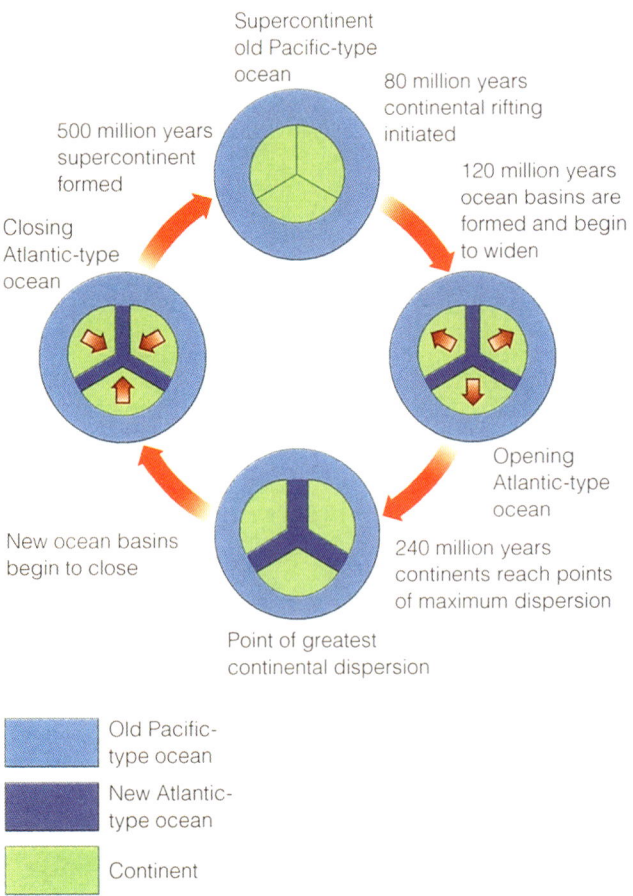

FIGURE 12.8 *Supercontinent Cycles. Modified after R. D. Nance, T. R. Worsley, and J. B. Moody. "The Supercontinent Cycle" in* Shaping the Earth, Tectonics of Continents and Oceans. Readings from Scientific American. *W. H. Freeman Co. 1990.*

opening and begin to close. After a total of about 500 million years, the continents collide to create a new supercontinent and the cycle repeats.

The last supercontinent cycle involved the breakup of Pangaea and the formation of the present continents and the Atlantic and Indian oceans. Note that according to the timing of events, the Atlantic and Indian oceans should soon stop opening and enter the closing phase of the cycle.

SPOT REVIEW

1. How does sonar determine the distance to the ocean bottom?
2. What are some of the ideas that have been proposed to explain the formation of the ocean basins?
3. What is a triple junction and how is it involved in the rifting of a continent?
4. Compare the geologic structures that characterize divergent and convergent plate margins.

LANDFORMS ON THE OCEAN BOTTOM

It may appear strange to speak of "landforms" on the ocean bottom. Historically, the term *landforms* has referred to surface features of the land. With our expanded knowledge of the ocean bottom, however, the realm of the physical geographer or *geomorphologist,* the scientist who studies landforms, has been extended beyond the limits of the land and into the ocean basins. This is consistent because the sea floor is simply that part of Earth's surface lying at the lowest elevations due to the higher density of the basaltic crust.

Oceanic Ridges

The **oceanic ridges** are the most prominent landforms on the ocean floor. It is said that if the water were removed from the ocean basins and Earth was viewed from space, the oceanic ridge system would be the most conspicuous surface feature.

Oceanic ridges are broad structures with gentle outer slopes. Their summits may rise as much as 2 miles (3 km) above the level of the surrounding ocean floor. Their slopes, stretching for 300 to 500 miles (500–800 km) on either side, are quite unlike the steep slopes of land-based mountains (Figure 12.9). Except for the East Pacific Rise, a rift valley occupies the summit of oceanic ridges. The East Pacific Rise lacks a summit rift valley due to its high rate of spreading and the large volume of lava erupted. In addition to the longitudinal fractures, the oceanic ridges are cut perpendicular to their trends by many fractures called transform faults that allow the plates to move over Earth's spherical surface (refer to Figure 12.9).

Even though oceanic ridges are the longest mountain ranges on Earth, they are, for the most part, obscured by the ocean waters. A well-known exception is the island of Iceland, which rises as a plateau along part of the Atlantic Oceanic Ridge (Figure 12.10). The icelandic Plateau formed along the Atlantic Oceanic Ridge over the last 16 million years because additional molten rock is being generated by a mantle plume (hot spot) located beneath the ridge. Iceland not only exhibits all the various kinds of volcanism associated with oceanic ridges, but it also exposes the summit rift valley (Figure 12.11).

The rocks generated at the oceanic ridges are mostly **pillow basalts** and basalt flows. Because the ocean basins cover 70% of Earth's surface, it can therefore be said that basaltic pillow lavas are the most common kind of rock exposed on the surface of Earth. Beneath the surface layers of pillow basalts,

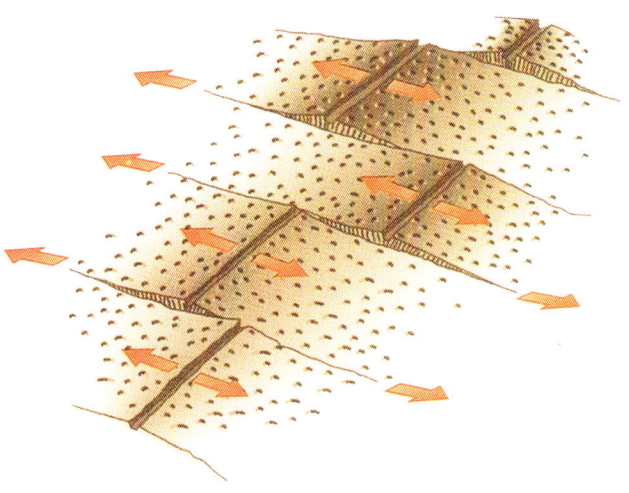

FIGURE 12.9 *The most dominant features of the ocean basins are the* oceanic ridges. *Interconnected throughout all of the ocean basins, they are the longest mountain ranges on Earth. Unlike steep-sided continental mountain ranges, the slopes of oceanic ridges are very gentle, and the trends of the ridges are cut with hundreds of* transform faults.

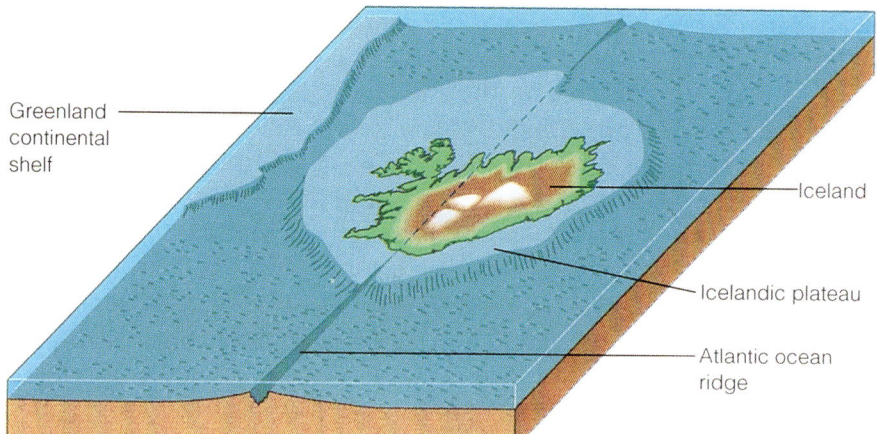

FIGURE 12.10 Iceland *is the exposed portion of the* Icelandic Plateau *which is in turn a part of the* North Atlantic Oceanic Ridge. *This portion of the oceanic ridge has been expanded to a plateau and built above sea level because of additional magma that is being provided by a hot spot located below the oceanic ridge.*

FIGURE 12.11 *The* rift valley *that runs along the summit of the Atlantic Oceanic Ridge is well exposed on Iceland. (VU/© Charles Preitner)*

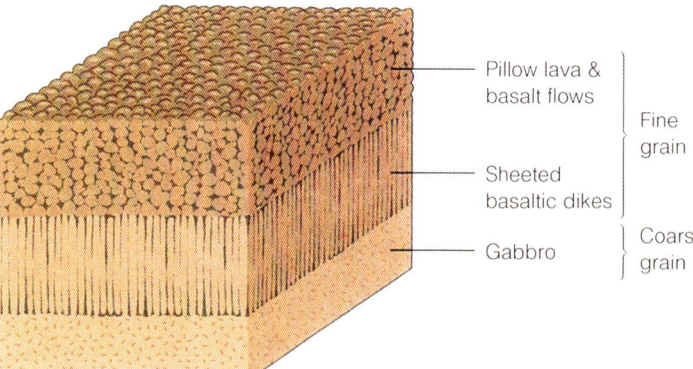

FIGURE 12.12 Oceanic crust *is a three part complex called an* ophiolite suite. *The oceanic crust is composed of* basaltic/gabbroic rocks. *The toplayer is largely* basaltic pillow lavas *with some* basaltic lava flows. *The middle layer consists of* basaltic sheet dikes, *and the lower layer, because it cooled more slowly, consists of* gabbro.

magma chamber (Figure 12.12). The entire complex of rocks is known as an **ophiolite suite** (Greek *ophis* = snake).

The elevation of oceanic ridges above the surrounding ocean floor is due to the buoyancy of the underlying, low-density, hot rocks. Away from the oceanic ridge, the ocean bottom slopes off gently as the newly formed lithospheric rocks move away from the ridge axis, cool, contract, and increase in density. As a result, the ocean basin progressively deepens away from the ridge with the depth to the ocean bottom being approximately proportional to the age of the crustal rock.

Abyssal Hills

Basalt peaks and irregularities in the ocean bottom protruding above a thin layer of fine sediment adjacent

molten basaltic rock solidifies within the fractures that are constantly being created along the divergent plate margins to form a layer composed of more or less vertical **sheeted dikes**. The sheeted dikes are in turn underlain by a layer of gabbro formed by the slow cooling of molten rock within the underlying

to the oceanic ridge make up the **abyssal hills.** Sediment fills the low areas between the abyssal hills. The sediment is largely composed of the remains of microscopic shelled animals and photosynthetic organisms (plankton) that lived in the sunlit waters from the surface down to about 260 feet (80 m), land-derived, windblown dust, and contributions of micrometeorites.

Abyssal Plains

The portion of the ocean bottom located between the margin of the continent and the abyssal hills is called the **abyssal plain.** Within the abyssal plain, the ocean floor is flat and nearly featureless. Under the abyssal plain, the rocks of the ocean bottom are covered with sediment, burying any topographic irregularities that may have existed in the basaltic bedrock surface. In addition to the accumulated remains of plankton, much of the sediment covering the abyssal plain is fine-grained sediment derived from the land and transported beyond the edge of the continental margin by various oceanic currents, which we will discuss shortly. The Pacific Ocean basin is generally devoid of abyssal plains because it is nearly surrounded by deep-sea trenches that serve as traps for any land-derived sediment. Except for limited areas such as off the northwestern coast of North America where deep-sea trenches do not exist, abyssal hills dominate the surface of the Pacific Ocean basin.

Deep-Sea Trenches

The deepest parts of the ocean bottom are the **deep-sea trenches** that form by the downwarping and consumption of the lithosphere at the zones of subduction. The dimensions of deep-sea trenches vary, but widths average about 60 miles (100 km), with lengths up to 2,800 miles (4,500 km) and depths up to 36,000 feet (13,000 m) below sea level.

Shield Volcanoes

Scattered across the abyssal ocean floor are thousands of shield volcanoes that formed over mantle plumes located beneath the moving oceanic crust. Most of these submarine volcanic peaks are **seamounts** while others rise above sea level to become **volcanic islands.** In time, the islands resubmerge, largely through cooling and sagging of the crust and mantle in response to the weight of the islands. In many cases, the volcano is truncated by weathering and erosion and sinks beneath the surface of the ocean to become a flat-topped seamount called a **guyot.**

CONTINENTAL MARGINS

The structure of a continental margin depends primarily on whether the margin is associated with an *opening* or a *closing* ocean basin. Because a continent adjoining an opening ocean is being carried passively as part of the lithospheric plate, the continental margin adjoining an opening ocean is called a **trailing edge** or **passive margin.** As the ocean continues to open, the fractured margin of the continent and the contact between the continental and the oceanic crust sinks and becomes buried under a wedge-shaped deposit of sediment largely derived from the land by stream erosion (Figure 12.13). Later, we will see that this wedge of sediment plays an important role in the formation of foldbelt mountains.

At some point, a convergent plate margin is created as the oceanic crust cools enough to become so dense that it breaks away from the adjoining continental margin and subsides into the underlying mantle. As the oceanic lithosphere subducts, the ocean basin closes (Figure 12.14). A continental margin adjoining a zone of subduction is called a **leading edge** or **active margin.**

The enormous amount of energy expended at a convergent boundary is expressed in various ways such as the uplift and folding of continental margin sediments, subduction zone volcanism forming island and continental arcs, and major earthquakes. From this brief description, we should expect leading and trailing continental margins to be very different.

Continental Trailing Edges

The eastern margins of North and South America are examples of continental trailing edges. The three basic elements of trailing continental margin are the (1) *continental shelf,* (2) *continental slope,* and (3) *continental rise* (Figure 12.15).

The Continental Shelf

The **continental shelf** extends outward from the land as a submerged continuation of the coastal plain. The widths of continental shelves associated with continental trailing edges vary greatly. The continental shelf off the southern tip of Florida is only a few miles wide while that off the coast of Arctic Siberia is more than a thousand miles wide. The average width of a continental shelf is about 40 to 50 miles (70 - 80 km), a width more typical of the continental shelf along the Atlantic margin of the United States.

The continental shelf generally exhibits a gently undulating surface with a slope averaging less than 1°.

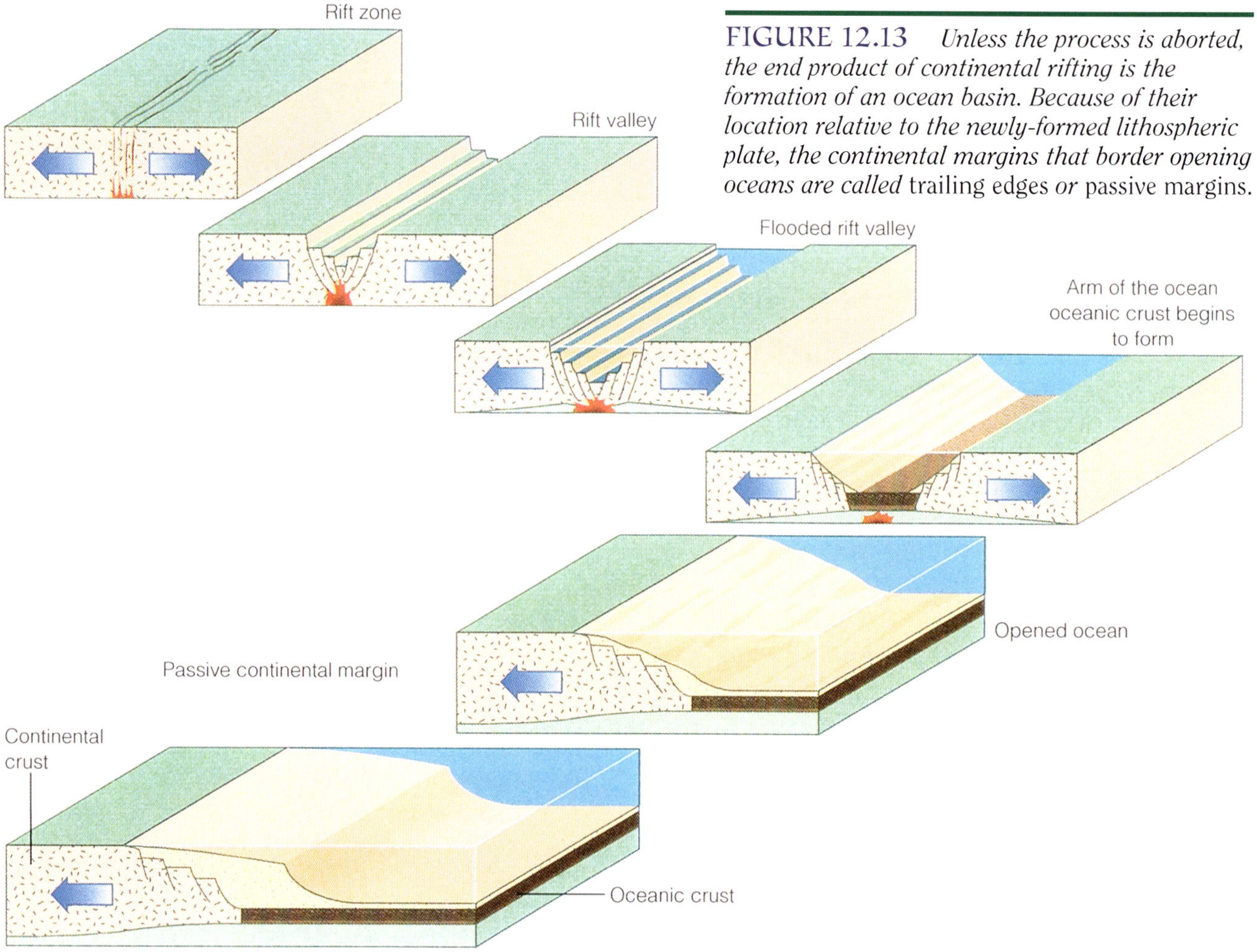

FIGURE 12.13 *Unless the process is aborted, the end product of continental rifting is the formation of an ocean basin. Because of their location relative to the newly-formed lithospheric plate, the continental margins that border opening oceans are called* trailing edges *or* passive margins.

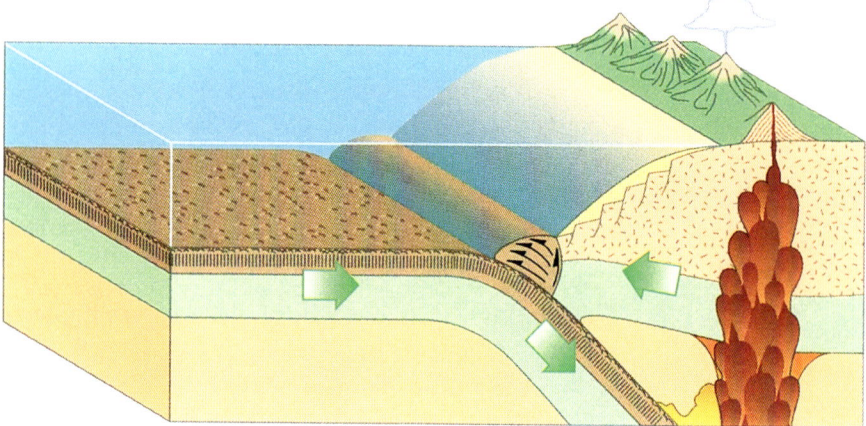

FIGURE 12.14 *To balance the* formation *of a new lithosphere at the* divergent plate boundary, *the lithosphere breaks and forms a* convergent plate boundary *where a portion of the lithosphere "subducts" and is consumed within the zone of subduction. The continental margin associated with a zone of subduction is called a* leading edge *or an* active margin.

With such a gentle slope, the continental shelf would appear to the eye to be perfectly horizontal if exposed as land.

Geologically, the continental shelves are important as the repository for much of the sediment stripped from the land and as the depositional environment of most sedimentary rocks including the thick sequences that are uplifted and deformed into mountains during the ocean-continent convergence or continent-continent collisions.

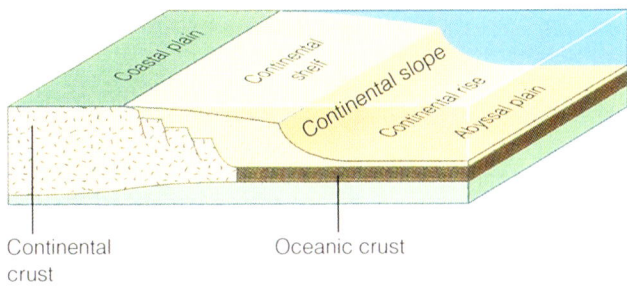

FIGURE 12.15 *A major component of a passive continental margin is a thick wedge of sediment that forms the* continental shelf, continental slope, continental rise *and a portion of the* abyssal plain.

FIGURE 12.16 *The sediments carried within a turbidity flow eventually settle to form* graded beds *on the abyssal ocean floor. When lithified, these sediments form a type of sedimentary rock called a* turbidite. *As a result of the deformation that characterizes convergent plate margins, turbidites, such as these near Monterey, California, may be uplifted and exposed.*

The Continental Slope   From the edge of the continental shelf, the continental margin slopes more steeply toward deeper water, forming the **continental slope.** Its inclination averages about 4° or 5°. Portions of the slope are devoid of loose sediment while others are covered with the fine silt- and clay-sized sediments that have been moved beyond the edge of the continental shelf. The coarse debris found on some slopes is a product of a glacial episode when rock debris was ice-rafted beyond the edge of the shelf.

Sediments accumulated on the continental slope are susceptible to sliding. From time to time, the sediments break loose and cause **turbidity** or **density currents,** which may travel far out onto the abyssal floor before the sediment is deposited. Deep-sea sediment deposited from turbidity currents forms **turbidites** that may eventually be lithified into a rock (Figure 12.16). We will discuss turbidity currents in more detail later in this chapter.

The Continental Rise   The base of the continental slope grades into the more gently sloping **continental rise,** which may continue basinward for hundreds of miles. Eventually, the continental rise merges imperceptibly into the *abyssal plains* bordering continental trailing edges.

Submarine Canyons   On many continental shelves, **submarine canyons** extend from the shore to the base of the continental slope (refer to Figure 12.17). Considerable disagreement exists over their origin. Most submarine canyons such as the canyon opposite the Hudson River appear to be seaward extensions of major river systems. It is believed that the tops of these canyons were incised into the surface of the shelf during the Pleistocene glacial episodes when much of the present-day continental shelf was above sea level. After the sediments accumulated at the mouth of the canyon, turbidity currents carried them down the continental slope onto the abyssal floor. Submarine erosion by the turbidity currents extended the canyon beyond the shelf and onto the slope. As the ice sheets melted and sea level rose, the submerged stream valleys became the upstream segments of the submarine canyons. In the case of the Hudson River, the sediments delivered to the ocean continued to be carried along the Hudson submarine canyon as a turbidity current to be deposited at the base of the slope. The continued transport of sediment along the submarine canyon explains the absence of a delta at the mouth of the Hudson River.

Some submarine canyons are not associated with river systems, however. Most geologists believe they were cut by turbidity currents generated at the upper edge of the continental slope. Subsequent tubidity currents carved the canyons landward by a process equivalent to headward stream erosion.

Deep-Sea Fans   Some **deep-sea fans** are sediments deposited at the mouths of submarine canyons while others form from sediments carried beyond the deltas of major rivers (Figure 12.17). The Mississippi fan builds into the Gulf of Mexico, the Saint Lawrence fan builds into the North Atlantic, and the Amazon and Congo fans build into the South Atlantic.

The two largest deep-sea fans, the Bengal and Indus fans, build into the Indian Ocean on opposite sides of the Indian subcontinent (Figure 12.18).

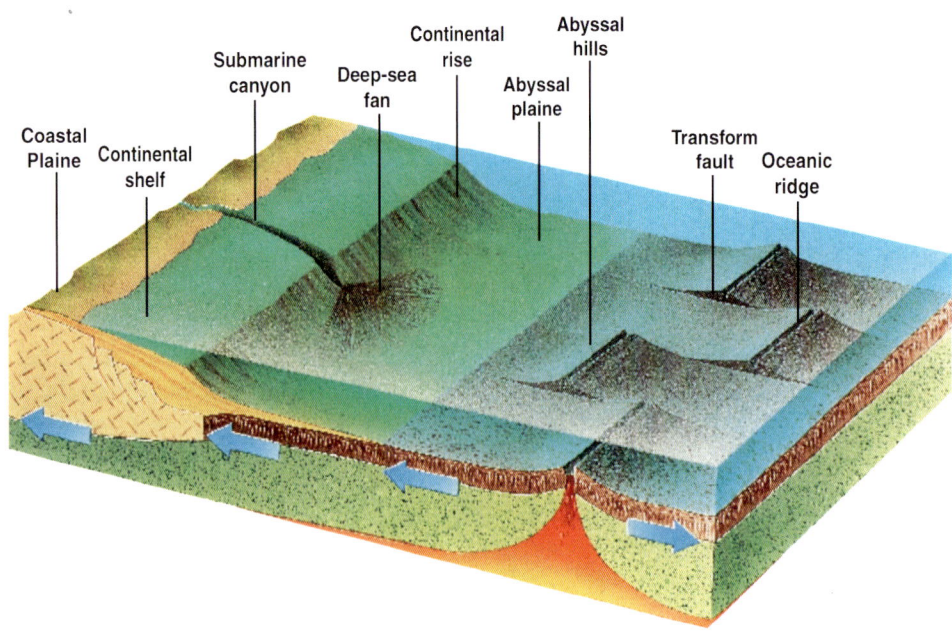

FIGURE 12.17 Deep-sea fans *are another example of deep-sea sedimentation. They form from sediments that have been carried by turbidity currents to the mouths of submarine canyons or beyond the edge of the continental shelf opposite major river deltas.*

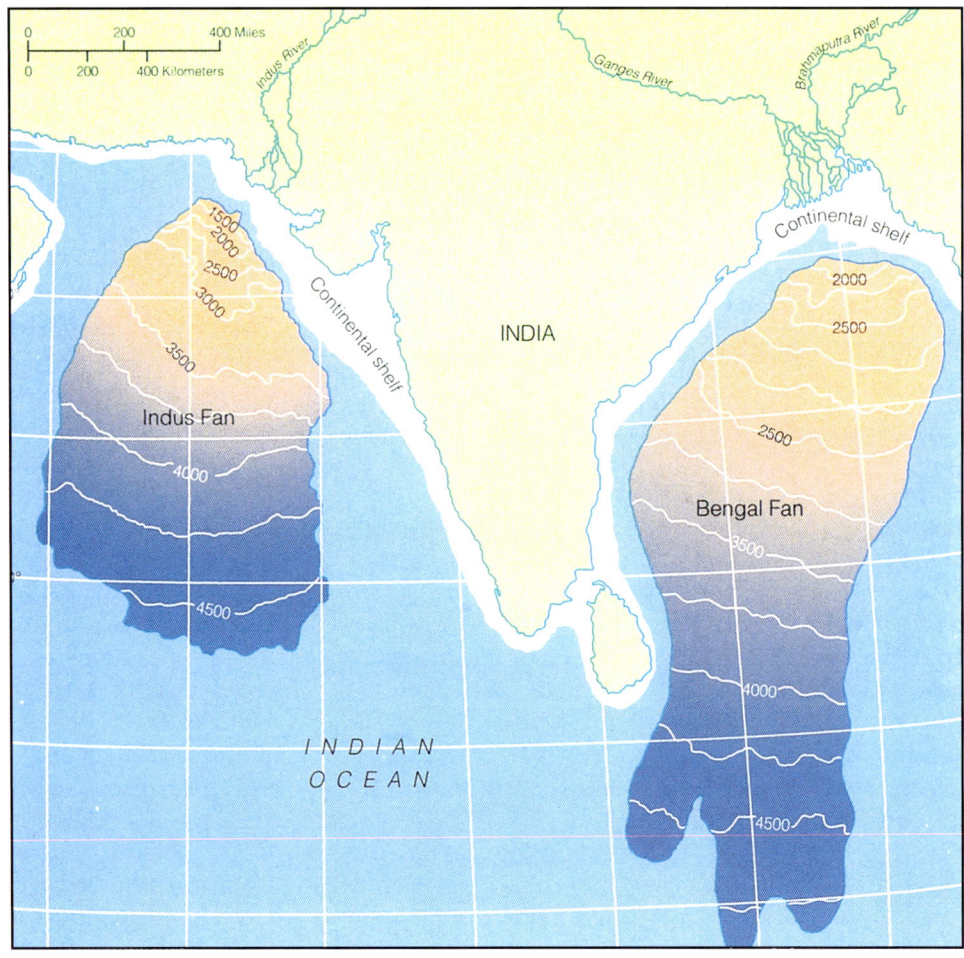

FIGURE 12.18 *The largest of Earth's deep-sea fans are those formed from sediments being stripped from the rising* Himalaya Mountains. *The sediments carried by the* Indus River *formed the Indus Fan while the combined sediments carried by the* Ganges *and* Brahmaputra Rivers *are accumulating in the Bengal Fan.*

Constructed of sediment derived from the rising Himalayas, these two fans represent 90% of the total volume of all the deep-sea fans currently forming.

Because of the presence of deep-sea trenches that capture land-derived sediment, the deep-sea fans are rare in the Pacific Ocean. The only deep-sea fans in the Pacific Ocean basin are in the Gulf of Alaska.

Continental Leading Edges

In contrast to the *tensional* forces that result in the formation of the continental trailing edge, *compressional* forces dominate the tectonics of *continental leading edges* such as the western margins of North and South America. These are areas of active mountain building, deep-seated metamorphism, intrusive igneous activity, volcanism, faulting, and folding. Depending on the location of the zone of subduction relative to the continental margin, volcanic mountains may either build from the ocean bottom as **island arcs** such as the Aleutian or Tonga Islands or develop on the edge of a continent as **continental** or **Andean-type arcs** such as the Cascade Range of the northwestern United States of the Andes Mountains of South America Figure 12.19).

(a) Continental arc volcanism
e.g. Cascade Mountains

FIGURE 12.19 *Volcanic* island arcs *and* continental arcs, *such as the Cascade Mountains, are the result of the enormous amounts of energy being released as the lithospheric plates converge along continental leading edges. (Photo courtesy of D. A. Swanson, MSH/USGS)*

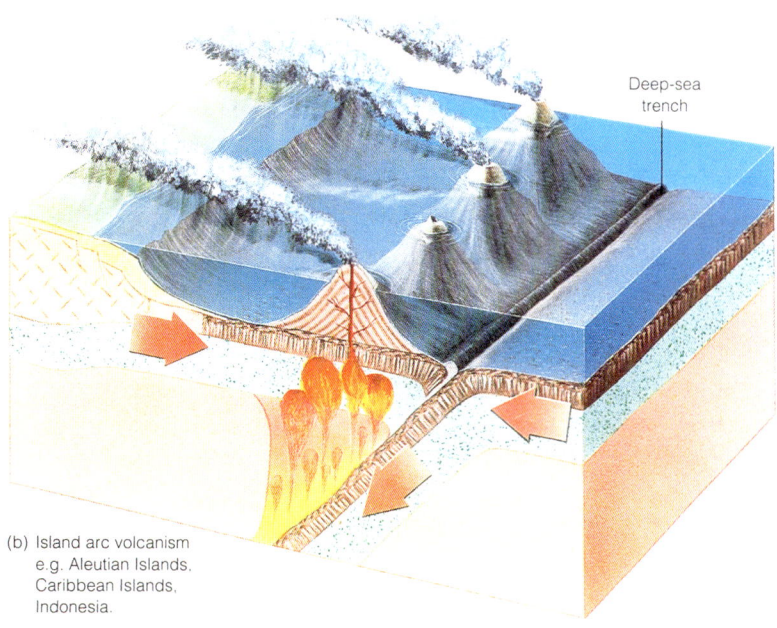

(b) Island arc volcanism
e.g. Aleutian Islands,
Caribbean Islands,
Indonesia.

Continental shelves along leading edges are either narrow or nonexistent. The continental margin slopes abruptly away from the shoreline down into the deep ocean or into a deep-sea trench, sometimes within a distance of a few hundred yards. Land-derived sediments are transported seaward into deep water with little accumulation on the continental margin. Because deep-sea trenches trap land-derived sediments, ocean bottoms offshore from continental leading edges are usually characterized by abyssal hills rather than abyssal plains.

SPOT REVIEW

1. How do abyssal hills and abyssal plains differ?
2. Why can the continental shelf be considered an extension of the coastal plain?
3. Why are abyssal plains and deep-sea fans rare in the Pacific Ocean basin?
4. Compare the geologic structure of continental leading and trailing edges.

MOVEMENT OF OCEAN WATER

The water of the oceans is in constant motion. Most of us have enjoyed watching the seemingly constant arrival of *waves* at the shoreline. Everyone is familiar with the *tides,* those comings and goings of the water along the coastline in response to the gravitational pull of the Moon and the Sun. *Longshore currents* carry sediment parallel to the shoreline and are responsible for many of the depositional features seen along some coastlines. Most important are the various *density currents* that exist within the vast expanse of the ocean.

Waves

Waves are the major agent of change along the shoreline. In some cases, the change is *destructive* as the waves carve and erode the rocks, constantly undercut the coast, drive sea cliffs landward, and remove the sediments to the ocean deep. In other cases where the waves build and modify beaches and offshore sand islands, the effect of the waves is *constructive.*

Waves are generated as stresses develop between the wind and the surface of the water, and energy is transferred from the wind to the water. Each wave is characterized by a *wavelength* and an *amplitude* (Figure 12.20) and travels with a *frequency* measured

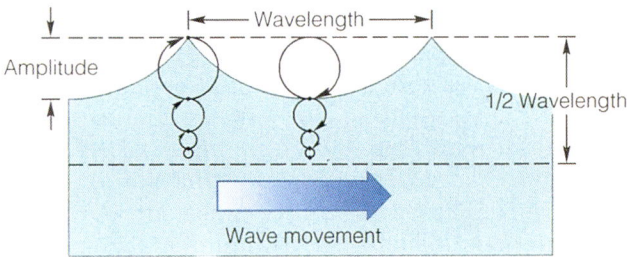

FIGURE 12.20 *In water deeper than one-half the wavelength, a passing waveform results in* no lateral motion *of the water; the water is moved in* circular paths that decrease in diameter with depth where, at a depth of one-half of the wavelength of the waveform, movement ceases.

by the number of wave *crests* or *troughs* that pass a given point in a given amount of time.

A passing wave will move water to a depth of about one-half its wavelength. The individual water particles move in orbital circles that decrease in diameter with depth (refer to Figure 12.20). The information presented in Figure 12.20 leads to two important points: (1) water deeper than half the wavelength is not moved by a passing wave, and (2) as the wave moves *laterally,* the *net* movement of water is *zero*—that is, there is *no* lateral movement of water.

When first formed by wind stress on the surface, waves have *minimum* wavelength and *maximum* amplitude. With their high amplitudes, the surface expression of newly created waves may be quite spectacular. Picture the surface of a stormy sea. However, because the wavelengths of storm waves are relatively short, the water is being moved to only a shallow depth, explaining why submarines need not dive very deep to avoid the turbulence of a rough sea.

As the wave moves away from its point of origin, the amplitude *decreases* and the wavelength *increases.* Eventually, the wave becomes a **swell** with *minimum* amplitude and *maximum* wavelength. While the surface of the ocean may appear to be perfectly calm, the water is being moved to the greatest depth.

Physical geologists are primarily interested in the wave energy available to erode and modify the coastline (in contrast to historical geologists who are more interested in the deep-ocean currents). The amount of energy available to modify the coastal rocks and sediments depends upon (1) the original energy of the wave and (2) the energy dissipated before the wave moves into shallow water.

As the wave moves into shallow water and the wavebase touches bottom, the *wavelength shortens,* the *amplitude increases,* and the surface waters *move laterally* (Figure 12.21). As the water begins to move *laterally,* the waves become increasingly asymmetric until they finally break over in the **surf.** The last surge of energy drives the water up to the beach; from there it flows back to the sea. At the point where the water reverses direction, the original, wind-derived energy has been dissipated. Gravity then takes over, and the water returns to the sea as **backwash.**

Whether a shoreline is subjected to dominantly erosive processes or to a combination of erosional and depositional processes depends primarily upon the amount of energy remaining as the wave breaks into the surf. In general, the longer and more gentle the offshore slope, the greater the amount of energy that will be extended in wave modification and bottom erosion as the wave makes its way toward the shoreline. As a result, *less* energy will be available in the surf zone for erosion and sediment transport. With steeper offshore bottom slopes, the distance between the point at which the wave "touches" bottom and the shoreline decreases, and *more* energy is available in the surf zone.

Tides

The **tides** are the result of the interaction of Earth and its companions in space, the Moon and the Sun. Most of us know that the tides are caused by the Moon, but few of us are aware that there is also a solar tide. According to Newton's laws, the gravitational attraction between any two bodies is *proportional* to the product of the masses of the two bodies and *inversely proportional* to the square of the distance between them. The Moon is many times less massive than the Sun, but due to its proximity to Earth, it has a gravitational effect on the tides approximately twice that of the Sun.

The question most often asked about the tides is why there are two tidal bulges rather than just a single bulge facing the Moon. The answer lies in the effect of gravitational attraction between two bodies. Consider the drawing in Figure 12.22. Points A and B represent unit masses of ocean water while points E and M are unit masses of material located at the centers of Earth and the Moon, respectively. The distance between Earth and the Moon, d_{EM}, is approximately 61 times the radius of Earth. The distance between the Moon and the water mass on the near and far sides of Earth, designed d_{AM} and d_{BM}, respectively,

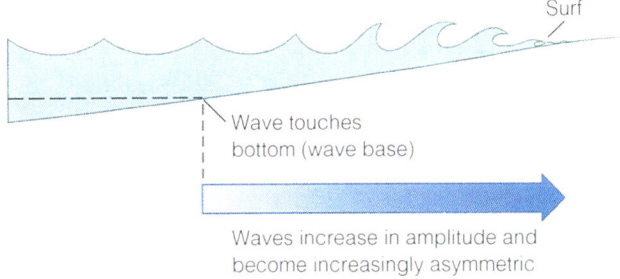

FIGURE 12.21 *As an approaching waveform touches bottom and moves landward, the wavelength shortens, the amplitude increases, and the wave profile becomes increasingly asymmetric as the mass of water moves laterally and collapses in the surf. Between the point of touchdown and the surf, the energy originally possessed by the water is consumed by water movement and erosion, the most important geologically being the energy available in the surf for the erosion of the shoreline.*

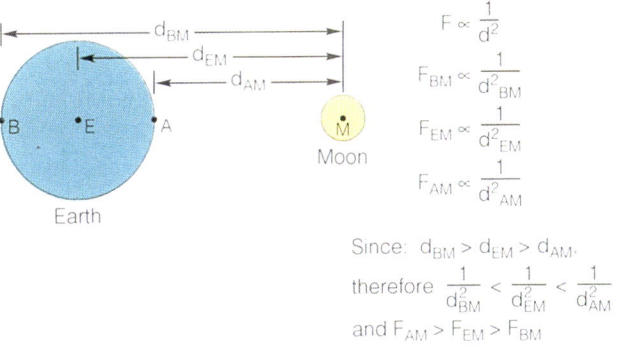

FIGURE 12.22 *The* tides *are largely due to the difference in the gravitational attraction between the Moon and (1) Earth (F_{EM}), (2) the oceanic water mass on the near side of the Earth (F_{AM}), and (3) the oceanic water mass on the far side of Earth (F_{BM}). The* two tidal bulges *exist because the Moon pulls the mass of oceanic water away from the near side of the Earth (F_{AM}) while at the same time pulling Earth away from the mass of oceanic water on the far side of Earth (F_{EM}).*

would therefore be 60 and 62 Earth radii. According to Newton's laws of gravity, the force between two bodies is inversely proportional to the square of the distance between them; that is,

$$F \propto \frac{1}{d^2}$$

Substituting the distances in radii between the Moon and water mass A, watermass B, and the center of Earth, you can see that $F_{AM} > F_{EM} > F_{BM}$. Therefore, the Moon simultaneously pulls water mass A away from Earth and pulls Earth away from water mass B.

Because Earth's hydrosphere is liquid and can flow with little resistance, the layer of water deforms into an oblate spheroid with one (tidal) bulge facing *toward* the Moon (the direction of maximum attraction, F_{AM}) while another (tidal) bulge develops on the opposite side of the Earth facing *away* from the Moon (the direction of minimum attraction, F_{BM}). The attraction of the Sun produces a similar but significantly lesser tidal effect.

Both high tides and low tides occur twice each day as Earth rotates through the two tidal bulges. Because of the changing orientation of the Sun, Earth, and Moon during the monthly lunar cycle, the *combined* effects of the Sun and the Moon change (Figure 12.23). When the Sun, Earth, and Moon are aligned, as they are at the new and full phases of the Moon, the effects of the solar and lunar tides are *additive*. At these times, Earth experiences **spring tides** where the tidal range is maximum, that is, the highest high tides and the lowest low tides. During the first and third quarter when the Moon and the Sun are at right angles, the forces due to the Sun and Moon tend to partially cancel each other. At these times, the difference between high and low tide is at a minimum, and the tides are referred to as **neap tides.**

The tidal range experienced along a particular coastline is dependent upon many variables, one of which is the slope of the offshore ocean bottom. In areas such as the Hawaiian Islands or along the Florida Keys, where the sea floor drops off steeply into deeper water, the tidal range is relatively small. On the other hand, the difference between low and high tide may be quite large. The greatest tidal range in North America is in the Bay of Fundy, Nova Scotia, Canada, where the tidal range can be as high as 50 feet (18 m) (Figure 12.24). The extreme tidal range is due to the gentle bottom slope within the bay combined with the fact that the waters are funneled into a restricted channel.

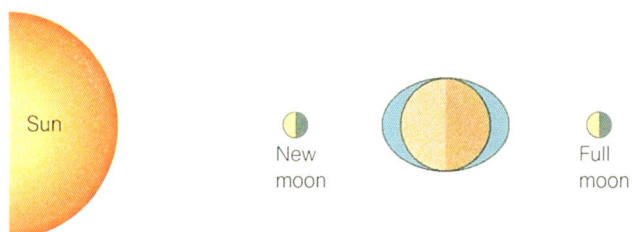

(a) Sun, Earth, Moon are aligned twice each month. Solar and lunar tides are additive to give the highest high tides and the lowest low tides—called spring tides.

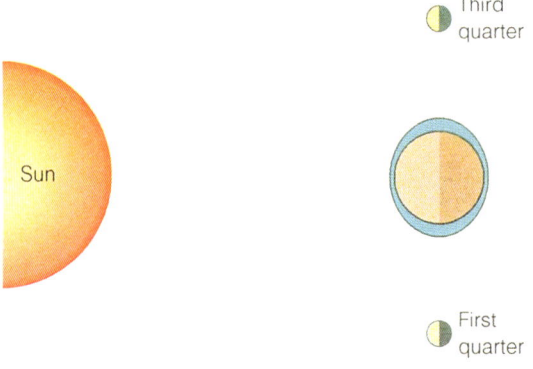

(b) At the first and third quarter of the month, the solar and lunar tides are nonadditive, producing the lowest high tides and the highest low tides—called neap tides.

FIGURE 12.23 *The monthly tidal range is the result of the changing relationship between the Sun, Earth, and the Moon. When all three bodies are aligned along the same axis, as they are during the new and full phases of the Moon, the tidal effects of the Moon and Sun are additive, resulting in the maximum tidal range, the so-called spring tide. When the Moon is either the first or third quarter phase, the tides are largely controlled by the Moon and the tidal range will be at its minimum, the so-called neap tide.*

FIGURE 12.24 *Because of its unique combination of a gentle offshore slope and confinement by the adjacent landmass, the* Bay of Fundy, *Nova Scotia, Canada, experiences the highest tidal range in North America. (Courtesy of Nova Scotia Tourism)*

FIGURE 12.25 *Because most waves approach the shoreline at an angle, they undergo refraction. As a result, the surf drives the water onto and along the shoreline, generating the longshore currents that are responsible for the lateral transportation of sand along the coastline. (VU/© John D. Cunningham)*

Longshore Currents

Along most coastlines, the waves approach at an oblique angle to the shoreline (Figure 12.25). As one end of the wave touches bottom, drags, and slows, the wave begins to bend or *refract* and becomes more nearly parallel to the beach. Even with refraction, the waves usually approach the shoreline at an angle. As a result, the surf drives the water both *onto* and *along* the beach before the water turns seaward. This lateral movement of water along the beach generates a **longshore current** moving parallel to the beach within the surf zone. The longshore current is primarily responsible for the continuous movement of sand along the beach, a process referred to as **longshore transport.**

Ocean Currents

Oceanic water undergoes constant, three-dimensional movement as currents are generated on and within the water mass. Ocean currents are classified as either *wind-driven, wind-induced,* or *thermohaline.* **Wind-driven currents** obtain their energy from the wind, move *horizontally,* and involve both the surface water and water masses to depths of about 300 feet (100 m). **Wind-induced currents** are vertical currents restricted to the upper 3,000 feet (1,000 m) of water and are generated by the displacement of surface water by wind-driven currents. **Thermohaline currents** are *density currents;* they are gravitationally driven, move *vertically,* and involve the entire ocean water column.

Wind-driven currents Within each ocean, surface, near-surface, and very deep waters move in large circular patterns called **gyres.** In the Atlantic and Pacific oceans, gyres exist in both hemispheres. Because most of the Indian Ocean is located in the Southern Hemisphere, it has only one gyre in the Southern Hemisphere (Figure 12.26).

Surface currents are set into motion as stresses develop between the wind and the ocean surface, and energy is transferred from the wind to the water. The winds primarily responsible for the generation of the gyres are the *trade winds,* which blow from the east toward the equator between 5° and 25° north and south latitude (Figure 12.27). The trade winds initiate two warm **equatorial currents** that move *westward* parallel to the equator separated by an **equatorial countercurrent** that flows *eastward* (see Figure 12.26). Eventually, the equatorial currents encounter the landmasses at the western edge of the ocean basin and are deflected by Earth's rotation; the *northern equatorial current* is deflected to the right and the *southern equatorial current* to the left.

As the deflected equatorial currents move *poleward* along the eastern margins of the landmasses, they eventually come under the influence of the *westerly winds,* which blow from the west between 35° and 60° north and south latitude, and are driven *eastward* (refer to Figure 12.27). Upon encountering the landmasses on the eastern margin of the ocean basin, the currents are deflected toward the equator where

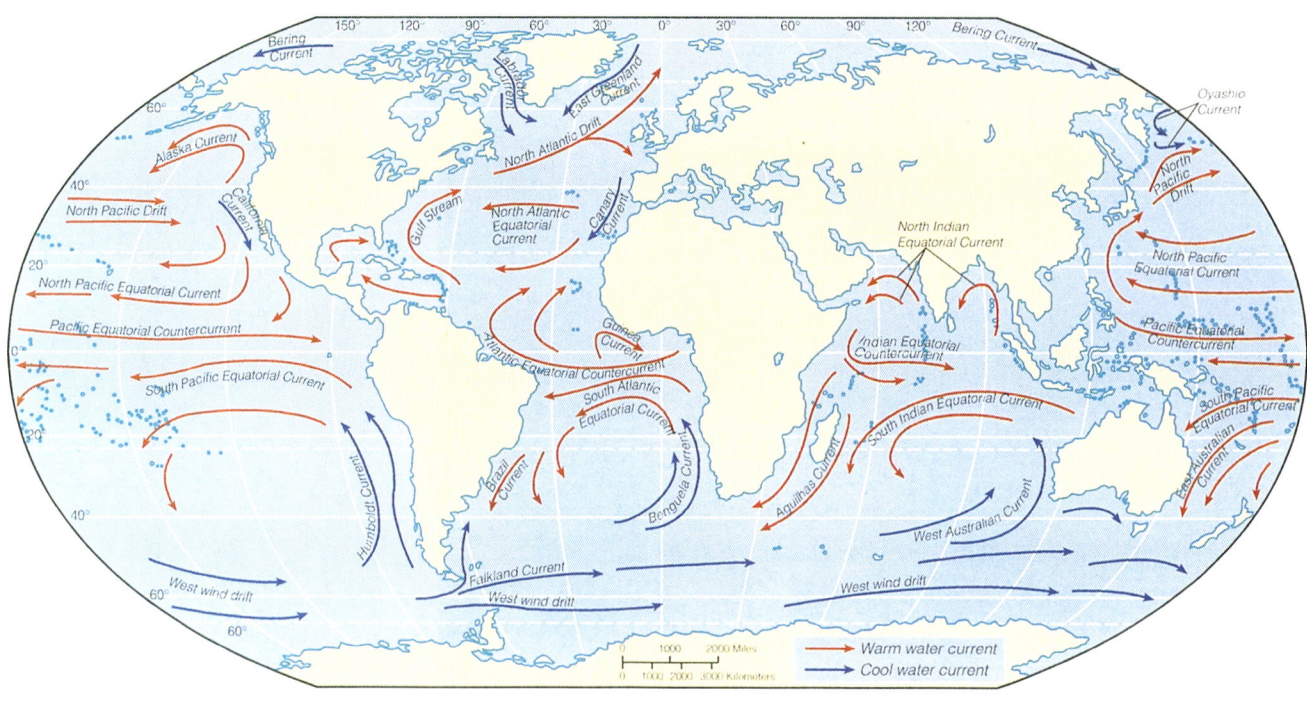

FIGURE 12.26 *Driven by energy provided by the winds, the surface waters of the ocean are driven in circular patterns called* gyres, *clockwise in the Northern Hemisphere and counter clockwise in the Southern Hemisphere. The Atlantic and Pacific oceans each have two gyres while the Indian Ocean, because it is almost totally located in the Southern Hemisphere, has only one.*

they join the equatorial currents to complete the gyre. By the time the currents are moving toward the equator, the temperature of the water has fallen, and the currents have become cold. In summary, as a result of the movement of water within the gyres, warm ocean currents flow along eastern continental margins while cold offshore ocean currents flow along western continental margins.

Because of the relative distribution of land and water, the Atlantic and Pacific countercurrents are located about 5° north of the equator and exist throughout the year. Within the Indian Ocean, however, the north equatorial current and the equatorial countercurrent exist only during the Northern Hemisphere winter months when the winds, called the *winter monsoon,* blow off the cold Asian continent toward the warmer waters of the Indian Ocean. During the summer months when the temperature of the Asian landmass becomes higher than that of the adjoining ocean water, air masses rising over the Asian continent draw air masses in from the ocean. During this period, the northeastern winds are replaced by an onshore wind called the *southwestern monsoon,* and the northern equatorial current and the equatorial countercurrent are eliminated (refer to Figure 12.26).

Because the equatorial countercurrent exists across the entire width of the Pacific Ocean, little or no exchange of surface waters occurs between the Northern and Southern Hemispheres. Within the Atlantic Ocean, however, some surface water moves from the Southern to the Northern Hemisphere as the south equatorial current is split by the landmass of Brazil (refer to Figure 12.26). The northern portion of the split current continues into the Caribbean and Gulf of Mexico where it joins with the northern equatorial current to become part of the North Atlantic gyre.

You may have learned in an elementary geography course that the relatively mild, but rainy, climate of Europe is the result of the landmasses being bathed in the heat brought by the North Atlantic gyre from more tropical climes. Two little known facts, however, are that the North Atlantic gyre was the first surface current to be charted and that the map was the cooperative effort of a former whaling captain, Timothy Folger, and the American statesman, scientist, journalist, and inventor, Benjamin Franklin.

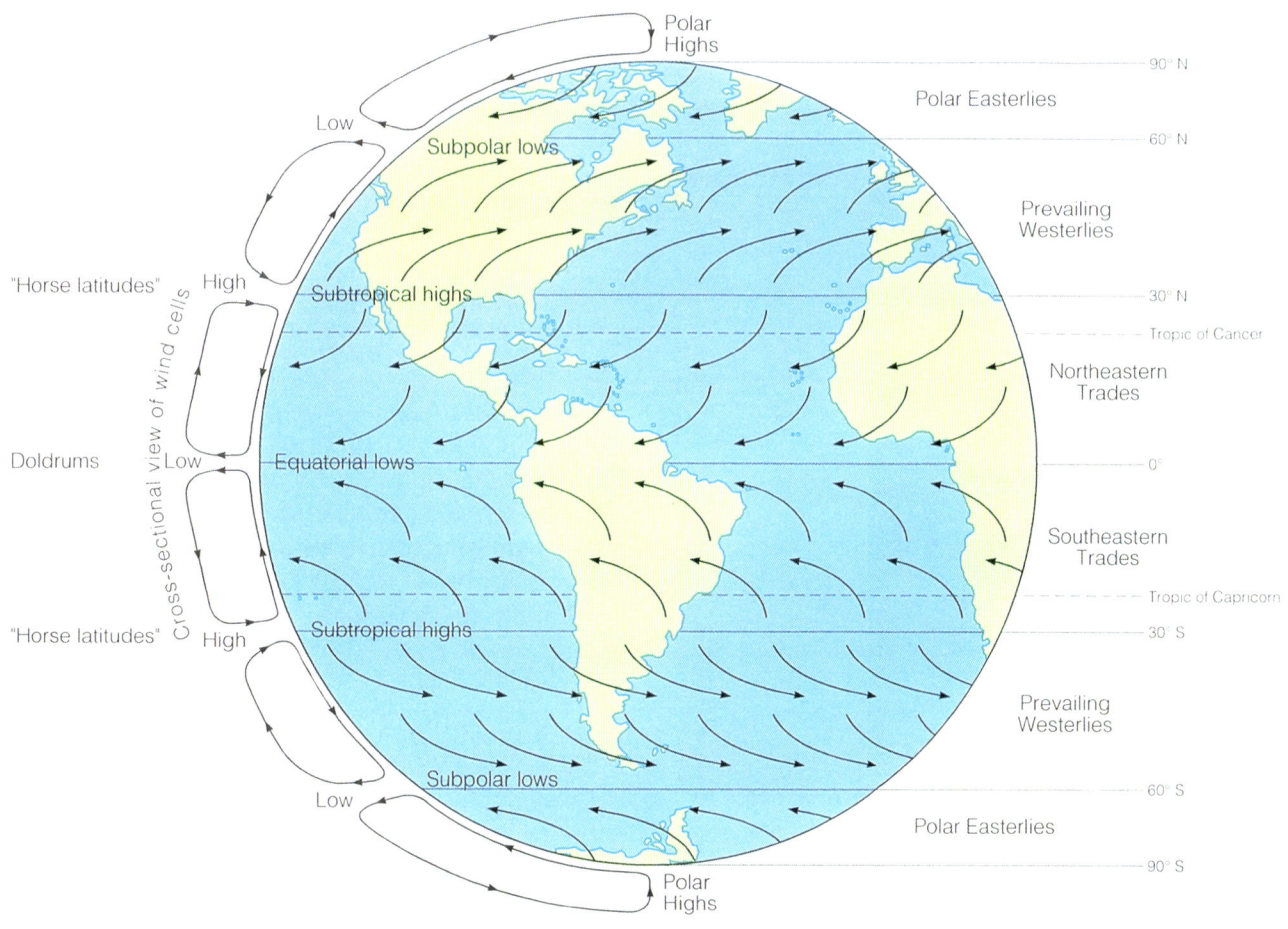

FIGURE 12.27 *Two dominant winds on Earth are the* easterly *trade winds and the* westerly *winds. The trade winds blow toward the equator from about 25° north and south latitude and are primarily responsible for the generation of the gyres. The westerly winds blow from the west between about 35° and 60° north and south latitude and deflect the gyres away from the eastern continental coastlines, across the ocean basins and toward the western continental coastlines.*

Wind-Induced Currents   Ocean water is subdivided vertically into two water masses: the *upper water mass,* which extends from the surface to a depth of about 3,000 feet (1,000 m), and the *deep water mass,* which extends from the bottom of the upper water mass to the ocean bottom. Both of these zones experience vertical water movements.

Vertical movements in the upper water mass are *wind-induced.* In areas where water is carried away by one surface current and not replenished by another, replacement water moves up from below. Although such *upwelling* can occur anywhere within the ocean, it is most prevalent in four areas: (1) where *equatorial currents* deflect water away from the equator, (2) *along western continental margins* where strong winds carry water away from the coastline, (3) where winds blow steadily in one direction and generate convection cells parallel to the wind flow (Figure 12.28), and (4) in the north Atlantic and Pacific Oceans where the combination of winds, heat exchange, and salinity differences cause global convective currents to be generated as described below.

Note that in the case of the coastal upwellings, the rising cold water serves to make the already cold coastal currents even colder. Examples are the Humboldt and Benguela currents that were previously cited as being responsible for the fog deserts of Chile and southwestern South Africa, respectively.

Thermohaline Circulation   Most of the vertical mixing of ocean water is due to **thermohaline circulation,** which affects the entire volume of ocean water. Thermohaline currents are *density currents* that originate in regions beyond 40° north and south latitude where masses of cold, dense, surface water sink.

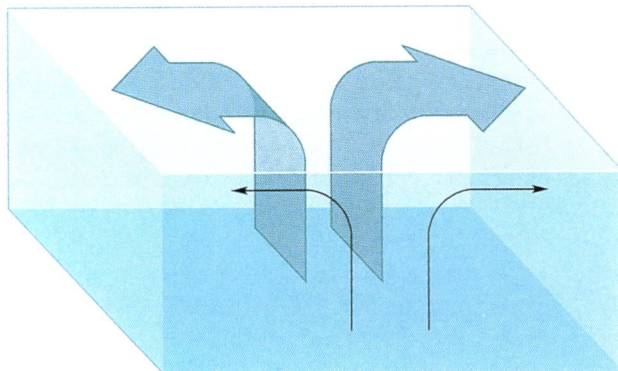
(a) Equatorial upwelling

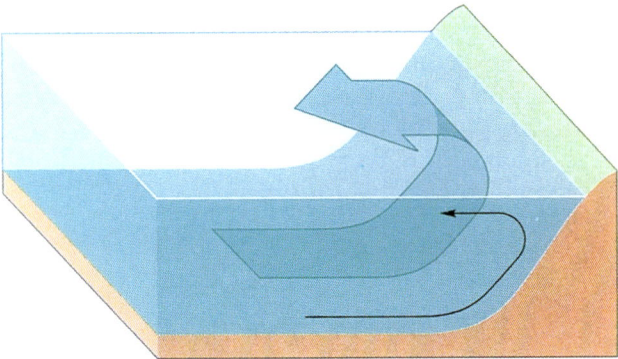
(b) Coastal upwelling

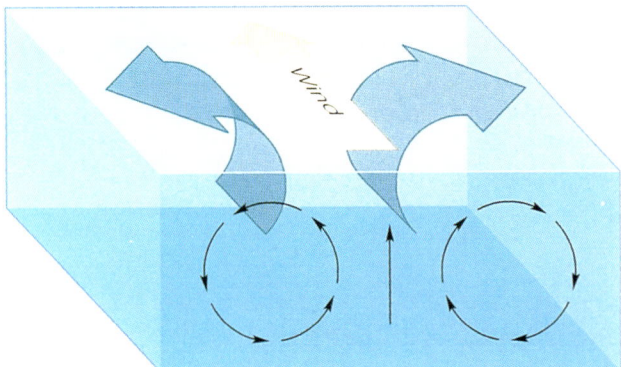
(c) Upwelling due to circulation cells set into motion by strong directional winds

FIGURE 12.28 Wind-induced currents *form as surface water is removed from a portion of the ocean by a wind-driven surface current without being replaced by water brought in by another surface current. Three places where they are commonly found are (1) along the equator, (2) parallel to western continental margins, and (3) anywhere at sea where winds blow steadily in one direction for long periods of time. A fourth scenario, not portrayed in the figure, involves areas in both the northern Atlantic and Pacific oceans where currents are generated by the complex interaction of the wind, thermal gradients, and salinity differences.*

Of the world's oceans, only the polar regions of the North and South Atlantic have surface water masses with densities high enough to sink. Within the Pacific and Indian oceans, the relatively uniform conditions of temperature and salinity that characterize the surface and near-surface water masses apparently prevent the formation of density differentials large enough to cause surface water to sink.

In the South Atlantic, very cold water (less than 33° F or 1° C) sinks off the Antarctic continental shelf, flows downslope, and moves along the ocean bottom as *Antarctic bottom water* to about 20° north latitude (Figure 12.29). Another cold Antarctic surface current sinks at about 60° south latitude. Because the temperature of the current is slightly higher and the density of the water therefore slightly lower than that of the shelf water (36° F or 2° C), it only sinks to about 0.6 miles (1 km). Called the *Antarctic intermediate water,* the current flows northward to about 20° north latitude.

In the North Atlantic off the coasts of Norway and southern Greenland, cold water (37° F or 3° C) sinks to the ocean bottom to become the *North Atlantic deep water* (refer to Figure 12.29). As the North Atlantic deep water moves southward, it eventually encounters and overrides the colder and denser Antarctic deep water. The current continues to flow southward between the Antarctic deep and intermediate water flows and eventually comes to the surface between 70° and 80° south latitude. The North Atlantic deep water acquires such a high level of nutrients during its travels that its surfacing masses are responsible for the high biological productivity of the waters off the coast of Antarctica. Because deep-ocean waters are not formed in either the Pacific or Indian oceans, masses of deep-ocean water found in the Pacific and Indian ocean basins originated in the North Atlantic and were eventually driven into the adjacent ocean basins by the Antarctic circumpolar current.

Other Density Currents Density currents may also result from salinity differences or from suspended loads.

Density Currents Due to Salinity Differences
Water density increases with increasing *salt content*. The current responsible for the continuous overturning of the water in the Mediterranean Sea is an example of a **density current resulting from salinity differences**. Because of the relatively dry, hot climate of the region, about 850 cubic miles (4,000

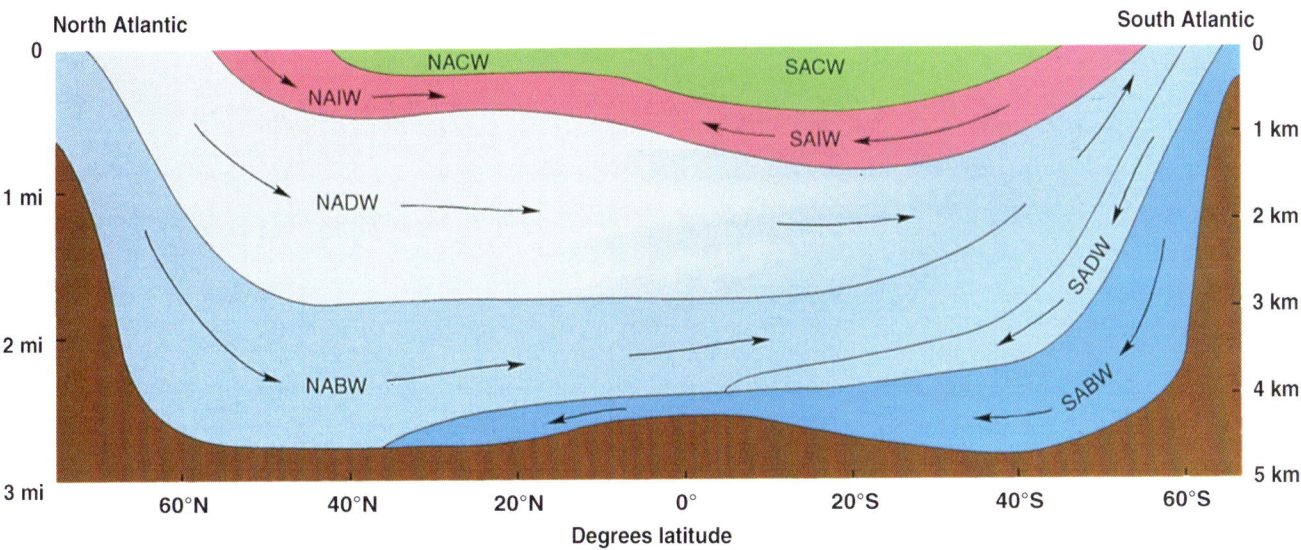

FIGURE 12.29 Thermohaline currents *are* density *currents generated in the Atlantic Ocean as masses of cold, dense, surface water sink in the higher latitudes of both the north and south Atlantic basins and move into the adjacent hemisphere. An important biological aspect of these currents is the well-known high biologic productivity of the Atlantic waters located between 70° and 80° south latitude. These masses of water originated as surface waters that sank off Norway and Greenland, moved southward along the ocean bottom, and rose to the surface laden with nutrients acquired in their journey.*

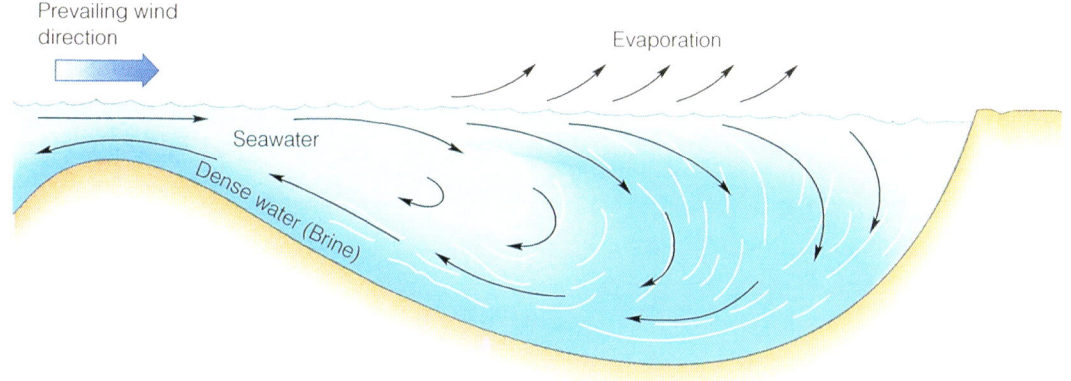

FIGURE 12.30 *Each year, nearly 800 cubic miles of water move into the Mediterranean Sea to replace water lost by evaporation and not replaced by runoff from the surrounding landmasses. This water movement, combined with subsequent evaporation due to the hot, dry climate, has generated a density current within the Mediterranean Sea and has established a stratification at the Strait of Gibraltar as normal-salinity, normal-density, oceanic waters move* into *the Mediterranean Sea while, at the same time, the denser, more saline waters generated by evaporation move* out *into the Atlantic Ocean.*

km^3) of water evaporate from the Mediterranean Sea each year. However, less than a hundred cubic miles of water are replaced by rain and direct runoff. The difference is made up by a surface current of normal-salinity ocean water that enters the Mediterranean Sea from the Atlantic Ocean through the Strait of Gibraltar (Figure 12.30). The denser Mediterranean surface waters descend to the bottom and *exit* through the Strait of Gibraltar *beneath* the incoming surface current from the Atlantic; thus, two distinct layers of water flow in opposite directions through the strait. The contact between the two layers is quite sharp. In fact, it is sharp enough to have reflected the sonar waves used during the early days of World War II to detect German submarines. Until the Allies realized that their sonar was not detecting the ocean bottom but rather the density interface between the two layers of water, the Germans were able to send submarines in and out of the Mediterranean Sea below the contact of the two layers.

EL NIÑO

Perhaps no ocean phenomenon has such far-reaching effects as the El Niño. An El Niño is the result of the complex, but not thoroughly understood, interaction of the Humboldt Current, the cold, upwelling waters along the west coast of South America, the South Pacific Equatorial Current, and the trade winds.

As we have already seen, the Humboldt Current is the eastern portion of the South Pacific gyre and brings cold water (50°F/10°C) from the coastal areas of Antartica northward along the west coast of South America (Figure 12.B1). Winds blowing seaward along the coast displace surface waters, which are then replaced by upwelling cold water from the deep ocean (refer to Figure 12.28). These cold upwelling currents bring nutrients from the deep ocean and produce a nutrient-rich zone of water about 6 to 12 miles (10–20 km) wide near the coast. The nutrients are responsible for the high primary productivity of the waters of the area and serve as the food supply for the large population of phytoplankton (microscopic floating plants) that is the base of the regional food chain.

Because the primary productivity of the waters is so high, populations of upper members of the food chain are also large. One member in particular, the anchovy, exists in such large numbers that the area is the world's largest fishery. In peak years, for example, the Peruvian fishing industry has harvested more than 12 million metric tons of anchovies from the coastal waters; the income from the anchovies is a major contributor to the economy of the country.

As the cold waters of the Humboldt Current approach the equator, they turn westward, become part of the South Pacific Equatorial Current (refer to Figure 12.B1), and are warmed by the tropical Sun. Under the influence of the intense trade winds, the warm surface waters are driven into a westward-thickening wedge as the contact between the warm surface water and the cold deep water is depressed simultaneously with an elevation of sea level (Figure 12.B2). As a result, sea level along the western portion of the Pacific Ocean is nearly a foot (30 cm) higher than normal. As long as the trade winds persist, this mass of warm water is maintained in the western Pacific. Meanwhile, along the western margin of South America, the combined effects of the Humboldt Current and the nutrient-rich, upwelling, coastal currents maintain the high biological productivity off the coast of Peru, a condition known as La Niña.

Once every five years or so, however, for unknown reasons, the trade winds diminish in inten-

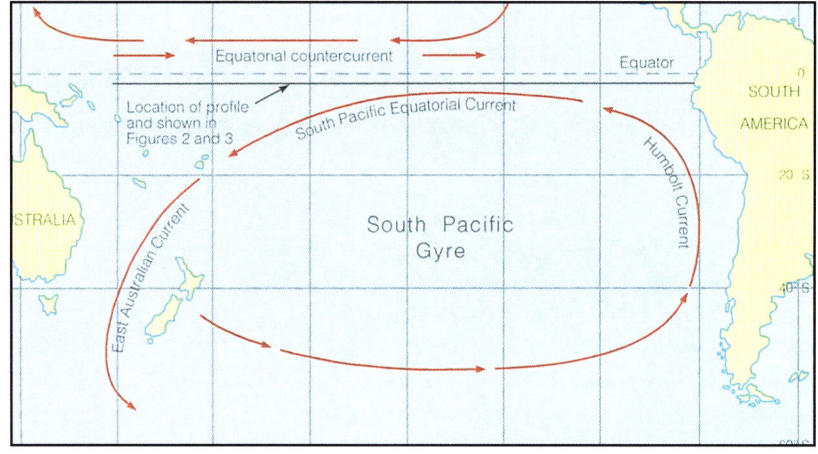

Figure 12.B1 *The Humboldt Current is the cold eastern portion of the South Pacific gyre. The displacement of surface waters by winds blowing seaward from the west coast of South America results in the upwelling of additional cold water from the deep ocean, making the waters of the Humboldt Current even colder and, at the same time, rich in nutrients. The high levels of nutrients in the more northerly portion of the Humboldt Current make the waters off the coast of Peru some of the richest fisheries on Earth.*

sity or cease to blow. With no force to contain it, the mass of warm water accumulated in the western Pacific begins to move westward as it collapses under the force of gravity. In about two months, usually in December, the warm water crosses the Pacific Ocean, reversing the orientation of the wedge of warm water (Figure 12.B3). The warm, tropical waters overun the Humboldt Current, drive southward along the coast of South America about 500 to 600 miles (800–950 km), elevate sea level about a foot (30 cm), and raise the temperature of the surface waters from 5°C to 10°C. Because of the overriding layer of nutrient-poor, warm water, the supply of nutrients provided by the upwelling cold currents is sharply reduced. As a result, the primary productivity of the waters rapidly declines. These combined events signal the beginning of an *El Niño*, a situation that may last as long as 18 months.

The immediate effects of the El Niño are felt throughout the food chain. During the most severe El Niño recorded, that of 1982–83, harvests of anchovies decreased from a peak of 12 million metric tons to less than 3 million metric tons. The effects on the economy of Peru were devastating. Some experts now feel that unless the Peruvian fishing industry prepares for future El Niños by reducing the yearly anchovy harvest, the effects of an El Niño, when added to the impact of overfishing and normal predation, may result in a population reduction from which the anchovy fishery may not recover.

El Niños also affect other members of the ecosystem. Because insufficient food is available following the hatch, the populations of young sea birds commonly experience sharp declines. In addition to young birds, large numbers of adult birds die from starvation as adequate supplies of food fish become more difficult to find. Similar mortalities have been documented in populations of seal pups as parents are forced to range so far for food that the young are unable to survive the long intervals between feedings.

In addition to the impact on life, El Niños have been blamed for worldwide climatic changes. During the 1982–83 El Niño, climatic changes were noted in every continent except Antarctica and Europe. For example, the removal of the warm water from the western Pacific was linked to droughts that ranged from Australia to India and as far away as South Africa. On the other hand, the increased moisture made available by the masses of warm water in the eastern Pacific resulted in severe storms along the Pacific coast of the United States and South America with widespread flooding and slope failures in the coastal mountains. Even as far east as the Gulf coast of the United States, the 1982–83 El Niño was blamed for torrential rainfalls. Only in the eastern United States, where the El Niño produced a milder than usual winter, was the effect not considered detrimental.

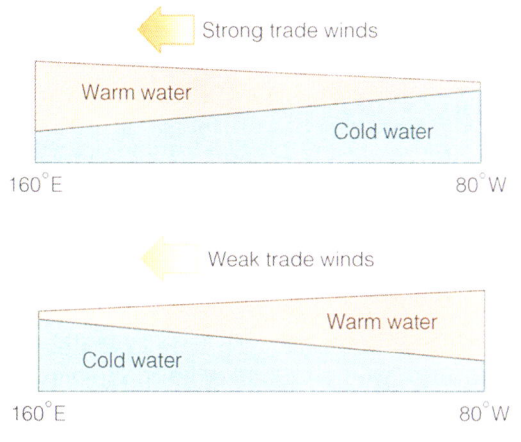

Figure 12.B2 *The strong easterly trade winds of the Pacific result in the formation of a westward-thickening wedge of warm water that both depresses the warm water–cold water interface and elevates the surface of the ocean in the western portion of the Pacific Ocean basin.*

Figure 12.B3 *For some unknown reason, about once every five years, the Pacific trade winds slacken. As a result, the mass of water that was being held in the western portion of the ocean basin moves eastward and produces a wedge of warm water that thickens toward the eastern portion of the basin, overruns the Humboldt Current, and results in a situation called* El Niño. *During the El Niño, the supply of nutrients in the surface waters is significantly reduced, severely impacting the fishing industry and economy of the area. Besides the biological impact, the El Niño is also thought to be responsible for far-reaching climatic effects.*

Another example of salinity-induced density currents can be found where low-density, freshwater streams enter the ocean and flow out across the surface of the higher-density, saline oceanic waters until they are intermixed and dispersed by the waves. (An example previously discussed is where the Mississippi River flows into the Gulf of Mexico.) In many coastal areas, for example, at the mouths of the Amazon and Saint Lawrence rivers, it is not uncommon for saline and brackish ocean water to flow *upstream* along the bottom during the rising tide. The lack of vegetation along the downstream banks of many coastal streams is due to the infusion of toxic salt water into the bank sediments.

Density Currents Due to Suspended Load
Density currents resulting from a high suspended load are called **turbidity currents**. Turbidity currents can commonly be seen wherever a silt- and clay-laden stream flows into the clear water of a lake or pond. The suspended materials, or *turbidity current,* can often be observed moving downward along the lake or pond bottom until currents within the larger body of water either dissipate the flow or the suspended material settles out (Figure 12.31). A density current can be modeled by pouring a spoonful of Kaopectate (a kaolinite-rich suspension) into a glass of water. A larger-scale example of this same process was referred to earlier when we discussed silt- and clay-sized sediments being dislodged from continental slopes. Once the sediments are dislodged, a turbidity current is generated that moves them downslope at speeds up to 60 miles per hour (100 kph). Such flows may travel for many miles out onto the abyssal plain before they finally come to rest. These flows were discovered when the first trans-Atlantic cable was broken following an earthquake. We have also seen that turbidity currents play a role in carving submarine canyons and in transporting land-derived sediment to the abyssal ocean floor.

FIGURE 12.31 As turbid, sediment-laden water enters clear water, the denser, turbid water forms a gravity-induced turbidity current that flows along the bottom of the larger body of water until the sediment settles and the turbidity flow is dissipated. (VU/© Ron Spomer)

SPOT REVIEW

1. How are waves formed?
2. How is wave energy consumed as it enters shallow water and moves toward the shoreline?
3. Why are there two tidal bulges due to the gravitational attraction between Earth and the Moon?
4. Why are there two high tides and two low tides each day?
5. Why do the tidal ranges of spring and neap tides differ?
6. What causes density currents?

COASTAL PROCESSES

No two coastlines are alike. They exhibit differences in structure, rock type, or process. For our purposes, however, we will consider coastlines as being of two basic types: (1) the **emergent** or **high-energy coastline** typified by most of the U.S. Pacific coast and portions of the northeast Atlantic coast and (2) the **submergent** or **low-energy coastline** found along most of the Atlantic and Gulf states.

Emergent or High-Energy Coastlines

Point Lobos, California, is a portion of the California coast that many consider exceptionally scenic (Figure 12.32). The scene at Point Lobos is repeated along most of the Pacific coast where the continental margin is rapidly rising because of the tectonic activity associated with converging plate margins. The offshore slope along coastlines such as Point Lobos is relatively steep. Because of the steep bottom slope, the waves feel bottom relatively near shore, wave amplitudes build rapidly, and relatively little of the wave energy is dissipated before the waves hit the surf. Consequently, strong forces are generated as the waves pound against the shore. In many cases, the clatter of boulder-sized rocks impacting each other can be heard above the sound of the surf. During storms, boulder-sized rocks are thrown against each other and against the cliff base like so many tennis balls, rounding the rocks and under-cutting the cliff face (Figure 12.33).

As the waves strike at the base of the cliff, water under great pressure is forced into fractures where mechanical weathering wears the rock away, opening **sea caves** that further undermine the **wave-cut cliff** (Figure 12.34). Eventually, the rocks are so weakened by weathering and undercutting that the cliff collapses into the sea. The rubble is broken into sand- and gravel-sized particles and quickly transported seaward and deposited in deeper water. As this process continues and the cliff retreats landward, a flat bedrock

FIGURE 12.32 Point Lobos, *California, an example of an* emergent, high-energy coastline, *is thought by some to be the most beautiful meeting of the land and the sea on Earth.*

FIGURE 12.33 *It is not uncommon for high-energy coastlines to be strewn with large, heavy boulders that one can hear being knocked together as they are moved about by an average-day surf. When one considers that during a storm, these same boulders are picked up and tossed against the cliff like so many marbles, the enormous amount of energy possessed by the sea can almost be comprehended.*

FIGURE 12.34 *The continuous pounding of the surf, especially the impact of storm waves, erodes the base of the cliff and carves out sea caves that undermine the overlying rock (a). Water, driven into sea caves under tremendous pressures by the surf, is often forced to the surface through fractures to form spectacular* water spouts *(b). Eventually, the cliff fails by slumping or rock fall, and the debris is quickly removed to the wave-built terrace (c). It is by the combined efforts of such processes that the sea claims the land.*

(a)

(a)

(b)

surface called a **wave-cut platform** is extended landward. The seaward edge of the **wave-cut platform** may be extended by the deposition of sediments as an offshore **wave-built platform** (Figure 12.35). The base of the wave-cut cliff is eventually carved back to just above the high tide line and represents the most landward transgression of storm waves.

Bedrock remnants rising from the wave-cut platform are called **sea arches** and **sea stacks** (Figure 12.36). These features can be seen all along the California, Oregon, and Washington coastlines. These rock monuments temporarily record a former, more seaward location of the coastline and stand in testimony to the incredible power of wave erosion. Should

Oceans and Shorelines 377

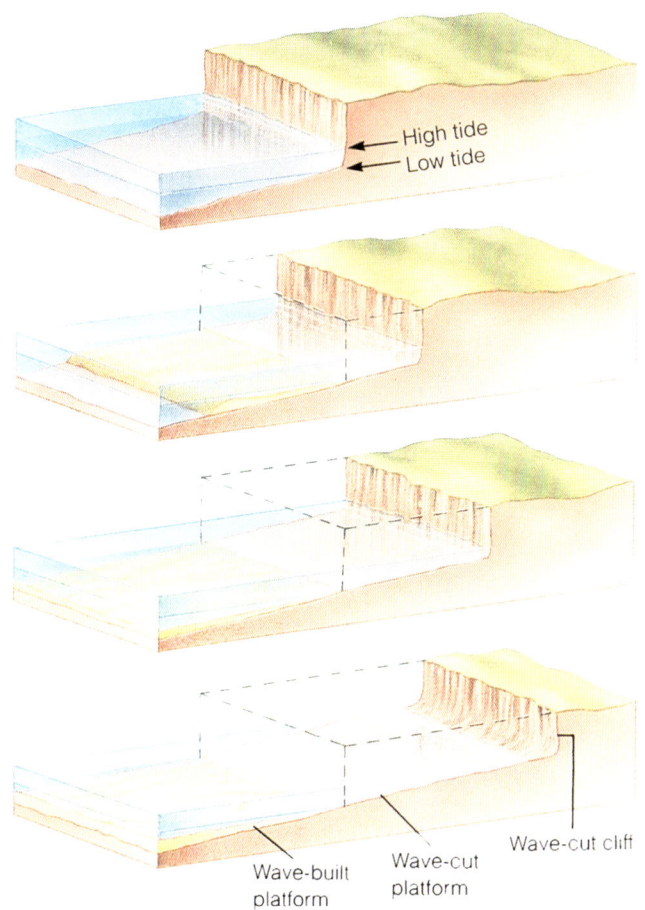

FIGURE 12.35 *The* wave-cut cliff *retreats as the energy provided by the surf erodes and undermines the cliff base. Eventually, the cliff will be driven to a point above high tide, which is the furthest transgression of storm waves. The generated rock debris is quickly transported seaward to construct the* wave-built platform, *leaving behind a flat rock surface called the* wave-cut platform *extending to the beach (a). The surface of the wave-cut platform is usually exposed at low tide (b).*

(a)

FIGURE 12.36 *As the wave-cut cliff retreats, rock remnants are commonly left behind, but only temporarily. Surrounded by the surf, (a)* sea stacks *and (b)* sea arches *are temporary reminders of the fact that the land once extended out to sea and was removed by the relentless erosion of the surf. (a Courtesy of Kara ZumBahlen)*

(b)

you ever visit the Pacific coast, a truly spectacular drive is along California Highway 1 where you will have ample opportunity to see the results of wave erosion and cliff retreat. You will probably not drive too far before a detour around a landslide or lost segment of highway will remind you of the efficiency with which the waves constantly reclaim the land.

Submergent or Low-Energy Coastlines

In contrast to the emerging, high-energy, Pacific-style coastlines, most of the Atlantic and Gulf coast shorelines are slowly subsiding. The slow subsidence of the Atlantic and Gulf continental margins is the result of the combined effects of crustal cooling away from the oceanic ridge and the accumulation of thousands of feet of sediment along the continental margin.

Along most of the Gulf and Atlantic coastlines, the gentle seaward slope of the coastal plain extends beyond the shoreline as the long, gentle offshore slope of the continental shelf. Because incoming waves touch bottom relatively far from shore, much of their energy is consumed by water movement and bottom erosion before they reach the shoreline. The amount of energy remaining in each wave upon arrival at the surf is commonly less than in waves approaching the typical Pacific coastline.

Slight topographic relief in the coastal plain is amplified into the irregularities commonly seen in Atlantic-type coastlines (Figure 12.37). Erosion concentrates at the seaward ends of promontories where the energy of the incoming waves is maximum. As the ends of the promontories retreat, the eroded materials are simultaneously deposited in adjacent areas (Figure 12.38). Longshore currents carry the materials parallel to the shoreline, deposit them between retreating headlands, and build **bay barriers** or **baymouth bars,** which create sheltered bays and lagoons. As sand moves parallel to the coast, long, narrow deposits called **spits** are created that are anchored to the receding headland at one end, extend parallel to

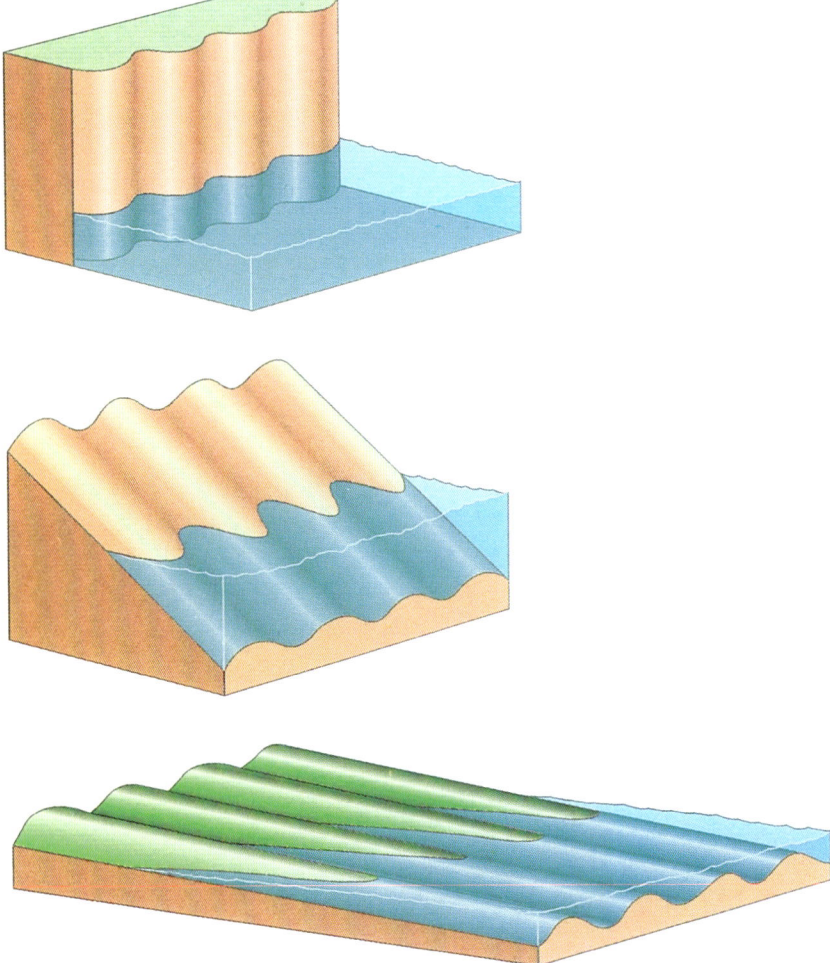

FIGURE 12.37 *Topographic variations that would result in a relatively* straight shoreline *along* emergent coastlines, *where the land enters the water at high angles, are amplified and produce a very* irregular shoreline *along a* submergent *coastline, where the land enters the water at a low angle.*

the shoreline, and usually terminate with a characteristic inward curl. The combined processes of erosion and deposition along Atlantic-type coastlines eventually convert an initially irregular coastal outline to a straight shoreline profile (refer to Figure 12.38).

Barrier Islands Perhaps the most dominant depositional forms observed along Atlantic and Gulf-style coastlines are **barrier islands.** Except for very large estuarian bodies of water such as Chesapeake Bay, barrier islands are found along nearly the entire Atlantic and Gulf coast. Excellent examples can be found continuously from just south of Ocean City, New Jersey, Miami, Florida, and along the Gulf coast from Brownsville, Texas, to the Florida Panhandle (Figure 12.39).

The origin of barrier islands is still debated. Some geologists believe they originate when storm-built sand ridges accumulate above high tide. This proposal may have some validity in that storm-generated sand ridges do develop in front of the breaker zone. Another theory suggests that barrier islands developed from near-shore depositional features such as bay barriers and spits that migrated landward as sea level rose.

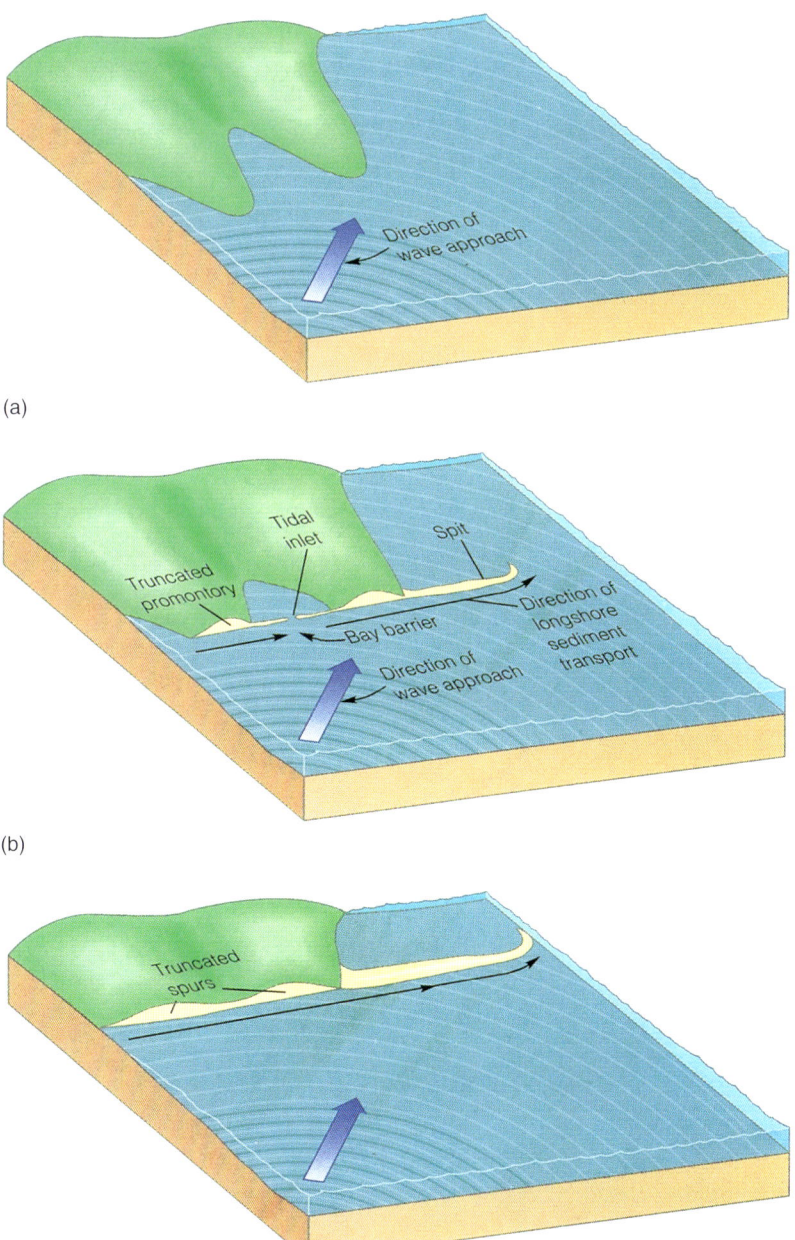

FIGURE 12.38 *Although submergent coastlines are initially irregular in outline, the shoreline straightens as waves erode the promontories and deposit the debris as* bay barriers *and* spits.

Other geologists, however, believe that the present system of barrier islands along the Atlantic and Gulf coastlines originated as sand beaches and beach ridges at the low stand of the ocean during the last glacial episode. As the ice melted and sea level rose, the beaches and beach ridges moved progressively landward and increased in height as the shoreline moved inland (Figure 12.40). Eventually, sea level rose high enough to create **lagoons** as sea-water flooded the low-lying areas landward of the ridges. As a network of lagoons developed and became interconnected, the original beach ridges became isolated as offshore barrier islands.

FIGURE 12.39 *One of the most common depositional features found along submergent or low energy coastlines is the* barrier island. *Composed of sand, barrier islands are found along much of the Atlantic and Gulf coasts where, because of their location, they provide partial, temporary protection for the land. Protection is partial in that storms may overwash the islands and enter the sheltered environments to landward. The protection is only temporary inasmuch as the islands will eventually be removed by storm erosion. At the present time, barrier island erosion along parts of the Louisiana coast is the highest in the United States (refer to our discussion of the Mississippi delta in Chapter 12). (Courtesy of S. J. Williams/ USGS)*

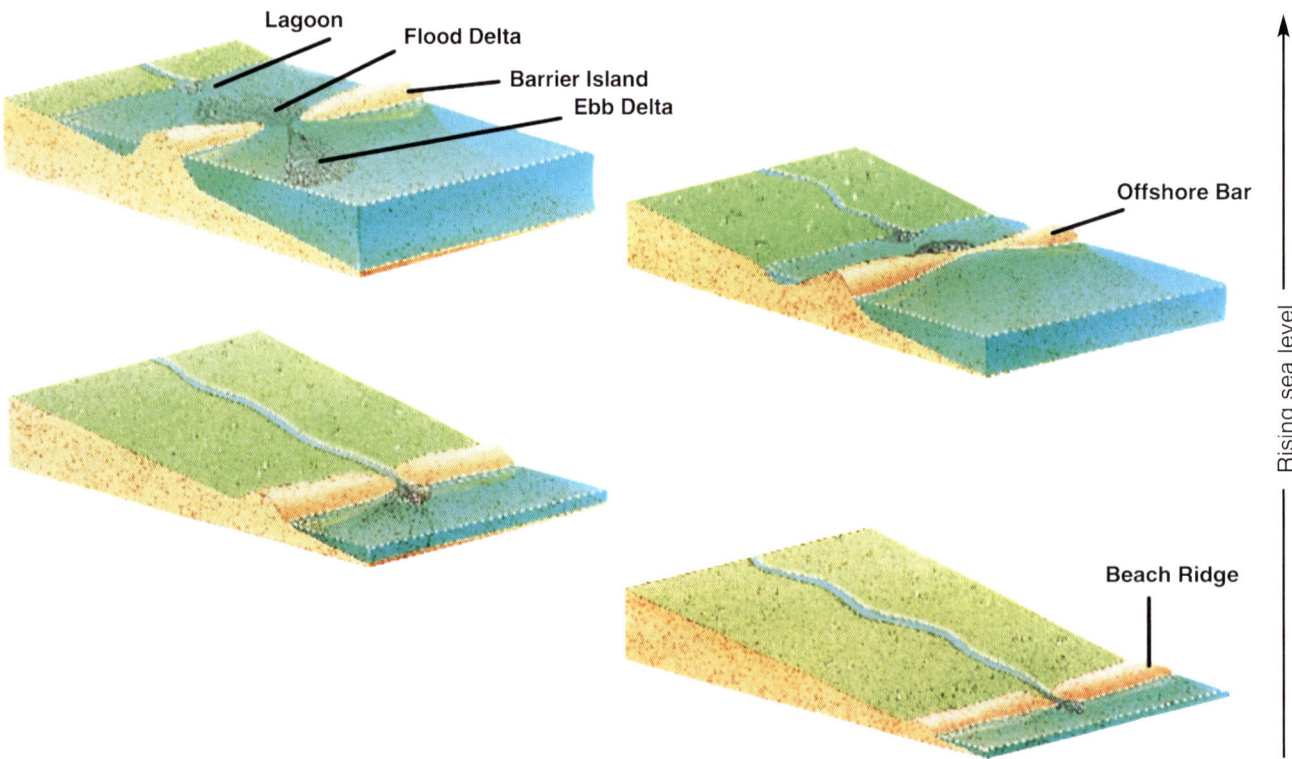

FIGURE 12.40 *Although barrier islands can form by a number of processes, most of those associated with the Atlantic and Gulf coasts are thought to have formed as sea level rose following the retreat of the last glacial episode.*

Once barrier islands are formed, along certain portions of the coastline, the shallow lagoons will begin to fill with sediment. Streams deliver sediments to the lagoons from the landward side while ocean waters, entering through **tidal inlets** during the incoming **flood tide**, carry materials from the seaward side of the islands and deposit them in the lagoon as a **flood delta**. During the outgoing **ebb tide**, some of the materials are carried back out to sea and redeposited on the seaward side of the tidal inlet as an **ebb delta**. Because the wave energy of the incoming tide is greater than that of the outgoing tide, the amount of material deposited on the flood delta is always greater than the amount redeposited in the ebb delta.

With time, any lagoon will gradually fill with sediment. As the water shallows, the bottom of the lagoon becomes exposed during low tide as a **mudflat** (Figure 12.41). Salt-tolerant grass begins to grow on the mudflat and traps sediment that normally would

FIGURE 12.41 *As barrier islands initially form, a portion of the ocean called a* lagoon *is isolated between the island and the land. With time, sediments provided from the land and from the ocean as tidal deltas begin to fill the lagoon. The first indication that the lagoon is being filled is the exposure of the bottom as a* mudflat *(a). As vegetation begins to grow on the mudflat, it is converted to a grass-dominated* marsh *(b). Eventually, as trees begin to grow on portions of the marsh, it is converted to a* swamp *(c). The end result is the development of the* marsh-swamp complex *that can be found along much of the Atlantic and Gulf coasts (d). (*a VU/ © John D. Cunningham; *b, c, d* Courtesy of S. J. Williams/USGS*)*

have been carried off. With the development of a continuous grass cover on the mudflat, sediment begins to accumulate at a higher rate, and the surface builds upward rapidly. Storm-deposited materials eventually elevate the surface of the grass-covered mudflat above high tide to create the beginnings of a coastal **wetland** or **saltwater marsh** (refer to Figure 12.41). With time, trees may begin to grow in portions of the marsh, transforming the wetland into a **marsh-swamp complex.**

The lagoon, however, never completely fills. The constant movement of tidal waters through the coastal wetlands maintains a network of interconnected streams (guts) and ponds throughout the system. Most of the seawater that enters through one tidal inlet returns to the sea through another. As we will discuss later, this circulation of marine water is vital for the survival of the wetland and of the offshore fisheries.

FIGURE 12.42 Coral *is an animal that secretes a shell of* $CaCO_3$ *for protection. Corals are among the most important animals contributing to the formation of* biochemical limestones. *(© Dave Fleetham/Tom Stack and Associates)*

REEFS

Many marine animals extract dissolved ions from seawater and use them to secrete protective shells. Most shell-bearing animals use calcium carbonate ($CaCO_3$) for this purpose. One of the most important of these rock-forming animals is **coral** (Figure 12.42).

The coral polyp consists of a barrel-shaped body surrounded by tentacles equipped with sting cells to paralyze its prey. The animal secretes a tube made of calcium carbonate in which it lives. Many corals are colonial in habit; that is, they join their carbonate living areas together and live as a colony. In some cases, many large colonies grow together to form a **reef** (Figure 12.43).

Corals have specific growth requirements. First, they need warm water. Reef-forming colonial corals will only grow in waters warmer than 65° F (18° C) so they are restricted to latitudes between about 35° north and south. There are so-called "cold-water" corals, but unlike reef-forming corals, their tissues do not contain algae to accelerate their growth, so they are usually small and rarely grow in sufficient numbers to create a reef.

Secondly, colonial corals grow in the **photic zone,** the zone of sunlit water. Although the depth of the photic zone varies, on average it is relatively shallow, about 260 feet (80 m). The colonial animals depend on photosynthetic algae contained within their bod-

FIGURE 12.43 *Large colonies of corals and associated animals grow together to form a* reef. *(Courtesy of S. J. Williams/USGS)*

ies to provide oxygen and to remove carbon dioxide, which enhances the production of $CaCO_3$.

Lastly, corals require *clear* water without suspended sediment. Sediment would both kill the animal and inhibit the growth of the algae by reducing the sunlight.

Kinds of Reefs

There are four basic kinds of coral reefs: (1) *patch reefs,* (2) *fringing reefs,* (3) *barrier reefs,* and (4) *atolls.* **Patch reefs,** a reference to their patchy or spotty distribution, can grow anywhere the necessary

FIGURE 12.44 *There are four different kinds of reefs:* patch reefs, fringing reefs, barrier reefs, *and* atolls. *Patch reefs, such as these growing in the shallow water behind the Great Barrier Reef, Australia, form anywhere the necessary growth requirements are met. (VU/© Edward Hodgson)*

conditions of sunlight, water temperature, and water clarity are provided (Figure 12.44). **Fringing reefs** grow in the shallow near-shore environment (Figure 12.45a). As sea level rises or, in the case of a volcanic island, as the island sinks, intergrowing coral colonies grow upward to maintain their position in the shallow water and become a **barrier reef**, separated from the land by a shallow, sheltered body of water or lagoon (Figure 12.45b). Many islands and larger landmasses are protected from the full energy of storm waves by barrier reefs. The largest barrier reef on earth is the 1,200-mile-long (1,900 km) Great Barrier Reef off the northeastern coast of Australia. Another example of a barrier reef parallels the Florida Keys about five miles offshore.

You will remember from our discussion of the formation of volcanic islands that as the oceanic lithosphere continues to move, it carries the island off the bulge formed by the mantle plume and into deeper water. As the volcanic islands sink below sea level and become seamounts, reefs that originally began as fringing reefs eventually are converted into coral islands called **atolls** (Figure 12.45c).

Charles Darwin was the first to suggest that the fringing reefs surrounding volcanic islands eventually developed into atolls. He also suggested that this change resulted from a rise in sea level. He was, however, unable to satisfactorily explain what caused the relative change in sea level. More than a hundred years passed before it was demonstrated that the sea-level changes involved were the result of the subsidence of volcanic islands due to the lateral movement of the oceanic lithosphere. Darwin's suggestion that atolls form by coral growing around volcanoes was proved correct when one of the two atolls used for the testing of the atomic bomb was drilled prior to the testing and the volcanic core was found.

SPOT REVIEW

1. Describe the processes that result in the formation of wave-cut platforms, wave-built platforms, and sea stacks.
2. Why do reef-forming coral grow only in the photic zone?
3. What are the requirements for the growth of coral?
4. Describe and compare the origins of the various kinds of reefs.

FIGURE 12.45 *Charles Darwin, in his studies of reefs associated with volcanic islands in the South Pacific, was the first to suggest that fringing reefs develop into barrier reefs and finally to atolls as a result of a rise in sea level, although he was not able to explain the change in sea level. When the islands are young,* fringing reefs, *such as these off Diamond Head, Hawaii, begin to develop practically in the surf zone (a). As plate movements carry the islands into deeper water, the reef structure builds vertically to keep in the photic zone, and as a result, the reef becomes isolated from the island and becomes a* barrier reef. *Earth's largest barrier reef is the Great Barrier Reef off the northeast coast of Australia (b). With time, the island sinks below sea level and the reef remains as a ring of coral islands called an* atoll. *(c). (Photo a VU/© Jeffrey Howe; b VU/© Kjell B. Sandred; c © Science VU/Visuals Unlimited)*

MODIFICATION OF SHORELINE PROCESSES

Natural coastal processes are often not to the liking of human inhabitants. Schemes are constantly being implemented to modify natural processes, usually with limited success. Along high-energy coastlines, for example, attempts are made to stop the erosion of coastal cliffs by placing large objects ranging from junked cars to steel and concrete structures in the breaker zone. The objective is to dissipate the energy of the waves before they reach the base of the cliff in much the same fashion as a *barrier reef* or a *barrier island* protects the mainland. The success of such attempts is temporary at best. Eventually, the energy of the waves will erode and remove these artificial barriers and attack the base of the cliffs.

Perhaps the most common attempts to control shoreline processes involve the manipulation of longshore sand transport. Structures called **jetties** are built at the mouths of harbors to prevent the deposition of sand in harbor entrances by confining the current and maintaining the velocity of water within the entrance (Figure 12.46). In many cases, the jetty is not completely successful, and sand must still be dredged from the harbor entrance.

Because much of the shoreline is used for recreation, attempts are made to ensure that supplies of beach sand are adequate. Barriers called **groins** are constructed perpendicular to the beach in an attempt to stop the longshore transport of sand (Figure 12.47).

In almost all cases, the groins are at best marginally successful. Invariably, additional sand does accumulate on the up-current side of the groin. However, the beach on the down-current side of the groin is denied a supply of sand while still undergoing the longshore removal of sand; as a result, the beach becomes sand-starved and shrinks. In some cases, the problem can be temporarily solved by bringing sand in from some distant source.

Shorelines are *constantly* undergoing change, and attempts to modify the natural processes of erosion and deposition are inevitably doomed to failure, usually after the expenditure of a great deal of money. Sooner or later, we may learn to enjoy the shoreline as it is and save money at the same time by not interfering with the shoreline processes.

ENVIRONMENTAL CONCERNS

Even though this is a text in geology, a point should be made concerning the great biological and ecological significance of the coastal wetland. The coastal wetland is literally a factory that produces the vast quantities of organic debris and organisms that occupy the bottom rungs of the marine food web. The outgoing tides flush these materials and organisms out to sea where they provide the basic food supply for a host of organisms living beyond the barrier island in the waters above the continental shelf and on the ocean bottom.

FIGURE 12.46 Jetties *are structures built at the entrances to harbors in an attempt to maintain current velocities high enough to prevent the deposition of sediment within the entrance.* (VU/© Bernd Wittich)

FIGURE 12.47 Groins *are structures built perpendicular to the shoreline in an attempt to control the loss of sand by longshore transport. Although additional sand does accumulate on the up-current side of the groin, it is at the expense of the sand on the down-current side, which becomes sand-starved. (VU/© Frank Hanna)*

At the present time, the very existence of the salt-water wetland environment is being threatened by pollution and by exploitations of various kinds. Marshes and swamps are being drained and filled to provide land for commercial developments, which in turn contaminate and overload the remaining wetlands with pollutants of every description. The building of roadways across the wetland and the indiscriminate filling of the marsh interfere with the circulation of the water upon which the life of the wetland depends.

The elimination of the coastal wetland environment seriously affects the entire offshore food web by decreasing the amount of biomass in the lower layers of the food web. In many coastal areas, the effect can already be seen in significant declines in populations of certain fish and shelled animals. Some species such as the bay scallop have all but disappeared from the mid-Atlantic coastal waters. Fishing fleets have to go further offshore to find commercial quantities of fish. How long will it be until each of us feels the effect of such foolish disregard for the natural system?

Another environmental concern is the progressive destruction of the barrier reef that protects the Florida Keys. An estimated 3,000 people a day visit the reefs within John Pennecamp Coral Reef State Park. Gasoline, oil, and diesel fuel from their boats pollute the water. The visitors leave behind aluminum cans, film wrappers, and styrofoam cups. Inexperienced operators drive boats onto the reef. Swimmers damage the coral by standing or walking on it. Although it is against the law, souvenir hunters take pieces of the coral. The near-shore waters are being polluted by pesticides, fertilizers, the runoff from streets, and septic system effluents from an increasing population. The tides eventually carry these polluted waters out to the reef where they destroy the coral. The high levels of nitrate and phosphate nutrients encourage the rapid growth of algae, which cover and eventually smother the living coral. The combined effect has been the destruction of extensive segments of the reef complex. With no living coral animals and thousands of years required to grow large coral heads, the reef will rapidly succumb to the impact of storms. Eventually, the barrier reef will be eliminated and once it is removed, the next more landward barrier, the Florida Keys themselves, will succumb to the energy of the storm waves. In other words, continued exploitation of the Keys and the offshore reef complex will eventually result in the destruction of both.

Perhaps the most serious environmental problem we face today with regard to the ocean is the dumping of waste. It is estimated that over half of the population of the United States lives within 50 miles (80 km) of the shorelines of the Atlantic and Pacific oceans, the Gulf of Mexico, or one of the Great Lakes. Since its passage in 1972, the Marine Protection Research and Sanctuaries Act has prohibited the dumping of certain toxic materials, such as radiological, chemical, and biological warfare agents and high-level radioactive waste, and has regulated the disposal of other materials. Nevertheless, the ocean is still the repository of a wide variety of pollutants ranging from the sediments dredged from harbors to raw sewage.

The impact on the marine environment has been profound. In many cases, the decomposition of organic wastes depletes the oxygen in the ocean waters surrounding the disposal sites. The oxygen content of the water is further reduced by the decomposition of algae whose growth was stimulated by the introduction of nutrient-rich wastes such as sewage sludge. Extensive areas of ocean bottom are now devoid of life as a result of the combined effects of the oxygen depletion.

The introduction of toxic materials has not only retarded both the growth and reproduction of marine organisms but in many cases has led to their extermination. Contaminated fish and shellfish have been identified as the source of infection and disease. It is estimated that pollution has resulted in the closure of more than 20% of the commercial fishing grounds in the United States.

The fallacy of the "out of sight, out of mind" concept has been dramatically demonstrated by recent reports of waste such as used syringes and hypodermic needles being washed up onto public beaches. Coastlines contaminated with petroleum residues and tar-coated sealife are poignant reminders of our dependence on oil as an energy source.

The disposal of waste is becoming a major problem nationwide. Solid waste disposal sites are being closed because of the potential of groundwater contamination. Those sites continuing in operation will eventually be filled and, increasingly, local statues prevent the opening of new sites. The potential of air pollution has limited the development of incineration as a method of disposal. With fewer options available for the land disposal of wastes, regulatory agencies are going to be under increasing pressure to allow more extensive ocean dumping.

SPOT REVIEW

1. What is the purpose of a jetty? Of a groin?
2. Why do most attempts to modify coastal processes fail?
3. What is the importance of coastal wetlands?

CONCEPTS AND TERMS TO REMEMBER

Hydrosphere
Deep-sea research
  determining depth of water
    weighted lines
    sonar
  submersibles
  deep-sea drilling
Origin of the ocean basins
  continental rifting
    triple junction
    rift zones
    rift valleys
      failed arms
Oceanic landforms
  oceanic ridges
    pillow basalts
    sheeted dikes
      ophiolite suite
  abyssal hills

abyssal plain
deep-sea trench
seamount
  volcanic island
  guyot
Continental margins
  trailing edge or passive margin
    continental shelf
    continental slope
      turbidity or density currents
      turbites
    continental rise
    submarine canyon
    deep-sea fan
  leading edge or active margin
    island arc
    continental or Andean-type arc

Movement of ocean water
  waves
    swell
    surf
    backwash
  tides
    spring
    neap
  longshore current
    longshore transport
  ocean currents
    wind-driven currents
      gyres
        equatorial currents
        equatorial countercurrents
    wind-induced currents
    thermohaline currents
    thermohaline circulation

CONCEPTS AND TERMS TO REMEMBER *continued*

    other density currents
        density currents due to salinity
        density currents due to suspended sediments
        turbidity currents
Coastal processes
    emergent or high-energy coastline
    sea caves
    wave-cut cliff
    wave-cut platform
    wave-built platform

    sea stack
    sea arch
    submergent or low-energy coastline
    bay barrier or baymouth bar
    spit
    barrier island
        lagoon
        tidal inlets
        flood tide and flood delta
        ebb tide and ebb delta
        mudflat

    wetland
        saltwater marsh
        marsh-swamp complex
Reefs
    coral
        photic zone
    patch
    fringe
    barrier
    atoll
Modification of coastal processes
    jetties
    groins

REVIEW QUESTIONS

1. Which of the following geographic features is commonly used to illustrate an "ocean in making"?
   a. Mediterranean Sea
   b. Red Sea
   c. Sea of Japan
   d. Black Sea
2. The most dominant topographic features on the ocean bottom are the
   a. oceanic ridges.   c. abyssal hills.
   b. seamounts.   d. deep-sea trenches.
3. The surface of the oceanic crust is composed of
   a. deep-sea sedimentary rocks.
   b. pillow basalts.
   c. granodiorite.
   d. gabbro.
4. Which of the following topographic features is not associated with continental trailing edges?
   a. continental shelf
   b. abyssal plain
   c. deep-sea trench
   d. submarine canyon
5. Wave erosion of the ocean bottom can occur
   a. at any depth of water.
   b. when the depth of water is equal to the amplitude of the waves.
   c. when the depth of water is equal to half the wavelength of the waveform.
   d. only in the surf zone.
6. The coral reefs associated with seamounts and guyots are
   a. atolls.   c. barrier reefs.
   b. fringing reefs. d. patch reefs.
7. What would be an example of a triple junction failed arm?
8. Why, in general, are the beaches along continental trailing edges not particularly good for the sport of surfing?
9. What landforms would you expect to find along a continental leading edge that you would not expect to find along a continental trailing edge?
10. How are the tidal ranges related to the changing positions of the Sun, Earth, and Moon?

THOUGHT PROBLEMS

1. All countries that adjoin the ocean place their international boundary out to sea. Initially, this was probably done to provide a protective buffer against naval attack. The boundary was commonly set at 3 miles (5 km), which was the range of ship-mounted cannon. In today's world of ballistic missiles, such a buffer would be meaningless. Although there is no international agreement as to the location of national marine boundaries, most countries recognize two types of boundaries. The innermost boundary designates "territorial waters" within which warships must have permission to pass. Beyond the territorial waters is an "economic boundary" within which the country has ownership of any exploitable material from fish to minerals. For the United States, the economic boundary extends out to 200 miles (320 km). Consider the problems that would be involved in any negotiations directed toward the placement of international marine boundaries.

2. One of the suggestions for the disposal of highly toxic radioactive and nonradioactive wastes is to encapsulate them and dispose of them in zones of subduction. Present arguments both for and against such a proposal.

FOR YOUR NOTEBOOK

Besides investigating the geology of your favorite beach, perhaps the most significant thing you can do for this particular chapter is to compile information concerning our attempts to modify, control, defile, and generally destroy the coastal environment. What is needed to bring such futile and destructive practices to a halt is an informed public, and there is no better way to become informed than to amass all the information you can about a subject.

CHAPTER OUTLINE

INTRODUCTION

CLASSIFICATION OF SEDIMENTARY ROCKS
- Size
- Composition
- Sorting
- Shape

DETRITAL SEDIMENTARY ROCKS
- Breccia
- Conglomerate
- Sandstones
- Siltstone
- Shales and Mudstones

NONDETRITAL SEDIMENTARY ROCKS
- Chemical Sedimentary Rocks
- Biochemical Sedimentary Rocks
- Evaporitic Sedimentary Rocks
- Organic Sedimentary Rocks

DEPOSITIONAL ENVIRONMENTS
- Continental Deposits
- Transitional Environments

MARINE ENVIRONMENTS

LITHIFICATION
- Compaction
- Cementation

SEDIMENTARY FEATURES
- Beds
- Ripple Marks
- Cross-Bedding
- Graded Bedding
- Mud Cracks
- Animal Trails

INTERPRETATION OF ANCIENT DEPOSITIONAL ENVIRONMENTS
- Walther's Law

ENVIRONMENTAL CONCERNS
- Sediment Pollution
- Engineering Problems

CHAPTER 13

Sedimentary Rocks

INTRODUCTION **Sedimentary rocks** are composed of the transported and subsequently redeposited insoluble and soluble products of weathering. The products of weathering, when deposited by the various agents of erosion, are referred to collectively as **sediment**. Sediments include the insoluble products of weathering such as *rock fragments, quartz,* residual *feldspar,* and *clay minerals* as well as minerals precipitated from materials originally in aqueous solution.

Sedimentary rocks are the most abundantly observed rocks at Earth's surface. Common examples of sedimentary rocks—*shales, sandstones,* and to a lesser degree *limestones*—appear in most road cuts, cliffs, and valley walls. This abundant display is due to the fact that sedimentary rocks cover 75% of Earth's exposed land surface. Though they constitute most of the surface rocks, sedimentary rocks represent only 5% of the volume of Earth's crust. The reason for this apparent discrepancy is that sedimentary rocks form a thin veneer covering the crystalline core of the continents in much the same fashion as a thin layer of icing covers the much larger volume of a cake. The thickness of sedimentary rocks over most of the continental surface averages a few thousand feet or meters with the thickest sections being found either associated with foldbelt mountains, where the total thickness of sedimentary rocks may measure tens of thousands of feet, or in deep basins.

The geologic and economic importance of sedimentary rocks is disproportionate to their limited representation within Earth's crust. Sedimentary rocks allow geologists to achieve one of the major goals of the science of geology, namely, to interpret the history of Earth. Sedimentary rocks literally record the geologic (and biologic) history of Earth. The kind of sedimentary rock, the physical features found within the rocks, the occurrence of fossils, and the minerals composing the rocks record many of the events and environmental conditions in existence at the time the sedimentary materials were deposited. Because sedimentary rocks contain the major clues to unraveling Earth's history, a large portion of a geologist's education is spent studying them to obtain the tools needed to interpret the historical record they contain.

Economically, sedimentary rocks contain many of the natural resources needed for modern society, not the least of which are the sources of energy. The primary sources of energy in the world today, oil and gas, are largely contained within the sedimentary rocks. Coal is also found as layers within sequences of sedimentary rocks. Our present supplies of uranium are primarily extracted from deposits contained in sedimentary rocks. In addition to supplies of energy, sedimentary rocks provide us with a wide variety of other valuable and necessary commodities such as iron and copper ore, bauxite for aluminum, basic building materials such as sand, clay, and stone, raw materials for the manufacture of cement, and essentials for life as phosphate, salt, and water.

CLASSIFICATION OF SEDIMENTARY ROCKS

Sedimentary rocks made from the insoluble products of weathering are called **detrital** rocks while those composed of minerals precipitated from solution or formed from biologic accumulations are called **nondetrital**. Nondetrital rocks are further subdivided as *chemical, biochemical,* or *evaporite* rocks depending upon the mechanism by which the minerals were removed from solution.

The name given to a detrital sedimentary rock depends primarily on the *size* and *composition* of the grains comprising the rock. Two other parameters, *sorting* and *shape,* although not specifically required for the classification of sedimentary rocks, provide important information about the origin of the particles that make up detrital sedimentary rocks. Grain size, sorting, and shape are collectively known as *texture*.

Size

Many of the terms used to describe the size of particles are common words such as *cobble, pebble,* and *sand.* To a geologist, these terms refer to specific sizes or, more accurately, to specific ranges of particle sizes. Table 13.1 summarizes the complete list of particle sizes.

Composition

The composition of the *rock fragments* generated by physical weathering obviously depends upon the kind of rock undergoing weathering and usually includes more than one kind of mineral.

The insoluble products of chemical weathering are primarily particles of *quartz* and *clay minerals* with lesser amounts of *feldspar.* Quartz is present because of its resistance to chemical weathering. The clay minerals are present because they are the stable silicate mineral formed by the chemical weathering of most of the aluminosilicate minerals. Feldspar is a residue of weathering because of its dominance in the rocks of Earth's crust. But since feldspar also has a relatively high susceptibility to chemical weathering, feldspar grains are not preserved in sediments except under conditions that minimize the amount of time the materials are exposed to the atmosphere before being buried. For this reason, significant concentrations of feldspar in a sedimentary rock indicate that the materials of which the rock is made were deposited and buried soon after formation and probably were not transported far from their point of origin.

TABLE 13.1
PARTICLE SIZES IN MILLIMETERS

Particle Size	Mean Diameter in Millimeters
Boulder	256
Cobble	64
Pebble	4
Granule	2
Sand	.06
Silt	.004
Clay	

Quartz and feldspar particles are dominantly sand and silt size while clay mineral grains are almost exclusively clay size. Usually, in fact, the great volume of clay-sized grains are clay minerals, which explains why the term *clay* is used to describe both a particle size and a kind of mineral. An alternative term frequently used for clay-sized material is *mud*.

Sorting

As sediment is transported by the various agents of erosion, the insoluble products of weathering are separated naturally by size. Most rock fragments are granule size and larger. As we have noted, most quartz and feldspar grains are sand size while the clay minerals are predominantly clay size. The process whereby materials are separated by size is called **sorting**. The more uniform in size the particles with sediment or rock become, the more *well sorted* the material is. Conversely, the wider the range of particle sizes, the more *poorly sorted* the materials.

Of the agents of erosion, water is the most efficient at sorting. Examples of the ability of moving water to produce *well-sorted* deposits are the fine-grained muds that cover a stream floodplain after a flood and the sand-sized quartz grains that constitute your favorite beach.

Wind is a unique sorting agent in that it is very selective with respect to the particle sizes it can transport. It is limited to sand-, silt-, and clay-sized materials. Sand-sized materials are moved within a few feet of the ground surface and accumulate in deposits called *dunes.* The finer silt- and clay-sized particles are lifted into the air as dust and may be carried great distances before being deposited as fine-grained deposits called *loess*.

Glacial ice is a relatively ineffective sorting agent, picking up and transporting whatever material it may encounter in its path. When the ice melts, the material is literally *dumped* with little discrimination by size. As a result, most glacial deposits are usually *poorly sorted* and consist of a mixture of particle sizes that may range from huge boulders to clay. Some glacial deposits are well sorted, but these are not deposited directly by the ice but by meltwaters emerging from the terminus of the glacier.

Shape

The two descriptors of particle shape are *roundness* and *sphericity*.

Roundness   The **roundness** of a particle is a measure of the sharpness of its corners or edges and is calculated by dividing the average diameter of the circles that can be inscribed into its corners by the diameter of the largest inscribed circle (Figure 13.1). A sphere represents perfect roundness with a roundness factor of 1.0.

Particles newly created by most means of mechanical weathering are characterized by sharp edges and corners and have *low roundness*. Particles with roundness values from zero to 0.15 are commonly referred to as being *angular*. Examples are the particles that accumulate in talus.

The presence of angular particles in sediment or in a sedimentary rock is usually interpreted as indicating that the material was transported only a short way from the point of origin before being deposited or converted into a sedimentary rock. Increasing roundness usually indicates that the particles have been transported greater distances from the point of origin and have undergone progressive rounding by mutual abrasion.

Sphericity   The **sphericity** of a particle is a measure of its three-*dimensional* shape; that is, it is a measure of the relationship of the length, width, and thickness of the particle. Sphericity is calculated as the ratio of the nominal diameter of the particle to the diameter of the largest possible circumscribed sphere (Figure 13.2). The nominal diameter of a particle is the calculated diameter of a sphere having the same volume as the particle.

The difference between roundness and sphericity is somewhat subtle. An equidimensional particle such as a cube has *high* sphericity but *low* roundness. On the other hand, a flat pebble found along a stream or shore may have *high* roundness but *low* sphericity. Figure 13.3 compares sphericity and roundness.

How closely the shape of a particle approaches a sphere is a complex product of crystal structure, mineral hardness, the type of transporting agent, the distance of transport, and the particle size. Most particles become rounded by the attack of chemical weathering and exfoliation (Figure 13.4) and by mutual abrasion as they are being transported either in streams or wave zones.

The process of rounding commonly affects only sand-sized and larger particles. Silt- and clay-sized particles transported by water or wind usually do not

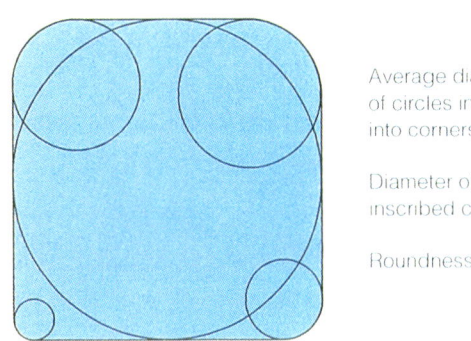

FIGURE 13.1   *Rock fragments are progressively rounded during transport as sharp edges and corners are worn away. Roundness is measured by dividing the diameter of the average inscribed circle by the diameter of the largest inscribed circle. A perfectly round particle would have a circular outline.*

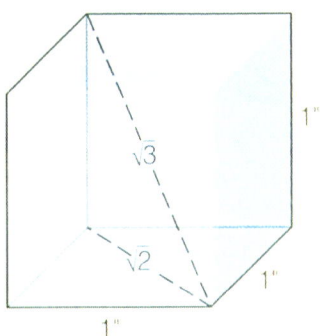

FIGURE 13.2   Sphericity *differs from* roundness *in that the former is a three-dimensional property while the latter is a two-dimensional property. The sphericity of a cubic particle 1 inch on a side is determined as follows: The diameter of the largest circumscribed sphere is equal to √3 or 1.73 inch. The diameter of a sphere with a volume of 1 cubic inch is 1.24 inch. Therefore the sphericity of the cube is or 0.72.*

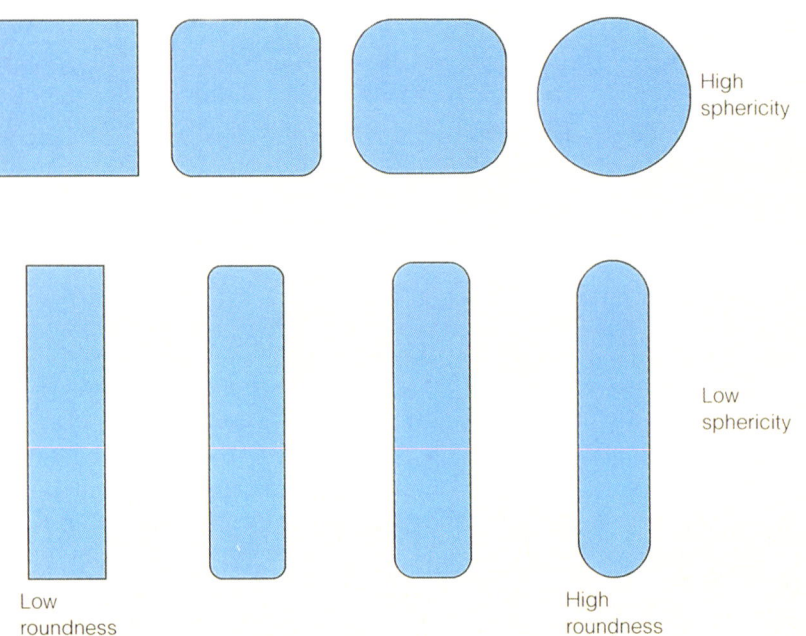

FIGURE 13.3 *Sphericity and roundness are not interrelated; that is, a particle may be highly rounded and yet not be spherical, or it may be highly spherical but not be rounded.*

FIGURE 13.4 *The end product of weathering is the rounding of rocks. (VU/© Doug Sokell)*

become rounded because the water or air provides a cushioning effect as the particles are transported in suspension. Consequently, the particles have little opportunity to experience mutual abrasion. The subrounded (almost round) silt-sized particles that have been observed in some glacially transported materials are the one exception to this rule.

Because nondetrital rocks consist of minerals precipitated from solution, the parameters of sorting, roundness, and sphericity have limited application to them. Nondetrital rocks are usually described solely on the basis of grain size and composition.

SPOT REVIEW

1. What accounts for the relatively high concentrations of quartz, clay minerals, and feldspar in sediments?
2. What is meant by "sorting"? How does the agent of erosion determine the sorting of sediments?
3. What is the difference between roundness and sphericity? What kind of information concerning the mode of sediment transport do roundness and sphericity provide?

DETRITAL SEDIMENTARY ROCKS

About 90% of the total volume of sedimentary rocks is made up of detrital sedimentary rocks. These rocks form from the insoluble products of weathering and are named primarily on the basis of the size and composition of their particles.

Breccia

A **breccia** is a sedimentary rock composed primarily of angular grains, granule size (greater than 2.0 mm) and larger, contained within a finer-grained matrix of sand, silt, or clay (Figure 13.5). For the most part, the individual grains are rock fragments. Most breccias form from rock fragments such as talus and the pyroclastic debris of volcanic eruptions that accumulate near the site of origin. Having been transported relatively short distances, the individual particles have had little chance to experience rounding by either chemical weathering or abrasion. Some breccias, called fault breccias, form within a fault zone from the rock that is broken or crushed by the movement of the rocks on opposite sides of the fault.

Conglomerate

A **conglomerate** is a sedimentary rock with the same size particles as a breccia, differing only in that the particles are rounded (Figure 13.6). Although some conglomerates may form from poorly sorted glacial deposits, most form from materials that accumulate in the channels of steep, highly turbulent streams and along rocky coastlines where wave action is intense.

Sandstones

A **sandstone** is a detrital rock consisting primarily of sand-sized (0.062 to 2.0 mm) grains, contained within a finer-grained matrix (Figure 13.7). Sandstones constitute about 20% to 25% of all sedimentary rocks. Because quarts is a major component, sandstones are generally more resistant to weathering and erosion than the enclosing rocks and, as a result, are usually the most conspicuous rock layers in outcrops such as cliffs and road cuts. This often leads beginning students of geology to overestimate the abundance of sandstones relative to other types of sedimentary rocks.

The term *sandstone* is commonly used without qualification to imply a quartz content of at least 85%. The proper term for such a rock is a **quartz sandstone**. Although quartz is the major mineral

FIGURE 13.5 *The angular rock fragments contained in a* breccia *indicate that the materials were not transported far from the point of origin before they were deposited. (VU/© Arthur Hill)*

FIGURE 13.6 *The rounded rock fragments in a* conglomerate *indicate that the materials were transported some distance from their point of origin before they were deposited. (VU/© John D. Cunningham)*

found in sandstones, some sandstones contain appreciable quantities of feldspar. The term **arkose** is used to describe a sandstone subtype composed of at least 25% angular to subangular sand-sized particles of feldspar. Arkoses commonly represent debris from rapidly disintegrating granite or granitic rocks that has undergone little transport or subsequent chemical weathering before being deposited and buried.

A third kind of sandstone is called a **graywacke**. The term is an old one, dating back to the early 1800s. There is little agreement about the specific definition of a graywacke and, in fact, some geologists

FIGURE 13.7a *The most common sandstone is a quartz sandstone. (VU/© A. J. Copley)*

FIGURE 13.7b *A photomicrograph of a sandstone cross section shows the individual quartz grains cemented with quartz. (Courtesy of Kathy Bruner)*

have recommended that the term be discarded. The term is still used, however, to describe a dark gray sandstone consisting of a mixture of poorly sorted, angular to subangular, sand-sized particles of quartz, feldspar, rock fragments, and significant concentrations (at least 15%) of clay minerals.

Siltstone

A **siltstone** is a rock composed primarily of silt-sized (0.004 to 0.062 mm) grains. As with sandstones, the composition of siltstones is not restricted to any mineral type. Most siltstones are intermediate in composition between the quartz-rich sandstones and the clay mineral–dominated shales and mudstones.

Shales and Mudstones

Shales and **mudstones** are the most abundant of all sedimentary rocks, constituting about 65% of the total. They are composed primarily of clay minerals, the major solid product of the chemical decomposition of igneous and metamorphic rocks, intermixed with silt- and clay-sized quartz and to a lesser extent feldspar. Shales and mudstones differ in the thickness of their individual layers. Shales characteristically split into thin layers due to the parallel alignment of the clay minerals while mudstones are more massive or thickly layered because the clay minerals are not aligned. Because of their more massive nature, mudstones tend to be more resistant to weathering and erosion. Of the two, shales are the more abundant.

In most exposures of rocks such as cliffs or road cuts, the dominance of shales is not at all obvious because they weather and erode at a relatively rapid rate. However, careful examination of most sites that expose a variety of sedimentary rocks will show that shales are the most abundant single sedimentary rock type.

NONDETRITAL SEDIMENTARY ROCKS

Nondetrital sedimentary rocks form from the soluble products of weathering and are subdivided into three types: (1) **chemical**, (2) **biochemical**, and (3) **evaporite**. The elements taken into solution in the greatest concentration as positive ions are *sodium* (Na^{1+}), *calcium* (Ca^{2+}), and *silicon* (Si^{4+}). The major negative ions resulting from chemical weathering include *sulfate* (SO_4^{2-}), *nitrate* (NO_3^{1-}), *phosphate* (PO_4^{3-}), *chloride* (Cl^{1-}), and the most abundant and important of all, the *bicarbonate* ion (HCO_3^{1-}). These ions combine to form the minerals that constitute the nondetrital rocks. In contrast to the detrital sedimentary rocks, which usually consist of a mixture of different minerals, nondetrital sedimentary rocks are usually *monomineralic;* that is, they are dominated by a single mineral. This characteristic explains why the classification of nondetrital sedimentary rocks is based primarily on composition.

Chemical Sedimentary Rocks

Limestone, the most abundant of all the nondetrital sedimentary rocks, is the major sedimentary rock type in both the chemical and biochemical categories.

rocks by volume and are made primarily of *calcium carbonate* ($CaCO_3$). The calcium carbonate in chemical limestones formed strictly by chemical precipitation from solution. The calcium carbonate in biochemical limestones, on the other hand, was produced by living organisms, plant or animal, although in practice it may be difficult to tell whether the origin was animal or plant.

Chemical Limestones Calcium carbonate is unusual in that it is more soluble in cold water than in warm water. Because the solubility of carbon dioxide increases with decreasing water temperature, cold water promoted the dissolution of $CaCO_3$ while warm water promotes its precipitation. This explains, at least in part, why shelled animals and the accumulation of shell remains are more abundant in warm waters than in cold. Although some limestones form in freshwater lakes, most limestones form from materials that accumulate in the ocean.

For the most part, the surface waters of Earth's oceans are saturated with $CaCO_3$. The materials for future limestones accumulate in the shallow offshore areas where the waters are warm. In some areas, cold ocean waters saturated with $CaCO_3$ rise from the ocean deep and flood out onto the warm, shallow ocean bottom. The combined effect of the decreasing depth and the warming of the waters results in the precipitation of $CaCO_3$ as a fine-grained carbonate mud. Should this material become a rock, it will be an example of a chemical limestone. In some areas where the carbonate mud is agitated by wave and tidal currents, the $CaCO_3$ precipitates in concentric layers onto the surface of tiny shell and mineral fragments to form spherical grains called *ooids*, which may give rise to *oolitic limestones*, another example of a chemical limestone (Figure 13.8). Another kind of limestone, *travertine*, forms by the precipitation of $CaCO_3$ from surface and groundwater. Travertine is found in fractures, on rock faces, as vein fillings, and perhaps most spectacularly, in limestone caves and caverns (Figure 13.9). Chemical precipitation, however, accounts for less than 10% of all limestones.

Dolostone Another common carbonate rock, **dolostone**, is made of the carbonate mineral *dolomite*, $CaMg(CO_3)_2$. Most dolomite forms when magnesium partially replaces the calcium in grains of $CaCO_3$. The carbonate sediments that initially accumulate are mostly composed of the mineral *aragonite*. Aragonite has the same composition as calcite, $CaCO_3$, differing

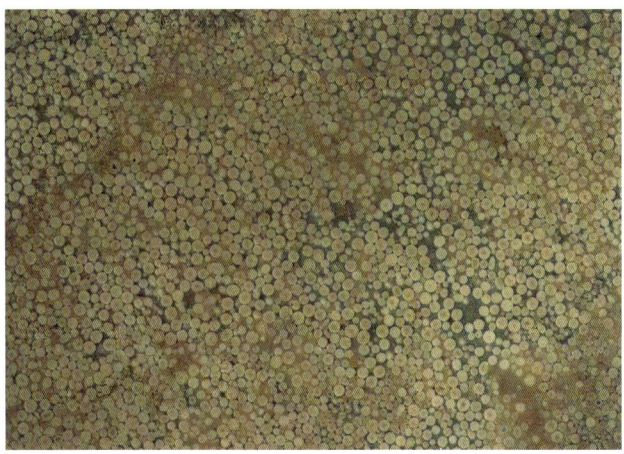

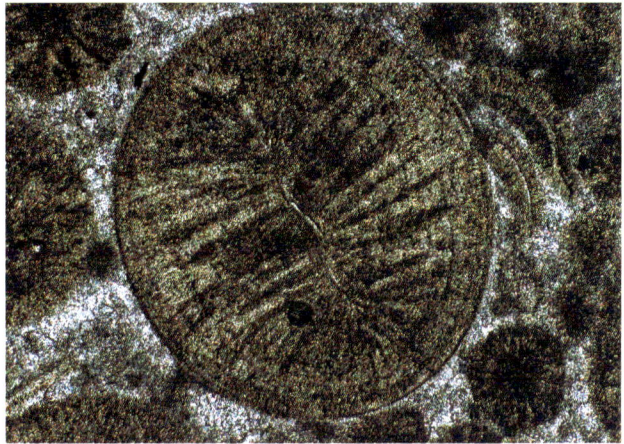

FIGURE 13.8 *An oolitic limestone tells a geologist that the waters at the time of accumulation were warm, shallow, and wave agitated. A magnification of one of the ooids (b) shows the concentric internal structure.* (a VU/© A. J. Copley, b © Ray Simons/Photo Researchers.)

only in crystal structure. If magnesium-rich solutions come in contact with aragonite, magnesium is emplaced into the carbonate lattice to form dolomite. The process is especially effective in arid regions where the evaporation of seawater preferentially increases the concentration of magnesium by precipitating calcium as gypsum ($CaSO_4 \cdot 2H_2O$). In some cases where leaching is limited to more porous sediments or where carbonate muds have been preferentially dolomitized due to their high surface area, the resultant rock will be a mixed limestone-dolostone. In situations where the leaching is intense or where the magnesium concentration is especially high, the process of dolomitization can convert all the carbonate grains to dolomite, forming a pure dolostone.

FIGURE 13.9 Speliothems, *the spectacular structures that adorn limestone caves and caverns, are made from $CaCO_3$ precipitated by the evaporation of carbonate-rich groundwater. (VU/© A. J. Copley)*

FIGURE 13.10 *A variety of microscopic siliceous shells accumulate as part of the fine* muds *or* oozes *that coat the deep ocean floor. (Courtesy of the Deep Sea Drilling Project/Scripps Institution of Oceanography)*

Biochemical Sedimentary Rocks

As we have pointed out, biochemical sedimentary rocks form from the remains of plant and animal organisms. Chert and biochemical limestone are among the most important.

Chert   Chert is a sedimentary rock composed entirely of *silica* (SiO_2). Most chert is described as being *cryptocrystalline* (Greek *kruptos*=hidden), which means that the individual crystals are so small that they are not visible under a light microscope and can only be viewed with the high magnification of an electron microscope.

Some chert forms from silica originally in solution. The dissolved silica may replace $CaCO_3$ to form the *nodules* that are commonly found within beds of limestone. In such a case, the chert would be chemical in origin. Most chert is biochemical in origin, however, and forms from the opaline shells of microscopic animals called *radiolaria*, microscopic plants called *diatoms*, and the siliceous spicules of *sponges* (Figure 13.10) that accumulate on the ocean floor. Eventually, they are transformed into *bedded chert*.

A common variety of chert, showing a more dense texture, black color, and perfect choncoidal fracture, is called *flint*. For hundreds of years, the term has been used to refer to any hard rock. Ever since the evolution of *Homo habilis* 2.5 million years ago, flint has been prized for the manufacture of tools such as knife blades and points for spears and arrows (Figure 13.11). Other varieties of chert are the red *jasper* and variously colored, banded *agate* where the coloration is due chiefly to impurities of iron oxide.

Biochemical Limestones Most limestones (about 90%) are **biochemical** in origin and form from $CaCO_3$ produced by the biological activity of animals and plants. Many marine animals such as *clams, oysters, snails,* and *coral* as well as many planktonic (floating) organisms secrete shells of $CaCO_3$. When they die, the shells accumulate on the ocean floor. Should this material enter the geological record as a limestone, it forms a biochemical limestone.

Many species of marine plants also secrete microscopic plates made of $CaCO_3$ on their outer surfaces to strengthen and protect their body parts. One common example found extensively on the bottom of Florida Bay is *Penicillus,* commonly called "shaving brush algae" because of their shape (Figure 13.12). Upon the death and decomposition of the plant, the tiny plates accumulate on the bottom of the bay as a fine-grained mud that may be converted into a biochemical lime-

FIGURE 13.11 *From prehistoric time, a form of chert called* flint *has been used to fabricate various kinds of points and blades. (VU/© John D. Cunningham)*

FIGURE 13.12 *A common alga growing in Florida Bay, the so-called shaving brush alga, secretes microscopic plates of $CaCO_3$ to strengthen its parts. When the algae die and decompose, the plates are released to become a component of the carbonate muds that coat the bottom of the bay. (Courtesy of Bill Grady)*

stone. Studies have shown that much, perhaps most, carbonate mud is derived from the breakdown of the fine skeletal materials of algae and bryozoans and the fragile frameworks of other organisms.

The limestone that forms from carbonate muds composed of microscopic algal materials is quite different in appearance from the limestone that is produced from shell debris. Although both are examples of biochemical limestones, the limestone that forms from the accumulated shell debris is *fossiliferous* with many of the shells and shell fragments identifiable within the rock. An extremely fossiliferous variety of limestone called **coquina** is composed almost entirely of loosely cemented shells and shell fragments (Figure 13.13). On the other hand, the rock formed from muds composed of microscopic algal plates appears as a dense, fine-grained limestone with no visible plant or animal remains.

Evaporitic Sedimentary Rocks

As the name implies, **evaporites** form by the evaporation of water. The two major evaporite minerals are **gypsum** ($CaSO_4 \cdot 2H_2O$) and **rock salt** or **halite** ($NaCl$). Most evaporites form in hot arid to semiarid regions where water may evaporate in seasonal environments such as desert lakes or along arid coastlines called *sabkhas* where seawaters continuously evaporate on shallow, nearly landlocked embayments. Most of the salt you use to season your food came from such deposits. Because high temperature and low relative humidity are required for their formation, evaporites are important indicators of paleoclimate.

Organic Sedimentary Rocks

An **organic** category is often included in a sedimentary rock classification scheme primarily to classify **coal**. Some textbooks omit coal altogether while others classify it variously as biochemical, nondetrital-biochemical, nondetrital-accumulated plant debris, or chemical-organic.

The problem of coal illustrates the inherent fallibility of all classification schemes; there will always be something that doesn't quite fit. By definition, coal is a rock containing 50% or more organic material by *weight*. Because of the difference in the densities of organic and inorganic materials, the *minimum* volume of organic material in coal is about 60%. Most coal—and all minable coal—forms from plant debris converted into *peat* within freshwater swamps and subsequently converted after burial into *lignite, subbituminous, bituminous,* or *anthracite* coal (Figure 13.14). Note that the major materials from which coal is made are *not* products of weathering but rather the remains of once-living plants. In the next chapter, we will see

FIGURE 13.13 *A coquina is an example of a biochemical limestone consisting almost entirely of cemented shell and coral fragments.* (VU/© Albert J. Copley)

FIGURE 13.14 *Coals form largely from the wood of trees growing in swamps dominated by fresh water.* (Courtesy of J. F. Elder/USGS)

that metamorphic rocks form by the solid-state recrystallization of an existing rock through the action of heat, pressure, and/or chemically active fluids. Although the transformation of peat into other types of coal takes place at significantly lower temperatures than are required for most metamorphic processes (less than 400°F or 200°C), and might not be considered metamorphism by a metamorphic petrologist, coal geologists regard this transformation as a metamorphic process. By their definition, coal is a metamorphic rock even though it is found associated with sedimentary rocks. Many geologists who are not involved in coal geochemistry would not classify coal as a metamorphic rock. Instead, because coal is obviously not an igneous rock, they would classify it as a sedimentary rock by default.

SPOT REVIEW

1. Why are shales the most abundant of all the various kinds of sedimentary rocks?
2. What possible differences in the origin of the sediments do the compositions of quartz sandstones, arkoses, and graywackes suggest?
3. In what way do quartz sandstones and cherts differ? In what way are they the same?
4. What different kinds of materials make up biochemical sedimentary rocks?

DEPOSITIONAL ENVIRONMENTS

The products of weathering are carried away by the processes of erosion to be deposited in a number of different depositional sites. Any site where the products of weathering can accumulate is a **depositional environment**. Depending on whether the sediment is deposited on land, at the margins of a continent, or in the ocean, the sites and the sedimentary rocks produced from them are categorized as one of three basic types: (1) *continental*, (2) *transitional*, and (3) *marine*.

Continental Deposits

Continental deposits vary depending upon the particular agent of erosion and the depositional site.

Fluvial Fluvial deposits are materials laid down directly from streams in a number of depositional

FIGURE 13.15 *Any sediment deposited by a stream is called* fluvial. *One of the most common examples of fluvial deposit, the* point bar, *is the sand that accumulates on the inside bend of a meandering stream.* (© E. R. Degginger)

sites including *floodplains, channels, point bars,* and *levees* (Figure 13.15). Except for deposits of desert streams, most fluvial deposits are well-sorted sands and exhibit well-rounded grains. Materials deposited by desert streams are usually less well sorted and more angular than most other fluvial deposits because they are transported relatively short distances and deposited rapidly.

Glacial Glacial deposits include materials that accumulate under ice sheets, at the margins of melting ice, and from glacial meltwaters (Figure 13.16). Except for glacial meltwater deposits, glacial deposits are almost always poorly sorted.

Eolian Eolian deposits are materials deposited by the wind. They are found both as sand-sized accumulations in *dunes* (Figure 13.17) and also as silt- and clay-sized materials in *loess* (Figure 13.18).

FIGURE 13.16 *Glacial deposits range from materials deposited directly from the melting ice to those deposited from meltwater streams. (Photos courtesy of D. R. Crandell/USGS)*

FIGURE 13.17 *Accumulations of windblown sand are called* dunes. *(Courtesy of E. D. McKee/USGS)*

FIGURE 13.18 *Accumulations of wind-transported silt and clay form a fine-grained deposit called* loess. *(VU/© John D. Cunningham)*

Some of the spectacular sandstone cliffs in the deserts of the southwestern United States, such as the Navajo and Entrada Sandstones, are examples of ancient dune deposits (Figure 13.19). Because of the extreme degree of grain-to-grain abrasion, wind-transported sand grains commonly become well rounded. The most extensive loess deposits on Earth are found in northern China where fine-grained, wind-transported materials coat the landscape with a blanket that is several hundred feet (meters) thick in places.

Lacustrine Lacustrine deposits accumulate in lakes. Geologically, all lakes are short-lived. Sediment introduced by streams will eventually fill the lake basin, converting the lake environment first to a marsh, next to a swamp, and then to a bog. With the complete infilling of the bog, the original lake is eliminated.

Paludal Paludal deposits accumulate in *marshes* and *swamps* (refer to Figure 13.18). Marshes are *wetlands* dominated by grasses and other nonwoody plants. Swamps, on the other hand, are wetlands dominated by wood-rich plants. Because of the slow movement of the water within wetlands and the chemistry of the waters, accumulating plant debris is partially preserved and converted into peat. Depending upon the relative concentrations of preserved organic material and mineral matter, the accumulated material will be the precursor of a range of rock types from organic-rich *black shales* to *coal*.

FIGURE 13.19 *Sandstones formed from dune deposits, such as those exposed in Zion National Park, commonly exhibit well-developed cross beds. (Courtesy of W. B. Hamilton/USGS)*

Transitional Environments

A number of **transitional** settings between the continental and marine depositional environments provide much of the material for marine sedimentary rocks. These environments include *deltas, beaches, barrier islands, coastal wetlands,* and *mudflats*.

Deltas A coastal **delta** accumulates as a wedge of stream-fed continental detrital sediment, extending the reaches of the land into the marine environment (Figure 13.20). At some point, the seaward development of the delta is terminated by ocean currents that become strong enough to pick up and redistribute the materials on the ocean floor. Deltas may also form where streams enter freshwater lakes.

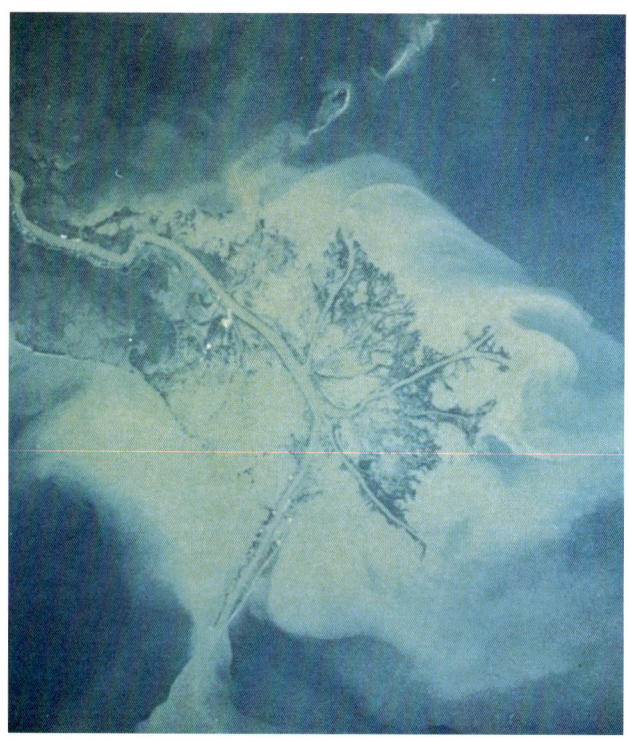

FIGURE 13.20 *Most of the present load of the Mississippi River is being deposited in the Balize delta lobe. Note the plume of suspended load as the lower density fresh river water spreads out over the surface of the denser saline water of the Gulf of Mexico. (Courtesy of NASA)*

Beaches Beaches are coastal accumulations of unconsolidated materials (Figure 13.21). The portion of the beach nearest the water is usually a relatively flat surface, sloping seaward. Inland from the beach, windblown sand may accumulate as dunes, or the land may rise more abruptly as a cliff. Along shorelines characterized by high-energy waves, the beach may consist of larger rock fragments, up to boulder size (Figure 13.22).

Barrier Islands Barrier islands are enlongated sand islands that parallel the shore and are separated from the mainland by a lagoon or a coastal wetland. The islands commonly have dunes at the center and a beach on the seaward side (Figure 13.23). The origin of barrier islands is a subject of debate. Some barrier islands form from land-derived sand transported along the coast by longshore currents. The most widely accepted idea is that most barrier islands originated as beaches and beach ridges during the last glacial episode when the ocean was at a lower level. As the ice age waned and sea level began to rise, the beach ridges migrated landward. Eventually, the ocean encroached into the low area behind the beach ridges isolating them from the mainland (Figure 13.24).

FIGURE 13.21 *The beach sands that accumulate along the continental margin are often the exposed edge of a wider belt of sand that parallels the shore. (Courtesy of S. J. Williams/USGS)*

FIGURE 13.22   *Along high-energy beaches, finer particles are removed by the waves, leaving behind particles that are often cobble size and larger.*

FIGURE 13.23   Barrier islands *can be found in many areas along the coastlines of the Atlantic Ocean and the Gulf of Mexico. (VU/© Frank M. Hanna)*

Coastal Wetlands and Mudflats   As the lagoons behind the barrier islands fill with sediment derived from the land and from the ocean by way of the tidal inlets, the water shallows and the bottom of the lagoon become exposed at low tide as a **mudflat** (Figure 13.25). Once the bottom is exposed, grass begins to grow on the mudflat and traps additional sediments. The sediment surface builds up rapidly and eventually emerges above high tide and develops into a *marsh*. As larger plants and trees begin to encroach into the wetland, the marsh changes into a *swamp*.

Marine Environments

The most important of all the depositional environments is the ocean bottom. Most of the products of weathering and erosion, including materials initially deposited in continental and transitional depositional environments, eventually accumulate on the ocean floor. Once deposited, the sediments are the source materials for *marine* sedimentary rocks, the most voluminous of sedimentary rocks.

The shallow waters of the **continental shelf** are the major depositional site for most of these materials, although some sediments are carried beyond the

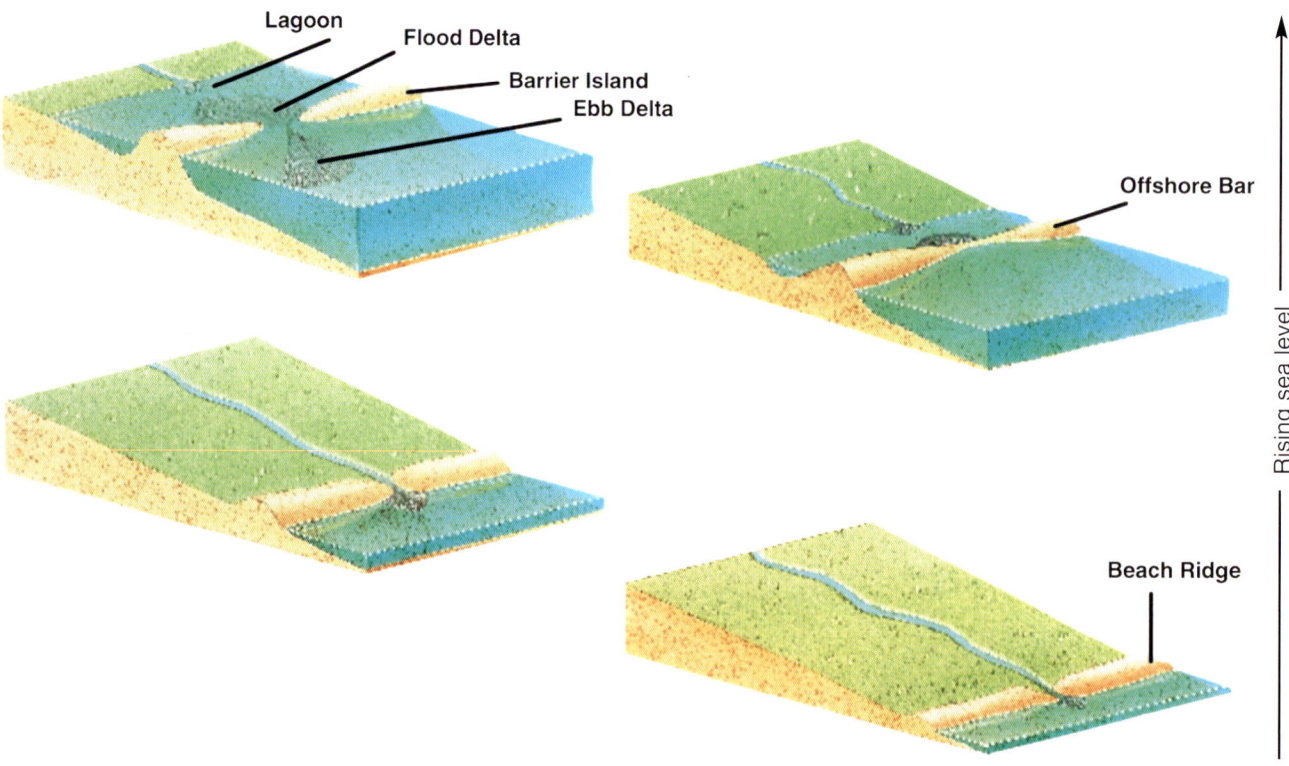

FIGURE 13.24 *Most of the barrier islands that exist along the Atlantic coast and at other places around the world are believed to have formed as sea level rose following the end of the Pleistocene Ice Age.*

FIGURE 13.25 *The process that culminates in the formation of coastal wetlands is initiated by the exposure of the near-shore ocean bottom as a* mudflat. *(VU/© John D. Cunningham)*

edge of the continental shelf to accumulate on the **continental slope** and on the **abyssal floor**. Ocean currents distribute the sediments on the surface of the continental shelf in bands parallel to the shoreline (Figure 13.26). The coarse materials, usually sand, accumulate nearest the land with the finer-grained silts and clays being carried farther seaward. Beyond the influence of the detrital sediments, carbonates may accumulate on the ocean bottom, chiefly where the waters are warm enough to promote the chemical precipitation of calcium carbonate and preserve the remains of shells. In some cases, the clay-sized sediments are transported seaward beyond the zone of carbonate accumulation.

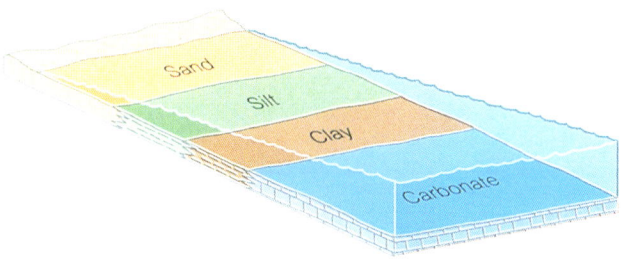

FIGURE 13.26 *The combined effects of the various coastal currents distributes the sediments in strips parallel to the coastline with the largest particles accumulating nearest the shore and decreasing in size seaward. When the water is warm, carbonates accumulate beyond the influence of the clastic sediments.*

LITHIFICATION

The process by which accumulated products of weathering are converted into a rock is called **lithification**. Sedimentary deposits normally do not lithify at Earth's surface although Sun-dried muds and newly deposited evaporite minerals can achieve a certain rock-like character. For the most part, the processes of lithification require the sediments to be buried and subjected to increased temperatures (up to a maximum range of 400°F to 570°F or 200°C to 300°C) and pressures of less than a kilobar (about 14,000 lb/in^2). Once buried, the sediments undergo lithification. The major processes of lithification are *compaction* and *cementation*.

Compaction

The initial change that takes place with burial is **compaction**, which forces the individual grains closer together, decreasing both the volume and the porosity of the sediment layer. When the layers consist primarily of materials that are sand size and coarser, the volume reduction due to compaction is relatively small because of the dominance of relatively rigid, equidimensional grains. If, on the other hand, the material is mud, which consists primarily of clay minerals with silt-sized quartz, the volume reduction can approach 50% as the water is squeezed from the mass of platelike clay mineral particles.

Compaction is the major mechanism by which muds are transformed into *mudstones* and *shales*. As the water is physically expelled from the layer, compacting the clay crystallites closer together, the individual clay particles begin to adhere to each other. In cases where mud is deposited along with larger grains, such as sand-sized quartz, compaction of the clay minerals may create a clay matrix that holds the larger grains together to form a sandstone.

Cementation

Cementation is the dominant process by which detrital materials consisting of sand-sized or larger particles are lithified. During cementation, minerals precipitate from solution in the pores between the grains. This precipitated material "cements" the grains together. The most common cements are *calcite* ($CaCO_3$), *quartz* (SiO_2), and *iron oxide* (Fe_2O_3), all provided by the chemical weathering of rocks.

Coarse-grained nondetrital materials such as shells and shell fragments are also lithified by cementation to form biochemical limestones. The cements that form most biochemical limestones are usually carbonate minerals such as *calcite* or *aragonite* (both calcite and aragonite consist of calcium carbonate, $CaCO_3$) and *dolomite*, $CaMg(CO_3)_2$. In some cases, *gypsum* $CaSO_4 \cdot 2H_2O$, is the cementing agent.

Fine-grained, nondetrital sediments such as those that would produce biochemical or chemical limestones and evaporites usually lithify by a combination of compaction and cementation. Lithification begins as the materials are buried and forced closer together with the subsequent expulsion of water and collapse of pore space. The sediments continue to lithify as additional minerals precipitate from solution and the individual mineral crystals grow and interlock. Biochemical rocks characteristically undergo lithification by cementation as calcite precipitates from solution and fills the pore spaces between the shells and shell fragments.

SPOT REVIEW

1. Which of the various sedimentary environments is responsible for the greatest volume of sedimentary rocks? Why?
2. Which depositional environments are considered "fluvial?"
3. What kinds of sediments lithify primarily by cementation? By compaction? By a combination of cementation and compaction?

SEDIMENTARY FEATURES

Many internal features associated with sedimentary rocks are the product of the environment under which the materials accumulated; therefore these features reveal a great deal about the conditions that existed within the depositional environment at the time of accumulation. Geologists are able to establish the history of Earth by studying the kinds and distributions of sedimentary rocks and the various sedimentary features contained within them.

Beds

All sedimentary rocks, regardless of type or mode of formation, are *layered* or *stratified* (Figure 13.27). The principal layers, which are called **beds** (strata), consist either of detrital rock and mineral fragments deposited by a stream, glacier, or the wind or of mineral grains precipitated from solution. According to the law of original horizontality, the sediments always accumulate in horizontal or near-horizontal layers, regardless of whether the accumulations occur on the bottoms of lakes, in stream floodplains, in desert basins, or, most commonly of all, in marine environments. If sedimentary rocks are observed in road cuts or cliffs with the beds tilted from the horizontal, it means that some force must have acted upon the original layers to change their original horizontality.

Other sedimentary features contained within the individual beds allow geologists to reconstruct certain aspects of the conditions that existed within the depositional basin when the sediments accumulated. Each detail that can be understood adds one more bit of information to the overall goal of establishing the history of Earth.

Ripple Marks

Currents of water or air moving across the surface of a recently deposited layer of sand-sized materials may produce standing waves or dunes on the surface (Figure 13.28). If the currents move in a single direction, as occurs with stream flow, the wind, or a current flowing on the ocean bottom, waves that develop on the surface of the sediment layer will be oriented perpendicular to the direction of the current and will be **asymmetrical** in cross section with the steep side in the direction of current flow (Figure 13.29). If this layer of sediment is quickly buried by the influx of another layer of sediment, the original waveforms may be preserved on the top surface of the bed as **current ripple marks**. Like cross-bedding, current ripple marks are used to interpret the direction of current flow.

When sediments in shallow coastal areas are exposed to the oscillating action of waves, the waveforms that develop on the surface of the sediments are **symmetrical** (Figure 13.30). Again, if an influx of sediment buries the first layer, the waveforms may be preserved as **oscillation ripple marks.** Oscillation ripple marks indicate that the materials accumulated in a shallow, near-shore environment.

Cross-Bedding

As sediments are moved by flowing water along the bed of a stream or are blown across a dune by the wind, grains constantly move along the top surfaces

FIGURE 13.27 *The one feature common to all sedimentary rocks is the bed. (VU/© Steve McCutcheon)*

FIGURE 13.28 *The wind forms asymmetrical ripple marks on the surface of sand. (Courtesy of E. D. McKee/USGS)*

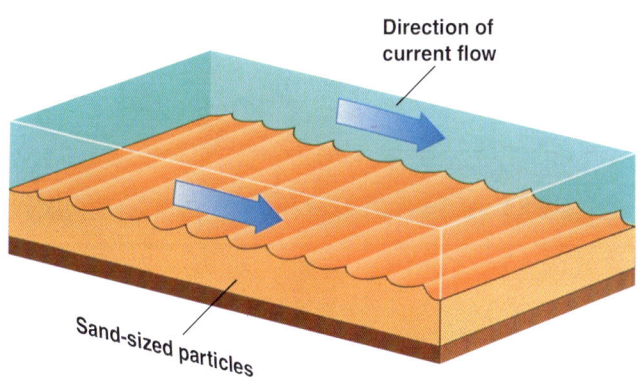

FIGURE 13.29 *Asymmetrical ripple marks are generated as water or wind moves across the surface of layers of sand. The ripples form perpendicular in the direction of flow with the steep side of the ripple in the downflow direction.*

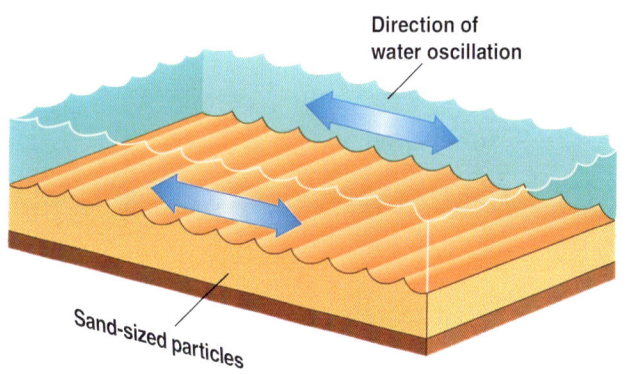

FIGURE 13.30 *Symmetrical ripple marks form where layers of sand are exposed to the oscillating currents in the shallow water near shore.*

of the bed and cascade down the leading edge of the deposit (Figure 13.31). The result is the progressive development of **cross-bedding** or layers that are oriented at an angle to the major bedding in the downstream direction. Because the cross-beds are contained within the main bedding and are an integral part of it, their formation does not defy the law of original horizontality.

The orientation of the cross-beds records the direction of water or wind movement within the original depositional environment. Spectacular examples of cross-bedding preserved in eolian sandstones record the direction of wind flow that created dune fields in some prehistoric desert (Figure 13.32). Because the wind commonly changes direction, the patterns of cross-bedding in eolian sandstones can be quite intricate as they record the changing wind directions. In similar fashion, the cross-beds within a fluvial sandstone can be used to determine the direction of stream flow.

Graded Bedding

When clastic sedimentary materials carried in moving currents settle out in standing water or in very slow moving water, the largest particles settle out first, followed by progressively finer-grained materials. The

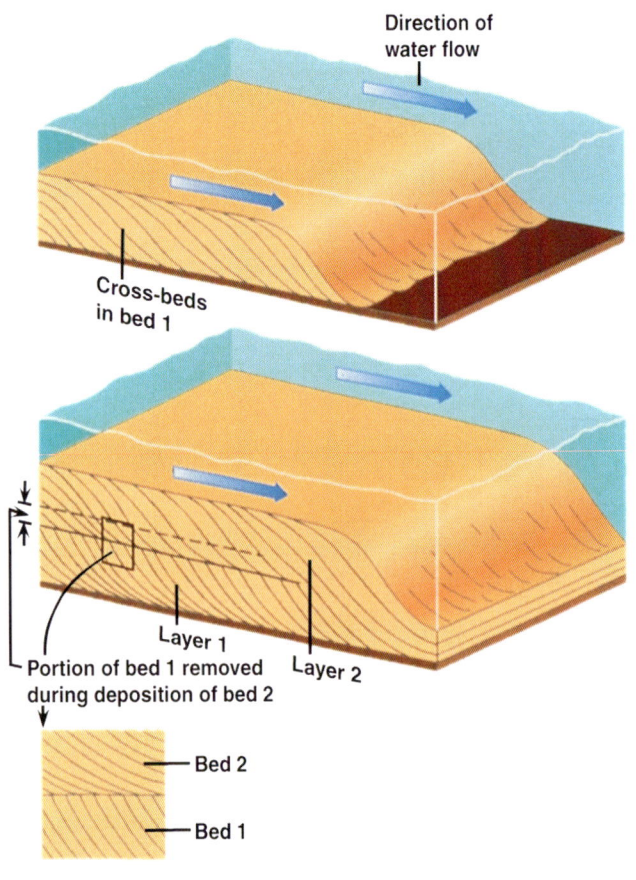

FIGURE 13.31 *The direction of movement of water or wind is recorded by the cross-beds that form within layers of sand.*

FIGURE 13.32 *The varying directions of the cross-beds in eolian sandstones record changes in the direction of wind flow in the ancient desert. (Courtesy of W. B. Hamilton/USGS)*

result is a feature called **graded bedding** in which the grain size changes from coarse at the bottom to fine at the top (Figure 13.33). Graded bedding is common in many sedimentary beds, but is especially well developed in certain kinds of deposits. An excellent example of graded bedding is found in the sedimentary rocks called *turbidites*. Sediments carried by ocean currents beyond the edge of the continental shelf are deposited on the continental slope. In many cases, these unconsolidated sediments rest nearly at the angle of repose. This precarious balance may be upset by a shock such as a local earthquake or by the gravitational pull of the deposit's own mass and set into motion downslope as a *turbidity flow*. The turbidity flow carries the sediment into deep water and onto the abyssal ocean floor. As the sediments carried by individual turbidity flow settle out, graded bedding develops in the layer of sediment. Once lithified, these trubidites may eventually be uplifted and exposed on land (Figure 13.34).

Another example of graded bedding commonly results from the seasonal accumulation of sediments within lakes located in temperate climates. During the spring and early summer when the discharge and loads of streams flowing into the lakes are high, the materials that accumulate on the lake bottom will be relatively coarse-grained. As the year progresses and the amount of rainfall lessens, usually during late summer and early fall, stream discharge declines, and the sediments brought into the lake and deposited decrease in both volume and size. During the winter when the streams and lakes may freeze over, the influx of sediments come to a halt. All that remains to settle out are the very finest materials still held in suspension within the lake waters. In some cases, there may be a period during late winter when no sediments remain in suspension and accumulation on the lake bottom ceases. With the spring thaw, the cycle begins again.

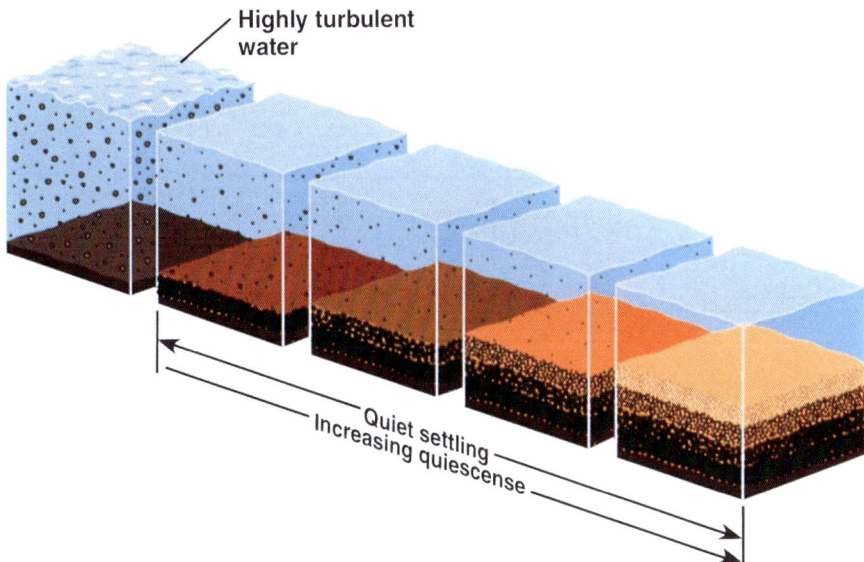

FIGURE 13.33 Graded beds *form as progressively smaller particles settle out of water with time.*

The result of the yearly cycle is a single graded bed called a *varve* (Figure 13.35). The age of the lake can be determined by counting the number of varves in a core taken from the lake bottom. The thickness of individual varves can also be used to derive other kinds of information such as variations in yearly rainfall throughout the life of the lake, and the pollen contained within the lake sediments can reveal changes in the vegetation around the lake.

Mud Cracks

Everyone has seen the polygonal, curling plates of semiconsolidated mud in the bottom of a dried mud puddle. If they are preserved by the rapid accumulation of another layer of sediment, they may appear in the rock record as **mud cracks** (Figure 13.36). Mud cracks are evidence that the area was once sufficiently shallow to undergo periodic flooding and evaporation, perhaps a flat coastal area or an inland seasonal lake, and usually suggest a dry, warm environment that promoted the drying of the sediment surface.

Animal Trails

Animal marks left in sediment range from the rather nondescript trails of worms crawling across the sediment surface or boring into the upper few inches or feet of the sediment layer to the more spectacular footprints of dinosaurs (Figure 13.37). Although **animal trails** are not the result of the process of sediment accumulation, they provide information not only about the depositional environment at the time

FIGURE 13.34 Graded beds, *called* turbidites, *which formed from materials that originally accumulated on the deep-ocean floor, may be uplifted and exposed at the surface following mountain-building episodes.*

FIGURE 13.35 *Graded beds that accumulate in lakes are called* varves. *Because each varve represents one year's accumulation, the varves can be used to calculate the age of the lake and to determine climate changes that occurred during the lifetime of the lake. (Courtesy of P. Carrara/USGS)*

FIGURE 13.36 *Clay mineral-rich sediments commonly shrink and crack upon drying to form* mud cracks, *which can in turn be preserved during lithification. (Photo courtesy of P. Carrara/USGS)*

of sediment accumulation, but also the kinds of creatures that existed in the area, the climate, and the means of locomotion of these long-extinct creatures.

SPOT REVIEW

1. Which sedimentary features can be used to determine the direction of sediment transport?
2. How do the conditions for the formation of symmetrical and asymmetrical ripple marks differ?
3. What is a "varve" and how does it form? What kinds of historical information can be obtained from the study of varves?

INTERPRETATION OF ANCIENT DEPOSITIONAL ENVIRONMENTS

A primary goal of geology is to determine the history of Earth. Considering their relatively small crustal volume, sedimentary rocks provide a disproportionately large amount of historical information. As each layer of sediment accumulates, it records information about the depositional environment, such as the kinds and relative abundances of minerals, sedimentary features, bed geometry, and the remains of plants and animals. The following examples illustrate how geologists utilize sedimentary rocks to determine Earth history.

Applying Hutton's concept of uniformitarianism as exemplified by the expression "the present is the

Sedimentary Rocks 415

FIGURE 13.37 *Some information about the depositional environment can be obtained from the trails left by animals that lived in the area. (a VU/© John D. Cunningham, b VU/© Scott Berner)*

key to the past," geologists have achieved considerable success in deciphering the history contained in sedimentary rocks. Fossils are one of the parameters that provide us with a great deal of information. By studying the habitats of modern plants and animals and assuming that comparable ancient organisms has similar affinities, we can use fossils to interpret the ancient depositional setting. For example, in modern marine environments, green plants, plant eaters, and animals that utilize plankton (floating organisms) as a food supply are abundant in water penetrated by sunlight, the *photic zone*. Below the photic zone, scavengers and sediment feeders become dominant. Filter feeders such as the colonial bryozoans that acquire their food by filtering seawater through their bodies can only live in clear water (Figure 13.38).

The shapes of animal shells reflect the degree to which the water is agitated. The shells of animals living in agitated waters such as clams are usually shorter, rounder, and thicker than those of animals living in quiet waters. Some colonial species modify the shapes of their colonies in response to water depth. For example, branching corals are usually found in shallow water. As the water deepens, the coral colonies become flat and platy while colonies in very deep water are small (Figure 13.39). When fossil remains with these shell or colonial forms are observed, geologists can assign the appropriate depositional setting to the layer of rock in which they are found.

FIGURE 13.38 *Animals such as scallops acquire their food by filtering seawater as it passes through their bodies. (VU/© Marty Snyderman)*

FIGURE 13.39 Corals *are among the most prolific rock-forming animals living in the ocean. (Courtesy of S. J. Williams/USGS)*

The specific rock lithology can often be used as an indicator of a certain depositional setting. For example, modern studies have shown that although carbonate materials are produced in most aquatic environments, limestones form primarily from materials that accumulate in marine environments where stream-borne detritus is excluded, such as the Florida Keys or the Bahamas. Within carbonate environments, lime muds preferentially accumulate in sheltered environments such as Florida Bay where water movement is restricted (Figure 13.40). Where waves, tides, and longshore currents agitate the water and winnow out the muds, such as on the seaward side of the Florida Keys, the sediment consists primarily of coarse carbonate particles. The limestones that would form from these two sediments would be quite different in appearance. Lithification of carbonate muds would produce a dense, fine-grained limestone while the materials accumulating in the more agitated waters would form a coarse-grained, highly fossiliferous limestone. Note that these two modern depositional environments (and the rocks that would form from them) are separated only by the narrow width of the Keys.

Sedimentary features provide important information about the depositional environment. In some cases, the same sedimentary feature can provide data at different scales. For example, the attitude of the cross-bedding in a single bed of a fluvial sandstone can be used to determine the direction of stream flow within the stream in which the sand accumulated. By combining the cross-bed orientation data for a large number of fluvial sandstones deposited within a

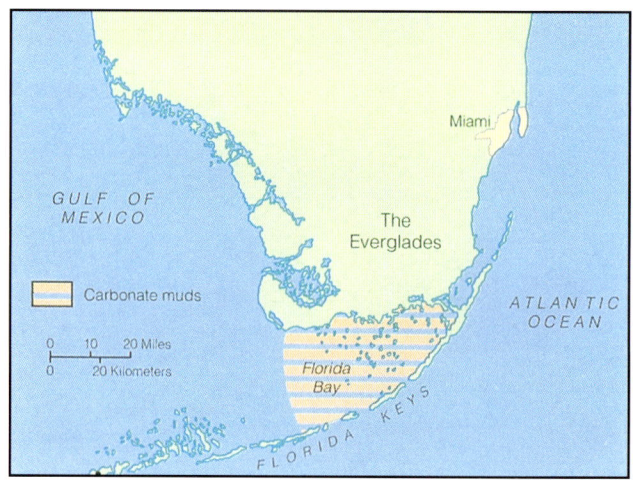

FIGURE 13.40 Florida Bay *is an excellent example of a marine environment accumulating carbonate muds. Because of the shallowness of the bay and its isolation by the Keys, the temperatures and salinities of the water in the bay are generally higher than normal for seawater. As a result, the kinds and numbers of both animals and plants living in the bay are significantly fewer than in the open Gulf of Mexico. (Photo courtesy of Bill Grady)*

drainage system, we can determine the extent of an ancient delta (Figure 13.41). Perhaps the most impressive examples of cross-bedding are exhibited in eolian sandstones such as those exposed in Zion National Park where the complex patterns of cross-bed orientation record the changing directions of sand transport within the dunes of an ancient desert (Figure 13.42).

Certain rocks, such as coal, can provide information about the chemistry of an ancient depositional environment. Coal consists primarily of partially preserved wood that initially accumulated as peat in a

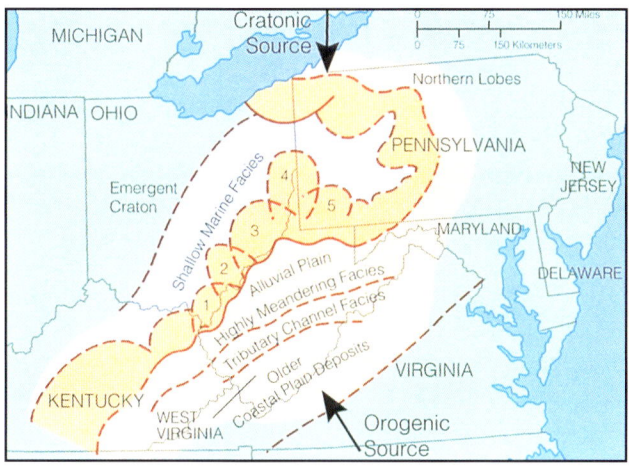

FIGURE 13.41 *Maps prepared by plotting the directions of stream flow based on the cross-beds found in fluvial sandstones have allowed geologists to recreate the patterns of deltaic deposition.*

swamp. As the wood decomposed, mineral matter originally contained in the wood accumulated within the peat. The relative amounts of preserved wood and mineral matter in the peat are a function of the pH of the swamp matter. At pH values above 5 to 6, the amount of mineral matter in the peat dominates over that of preserved wood. When such a peat lithifies, it forms an organic-rich black shale. As the swamp water pH decreases, the relative portion of preserved wood increases, eventually resulting in the formation of a coal-forming peat. Initially, however, the high mineral matter content results in the formation of a poor-quality coal. As the pH of the swamp water decreases, the relative amount of preserved wood increases, and the quality of the resultant coal improves. High-quality coals with organic contents in excess of 90% by weight form when swamp waters have a pH lower than about 2.5

Walther's Law

The change in rock lithology within a vertical sequence of rocks may record an event such as a change in sea level. A principle called **Walther's law** states that the vertical sequence of rocks observed at any locality reflects the lateral sequence of depositional environments that existed side by side perpendicular to the shoreline at any one time. To illustrate, refer back to Figure 13.26, which is an idealized illustration of the distribution of depositional environments parallel to a coastline. Sand dominates the sediments in the environment immediately adjacent to the coastline. Seaward, the sand-sized materials grade (change) laterally to the silt- and clay-sized sediments that dominate the intermediate environment. These materials, in turn, grade seaward into the carbonates that usually characterize the outermost environment. It is important to note that these three different kinds of sediments are accumulating *simultaneously* in a *single* layer on the ocean floor. It is equally important to note that when these sediments are eventually lithified, the rocks will be in *lateral continuity* within a single layer; that is, they will grade laterally into each other. The sandstone will grade laterally into shale, which will in turn grade into limestone. Rock bodies that differ in lithology from other beds that are deposited at the same time and in lateral continuity are called *facies*.

FIGURE 13.42 *Because the wind can change direction over relatively short periods of time, the orientations of cross-beds in eolian sandstones are more variable than those found in stacked fluvial sandstones.*

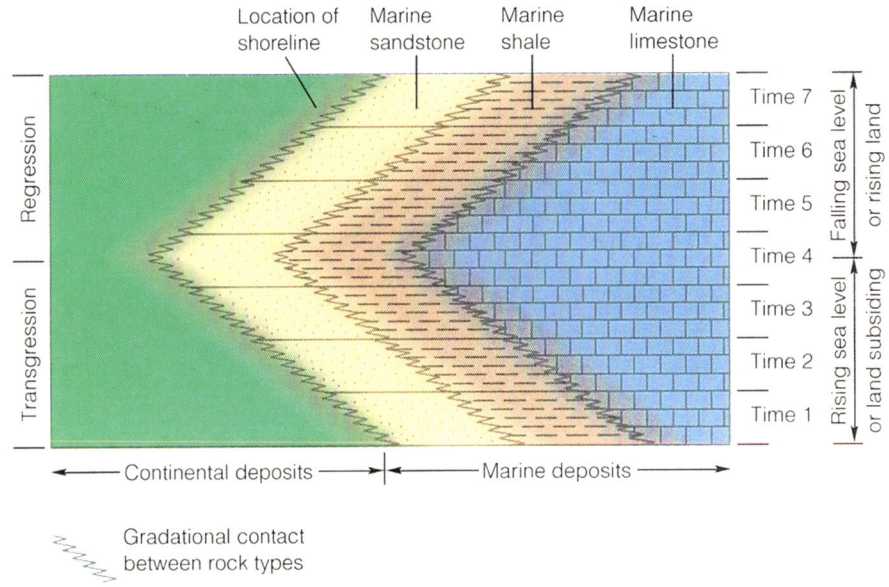

Figure 13.43 *The stratigraphic (vertical) sequence of marine rocks exposed in an outcrop provides information about the relationship between the elevation of the land surface and sea level during the period when the sediments were accumulating.*

During a period of rising sea level, the shoreline transgresses or moves inland, and the depositional environments parallel to the shoreline move inland correspondingly. As a result of the transgression, new carbonate deposits that will in time give rise to limestone are positioned over the shales and sandstones of the original depositional environments. During a period of falling sea level, when the sea transgresses or withdraws from the land, the opposite process occurs. Thus, during a transgression, new sediments that will eventually give rise to sandstones are deposited over the shales and limestones of the earlier depositional environments.

Consider what would happen if sea level rose due either to a subsidence of the land or to an actual rise in sea level (Figure 13.43). As the coastline moved inland, each of the three depositional environments would follow or *transgress,* with each environment moving *over* the adjacent landward environment. Note that, in time, the carbonate depositional environment could eventually be positioned above the original location of the sand-dominated environment.

The transgression would be recorded by a vertical sequence of rocks with more seaward rocks positioned over more landward rocks. The sequence from bottom to top would be

sandstone → shale → limestone

Note that as Walther's law predicts, the sequence duplicates the distribution of depositional environments from the land seaward. Should the process reverse and the sea *withdraw* from the land, a process called *regression,* the sequence of rocks recording the regression would also reverse and form a regressive sequence with more landward rocks positioned over more seaward rocks (refer to Figure 13.43).

SPOT REVIEW

1. How is the concept that the present is the key to the past used to understand the historical significance of sedimentary rocks?
2. Give an example of the application of Walther's law.
3. What kinds of features within a sedimentary rock can provide information about the climate in the area of the depositional environment at the time of sediment accumulation?
4. What kinds of information contained within sedimentary rocks allow geologists to determine whether a particular sequence of rocks was of continental or marine origin?

ENVIRONMENTAL CONCERNS
(Sediments)

Sediment Pollution

Sediments are often the cause of serious pollution problems. Any disruption of the natural land surface has the potential to produce sediments in such volume that they may pollute adjacent land surfaces and water bodies. For example, denuding the land of vegetation during construction, forest clear-cutting, or the conversion of land to agricultural use have all been cited as sources of sediment that pollute streams and lakes. The high suspended loads that are produced as the sediment enters the water body may harm or destroy aquatic life. The useful life span of reservoirs and dams constructed for the generation of hydropower or flood control may significantly shorten as they fill with increased volumes of sediment, while sediment flushed into drainage canals and aqueducts may clog the systems and render them ineffective.

In cases where the generation of sediment cannot be prevented, control measures can be initiated to minimize the potential for pollution. The use of contour plowing significantly reduces the amount of soil lost from agricultural land. The replanting of clear-cut forests and the judicious placement of access roads can reduce the sediment load generated by logging operations. Many urban areas have enacted legislation that requires the establishment of diversion channels and sediment collection ponds in areas undergoing construction. Similar laws have been applied to surface mining operations where large areas are disturbed and large volumes of sediment can be generated when piles of unconsolidated rock materials are exposed to heavy rains and snow melts.

It seems unlikely that the production of sediment by a wide range of human activities can be totally eliminated, and it must therefore be accepted as part of our daily activities. The production of sediment can be controlled, however, and the potential environmental impacts of sediment pollution can be minimized.

Engineering Problems

Sedimentary rocks, especially shales and limestones, may potentially present several engineering problems, including difficulties at construction sites and surface subsidence.

Shales Of the various kinds of sedimentary rocks, shales are, environmentally, the most troublesome. Shales are so prone to failures of various kinds that a design and construction engineer views even the presence of shales in a construction site as a sign of potential trouble. Unfortunately, shales are the most abundant sedimentary rock type.

Most engineering problems arise when poorly cemented shales and mudstones are cyclically wetted and dried, especially if they contain expandable clay minerals. Such rocks can repeatedly absorb and release water, causing the rock to cyclically swell and shrink. Should shale be exposed in the foundation for a building, bridge, or dam, the foundation may be exposed to stresses of sufficient magnitude to result in structural failure.

Road cuts containing expandable shales are prone to slumps and rock slides as the layers of shale absorb water, lose cohesion, and develop slippage planes. The situation is compounded when the bedding dips into the roadway (refer to Figure 11.33). In fact, good engineering practice usually dictates that, if at all possible, road cuts be oriented such that the bedding does not dip toward the roadbed.

The shales associated with coal often contain appreciable amounts of calcite ($CaCO_3$) and pyrite (FeS_2). When exposed to weathering, products of the chemical decomposition of these two minerals combine to form $CaSO_4$, which subsequently hydrates to gypsum ($CaSO_4 \cdot 2H_2O$). As the gypsum forms, the rock expands. Should the shale layer be located under the foundation of a building or other structure, the swelling of the shales may subject the foundation to stress sufficient to cause serious structural damage.

Limestone Limestones are common sources of environmental problems, especially in humid areas where subsurface dissolution of the beds results in the formation of caves and caverns and the development of karst topography (we will discuss the development of karst in Chapter 14). During times of drought, groundwater that normally fills the cave system and serves to support the overlying rocks may be removed. Depending on the size of the cave and the thickness of the overlying rock layers, the cave roof may collapse, resulting in the formation of surface depressions called sinkholes (refer to Figure 14.23). Such surface subsidence can not only result in the damage or destruction of surface structures but may also significantly alter the utilization of the surface.

Sediments and Sedimentary Rocks as Conduits of Water Water moving underground passes through sediments and sedimentary rocks called aquifers. In many cases, domestic and industrial pollutants that have entered the groundwater from the surface are removed as the water makes its way through the layers of sediment or rock. The sedimentary materials act, therefore, like a purifier. The sediments surrounding a septic field, for example, remove noxious materials by a combination of filtration, adsorption to the clay minerals, and retarding the movement of the water so that chemical processes such as oxidation are given sufficient time to destroy harmful components.

There is, however, a limit to the ability of sedimentary materials to cleanse groundwater. For example, if the movement of the water through the sediments is impeded, effluents from a septic system may rise to the surface before they have undergone adequate treatment and create both an environmental and a health problem. A similar situation may arise where septic systems are located on steep slopes. Unless measures are taken to route the drainage parallel to the contours of the slope in order to increase the time of contact between the water and the sediments, the effluents may come to the surface downslope before they have been properly treated.

CONCEPTS AND TERMS TO REMEMBER

Sedimentary rocks
Sediment
Classification of sedimentary rocks
  detrital
  nondetrital
Sorting
  roundness versus sphericity
Kinds of detrital rocks
  breccia
  conglomerate
  sandstone
    quartz sandstone
    arkose
    graywacke
  siltstone
  shale versus mudstone
Kinds of nondetrital rocks
  chemical rocks
    chemical limestone

dolostone
biochemical rocks
  chert
  biochemical limestone
    coquina
evaporite rocks
  gypsum
  rock salt or halite
organic rocks
  coal
Depositional environments
  continental environments/deposits
    fluvial
    glacial
    eolian
    lacustrine
    paludal
  transitional environments
    delta

    beach
    barrier island
    coastal wetland and mudflat
  marine environments
    continental shelf
    continental slope
    abyssal floor
Processes of Lithification
  compaction
  cementation
Sedimentary features
  beds
  ripple marks
    asymmetrical/current
    symmetrical/oscillation
  cross-bedding
  graded bedding
  mud cracks
  animal trails
Walther's law

REVIEW QUESTIONS

1. Which of the major minerals found in sedimentary rocks is usually nondetrital in origin?
   a. quartz
   b. calcite
   c. feldspar
   d. clay
2. A well-sorted rock is composed of
   a. one dominant mineral.
   b. a narrow range of particle sizes.
   c. a mixture of particle sizes.
   d. particles that all show the same degree of roundness.
3. A cube would be described as showing
   a. high sphericity but low roundness.
   b. both high sphericity and high roundness.
   c. low sphericity but high roundness.
   d. both low sphericity and low roundness.
4. The difference between a conglomerate and a breccia is
   a. composition.
   b. bedding thickness.
   c. particle shape.
   d. particle size.
5. Fluvial deposits are laid down
   a. by the wind.
   b. on the ocean bottom.
   c. by streams.
   d. in wetlands such as marshes and swamps.
6. An arkose in a sandstone that contains an appreciable concentration of
   a. feldspar.
   b. calcite.
   c. clay minerals.
   d. rock fragments.
7. Compaction is the major lithification process in the formation of
   a. limestones.
   b. sandstones.
   c. conglomerates.
   d. shales.
8. Compare the various processes of erosion in terms of their relative ability to sort clastic materials.
9. Why are arkoses not very abundant?
10. What specific sedimentary features would enable you to determine each of the following:
    a. the paleoclimate
    b. the direction of water or wind flow
    c. whether the materials accumulated on land or on the ocean bottom
    d. the specific kind of continental deposit

THOUGHT PROBLEMS

1. In what ways would the assemblage of sedimentary rocks that forms in a hot, dry climate differ from the assemblage that forms in a warm, moist climate?
2. Which detrital sedimentary rock could best be used to determine the kinds of rock that were present in the highland area supplying the sediment?
3. Assume that you are observing a sequence of marine sedimentary rocks in an extensive cliff exposure such as the Grand Canyon. What kind of evidence would you seek to determine the direction to the ancient shoreline, the depth of water at any point in time, and the possible presence of hiatuses in the record?

FOR YOUR NOTEBOOK

Most of us live in areas underlain by sedimentary rocks. The distribution of the various rock types can be obtained from a geologic map of your area. Relate the topography in your area to the distribution of the various kinds of sedimentary rock. Identify the individual rock layers that are responsible for the hills and valleys in your area.

Sedimentary rocks are commonly used for various construction purposes from building stone to crushed stone for road construction. Make a survey of your area and indicate whether sedimentary rock is being quarried and, if so, the purpose for which it is being used. At the same time, a field trip through your town may discover a number of examples of sedimentary rocks being used for construction. Pay particular attention to the kinds of sedimentary rock used for decorative and trim stone.

You might also investigate the various kinds of depositional environments in your area where new sediments are accumulating. Attempt to determine what kinds of rocks will most likely result from these materials.

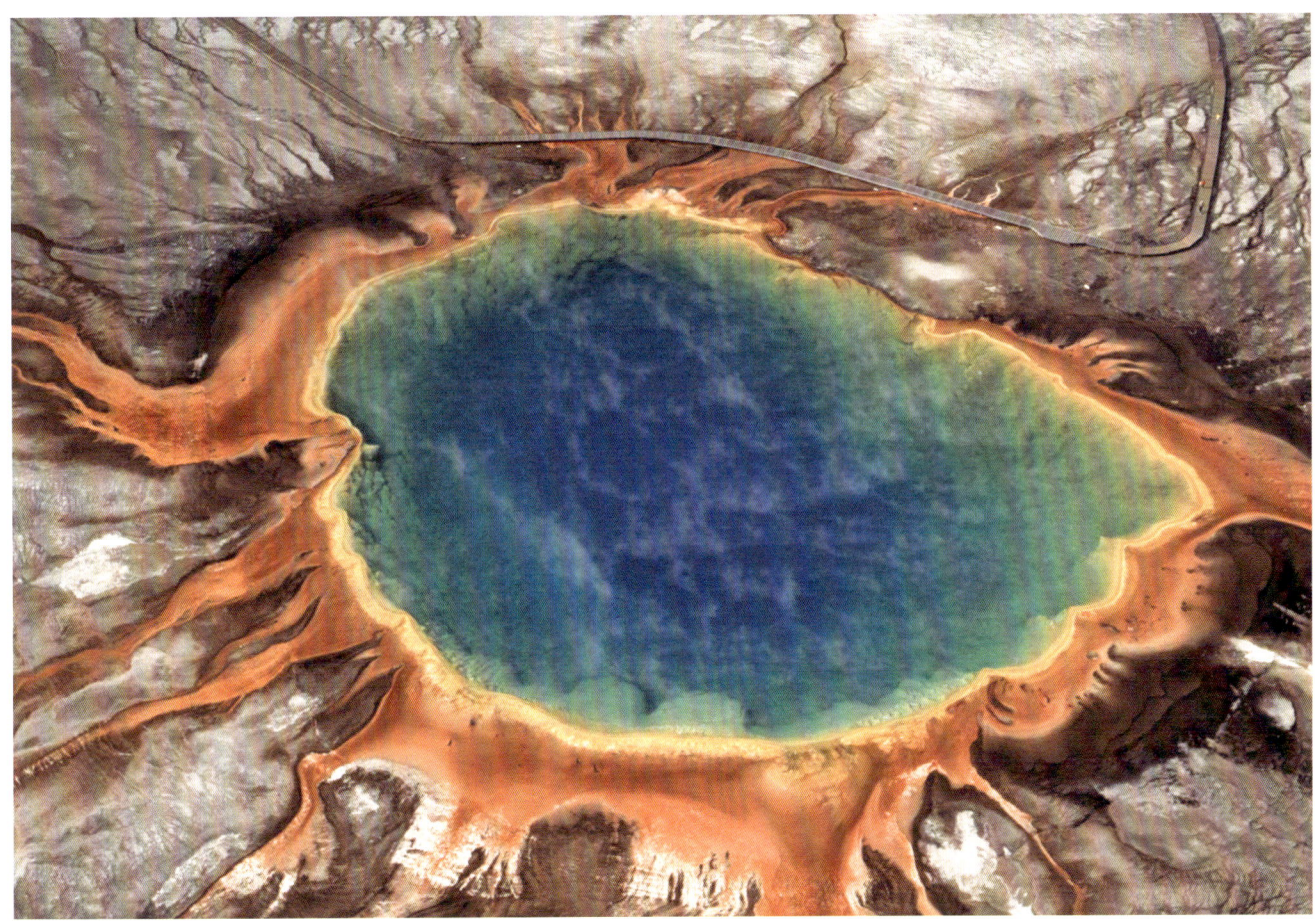

CHAPTER OUTLINE

INTRODUCTION

POROSITY AND PERMEABILITY
  Porosity
  Permeability
  Aquifers, Aquitards, and Aquicludes

THE WATER TABLE
  Regional Water Tables
  Perched or Hanging Water Tables

WATER WELLS
  Confined Aquifer
  Municipal Water Supplies

KARST TOPOGRAPHY
  Dissolution of Carbonate Rocks
  Sinkholes
  Surface Drainage
  Subsurface Drainage
  Caves and Caverns

HOT SPRINGS, FUMAROLES, AND GEYSERS

ENVIRONMENTAL CONCERNS
  Groundwater Pollution
  Dewatering
  Saltwater Encroachment

CHAPTER 14

Groundwater

INTRODUCTION An adequate explanation of groundwater defied the best minds for centuries. The ancient Greeks were convinced that all of the water that fell as rain ran off as streams and rivers. As a result, they had to employ extraordinary processes to explain the water they saw emerging from the ground. Their explanation of springs, for example, was that water was drawn up through the rocks from the ocean basin. However, their theory failed to explain how the salt content of the seawater was lost during the trip. Since few places in Greece are far from the ocean, it is easy to see how the Greeks could visualize such a process, but it might have been interesting to hear how they would have explained the presence of a spring thousands of miles inland from the sea.

Aristotle had his own version of the origin of springs. He thought that Earth was hollow with a cavity so large that it had its own internal weather systems. His explanation for springs was that they were the "overflow" from torrential downpours that occurred inside Earth.

The Romans apparently were not aware of the existence of groundwater, otherwise they would not have expended enormous amounts of energy and money to build aqueducts to transport water great distances (Figure 14.1). Had they known that a store-house of water lay beneath their feet, they would have dug wells to supply all the water they needed for a much smaller investment of both energy and capital.

It was not until the seventeenth century that our present understanding of groundwater began to evolve. A French physicist, Pierre Perrault, was the first to prove that the water contained within Earth was not drawn up from the oceans but rather was provided by rainfall and snow melt. Perrault kept accurate records of precipitation and discharge for a portion of the Seine River drainage basin over a period of several years and showed that the amount of precipitation *far* exceeded the discharge of the streams. He concluded that a significant portion of the water that falls on Earth's surface sinks in to become groundwater.

Perrault's conclusions were indeed correct, as a brief review of the distribution of water on Earth indicates. About 97% of all water is contained within the ocean basins.

FIGURE 14.1 *Many of the aqueducts constructed by the Romans to transport water to their cities are still used today. (VU/© Bayard H. Brattstrom)*

Of the 3% outside the ocean basins, nearly 80% is contained within glaciers and polar ice with only 0.7% being represented by the more visible surface accumulations of water. Many individuals are surprised to discover that nearly 20% of all the water outside the ocean basins resides underground. Hydrologists estimate that more than 1.92million cubic miles (8 million km³) of water exist below Earth's surface.

POROSITY AND PERMEABILITY

The amount, the availability, and the movement of groundwater are largely dependent upon two physical parameters of rocks: *porosity* and *permeability*.

Porosity

The question that immediately arises is, how can "solid" rocks even contain water let alone allow the passage of water back to the sea? Explaining how *solid* rocks can hold fluids such as water, oil, and gas has led many an instructor to compare rocks to sponges. Although rocks and sponges may at first seem to have little in common, there are some valid comparisons. Few rocks are completely solid but rather contain spaces between the individual grains called **pores**. In addition, most rocks are broken by **fractures** that generate additional space within the rock (Figure 14.2). In carbonate rocks, fractures are opened further by dissolution to form *solution channels*. The combined space represented by pores, fractures, and solution channels is called **porosity**. Porosity is expressed as a percentage of the total rock volume and is subdivided into the space represented by the pores, called **intergranular porosity**, and the space represented by the fractures, called **fracture porosity**.

Porosity determines how much water (or oil and gas) a rock can hold. For example, in the Appalachians the Oriskany Sandstone, an important source of gas and oil, often has a porosity as high as 30%, which means that a third of the total rock volume is open space. At least in this respect, the rock certainly is like a sponge.

Different kinds of rocks exhibit quite different porosities. Because of the interlocking nature of the mineral grains, intergranular porosity is low to nonexistent in fresh (unweathered) igneous, metamorphic, and nondetrital sedimentary rocks. Unless these rocks are subjected to chemical weathering, which induces pore spaces by preferentially removing the more reactive mineral grains or portions of the rock, initial fracture porosity accounts for the little porosity the rocks possess.

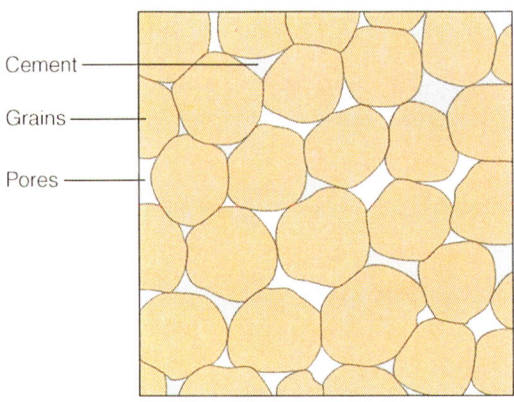

(a) Intergranular porosity, e.g., sandstone

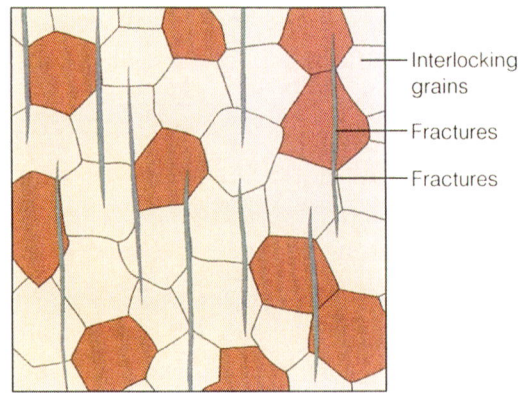

(b) Fracture porosity, e.g., unweathered igneous rock

FIGURE 14.2   The porosity *of a rock is determined by the combined space between mineral grains, called* intergranular porosity, *and within fractures, called* fracture porosity.

Intergranular porosity is much more important in detrital sediments and sedimentary rocks where the amount of porosity is a function of grain shape, the degree of sorting, and the extent of cementation. The intergranular porosity of unconsolidated sediments will reach a maximum in an ideal (but rarely seen) deposit that shows *perfect* sorting of *perfectly* spherical grains accumulated in an open, closely packed arrangement (Figure 14.3). Note that in such an ideal deposit, particle size does not affect porosity. A layer of packed, spherical, sand-sized grains will have the *same* percentage of intergranular porosity as a similar layer where the particles are cobble (or any other) size.

Most sediments and grains within sedimentary rocks do not exhibit perfect sorting and shape. Porosity declines as the degree of sorting decreases and smaller particles begin to fill the spaces between the larger grains. For this reason, a well-sorted beach sand will generally have a higher porosity than a poorly sorted glacial gravel deposit. Similarly, the average porosity of a sandstone will be higher than that of a conglomerate. Porosity also decreases as the individual particles become more angular and projections of individual grains begin to encroach into adjoining pore spaces (Figure 14.4)

A special situation arises with fine-grained materials dominated by clay minerals. Because of the platy crystal habits of the clay minerals, the porosity will depend upon the orientation of the clay mineral platelets within the deposit (Figure 14.5). In deposits where the platelets prop against each other or "bridge," the porosity will be significantly higher than where the platelets are parallel to each other. As a result, sedimentary deposits or rocks dominated by clay-sized particles can have porosities ranging from as low as 10% to as high as 60%.

Porosity is reduced as sediments undergo lithification. Lithification by compaction physically reduces the pore spaces as the particles are forced close together while cementation reduces porosity as growing mineral crystals fill the pore spaces.

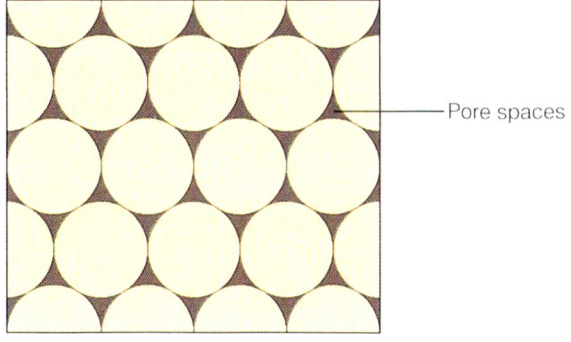

FIGURE 14.3 *The pore volume between spherical, close-packed particles is independent of particle diameter. Thus, regardless of the diameter assigned to the particles in the figure, the percentage of the total area represented by pore spaces would remain unchanged.*

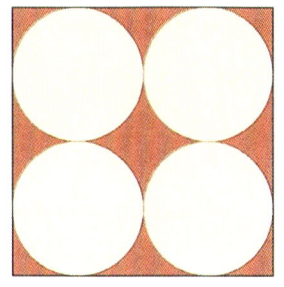  

Roundness = 1.00
Open area = 21.46% of total

Roundness = 0.50
Open area = 5.36% of total

Roundness = 0.00
Open area = 0.00 % of total

FIGURE 14.4 *As sediments become more poorly sorted and the particles become more angular, the porosity decreases as the original pore spaces become filled with smaller particles and the projections of irregular-shaped particles.*

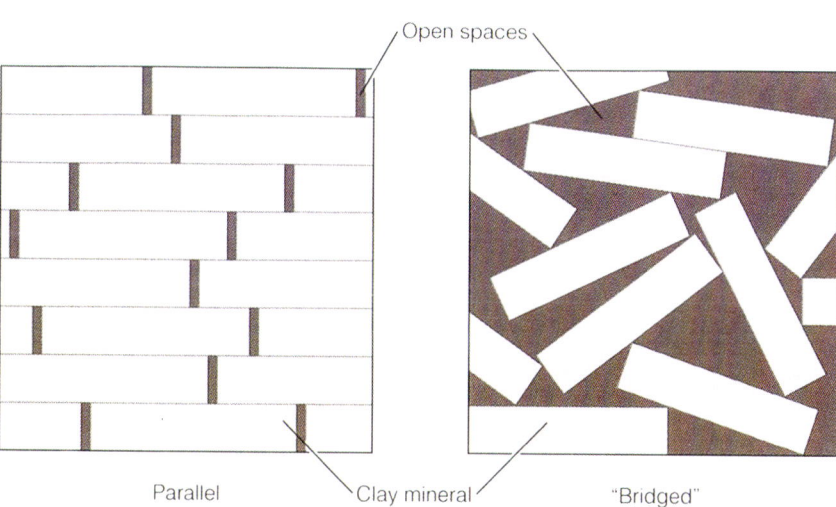

FIGURE 14.5 *The porosity of sediments dominated by clay minerals depends on the orientation of the clay particles. As the bridging between particles decreases and their orientation becomes more parallel, the porosity decreases.*

Permeability

Although porosity is important in determining how much water a deposit or rock can hold, unless the pores and fractures are interconnected, the water will not be able to make its way back to the sea or to a well or a spring. The ability of rocks to transmit fluids is called **permeability** and is usually measured in feet or meters of flow per day. The rate at which a rock will transmit water (or oil and gas) is determined by both the degree to which the spaces within the rock are interconnected and the physical dimensions of the passageways. It is important to note that porosity and permeability are not always directly related. While low porosity always ensures low permeability, high porosity does not ensure high permeability. For example, in a vesicular basalt flow, the vesicles may represent significant porosity but are largely unconnected; therefore the rock has low permeability.

The movement of water through sediments or rocks is significantly affected by the molecular attraction of water for the surfaces of unconsolidated particles or grains within rocks. A layer of water one molecule thick (monomolecular), tenaciously held to the surface of each particle and grain, serves to retard water movement under certain conditions. For example, the porosity of a poorly sorted glacial gravel deposit may average only 25% while the average porosity of a well-sorted beach sand may be nearly 40%. However, because the space between the grains in the glacial deposit are both larger in cross section and less tortuous than the spaces in the sand deposit, the absorbed water layer in the glacial deposit is thinner relative to the volume of the space between the particles. As a result, the retardation of water flow due to the molecular attraction of water to the surfaces of the individual particles will be significantly less in the glacial gravels than in the sand deposit. It is not uncommon, therefore, for glacial gravel deposits to have permeabilities of several thousand feet per day while sand deposits may transmit water at rates of only 30 feet (10 meters) per day or less. The retardation of water flow by molecular attraction in deposits of clay-sized materials is so high that, although they may have porosities as high as 50% or 60%, the permeabilities are very low.

Aquifers, Aquitards, and Aquicludes

A sedimentary deposit or rock that is sufficiently permeable to transmit economically significant quantities of water is called an **aquifer**. Because the permeability of rocks varies, some rocks make better aquifers than others. Rocks that have porosities and permeabilities so low that they significantly retard the flow of water to or from an adjacent aquifer are called **aquitards**. Rocks that do not contain or transmit water are called **aquicludes**. Since most rocks contain and transmit some water, although very slowly, true aquicludes are not very common. We have already commented on the general low porosity and permeability of fresh igneous and metamorphic rocks. Unless they exhibit interconnected pore spaces (produced by the removal of certain mineral grains during chemical weathering) or contain fractures (included by physical weathering, cooling, faulting, or unloading), igneous and metamorphic rocks are aquitards. An exception to this generalization is found in certain lavas such as Thousand Springs along the Snake River (Figure 14.6) where the presence of lava tubes and rubble zones between successive lava flows produces exceptional permeability.

FIGURE 14.6 *Although the porosity and permeability of basalt are normally quite low, lava tubes formed as basalt flows cool and rubble trapped between successive lava flows may become conduits for groundwater. One of the best-known examples of groundwater being produced from basalt is* Thousand Springs *along the Snake River near Twin Falls, Idaho. (Courtesy of G. F. Lindholm/USGS, Boise, Idaho)*

RATES OF GROUNDWATER FLOW

In the mid-nineteenth century, Henry Darcy, the engineer in charge of public works in Dijon, France, reasoned that the rate of movement of groundwater depended upon the permability of the aquifer and the slope of the water table (Figure 14.B1). The rate at which water moves through an aquifer depends on the **hydraulic head** or the difference in elevation over which the water flow's, $E_1 - E_2$. The hydraulic head divided by the horizontal distance over which the water moves, D, is called the **hydraulic gradient**, the driving force for the movement of groundwater.

Darcy reasoned that the amount of water that would flow through an aquifer, V, would be determined by the driving force, or hydraulic gradient, $(E_1 - E_2)/D$, the cross-sectional area of the aquifer through which the water moves, A, and a parameter, K, that takes into account the complex interaction of a number of factors including gravity, the porosity and permeability of the aquifer, the viscosity of the water, and the molecular attraction between the water and the rock surfaces adjoining the pores and passageways. The relationship is expressed as Darcy's law:

$V = AK(E_1 - E_2)/D$

In extremely porous and permeable sands and gravels, the water may move at rates of 5 or 6 inches (12–15 cm) per day. At the other extreme, in relatively poor aquifers such as fine-grained materials, water may flow at rates of a few inches (centimeters) per year, especially if the slope of the water table is low. On average, groundwater moves through most aquifers at a rate of an inch or so per day. One result of such slow movement is that the water that you drink from a well or spring may be hundreds and perhaps thousands of years old. More practically, the rate of flow determines the time required for a water well to refill after being pumped dry.

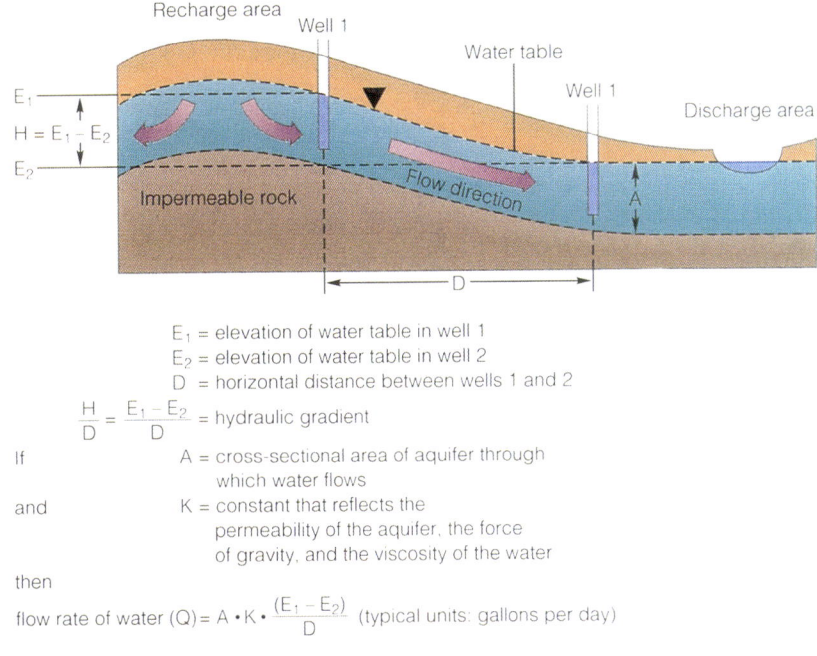

E_1 = elevation of water table in well 1
E_2 = elevation of water table in well 2
D = horizontal distance between wells 1 and 2
$\frac{H}{D} = \frac{E_1 - E_2}{D}$ = hydraulic gradient

If
A = cross-sectional area of aquifer through which water flows
and
K = constant that reflects the permeability of the aquifer, the force of gravity, and the viscosity of the water
then
flow rate of water (Q) = $A \cdot K \cdot \frac{(E_1 - E_2)}{D}$ (typical units: gallons per day)

Figure 14.B1 *The volume of water that will flow through an aquifer system from the recharge area to the discharge area will depend on the hydraulic gradient (similar to the gradient of a stream), the thickness of the aquifer, and a constant that takes into account the force of gravity, the permeability of the aquifer, and the viscosity of water.*

Sedimentary rocks, on the other hand, are generally good aquifers. Of the various kinds of sedimentary rocks, sandstones are generally the best aquifers because they commonly have high permeabilities. Limestones vary; those that are relatively unweathered are aquitards whereas others where solution has opened cavernous underground passageways are exceptional aquifers.

The extremely small dimensions of the pores and passageways in shales and mudstones, combined with the molecular attraction of the water to the surfaces of the grains, nearly preclude the flow of water. Because of their very low permeabilities, most fine-grained sedimentary rocks are aquitards.

THE WATER TABLE

Upon sinking into the ground, water is attracted by molecular forces to individual particles of the soil or regolith or to the exposed surfaces of the grains within the rocks. After all the rock particles or mineral grains have been coated with water, additional water percolating from the surface begins to move downward.

As the water moves deeper beneath the surface, it begins to displace the air in the pores and fractures within the rocks or buried sediments. Eventually, below a certain depth, all the pores and fractures will be filled with water, creating a **zone of saturation**. The layer of material from the top of the zone of saturation to the surface is called the **zone of aeration**; here spaces within the sediments or rocks contain both water and air. The boundary between the zone of aeration and saturation is called the **water table** (Figure 14.7). Within the zone of aeration, water moves vertically, whereas below the water table, the water begins to move laterally down a slope (the hydraulic gradient), usually in the same general direction as the surface waters move.

The depth to the water table is a function of rainfall, topography, slope, and orientation to the Sun, with rainfall being the most important. In general, as rainfall increases, the depth to the water table decreases. In a region receiving 30 to 40 inches (75–100 cm) of rainfall per year, the average depth to the water table may be measured in tens of feet (meters), while in a desert area receiving less than 10 inches (25 cm) of rain, the water table may lie hundreds of feet beneath the surface. In any region, the depth to the water table varies seasonally, with the water table rising during the rainy season and falling during the months of low rainfall.

While the water table is relatively well defined, the lower limit of the zone of saturation is a little more difficult to determine. With the exception of some lavas, unweathered igneous and metamorphic rocks normally have low porosities and permeabilities and are usually considered to be dry. Therefore, in regions underlain by igneous and metamorphic rocks, the bottom of the zone of weathering is the base of the zone of saturation. In regions where igneous and metamorphic rocks are covered with thousands and perhaps tens of thousands of feet of sedimentary rocks, porosity and permeability decrease with depth as the pore spaces and fractures collapse under the weight of the overlying rocks. Although deep wells drilled into sites of modern deltaic accumulations such as the Mississippi delta still encounter porous rocks at depths of nearly 30,000 feet (9,100 m), measurements have shown that the porosity and permeability of rocks at depths of 12,000 to 14,000 feet (3,600–4,300 m) are quite low. In general, by the time sedimentary rocks are buried to depths of 15,000 to 20,000 feet (4,600–6,100 m), porosity is virtually eliminated by compaction due to the weight of the overlying rocks and sediments. We can therefore assume that below such depths, nearly all rocks, including sedimentary rocks, would be dry. Although some groundwater may be found at such great depths, most is located within about 2,500 feet (760 m) of the water table.

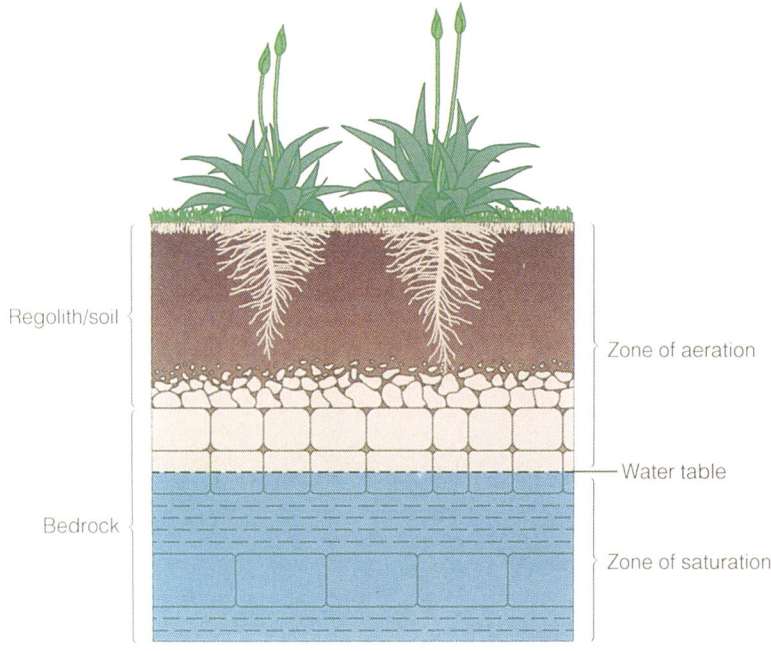

FIGURE 14.7 *The* water table *is the contact between the zones of* aeration *and* saturation.

Regional Water Tables

The water table that exists nearly everywhere beneath the surface is called the **regional water table**. Because the hydrostatic pressure (pressure that exists at any point within a body of water) and the atmospheric pressure are equal at the water table, the water table generally follows the surface topography except that the slopes are usually not so steep (Figure 14.8). Surface accumulations of water such as streams, lakes, ponds, and swamps occur where the water table intersects topographic lows. In each case, the surface of the water is the location of the water table. Have you ever wondered why the water stays in a stream or a pond or a lake? Why doesn't it sink into the ground? The answer is that rather than the water moving *downward* from the stream or lake to become groundwater, the stream or lake exists because it is actually being fed with water moving *upward from* the groundwater reservoir.

During times of drought, we say that streams, lakes, and ponds "dry up." "Dry up" is an unfortunate choice of words, however, because it implies that the disappearance of the water was primarily due to evaporation. Undeniably, some water evaporates during dry seasons, but the phrase "dry up" misrepresents what actually happened to the water. Because rainfall is a major factor controlling the location of the water table, a seasonal decrease in rainfall will result in the lowering of the regional water table, which in turn will cause the level of the water in streams, lakes, and ponds to fall. Should the water table drop below the bottom of the stream channel, lake, or pond, the water will disappear, not by *drying up* but by *sinking in* (Figure 14.9). During times of drought, it is not

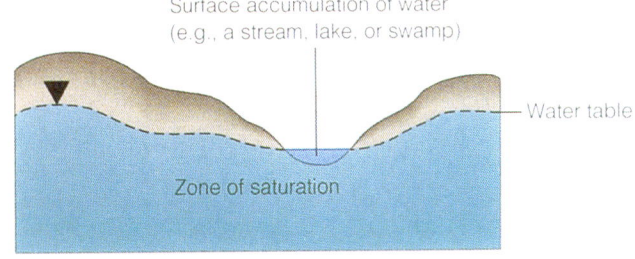

FIGURE 14.8 *In general, the regional water table follows the surface topography at a depth largely determined by the regional rainfall.*

FIGURE 14.9 *Surface accumulations of water such as lakes and streams exist where the water table comes to the surface and crosses a topographic low. The level of water in a stream or late depends on the location of the water table; should the water table drop below the surface, the body will "dry up."*

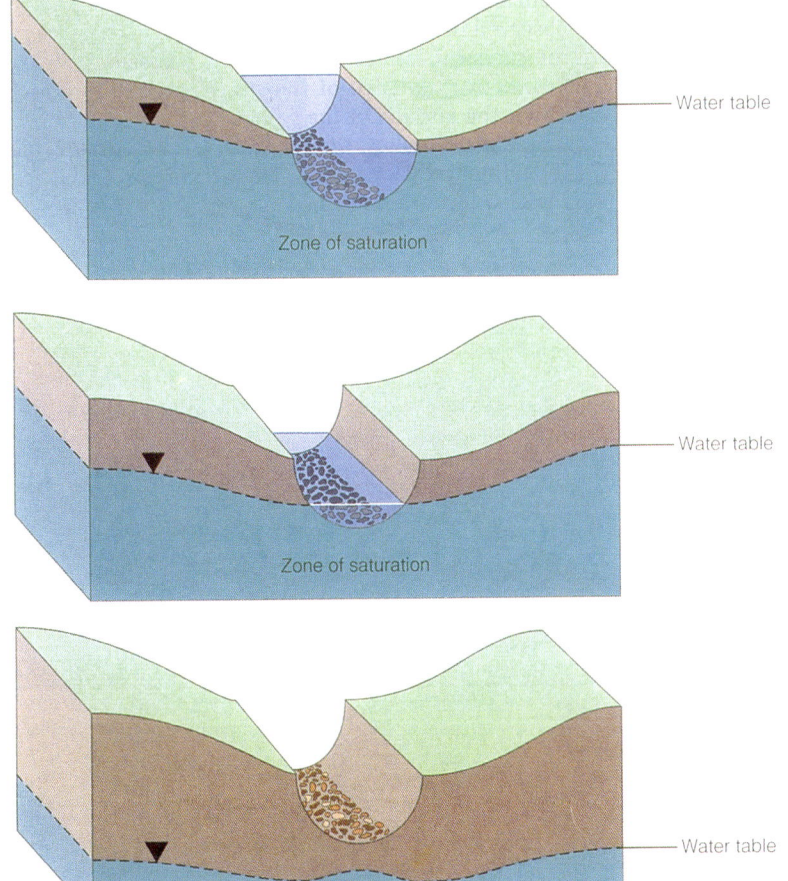

uncommon to see animals digging in dried streambeds or lake bottoms in an attempt to find water by reaching the water table.

After a dry period ends and the stream channels fill with rainwater draining directly from the land surface, water moves from the stream channel into groundwater storage (Figure 14.10). This is one of the ways that groundwater supplies are replenished. Such streams are called **losing** or **influent** streams. Once the water table rises above the bottom of the stream channel, the water level in the stream is maintained between rainfalls by water moving from groundwater storage into the stream channel (Figure 14.11). Such streams are called **gaining** or **effluent** streams.

Perched or Hanging Water Tables

In addition to the regional water table, numerous water tables of limited extent called **perched** or **hanging water tables** exist above the regional water table (Figure 14.12). Perched or hanging water tables are especially common in regions of sedimentary rocks where aquitards, most likely shales, located in the rock section above the regional water table intercept the downward-moving water. Where the aquitard intersects the slope of a hill, "leakage" of the water from the perched water table commonly forms springs (refer to Figure 14.12). Water leaking from an intersected perched water table can often be observed in road cuts. During the rainy season when the zones of saturation are filled to capacity, water flows over the edge of the truncated aquitard and down the surface of the road cut (Figure 14.13). In temperate climates, the flow of groundwater during the winter may feed a growing mass of ice that coats the cliff face (Figure 14.14).

SPOT REVIEW

1. What rock parameters determine porosity?
2. What is the difference between porosity and permeability? Are they related in any way?
3. What factors determine the depth of the water table?
4. What role does the water table play in determining whether a stream will be influent or effluent?
5. What is the difference between an aquifer and an aquitard?

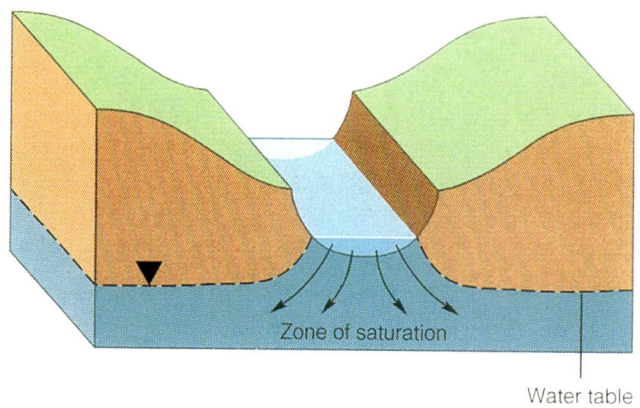

FIGURE 14.10   *After periods of low rainfall, rainwater diverted overland into a stream channel moves from the channel into groundwater storage to replenish the groundwater supply.*

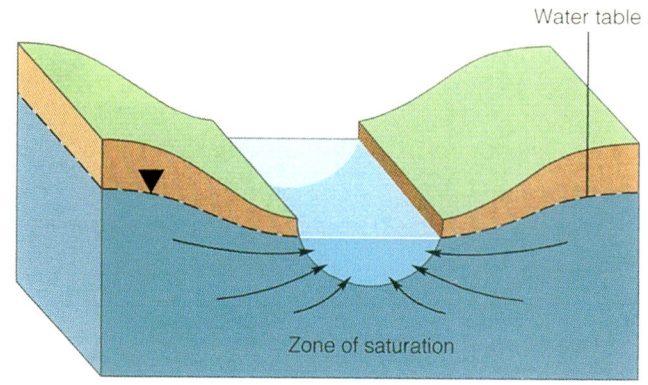

FIGURE 14.11   *Following a period of rainfall when the elevation of the water table rises above the bottom of the stream channel, the stream discharge is maintained between subsequent rain events by water moving from groundwater storage into the stream.*

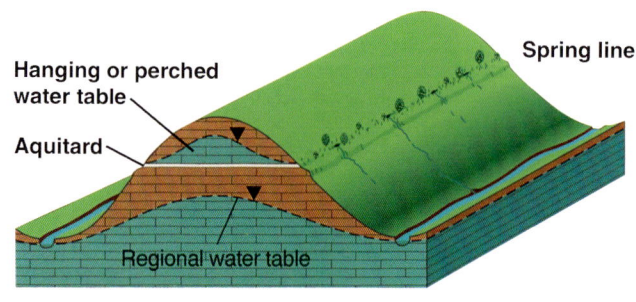

FIGURE 14.12   *Although there is only one regional water table, there can be any number of* perched *or* hanging water tables, *depending on the number of aquitards that exist in the rock section above the regional water table. Spring lines along hillsides commonly mark the intersection of a hanging water table and the surface.*

FIGURE 14.13 *The water observed flowing down the face of a cliff or road cut originates from the edges of an exposed perched or hanging water table. (Courtesy of WVGES)*

WATER WELLS

A **water well** is a hole either dug or drilled into the ground that allows water to flow to the surface or provides water that can be pumped to the surface. No one knows who first discovered that if a hole was dug deeply enough, the bottom would fill with water, and that if the water was removed, more water would flow into the hole. Before the discovery of the well, towns and cities had to be located at the margins of rivers or lakes in order to guarantee a supply of water for domestic and agricultural needs. Consider the "Cradle of Civilization" where the earliest known settlements were established along the banks of the Tigris and Euphrates rivers in what is now Iraq.

With the discovery of the water well, people could move away from bodies of water and settle the vast interiors of the continents. A town could be established anywhere a supply of water could be obtained by digging a well. The town then grew up around the well. When the growing population outstripped the ability of the first well to supply the needed water, another well was dug on the outskirts of the town, and the community continued to grow. This pattern of wells survives in the town plazas that grace many towns and cities throughout the world. No longer needed as a source of domestic water, many of the wells have been converted into decorative fountains.

Originally, wells were literally **dug wells**. A relatively large hole was dug until the bottom began to fill with an adequate supply of water; then the hole was lined, usually with brick or stone, and equipped with a windlass and a bucket for removing the water (Figure 14.15). Today, wells are not so picturesque. A modern well consists of a hole drilled into the ground, perhaps 6 inches (15 cm) or so in diameter, lined with a piece of perforated pipe, and equipped at the bottom with an electric pump to lift the water to the surface.

FIGURE 14.14 *During the winter, ice hanging from outcrops marks the location of an exposed perched or hanging water table. (Courtesy of W. B. Hamilton/USGS)*

FIGURE 14.15   *As the name implies, a* dug well *is one that has literally been* dug *to below the water table. (Courtesy of E. D. McKee/USGS)*

Regardless of the kind of well, the initial question to be answered is how deep must the well be to provide the required volume of water? From our previous discussions, you can see that unless a well penetrates the zone of saturation of either a regional or a hanging water table, it will produce no water at all. Once below the water table, the well must intersect the number of aquifers necessary to provide the amount of water desired. The average family in the United States uses about 2,000 gallons (7.600 l) of water per week for its domestic needs. To produce this weekly requirement, most wells must intersect several aquifers. Rarely will a single aquifer produce enough water to meet the demand.

In wells drilled into the zone of saturation associated with regional or hanging water tables, the water level will rise to the water table. As the well is pumped, the water drains from a cone-shaped volume of rock surrounding the well called a **cone of depression** (Figure 14.16). The size of the cone of depression is determined by the rate at which the water is being removed by pumping relative to the rate at which the water is being returned by **recharge** from aquifers. Once pumping stops, the cone of depression will shrink as water flows back into the rocks from the surrounding aquifers; given enough time to allow complete recharge, the cone of depression will be eliminated altogether.

If the rate of pumping is higher than the rate of recharge from the aquifers, the diameter of the cone

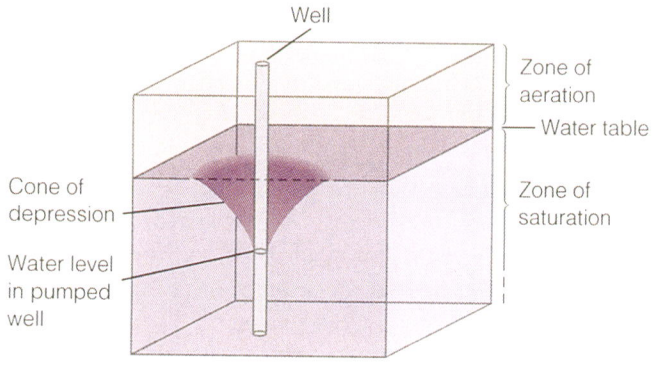

FIGURE 14.16   *As water is pumped from a well, the water table surrounding the well is depressed in the form of a cone called the* cone of depression.

of depression may become so large that the apex of the cone extends to the bottom of the well (Figure 14.17). At this point, the well will be pumped dry. If a well is repeatedly pumped dry before the needed supplies of water have been obtained, the demands on the well are exceeding the capacity of the aquifers. To provide the required volume of water, the well must be deepened to intersect additional aquifers.

Confined Aquifers

Most shallow wells drilled below either a hanging or a regional water table will fill with water up to the level of the water table. In some wells, however, the

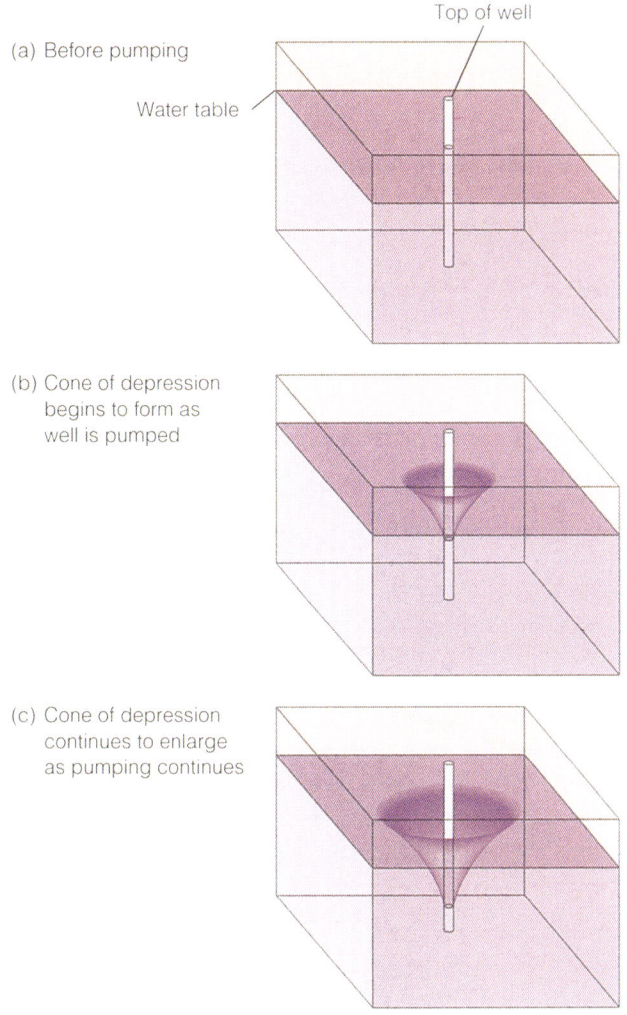

FIGURE 14.17 *Should water be withdrawn from a well faster than it can be replaced, the cone of depression will increase in diameter. When the apex of the cone reaches the bottom of the well, the well will have been pumped dry.*

water rises to a level unrelated to the water table and may actually spout above the surface to the ground. Such wells produce water from a **confined aquifer**.

The requirements for a confined aquifer are very specific (Figure 14.18). First, it is nearly always located in a region where a highland area adjoins a lowland. The aquifer underlies the lowland and comes to the surface in the highland. Most importantly, the aquifer is overlain by an aquitard. As Figure 14.18 illustrates, rainfall in the lowland area cannot gain access to the aquifer because of the overlying aquitard. Water can only enter the aquifer at the *recharge area,* the place where it comes to the surface in the highland.

Upon entering the aquifer, the water percolates down through the rock layer. Once in the aquifer, the water cannot be drawn to the surface by capillary action because of the overlying aquitard. Thus, as the name implies, once the water enters the aquifer, it is *confined*.

The major feature that distinguishes an **unconfined aquifer** from a confined aquifer is that the water in a confined aquifer is under *pressure*. In any water-filled container, the pressure increases with depth beneath the water surface. For example, when you dive into a swimming pool, the pressure increases on your body by 62.4 pounds per square foot (30.5 gm/cm^2) of body area for each foot you descent. A comparable situation exists within a confined aquifer. Theoretically, the pressure that exists at any point within a confined aquifer is the combined result of the height of the overlying column of water and the pressure exerted by the weight of the overlying rocks. If a well were to penetrate the overlying aquitard and

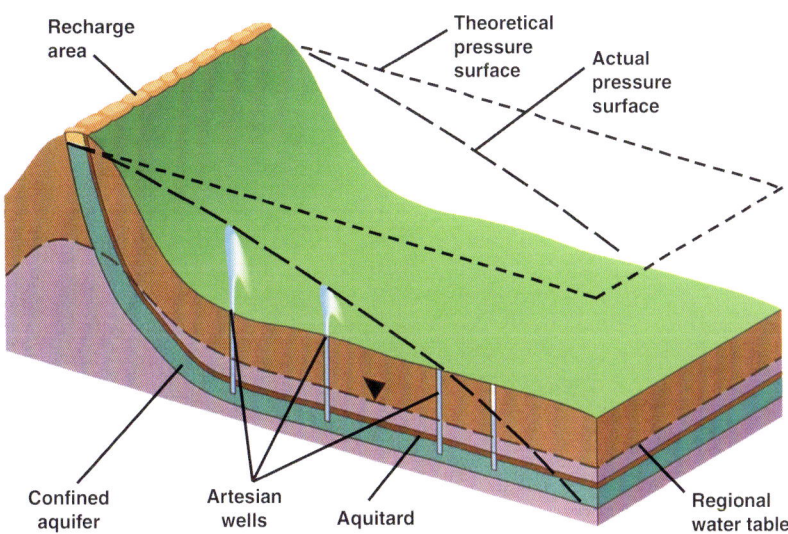

FIGURE 14.18 *Wells drilled into a confined aquifer produce water under pressure. Sometimes the pressure is sufficient to allow the water to flow freely at the surface, negating the need to pump the well.*

enter the aquifer at any point, the pressure should lift the water to the elevation of the recharge area, a horizontal surface indicted in Figure 14.18 as the *theoretical pressure surface.* However, the pressure within the aquifer decreases away from the recharge area because energy is lost due to the resistance to water flow. As a result, rather than rising to the theoretical horizontal pressure surface, the water will rise to an *actual* **pressure surface,** which slopes away from the recharge area (refer to Figure 14.18). If the well is drilled at a locality where the pressure surface is *above* the surface of the ground, the water will spout into the air. If the pressure surface at the well site is *below* the level of the ground, the water will still rise to the pressure surface but will not spout into the air. Any well producing water from a confined aquifer is called an *artesian well,* whether the water rises above the surface or not. Note that the rise of water in wells drilled into confined aquifers has *nothing* to do with either a regional or a hanging water table.

One of the best-known artesian systems in the United States is the Dakota aquifer system that underlies the Great Plains. The recharge areas for the Dakota Sandstone, a highly porous and permeable confined aquifer, are exposures in the Black Hills in southwestern South Dakota and along the front ranges of the Bighorn and Rocky mountains to the west. The Dakota Sandstone extends eastward beneath the surface of the Great Plains. In the early days of settlement, almost every well drilled into the Dakota aquifer produced water above the surface of the ground. The advantages of such wells are obvious. Unfortunately, tens of thousands of wells have been drilled into the aquifer, and large volumes of water have been removed, primarily for irrigation. Today, more water is being removed from the aquifer than is being replenished in the recharge areas. Overproduction from a confined aquifer leads to an additional decline in pressure within the aquifer with a subsequent lowering of the pressure surface. Throughout most of the Great Plains today, the pressure surface of the Dakota aquifer is below ground level. Although still artesian, the wells producing water from the Dakota aquifer no longer produce water at the surface and must be pumped like wells drilled into unconfined aquifers. A more serious problem is that a certain, and increasingly large, portion of the water being removed from the Dakota aquifer will not be replaced in anything shorter than geologic time. The water has become a *nonrenewable* resource.

Municipal Water Supplies

An example of an artificial confined aquifer that most of us utilize every day is the municipal water system (Figure 14.19). If asked why water emerges from the household faucet, most people would respond that the water is "pumped" to the house, a belief that is only partially correct. A pump is employed to obtain the water from the source, most likely a local river or reservoir. After treatment, another pump is used to transfer the water to a water tank or tower. From that point, the water simply flows by gravity to the customers. The force with which water emerges from a faucet is a function of the location of the faucet relative to the pressure surface extending outward from the water tank. The faucet is comparable to an artesian well that spouts water into the air except that the tap directs the water downward for obvious reasons.

The pressure at a tap does not depend on the proximity of the tap to the tank but rather is determined by the vertical distance between the tap and the pressure surface. The distance from the tank, however, may affect the pressure at certain taps and may present problems. At times of high water demand such as in summer, individuals living at the fringes of the pressure surface, where the pressure surface is approaching the surface of the ground, may experience low water pressure (Figure 14.20). As in the case of overproduction of water from a natural confined aquifer, overdemand for water from a

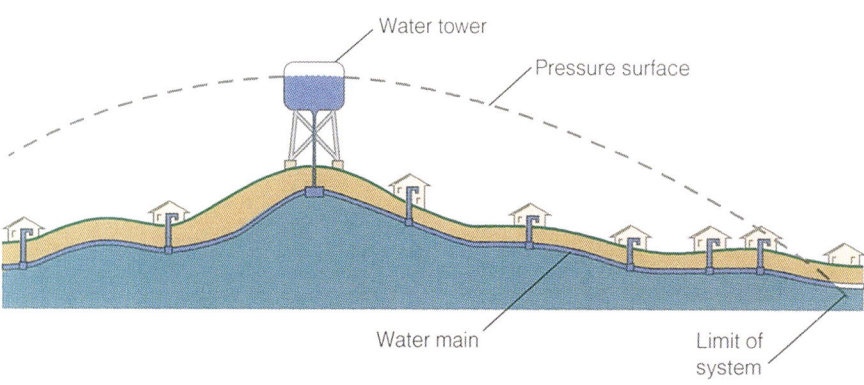

FIGURE 14.19 Municipal water systems *are artificial confined aquifers that operate under the same principles that govern natural confined aquifers.*

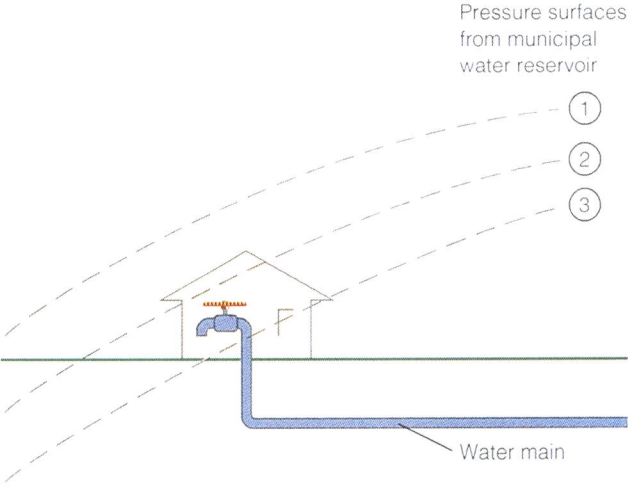

FIGURE 14.20 *Individuals located at the "fringe" of a municipal water system where the pressure surface is approaching ground level may experience low water pressure and interruptions of water supply during times of high water demand.*

municipal water system will cause a lowering of the pressure surface. If the pressure surface at the extremities of the system drops below the level of the tap, no water will flow at all. For this reason, to ensure that all customers are located beneath the umbrella of one or more pressure surfaces, as municipalities expand, they construct additional water towers, all connected to the same delivery system. To supply tall buildings where the top floors are obviously higher than any nearby water tower, the water is usually pumped to the top of the building and stored in a reservoir. The water is then fed by gravity to the building. In other words, the building has its own confined aquifer system.

SPOT REVIEW

1. What is the cone of depression? How does it form, and what determines its dimensions?
2. What is the difference between a confined and an unconfined aquifer?
3. What is the source of pressure that raises the water in an artesian well above the aquifer?
4. What is the source of pressure in a municipal water system?

KARST TOPOGRAPHY

Groundwater is responsible for a number of geologic features formed either by the dissolution of the surrounding rock or by the deposition of minerals originally carried in solution. Of these, none are more spectacular than those exhibited in areas characterized by *karst topography.*

In regions underlain by carbonate rocks, and to a lesser extent by evaporites, the processes of weathering and erosion produce a landform called **karst.** The term *karst* is a German word taken from the Italian word *carso,* which means "a bleak, waterless place." The original reference described a region on the border between Italy and Croatia. Karst topography is characterized by circular to elliptical surface depressions and a unique, disrupted type of surface drainage. In the United States, karst is especially well developed throughout the Valley and Ridge Province of the Appalachians, in central Florida, from central Indiana to western Kentucky, in central Missouri, and throughout the Edwards Plateau of Texas (Figure 14.21).

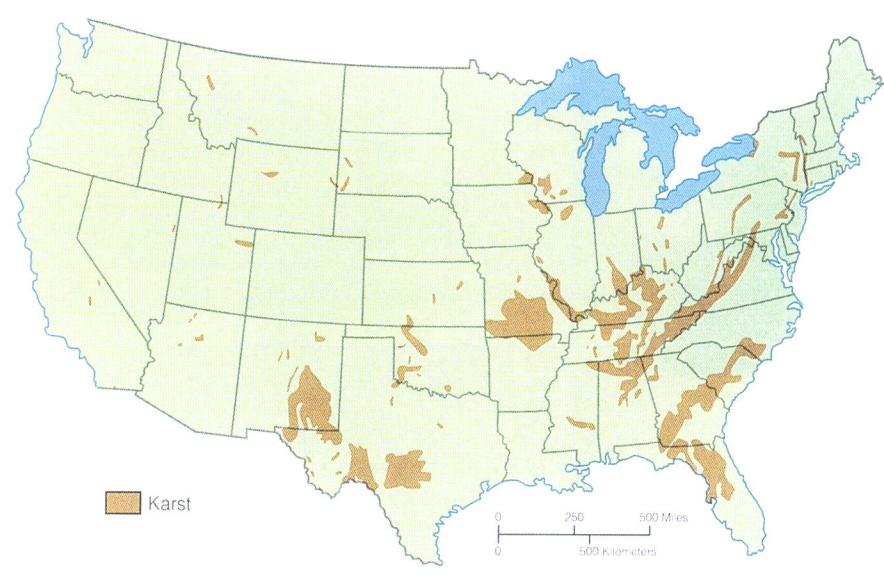

FIGURE 14.21 Karst topography *in many areas throughout the United States where relatively pure limestones are located immediately beneath the surface.*

Dissolution of Carbonate Rocks

The characteristic of all karst areas is that they are underlain by highly soluble rocks. In most cases, the rocks are carbonates, although karst can develop in areas underlain by evaporites such as gypsum and halite. The most common minerals in carbonate rocks are *calcite,* $CaCO_3$, and *dolomite,* $CaMg(CO_3)_2$. Carbonate rocks by dominated calcite (greater than 50% by volume) are called *limestones* while carbonate rocks dominated by dolomite are called *dolomites* or *dolostones.* Of the two minerals, calcite is more soluble in water.

Studies have shown that karst will not develop in carbonate rocks that are less than 60% calcite; for extensive development of karst, the rock must consist of at least 90% calcite. In other words, well-developed karst is limited to areas underlain by high-purity limestones.

Karst develops by the dissolution of limestone by moving groundwater. As you now know, the volume of groundwater depends on the porosity and permeability of the aquifer. Of the two types of porosity, fracture porosity is more important for the development of karst. Later in Chapter 15, we will see that fractures occur in sets that usually intersect at nearly right angles.

The development of karst also requires large quantities of water to dissolve the rock and form underground passages. For this reason, karst develops best in humid regions. The dissolution of limestone results from the reaction between calcite and *carbonic acid:*

$$CaCO_3 + H_2CO_3 \rightarrow Ca^{2+} + 2HCO_3^{1-}$$

The amount of carbonic acid contained in the groundwater depends on the availability of *carbon dioxide:*

$$H_2O + CO_2 \rightarrow H_2CO_3$$

As more carbon dioxide dissolves, the water becomes a more effective solvent. One of the major sources of carbon dioxide is the microbial decay of plant remains. Because large amounts of carbonic acid are needed, karst develops best in areas of high rainfall and lush vegetation (the tropics) and is least developed in dry (desert) and frozen (polar) areas.

Sinkholes

The most common feature of karst topography is the **sinkhole** (Figure 14.22). Sinkholes are circular to elliptical depressions with diameters ranging from a few feet to hundreds of feet. Most sinkholes are **solution sinkholes** that form by dissolution of the underlying limestone, especially at the intersection of fracture sets (Figure 14.23). The enlargement of the fractures by the downward flow of water, the dissolution of the surrounding rock, and the flushing of surface materials into the underground drainage all combine to create a

FIGURE 14.22 *Among the most obvious characteristics of karst topography are circular to elliptical surface depressions called* sinkholes. *Sinkholes are commonly aligned because of their association with rectangular fracture patterns in the underlying rock. (a VU/© Albert J. Copley, b VU/© Martin G. Miller)*

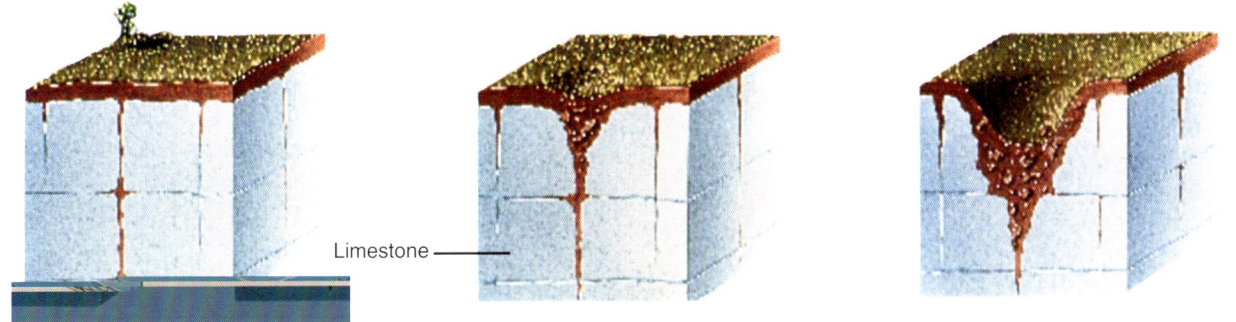

FIGURE 14.23 *The most common type of sinkhole is the* solution sinkhole *that forms by the progressive dissolution of the limestone in the immediate vicinity of a set of intersecting fractures called joints.*

closed surface depression. Because most sinkholes are located at the intersection of fractures, they are commonly aligned in a more or less rectangular pattern (refer to Figure 14.23). As sinkholes enlarge, the land surface between adjacent sinkholes become a divide that diverts surface water toward the center of the depression, the *reversed radial* drainage pattern described in Chapter 9.

In tropical areas, as the rock in the divide dissolves, increasing the diameter of the sinkholes, the surface takes on a dimpled appearance, called **cockpit karst,** not unlike the surface of a gigantic golf ball. In time, portions of the divides are isolated by erosion and form **karst towers,** which are completely surrounded by flat ground, usually the top surface of the insoluble rock layer below the limestone. The steep-sided towers are all that is left of the once-continuous layer of limestone. Some of the most spectacular karst towers are found near Kuei-lin, China (Figure 14.24).

Surface Drainage

The development of sinkholes, the subsurface drainage, and the diversion of surface water have a profound effect on the surface drainage. Karst areas characteristically have few, if any, continuous streams because surface drainage is eventually diverted into the underground system, the disappearing streams discussed in Chapter 9 (refer to Figure 9.16). Typically, disappearing streams lose water through **swallow holes** within their channels. During periods of low flow, the entire discharge of the stream may be *swallowed* by the underground system.

Subsurface Drainage

The subsurface movement of water in karst areas is very complex, and hydrologists disagree about its relationship to the water table. Some geologists believe that karst areas have a classic water table, meaning that all space below the water table is filled with water, including pores and microscopic fractures within the rock and any larger passageways that have been created by the dissolution of the rock adjacent to the fractures. According to this hypothesis, the groundwater contained within pores or microscopic fractures flows slowly through the rock while most of the groundwater moves much faster through the larger passageways created by dissolution.

Other hydrologists deny the involvement of a water table, at least in the classic sense. In their view, although a water table may exist, *all* of the groundwater below the water table moves through the larger passageways opened by dissolution and *no* water exists within the pores and microscopic fractures of the adjacent rock. In this interpretation, the passageways constitute a confined aquifer system that is *totally independent* of any water table.

As is often the case, another group of hydrologists have suggested a compromise that draws on both concepts. According to the compromise model, during the *early* stages of development of the underground system, the layer of limestone is below the water table, and the water moves through this layer as *diffuse flow,* largely controlled by the initial fracture porosity of the rock. As dissolution continues, the passages are enlarged, and extensive conduits, cavities, and eventually caves are formed; subsequently, the volume of water increases and the flow becomes increasingly *confined.* As the system of passages grows larger, the water table becomes less important. Eventually, the water flows freely, often turbulently, through the underground passages with little or no control by the water table.

During this latter stage, the large volumes of water moving underground eventually return to the surface as springs. Depending on the geology of the

FIGURE 14.24 *In the later stages of karst development, all that remains of the original layer of limestone are pillars of limestone called* karst towers. *Perhaps the most famous karst towers in the world are located in Kuei-lin, China. (Courtesy of Eleanor Renton)*

area, the flow of water to the spring may be strictly gravity controlled whereas in other cases, pressures may develop in the confined system and force the water to the surface as an artesian spring.

Caves and Caverns

Some of the most spectacular features resulting from the development of karst are **caves** and **caverns** and the features that eventually adorn them. Caves evolve in two phases. The **cave development phase** begins when groundwater starts to move through the fracture system. As the fractures widen and the volume of water increases, the limestone dissolves at a faster rate. With time, passages are created within the rock following the fracture pattern. The shapes and dimensions of the underground passages are controlled by the composition of the rock, structures within the rock, and the hydraulics of water flow. The passages are usually narrow, vertical openings called *canyons*. As the cross-sectional shape becomes more elliptical or circular, the passageways are referred to as *tubes*. *Rooms* form where canyons or tubes intersect.

During the millions of years that are required for passages to develop within the limestone layer, the rock above the limestone is being weathered and eroded. As the ground surface is lowered and the water table drops, the passages enter the zone of aeration and become a network of caves. As the water table continues to fall, more of the cave system becomes exposed above the water table. The rules that govern the relationship between surface streams and the position of the water table also hold true underground. In those portions of the cave system where the cave floor remains below the water table, the passages will be the sites of underground streams, ponds, or lakes. Where the cave floor is above the water table, the cave will be dry. As the water table drops below the cave system, the cave development phase comes to an end.

As the rock in the cave roof continues to be worn down by weathering and erosion at the surface, portions of the roof will collapse to form *collapse sinkholes* (Figure 14.25). Like solution sinkholes, collapse sinkholes align in rectangular patterns. A cave

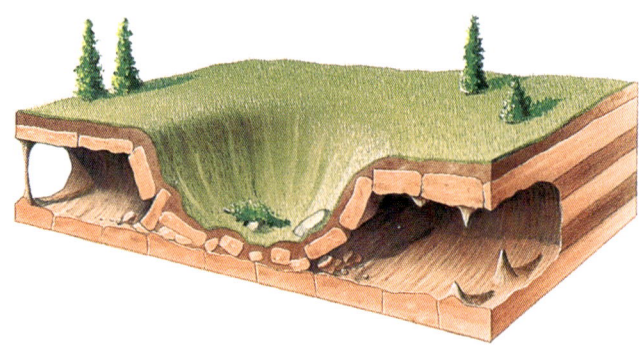

FIGURE 14.25 *Another type of sinkhole and potentially the most destructive, is the* collapse sinkhole *that forms when the roof of a cave collapses. In some cases, the collapse is due to the inability of the roof rocks to support their own weight. More often, however, the collapse follows a period of drought during which the regional water table dropped and groundwater that normally filled the cave and supported the cave roof was removed.*

system may be entered through the bottom of a collapse sinkhole, the mouth of a spring, or the point where a cliff or road cut has intersected a cave. The most spectacular entry, however, is a *chimney* or *shaft*, which may extend downward vertically from the surface for hundreds of feet or meters (Figure 14.26).

Once the cave system is located above the water table, the **cave construction phase** begins. All of the incredibly beautiful structures seen in limestone caves form as water continues to move downward from the surface, attempting to reach the underlying water table. As water percolates downward from the surface and passes through the layers of limestone remaining in the cave roof, it becomes saturated with calcium carbonate. The water beads on the roof of the cave and drops to the cave floor. The tiny bit

FIGURE 14.26 *Caves and caverns can be entered in a variety of ways including (a) the bottom of collapse sinkholes, (b) at points where underground streams return to the surface, (c) where a passageway is exposed along a cliff face, and (d) perhaps the most dangerous of all, by way of* chimneys or shafts *that may extend vertically hundreds of feet below the surface. (a VU/© John D. Cunningham; b and c VU/© Bill Palmer; d VU/© Albert J. Copley)*

of water remaining on the roof of the cave evaporates and deposits a tiny bit of calcite called **dripstone**. As the drop hits the floor of the cave and spatters, the water evaporates, depositing another tiny bit of calcite on the cave floor immediately below. Given millions of years, this drop-by-drop process will produce a pendulous buildup of dripstone hanging from the cave roof, called a **stalactite**. Simultaneously, a comparable conical structure called a **stalagmite** builds upward from the cave floor. In time, the two structures may meet to form a **pillar** or **column** (Figure 14.27). More complex

FIGURE 14.27 *The structures that adorn the interiors of limestone caves literally form* drop by drop. *Pendulous (a)* stalactites *drop from the ceiling while (b)* stalagmites *build upward from the floor to meet stalactites and form (c)* pillars. *(d) Sheets of dripstone descend from fractures in the ceiling to form even more spectacular formations. (a and d VU/© Albert J. Copley; b VU/© Daniel W. Gotshall; c VU/© Bill Kamin)*

FIGURE 14.28 *One of the best-known cave systems in the United States is* Carlsbad Caverns *in New Mexico. (Courtesy of W. B. Hamilton/USGS)*

structures of dripstone, variously referred to as *bacon rind* or *bridal veils,* move down from overhead fractures. These beautiful features, created during the later constructional phase of the cave system, attract millions of visitors each year to such places as Mammoth Caves, Kentucky, and Carlsbad Caverns, New Mexico (Figure 14.28).

HOT SPRINGS, FUMAROLES, AND GEYSERS

In our discussion of volcanism, we introduced some of the more spectacular manifestations of groundwater, namely, *hot springs, fumaroles,* and *geysers.* Groundwater, heated by magmas or the heat contained in newly formed lavas and pyroclastic rocks, commonly rises to the surface to produce **hot springs** and **fumaroles.** A very special situation where a hot spring periodically erupts hot water and steam in a fashion similar to a fumarole is called a **geyser.** Perhaps the best-known area where all of these features can be seen is Yellowstone National Park in Wyoming (Figure 14.29).

Not all hot springs obtain their heat from magmatic sources. Along the western margin of the Appalachian Mountains are a series of thermal springs that are the result of groundwater penetrating deep enough within Earth's crust to be heated by the geothermal gradient. The waters rise and create springs in sites such as Warm Springs and Hot Springs, Virginia (Figure 14.30). In most cases, the waters are not exceptionally hot. Even in Hot Springs, Virginia, the water temperature is about body temperature. Although not as spectacular as some of the western thermal springs, the eastern springs have attracted visitors for a long time.

SPOT REVIEW

1. What are the characteristics of an area exhibiting karst topography?
2. How do limestone caves and caverns form?
3. When and how do limestone cave features such as stalagmites and stalactites form?

ENVIRONMENTAL CONCERNS

The problems associated with groundwater are legion, and textbooks have been written on the topic. The intent here is to give you a general idea of the kinds of problems that may be encountered when groundwater is used as a source of water.

(a)

(b)

(c)

FIGURE 14.29 *One of the bests-known areas in the world exhibiting groundwater emissions of various kinds is* Yellowstone National Park *in Wyoming where visitors can see a variety of thermal features from (a) hot springs to (b) fumaroles to (c) geysers. (a and b VU/© Richard Thom, c courtesy of J. R. Stacy/USGS)*

FIGURE 14.30 *The octagonal buildings enclosing the thermal springs at* Warm Springs, Virginia *were designed by Thomas Jefferson. Unlike the magmatically heated springs at Yellowstone Park, the water in the thermal spas throughout the Appalachian area is warmed by geothermal heat contained within the upper portion of the crust. (Courtesy of The Homestead, Hot Springs, VA)*

Groundwater Pollution

The purity of groundwater is normally considered to be equal to that of rainwater. Except for groundwater produced from water-soluble limestone aquifers, most groundwater contains a low level of dissolved materials. Unfortunately, since groundwater ultimately comes from the surface of Earth, contamination can be introduced from a wide variety of sources. With few exceptions, most pollution stems from human activities. In rural areas, for example, where groundwater provides water for domestic and farm use, major sources of contamination are human sewage and the effluents from livestock. The effluents from the drainage fields of *septic systems* and from *barnyards* and *feedlots* contain bacteria and organic contaminants that are potentially injurious to human health.

Fortunately, sediments and rocks contained in the zones of aeration and saturation have an impressive ability to remove such materials from even the most contaminated waters. The molecular attraction of the water to the surfaces of both particles within the regolith and mineral grains within the rocks slow the movement of the water and allows the bacteria to be physically filtered from the water and the organisms and organic chemicals to be chemically decomposed into harmless by-products. Although the efficiency of the process depends upon the rate of water movement, purification can be accomplished in most sand-sized materials within a few hundred feet (meters) of the point of entry. Nevertheless, there is a limit to what the natural purification system can do. Therefore, in locating wells to be used as water supplies for human consumption, the subsurface geology and the locations of potential sources of pollution must be considered (Figure 14.31).

Certain sources of groundwater contamination can be controlled if the potential for pollution is recognized before the source is established. A common example in temperate areas is the storage of salt-rich aggregate materials to be applied to snow- and ice-covered winter roads. Not only should such materials be stored under cover to prevent leaching due to rain or melting snow, but a water barrier such as a plastic sheet or packed clay should be placed beneath them to prevent salt water from infiltrating the groundwater.

The same potential problem exists with the ill-named **sanitary landfill** where effluents, generated with the fill as the materials decompose, may seep

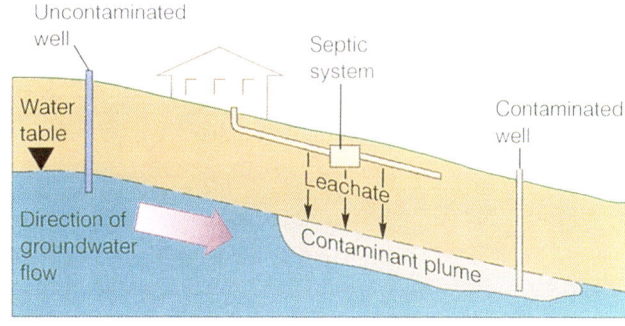

FIGURE 14.31 *A potential source of contamination in rural areas is the effluent from* septic systems. *To prevent the contamination of wells, septic systems should preferably be placed "downstream" from the well or beyond the influence of the septic system drainage field.*

downward and contaminate the groundwater supply (Figure 14.32). Although many local codes require the installation of either a water barrier to contain these effluents or a treatment system to purify the solutions draining from the fill before they are released into the environment, such legislation is not universal.

Dewatering

Another common problem encountered with wells and aquifers is **dewatering**, which simply means the loss of water. Again, the mechanisms of dewatering are many. One example will be presented to make a point. In figure 14.33, a perched water table is supplying water to several households along a ridge. The figure shows the same ridge after highway construction has removed a portion of the hill and, along with it, a significant portion of the aquitard that was responsible for the original perched water table. The position of the water table and the volume of stored water will readjust to the smaller area of the underlying aquitard. The resultant drop in the water table repositions its surface below the bottoms of the wells. Now located within the zone of aeration, the wells become dry. They have been "dewatered." If the reduced volume of water in the new perched water table system is inadequate to provide all the households with water, the only course of action is to drill new wells through the aquitard to the regional water table below, with the immediate expense of drilling and the longer-term increased cost of pumping the

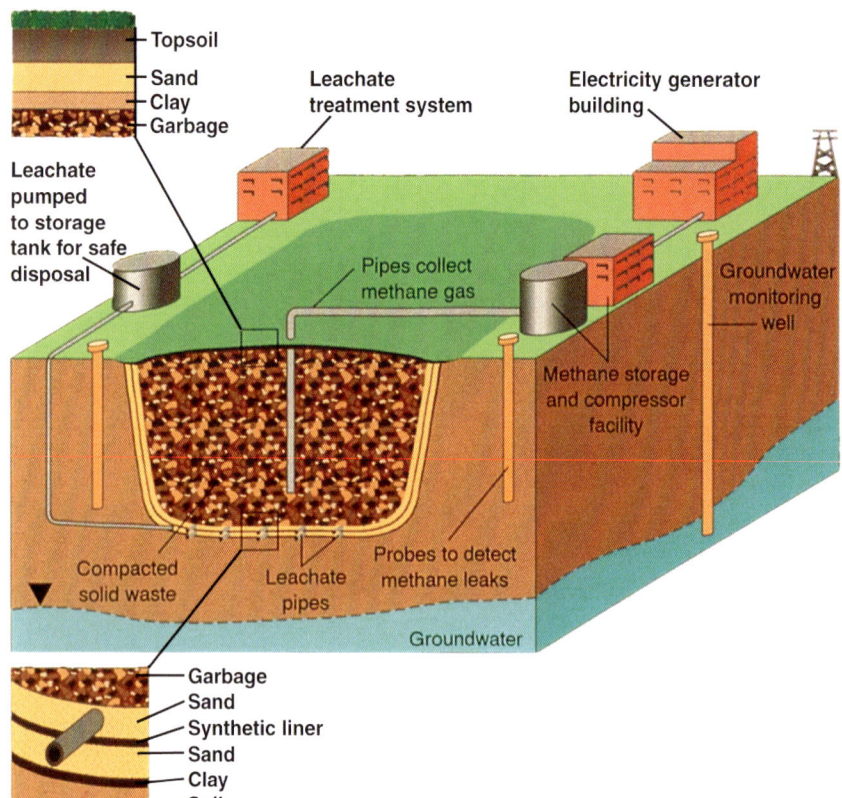

FIGURE 14.32 *Modern sanitary disposal systems are designed not only to collect and treat effluents so as to prevent groundwater contamination, but also to utilize the methane generated within the decomposing material as a source of energy.*

water from a deeper source. The point is that large volumes of rock cannot be removed from Earth, whether by excavating a road cut or opening a mine or a quarry, without potentially having serious effects upon the availability of groundwater in the immediate area.

Saltwater Encroachment

Those who live in coastal areas may experience special groundwater problems. Figure 14.34 illustrates a typical situation on any barrier island or beach area adjoining the ocean. Fresh water infiltrating from above displaces salt water that encroaches into the sediment below sea level. As a rule of thumb, for every foot that the water table rises *above* sea level, the fresh water-salt water contact is *depressed* 40 feet (12 m). As in any unconfined system, overproduction of water from the aquifer will cause the water table to be lowered. In this case, however, each *drop* in the elevation of the water table will be accompanied by a 40 times greater *rise* in the fresh water-salt water interface and subsequent encroachment of salt water. Eventually, salt water will seep into wells located at the fringes of the system making them unusable. Relief from saltwater encroachment has been achieved in some areas by pumping industrial waste

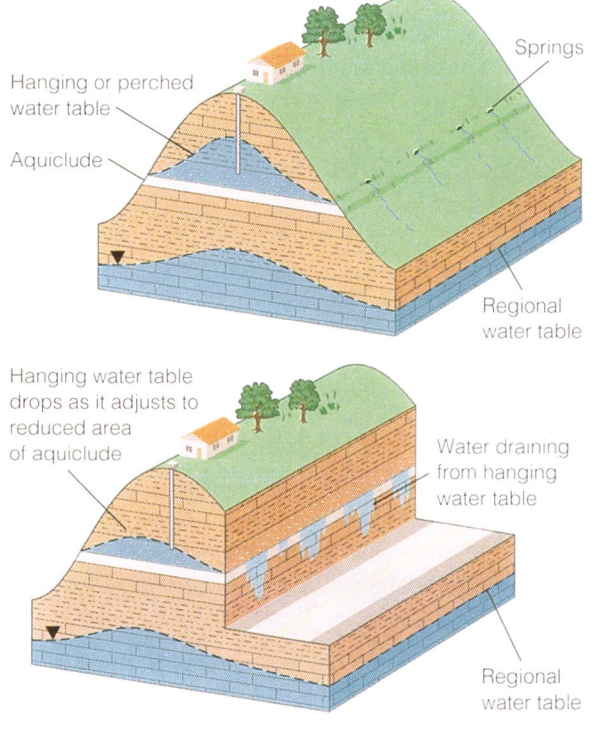

FIGURE 14.33 *Dewatering of aquifers, especially perched or hanging water tables, is always a potential problem when large volumes of rock are removed by construction or mining.*

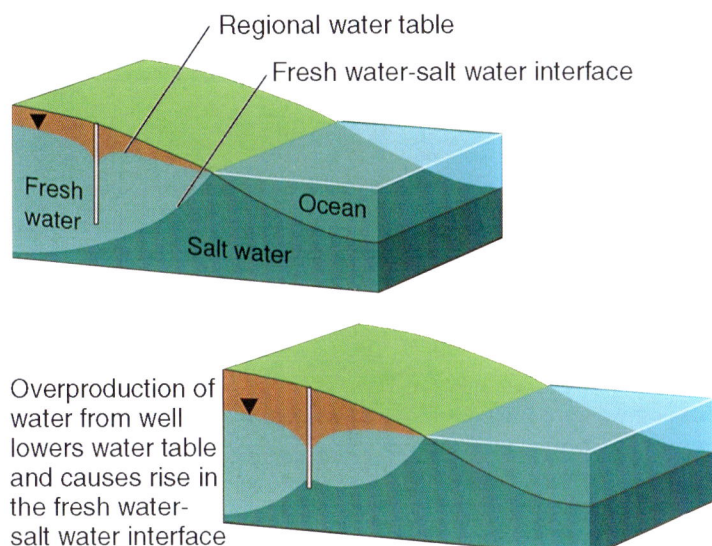

FIGURE 14.34 Saltwater encroachment *into* wells is a problem unique to coastal areas that may result when groundwater supplies are consumed faster than they are being recharged.

water into the aquifer system or by constructing surface containments that collect and divert rainwater into the groundwater supply to replenish the water lost by consumption (Figure 14.35).

The removal of more water from aquifers than is being replenished by natural or induced recharge is a general problem that is beginning to be experienced in many parts of the country. In areas of Arizona, New Mexico, and parts of California where water problems are particularly acute, the apparent solution has been to divert water from other parts of the country where surface waters are relatively plentiful. Unfortunately, the residents of water-rich states are coming to realize that they too may experience water shortages in the not-too-distant future and are becoming increasingly reluctant to allow their water to be transferred to other states.

SPOT REVIEW

1. Describe several ways in which an aquifer can be "dewatered."
2. What is the mechanism by which coastal wells are contaminated by salt water?

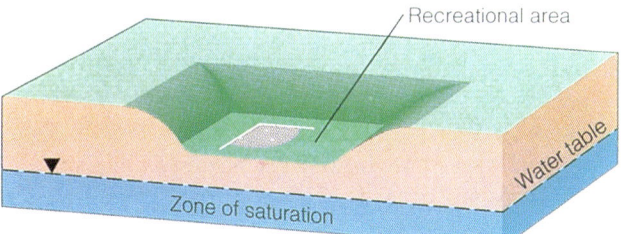

(a) During "dry" season

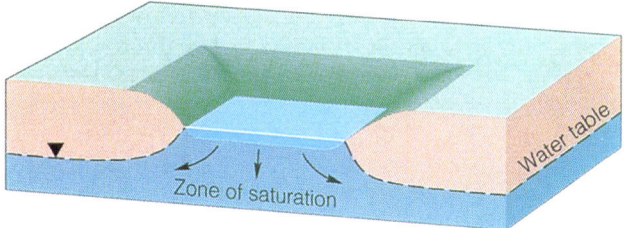

(b) During times of high surface runoff, water is collected and diverted to groundwater storage

FIGURE 14.35 *One way municipalities are attempting to ease the overproduction of groundwater is to construct catchment basins that collect rainwater that is subsequently diverted into groundwater storage. These basins are commonly large enough that, during times of low rainfall, they may serve as recreational areas such as ball fields.*

CONCEPTS AND TERMS TO REMEMBER

Development of porosity and permeability
  pores
    porosity
      intergranular porosity
  fractures
    fracture porosity
  permeability
  aquifer
  aquitard (aquiclude)
Rates of groundwater flow
  hydraulic head
  hydraulic gradient
The water table
  zone of saturation
  zone of aeration

regional water table
  losing (influent) stream
  gaining (effluent) stream
  perched or hanging water table
Water wells
  dug well
  cone of depression
  recharge
  confined aquifer
  unconfined aquifer
    pressure surface
Karst topography
  sinkhole
    solution sinkhole
    cockpit karst
    karst towers

swallow holes
caves and caverns
  cave development phase
  cave construction phase
    dripstone
      stalactite
      stalagmite
      pillar or column
Other groundwater features
  hot springs
  fumaroles
  geysers
Environmental concerns
  sanitary landfills
  dewatering
  saltwater encroachment

REVIEW QUESTIONS

1. Which of the following combinations is most improbable?
   a. high porosity-high permeability
   b. high porosity-low permeability
   c. low porosity-low permeability
   d. low porosity-high permeability

2. Which of the following rocks is most likely to be a good aquifer?
   a. sandstone   c. limestone
   b. shale   d. granite

3. The water level in a well producing from a confined aquifer will always rise up to the
   a. top of the aquifer.
   b. regional water table.
   c. surface of the ground.
   d. pressure surface.

4. In order for karst topography to develop, the area must
   a. be underlain by sandstone.
   b. be underlain by limestone.
   c. be associated with a confined aquifer.
   d. receive at least 40 inches of rainfall per year.

5. A pendulous dripstone structure hanging from a cave roof is called a
   a. stalagmite.   c. stalactite.
   b. pillar.   d. hoodoo.

6. The ability of water to move through an aquifer is largely dependent on the rock's
   a. composition.   c. porosity.
   b. thickness.   d. permeability.

7. How does particle size affect permeability?

8. How does sorting affect the porosity and permeability of a sediment or a rock?

9. Why are shales aquitards even though they may have porosities as high as 50% to 60%?

10. What is responsible for the pressures developed within confined aquifers?

11. Exactly what happens when streams "dry up" during periods of drought?

12. Why do customers in the fringes of municipal water systems find themselves with low water pressure and, at times, no water at all during the summer months?

13. How may the saltwater encroachment problem commonly encountered in coastal areas be alleviated?

THOUGHT PROBLEMS

1. Assume that you have been hired to locate the septic system and water well for a rural home. What factors must you take into account to ensure that the effluents from the septic system do not contaminate the well?

2. Consider an area of karst and the problems inherent in obtaining domestic water supplies and disposing of waste of all kinds. What problems do you anticipate for the rural inhabitants in the area? What particular problems do you anticipate for any urban development?

FOR YOUR NOTEBOOK

For most of us who live in urban areas, contact with groundwater and the problems associated with it are limited. In some communities, however, the municipal water supply is obtained from wells. Inquire about the source of water for your community. Information concerning all aspects of groundwater can be acquired from the Groundwater Branch of the U.S. Geological Survey (USGS) or from your state geological survey. If neither organization has an office in your immediate area, ask your instructor for the address of the state or USGS office. Ask about groundwater problems that are specific to your area and actions that have been taken to remedy them.

In most areas, a growing and very serious problem is the disposal of waste and the possibility that waste disposal sites will contaminate the local groundwater supply. Investigate how domestic and industrial wastes are disposed of in your area. Do any municipal regulations govern or limit solid waste disposal?

Another area you may investigate is the deep disposal of toxic liquid wastes and the disposal of low- and high-level radioactive wastes, all of which have the potential to seriously affect the groundwater in the region.

CHAPTER OUTLINE

INTRODUCTION

STRESS, DEFORMATION, AND THE STRENGTH OF MATERIALS
  Kinds of Stress
  Kinds of Deformation

THE STRENGTH OF ROCKS

ROCK DEFORMATION
  Elastic Deformation
  Plastic Deformation
  Brittle Deformation

THE CONCEPT OF STRIKE AND DIP

GEOLOGIC STRUCTURES
  Folds
  Faults
  Faulting after Initial Folding
  Joints

DETERMINING THE DIRECTION OF ROCK MOVEMENT
  Rock Transport
  Actual Direction of Fault Movements
  Relative Direction of Fault Movements

CHAPTER 15

Rock Deformation and Geologic Structures

INTRODUCTION In earlier discussions, we saw that sedimentary rocks are originally deposited as horizontal layers. If you see horizontal sedimentary rocks in a hillside or road cut, it means that the rocks were not tilted as they were uplifted (Figure 15.1). Unless they are exposed in an outcrop such as the Grand Canyon, horizontal sedimentary rocks are not exceptionally awe inspiring, even to a geologist. The surrounding scenery is likely to be more eye-catching, whether it is the rolling hills of the Appalachian Plateau or the rugged beauty of the Colorado Plateau. When tipped on end, however, soaring vertically for hundreds of feet, or intricately folded and contorted, sometimes broken and thrust one over another, sedimentary rocks capture the imagination like no other kind of exposure (Figure 15.2). How did they get that way? What force could bend and twist these "rock-hard" materials like so much taffy?

These bend and broken rocks exhibit the basic geologic structures of *folds*, *faults*, and *joints*. The forces that are responsible for making geologic structures and deforming the rocks are the topics of this chapter.

FIGURE 15.1 *Horizontal rocks observed in outcrop, such as these at Dead Horse Point, Utah, indicate that they were uplifted and exposed by erosion without being either folded or broken. (VU/© Albert Copley)*

FIGURE 15.2 *Sedimentary rocks such as these exposed near Pinto, West Virginia, were folded during the continent-continent collision that formed the Appalachian Mountains and created Pangaea. Keep in mind that folded rocks such as these were deformed deep within a zone of subduction and are now exposed only after thousands of feet of rock were removed by erosion. (Courtesy of Robert Behling)*

STRESS, DEFORMATION, AND THE STRENGTH OF MATERIALS

This chapter is about *stress, deformation,* and the *strength* of materials. **Stress** is *force per unit area,* whether it is the stress you apply to mold a piece of modeling clay or the stress that creates mountains. The response to stress, called **strain** or **deformation**, is the *change* in either *shape* or *volume* of the material undergoing stress. Deformation may or may not be permanent. The breaking of a window by the stress of an impacting rock is permanent deformation while the flattening of a tennis ball by the impact of a racket is not.

Every solid material has an inherent **strength**, which is the ability of the material to withstand stress. Once stress *exceeds* the strength of the material, the object will *deform;* that is, it will *change* in either shape or volume. It is important, however, to realize that the amount of deformation may not be sufficiently great to be detected by the unaided eye. For example, the deformation in rocks before they rupture to cause an earthquake can only be detected by very sensitive instruments.

Kinds of Stress

There are two fundamentally different kinds of stress: (1) *compression* and (2) *tension.*

In **compression**, the forces act *toward* each other. In some cases, the forces are *directly opposed* to each other, and the material is squeezed (Figure 15.3). An example would be the stress a pencil experiences as you hold it between your fingertips. In other cases, the forces are directed toward each other, but along *parallel, nonopposing* paths (Figure 15.4). These **shear** forces, as they are called, can be modeled by pushing on the diagonal edges of a deck of cards or by turning a doorknob (Figure 15.5).

The stresses involved in **tension** also act directly opposite each other and along parallel paths, but they act *away* from each other (Figure 15.6). In other words, tension is the *opposite* of compression. The stretching of a rubber band is an example of a tensional extension.

The stresses that deform the rocks of Earth's crust have many sources. The greatest stresses are associated with the plate margins. Tensional stresses are generated at divergent margins while enormous compressional stresses are created at convergent margins. Other more localized, and significantly less intense, stresses are generated by a number of

FIGURE 15.3 *The forces involved in* compression *act* toward *and* directly opposite *each other. With few exceptions, this is the force you apply whenever you pick up an object.*

FIGURE 15.4 *The forces in* shear *also are directed* toward *each other, but are not directly opposed. This is the kind of force you apply when you take the cap off a bottle or turn a doorknob.*

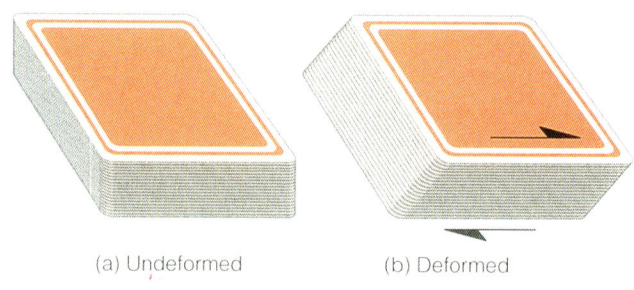

(a) Undeformed (b) Deformed

FIGURE 15.5 Shear Forces *can be modeled using a deck of cards. Applying forces directed toward but not directly opposite each other causes the cards to move over one another (shear), all in the direction of the applied force.*

FIGURE 15.6 *The forces involved in* tension *operate* directly opposite *and* away *from each other. Anytime you pull on an object, be it a drawer or a rubber band, you are applying tensional forces.*

processes including frost wedging, the near-explosive impact of waves at the base of a wave-cut cliff, and the force of gravity, which drives mass wasting, bows Earth's surface down under huge masses of glacial ice, and causes the lithospheric plates to move.

Kinds of Deformation

We will consider three different kinds of deformation: (1) *elastic*, (2) *plastic*, and (3) *brittle*. A fourth type of deformation, *viscous*, is experienced by liquid rock such as intruding magmas and mobile rocks such as those of the asthenosphere. Because our discussions will involve only deformation of the crust, we will not consider viscous deformation further.

Elastic Deformation   A generalized stress-strain relationship is illustrated in Figure 15.7. The first response to stress is **elastic**. The characteristic of elastic deformation is that when the stress is *released*, the object will *return* to its original shape; that is, it is not permanently deformed. Think of squeezing a rubber ball or stretching out a rubber band. The importance of the elastic response is that while the material is being stressed, some of the applied energy is *stored* within the material and is *released* when the stress is removed. It is also important to realize that the released energy can be used. For example, the impact of a tennis ball and a racket results in the elastic deformation of both the ball and the racket strings. As the tennis ball and the racket strings return to their original shape, the combined energy is released and drives the tennis ball over the net.

At some point during elastic deformation, deformation will begin to increase nonlinearly with stress, signaling that the material has reached its **elastic limit** (point 1 in Figure 15.7). If the material is stressed beyond the elastic limit, it will undergo permanent deformation, either plastic or brittle.

Plastic Deformation   If stresses are applied slowly, especially under conditions of high confining pressure and high temperatures, the material will begin to undergo **plastic deformation**. During plastic deformation, the energy applied to the object is *absorbed* internally by friction and is used to rearrange the makeup of the material as it flows to take on a new *shape*. An example would be the response of a ball of modeling clay as you squeeze it in your fist. The modeling clay *remains* deformed when the stress is released. The clay cannot go back to its original shape because there is no stored energy to

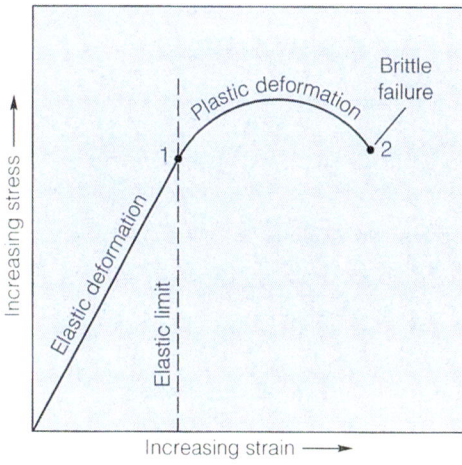

FIGURE 15.7   Strain *or* deformation *is the response to* stress. *These are responses that are part of our everyday life. When subjected to tensional forces, a rubber band deforms by* elastic deformation *and stretches, but returns to its original shape when the stress is released. The butter you spread on your toast deforms by* plastic deformation. *Once it is spread, it stays spread. Finally, we are all familiar with the common example of* brittle deformation, *the breaking of a glass as it hits the floor.*

generate a stress to make it go back. Structural geologists would say that the material "has no memory." Except for the tiny amount of energy involved in elastic strain, nearly all of the applied energy was absorbed and *used* to deform the clay.

Brittle Deformation   When some materials are subjected to high rates of stress application, especially under low confining pressures and low temperatures, they break with *little* or *no* plastic deformation beyond the elastic limit (point 2 in Figure 15.7). Such materials are said to undergo **brittle deformation**. As the material fails, the energy stored during elastic deformation is released. Following brittle failure, the material may once again undergo elastic deformation.

A few comments concerning the significance of breaking are in order. Solid materials release energy from their surface. The sound of a ringing bell or a crystal goblet struck with the fingernail is the response of your ear to the energy released from the surface as the bell or glass deforms elastically. Consider now striking the glass a little too hard or perhaps dropping it. The amount of energy that can be released during any interval of time is limited by the surface area of the object. Because the amount of energy applied to the glass by the impact is more than can be absorbed and released from the *existing* surface of the glass, the glass breaks in order to generate the *additional* surface area needed to release the excess energy. The number of pieces into which the glass will break depends upon the amount of energy that must be released. In contrast to the melodious ringing of the crystal goblet carefully struck with a fingernail, the familiar, discordant sound of breaking glass is the result of the mixing of the sounds from a number of different-sized and odd-shaped "bells."

The Effect of Time Whether a solid material deforms as a plastic or as a brittle solid is often determined more by the *rate* at which the energy was applied than the *amount* of energy. Consider, for example, the response of glass to stress. Most individuals would consider glass to be a brittle solid as evidenced by the ease with which it breaks when impacted by a rock. As a rock strikes a pane of glass in a window, the rate of energy application is very high. Within a fraction of a second, the strength of the glass

THE STRAIN ELLIPSOID

A helpful tool used by geologists to portray the response of rocks to stress (strain) and to illustrate the various strain orientations represented by the geologic structures is the **strain ellipsoid**. An ellipsoid is a geometric figure where the maximum, intermediate, and minimum diameters are mutually perpendicular and the corresponding sections are ellipses (Figure 15.B1). A sphere is a special case where all axes are equal in length and the corresponding sections are circles.

To illustrate the use of the strain ellipsoid, consider Figure 15.B2. In Figure 15.B2a, the sphere represents the distribution of strain within a mass of buried rock that is being subjected to compressive forces equally from all directions. Such forces would be the confining forces to which rocks are subjected at depth. Now consider subjecting the rock mass to a *directional* compressive stress. The sphere will be deformed into an ellipsoid as the three axes change length (refer to Figure 15.B2b). The axis parallel to the direction of the maximum compressive stress experiences the greatest shortening (strain)

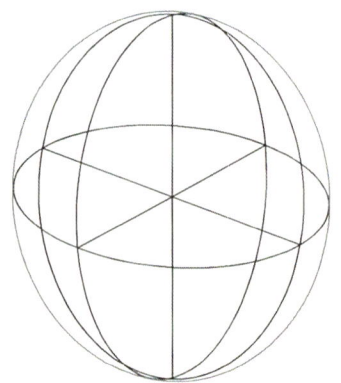

Figure 15.B1 *An* ellipsoid *is a geometric figure with three mutually perpendicular axes of different lengths and corresponding elliptical cross sections. A sphere is a special case where the three mutually perpendicular axes are equal in length and all sections are circles.*

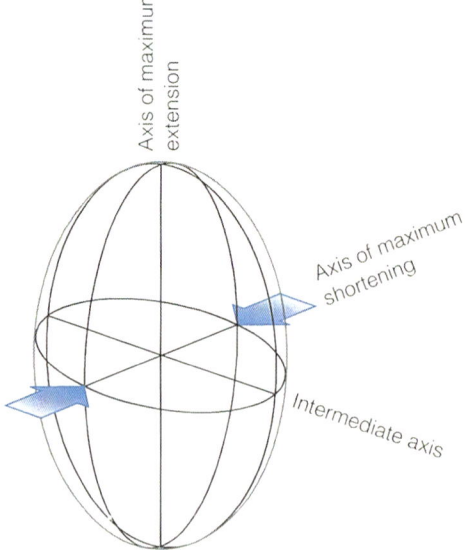

Figure 15.B2 *An* ellipsoid *forms as one axis of a sphere becomes the direction of maximum directional compressive stress and shortens. In order to maintain a constant volume and remain symmetrical, one of the other axes becomes elongated while the length of the third remains essentially unchanged. In terms of the strain ellipsoid, the three axes are called the* axis of maximum shortening, *the* axis of maximum extension, *and the* intermediate axis, *respectively.*

is exceeded, the applied energy is far beyond the amount that can be absorbed by elastic deformation, the glass responds as a brittle solid, and it breaks.

Consider, on the other hand, the same pane of glass standing on the end in the door frame of an antique china closet, subjected to the *same* amount of energy but this time in the form of gravity pulling downward on the glass over a period of a hundred or more years. The rate of energy application will be very low. During a century or more of confinement within the frame of the china closet door, the glass will have sufficient time to absorb the energy, make internal molecular rearrangements, deform plastically, and flow downward. Variations in thickness in the glass will be created in the form of horizontal wavy lines across the pane while vertical shortening of the pane will be evident at the upper portion of the door frame. Antique hunters look for these results of plastic deformation as evidence for the age of glass in considering potential purchases.

and is therefore called the **axis of maximum shortening**. Of the two remaining axes, one will undergo maximum extension (strain) and is called the **axis of maximum extension**. The remaining axis is the **intermediate axis**.

The response of rocks to stress can be represented by three different orientations of the strain ellipsoid (Figure 15.B3). Note that the strain ellipsoids in Figures 15.B3a and 15.B3c both characterize the response of rocks to horizontal compressive forces such as those that exist at convergent plate margins. The axis of maximum extension of the strain ellipsoid shown in Figure 15.B3b, on the other hand, is horizontal, and the axis of maximum shortening is vertical. This is the strain orientation characteristic of areas undergoing extension, such as the margins of divergent plates.

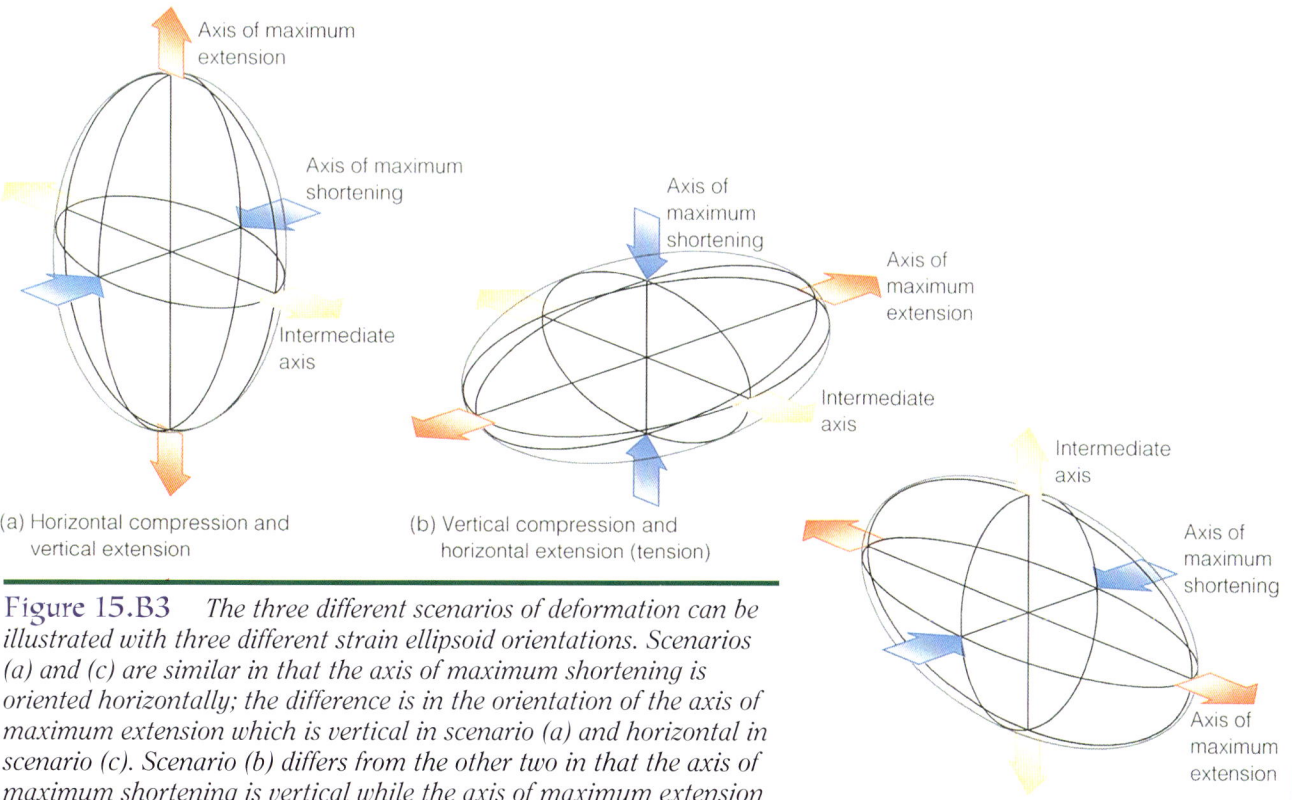

Figure 15.B3 *The three different scenarios of deformation can be illustrated with three different strain ellipsoid orientations. Scenarios (a) and (c) are similar in that the axis of maximum shortening is oriented horizontally; the difference is in the orientation of the axis of maximum extension which is vertical in scenario (a) and horizontal in scenario (c). Scenario (b) differs from the other two in that the axis of maximum shortening is vertical while the axis of maximum extension and the intermediate axis are horizontal.*

An example of a natural material that may respond by brittle failure under one set of conditions but deform plastically under another is ice. In our discussions of glaciers, we pointed out that ice, although a brittle solid when responding to a short-term stress as a jab of an ice pick, deforms plastically when it accumulates to thicknesses in excess of 150 feet (46 m) and begins to flow under the force of gravity.

The Effect of Temperature and Pressure Increasing temperatures usually favor the plastic response of solids, an example being the ease with which glass rods bend when heated to redness over a Bunsen burner. Increased confining pressure increases the ultimate strength of most solids by extending the range of plastic deformation. As a result of the combined effects of temperature and pressure, rocks under stress at or near Earth's surface usually behave as brittle solids and break, while with increased depth of burial and the resulting increase in both temperature and pressure, rocks become increasingly plastic and tend to fold or flow.

Effect of Rock Composition The composition of rocks may critically control their response to stress. For example, under similar conditions a sandstone composed of brittle quartz grains may fail by brittle fracture while an adjoining shale made up primarily of clay minerals may deform plastically and flow. Water contained within rocks usually promotes a plastic response by enhancing the dissolution of the more water-soluble minerals while at the same time reducing cohesion and friction at the contacts of others.

SPOT REVIEW

1. What is stress?
2. How are the strength of a material, stress, and deformation related?
3. What is the difference between compression and shear?
4. What is the elastic limit of a material?
5. How do plastic and brittle deformation differ? In what way(s) are they similar?
6. How do time, pressure, and temperature affect the deformation response of a material?

THE STRENGTH OF ROCKS

Rocks are very strong under compression but weak under tension. Perhaps a historical example would serve to demonstrate. When one compares the architectural styles of the ancient Greeks and Romans, the major difference is that the Greeks made prolific use of **columns** whereas the Romans used **arches** (Figure 15.8). The Greeks were master column makers, not so much because they wanted to be, but because they never discovered how to overcome the basic weakness of rocks under tensional forces. The Greeks spanned the distance between columns with flat rock slabs. They found that when they attempted to span too great a distance, the rock slab would sag and eventually break when loaded from above. The rock slabs failed due to *tensional* forces that are generated within any slab supported at its ends. A load placed on the slab increases the amount of tensional stress. As a result, the Greeks had to limit the distances to be spanned by using many columns spanned with short rock slabs, all of which provided a lot of work for column makers.

The Romans, on the other hand, invented the Roman arch. The secret of the Roman arch is the **keystone** (Figure 15.9). When loaded from above, the wedge-shaped keystone directs the forces outward against the first blocks within the arch. The contact between the keystone and the first block of the arch is under compression, as are all the contacts between successive blocks down to the supporting columns and continuing through the columns to the ground surface. No parts of the arch or the supporting columns are under tension. Because rocks are *very* strong under compression, the arch and supporting columns can support enormous weight. The concept of the Roman arch has been responsible for all the beautiful vaulted arches constructed throughout history (Figure 15.10). It is this inherent strength under compression that allows a volume of rock at depth to support the weight of all the overlying rocks.

ROCK DEFORMATION

Once the strength of a rock is exceeded, its response depends upon (1) the kind of stress, (2) the rate of application of the stress, (3) the conditions of confining pressure and temperature, (4) the composition of the rock, and (5) the presence of fluids.

FIGURE 15.8 *One reason the ancient Greeks used columns so prolifically was that they never learned how to overcome the inherent weakness of rocks under tension. (Courtesy of M. M. Reddy/USGS)*

FIGURE 15.9 *The Romans discovered that positioning a wedge-shaped rock at the top of an arch allowed it to support enormous loads. (VU/© Bayard H. Brattstrom)*

FIGURE 15.10 *We can thank the Romans for all of the beautiful vaulted ceilings in the buildings around the world. (VU/© Bruce Berg)*

Elastic Deformation

The initial response of rocks to stress is elastic. Some examples of elastic deformation involve large volumes of Earth's crust. During the Great Ice Age, for example, a large portion of the North American continental crust centered over what is now Hudson Bay and a comparable region underlying the Scandinavian Peninsula were depressed thousands of feet by the weight of the overlying ice mass, displacing the underlying asthenospheric rocks. Since the ice melted about 10,000 years ago, these two areas have been rising at the rate of about 3 feet (1 m) per century. Because the lithosphere is physically joined to the asthenosphere, the rate of elastic rebound is being controlled by the rate at which the asthenospheric rocks flow back under the crust. As the crust returns to its original elevation, the water will drain from Hudson Bay and the Baltic Sea and return to the North Atlantic Ocean. As is the case with all elastic deformation, once the crust returns to its pre-Ice Age position, no indication will remain that the rocks were ever deformed.

Another example of the elastic response of large volumes of rocks, but on a totally different scale in terms of both time and amount of rock movement, involves the passage of *seismic* (earthquake) waves through Earth's interior. As seismic waves propagate, the rocks are subjected to stress for a period of a few seconds or minutes and are moved millimeters or fractions of millimeters. The rocks respond elastically and are neither broken nor permanently deformed but return to their pre-earthquake positions after the seismic wave passes.

The elastic response of relatively small volumes of rock, such as that exposed in a single cliff face or road cut, is so small that it can be detected only by very sensitive instruments called strain gauges. Yet, the combined amount of energy stored within the rocks may one day be released as a destructive earthquake.

Plastic Deformation

In our discussion of plate tectonics, the asthenosphere was described as being *plastic*. You will remember that it is the heat-driven plastic flow of rocks within the asthenosphere that reduces the frictional contact between the asthenosphere and the lithosphere to the point where the lithospheric plates can move over the underlying asthenosphere. Because of the high temperatures and pressures within the asthenosphere and mantle, the rocks tend to respond to stress as plastic solids and do not break.

As foldbelt mountains are created by the collision of lithospheric plates, the conditions of temperature and pressure within the cores of the mountains are oftentimes sufficient to metamorphose and cause the plastic deformation of rocks. The swirling patterns of color commonly seen in marble are an example.

Brittle Deformation

The outermost layer of Earth, the lithosphere, is made up of the *brittle* crust and the outermost *brittle* portion of the mantle. When stress is applied at rapid rates or in great amounts, the lithosphere will tend to break. Indeed, it has. The lithosphere is broken into the *plates* that move relentlessly over the surface of Earth.

Not all rock deformation within the lithosphere is by brittle failure, however. Within the volume of rock involved in the creation of a mountain range, for example, localized stress conditions generated by the movements of the lithospheric plates may result in either plastic deformation or brittle failure of the rocks. The folds that characterize the great mountain ranges of Earth are examples of the plastic deformation of the lithosphere (Figure 15.11) while the generation of faults is an example of brittle deformation (Figure 15.12).

FIGURE 15.11 *The folds seen in this satellite photograph of the northern Appalachian Mountains are a spectacular example of plastic deformation of the lithosphere. For reference, note the Potomac River in the center near the bottom and Harrisburg, Pennsylvania to the middle right. (Courtesy of NASA)*

Rock Deformation and Geologic Structures   461

FIGURE 15.12 *This linear fault trace is an example of the breaking of the brittle crust in response to deformational forces. (Courtesy of R. E. Wallace/USGS)*

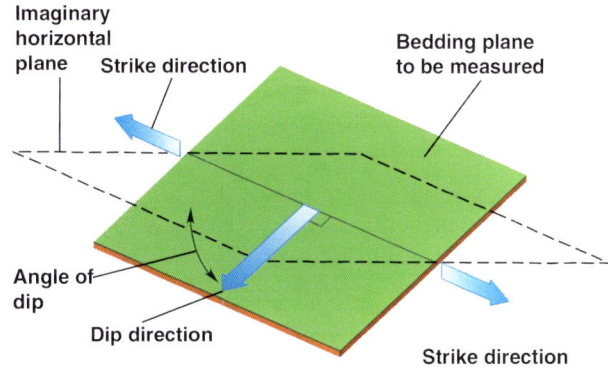

FIGURE 15.13 *The two measurements of* strike *and* dip *allow a geologist to describe a planar surface in three-dimensional space. Note that the directions of strike and dip are mutually perpendicular.*

Largely because no effect of the elastic response to stress remains after the stress is released, most geologists, except for seismologists (those who study earthquakes), tend to be concerned only with brittle or plastic deformation.

THE CONCEPT OF STRIKE AND DIP

The features we will soon discuss, folds, faults, and joints, all have *planar* (two-dimensional) components. The method used by geologists to describe the three-dimensional orientation of these planar surfaces employs two parameters: (1) *strike* and (2) *dip* (Figure 15.13).

Strike is defined as the *direction* of the *line* of intersection between the plane and a horizontal plane. The definition makes two important points: (1) strike is a *direction* and (2) strike is a *horizontal line*. Although the direction of either end of the line will suffice to describe the direction of the line, we may want to specify the use of one end rather than the other. By convention, for example, the direction of strike is always noted relative to the *north magnetic pole*.

Dip is defined as the *angle* the plane makes with the horizontal and is always measured *downward* from the horizontal. Although not implicit in the def-

inition, the *direction* of the dip must also be given because a plane may dip in two different directions (to either side of the line of strike).

Strike and dip directions are mutually *perpendicular* (refer to Figure 15.13). Once the strike of a plane has been determined, the direction of dip can be only in one of two directions. For example, if the strike of a plane is to the north, the dip must be to the east or west.

GEOLOGIC STRUCTURES

Geologic structures are (1) *folds,* (2) *faults,* and (3) *joints*. The plastic response of rocks to stress produces folds. Faults and joints form by brittle failure.

Folds

Folds form primarily by compressional stress. This can be easily demonstrated with a sheet of paper. You can form folds easily by pushing any two opposite edges of the sheet *directly* toward each other. Folds may also form by shear forces where compressive forces act along parallel, nonopposing paths. This can be demonstrated by moving opposing edges of a sheet of paper toward each other while simultaneously moving them in opposite directions. Although folds can be formed by nonopposing compressive forces, most folds form by directly opposing forces.

Basically, there are three different kinds of folds: (1) *monoclines,* (2) *anticlines,* and (3) *synclines.* The simplest fold is the **monocline** in which the layers are tilted in *one direction*. Monoclines commonly form around the margins of local or regional uplifts and in the strata above faults (Figure 15.14).

Rocks may be arched either *upward* to form an **anticline** or *downward* to form a **syncline**. Normally, anticlines and synclines form together as is illustrated in Figure 15.15a.

Anticlines and synclines are described using (1) the strike and dip of the *limbs* relative to the *axial plane*, (2) the direction of the fold *axis*, (3) the *plunge* of the fold axis, and (4) the *amplitude* of the fold.

The **axial plane** is an imaginary plane that parallels the length of the fold and divides the cross section of the fold in half. The **limbs** are the flanks of the fold, which make an angle with the axial plane. The **axis** of the fold is the line of intersection between the axial plane and the limbs. The **plunge** is the angle between the axis and the horizontal. The **amplitude** of the fold is the distance from the top of the anticline to the bottom of the adjacent syncline (refer to Figure 15.15a).

Based on the angle and direction of the dip of the limbs relative to the axial plane, folds are classified as (1) *symmetrical*, (2) *asymmetrical*, (3) *overturned*, or (4) *recumbent*. If the limbs dip *away* from the axial

FIGURE 15.14 *The simplest kind of fold, the* monocline, *is illustrated by the inclined outcrop of rocks that surrounds the San Rafael Swell, Utah.*

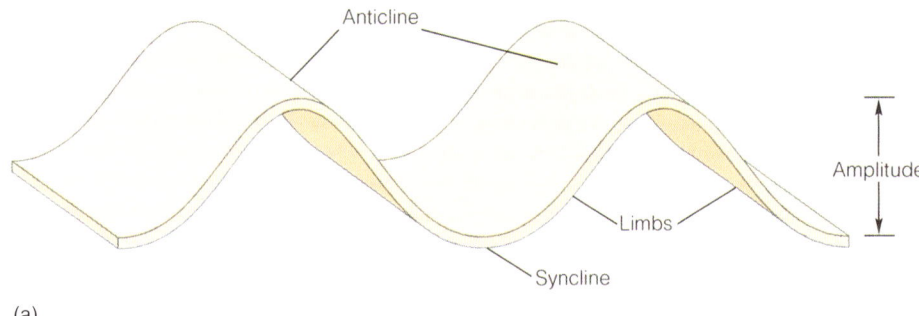

(a)

FIGURE 15.15 *The most commonly observed folds are anticlines and synclines. For simplicity, only one bed has been shown in the drawing. Anticlines and synclines are common sights in road cuts in areas of foldbelt mountains. (Photo b courtesy of Tom Kammer, c courtesy of Maryland Department of Highways)*

(b)

(c)

plane in *opposite* directions at *equal* angles, the fold is described as **symmetrical** (Figure 15.16a). Folds whose limbs dip *away* from the axial plane in *opposite* directions but at *different* angles are called **asymmetrical** (Figure 15.16b). If both limbs and the axial plane dip in the *same direction*, the fold is referred to as being **overturned** (Figure 15.16c). When rocks are subjected to extreme compression, the folds may assume the attitude shown in Figure 15.16d, where the axial plane *approaches the horizontal*. Such folds are called **recumbent**. Symmetrical, overturned, or recumbent folds where the limbs are parallel are described as being **isoclinal**.

Along most of its length, the axis of a fold is approximately horizontal; that is, the plunge is zero. Eventually, however, every fold comes to an end by plunging as the axis dips away from the horizontal and the amplitude of the fold diminishes (Figure 15.17).

Plastic deformation of rocks usually requires such high confining pressures and temperatures that they cannot be achieved anywhere near Earth's surface. Because folded rocks are a common sight at Earth's surface, we tend to forget that we see them only after they have been uplifted from their place of origin deep within Earth and exposed by millions of years of weathering and erosion.

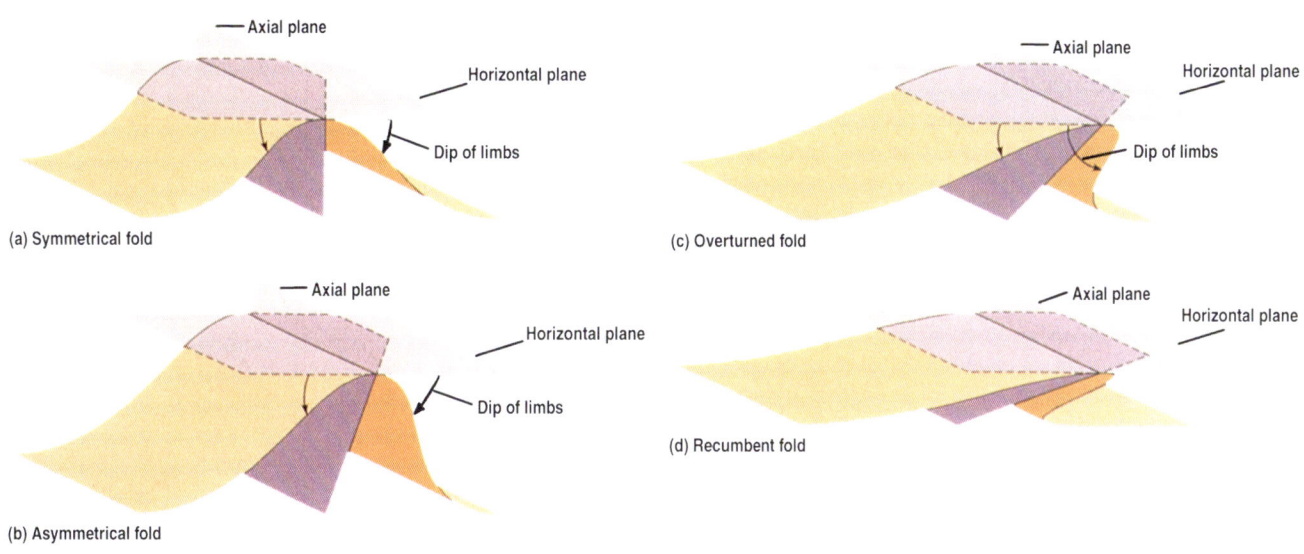

FIGURE 15.16 *Folds are described as* symmetrical, asymmetrical, overturned, *or* recumbent *depending on the relationship of the* axial plane *and the* limbs *relative to the* horizontal.

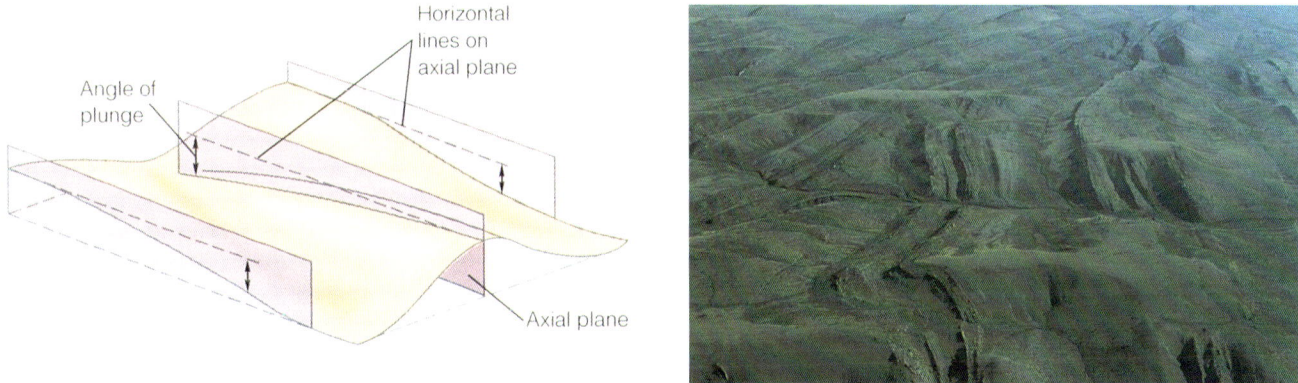

FIGURE 15.17 *Folds eventually come to an end by* plunging *as the amplitude of the fold decreases along its length. As erosion removes the axial region of an anticline, the eroded limbs dip away from the axis of the fold and wrap around the plunging structure as is shown in the photo. Refer also to Figure 18.11 to observe the plunge of folds in the satellite photograph of the Appalachian Mountains. (Photo VU/© Martin G. Miller)*

FOLDS AND THE STRAIN ELLIPSOID

Each of the different types of folds has a characteristic orientation of the strain ellipsoid. Consider the symmetrical folds illustrated in Figure 15.B4 and note the orientation of the strain ellipsoids. In both cases, the axis of maximum extension is vertical, the axis of intermediate extension is parallel to the axis of the fold, and the plane of the axes of intermediate and maximum extension is parallel to the axial plane of the fold. The greater amplitude of the fold in Figure 15.B4b is represented by the increased length of the axis of maximum extension while the increased deformation that produced the higher-amplitude fold is represented by the shorter axis of maximum shortening.

Consider now the nonsymmetrical folds in Figure 15.B5. The nonsymmetrical folds form by the application of horizontal, nonopposing, compressive forces.

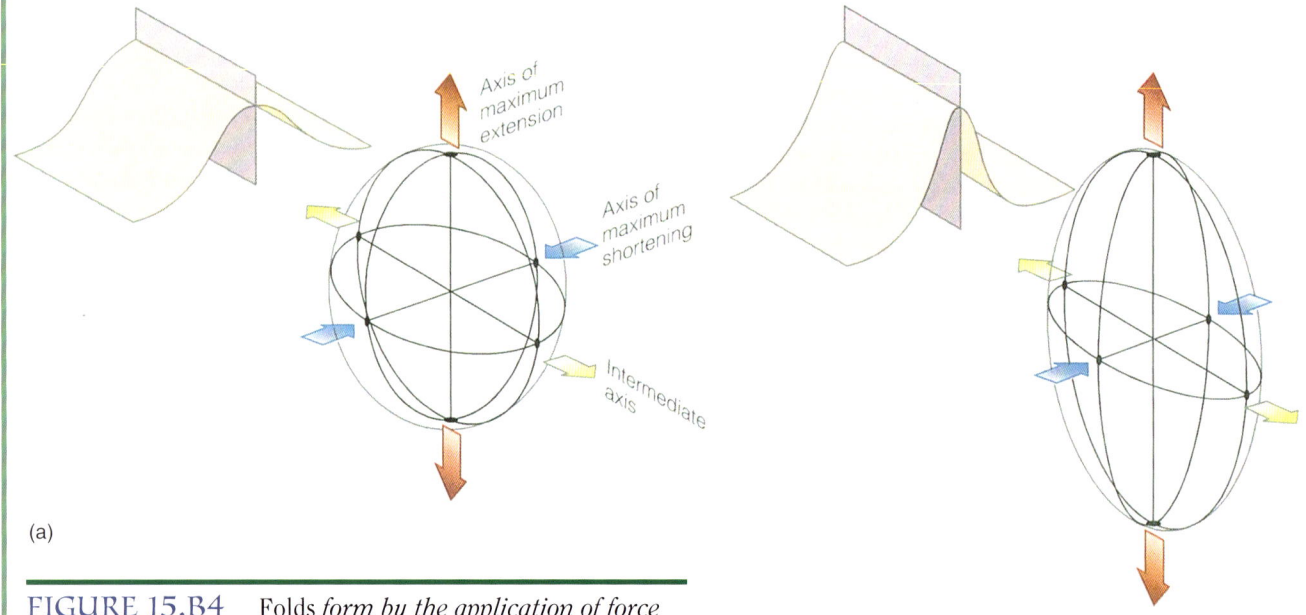

FIGURE 15.B4 Folds *form by the application of force described by the strain ellipsoid orientation shown in scenario (a) of Figure 15.B3. In all cases, the intermediate axis is parallel to the axis of the fold.* Symmetrical folds *form by the application of compressive forces such that the axis of maximum shortening is* horizontal *and the axis of maximum extension is* vertical. *As a result, as stress is applied, the crustal segment is shortened and the rocks arch* upward *into anticlines and/or* downward *into synclines. Continued application of force results in increased fold* amplitude *and additional shortening of the crust.*

SPOT REVIEW

1. Give an example of rocks undergoing elastic deformation.
2. Why would you not expect rocks to undergo plastic deformation at Earth's surface?
3. What is strike? How are strike and dip related?
4. What is the axial plane of a fold? How does the orientation of the axial plane differ among the various kinds of folds?
5. How do the forces that produce symmetrical folds differ from those that produce nonsymmetrical folds?
6. What orientation of the strain ellipsoid is common to all folds?

Faults

Faults are relatively planar breaks in the lithosphere along which the rocks have been *offset*. In other words, all faults form as a result of *shear*. The frequency of faulting is highest near Earth's surface

The effect of such forces is to rotate the strain ellipsoid around the intermediate axis. Note that, as was the case with the symmetrical folds, the intermediate axis is always parallel to the axis of the fold and the axial plane parallels the plane of the axes of intermediate and maximum extension. The three types of folds differ mainly in the change in the orientation of the axial plane as increased deformation rotates the strain ellipsoid.

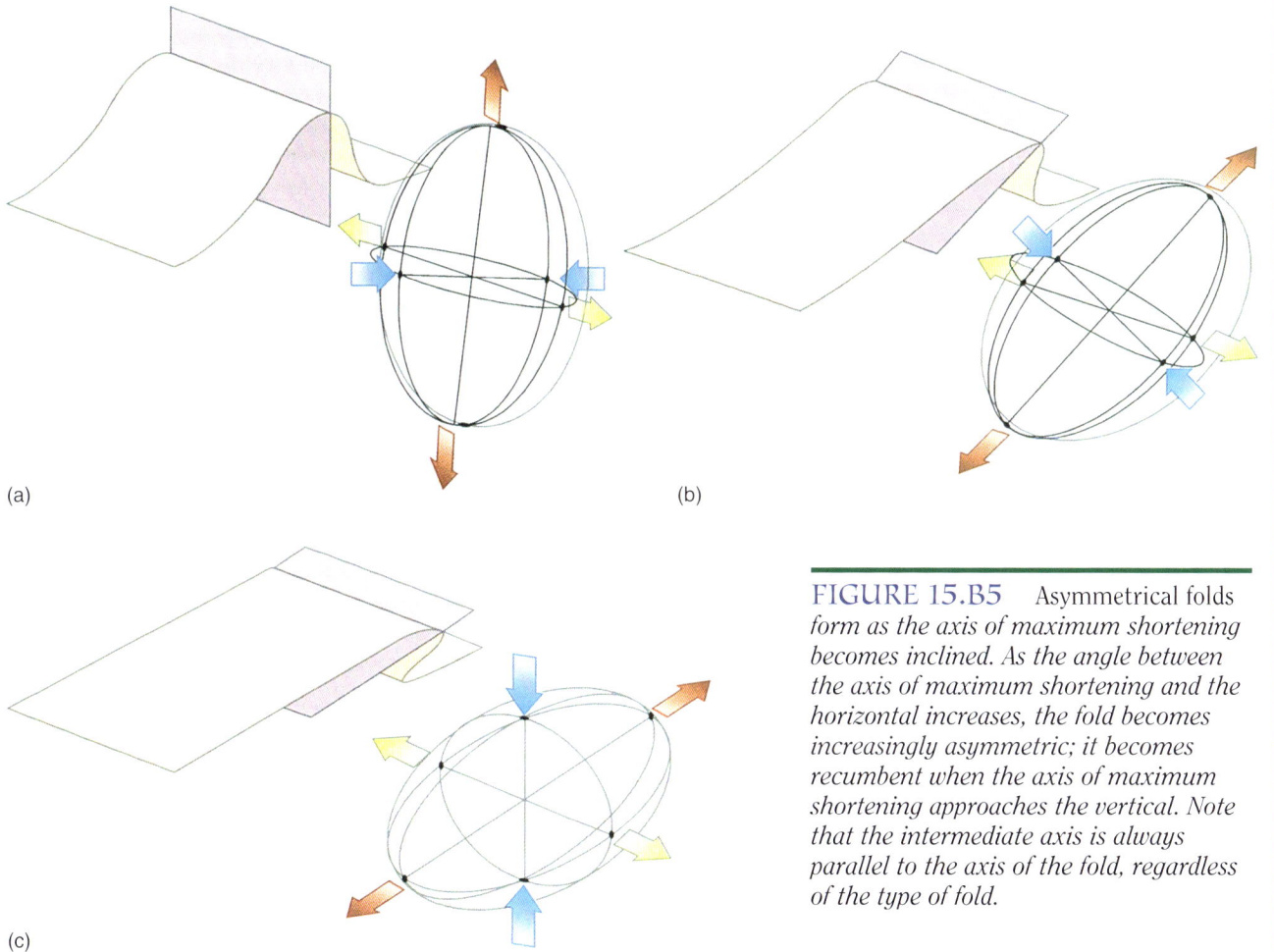

FIGURE 15.B5 *Asymmetrical folds form as the axis of maximum shortening becomes inclined. As the angle between the axis of maximum shortening and the horizontal increases, the fold becomes increasingly asymmetric; it becomes recumbent when the axis of maximum shortening approaches the vertical. Note that the intermediate axis is always parallel to the axis of the fold, regardless of the type of fold.*

where the rocks are most brittle and decreases with depth as the rocks become increasingly plastic due to the increase in temperature, pressure, and the presence of fluids. Deep-focus earthquakes disappear with depth along subducting plates, indicating that below a depth of about 450 miles (725 km), rocks respond totally as plastic solids and faults do not occur. Within continental crust, the brittle-plastic transition occurs at a depth of about 10 miles (15 km).

Most earthquakes result either from movement along existing faults or from a rupture that initiates new faults. Before rocks can be deformed by faulting, they must first undergo elastic deformation. The energy that destroys buildings and triggers the mass wasting during an earthquake is some of the energy that was stored in the rocks as they underwent elastic deformation over a period of perhaps hundreds or thousands of years. Because the elastic response of rocks under localized conditions is *very* small, there is generally no visual evidence that energy is being stored within the rock. Yet all the while the rock is silently undergoing strain. When the ability of the rock to store additional energy is exceeded, it breaks or slips along a new or preexisting break, and the stored energy is released. Some of the energy is used to move the rocks on opposite sides of the fault. The remaining energy is released as seismic waves. A seismic wave has many of the same characteristics as the waveform that is generated by a breaking stick and is converted by our ear to a sound. We will learn more about seismic waves in Chapter 18.

The three kinds of stress conditions mentioned earlier result in the generation of different kinds of faults.

Reverse or Thrust Faults   Figure 15.18 shows a block of rock that has failed under compression where the forces were acting directly opposite each other. Upon fracturing, the rock is displaced (sheared) along a planar surface that is inclined to the direction of compression and at an angle to the horizontal. Some of the energy stored in the rock during the elastic phase of deformation is used to move the blocks of rock on opposite sides of the fault plane in opposite directions. The block *above* an inclined fault plane is called the **hanging wall**, and the one *below* the fault plane is called the **foot wall**. Miners coined these terms many years ago as a result of following mineral veins that are commonly emplaced along faults. When the miners were at work, one block *hung* over their heads while the other was below their feet (refer to Figure 15.18).

The fault shown in Figure 15.18, which formed by compression with the hanging wall moving *up* relative to the foot wall, is a **reverse fault**. A reverse fault where the dip of the fault plane is less than 45° is often referred to as a **thrust fault**.

Three measurements are made on the most faults: (1) *displacement,* (2) *throw,* and (3) *heave* (refer to Figure 15.18). The **displacement** is the amount of offset measured along the fault plane. The **throw** and **heave** are the apparent *vertical* and *horizontal* components of movement, respectively. In the case of a reverse or thrust fault, the throw is the amount of crustal *uplift* generated by the fault, and the heave is the amount of crustal *shortening* the fault has created. All of these measurements may vary in magnitude along the fault plane and will diminish to zero at the ends of the fault.

A major thrust fault in the eastern United States underlies the Blue Ridge Mountains of the Appalachians and the rocks to the east. Although the actual displacement is neither known or agreed upon, it is estimated that during the formation of the Appalachians, the rocks that make up the Blue Ridge Mountains and the Piedmont to the east were moved along this thrust fault at least 50 miles (80 km) westward from their original location.

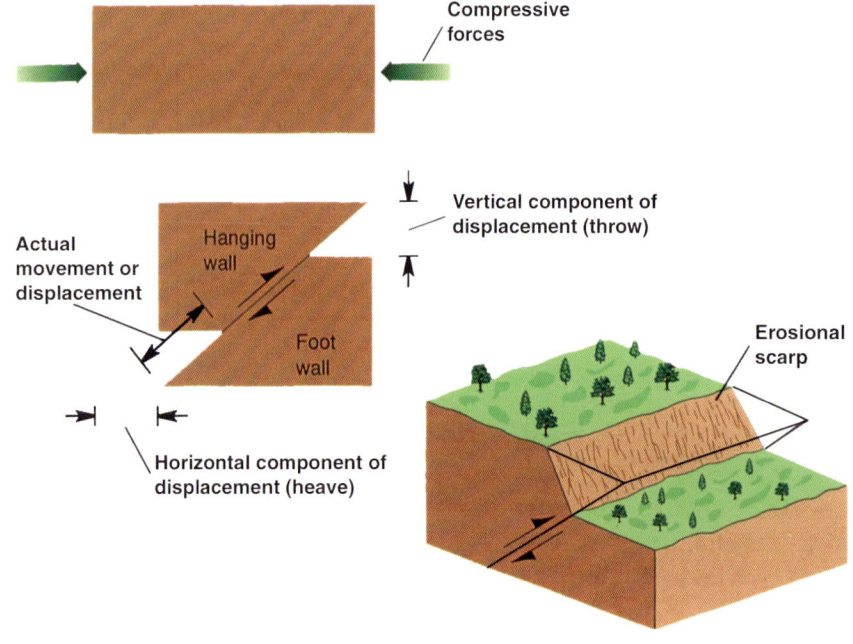

FIGURE 15.18   Reverse *or* thrust faults *form by the brittle failure of rocks under compression. As a result of the movement, the length of Earth's crust is* shortened *and the crust is* locally thickened.

Rock Deformation and Geologic Structures   467

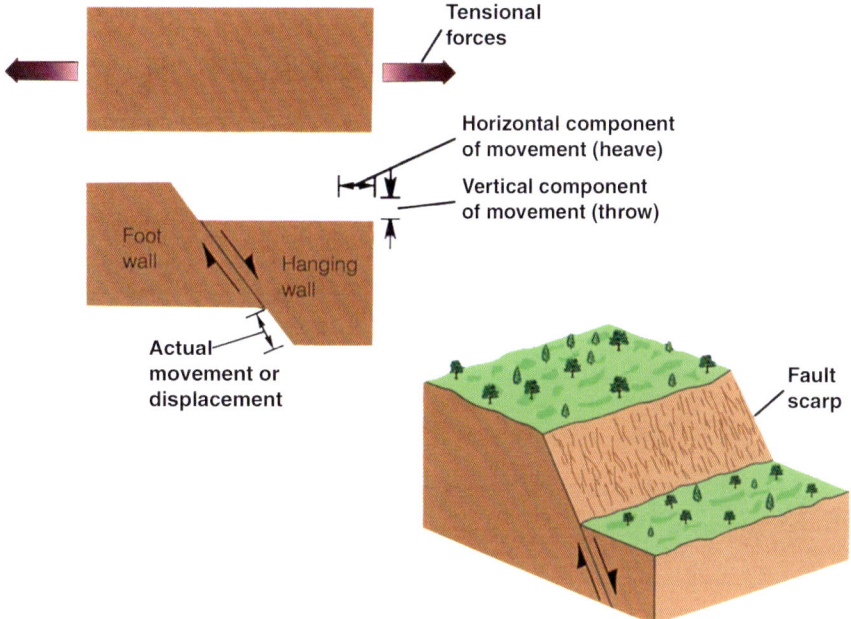

FIGURE 15.19 Normal faults form by the brittle failure of rocks under tension. As a result of the movement, the length of Earth's crust is lengthened. The three-dimensional drawings in Figures 18.18 and 18.19 illustrate the potential difficulty of identifying the type of fault based on the surface expression. Even though the forces and rock movements involved in reverse and normal faulting are exactly opposite, the surface scarps may be identical in appearance. Note, however, that one scarp is an actual fault surface while the other is erosional.

Normal Faults As previously indicated, rocks are weakest under tensional stress and will usually fail by brittle fracture. Figure 15.19 shows a block of rock before and after failure under tensional stress. The fault, where the hanging wall moves down relative to the foot wall, is called a **normal fault**. Displacement, throw, and heave for a normal fault are measured in the same way as for a reverse fault.

Note that the orientation of forces that produce normal faults is opposite to the orientations of forces that produce reverse faults. As a result, certain characteristics of the two fault types are also opposites. For example, the relative movements of the hanging and foot walls are reversed. Note also that normal faults result in crustal *extension* or *elongation* while reverse faults produce crustal *shortening*.

In cases where the dip of reverse or normal faults decreases with depth, the faults are referred to as being **listric** (Figure 15.20).

Strike-Slip Faults When a segment of Earth's crust breaks as a result of compression where the shear zone is oriented vertically, the fracture surface will be vertical or near-vertical and parallel to the direction of the compressive stress. The blocks on either side of the fault move horizontally in opposite directions and produce a **strike-slip fault**. The name arises from the fact that the movement (the *slip*) is

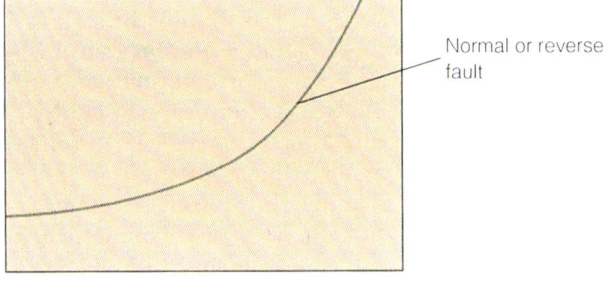

FIGURE 15.20 *The term* listric *refers to normal or reverse faults that decrease in angle with depth. For examples of listric faults, refer to Figures 19.10 and 19.12.*

parallel to the *strike* (the compass direction) of the fault plane (Figure 15.21).

When describing strike-slip faults, the terms foot wall, hanging wall, throw, and heave have no meaning because there is no vertical movement of the rocks. Displacement, however, is measured as before.

Strike-slip faults are described as being either **right-lateral** or **left-lateral** faults depending upon the relative direction of movement of the two blocks (Figure 15.22). The best-known strike-slip fault in the United States, the San Andreas Fault, is responsible for many of the earthquakes that shake the west

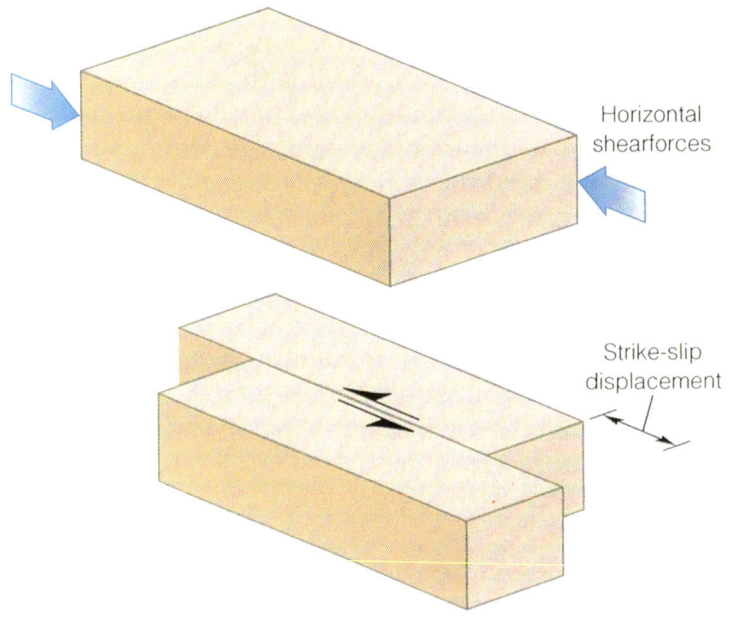

FIGURE 15.21   Strike-slip faults *have only horizontal displacement. The best-known strike-slip fault in North America, the San Andreas Fault, is the boundary between the North American and Pacific lithospheric plates and the source of most earthquakes in the United States.*

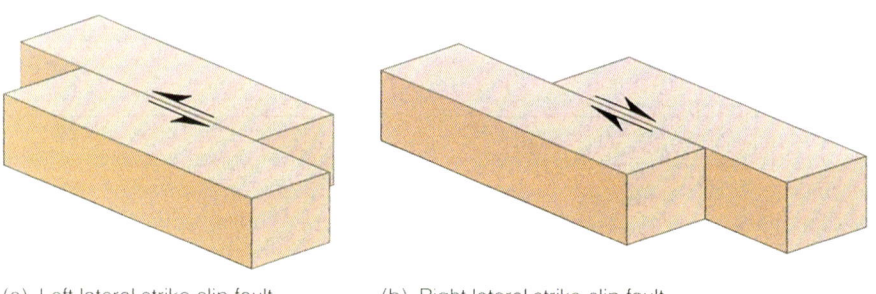

FIGURE 15.22   *Strike-slip faults are described as* right-lateral *or* left-lateral *depending on the relative movements of the rocks on opposite sides of the fault. Picture yourself walking toward the fault and asking, "Which way do I have to turn to reach the same point on the other side of the fault?" The San Andreas Fault is an example of a right-lateral strike-slip fault.*

coast. The San Andreas Fault is also the plate boundary between the North American and Pacific plates.

Oblique-Slip or Diagonal-Slip Faults   In some cases, the stress orientation results in two components of rock movement. One component will be horizontal, within the plane of the fault and *parallel* to the strike of the fault plane, while the other will be within the plane of the fault and *perpendicular* to the strike of the fault plane (Figure 15.23). The resultant movement will angle across the fault surface and produce an **oblique-slip** or **diagonal-slip fault**.

Faulting after Initial Folding

Folding and faulting can develop during the same deformational episode if the rocks break after a period of plastic deformation. The onset of deformation initiates the formation of folds within a horizontal shear zone. As the folds develop to the highly asymmetrical or overturned stage, the volume of rock simultane-

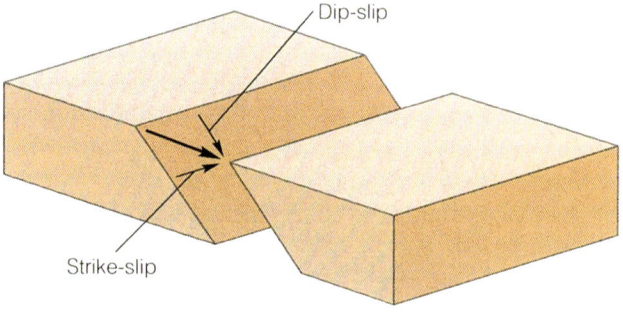

FIGURE 15.23   Oblique-slip *or* diagonal-slip faults *are those that have both a dip-slip and a strike-slip movement.*

ously *shortens* and *thickens* in order to maintain a constant volume. Think of squeezing a stick of modeling clay between your hands. As the rock section thickens, the strength of the rock mass increases. Eventually, the rocks begin to resist continued folding and become unable to consume additional energy by plastic deformation. At this point, the rocks will break and develop reverse faults. The faults will invariably break the steeper or overturned limb of the anticlinal folds. Deformation then proceeds with additional shortening and uplift as the folds are thrust one on top of another (Figure 15.24). This type of deformation is common during the formation of foldbelt mountains where thick sequences of sedimentary rocks are involved. The process will be discussed in more detail in Chapter 19 dealing with mountain building.

Joints

A **joint** is one of a set of fractures in Earth's crust where only opening displacement has occurred with no appreciable displacement parallel to the fracture surface. Joints are the most common geologic structures, being found in all exposed rocks.

Joints occur in sets within which the individual joints are parallel or subparallel to each other. There are two major types of joints depending on the orientation of the fractures relative to the direction of maximum stress (Figure 15.25). Joints that are not parallel but form at an angle to the direction of stress are called **shear joints**. Shear joints usually occur as two equally well developed sets intersecting at an *acute* angle of about 80° that is bisected by the axis of greatest stress (Figure 15.26). In **tensional joints**, the fractures are aligned *parallel* to the direction of maximum compressive stress (Figure 15.27). They are called tensional joints because the fracture opens by tension. Think of how the cards in a deck of cards open when you compress the deck along the long dimension of the cards.

Certain types of joints are specifically associated with igneous rock bodies. It is common to find joints parallel to the surface of an igneous rock body, apparently formed by the release of pressure as the

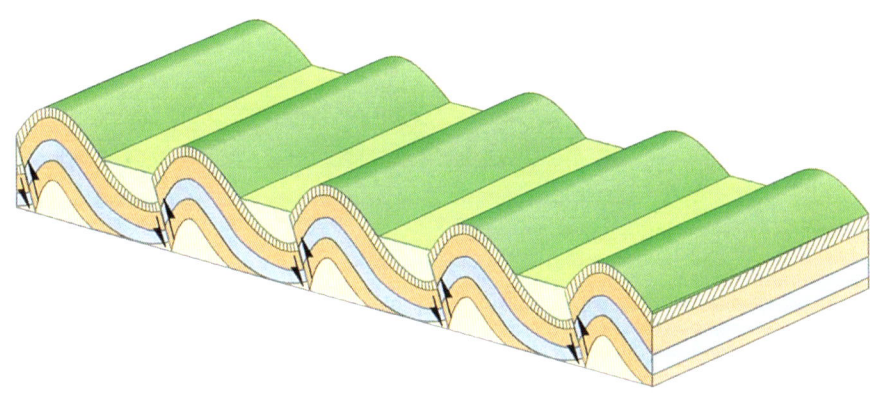

FIGURE 15.24 *A common type of deformation, especially in foldbelt mountains, is asymmetric or overturned folds that have been broken by reverse faults on the oversteepened limb. Refer to Figure 18.31 to see combined folding and thrust faulting in the structure of the Appalachian Mountains.*

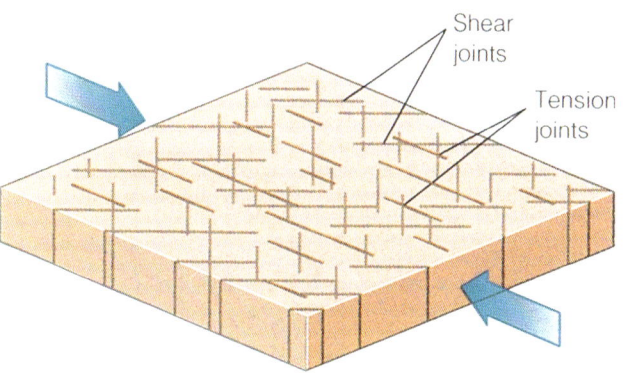

FIGURE 15.25 *Of all the geologic structures, joints are the most common. There are two types of joints;* shear joints *which form at an angle to the direction of applied stress, and* tension joints *which form parallel to the direction of applied stress.*

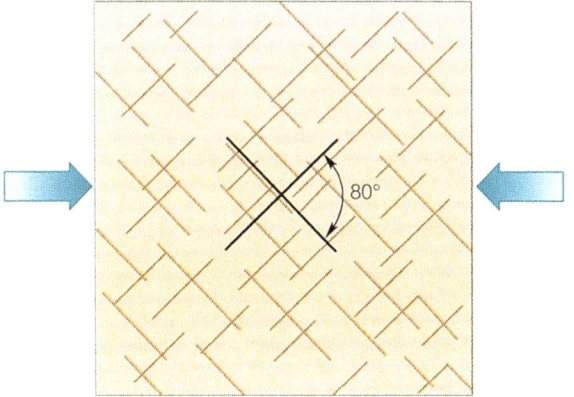

FIGURE 15.26 *Shear joints consist of two sets of fractures that intersect at an angle of about 80° to the direction of applied stress. Geologists use this relationship to determine the direction of applied stress.*

FIGURE 15.27 *The well-developed joints oriented parallel to the pencil are the tension joints. The ruler is approximately parallel to one set of shear joints. Note the other set of shear joints oriented at nearly right angles. (VU/© Martin G. Miller)*

FIGURE 15.28 *The sheeting exhibited by the plutonic rocks exposed in Yosemite Park is the result of joints that formed parallel to the surface of the igneous body when pressure was released as erosion removed the overlying rocks. (Courtesy of N. K. Huber/USGS)*

FIGURE 15.29 Columnar jointing, *such as that exposed at Devil's Post Pile, California, is a common type of jointing that forms in many types of igneous intrusions as the rock mass shrinks during cooling and solidification. (© Tony Freeman/Photo Edit)*

overlying rocks are removed by weathering and erosion (Figure 15.28). A type of joint found in a variety of igneous rock bodies is called **columnar jointing**. Columnar joints typically break the rock into elongate hexagonal columns, generally perpendicular to the cooling surface; they are formed by the tensional stresses that are generated as the rock mass shrinks upon cooling (Figure 15.29).

In some cases, geologists agree on neither the time nor the mode of formation of joints. We normally think of the formation of joints as the brittle response of an existing rock. Field evidence, however, indicates that in sedimentary rocks some joints may form not long after burial while the sediments are still consolidating. Studies of jointing in some folded sedimentary rocks show that, regardless of the orientation of the bedding, the fractures are always nearly perpendicular to the bedding, indicating that the joints formed before folding began (Figure 15.30). Based on our previous discussions that indicated that the initial response to stress is elastic followed in order by plastic and brittle deformation, you might question why rocks would fail by brittle deformation

Rock Deformation and Geologic Structures

even before they began to deform plastically. You will remember, however, that under conditions of low temperature and low confining pressure, as are found near Earth's surface, rocks can go directly from elastic deformation to brittle failure with little or no plastic deformation.

SPOT REVIEW

1. What fault movements are measured by heave, throw, and displacement, respectively?
2. How does the orientation of the strain ellipsoid differ between reverse and normal faults? What do the different strain ellipsoid orientations say about the difference in the forces that produced the faults?
3. How would you determine whether a strike-slip fault was right or left lateral?
4. How do the relative movements of the hanging and foot walls differ between reverse and normal faults?
5. What are the differences between shear and tensional joints?
6. How are the orientations of shear and tensional joints used to determine the direction of maximum compression?

FIGURE 15.30 *Joints oriented perpendicular to the bedding of folded sedimentary rocks, regardless of the angle of dip, indicate that the joints formed before the rocks were deformed. (VU/© Doug Sokell)*

FAULTS AND THE STRAIN ELLIPSOID

As was the case with folds, the stresses involved in the formation of the various types of faults can be portrayed by the strain ellipsoid. When materials are stressed to the point of brittle failure, fractures develop at angles of about 30° to the direction of maximum compressive stress. Consider Figure 15.B6. Figure 15.B6a illustrates the orientation of the strain ellipsoid for a reverse fault. Note that the intermediate axis parallels the strike of the fault, the crustal shortening is in the direction of the axis of maximum shortening, and the crust thickens as a result of the fault movement in the direction of the axis of maximum extension. Note also that the fault makes an angle of about 30° with the direction of maximum shortening.

As we have observed, normal faults are the opposite of reverse or thrust faults. Compare the strain ellipsoid orientations in Figures 15.B6a and in the strain ellipsoid in Figure 15.B6b has been rotated about the intermediate axis by 90°, with the result that the directions of maximum shortening and extension are opposite to those shown for the reverse fault. Note also that the fault plane makes an angle of about 30° with the vertical, which explains why most normal faults are high angle.

Figure 15.B6c illustrates the orientation of the strain ellipsoid for a strike-slip fault. As was the case with the reverse and normal faults, the intermediate axis is within the fault plane. Because both the axes of maximum shortening and extension have a horizontal orientation, however, the intermediate axis does not parallel the strike of the fault.

(continues on next page)

FAULTS AND THE STRAIN ELLIPSOID *continued*

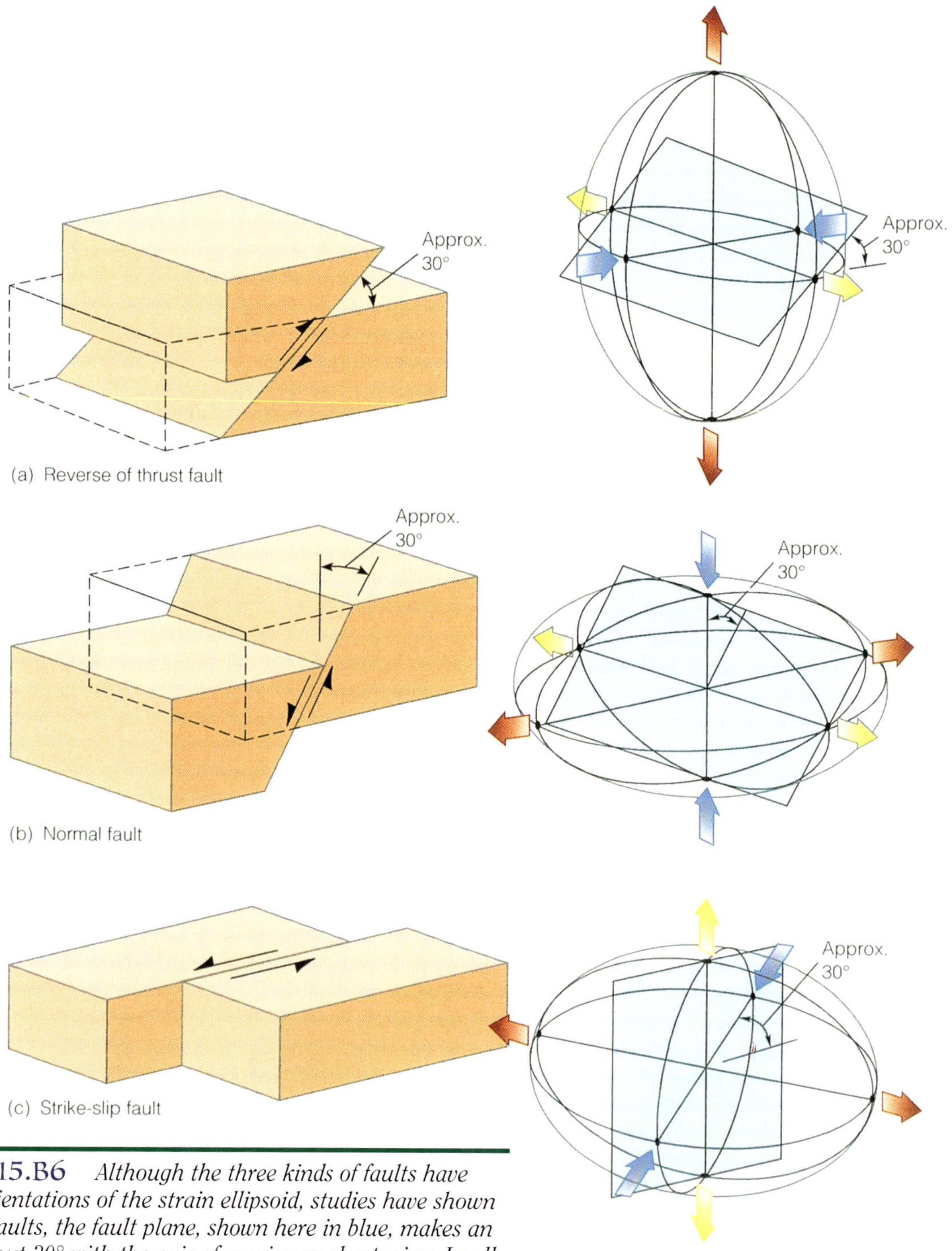

(a) Reverse of thrust fault

(b) Normal fault

(c) Strike-slip fault

FIGURE 15.B6 *Although the three kinds of faults have different orientations of the strain ellipsoid, studies have shown that in all faults, the fault plane, shown here in blue, makes an angle of about 30° with the axis of maximum shortening. In all cases, the intermediate axis is located within the fault plane.*

Reverse faults *result in the vertical stacking of rocks in response to the vertical orientation of the axis of maximum extension.* Normal faults, *on the other hand, result in an orientation of the axis of maximum extension.*

In strike-slip faults, *the intermediate axis is oriented vertically within the fault plane with the axis of maximum shortening oriented horizontally and at an angle of about 30° to the fault plane. The orientation of the axis of maximum shortening relative to the fault plane determines whether the fault will be right-lateral or left-lateral. The fault shown in Figure 18.B6c is left-lateral. Note, however, that by rotating the axis of maximum shortening 60° clockwise, the fault movement becomes right-lateral.*

DETERMINING THE DIRECTION OF ROCK MOVEMENT

Figures such as those used earlier to demonstrate folding and faulting might give the impression that the opposing rock masses were both moving; in other words, that the movement of the opposing rock masses is comparable to the head-on collision of two moving automobiles. This, however, is not always the case. As rocks undergo deformation, the stresses may be the result of one rock mass impinging on another in much the same fashion as a car collides with a wall. An example is the collision of India and Asia that gave rise to the Himalayas. In this case, we are fairly certain that India was moving northward while Asia was relatively stationary. In many other cases, however, we are unable to determine the actual movement of two impinging plates based on the observed deformation. Consider that the deformation experienced by the head-on collision of two automobiles each traveling at 30 miles per hour (50 kph) will be the same as that resulting from the collision of an automobile traveling at 60 miles per hour (100 kph) with one that is parked.

Rock Transport

Although we may not be able to determine the *actual* direction of rock movement, we can make judgments as to the *relative* directions of the rock movement initiated by the deformation. This is accomplished by determining the direction of **maximum rock transport** observed in the resultant structures. Figure 15.31 illustrates the generalized structures observed in the Appalachian Mountains. The continental collision that

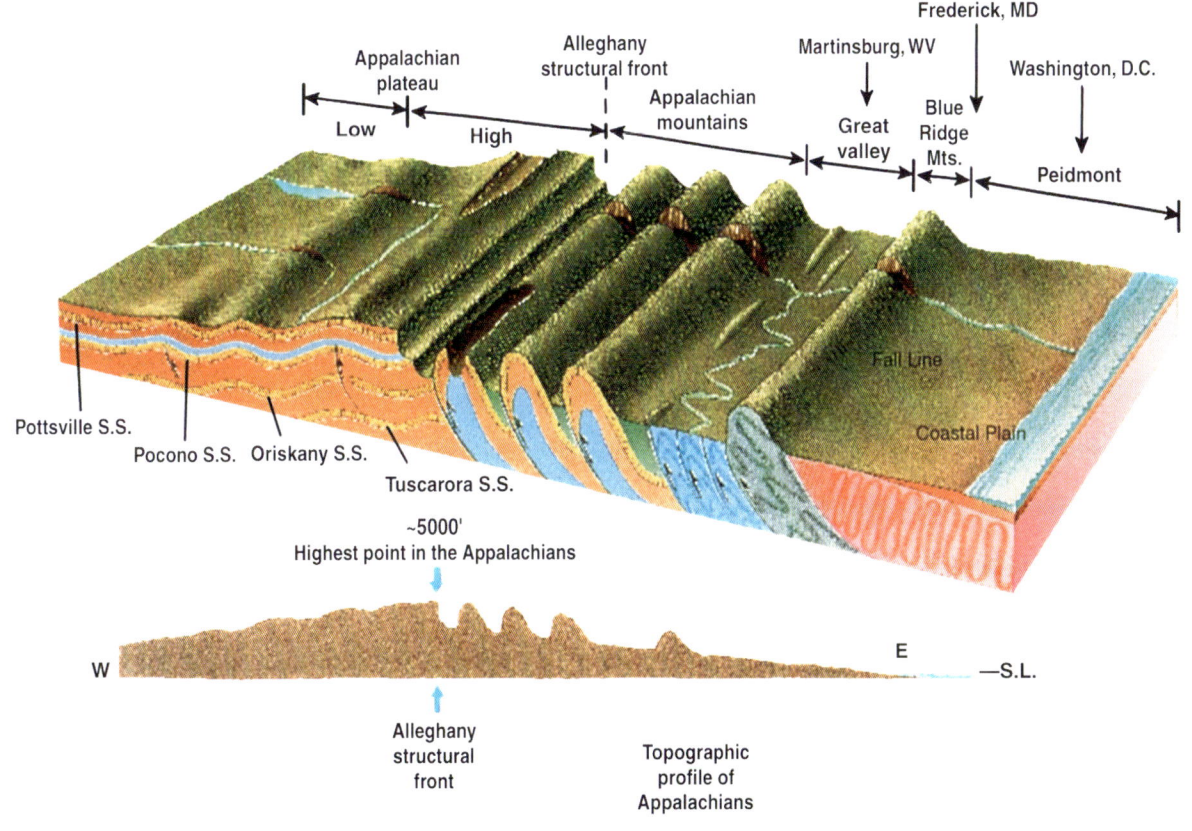

FIGURE 15.31 *The structures of the Appalachians illustrate how geologists determine the direction of* maximum rock transport. *Note that from east to west (a) the folds become increasingly symmetrical and lower in amplitude; (b) the displacements of the thrust faults progressively decrease until, within the Appalachian Plateau, they can no longer be recognized; and (c) the overall degree of rock deformation decreases from the metamorphosed and intruded rocks underlying the Piedmont to the essentially undeformed sedimentary rocks of the Appalachian low plateau. These observations indicate that as the continents collided to form Gondwana, the forces that formed the Appalachians were directed from east to west. In the same fashion, the forces that affected the continent of India were directed from north to south as India collided with Asia to form the Himalaya Mountains. The eastward-dipping thrust faults and steepening of the western limbs of folds indicate that the deformation resulted in the westward movement of rock masses from their original locations.*

formed the Appalachian Mountains drove the rocks now seen in the Appalachians *westward* by way of thrust faults and by the formation of folds. Note that in Figure 15.31, the folds become increasingly asymmetrical toward the east and that both the fault planes and the axial planes of the nonsymmetrical folds all dip toward the east. As the easternmost rocks were deformed beyond their ability to respond plastically, they broke to form thrust faults with the hanging walls being moved upward and westward. Note that in the development of the faults, the rock mass of the hanging wall actually moves *laterally* westward relative to the foot wall as the hanging walls of the faults move up and over the foot walls. The development of asymmetry in the folds with the axial planes tipping to the west also results in a westward movement of the rock mass.

Structures such as those observed in Figure 15.31 have been reproduced experimentally using models where layers of clay represent the rock layers. These model layers are deformed by the same kind of stress conditions that have been suggested for the formation of the Appalachian Mountains (Figure 15.32). We can use our belief about how major mountain ranges have formed, coupled with carefully conducted laboratory experiments, to create some basic rules for interpreting the observed geologic structures. In this case, both the experimental and the field data indicate that when rocks are deformed by compressive stresses, the axial planes of nonsymmetrical folds and the planes of thrust faults usually dip in a direction *opposite* to both the direction of the maximum applied stress and the direction of maximum rock transport. The structures illustrated in Figure 15.31 indicate that the Appalachian Mountains formed by compressive stresses and that the direction of maximum rock transport was toward the west.

Actual Direction of Fault Movements

The actual movements involved in the formation of faults are not always easily determined. Experiments producing reverse faults by subjecting clay cakes to compressional forces indicate that most of the movement is experienced by the hanging wall as it moves up and over a stationary foot wall. However, this does not preclude the possibility of *underthrusting* in which the foot wall is driven under the overlying rock mass. The same kinds of experiments also show that in normal faulting, most of the movement is experienced by the hanging wall. Consider our discussions of the East African Rift Valley where the valley formed as the adjoining continental masses moved away from each other. A similar scenario was used to describe the downfaulting along the crest of the oceanic ridges as the oceanic plates move away from each other (refer to Figure 15.33). In either case, the formation of the valley is interpreted as the result of the *downdropping* of the rock mass between normal faults located on opposite sides of the rift. It must be pointed out, however, that in both scenarios the valley margins are usually uplifted during rifting due to the buoyancy of the underlying rocks. Did the rift valleys form as the rocks of the valley floor moved

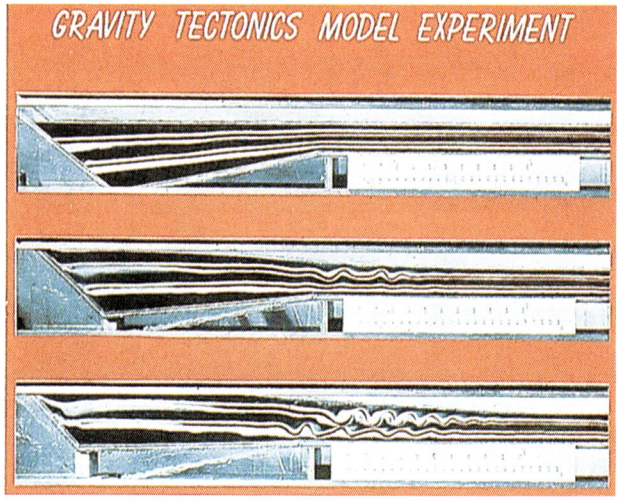 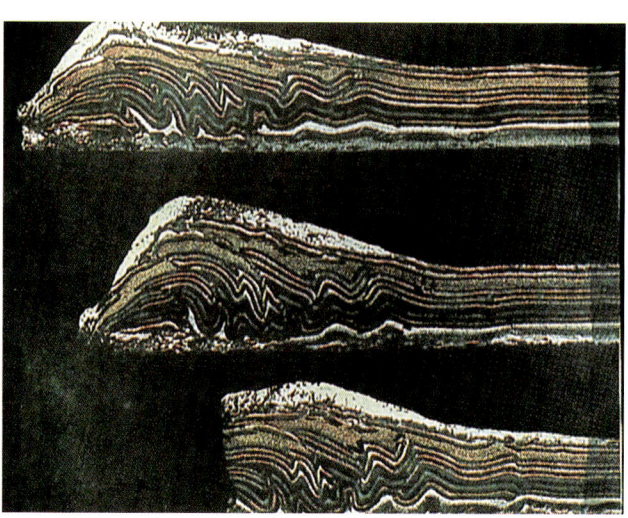

FIGURE 15.32 *Much of our understanding of rock deformation has come from laboratory experiments that subject plastic materials such as clay and wax to compressional forces generated by gravity sliding or in a squeezebox. The experimental structures shown in the two photographs were generated under the direction of one of the pioneers in structural modeling, Walter Bucher. (Courtesy of Robert Shumaker)*

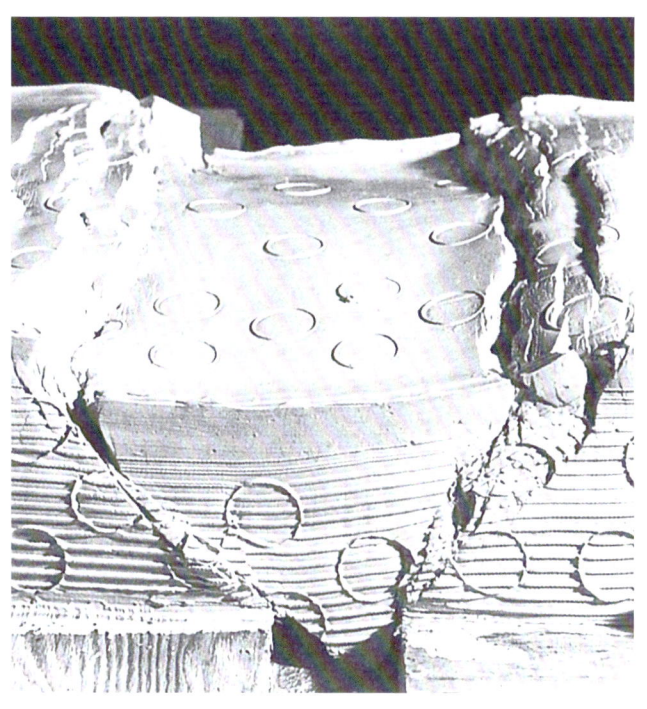

FIGURE 15.33 *Normal faulting is studied experimentally by subjecting clay cakes to tensional forces. The circles impressed into the clay are used to evaluate the degree of elongation of the clay cake under tension, and the stripes are used to simulate bedding. The photo shows the result of an experiment designed to model the formation of a rift valley. Note the surface fractures and the steps along the sides of the main valley formed by normal faulting and the curving of the fault surface with depth. Compare the results of this experiment with the drawing in b. (Photo courtesy of Robert Shumaker)*

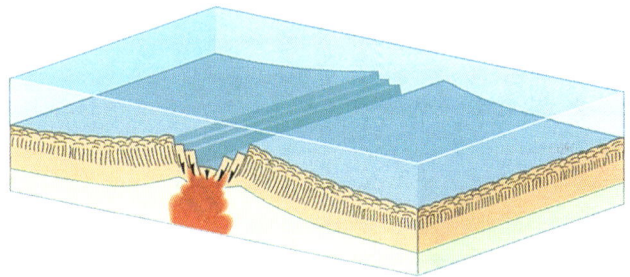

down, did they form as the rocks along the valley margins moved *up,* or was it a combination of both? Because of the difficulty in determining the *actual* movement along normal faults, the question cannot be answered with certainty.

Evidence of the direction of maximum rock transport involved in strike-slip faults is usually not present. Only in cases such as the right-lateral strike-slip movement of the San Andreas Fault, where we understand the plate movements that are responsible for the creation of the fault, can we make a statement relative to the direction of maximum rock transport. We can say with some certainty that the rock mass west of the San Andreas Fault is moving northwestward while the North American continental plate east of the fault either remains relatively stationary or is moving northwestward at a slow rate.

Relative Direction of Fault Movements

A problem commonly confronted by a geologist studying a particular fault is how to determine the *relative* direction of the fault movement. If the fault cuts across the rock layers and there are distinctive layers that can be identified on both sides of the fault, determining the relative direction in which each side has moved is an easy task (Figure 15.34). Without such marker beds, however, the direction of movement may not be apparent. Without some evidence as to the directions of movement, a geologist may not be able to determine whether a particular fault is a normal or reverse fault. Another situation where a geologist may not be able to determine the relative directions of movement is where the fault parallels the rock layers.

Some minor structures, however, may be used to indicate the relative direction of movement. During the movement of one rock mass by the other, less competent rocks such as shales immediately adjacent to the fault surface are deformed into small folds called **drag folds** (Figure 15.35). The direction in which the broken ends of the beds are bent indicates the relative direction of the movement and allows the type of fault to be determined.

Another useful feature, called a **slickenside**, is commonly generated as a result of the two rock masses grinding against each other (Figure 15.36). The high polish of the opposing rock surfaces is an indication of rock movement. Close examination of the polished surface will invariably show **striations** that parallel the direction of movement. In some cases, the relative direction of movement of the blocks on opposite sides of the fault can be determined by running one's hand over the slickensided surface parallel to the directions of the striations. Whichever direction feels the smoother is the direction the block *above* the slickensided surface was moving. Field geologists learn to employ a great number of such techniques in the conduct of their everyday work.

FIGURE 15.34 *Having a recognizable bed such as a coal bed, sill, or distinctively colored sandstone or shale on opposite sides of the fault allows the direction of fault movement to be determined. (Photos VU/© Albert Copley)*

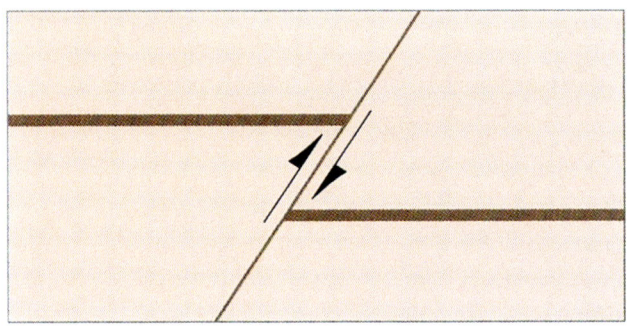

FIGURE 15.35 *The bending of incompetent layers, such as shales, into drag folds allows the relative motion of the rocks on opposite sides of a fault to be determined. (Photo VU/© Martin G. Miller)*

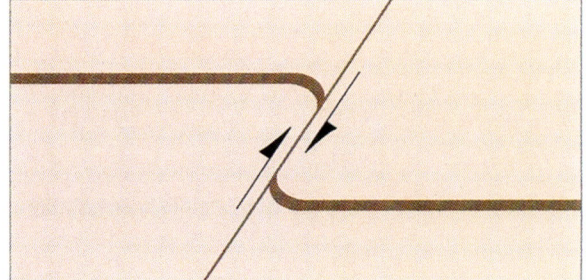

SPOT REVIEW

1. In terms of deformation, what is meant by "rock transport"?
2. What kinds of evidence can be used to determine the most probable direction of rock transport within a deformed area?
3. What kinds of features can be used to determine the relative directions of movement of the rocks on opposite sides of a fault?

FIGURE 15.36 *Slickensides allow a geologist to determine the relative direction of fault movement and in some cases, where friction between the moving rock masses form steps on the slickensided surface, the actual direction of movement can be determined. The striations on the fault surface shown in the figure indicate that the relative direction of movement was parallel to the pencil. The stepped surface indicates that the block that overlaid the fault surface moved from bottom right to top left. (VU/© Martin G. Miller)*

CONCEPTS AND TERMS TO REMEMBER

Strength
Stress
  compression
    shear
  tension
Deformation or strain
  elastic
    elastic limit
  plastic
  brittle
Strain ellipsoid
  axis of maximum shortening
  axis of maximum extension
  intermediate axis
Strength of rocks
  columns
  arches
  keystone
Orientation of planar features
  strike
  dip

Geologic structures
  folds
    monocline
    anticline
    syncline
    descriptive elements
      axial plane
      limbs
      axis
      plunge
      amplitude
    classification of folds
      symmetrical
      asymmetrical
      overturned
      recumbent
      isoclinal
  faults
    descriptive elements
      hanging wall
      foot wall

      displacement
      throw
      heave
    kinds of faults
      reverse or thrust
      normal
      listric
      strike-slip
        right lateral
        left lateral
      oblique-slip or diagonal-slip
  joints
    shear
    tensional
    columnar jointing
Rock movement
  maximum rock transport
  drag folds
  slickensides
  striations

REVIEW QUESTIONS

1. The force that you apply when opening a drawer is an example of
   a. rotational compression.
   b. shear.
   c. tension
   d. nonrotational compression.
2. Which of the following kinds of strain is not permanent?
   a. elastic   c. brittle
   b. viscous   d. plastic
3. Rocks are weakest under
   a. nonrotational compression.
   b. shear.
   c. rotational compression.
   d. tension.
4. An anticline that has an inclined axial plane with limbs dipping in opposite directions is an example of a (an) _____ anticline.
   a. overturned   c. asymmetrical
   b. recumbent   d. symmetrical
5. A fault in which the hanging wall is down relative to the foot wall is a _____ fault.
   a. strike-slip   c. thrust
   b. normal   d. reverse
6. If, in a series of folds, the axial planes all dip toward the east, the direction of rock transport can be interpreted as having been from the
   a. east toward the west.
   b. west toward the east.
   c. north toward the south.
   d. south toward the north.
7. The most common of all the geologic structures are
   a. thrust faults.   c. folds.
   b. normal faults.   d. joints.
8. How are stress, strain, and strength related?
9. How are the various kinds of strain related to the depth below the surface of the crust?
10. How can folds and faults be used to determine the direction of rock transport resulting from deformation?
11. What kinds of evidence can be used to indicate the direction of fault movement?

THOUGHT PROBLEMS

1. How do geologic structures control the processes of weathering and erosion?
2. How can you utilize topography to determine the structures present in subsurface rocks?

FOR YOUR NOTEBOOK

If you live in an area where various kinds of structures are exposed, certainly a photo-taking field trip is in order to document the various kinds. Even though the more spectacular folds and faults may not be exposed, joints are always present. In the case of joints, note the spacing of the joints relative to the bed thickness and to the competence of the individual rock layers.

Should the geologic structures not be readily available, this might be a time to further investigate the various kinds of stress and strain by observing everyday activities and objects. List common activities, from closing drawers to opening twist-top bottles, that are examples of the various kinds of stress.

Simple experiments using sheets of paper or thin layers of clay to reproduce the various kinds of folds and faults could also be performed. Perhaps the geology department has a clay-cake table that you may use, or perhaps structural geology students may be conducting clay-cake experiments that you may be able to observe.

CHAPTER OUTLINE

INTRODUCTION

METAMORPHISM

METAMORPHIC TEXTURES AND CLASSIFICATIONS
- Foliated Texture
- Nonfoliated Texture

KINDS OF METAMORPHISM
- Contact Metamorphism
- Dynamo-Thermal Metamorphism
- Hydrothermal Metamorphism

METAMORPHIC GRADE

METAMORPHIC FACIES

METAMORPHISM AND PLATE TECTONICS

ENVIRONMENTAL CONCERNS

CHAPTER 16

Metamorphic Rocks

INTRODUCTION Simply stated, metamorphism is *change*. We have talked about change before: *crystallization* is the change of magma or lava into an igneous rock, *weathering* is the change from one mineral assemblage to another resulting from the chemical attack of oxygenated or carbonated waters, and *lithification* is the change of sediment into sedimentary rocks.

Two points set the scene for our discussion of metamorphism and metamorphic rocks: (1) in all of the transformation studied so far, the change was a response to the combined effects of *heat, pressure,* and *chemically active fluids,* primarily water; and (2) transformations take place when materials created under one set of conditions are exposed to a new set of conditions.

In the case of sedimentary rocks, calcium carbonate crystallizes from solution as the water temperature rises. Mud, accumulated at Earth's surface, is converted into shale by the increased pressure of burial. Various cements that convert loose sediment such as beach sands into a sandstone precipitate from groundwater.

In the case of igneous rocks, heat, pressure, and water all contribute to the creation of magma. Heat breaks bonds within crystal lattices and allows the rocks to melt. The decrease in pressure as convection-driven mantle rocks approach the top of the asthenosphere results in their melting. The waters provided by the dehydration of hydrous minerals and the hydrothermal solutions released from crystallizing magmas aid in the formation of magma. Thus, to one degree or another, heat, pressure, and chemically active fluids in various combinations play critical roles in the formation of both sedimentary and igneous rocks.

Most minerals are stable unless exposed to conditions different from those under which they formed. Minerals within plutons are stable as long as they remain deep within Earth. When igneous rocks are exposed at the surface, however, they become unstable and are changed by the process of weathering into a new assemblage of materials that are stable. Similarly, the products of weathering remain relatively unchanged at Earth's surface. If they are buried at depths of a few thousand feet or meters, however, where they are subjected to increased temperatures and pressures and exposed to mineral-laden waters, the sediments change into rocks that are stable under the new conditions.

Now we are about to introduce the third kind of rock, **metamorphic rock**. Once again, we will see that heat, pressure, and chemically active fluids play important roles. We will see that the process of metamorphism involves changes that take place when rocks are subjected to conditions *different* from those under which they formed. In this respect, the process of metamorphism can be viewed as the inverse of weathering in which minerals that formed at conditions of high pressure and temperature are exposed to the atmosphere and converted to a mineral assemblage stable at atmospheric conditions.

METAMORPHISM

Metamorphism is a process of *transformation* by which the mineral composition, texture, or both of an existing rock are changed, creating a new rock through the application of (1) *heat,* (2) *pressure*, and/or (3) *chemically active fluids.* Of the three, heat is the most important, being involved even when pressure and chemically active fluids are the dominant agents. For the most part, metamorphism takes place deep within the confines of active mountain belts at the convergence of lithospheric plates, especially during continent-continent collisions.

A requirement of the process of recrystallizataion involved in metamorphism is that it proceed in the *solid state.* During metamorphism, the mineral assemblage of the original rock recrystallizes into a new mineral assemblage without melting. In some cases, the new minerals are identical in composition to the original minerals, differing only in grain size. The best example is the transformation of limestone to marble. Both rocks are made of calcite ($CaCO_3$). The original limestone may have been a chemical sedimentary rock made up of fine-grained calcite or biochemical limestone consisting of a mixture of fossils, shell fragments, and chemically precipitated carbonate cement. The process of recrystallization destroys the fossil remains and the original calcite grains and replaces them with new crystals of calcite, all of which will be relatively uniform in size.

In other cases, the recrystallization process results in both a change in texture and the creation of new and different minerals. A common example is the metamorphism of shale to produce *slate*. During metamorphism, the clay minerals originally present in the shale are changed into minerals such as mica and chlorite.

These examples illustrate a general characteristic of most metamorphic processes, namely, that the transformations take place with little, if any, addition to or removal of material from the original rock. The original rock and the subsequent metamorphic rock has the same *elemental* composition. Only the texture or the mineralogy has been changed.

Some metamorphic transformations do involve changes in elemental composition, however. In some cases, new elements are *added* to the rock while in other cases, elements are *removed* or *replaced* with different elements. In such cases, mineralogy of the metamorphic rock will always be different from that of the original rock.

As these brief descriptions of the process of metamorphism illustrate, metamorphism is simply another example of change involving the basic agents of heat, pressure, and chemically active fluids, which are also needed for the formation of igneous and sedimentary rocks. These three conditions can be considered as lying along an intensity continuum. As we discuss metamorphism, it will become evident that the conditions required to produce metamorphic rocks are usually *more intense* than those involved in the formation of sedimentary rocks and *less intense* than those involved in the generation of a magma.

Most geologists believe that the transformation from sedimentary to metamorphic conditions occurs at temperatures of about 400°F (200°C) and pressures equivalent to 6,000 to 9,000 feet (2,000–3,000 m) of burial. At some point, the conditions of temperature, pressure, and solution activity that convert sediments into sedimentary rocks may result in mineralogic or textural changes within the rock that go beyond those attributable to lithification. A sedimentary rock that shows evidence of having been subjected to metamorphism is called a *metasediment.*

Because of the transitional relationship that may exist between igneous and metamorphic conditions, some rocks may partially melt as they undergo metamorphic changes. The rock that is produced is called a **migmatite** (meaning "mixed rock") and is considered to be both igneous and metamorphic. Some metamorphic petrologists believe that migmatites can also form by the reaction between the host rock and chemically active fluids, a process called *metasomatism,* to be discussed shortly. The existence of migmatites notwithstanding, however, it is important to remember that the conditions cannot result in the melting of the original rock. Should melting occur, the newly formed rock will be igneous rock, not a metamorphic rock. The conditions may cause the rocks to react plastically and flow, but all metamorphic changes must take place in the solid state.

METAMORPHIC TEXTURES AND CLASSIFICATIONS

Metamorphic textures are of two types: (1) *foliated* and (2) *nonfoliated.* Foliated rocks exhibit more or less pronounced parallelism of platy minerals (those having sheet structures) and commonly split along planes of weakness into layers, sheets, or flakes, a property called **rock cleavage**. The rock cleavage responsible for foliation results from the orientation of platy minerals that exhibit excellent mineral cleavage, such as the micas and chlorite.

Foliated Texture

Foliated rocks develop primarily as a result of dynamo-thermal metamorphism (described later in the chapter). In dynamo-thermal metamorphism, stress can be applied in two different ways, by *direct compression* and by *shear*. In direct compression, the forces act toward and directly opposite each other in much the same fashion you would grip a pencil. In shear, the forces act toward each other, but not directly opposite. As the rocks are exposed to the directional stress of direct compression and the platy minerals begin to crystallize, they grow preferentially in the direction of *least* stress, that is, *perpendicular* to the direction of *maximum* stress (Figure 16.1b). The rock develops cleavage with the plane of cleavage parallel to the plane of mineral cleavage and perpendicular to the direction of maximum stress (Figure 16.2). Note in Figure 16.2 that the cleavage cuts across the bedding in the folded strata. When recrystallization takes place under shear, the platy minerals are parallel to each other and at a small angle to the direction of stress, as shown in Figure 16.1c.

Slaty Cleavage Depending upon the relative abundance and preferred orientation of the platy minerals and the perfection of the cleavage, the rocks can exhibit varying degrees of foliation. The fullest development of foliation is called **slaty cleavage** because it is exhibited by slate where the orientation of the tiny platy minerals is so perfect that the resultant cleavage allows the rock to be split into

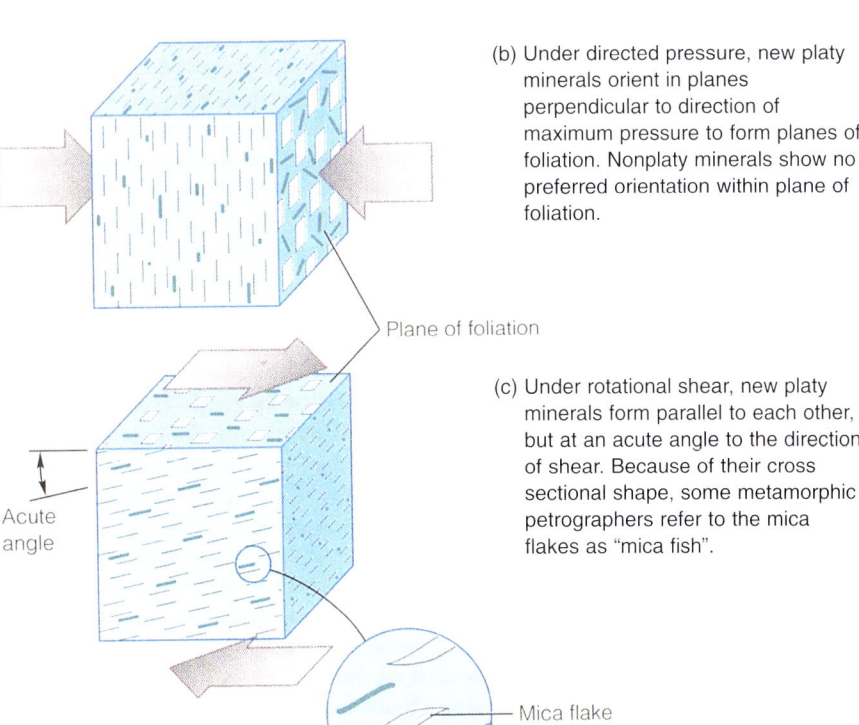

FIGURE 16.1 *When rocks enriched in clay minerals undergo metamorphism, the clay minerals recrystallize to form new types of platy minerals, such as chlorite. The orientation of the new minerals within the rock depends on the kind and direction of the applied force.*

FIGURE 16.2 *As new-formed metamorphic rocks fold, the orientation of the cleavage planes is maintained relative to the forces of deformation. As a result, the cleavage planes cut across the folded beds and are oriented approximately parallel to the axial planes of the folds.*

FIGURE 16.3 *Because of the development of nonplaty minerals within the rock, the surfaces of phyllitic cleavage planes are more irregular than the smooth, flat surfaces that characterize slaty cleavage. (VU/© Martin G. Miller)*

slabs of uniform thickness with smooth surfaces. Before the days of painted chalkboards, slate was a common blackboard material as well as being used for roofing tiles and paving stones.

Phyllitic Cleavage Continued metamorphism of slate produces the metamorphic rock, **phyllite**. Phyllite resembles slate, but close examination reveals coarser mica crystals, as well as nonplaty minerals such as garnets, quartz, and feldspars, which serve to decrease the degree of perfection of the cleavage. Rather than splitting in the parallel-faced slabs characteristic of slate, phyllite splits into sheets with more irregular surfaces that exhibit a characteristic sheen (Figure 16.3). Because it is characteristic of phyllite, this type of cleavage is referred to as **phyllitic cleavage**.

Schistose Cleavage More intense conditions of metamorphism result in the coarsening of micas and nonplaty minerals and the formation of **schist**. Schists can form from a variety of predecessors and, as a result, vary widely in composition. Due to the larger grain size, schist normally splits into fragments with irregular, shiny faces rather than the sheets characteristic of slate and phyllite (Figure 16.4). This grade of cleavage is referred to as **schistose cleavage** or **shistosity**.

Gneissic Banding The poorest development of cleavage is found in the metamorphic rock called **gneiss** where the minerals have segregated into bands of dark- and light-colored minerals (Figure 16.5). In fact, some gneisses show no rock cleavage at all.

Gneiss is the most abundant of all metamorphic rocks, forming from a wide variety of premetamorphic rocks and under a wide range of metamorphic conditions. Generally, however, gneisses are the result of high temperatures and pressures. Because gneiss consists largely of nonplaty minerals such as quartz, feldspars, amphiboles, and pyroxenes, it exhibits minimal development of foliation, referred to as **gneissic banding**.

Nonfoliated Texture

Nonfoliated metamorphic rocks consist primarily of interlocking, equidimensional grains. These rocks commonly form under metamorphic conditions where little directional stress is present to orient the newly formed minerals and also where the original rock lacks layered silicate minerals. A well-known example of a nonfoliated metamorphic rock is marble. Sculptors seek specimens of marble that show exceptional uniformity in nonfoliated texture for their work.

FIGURE 16.4 Schistose cleavage *results in the rock splitting into relatively small, irregularly shaped rock fragments. (VU/© Martin G. Miller)*

In general, monomineralic rocks have less tendency to layer preferentially, and the layering in those that do is difficult to see. Examples include *metaquartzites,* which consist of interlocking quartz grains; *marble,* which is made up of recrystallized calcite; and *serpentinite,* which consists of various serpentine minerals including antigorite, chrysotile, and lizardite (Figure 16.6). In some cases, impurities such as clay minerals in the original rock result in the development of a degree of foliation in rocks such as marble and metaquartzites. Any foliation in marble produces zones of weakness and precludes its use for sculpting.

FIGURE 16.5 *Exhibiting the poorest development of rock cleavage, gneisses are characterized by the separation of light and dark minerals that form the characteristic* gneissic banding. *(VU/© Martin G. Miller)*

FIGURE 16.6 *During the transformation of limestones and dolostones, the original carbonate grains recrystallize to form crystallites of relatively uniform dimensions that produce the sugarlike texture of marble. (© E. R. Degginger)*

SPOT REVIEW

1. What is a "migmatite" and what is its significance?
2. At what point do the conditions of temperature and pressure that convert sediments to sedimentary rocks become sufficiently intense to produce metamorphic changes?
3. Why is the orientation of platy minerals more important than their abundance in the development of foliation?

KINDS OF METAMORPHISM

There are three primary kinds of metamorphic processes: (1) *contact metamorphism* where *heat* is the dominant agent, (2) *dynamo-thermal metamorphism* where both *pressure* and *temperature* are important, and (3) *hydrothermal metamorphism* where the dominant agents are *chemically active fluids* (largely superheated water) derived from the crystallization of magmas.

Contact Metamorphism

Contact metamorphism is primarily associated with magmatic intrusions. The host rocks in contact with the molten mass are subjected to heat from the magma and are literally "baked," as you would bake a ceramic piece in a kiln. Just as the ceramic undergoes chemical and physical changes as it bakes, changes take place within the host rocks as they adjust to the elevated temperatures to which they have been subjected. The zone of metamorphic rocks that forms around an intrusive body is called a **halo** or an **aureole** (Figure 16.7).

A metamorphic rock formed by contact metamorphism of an argillaceous (containing clay minerals) rock is referred to as a **hornfels**. Hornfelses are usually harder, denser, and darker in color than the original rocks. The extent of the changes that take place and the volume of host rock affected depend upon a number of factors including (1) the volume and temperature of the intruding magma body, (2) the amount of water in the host rock, and (3) the grain size and composition of the intruded rock. For example, if a shale is intruded by a relatively small volume of magma such as the amount that would produce a dike or sill, metamorphism of the host rock will be limited to a relatively narrow zone where

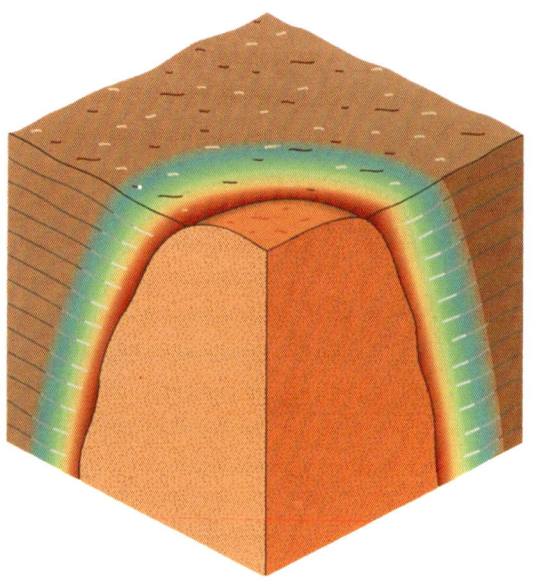

FIGURE 16.7 *The host rock immediately adjacent to an intruding mass of magma is altered to form a zone of metamorphism called a* halo. *The thickness of the halo depends on the amount of heat available from the mass of the intruding magma.*

the clay minerals are altered to mica and the quartz grains are recrystallized to form larger grains. If the shale in intruded by a volume of magma of batholithic dimensions, however, the amount of heat will be so great that the metamorphic process will involve a relatively thick halo. Within the halo, the clay minerals and the quartz will react with each other to form a totally different assemblage of minerals including micas, garnet, and andalusite/sillimanite. In both cases, the rock is called a hornfels, but the garnet, andalusite, and sillimanite in the latter rock record its formation at significantly higher temperatures. Note that the minerals that form are those that are stable at the new, higher temperatures.

Water tends to accelerate most metamorphic processes by serving as a transport medium for ions within the rock mass. As a result, the volume of rock within the aureole will be larger if the intruded rock was wet. Because the rate at which rocks undergo metamorphism increases with decreasing grain size, the extent of metamorphism beyond the magma-host rock contact for a given set of metamorphic conditions will be greater for fine-grained host rocks.

Dynamo-Thermal Metamorphism

The primary agents in **dynamo-thermal metamorphism** are pressure and heat. One type of dynamo-thermal metamorphism is **regional metamorphism**.

Regional metamorphism is a general term for the metamorphism of rocks that are exposed over an extensive area as opposed to *local* metamorphism where the rocks of a limited area are affected. Metamorphic rocks formed by dynamo-thermal metamorphism are most commonly found in the cores of foldbelt mountains where the rocks have been subjected to the intense pressures generated by plate convergence, especially continent-continent collisions. A variety of foliated rocks is formed by dynamo-thermal metamorphism. A typical and very common example is *schist* (Figure 16.8). Schists are strongly foliated metamorphic rocks and can be of variable composition.

In some cases, the conditions of pressure and temperature involved in dynamo-thermal metamorphism are so intense that the newly formed rock flows plastically and creates the familiar flowage patterns often seen in marble (Figure 16.9). Metamorphic rocks can also be formed by dynamo-thermal metamorphism at lower temperatures. In major fault zones, for example, a dense, fine-grained metamorphic rock called *mylonite* forms when rocks, caught between the two moving fault blocks, are pulverized and sheared.

Hydrothermal Metamorphism

In both contact and dynamo-thermal metamorphism, the recrystallization usually takes place with little addition or loss of material. With the exception of some loss of the more volatile components such as water, the process simply rearranges the original elements into a new assemblage of minerals. As a result, although the *mineralogy* may change, the *elemental* composition of the metamorphic rock is the same as that of its predecessor.

In **hydrothermal metamorphism**, on the other hand, materials may be both *added to* and *removed from* the original rock composition by the intervention of "chemically active fluids." As the name *hydrothermal* implies, the fluid responsible for the movement of ions is hot water. The water may come from several sources including ocean water drawn down into the zones of subduction and the thermal decomposition of hydrated minerals originally contained within subducted oceanic sediments.

A major source, however, is the water that concentrates during the final stages of magma crystallization. Water is one of the ingredients dissolved in molten rock. As the magma crystallizes, little water is needed for the formation of the major silicate miner-

FIGURE 16.8 *Individual kinds of metamorphic rocks can form from a wide variety of precursors depending on the composition of the original rock and the conditions of metamorphism. A common metamorphic rock that can form from a wide variety of precursor rocks is* schist. *(VU/© Albert J. Copley)*

FIGURE 16.9 *The intricate bedding patterns often seen in metamorphic rocks indicate that the conditions of temperature and pressure were so high that the rock actually flowed plastically. (VU/© Albert J. Copley)*

als; this is apparent from their formulas, few of which contain either waters of hydration indicated by ($\cdot n H_2O$) or the hydroxyl radical (OH^{1-}). Because water is not needed for the formation of the early formed silicate minerals, it becomes increasingly concentrated in the remaining magma as the minerals precipitate. Eventually, crystallization progresses to the point where what remains is no longer molten rock but rather a hot water (hydrothermal) solution containing the remaining elements. This hydrothermal solution then permeates into and reacts with the host rock to produce new assemblages of minerals.

An example of hydrothermal metamorphism that simply involves the reaction of water with the intruded rock is the conversion of an olivine-rich dunite into serpentine and brucite by the reaction:

$$2Mg_2SiO_4 + 3H_2O \rightarrow Mg_3Si_2O_5(OH)_4 + Mg(OH)_2$$
olivine + water → serpentine + brucite

Serpentine is a deep green color. When shot through with fracture-filling white calcite, the rock is the beautiful green stone known in the industry as *verde antique,* a rock that is widely used as a decorative stone. Cut and polished, serpentine has been used for years as a decorative trim stone and can be found in many older public buildings. The columns around the rotunda of the National Gallery of Art in Washington, D.C., are spectacular examples (Figure 16.10).

Metasomatism In a special type of hydrothermal metamorphism, called **metasomatism,** the original mineral assemblage of a host rock is replaced atom-for-atom with a new mineral assemblage.

Another, somewhat more controversial example of metasomatism is *granitization.* In our discussions if igneous rock, we indicated that granite is the most common intrusive igneous rock, emplaced either forcefully or by the process of stoping. However, certain granite bodies can be found where features within the host rock such as bedding can be traced into the granite body. The presence of the residual bedding appears to exclude both forceful entry and stoping as the means of emplacement.

Several possible scenarios could result in relic host rock structures being found within the granite body. The host rock in contact with the magma could have melted in place with the melt moving little before solidification; structures of the host rock would then be preserved in the outer portion of the igneous rock body. On the other hand, the mineral assemblage of the original rock could have been subjected to metasomatic replacement with the original rock being replaced literally atom-for-atom by solutions containing the granitic assemblage of minerals. In this case, the granite would be metamorphic rock, not igneous rock.

Although some geologists once believed that the majority of granite bodies were the result of metamorphic processes, the concept of granitization has now been largely discredited. Most geologists, although agreeing that some granite bodies can form by metasomatism, believe that the vast majority of granites result from the crystallization of magmas.

SPOT REVIEW

1. How do the grain size and the presence or absence of water within the host rock affect the volume of rock modified by contact metamorphism?
2. What are the major sources of water in hydrothermal metamorphism?
3. What is unique about the process of metasomatism?

FIGURE 16.10 *Because of their colors and internal structures, a variety of metamorphic rocks find use as decorative stone. An example consisting largely of serpentine shot through with calcite is* verde antique, *which is used effectively in the National Gallery of Art in Washington, D.C. (Courtesy of Washington D.C. Convention and Visitors Bureau)*

METAMORPHIC GRADE

When rocks produced by dynamo-thermal or regional metamorphism are exposed by erosion, the regional variation in the mineral assemblages records information about the intensity of the heat, pressure, and deformation experienced by the rocks. Most often, these rocks represent the core of an ancient mountain range, now exposed by erosion. An area showing the effects of regional metamorphism is the *Canadian Shield,* which is the exposed craton of the North American continent (Figure 16.11). A bewildering array of igneous, metamorphic, and sedimentary rocks that have been subjected to metamorphism are found within the Canadian Shield, some of which date back to some of the earliest mountain-building episodes to occur on Earth. Some of the metamorphic rocks represent rocks that formed deep within collision zones at the extreme metamorphic conditions of heat and pressure. These rocks grade laterally into other assemblages or metamorphic rocks, which, based upon their mineral contents, are interpreted to have formed under less severe conditions (Figure 16.12).

The **metamorphic grade** of a rock is determined by identifying its **index minerals,** which are silicate minerals that form only under specific metamorphic conditions. The *micas* and *chlorite* are indicators of **low-grade** metamorphism. **Intermediate-grade** metamorphism is indicated by the presence of *garnets* and *staurolite*. Fibrous crystals of *kyanite* and *sillimanite* indicate **high-grade** metamorphism, the most intense metamorphic conditions.

METAMORPHIC FACIES

Metamorphic rocks that form under the same conditions of temperature and pressure (depth of burial) are said to belong to a **metamorphic facies.** It is important to remember that most metamorphism does not involve a change in the elemental composition of the original rock. As a result, the kind of metamorphic rock contained within a particular metamorphic facies depends on the composition of the original rock. Although a suite of metamorphic rocks may differ in kind, they will exhibit comparable index minerals if they formed under identical conditions of metamorphism. Thus, a geologist can determine that the rocks formed under the same conditions by examining the index minerals.

It is also important to remember that metamorphism is a dynamic process. Consequently, the mineral assemblage of a rock being subjected to changing conditions of temperature and pressure will change as it approaches equilibrium with the new conditions. When the mineral composition changes in response to *increasingly intense* conditions of temperature and pressure, such as deeper burial, the process is called **prograde metamorphism.** On the other hand, if a metamorphic rock is exposed to new conditions of temperature and pressure that are *less intense* than those under which the rock formed, such as when burial depth decreases as erosion removes the overlying rocks, the original minerals may change to an assemblage that is stable at the lower conditions of temperature and pressure. Such a change is called **retrograde metamorphism.**

FIGURE 16.11 *The most extensive exposure of rocks showing regional metamorphism in North America is found in the* Canadian Shield. *These rocks formed in the cores of some of Earth's earliest mountains. (Landsat imagery courtesy of the Canada Centre for Remote Sensing, Energy, Mines, and Resources)*

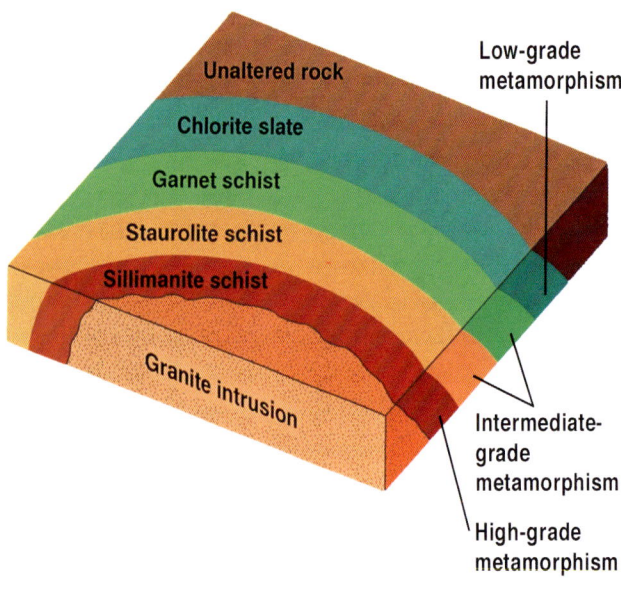

FIGURE 16.12   *In regional metamorphism, the kinds and compositions of the metamorphic rocks record a* metamorphic gradient *with the grade of metamorphism progressively decreasing from a central locale, where conditions were most intense, outward to rocks that were unaffected.*

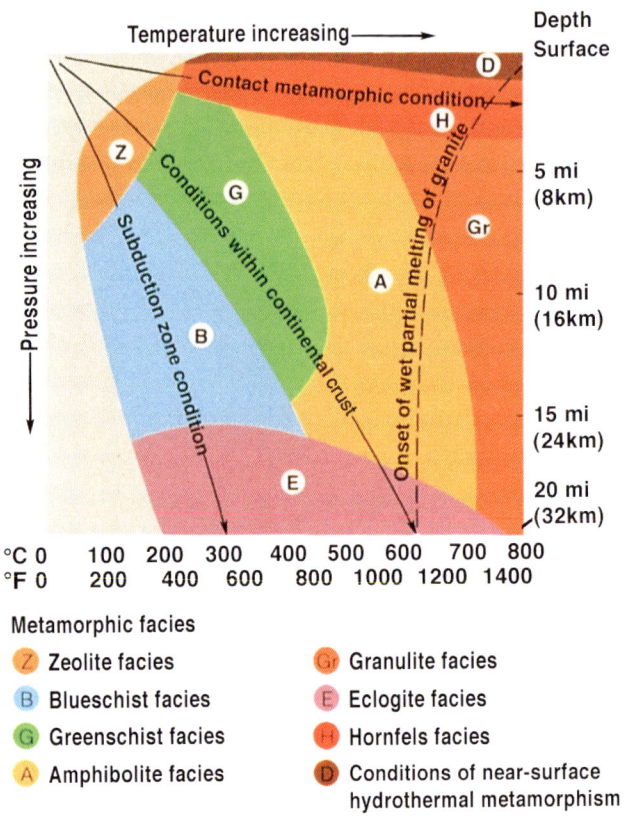

FIGURE 16.13   *Every metamorphic rock belongs to a particular* metamorphic facies *depending on the conditions of temperature and pressure under which it formed.*

METAMORPHISM AND PLATE TECTONICS

Perhaps the major contribution that plate tectonics has made to our understanding of metamorphism is that it has allowed geologists to explain the spatial distribution of metamorphic facies within regional metamorphism. Figure 16.13 illustrates the various metamorphic facies and their temperature/pressure boundaries. Throughout this discussion, the circled letters in the text refer to the corresponding letters in the figure.

Although some metamorphic rocks form at divergent plate boundaries and hot spots where host rocks are intruded by magmas and within some fault zones, most metamorphic rocks are formed at convergent plate boundaries within the overriding plate (Figure 16.14). Within the overriding plate, the temperature, pressure, and availability of chemically active fluids depend on location. For example, near-surface rocks within the continental crust may be modified by hydrothermal metamorphism as they are permeated by solutions being released from crystallizing magma bodies Ⓓ. Many of our rich metal deposits are formed by these processes.

Within a deep-sea trench, land-derived sediment combines with deep-sea sediments and fragments of oceanic basalts stripped from the subducting oceanic plate to form a *mélange* (French for mixture), which may be subsequently transformed into zeolite facies metamorphic rocks under relatively low conditions of temperature and pressure Ⓩ. On the other hand, as rocks are subducted to depths of 10 to 20 miles (15–30 km), they are subjected to pressures in excess of 5 kilobars (37 tons/in^2). Because the oceanic rocks are initially cold, however, and warm only slowly as they descend into the mantle, the temperatures at these depths reach only moderate levels Ⓑ. Under these conditions of high pressure and relatively low temperature, the rocks are converted to the blueschist/eclogite facies Ⓑ–Ⓔ. Because of our understanding of plate tectonics, geologists now use the presence of blueschists to identify a sequence of rocks that formed in an ancient zone of subduction. An example of such a rock sequence is the Franciscan Formation that makes up much of the Coastal Range mountains of California (Figure 16.15).

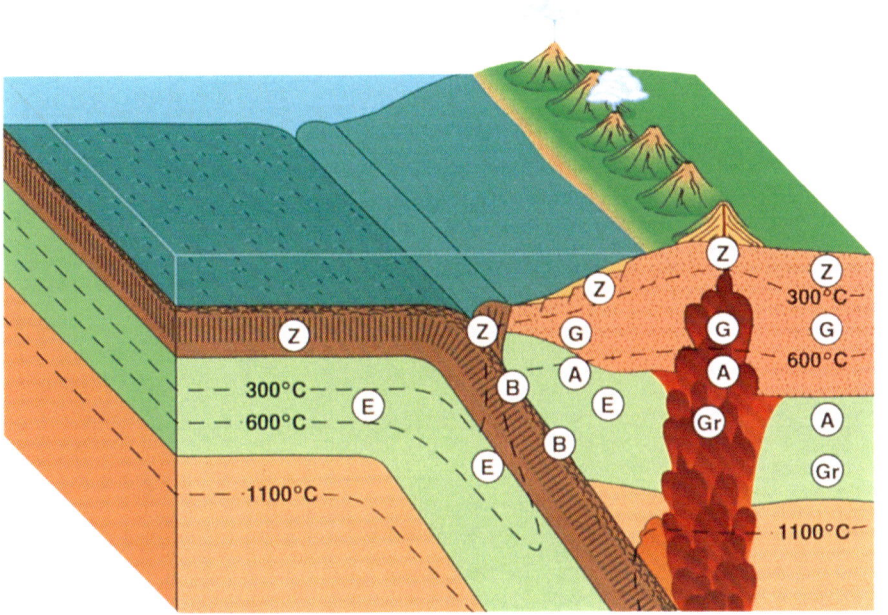

FIGURE 16.14 *Most metamorphic rocks form at the convergence of lithospheric plates where the conditions of temperature and pressure necessary for extensive metamorphism are achieved.*

Metamorphic Facies

- (Z) Zeolite
- (B) Blueschist
- (A) Amphibolite
- (G) Greenschist
- (E) Eclogite
- (Gr) Granulite

FIGURE 16.15 *The metamorphic rocks that make up most of the* Coastal Range *mountains of California were formed deep within an ancient zone of subduction. (VU/© LINK)*

As the plates converge, the rocks of the continental crust are subjected to compression that generates mountains along the margin of the continental crust. As the continental crust thickens, the rocks within the core of the developing mountains are buried deeper and therefore are subjected to increasing pressures. At the same time, temperatures rise as the core rocks are intruded by magmas. As a result of these high temperatures and pressures, the rocks are converted to metamorphic rocks of the greenschist/amphibolite (G)–(A) facies. At the present time, vast volumes of greenschist/amphibolite facies rocks are being created within the core of the Himalayas as the collision of the Asian and Indian plates continues.

SPOT REVIEW

1. What is meant by prograde and retrograde metamorphism?
2. How can a geologist determine the intensity of conditions under which a particular metamorphic rock formed?
3. Where on Earth are the greatest volume of metamorphic rocks formed?

ENVIRONMENTAL CONCERNS

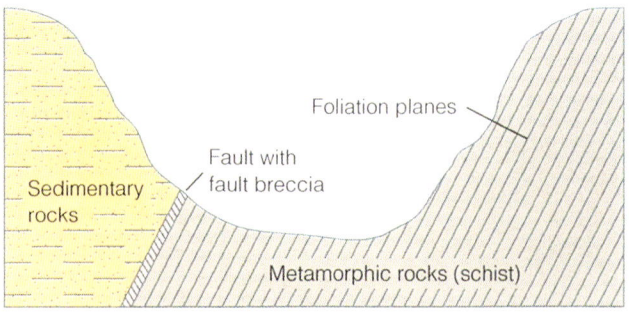

FIGURE 16.16 *The geology of the construction site of the St. Francis dam in San Francisquito Canyon could not have been more unfavorable for the location of a dam.*

Like igneous rocks, most metamorphic rocks undergo relatively high rates of chemical weathering due to the large difference between the conditions of temperature and pressure under which they form and atmospheric conditions. As a result, thick layers of weathering products quickly develop on exposed surfaces. Should the rocks crop out on steep slopes, the products of weathering are prone to slumps and landslides. Should the rocks exhibit foliation, rock slides are common as water penetrates the planes of foliation, promotes weathering that subsequently decreases the cohesion between the foliation surfaces, and generates planes of slippage.

A good example of the type of engineering problems that are commonly associated with metamorphic rocks, and an excellent example of how preconstruction geologic investigations can prevent deadly and costly structural failures, is the collapse of the St. Francis Dam near Saugus, California.

The St. Francis Dam was built in a narrow section of the San Francisquito Canyon. The geology of the site is illustrated in Figure 16.16. A worse geologic setting for the construction of a dam could not be imagined. The floor and east wall of the canyon were underlain by a schist with planes of foliation oriented parallel to the length of the canyon and dipping westward toward the canyon floor. Numerous landslide scars on the slopes of the eastern canyon wall attested to the inherent instability of the metamorphic rocks. The west wall of the canyon was composed of what appeared to be coherent sedimentary rocks. Unfortunately, strength tests conducted *after* the failure of the dam showed that, like many rocks that appear competent under arid conditions, the sedimentary rocks became unstable when continuously exposed to water. To make the situation worse, the sedimentary and metamorphic rocks were separated by a fault zone containing about 5 feet (1.5 m) of fault breccia. Ironically, the presence of the fault was known before construction began, but the builders apparently chose to ignore it. The setting was ready-made for disaster.

When completed in 1927, the dam was 190 feet (63 m) high, 650 feet (214 m) long, and held 62 million cubic yards (47 million m^3) of water. The pressure at the base of the dam was about 5.7 tons per square foot.

Within one year of the completion of the dam, water had weakened the sedimentary rocks anchoring the west end of the dam. Water penetrated the fault, moved under the dam, and resulted in the removal of crushed rock from the dam foundation. Some believed, however, that the major cause of the dam failure was the development of slip planes in the schists anchoring the east end of the dam as water permeated the rock foliation planes.

No one agreed on the exact sequence of events that preceded the failure of the dam on March 12, 1928. The torrent of water that surged down San Francisquito Canyon killed more than 500 people and caused an estimated $10 million in property damage (Figure 16.17). Had modern preconstruction geologic investigations been applied to the dam site, the St. Francis Dam would never have been built.

Metamorphic Rocks 495

FIGURE 16.17 *The failure of the St. Francis dam could have been predicted by a preconstruction geologic evaluation of the site; unfortunately an evaluation was never made. (Courtesy of the California Department of Water Resources)*

CONCEPTS AND TERMS TO REMEMBER

Metamorphic rock
Metamorphism
  migmatite
Metamorphic textures
  foliated texture
    rock cleavage

slaty cleavage
phyllitic cleavage/phyllite
schistose cleavage or schistosity/schist
gneissic banding/gneiss
nonfoliated texture

Kinds of metamorphism
  contact metamorphism
    halo or aureole
    hornfels
  dynamo-thermal metamorphism
    regional metamorphism

CONCEPTS AND TERMS TO REMEMBER continued

hydrothermal metamorphism
  metasomatism
Metamorphic grade
  index minerals
  low/intermediate/high grade
Metamorphic facies
  prograde metamorphiism
  retrograde metamorphism

Important Minerals

calcite	$CaCO_3$
chlorite	$(Mg, Fe^{2+}, Fe^{3+})_6AlSi_3O_{10}(OH)_8$
micas	$(K, Na, Ca)(Mg, Fe, Li, Al)_{2-3}(Al, Si)_4O_{10}(OH, F)_2$
garnet	$(Ca, Mg, Fe^{2+})_3(Al, Fe^{3+}, Mn, V)_2(SiO_4)_3$
andalusite	Al_2SiO_5
sillimanite	Al_2SiO_5
magnetite	Fe_3O_4
staurolite	$(Fe, Mg)_2Al_9Si_4O_{23}(OH)$

REVIEW QUESTIONS

1. Heat is the major agent of metamorphism in
   a. contact metamorphism.
   b. dynamo-thermal metamorphism.
   c. hydrothermal metamorphism.
   d. metasomatism.
2. The most abundant metamorphic rock is
   a. schist.
   b. gneiss.
   c. migmatite.
   d. hornfels.
3. Foliation forms primarily as a result of
   a. dynamo-thermal metamorphism.
   b. contact metamorphism.
   c. hydrothermal metamorphism.
   d. metasomatism.
4. Most metamorphic rocks form in association with
   a. divergent plate margins.
   b. convergent plate margins.
   c. volcanic hot spots.
   d. rift zones.
5. In what ways are the processes of metamorphism and weathering similar?
6. How can a geologist determine the relative intensity of the metamorphism that created a particular assemblage of metamorphic rocks?
7. Explain how gneiss can form from rocks as diverse as granite and argillaceous sandstones.
8. Although the theory is now discredited, what led geologists to consider that most granites had formed by hydrothermal metamorphism?
9. Why do metamorphic rocks such as marble show no foliation?

THOUGHT PROBLEMS

1. How would you distinguish between the conditions of burial that result in the conversion of sediments into sedimentary rocks and those that produce metamorphic rocks?
2. Although the conditions of temperature and pressure required for metamorphism are usually found only at depths of tens of hundreds of miles (kilometers) below the surface, what surface or near-surface situations could result in the metamorphism of rocks?
3. Consider all of the examples of landslides used in the text. Can you ascertain certain pre-failure conditions that seem to be associated with most rock mass movements?

FOR YOUR NOTEBOOK

The first step is to investigate a geologic map to see whether metamorphic rocks are exposed in your area. If they are, note the kinds of rocks and the metamorphic grade. From these data and the general geology of your area, attempt to determine the type and extent of deformation that produced the regional geology. Because such a study could be quite complex, you may want to consult your instructor for some help and advice.

For the greater percentage of us who live in areas where metamorphic rocks are not exposed, a field trip through town will often produce examples of metamorphic rocks. As the chapter text noted, metamorphic rocks are favorite materials for decorative stone. Historically, for example, banks have often utilized various kinds of metamorphic rock in their interior decor. Once again, a short conversation with a local architect may prove useful, not only about the use of various kinds of metamorphic rock but also in locating specific examples of their use. A visit to the local cemetery will also provide examples of metamorphic rocks.

CHAPTER OUTLINE

INTRODUCTION

DEFINITION OF A MOUNTAIN

KINDS OF MOUNTAINS
- Volcanic Mountains
- Domal and Block-Fault Mountains
- Foldbelt Mountains

OROGENIC STYLES
- Ocean-Continent Orogenesis
- Ocean-Island Arc-Continent Orogenesis
- Continent-Continent Orogenesis

CHAPTER 17

Mountain Building

INTRODUCTION No geologic feature captures the imagination quite like mountains. Many people are attracted by their overwhelming beauty. Who would not be impressed by the towering peaks of the Tetons or the majesty of Fujiyama? Others are attracted by the serenity where they can escape from the urban turmoil and the drone of city life. Still others are attracted to the danger that a mountain may present and risk their lives to climb sheer rock faces to feel the exhilaration of triumph as they stand literally on top of the world.

Although most geologists are not willing to risk their lives scaling a sheer rock face, they still are attracted to mountains. According to an old saying, the best geologist is the one who has seen the most rocks. The simple fact is that more rocks are exposed in mountains than anywhere else on Earth. You will also remember from earlier discussions that one of the major objectives of geology is to decipher the history of Earth. Because of the extensive exposures of rock, the greatest segments of Earth's history are to be found recorded in the mountains.

Some geologists are particularly interested in studying and observing basic geologic processes, many of which operate at accelerated rates in the mountains. Relentless freeze-thaw cycles rip at the rocks and transform them into the fragments that commonly litter the ground. Glaciers ply their methodical trade and sculpt mountain peaks and ridges to breathtaking rugged beauty. Turbulent mountain streams crash down steep slopes, carving their channels ever deeper as they attempt to undo the process of mountain building.

Everywhere in the mountains you are surrounded by the feeling of power. Deformed rocks and lofty peaks attest to the enormous forces that brought the mountains into being, while all around the energy of the basic erosional processes works to return the rocks to the sea.

Mountains are also the source of many of the world's mineral deposits. We will discuss mineral resources in Chapter 20.

DEFINITION OF A MOUNTAIN

How does one define a mountain? Someone born and raised in the shadow of the Alps or the Rocky Mountains might wonder how the low ridges of the Appalachians could possibly qualify as mountains. Similarly, someone raised in Appalachia might wonder why a small hill that barely rises a few hundred feet above the flat coastal plain of North Carolina is called "Mount." Nevertheless, that is indeed the definition of a mountain: "a part of Earth's crust sufficiently elevated above the surrounding land surface to be considered worthy of a distinctive name." What "sufficiently elevated" means in terms of feet or meters is not indicated. In like fashion, a mountain chain is defined as "a series of mountains whose bases are continuous." Again, there is no indication of how high the mountain crests must be. Obviously, what constitutes a mountain or mountain chain is in the eye of the beholder. Generally, however, most geologists restrict the term *mountain* to a topographic feature that projects at least 1,000 feet (300 m) above the surrounding land. Some elevated features such as buttes and mesas are usually excluded. It is interesting to note that the government of Nepal restricts the term *mountain* to those topographic features higher than 26,000 feet (8,000 m) above sea level.

Because the definition of mountains is so general, one would expect the classification of mountains to be a difficult task. Actually, mountains can be grouped, at least in general terms, according to their mode of origin and internal structure. Their size or surface appearance is really of little concern. Nevertheless, as we attempt to classify mountains in the discussion that follows, keep in mind that although one mountain range may share some basic qualities with another, each range has a unique geographic and geologic setting and history.

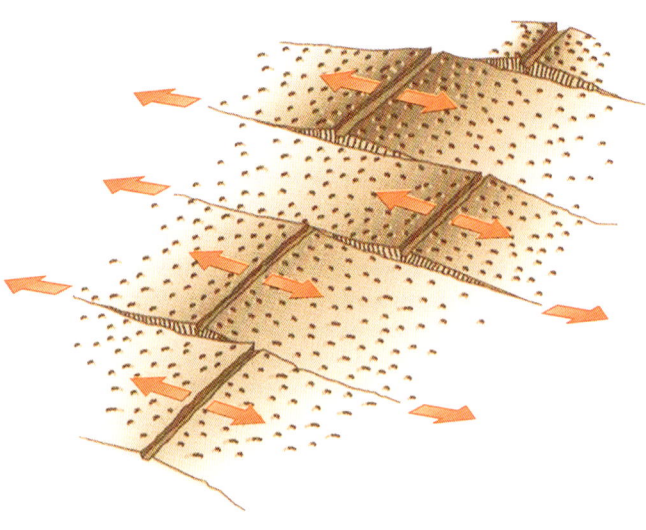

FIGURE 17.1 *Unlike continental mountain ranges with their steep, often precipitous slopes, oceanic ridges are very broad structures with gentle outer slopes.*

KINDS OF MOUNTAINS

Mountains are the result of the application of enormous amounts of energy acting on and within Earth's crust. Energy sufficient to result in the formation of mountains is available from three different sources: (1) **volcanism** resulting from thermal energy associated with rising asthenospheric plumes and mantle convection cells; (2) **epeirogenic forces** (Greek *epeiros* = continent) involving either (a) local or regional upwarping of Earth's crust *without* faulting or (b) uplift accompanied by extension and normal faulting of segments of Earth's crust; and (3) **orogenic forces** (Greek *oros* = mountains), which involve horizontal compressive forces generated by the convergence of lithospheric plates. The term **diastrophism** (Greek *diastrophe* = twisting or distortion) refers to all tectonically induced crustal movements and includes both epeirogenic and orogenic forces.

The combined effects of these forces result in the creation of four basically different kinds of mountains: (1) *volcanic mountains,* (2) *domal mountains,* (3) *block-fault* (or *fault block*) *mountains,* and (4) *foldbelt mountains.*

Volcanic Mountains

Volcanic mountains are found either in areas where rising mantle convection cells or mantle plumes impinge upon the bottom of the lithosphere or in areas where the thermal energy is being released within zones of subduction. Fujiyama in Japan and Mount Vesuvius in Italy are well-known examples of individual volcanic mountains, while the Cascade Mountains, the Aleutian Islands, and portions of the Andes Mountains are examples of volcanic mountain ranges.

Continental Rift Zones Upwelling mantle convection cells located beneath continental lithosphere initiate the **rifting** of the lithospheric plate. As the fractures develop, volcanism begins, first in the form of hot springs and gaseous emanations; later these are followed by outpourings of lava and finally by the building of cones. An example of rift zone volcanism in the United States is the Rio Grande Rift Zone extending from the Mexico-New Mexico border to Colorado. Cinder cones, small lava cones, and lava flows are common throughout the rift zone. One of the world's most spectacular volcanoes, Mount Kilimanjaro, is located along the East African Rift Valley.

Oceanic Ridges In the past, **oceanic ridges** were not included in discussions of mountains mainly because they are not on land. Since the recognition of the concept of plate tectonics, however, and our increased understanding of the ocean bottom, fewer geologists are making a distinction between elevated features of the continental crust and those of the oceanic crust.

Encompassing more than 40,000 miles (65,000 km) in length, the oceanic ridges are the longest mountain ranges on Earth. They are composed of basalt erupted along oceanic rift zones. Along most oceanic ridges, the summit area is downfaulted forming a long, narrow valley similar in structural form to continental rift valleys. The oceanic ridges are mostly submarine features, but isolated segments do occasionally come to the surface. The best-known example, Iceland, is part of the Atlantic Oceanic Ridge and is emergent only because additional magma is being provided by a hot spot under the oceanic ridge.

For the most part, oceanic ridges do not exhibit the steep slopes characteristic of most terrestrial mountain ranges but are rather broad ranges with gentle outer slopes. In some areas where the height of the ridge is less than 2 miles (3 km), the ridges may be more than 1,000 miles (1,600 km) wide (Figure 17.1).

Hot Spot Volcanoes Localized mantle plumes rising beneath oceanic lithosphere result in the generation of **shield volcanoes** building up from the ocean floor (Figure 17.2). Once the rising magma penetrates the lithosphere, the volcanic cone builds rapidly. In some cases, the peaks remain submarine and are called seamounts. In other cases, the peaks build

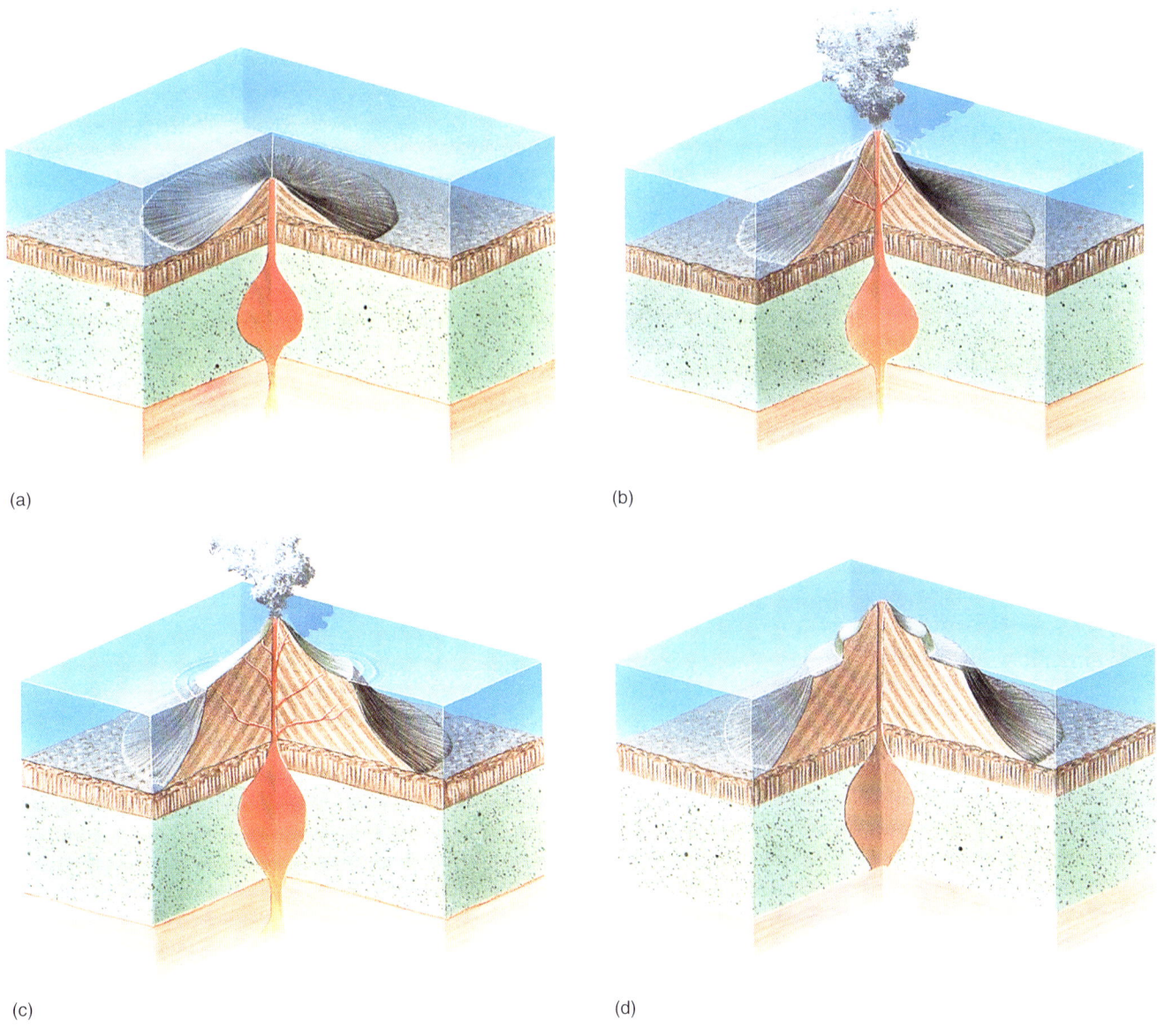

FIGURE 17.2 Shield volcanoes, *formed over stationary hot spots, dot the ocean basins, especially in the Pacific Ocean. Many, called* seamounts, *are submarine while others, such as the* Hawaiian Islands, *build above sea level, only to be eroded and, in time, sink below sea level to once again become seamounts or* guyots.

above sea level to form volcanic islands such as the Hawaiian Islands. As the hot spot goes through as yet unexplained cycles of activity and nonactivity, chains of shield volcanoes—some islands, others seamounts and guyots—form along the ocean floor, aligned in the direction of plate movement.

Subduction Zone Volcanoes   The volcanic mountains that form on the edge of the overthrust plate at convergent plate boundaries differ from oceanic ridge and hot spot volcanoes in that they are primarily constructed of andesitic lavas although basalts and rhyolites also occur. Because silicic magmas are more viscous, subduction zone volcanism is often explosive and generates large volumes of pyroclastic materials in addition to outpourings of lava (Figure 17.3). Accumulations of alternating lava and pyroclastic layers construct the stratovolcanoes (composite volcanoes). These volcanoes have steeper slope angles than the shield volcanoes associated with hot spots. Subduction zone volcanoes occur either as **island arcs**, such as the Aleutian or Tonga islands, or as **continental arcs**, such as the Cascade Mountains or the northwestern United States and the western portion of the Andes Mountains.

Domal and Block-Fault Mountains

Of the various kinds of mountains, those produced by the epeirogenic forces are in many ways the simplest both geologically and structurally. Mountains forming as a result of epeirogenic forces fall into two categories: (1) **domal mountains** where the relief is the result of erosion following either (a) *local doming* of the continental crust, where the affected area may be a hundred or so miles in diameter, or (b) *regional doming*, where large portions of the continental crust are affected by broad, *regional upwarping;* and (2) **block-fault** (or **fault-block**) **mountains**, which result from *extension* and normal faulting of large segments of the continental crust, usually associated with vertical uplift.

Of all the mountain-building forces, the epeirogenic uplift is the most difficult to explain. In many cases, because the regions undergoing uplift are located far from plate boundaries, the forces do not appear to be directly associated with plate tectonics. Some geologists suggest that the forces originate when heat accumulates under the lithosphere, resulting in the increased buoyancy of the underlying asthenospheric rocks.

FIGURE 17.3   *Because of the higher viscosity of the andesitic magmas involved, volcanic eruptions associated with subduction zones are invariably violent and expel large volumes of pyroclastic materials. (Courtesy of T. J. Casaderall/USGS)*

Local Doming   Except for high-angle faults and monoclinal folds that may develop along the margins of the domal uplift, the upwarping usually occurs without any associated rock deformation. The Black Hills of South Dakota are an example of mountains produced by local doming (Figure 17.4). Located on South Dakota's southwestern border with Wyoming, the Black Hills are carved from a domal upwarp approximately 100 miles (160 km) long by 50 miles (80 km) wide. Streams rejuvenated by the doming stripped the sedimentary rocks from the central portion of the up-arched crystalline continental core, producing monoclinal ridges of sedimentary rocks that dip away under the adjoining Great Plains.

Mountain Building   505

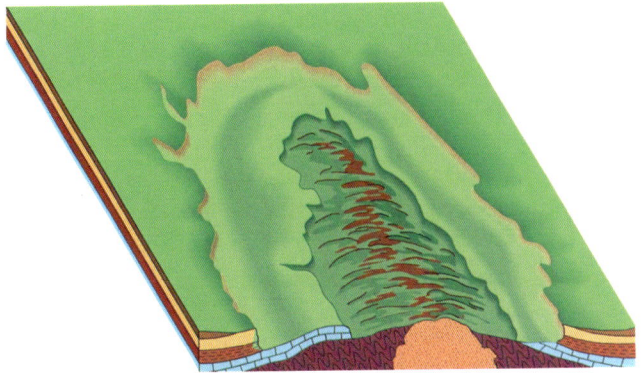

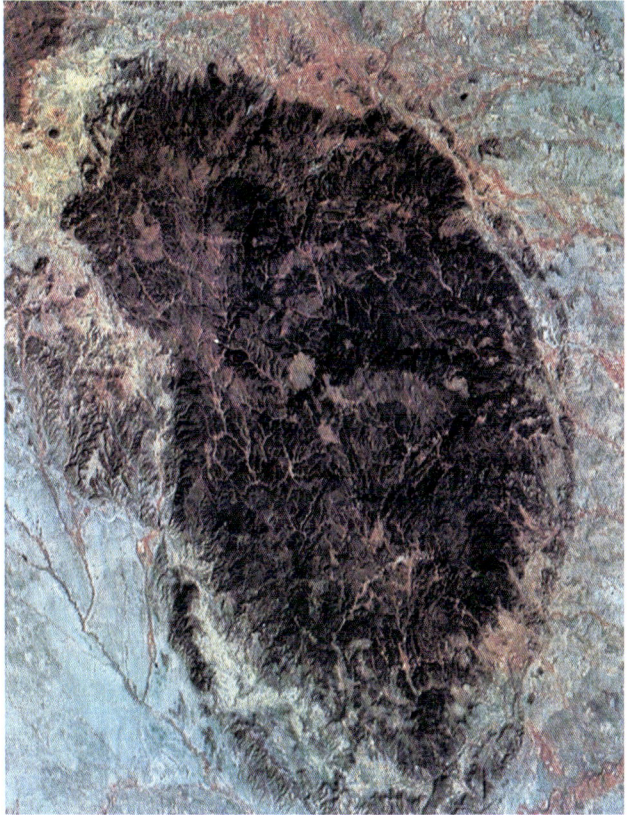

FIGURE 17.4 Local doming *followed by erosion is responsible for the* Black Hills *on the border between South Dakota and Wyoming. The major site of gold production in the United States until the discovery of the California deposits, the Black Hills are now best known for the location of Mount Rushmore on which the faces of several past presidents have been carved. (Photo courtesy of USGS)*

Geologists still do not agree on the source of the energy that created the Black Hills doming. Some argue that the uplift was part of the uplift and later Tertiary rejuvenation of the northern Rocky Mountains to the west.

Regional Doming Epeirogenic forces are also responsible for the uplift of large portions of lithosphere. An example is the uplift that affected a large area in the southwestern United States centered over the state of Utah and including parts of Colorado, New Mexico, and Arizona (Figure 17.5). Once again, the origin of the epeirogenic forces is not clear. Some geologists believe the upwarping was due to a mantle plume hot spot located beneath the North American lithosphere. Others believe that the uplift was the result of the North American lithosphere overrunning the Pacific spreading center. Whatever the source of the forces, the total uplift of the region amounts to about 1 to 1.2 miles (1.5–2 km) with most of the uplift occurring during the last 15 to 20 million years. Except for some normal faulting and monoclinal folding that developed along the margins, the eastern portion of the uplift, referred to as the

FIGURE 17.5 *One of the most spectacular results of epeirogenic forces in the United States is the* Colorado Plateau, *which has been elevated more than a mile. Erosion of the uplifted rocks has given us many of the awe-inspiring vistas that grace the Southwest.*

Colorado Plateau, was elevated to its present level with minimal deformation. Consequently, it is characterized by essentially horizontal, undeformed rocks. Rejuvenation of the streams throughout the Colorado Plateau resulted in the formation of some of the most spectacular scenery in the United States including the Grand Canyon (Figure 17.6), Monument Valley (Figure 17.7), Bryce Canyon (Figure 17.8), and Zion National Park (Figure 17.9).

While the eastern portion was upwarped with little or no deformation, uplift in the western portion was accompanied by tensional forces and lateral extension of the continental crust by normal faulting (Figure 17.10). An obvious question is why the crustal extension did not affect the rocks of the Colorado Plateau. Two explanations have been suggested: (1) tensional forces were present but the rocks of the Colorado Plateau were too strong to break, or (2) the temperature of the underlying mantle rocks was not high enough to initiate the lateral movement necessary to produce the tensional forces. Whatever the explanation, the crustal extension of

FIGURE 17.7 *Monument Valley is truly a monument to the ability of desert streams to erode and remove enormous volumes of rock. The mesas, buttes, and needles that dot the landscape are the remnants of the sandstone body that once extended throughout the area. (Courtesy of E. D. McKee/USGS)*

FIGURE 17.6 *Certainly one of the best-known results of the rejuvenated erosion of the uplifted Colorado Plateau is the* Grand Canyon *of the Colorado River, which has been incised a mile into the rocks of the plateau. The Colorado River is one of the few rivers of the world that flows completely through a desert region to the sea. (Courtesy of E. D. McKee/USGS)*

FIGURE 17.8 *To be truly appreciated, the beauty of* Bryce Canyon *must be experienced firsthand. In the light of the rising or the setting Sun, the colors of the rocks are magnificent.*

FIGURE 17.9 *The sandstones within the soaring cliffs of* Zion National Park *record the presence of an ancient desert and the dunes that swept across its surface.*

the western portion of the uplift resulted in the development of north-south trending normal faults and the subsequent development of the *block-fault mountains* that characterize the area.

Block-Fault Mountains The faulting associated with block-fault mountains occurs in two different styles that are found together (refer to Figure 17.10). In the first, the strikes of the fault planes are *parallel,* and the fault planes dip in the *same* direction. Because of the listric nature of the faults, the relative motion results in a rotation of the block between each set of parallel faults. As one edge of the block moves *upward* to form the edge of a ridge, the other edge of the block moves *downward* to form a parallel valley. The steeper flank of the ridge is the **scarp** of the exposed fault plane.

In the second tectonic style, the strikes of the fault planes are *parallel,* but the dip directions of the fault planes *alternate.* The relative motion now results in the formation of uplifted blocks called **horsts** and downthrown blocks called **grabens**. Note that, in this case, both flanks of the ridges are occupied by fault plane scarps.

In western North America, the combination of both faulting styles has given rise to a particularly rugged topography. Because of its unique geology, the region has been designated as a separate geological province called the **Basin and Range**. The Basin and Range Province extends from southernmost Oregon to Mexico with the best development of the structural style visible in Nevada. A quick scan of a Nevada road map reveals one effect of the Basin and Range topography. Because of the northeast-southwest trending block-fault mountains, east-west roadways are scarce.

Difficult though east-west travel within the Basin and Range is today, crossing Nevada was even harder for the early pioneers who found the parallel ridges and adjoining desert valleys to be impenetrable barriers. Westward travel to the Pacific coast followed either a southerly route across Arizona into southern California, roughly the route of old U.S. Route 66, or a more northerly route, used today by the railroad, from Denver, Colorado, to Salt Lake City, Utah, through Reno, Nevada, and across the Sierra Nevada to the Pacific coast. Reno, Nevada, developed as a waiting station for the seasonal crossing of the high Sierra. The crossing is now made through the Donner Pass, named after the Donner party who attempted a late crossing and were stranded by an early snowfall. Most of the party perished in the intense cold of a high Sierra winter.

Foldbelt Mountains

Foldbelt mountains are generated at convergent plate margins. The folds and faults that characterize foldbelt mountains are typically those associated with horizontal compressive stress.

Foldbelt mountains usually consist of two parallel or subparallel linear components: (1) a complex core composed of deformed igneous, metamorphic, and volcanic rocks and (2) an adjoining system of folded and faulted sedimentary rocks (Figure 17.11). A chain of volcanic mountains usually is associated with the complex core either in the form of an island arc separated from the mainland by a shallow sea or as a continental arc paralleling the coast.

Most foldbelt mountains develop near the edges of continents. The two components generally parallel the continental margin with the complex core seaward and the folded mountains landward of the continental margin.

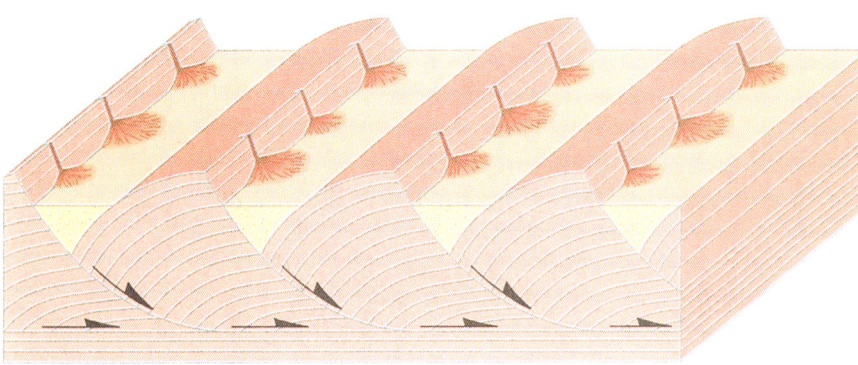

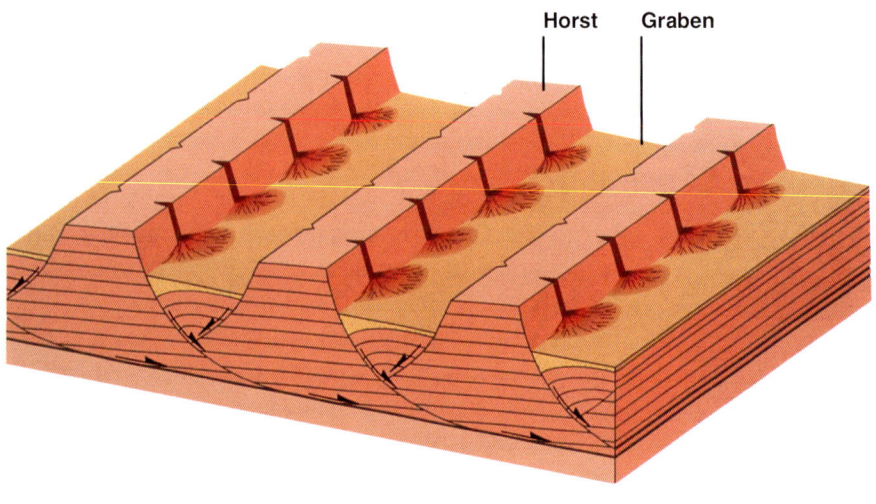

FIGURE 17.10 *The* Basin and Range Province *west of the Colorado Plateau was part of the epeirogenic uplift that affected the entire southwestern portion of the United States. Rather than being uplifted with little deformation, however, the region of the Basin and Range Province was subjected to* extensional forces *that formed two different styles of north-south trending faults. The results is an extremely rugged topography consisting of north-south trending mountain* ranges *separated by* basins *now occupied by deserts.*

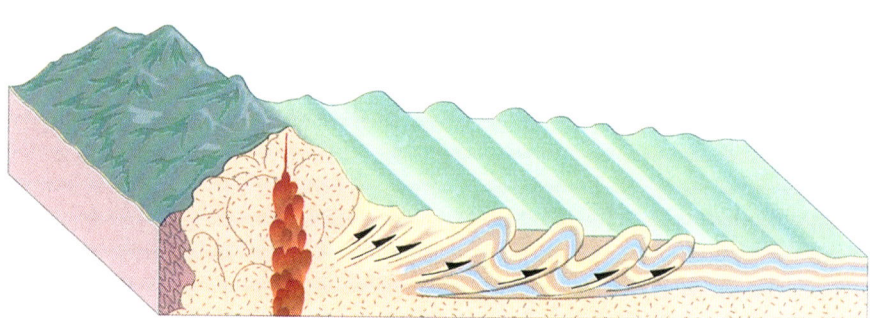

FIGURE 17.11 *Most of the great mountains of the world are* foldbelt mountains, *which consist of two basic components: a* complex core *of igneous, metamorphic, and volcanic rocks and a sequence of folded and faulted sedimentary rocks. The mountains are usually located along the leading edge of a continent with the complex core located on the seaward side.*

As the lithospheric plates converge, energy is partly converted to massive compressional forces, which are used to form, deform, and uplift rocks to awe-inspiring heights. Uplift occurs as the result of two processes: (1) crustal thickening due to both igneous intrusion and crustal shortening and (2) a reduction in density resulting from heat and the generation of low-density siliceous magmas. The great mountains of Earth, including the Alps, the Andes, the northern Rockies, and the Himalayas, are examples of foldbelt mountains. The Appalachians are also foldbelt mountains, though greatly subdued due to millions of years of erosion (Figure 17.12).

The Appalachians may be the most studied foldbelt mountains on Earth. Indeed, our present understanding of how the major components of foldbelt mountains are created stems from an early study of the folded sedimentary portion of the Appalachian Mountains.

In the mid-1800s, an American paleontologist, James Hall, made two observations concerning the sedimentary rocks of the folded Appalachians: (1) most of the sedimentary rocks within the folded

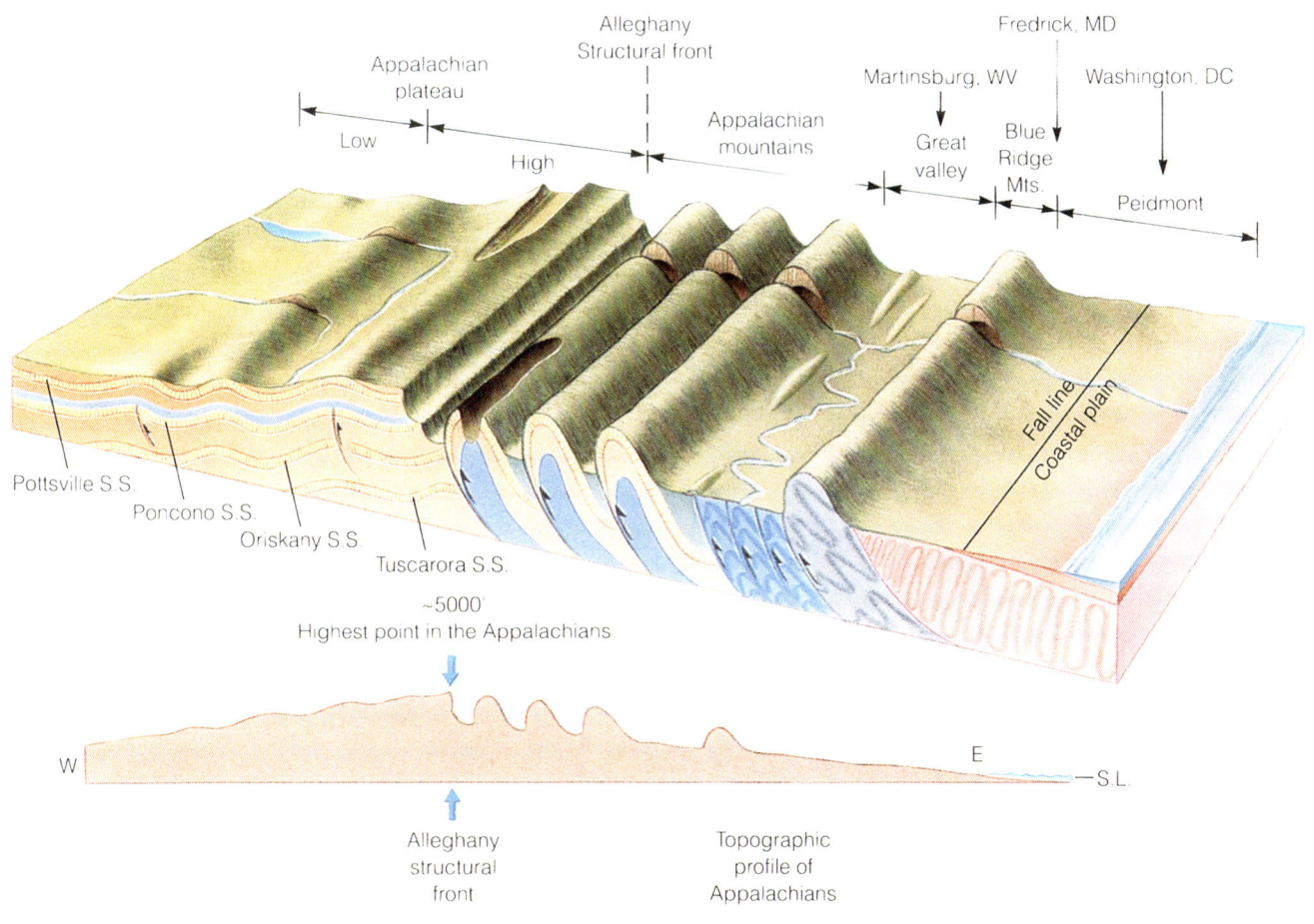

FIGURE 17.12 *One of the first examples of foldbelt mountains to be studied in detail, the* Appalachian Mountains *were formed during the continental collision that created Gondwana. Since their formation, the Appalachians have been reduced by erosion, perhaps to the featureless surface proposed by Davis called a peneplane, and reuplifted by epeirogenic forces that affected the entire eastern portion of North America. The present topography of the Appalachians is the result of erosion by the rejuvenated streams.*

Appalachians were of shallow marine origin and were the same types found within the continental interior, and (2) the sedimentary rocks in the folded Appalachians were many times thicker than the sedimentary rocks that covered the crystalline basement of the continent outside the mountains. The sedimentary rocks over the continental interior were estimated to be a few thousand feet or meters thick, but within the folded Appalachians, Hall found the total thickness of sedimentary rocks to be in the *tens* of thousands of feet. Hall was faced with the problem of explaining how tens of thousands of feet of marine sedimentary rock could accumulate along a continental shelf where the water depth rarely exceeded 500 or 600 feet (150–180 m). He concluded that as the sediments accumulated, the surface of the continental shelf must warp downward into a broad synclinal trough parallel-ing the continental margin (Figure 17.13). Hall's con-temporary, the mineralogist James D. Dana, coined the name **geosyncline** to describe the structure. This term was intended to reflect the great size of these basins and should not be confused with a *structural syncline*. Hall and Dana agreed on the existence of the geosyncline, but developed different hypotheses as to how the structure formed. Hall was of the opinion that the geosyncline formed as the weight of the sediments downwarped the surface of the continental shelf. Dana argued that the accumulating sediments alone could not cause the downwarp of the shelf because the density of the sediments was too low to cause the displacement of the higher-density mantle rocks below. Dana was a proponent of the idea that all the deformation exhibited by Earth's crust, including the formation of geosynclines, could be attributed to the fact that Earth was cooling; as it cooled, the crust supposedly was wrinkling like the skin of a cooling baked

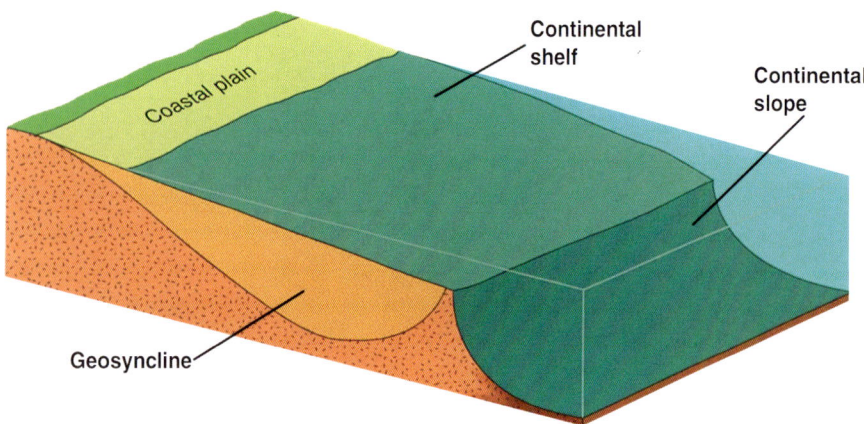

FIGURE 17.13 *As originally conceived, the* geosyncline *was a synclinal basin that formed along the trailing continental margin and continued to downwarp as sediments were introduced from the land.*

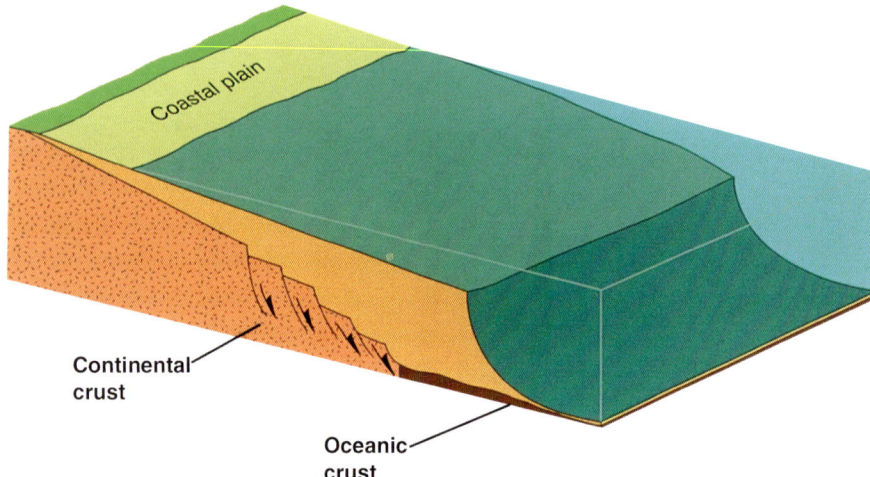

FIGURE 17.14 *With the advent of plate tectonics, combined with seismic data that have allowed geologists to "see" within the rocks and sediments of the continental margin, our structural picture of the passive continental margin has changed. Rather than a synclinal basin, we now recognize a thick wedge of sediment, called a* geocline, *that extends out from the edge of the passive continental margin.*

apple. This "baked apple theory" or "shrinking Earth hypothesis" dominated the thinking of geologists for many years but eventually gave way to the theory of plate tectonics.

The advent of plate tectonics drastically changed our picture of the formation of the thick sedimentary wedge that characterizes continental trailing edges. Because the wedge is not synclinal, some geologists have suggested that it should be called a **geocline** rather than a geosyncline. Although our ideas about the formation of the thick wedge of sediment have changed, the term *geosyncline* is still used today to describe the overly thick accumulations of sediments that are considered to be the precursors to the sedimentary rocks associated with foldbelt mountains.

As the plates move, the continents are carried along as components of a lithospheric plate. Trailing or passive continental margins, such as the eastern margins of North and South America, develop a slowly downwarping continental shelf, which initiates the development of the geocline. The downwarping is thought to be due to crustal thinning and subsidence resulting from continued movement along the high-angle normal faults that formed as a result of extensional forces induced by the former plate movements (Figure 17.14). As the geocline continues to evolve, sediments derived from the weathering and erosion of the adjoining landmass continuously accumulate on the surface of this downwarping margin in the form of a thick sedimentary wedge (refer to Figure 17.14). As the earlier layers of sediment become buried within the downwarping basin, lithifications transforms the sediments into marine sedimentary rocks.

With the advent of plate tectonics, we have come to recognize other settings in which thick sequences of sediments may accumulate. For example, a **back-arc basin** forms between a volcanic island arc and the continental margin (Figure 17.15). In some cases, a mantle convection cell may develop below the back-arc basin, rift the rocks by extension, and pour basalt onto the ocean floor as happens at an oceanic ridge.

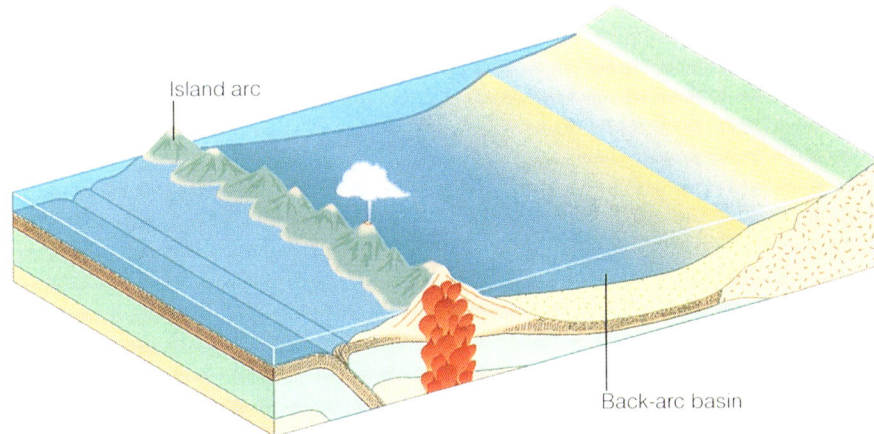

FIGURE 17.15 *The sediments accumulating in geoclines are one source of the sediments from which sedimentary rocks form. Another major source is the* back arc basin *that exists between the island arc and the continental margin. Accumulating sediments from both the continent and the island arc, the back arc basin may develop into a* marginal sea *if a mantle plume develops beneath the underlying oceanic crust. An example of a marginal sea is the* Sea of Japan.

A back-arc basin expanded by tensional forces is called a **marginal sea**. The Sea of Japan located between the Japanese Islands and the Asian mainland is an example of a marginal sea.

The back-arc basin and marginal sea differ from the geocline associated with a continental trailing edge in that the accumulated sediment in the back-arc basin is introduced from *both* the mainland and the island arc. With the passage of time, as the mainland is reduced in elevation by weathering and erosion, it becomes less important as a source of sediment. Because the eroded volcanic mountains are "rejuvenated" by fresh eruptions, the island arc eventually becomes the major source of the sediment and will provide sediment more or less continuously for the entire life of the basin.

SPOT REVIEW

1. What is the definition of a mountain?
2. What is the difference between epeirogenic and orogenic forces?
3. Compare and contrast three different scenarios under which volcanic mountains form.
4. What are the basic characteristics of epeirogenic mountains?
5. What are the characteristics of block-fault mountains? Where in the United States are block-fault mountains best developed?
6. What are the basic characteristics of foldbelt mountains? What are some examples of foldbelt mountains?
7. What are the possible depositional sources for the sediments that make up the sedimentary rocks of foldbelt mountains?

OROGENIC STYLES

Three basic convergent plate scenarios can be used to explain most of Earth's foldbelt mountain systems: (1) *ocean-continent,* (2) *ocean-island arc-continent,* and (3) *continent-continent* collisions.

Ocean-Continent Orogenesis

The **ocean-continent** orogenic style, illustrated in Figure 17.16, shows the process of mountain building beginning with a passive continental margin and the associated geoclinal accumulation of sediments. An example is the present west coast of South America or North America. If we go back about 200 million years to the time when our present-day continents were still joined together in Wegener's supercontinent of Pangaea, Figure 17.16a would represent the western part of Gondwana, that portion of the supercontinent that would soon break away to become South America. As South America broke away from Africa and began to move westward against the oncoming Pacific lithospheric plate, a zone of subduction formed seaward of the geoclinal deposits (Figure 17.16b), transforming the western edge of the South American plate from a *passive* to an *active* margin.

As subduction continued, compressional forces developed, initiating the following sequence of events. The downward-moving oceanic lithosphere carried ocean bottom sediments and water with it. The water lowered the melting temperature of rocks within the subduction zone as the system converted from dry to wet, as discussed in Chapter 5. Intense heat and pressure metamorphosed the descending sediments and existing crystalline rocks. Thrust faults formed in the overriding crustal block and brought a mixture of metamorphic rocks, slivers of

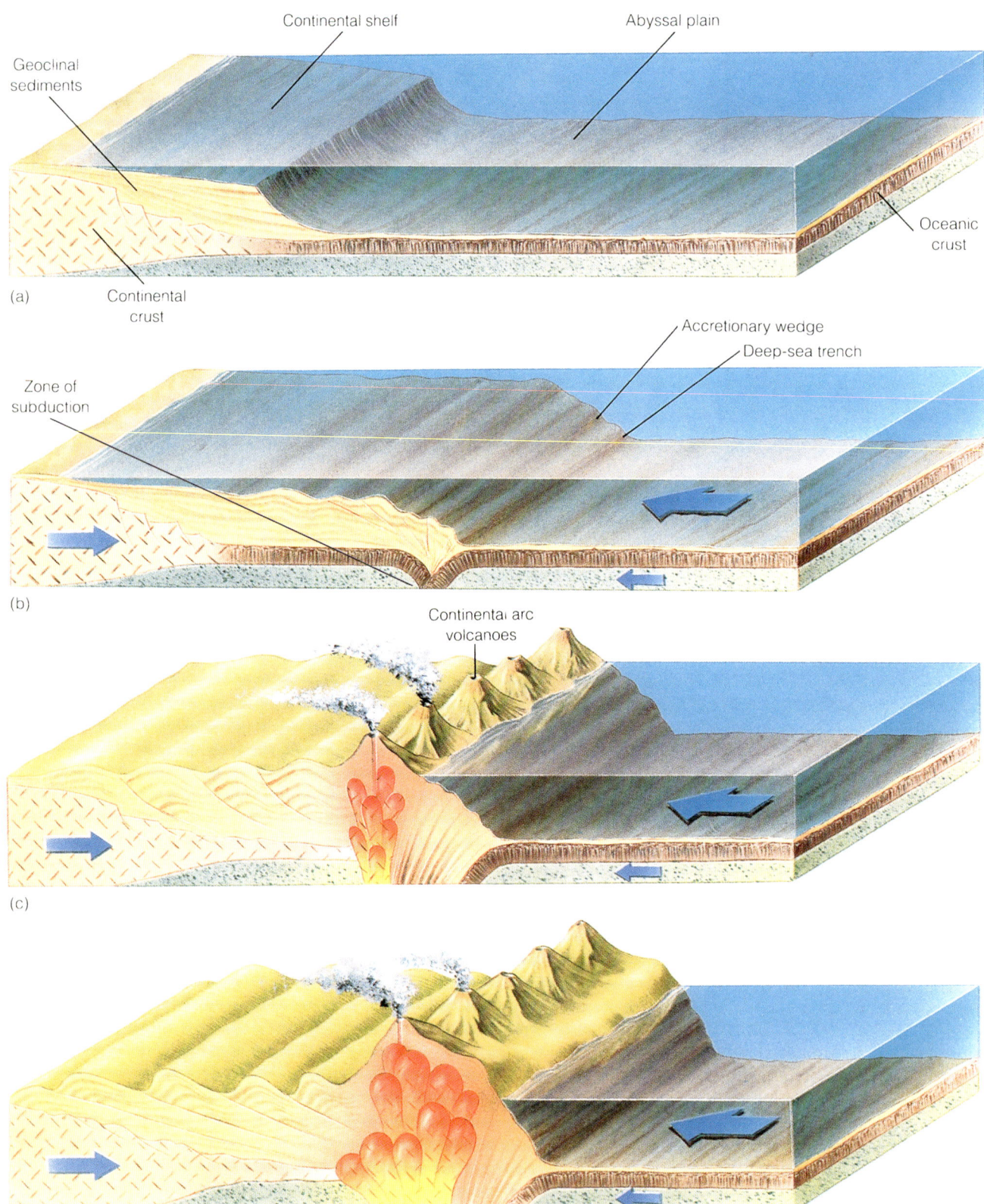

FIGURE 17.16 *Opening oceans* form by sea-floor spreading as new oceanic lithosphere is formed at an oceanic ridge (a). At some point in time, the cooling oceanic lithosphere breaks and sinks into the underlying asthenosphere, forming a zone of subduction (b). As the oceanic plate continues to subduct, an accretionary wedge forms at the edge of the overriding plate; granitic magmas intrude into the edge of the continent while less viscous andesitic magmas are extruded to form a continental arc; and the sediments of the geocline are folded and thrust onto the continent (c). The result of this orogenic process is a mountain range such as the *Andes Mountains that form the spine of South America (d).*

the oceanic crust, igneous rocks, and metasediments to the surface. This assemblage of different rock types, called a **mélange** from the French word for mixture, formed a structure called an **accretionary wedge** (Figure 17.16c). As each new thrust sheet formed and was emplaced *under* the previous sheet, the previously deformed materials were lifted upward as a growing rock mass, eventually forming a coastal mountain range. An example of a mélage formed within an accretionary wedge is the Franciscan Formation, the rocks that make up most of the mountain ranges along the California coast (Figure 17.17).

Massive intrusions of granitic magmas began to be emplaced into the edge of the continental crust while andesitic magmas reached the surface where they created the chain of continental arc stratovolcanoes associated with folded mountain belts. With time, the sediments of the geocline began to respond by plastic deformation as thrust and reverse faults carried the folded sediments eastward onto the continental surface (refer to Figure 17.16d).

The deformation produced the structures that we see today along the mountainous backbone of South America, the **Andes Mountains** (Figure 17.18). The mélage is exposed in the coastal areas of South America and grades eastward into complex exposures of granites and granodiorites capped by the volcanic mountains of the Andes. The crystalline mass makes up the complex core of the Andes mountain chain. The folded portion of the mountain chain is represented by the folded and faulted sedimentary rocks that make up the eastern flanks of the Andes Mountains. The level of present-day volcanism and earthquake activity along the western margin of the South American continent tells us that the Andes are still growing as South America continues to ride over and consume the Pacific ocean floor.

An important aspect of orogenesis is that the batholithic bodies of granite and granodiorite that are generated within the subduction zone through its history are eventually *added* to the continental crust, compensating for losses of continental rocks by weathering, erosion, and sediment transport to the deep sea, followed by subduction. The continental crust that we see today has grown to its present mass and size as a result of repeated orogenic events since the origin of Earth.

FIGURE 17.17 *Most of the rocks seen in the Coastal Range of California, Oregon, and Washington accumulated in an* accretionary wedge *that has been uplifted above sea level.* (Courtesy of R. E. Wallace/USGS)

FIGURE 17.18 *Earthquakes and volcanic activity are evidence that the* Andes Mountains *are still forming as the Pacific plate subducts beneath the edge of South America.* (VU/© 1993 David Matherly)

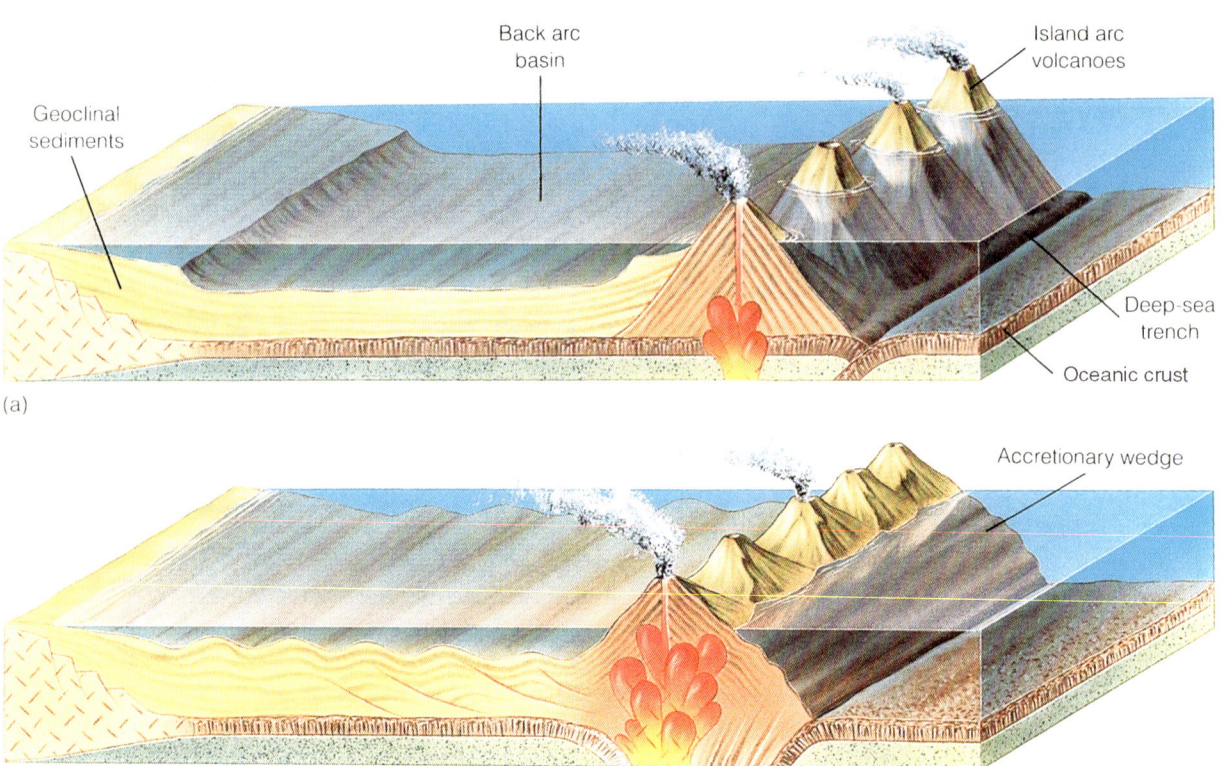

FIGURE 17.19 *An* ocean-island arc-continent orogeny *begins with the formation of a chain of andesitic, volcanic islands, such as the* Aleutian Islands, *along a zone of subduction separated from the mainland by a back arc basin (a). As convergence continues, granitic magmas intrude into the island arc, creating an ever-increasing core of continental-type rocks within the chain of islands. The* Japanese Islands *are an example. At the same time, the sediments within the back-arc basin begin to be deformed (b). Eventually, the sediments of the back-arc basin will be thrusted and folded above sea level and become a part of the continent (c). The orogenic event will end as the original island arc complex is welded to the edge of the continent and the former back-arc basin sediments are stacked onto the surface of the continent along low angle thrust faults (d). The northern Rocky Mountains from Wyoming to Alaska are thought to have formed by such an orogenic event.*

Ocean-Island Arc-Continent Orogenesis

The basic scenario of **ocean-island arc-continent** orogenesis is illustrated in Figure 17.19. The events of the early phases of subduction are the same as those for ocean-continent orogenesis except that an island arc develops. Once again, an accretionary wedge, with a mélange, forms along the leading edge of the overriding oceanic plate. Intermediate and granitic magmas are generated and emplaced into the overriding plate, in this case, oceanic lithosphere, which progressively thickens and increases in mass. As a result, the oceanic lithosphere becomes increasingly more continental in character. An example would be the Japanese Islands.

A back-arc basin—and, in some cases, a marginal sea—develops between the growing island arc and the adjacent continent. Sediment derived from both the continent and the island arc pours into the basin (Figure 17.19a). Tensional forces extend the marginal sea, if developed, allowing the continued accumulation of sediment. As the plates continue to converge, the sediments in the back-arc basin or the marginal sea begin to be folded and moved landward along low-angle, high-displacement thrust faults (Figure 17.19b). In time, as the entire island arc complex with its core of granitic rocks is thrust and fused to the edge of the continental crust, the deformed sediments of the back-arc basin or marginal sea are thrust onto the continent and stacked, one massive thrust sheet on top of another (Figure 17.19c). Thrust sheets thousands of feet thick are moved laterally with displacements as large as 10 to 20 miles (16–32 km) (Figure 17.19d). An example of a mountain range created by such an orogenic event is the **North American Cordillera** represented by the northern Rocky Mountains extending from Alaska south to Wyoming (Figure 17.20).

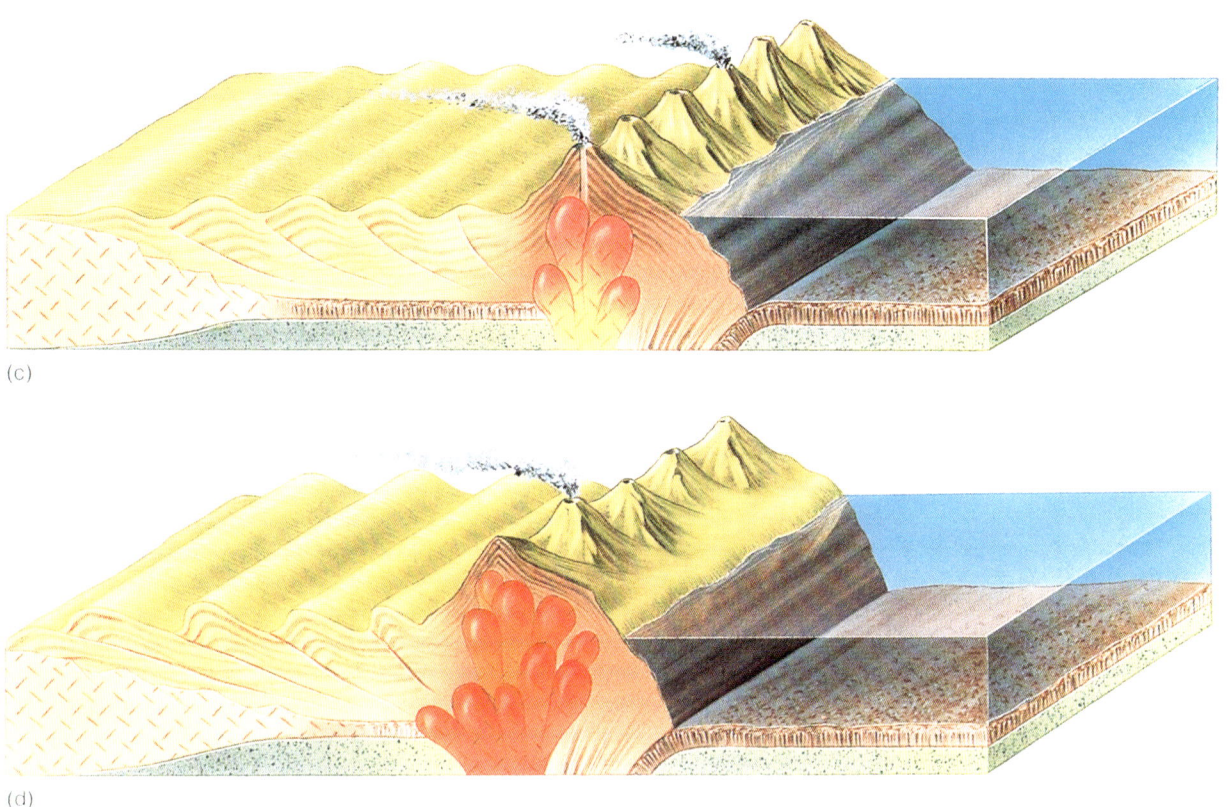

(c)

(d)

FIGURE 17.19 continued

FIGURE 17.20 *The northern Rocky Mountains originally formed along the western margin of North America. Mountain-building events since the formation of the Rocky Mountains have added the portion of the North American continent that extends from the present mountains to the Pacific coast.*

Continent-Continent Orogenesis

Some of the most spectacular mountains are the result of **continent-continent** collisions. The continent-continent scenario differs from the ocean-island arc-continent scenario only in that the oceanic lithosphere brings with it a passive continental margin with associated geoclinal sediments.

To demonstrate the mountain-building process, let us return to Gondwana, this time to the eastern side of the continent (Figure 17.21). At the time continental rifting took place, Gondwana had developed a geoclinal basin along its eastern margin. As the supercontinent broke up, the Antarctic continent moved southward, Australia headed eastward, and the fragment that was to become India rifted from

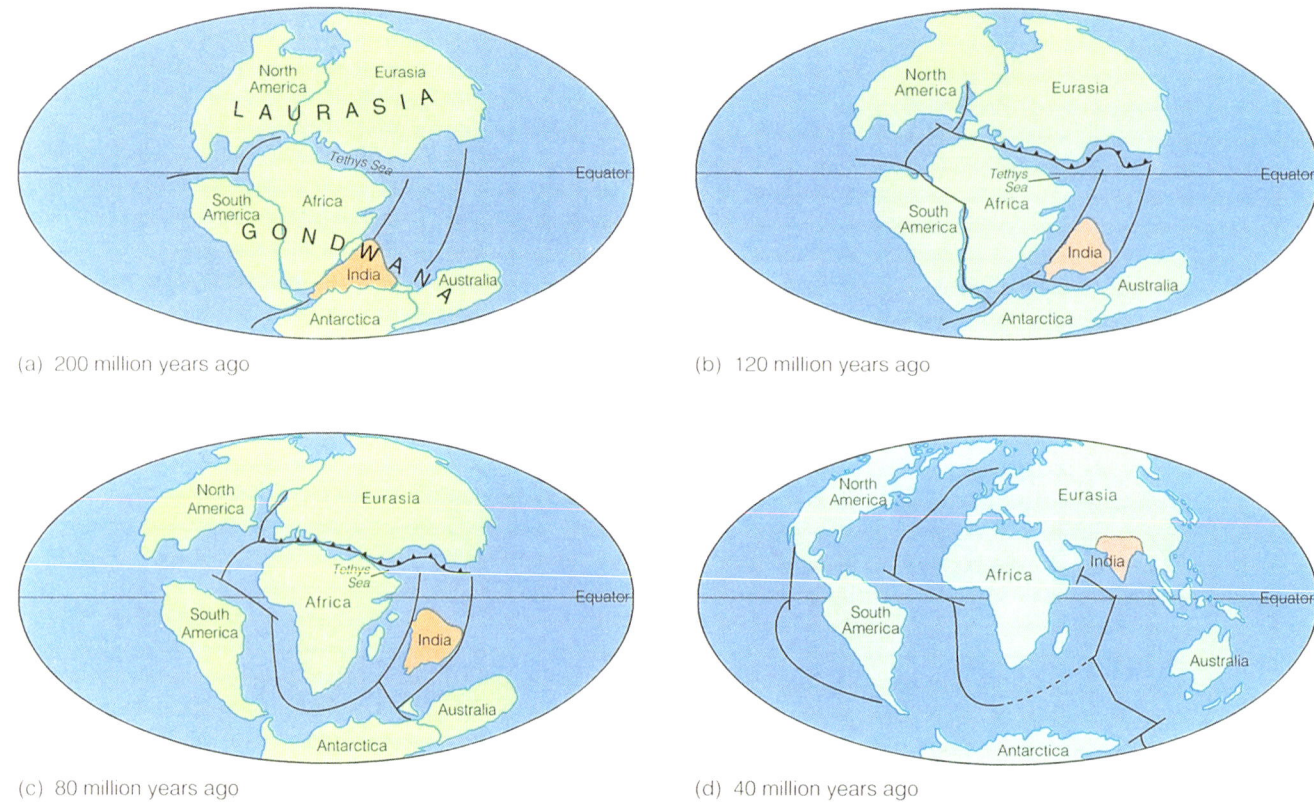

FIGURE 17.21 *The continents that are currently involved in* continent-continent *collisions and those that will be involved in the near geologic future were formed during the breakup of Pangaea about 200 million years ago.*

the main continental mass and began its journey northward, taking with it a portion of the geoclinal basin with its contained sediments (Figure 17.22).

Across the Tethys Sea, Asia waited with a chain of island arc volcanic mountains and a sediment-filled back-arc basin (Figure 17.22b). As India approached Asia and the Tethys Sea narrowed, the sediments of the Asian back-arc basin began to be folded and thrust northward onto the Asian continent (refer to Figure 17.22). With continued narrowing of the back-arc basin, the island arc complex collided with the continental crust of the Asian mainland and welded to it (Figure 17.22c). As the Tethys Sea narrowed and India approached, the scene was set for the creation of the **Himalayas**.

About 40 million years ago, India collided with Asia. As the continents collided, the geoclinal sediments located along the northern margin of the Indian continental fragment began to deform as the oceanic lithosphere in front of the Indian subcontinent was subducted below the Asian continental lithosphere. As the northern edge of the Indian continental lithosphere plunged downward beneath the southern edge of Asia, sheets of deformed geoclinal sediments and parts of the crystalline crust of India broke away and were thrust *southward* up over the Indian continental crust along major low-angle thrust faults (Figure 17.22d).

Unlike the dense basaltic oceanic lithosphere, the Indian continental lithosphere with its lower-density granitic crustal rocks resisted subduction. The breathtaking heights of the Himalayas that we see today are the result of the buoyancy of the Indian lithosphere as it rams beneath the edge of the Asian plate and lifts the overlying rocks of the Asian margin (refer to Figure 17.22).

Continued movement of the Indian continent northward beneath Asia resulted in a second episode of major low-angle thrusting that drove the Indian geosynclinal sediments farther southward onto the Indian continent (Figure 17.22e). At the same time, some of the folded sediments were thrust back *northward* over the rising edge of the Asian plate. Eventually, these folded and thrust-faulted sediments were uplifted to the very highest elevations in the Himalayas, even above those of the crystalline complex. One result of this uplift is that Mount Everest with its summit at 29,696 feet (9,051 m) is capped by a marine limestone that once lay many thousands of feet below at the bottom of the Tethys Sea!

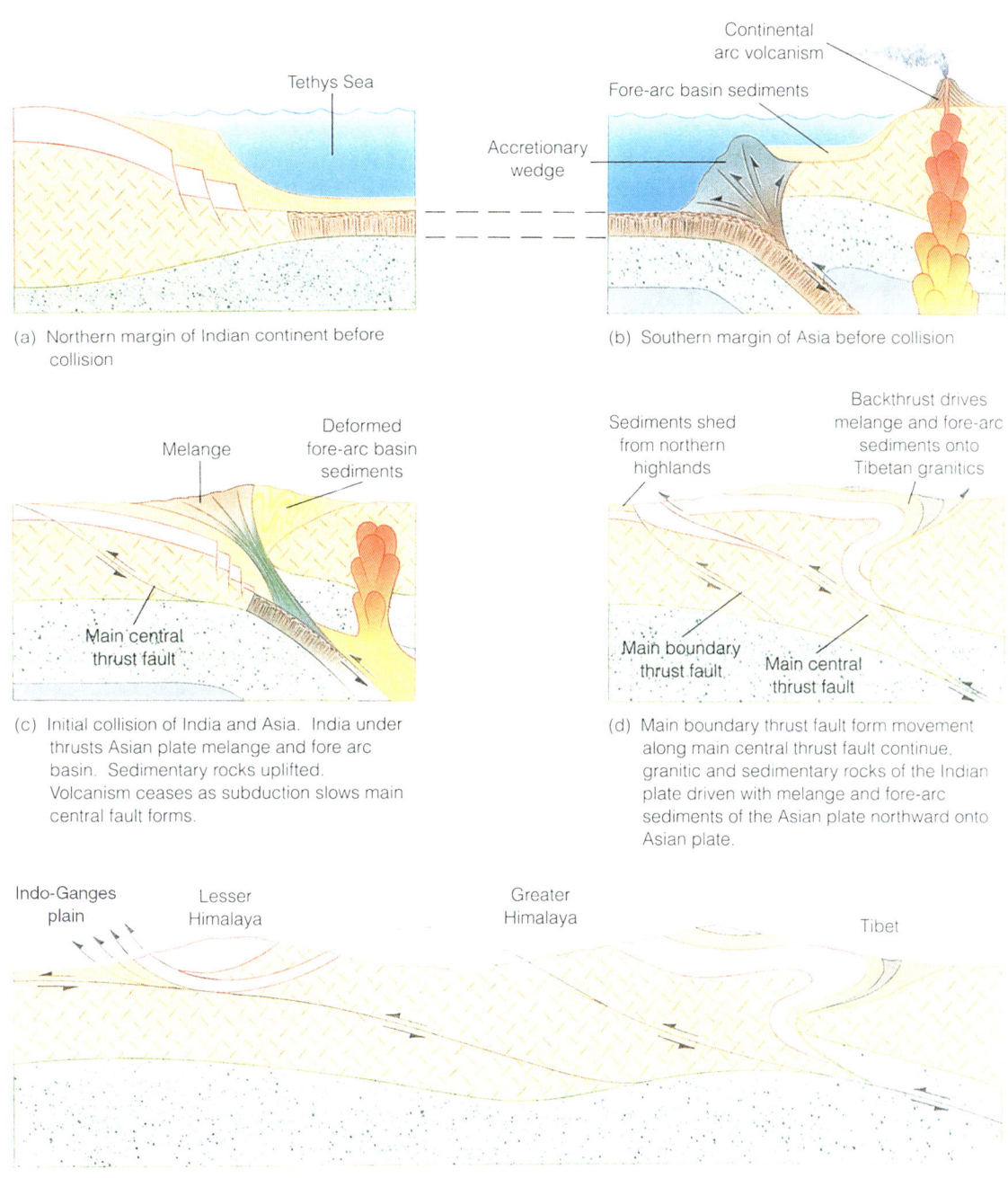

FIGURE 17.22 *The formation of the Himalaya Mountains by the collision of India and Asia about 40 million years ago is an example of a* **continent-continent orogeny.** *The orogenic event began with the rifting of Gondwana and the closing of the Tethys Sea. India, with its geoclinal wedge of sediments, lay to the south (a). To the north, along the southern margin of Asia, a zone of subduction existed with an accretionary wedge and a continental arc (b). The drawing shows a* **fore-arc basin** *that commonly develops where an accretionary wedge builds upward fast enough to trap sediment eroded from the continental margin. Eventually, the two continental masses collided. As the Indian continental crust was subducted beneath the edge of Asia, the sediments of the Indian geocline and part of the Asian accretionary wedge were thrust southward onto the Indian continent. At the same time, the sediments that had accumulated in the Asian fore-arc basin, plus a portion of the accretionary wedge, were thrust onto the Asian continent (c). Because of its low density, the Indian continental crust resisted subduction. As convergence continued, the buoyancy of the low-density rocks of the Indian continental crust resulted in the uplift of the region. The complex folding of the entire rock complex (d & e) eventually elevated a limestone that had originally formed on the bottom of the Tethys Sea to the top of Mount Everest. Although volcanic activity was terminated as subduction and the subsequent generation of magma slowed, seismic activity indicates that India is still moving northward. No one knows how long it will be before the orogenic episode is concluded.*

The formation of the Himalayas is not yet over. Earthquakes indicate that India is still thrusting northward under the Asian plate at an average rate of 2 inches (5 cm) per year. As a result of India's northward movement, the Himalayas are being lifted each year at a rate *greater* than the rate at which the rocks are being removed by erosion. As grand as the Himalayas are, they will become even grander.

Another continent-continent collision that is well underway is that between Africa and Europe. The relationship between Africa and Europe is similar to that between India and Asia before the two continents collided. The volcanism in the Mediterranean area and the rise of the Alps are the result of the forces being generated as the two continents converge. Eventually, when the collision is complete, the Mediterranean Sea—the last large vestige of the Tethys Sea—will be eliminated, and the Alps will probably be as grand as the present-day Himalayas.

It is important to note that the addition of continental lithosphere to the margin of another continent results in the formation of mountains located *within* the continental mass. The Ural Mountains, which are located well within the Eurasian continent, are an excellent example. The Urals were formed when smaller continents were "sutured" together during a continent-continent collision to form the huge Eurasian continent. In fact, the Ural Mountains are usually considered to be the boundary between Europe and Asia.

Before we leave the topic of continent-continent collisions, let's go back to the time before the supercontinent Gondwana was formed by the collision of whatever continents existed at that time. Referring once again to Figure 17.22, consider this time that the continent to the left is the one that will become North America and the one on the right will become Africa. Picture these two masses of continental lithosphere colliding about 250 million years ago, forming Pangaea and creating a mountain complex that was perhaps every bit as grand as today's Himalayas.

About 20 million years later, Gondwana began to break up. We have already discussed the events that occurred as South America broke away from west Africa. A similar rifting formed North America. When North America broke away from northwestern Africa and Europe, the break occurred somewhere within the crystalline complex of the newly formed mountain range. The northern portion of the Atlantic Ocean was created as the two parts of this once grand mountain range separated.

The western part of the original orogenic system became the rejuvenated Appalachian Mountains of North America. The folded sedimentary rock component of that pre-Atlantic mountain range (comparable to the sediments that were thrust southward onto the Indian continent) now makes up the Appalachian Plateau, the Valley and Ridge, and the Great Valley portions of the Appalachians. A part of the original crystalline core of that early mountain range can be found today in the rocks of the Blue Ridge Mountains and the Piedmont (see Figure 17.12). However, much of the original core complex is buried beneath the modern coastal plain and continental shelf. More extensive exposures of the crystalline complex can be seen today in northern New England, Nova Scotia, and Newfoundland.

As rifting took place and the two parts of this mountain complex separated, streams methodically reduced this once impressive mountain range to a flat erosional surface near sea level. Within the last 20 million years, however, epeirogenic forces in the form of glacio-eustatic rebound upwarped the entire eastern portion of the North American continent. As streams were rejuvenated throughout the area, the increased erosion not only produced the present topography of the eastern United States and southeastern Canada, but also reexposed the structures and rocks of that ancient orogeny. We see them today, perhaps not with the grandeur they once had, but certainly with sufficient clarity to attest to their proud history.

An interesting question might be, what happened to the rocks of the "other half"? On the western side of the Atlantic Ocean, the rocks of the crystalline core complex now lie buried beneath the sediments of the modern Atlantic coast geocline of North America and are exposed as the rocks of Nova Scotia and Newfoundland. On the "other side" of the Atlantic, the rocks of that same crystalline core complex can be seen in the rocks along the western coastlines of Ireland and Scotland and throughout the Scandinavian Peninsula. The folded mountains that existed east of the original central core complex as the "twin" to the Appalachians can now be found in the Mauritanide Mountains of northern Africa, a mountain range structurally quite similar to the Appalachians.

SPOT REVIEW

1. Compare passive and active continental margins. Give an example of each.
2. How can a passive continental margin be converted into an active margin? Give an example.
3. What is a melange and how does it form?
4. Compare the mountain-building scenarios associated with ocean-continent, ocean-island arc-continent, and continent-continent collisions.
5. Where would you look for companion rocks and structures to those exposed in the Appalachian Mountains?

CONCEPTS AND TERMS TO REMEMBER

Mountain-building forces
  volcanic
  epeirogenic
  orogenic
  diastrophism
Kinds of mountains
  volcanic mountains
    rift zones
    rift valleys
    oceanic ridges
    hot spot shield volcanoes
    island arc stratovolcanoes

continental arc
  stratovolcanoes
epeirogenic mountains
  domal mountains
  block-fault mountains
    scarp
    horst
    graben
    Basin and Range Province
orogenic mountains
  foldbelt mountains
    geosyncline or geocline

back-arc basin
  marginal sea
Orogenic styles
  ocean-continent collision
    mélange
    accretionary wedge
    Andes Mountains
  ocean-island arc-continent collision
    North American Cordillera
  continent-continent collision
    Himalaya Mountains

REVIEW QUESTIONS

1. The Black Hills of South Dakota are an example of _____ mountains.
   a. block-fault   c. erosional
   b. domal   d. foldbelt
2. The faults associated with block-fault mountains are
   a. thrust.   c. normal.
   b. strike-slip.   d. reverse.
3. A modern example of a geocline or geosyncline is now developing
   a. within the Sea of Japan.
   b. along the east coast of North America.
   c. along the west coast of North America.
   d. along the trend of the Aleutian Islands.
4. The forces associated with epeirogenic mountain building are dominantly
   a. horizontally directed nonrotational compression.
   b. volcanic.
   c. vertical uplift with or without tensional extension.
   d. horizontally directed rotational compression.
5. Which of the following features is primarily the result of epeirogenic uplift?
   a. Cascade Mountains
   c. Hawaiian Islands
   b. Grand Canyon
   d. Sierra Nevada

REVIEW QUESTIONS *continued*

6. Why is there no volcanism associated with the Himalayas?
7. Why do geologists find it difficult to explain the existence of epeirogenic forces?
8. What role do geoclines or geosynclines and back arc-basins play in the formation of foldbelt mountains?
9. What similarities exist between the Appalachian Mountains and the Himalayas?

THOUGHT PROBLEMS

1. Although most foldbelt mountains are located near the margins of continents, the Rocky Mountains of North America and the Ural Mountains of Russia are located inland, but for different reasons. How are the inland locations of these two foldbelt systems explained?
2. Based on what you know about the origin of mountains and about plate tectonics, where do you predict the next ranges of foldbelt mountains will form on Earth? What scenario(s) will bring them to being?

FOR YOUR NOTEBOOK

If you live in an area of deformed rocks, a field trip is definitely in order to describe and perhaps photograph some of the more impressive outcrops. You will also want to investigate the geology of your area in more detail to see how it fits into the larger mountain system.

Another possible exercise, especially if there are no mountains in your area, is to investigate in detail some other mountainous areas that were not treated in the text. In the northeast, there are the White Mountains, the Berkshires, and the Adirondacks of upstate New York. In the mid-continent, there are the Ozark and Arbuckle mountains. In the west Texas, the structure and lithologies of the Guadalupe mountains are unique. Another interesting study would be to compare the structure and stratigraphy of the southern and northern Rocky Mountains.

CHAPTER OUTLINE

INTRODUCTION

THE DISTRIBUTION OF EARTHQUAKES

THE FREQUENCY AND LOCATION OF EARTHQUAKES
    Focus
    Epicenter

MEASURING EARTHQUAKE INTENSITY AND MAGNITUDE
    Intensity
    Magnitude

DAMAGE FROM EARTHQUAKES
    Tsunami

EARTHQUAKE (SEISMIC) WAVES
    Body Waves
    Surface Waves

THE SEISMOGRAPH

LOCATING EARTHQUAKES FROM SEISMIC DATA

WORLDWIDE SEISMIC NETWORK

EARTHQUAKE PREDICTION

CHAPTER 18

Earthquakes and Seismology

INTRODUCTION

Those of us whose daily lives are unlikely to be touched by earthquakes often find it difficult to imagine how people living in earthquake-prone areas such as parts of California can tolerate knowing that at any moment of any day they may become victims of a major earthquake. Unlike tropical storms and most volcanic eruptions, earthquakes seldom, perhaps never, give any warning of their coming. Within a matter of seconds, a quiet, stable landscape is converted into a quaking surface with fearsome possibilities of the destruction of both property and life. Then, as quickly as it began, the quake is over, except for the damage and suffering that it leaves behind.

Records of earthquakes have been kept for some time. The Chinese records go back nearly 3,000 years. The ancient Greeks were the first to attempt a "scientific" explanation of earthquakes. The philosopher Strabo noted that the frequency of earthquakes decreased as one moved inland from the coast. Aristotle thought earthquakes were caused by strong winds blowing through Earth's interior. (You will remember from our discussions of groundwater that Aristotle thought that the interior of Earth was hollow.) Some scholars have interpreted portions of the Bible as describing the aftermaths of earthquakes. Foremost among these are the collapse of the walls of Jericho about 1100 B.C. and the destruction of the cities of Sodom and Gomorrah.

The first recorded earthquake in the Western Hemisphere occurred in Massachusetts in 1636—reportedly, chimneys were knocked down. Nearly three decades later, in 1663, an earthquake was recorded in the lower Saint Lawrence River Valley, a site of frequent earthquakes even today.

Parts of California have the highest frequency of earthquakes in the continental United States. A relatively complete record of earthquakes for the area is available, dating back to the early 1800s. The record is largely the result of diaries kept by the Franciscan missionaries who established a chain of missions along the west coast. The missions were unknowingly located near faults because of the presence of fault-controlled springs.

THE DISTRIBUTION OF EARTHQUAKES

A map plotting the locations of recent earthquakes shows a pattern similar to that of active volcanoes (Figure 18.1). Before the advent of plate tectonics, the similarity in the distribution of earthquakes and active volcanoes led geologists to believe that one was the cause of the other. Although the rise of molten rock does cause swarms of earthquakes as it shoves rock out of its path, the earthquakes are usually weak and rarely result in damage or loss of life. The spatial relationship that exists between volcanoes and earthquakes occurs not so much because one causes the other as because both of these geological phenomena are related to a common feature, the **plate margins**.

Most earthquakes occur along the margins of the plates with only a few originating within the plates. Ironically, the highest magnitude earthquake ever to occur in North America was an intraplate earthquake;

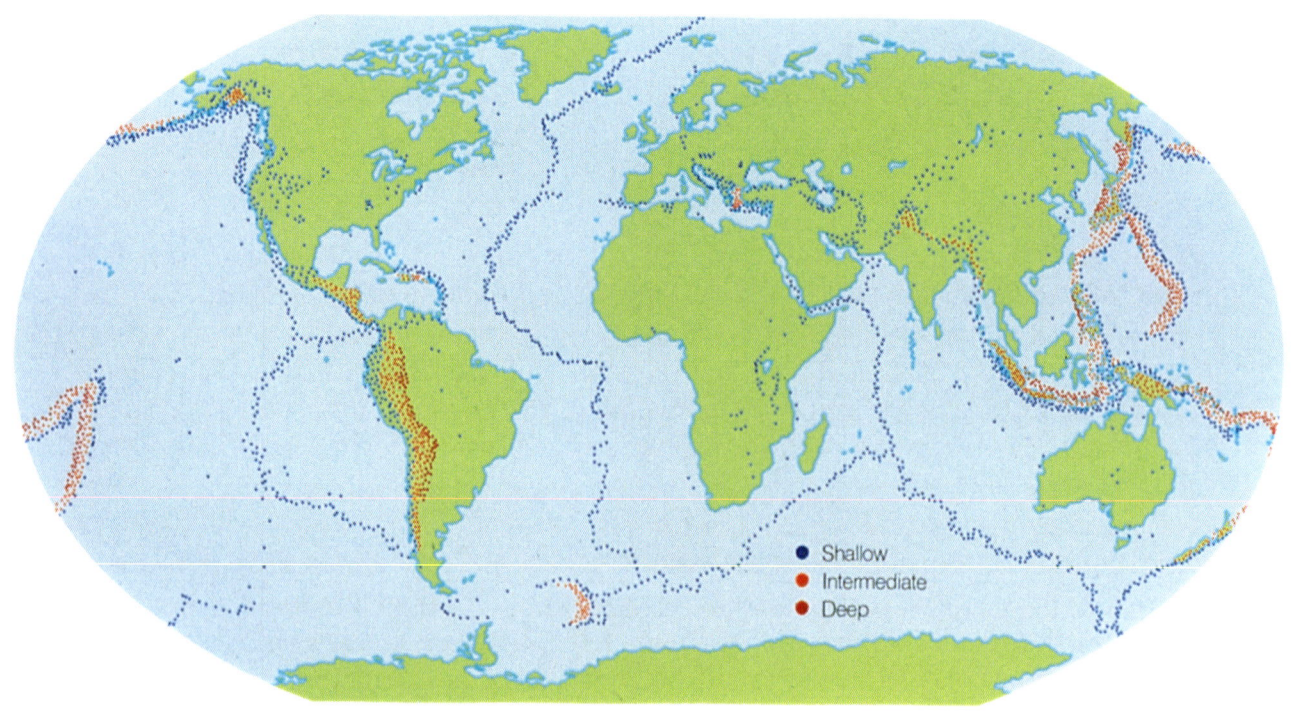

FIGURE 18.1 *With the advent of plate tectonics, the explanation for the* distribution *of earthquakes became apparent. Most of the energy released in the form of earthquakes around the world, and all of the* intermediate- *and* deep-focus *earthquakes that occur, are associated with the convergent plate boundaries.*

it occurred in New Madrid, Missouri, on December 6, 1811. Although the Richter Scale did not exist at the time of the earthquake, seismologists have estimated that the New Madrid earthquake would have been between 8.1 and 8.7 on the Richter Scale. We will discuss the Richter Scale later, but an earthquake with a Richter magnitude greater than 8.0 is considered to be major. Geologists believe the New Madrid earthquake was the result of fault movements along a failed rift zone that dates back to the Pre-Cambrian. Two other well-defined earthquake zones within North America are not associated with modern plate boundaries: one is located along the Saint Lawrence River, and the other forms a belt trending from Charleston, South Carolina, to Roanoke, Virginia.

THE FREQUENCY AND LOCATION OF EARTHQUAKES

Earthquakes are caused by the sudden rupture of rocks and the subsequent movement along faults. In our discussions of rock deformation in Chapter 18, we saw that when stressed, rocks will store energy up to the elastic limit; after the limit is reached, they will either react plastically and bend (fold) or react as brittle solids and break. If they bend, the stored energy is consumed as the rock is contorted into various kinds of folds. If the rocks break, however, the stored energy is released. Some of the released energy is used to move the rocks on either side of the fault. The remainder is released as an earthquake or **seismic wave** in much the same way that the energy released from a breaking stick produces a snapping sound. The released energy causes surface rocks to move or "quake." As the seismic wave moves away from the fault into the surrounding rocks, it may spread the impact of the earthquake and its potential for destruction over a large area.

The severity of earth movement depends upon the amount of energy released. Although only a small amount of energy may be stored in any cubic foot or meter of rock during elastic deformation, the total volume of rock involved in an average fault movement may measure miles wide and deep and tens of miles long; thus, the total amount of potential energy available to be released can be awesome.

Because rocks are stronger under compression than they are under tension, a mass of rock undergoing compression can store more energy before it finally fails than a mass of rock being subjected to tension. As a result, 90% of all earthquakes—and 95% of all the seismic energy released—are associated with **subduction zones** where all types of stress occur. This relationship explains why the area around the Pacific Ocean basin experiences the highest frequency of destructive earthquakes.

Most earthquakes occur near the surface, where the rocks are most brittle; the number of earthquakes decreases exponentially with depth as the rocks become increasingly plastic. The deepest earthquakes are recorded at about 450 miles (725 km). At depths greater than 450 miles (725 km), all rocks respond as plastic materials and will not break regardless of the amount or the rate of increase of the applied stress.

The 450-mile (725 km) depth limit of earthquakes represents the deepest penetration of subducting oceanic lithosphere into the mantle. The swarms of earthquakes that are associated with zones of subduction constitute the **Benioff Zone** (Figure 18.2).

Focus

The initial rupture point causing an earthquake, and also the point where the energy is first converted into a seismic wave, is called the **focus** (or **hypocenter**) of the earthquake. Earthquakes that occur from the surface to depths of about 40 miles (65 km) are referred to as **shallow-focus** earthquakes. As we have noted, the frequency of earthquakes decreases with depth. Nearly 75% of all earthquakes, including all major earthquakes, are shallow focus. **Intermediate-focus** earthquakes occur within the subducting plate from 40 miles down to about 185 miles (300 km) while those that occur from depths of 185 miles down to the maximum depth of about 450 miles (725 km) are called **deep-focus** earthquakes (Figure 18.3). Some geologists believe that rapid changes in the crystal structures of the minerals contained within the rocks rather than faulting may cause some of the deeper earthquakes.

Epicenter

The point on the surface directly above the focus of the earthquake is called the **epicenter** (Greek *epi* = upon, over). Note that the focus and the epicenter of an earthquake are *not* synonymous. The focus of an earthquake may be located anywhere from the surface down to a depth of about 450 miles, but the epicenter is *always* on the surface.

The significance of the epicenter and its relationship to the focus are illustrated in Figure 18.4. The energy released from the earthquake radiates from the focus in all directions. The decrease in energy

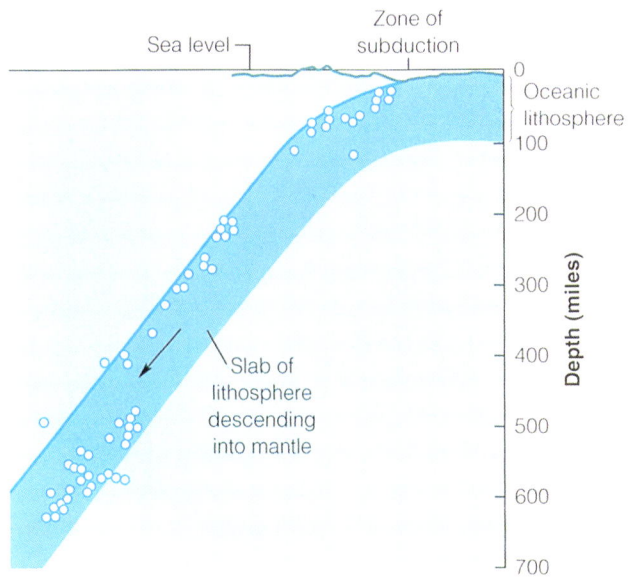

FIGURE 18.2 *Below the zones of subduction and dipping toward the continents at an angle of about 45° is a plane called the* Benioff Zone *along which abundant earthquake foci are located.*

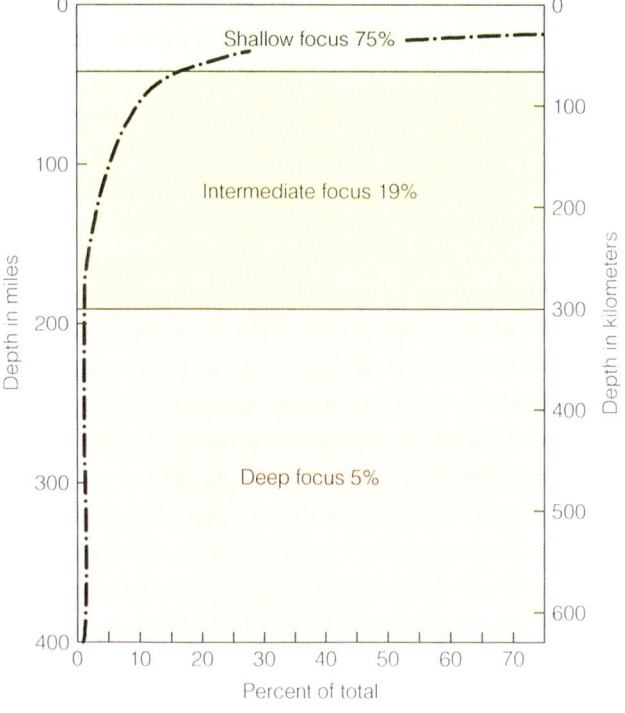

FIGURE 18.3 *The increase in the number of earthquakes above depths of 200 miles (300 km)—and especially the rapid increase above a depth of about 40 miles (65 km)—is due to the increasingly brittle nature of the rocks approaching Earth's surface.*

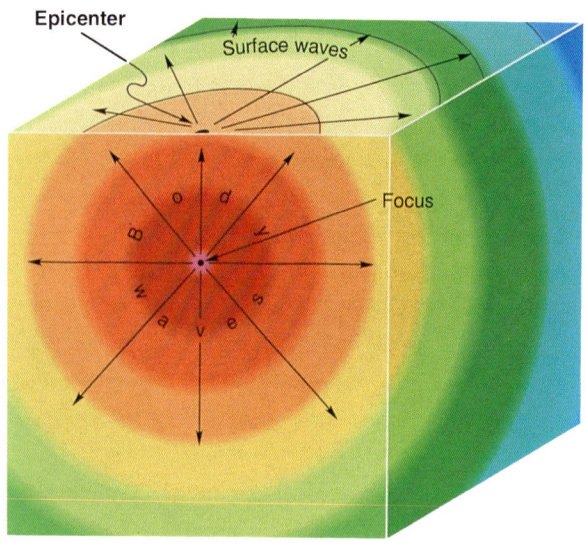

FIGURE 18.4 *The* focus *of the earthquake, the point where the seismic energy was released, can be located anywhere from Earth's surface to a depth of about 450 miles (725 km). The epicenter, on the other hand, is always located at the surface directly above the focus. Note that if the focus is at the surface, it and the epicenter are located at the same point.*

away from the focus is represented by a set of nested spheres with successive outer spheres representing less energy. Because Earth's surface cuts across the set of nested spheres, as a knife would cut across an onion, each sphere is represented at the surface by one of a set of concentric circles with each successive inner circle representing a higher-energy sphere. The epicenter, located at the center of the concentric circles, is the point on the surface closest to the focus and therefore the point on the surface where the earthquake energy is the greatest. In practical terms, the concentric circles can be considered zones indicating potential damage and destruction resulting from the earthquake, which decrease with distance from the epicenter. Although the concentric zones in Figure 18.4 are perfect circles, in reality their outlines would be irregular because of variations in both rock composition and structure at or near Earth's surface.

SPOT REVIEW

1. What is the source of energy for earthquakes?
2. Why are the distribution patterns for recent earthquakes and active volcanism similar?
3. Why does the frequency of earthquakes decrease with depth?
4. What is the difference between the focus and the epicenter of an earthquake? Can they ever be located in the same place?

MEASURING EARTHQUAKE INTENSITY AND MAGNITUDE

Seismologists use two different measurements to describe the strength of an earthquake. The *intensity* is a qualitative assessment of the effects of the earthquake. The magnitude is a quantitative measurement of the energy released by the earthquake.

Intensity

The **intensity** of an earthquake is determined by the effects observed (such as swinging chandeliers) and the kinds of **damage** caused. The first of several scales used to measure earthquake intensity was devised in 1873 by *M. S. de Rossi* of Italy and *François Forel* of Switzerland. A more detailed intensity scale was introduced in 1902 by an Italian volcanologist and seismologist, *G. Mercalli*. The **Modified Mercalli Scale** presented in Table 18.1 is the result of a revision of the original Mercalli Scale in 1931 by *H. O. Wood* and *F. Neumann* at the California Institute of Technology Seismological Laboratory.

An earthquake intensity scale is in part a stepwise verbal description of what one would expect to experience and observe at the time of the earthquake. This aspect of the scale is based on the comments of individuals who have experienced the particular earthquake combined with the direct observations of the earthquake's effect. Persons who have experienced low-intensity earthquakes typically give such responses as "dishes rattled," "bells rang," and "chandeliers began to swing."

Note, however, that above an intensity of about VII on the Modified Mercalli Scale, the major criterion for establishing the level of earthquake intensity is the amount of *observed* damage. Significant damage, including cracked masonry walls, shattered windows, and the collapse of tall brick smokestacks, begins to occur at intensity readings of VIII. At an intensity of IX, most structures such as buildings, interstate overpasses, and dams could be heavily damaged. At intensities of X, few structures will survive without extensive damage, and mass movements such as rock slides and snow avalanches will be common in areas of high relief. At an intensity of XI, the surface of the ground undulates visibility as the seismic waves pass by. At an intensity of XII, most humanmade structures will be destroyed, trees will be uprooted, and streams will be diverted from their channels. Seismologists estimate that the earthquake that struck New Madrid, Missouri, in 1811, had an intensity of XII at its epicenter.

TABLE 18.1
Modified Mercalli Intensity Scale of 1931 (Abridged)

I. Not felt except by a very few under especially favorable circumstances.

II. Felt only by a few persons at rest, especially on upper floors of buildings.

III. Felt quite noticeably indoors, especially on upper floors, but many people do not recognize it as an earthquake. Vibration like passing of truck.

IV. During the day felt indoors by many, outdoors by few. At night some awakened. Dishes, windows, doors disturbed; walls make cracking sound. Sensation like heavy truck striking building.

V. Felt by nearly everyone; many awakened. Some dishes, windows, etc., broken; a few instances of cracked plaster; unstable objects overturned. Disturbance of trees, poles, and other tall objects sometimes noticed. Pendulum clocks may stop.

VI. Felt by all; many frightened and run outdoors. Some furniture moved; a few instances of damaged chimneys. Damage slight.

VII. Everybody runs outdoors. Damage *negligible* in buildings of good construction; *slight* to moderate in well-built ordinary structures; *considerable* in poorly built or badly designed structures.

VIII. Damage *slight* in specially designed structures; *considerable* in ordinary substantial buildings; *great* in poorly built structures. Fall of chimneys, factory stacks, columns, monuments, walls.

IX. Damage *considerable* in specially designed structures; well-designed frame structures thrown out of plumb; *great* in substantial buildings, with partial collapse. Ground cracked conspicuously. Underground pipes broken.

X. Some well-built wooden structures destroyed; most masonry and frame structures destroyed; ground badly cracked. Considerable landslides from river banks and steep slopes.

XI. Few, if any, (masonry) structures remain standing. Bridges destroyed. Broad fissures in ground. Underground pipe lines completely out of service. Earth slumps and land slips in soft ground.

XII. Damage total. Waves seen on ground surface. Lines of sight and level distorted. Objects thrown upward into the air.

Magnitude

In contrast to the descriptive nature of earthquake intensity scales, the **Richter Scale** is a measure of the **magnitude** of an earthquake, which is an evaluation of the actual amount of **earth movement** and **energy released**. In 1935, *Charles Richter,* a professor at California Institute of Technology, devised a method of calculating the amount of energy released during an earthquake based upon actual measurements of Earth movement made by a sensitive instrument called a **seismograph** (refer to Figures 18.14–17). Richter based his method on experiments that subjected rocks to compressive forces up to the breaking point while he carefully measured the amount of energy applied and the amount of subsequent rock deformation. Each step on the Richter Scale represents a 10-fold increase in the amount of shaking (amplitude) and an approximately 30-fold increase in the amount of energy released. An earthquake of magnitude 7 therefore involves 100 times (10 times 10) the amount of rock movement and 900 times (30 times 30) the amount of energy released as an earthquake of magnitude 5. The Richter Scale is now used throughout the world to evaluate earthquake magnitude.

Table 18.2 shows the frequency of earthquakes at each level of the Richter Scale in any given year. Fortunately, most earthquakes are low magnitude and can only be detected by sensitive instruments. These earthquakes cause no damage of any kind.

TABLE 18.2
FREQUENCY AND ENERGY RELEASE OF EARTHQUAKES OF VARIOUS MAGNITUDES

Richter Magnitude	Number Per Year
Over 8.0	1 to 2
7.0–7.9	18
6.0–6.9	120
5.0–5.9	800
4.0–4.9	6,200
3.0–3.9	49,000
2.0–2.9	300,000

Frequency data from *Earth* by F. Press and R. Seiver. Copyright © 1986 by W.H. Freeman and Company. Reprinted with permission

Note: For every unit increase in richter magnitude, ground displacement increases by a factor of ten, while energy release increases by a factor of thirty. Therefore, most of the energy released by earth quakes each year is released not by the hundreds of thousands of small tremors, but by the handful of earthquakes of magnitude 7 or larger, the so-called major or great earthquakes. For comparison, an earthquake of magnitude 6 releases about as much energy as a 1-megaton atomic bomb.

THE GREAT SAN FRANCISCO EARTHQUAKE AND FIRE

At the turn of the century, San Francisco was the largest and fastest-growing city on the west coast. It was, for its time, a modern city with many innovations to which its citizens pointed with pride. Although "skyscrapers" were being built around steel frameworks, most of the buildings in the city were still built of wood. To make them look more substantial, the buildings were constructed with masonry facades facing the streets.

San Franciscans were also proud of their modern firefighting system. The city streets were equipped with a system of fire hydrants fed with water from central storage tanks, which eliminated the need to use horse-drawn wagons to haul water and steam-powered pumps to the sites of fires.

Some of the streets were illuminated by the recently invented electric light, another source of pride for San Franciscans. The main sources of light throughout the city were gas- or kerosene-fueled lamps, however.

Much of the city was built on steep slopes covered with the thick regolith that characterizes the coastal regions of California, setting the scene for disaster. At 5:12 A.M. on April 18, 1906, a segment of the San Andreas Fault to the north of the present location of the Golden Gate Bridge moved about 18 feet (5.5 m) (Figure 18.B1). As is usually the case, buildings with foundations firmly anchored into bedrock survived the quaking with little damage, but many of the buildings and homes built on the unconsolidated regolith experienced substantial movement. Kerosene lamps crashed to the floor and set the wooden buildings on fire. Downslope movement of the regolith broke gas lines, which released gas that fueled the fires. Firefighters attempting to reach

Figure 18.B1 *The earthquake that devastated San Francisco in 1906 was one of the most severe to occur in North America in historic time. You will recognize the statue with its head stuck in the pavement after being toppled from its perch as being that of Louis Agassiz, the father of the Great Ice Age. Considering the severity of the earthquake, it is surprising that only about 600 people lost their lives. In this classic photo, citizens watch the raging fires that destroyed nearly 90% of the city before they were finally extinguished. The debris generated by the cleanup was dumped into San Francisco Bay where it formed the land on which the Marina was built. Little did the inhabitants realize at the time that they were setting the scene for most of the damage that would occur during the 1989 earthquake. (a Courtesy of W. Mendenhall/USGS; b courtesy of T. L. Youd/USGS)*

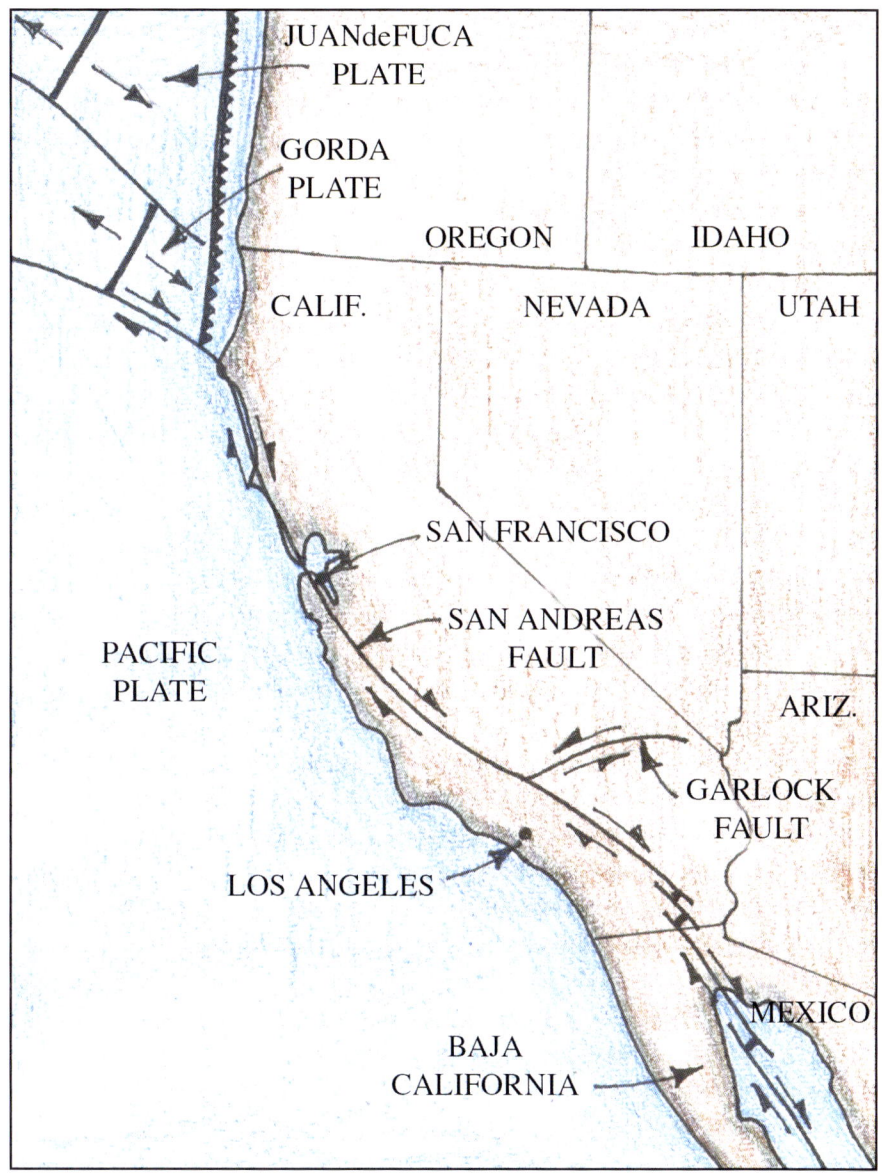

Figure 18.B2 *Many of the major earthquakes that occur in California are associated with the* San Andreas Fault. *The San Andreas Fault is a right-lateral strike slip fault and represents the boundary between the Pacific Ocean and North American lithospheric plates. The earthquakes that occurred in San Francisco in 1906 and in Loma Prieta in 1989 were the result of movements along segments of the San Andreas Fault.*

the fires found that many of the masonry building facades had fallen into the streets and made them impassable. When firefighters were able to arrive at the site of a fire, they found that the same earth movements that had severed the broke gas lines had also broken the water mains that fed the hydrants and the water hydrants were dry. The city was soon ablaze (Figure 18.B1). Before the fires were finally brought under control several days later by dynamiting entire city blocks, nearly 90% of the city lay in ruins.

DAMAGE FROM EARTHQUAKES

Although most earthquakes are of low magnitude on the Richter Scale, a single high-magnitude earthquake can kill thousands of people and cause millions of dollars worth of damage. The greatest loss of life due to a single earthquake occurred in 1556 in Shensi Province, China, where an estimated 800,000 people died. As recently as 1976, at least 650,000 people died in Tangshan, China, when an earthquake with a Richter Scale magnitude of 7.6 caused dwellings carved into loess cliffs to collapse, trapping the inhabitants.

Some damage such as the offsetting of roadways, fence lines, and the buckling of pavement is a direct result of the fault movement (Figure 18.5). Most earthquake damage is not directly related to fault movements, however, but results from events caused by the seismic energy, with fire being a major source of damage. An example is the earthquake that destroyed nearly 90% of San Francisco in 1906 where a combination of factors conspired to initiate a fire that caused 10 times more damage than was incurred by the quaking of the earth.

In 1989, San Francisco experienced another severe earthquake. Its epicenter was located in the Santa Cruz Mountains approximately 75 miles (120 km) to the south at Loma Prieta, California. Significantly, the Marina area of San Francisco, which suffered the greatest damage during the 1989 earthquake, was built atop the debris removed from the city after the 1906 earthquake (Figure 18.6). The material had been dumped into San Francisco Bay and partially compacted to provide additional building space.

FIGURE 18.5 *Although some earthquake damage is due to fault-induced vertical or horizontal offsets of the surface, most earthquake damage is only indirectly the result of fault movements. Fire is actually the most common cause of damage. (Courtesy of W. B. Hamilton/USGS)*

Thus, in a sense, the aftermath of the 1906 earthquake laid the foundation for some of the destruction suffered by the city in the 1989 earthquake. Unfortunately, most of the damage experienced by the Marina area occurred for the same reasons as the destruction in 1906. For one thing, the extensive structural damage experienced by buildings was directly the result of having been built on loose, unconsolidated materials. As in 1906, fire fueled by gas leaking from broken gas mains was a major cause of damage in the Marina area until shutoff valves could be located beneath the debris. The collapse of the Nimitz Freeway overpass, crushing motorists in their vehicles, accounted for most of the lives lost during the 1989 earthquake (Figure 18.7). Portions of the roadway failed, at least in part, because they had been constructed on poorly consolidated bay mud.

As the 1989 Loma Prieta earthquake clearly showed, the greatest amount of damage resulting from an earthquake is not necessarily in the area nearest the epicenter. An even better example of this relationship occurred on September 19, 1985, when

FIGURE 18.6 In general, structures anchored in bedrock have the best chance at surviving an earthquake, while those built on unconsolidated materials generally suffer the most damage. A case in point is the effect of the 1989 earthquake in Loma Prieta, California, on San Francisco. In the downtown area where the foundations of the buildings are in bedrock, there was little damage. Most of the damage occurred in the Marina area where the structures were built on land formed by dumping the unconsolidated debris of the 1906 San Francisco earthquake along the edge of the bay. (a Courtesy of J. K. Nakata/USGS; b courtesy of C. E. Meyer/USGS)

FIGURE 18.7 Most of the lives that were lost in the 1989 Loma Prieta earthquake were the result of the collapse of the Nimitz Freeway. Although the experts still do not agree on the exact cause of the structural failure experienced by the overpass, the fact that portions of the freeway that collapsed were located on unconsolidated bay sediments may have contributed to its failure. (Courtesy of H. Wilshire/USGS)

movement along the subducting Cocos plate caused an earthquake of Richter magnitude 8.1 on the Pacific coast of Mexico. The greatest amount of destruction, however, was experienced in Mexico City, more than 200 miles (300 km) to the east. The earthquake resulted in more than 8,000 deaths, 30,000 injuries, and the destruction of about 500 buildings. Most of the deaths occurred in ghetto areas where the buildings were poorly constructed and unreinforced. The total damage was estimated to exceed $4 billion (Figure 18.8).

How could an earthquake 200 miles away possibly cause such extensive damage? The answer is not yet fully understood, but certainly the local geology played a major part. Much of Mexico City is built on the bed of ancient Lake Texcoco, which was drained by the Spaniards to allow for the expansion of the city after they defeated the Aztecs. Most of the buildings destroyed during the earthquake were in the portion of the city built on the soft, water-saturated, unconsolidated sediments of the old lake bed. Buildings in adjacent portions of the city constructed on the bedrock surrounding the old lake shoreline experienced little or no damage from the seismic waves even though the amount of earth movement was greater than in the devastated portion of the city. Seismologists and engineers examining the damage concluded that upon reaching the old lake sediments, the seismic waves caused the sediments to begin vibrating like a bowl full of Jello. The vibration of the lake sediments was then transferred to and amplified by the buildings. Buildings with the same natural vibration frequency as the underlying sediments began to vibrate in harmony with the sediments and literally ripped themselves apart. Other buildings whose dimensions and construction rendered them "out of tune" with the vibrating sediments experienced less damage. All the while, buildings with foundations firmly anchored in bedrock survived with little or no damage.

Tsunami

The ultimate in extensive damage caused by a distant earthquake involves the impact of a **tsunami**. Commonly, but incorrectly, referred to as a "tidal wave," a tsunami is a gigantic seawave that has nothing to do with the tides but is usually generated by energy released during a submarine earthquake. Because most major earthquakes are associated with the zones of subduction surrounding the Pacific Ocean, the Pacific Ocean basin suffers more tsunamis than any of the other ocean basins, with Japan and Hawaii experiencing the most tsunamis (Figure 18.9). The worst tsunami to hit Japan occurred on June 15, 1896, when a wall of water estimated at 75 to 100 feet (23–30 m) high crashed onto the eastern shore of Honshu, sweeping away more than 10,000 homes and 26,000 people.

FIGURE 18.8 *Another example of indirect earthquake damage is the 1985 Mexico City earthquake where the fault movement occurred more than 200 miles (300 km) away. As was the case in the Marina area of San Francisco, most of the damage was located in areas of the city underlain by unconsolidated materials, in this case, lake sediments. The observed variation in the magnitude of the damage experienced by individual buildings during the Mexico City earthquake has led engineers to hypothesize that the extent of damage could be due to the specific vibrational frequency of the buildings relative to the seismic vibrations of the underlying rocks and sediments. (Courtesy of M. Celebi/USGS)*

FIGURE 18.9 *With major earthquakes occurring around nearly the entire perimeter of the Pacific Ocean basin, the Hawaiian Islands located in mid-Pacific are especially prone to destructive* tsunamis. *(Courtesy of USGS/HVO)*

The occurrence of earthquakes within the ocean basins is not the only cause of tsunamis. You will remember from our discussions of volcanoes in Chapter 6 that energy released during the eruption of Krakatau in 1883 produced a tsunami that overwhelmed all the low-lying parts of the surrounding islands and carried 36,000 people to their deaths.

With an average amplitude of only a few feet, a tsunami crossing the open sea would pass undetected, even by a relatively small ship. However, with an average wavelength of 60 to 70 miles (100–110 km) and speeds of 500 miles per hour (800 kph) not uncommon, the energy of a tsunami is almost unimaginable! To visualize the enormous destructive power of a tsunami, one need only recall our discussion of the generation of seawaves in Chapter 12 and review Figure 12.21, which graphically described the sequence of events as a waveform encounters the shoreline. When a tsunami with an amplitude of a few feet at sea moves onto the shoreline, the amplitude may increase to 100 feet (30 m) or more and cause enormous damage.

The **Seismic Sea Wave Warning System (SSWWS)** was established in 1946 to warn the inhabitants of Pacific coastal areas of potentially destructive tsunamis. Earthquakes occurring around the Pacific Ocean basin or on the ocean floor are constantly monitored at the headquarters in Honolulu, Hawaii. Warnings of potential tsunamis are broadcast to stations throughout the Pacific Ocean basin where strategically located sirens warn inhabitants and provide adequate time for protective action to be taken.

SPOT REVIEW

1. What is the difference between the Modified Mercalli Scale and the Richter Scale? In what ways are the two scales similar?
2. What factors have we learned from experience are most important in determining the amount of damage that results from an earthquake?
3. Why are tsunamis so destructive?

EARTHQUAKE (SEISMIC) WAVES

Whether generated by an earthquake, an explosion, or a breaking stick, waves are of two kinds: (1) **shear waves**, in which the material conducting the wave is moved back and forth *perpendicular* to the direction in which the wave is moving, and (2) **compressional-extensional (compression) waves**, in which the material moves back and forth *parallel* to the direction of propagation (Figure 18.10).

A shear wave can be visualized as the ripple that moves from the handle to the tip of a whip as it is snapped and can be illustrated as a sine wave (Figure 18.10a). A compressional-extensional wave is somewhat more difficult to represent graphically but is commonly illustrated as a "Slinky" toy, which, you may remember, is a spring along which a series of coils open and close, forming a wave that moves from one end of the spring to the other and allows the Slinky to "walk" downstairs and do other tricks.

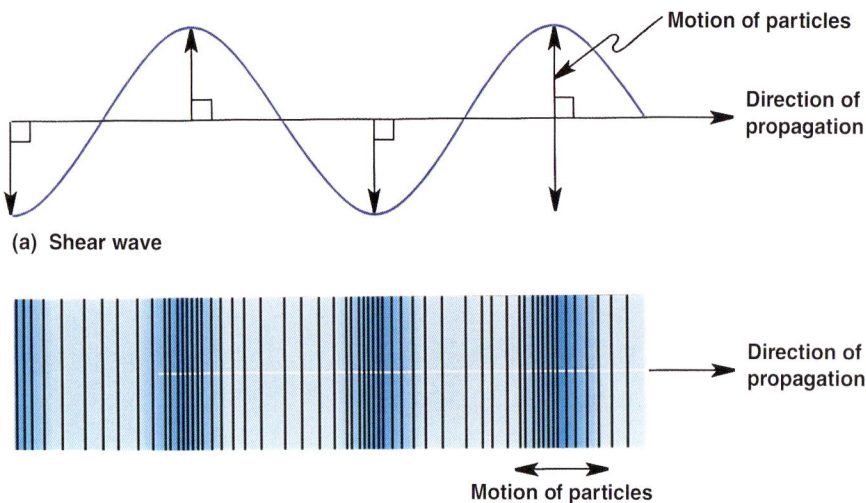

FIGURE 18.10 *Waves are of two kinds:* shear waves *where the movement of the material through which the wave is propagating (moving) is* perpendicular *to the direction of propagation (a), and* compression waves *where the movement of the material is* parallel *to the direction of propagation (b). In the case of shear waves, it is important to keep in mind that the material moves perpendicular to the direction of propagation in* all *directions, not simply in the vertical orientation illustrated in the drawing.*

An important distinction between the two kinds of waves is that shear waves can only be transmitted through *solids* while compression waves can be transmitted through matter in *any* state—solid, liquid, or gas. Shear waves cannot be transmitted through a liquid or a gas because neither has the rigidity necessary to allow it to be sheared (cut). All materials, on the other hand, can be compressed and therefore can transmit compressional-extensional waves.

Body Waves

As movement occurs along a fault and the energy stored in the rock during the elastic deformation is released, shear- and compression-type seismic waves are generated simultaneously and move out in all directions from the focus (refer to Figure 18.4). The seismic waves that move through Earth, called **body waves**, are both of shear and compression type (Figure 18.11). The shear body waves are called **S waves** (*secondary waves*) while the compression-type body waves are called **P waves** (*primary waves*). An important characteristic of both types of body waves is that they are of low amplitude; that is, they do not generate much shaking or vibration in the rock mass as they pass through Earth's interior. Of the two body waves, the P wave has the higher velocity and is therefore referred to as primary.

Surface Waves

Upon reaching Earth's surface, the seismic energy generates another set of waves called **surface waves**, which are of two types: (1) **Love (LQ) waves** and (2) **Rayleigh (LR) waves**. Love waves are shear waves

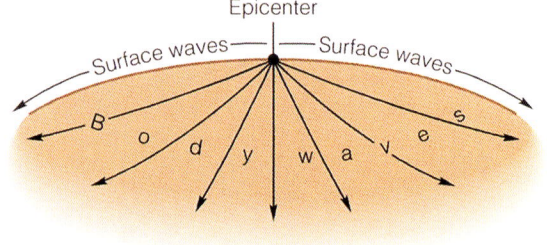

FIGURE 18.11 *The* body waves *that move through the* body *of Earth consist of both shear waves (*S waves*) and compression waves (*P waves*). Body waves are lower in amplitude and travel with higher velocities than surface waves. Of the two types of body waves, the P waves have the higher velocity.*

that move Earth's surface back and forth *horizontally* perpendicular to the direction of propagation (Figure 18.12a) while Rayleigh waves are a combination of compression and shear waves. They produce an elliptical, rolling motion that moves the rocks in a direction opposite to the direction of propagation (Figure 18.12b). The combined effect of these waves results in the *vertical* and *horizontal* movement of Earth's surface. The surface waves radiate from the epicenter of the earthquake.

Surface and body waves differ in several important respects. Surface waves are slower and have much higher amplitudes than body waves. The amplitude of the body waves is too low to cause any damage upon reaching the surface. The surface waves with their higher amplitudes are therefore responsible for the destruction resulting from an earthquake.

Earthquakes and Seismology 535

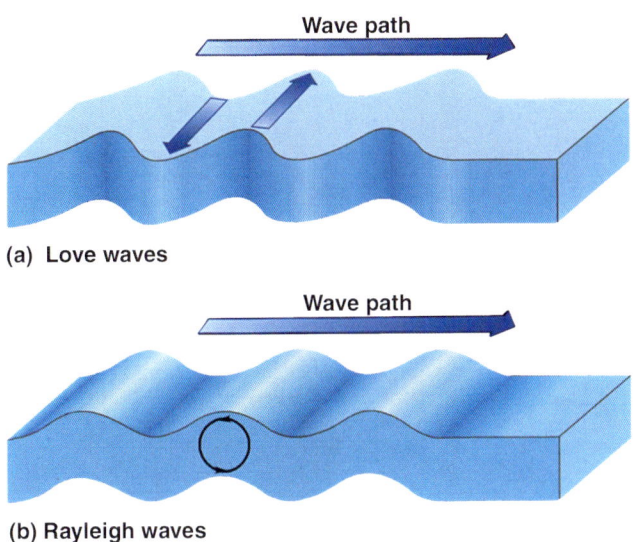

FIGURE 18.12 *There are two types of* surface waves: *(a) the shear-type* Love wave, *which moves the surface in a* horizontal *plane, and (b) the* Rayleigh wave, *which is a combined shear and compression wave that moves the surface in a* rolling *motion opposite to the direction of propagation. Studies have indicated that most of the damage caused by an earthquake is the result of the horizontal motion of the Love waves.*

FIGURE 18.13 *The first instrument designed to detect the arrival of seismic waves was invented by the Chinese scholar Chang Heng about A.D. 132. The instrument consisted of balls balanced in the mouths of eight dragons arranged around the perimeter of a large jar. An earth movement would dislodge the balls, which would fall into the mouths of waiting copper frogs. Because the instrument only detected the seismic event and did not provide a record, it is called a seismometer. (Reproduced by permission of the Trustees of the Science Museum, London).*

Engineering studies have shown that the horizontal motion induced by the Love waves is more destructive to structures such as buildings, bridge supports, and dams than the vertical motion caused by the Rayleigh waves. Consider holding a stack of dominoes in your hand. Moving your hand vertically may not disturb the stack, but moving your hand sideways is more than likely to cause the stack to fall.

As we will see later in our discussions of seismology, the magnitude of an earthquake on the Richter Scale is calculated from the largest measured amplitude of the surface waves.

THE SEISMOGRAPH

The **seismograph** is an instrument designed to *record* the movements of Earth's surface generated by a distant earthquake. The Chinese were the first to devise an instrument to *detect* the arrival of a seismic wave. The first true seismic instrument was invented by the Chinese scholar Chang Heng about A.D. 132 (Figure 18.13). The instrument was a large jar with dragon heads evenly spaced around it. An earthquake would cause balanced balls to drop out of a dragon's mouth into the mouth of a waiting copper frog. Assuming that the instrument worked according to plan, the early Chinese seismologists could both detect the arrival of a seismic wave and tell the direction of approach. However, they could tell very little else. The instrument could not, for example, measure or record any information concerning the *amount* of earth movement. Nevertheless, with all of its limitations, many centuries would pass before anyone designed a better instrument.

Consider the difficulty of inventing a *"black box"* that could be set anywhere on the surface of Earth and measure the amount of earth movement. The problem is that as the surface of Earth moves, so does the box. How can the box detect that Earth is moving, let alone the direction and amount of movement?

Any instrument used to measure the movement of another object must possess a component that either (1) moves in a known direction at a known velocity or (2) remains immobile as everything around it moves. It was the latter requirement that was finally satisfied by the first seismograph, which was simply a pendulum suspended from an arm attached to a ringstand. A pen tip extended from the bottom of the pendulum, just barely touching a pad of paper mounted on the base of the ringstand (Figure 18.14). As a seismic wave passed, it moved the base of the ringstand with the attached pad of paper, but the pendulum, because of its large mass and consequent *inertia,* remained *stationary.* The pen attached to the immobile pendulum drew a line on the moving paper that was proportional in length to the amplitude of the seismic wave and was oriented perpendicular to the direction of propagation (Figure 18.15). However, the instrument could not tell from which of the two directions the seismic wave arrived.

The vertical movement of the Earth's surface was measured by a similar instrument in which the pendulum was suspended by a spring rather than by a string (Figure 18.16). With its ability to measure the seismic wave amplitude and determine the direction of the line along which the seismic wave approached, the pendulum seismograph was a definite improvement over the Chinese design. The major shortcoming of both the Chinese and the early pendulum instruments was that neither could record *time*-related parameters. As a result, parameters such as the *frequency* and *time* of arrival of the seismic wave, the *duration* of the quake, or any changes in either wave amplitude or frequency with *time* could not be evaluated.

Seismograph design did not undergo another major change until the mid-1800s. John Milne, an English engineer, had gone to Japan shortly after its leaders opened the country to westerners. Milne

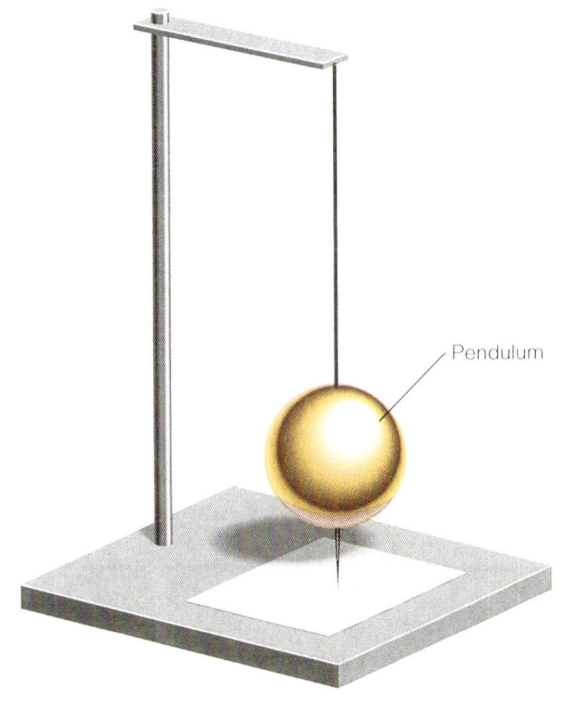

FIGURE 18.14 *The first real improvement on Heng's instrument was a vertically suspended pendulum. Because of its* inertia, *the mass of the pendulum would remain stationary as the remaining portion of the instrument was moved by the passing seismic waves. As we will see below, the instrument can generate a* record *of the seismic event called a* seismogram. *Any instrument that generates a record is identified by the suffix* -graph. *As a result, the pendulum instrument is a seismo*graph.

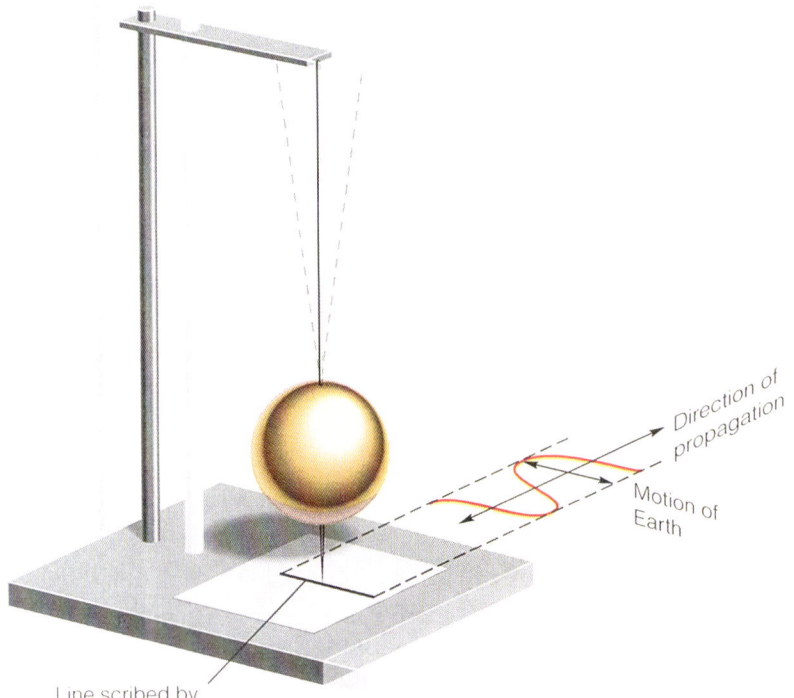

FIGURE 18.15 *As a* Love *wave arrived at the pendulum instrument, the base would be moved horizontally a distance equal to the amplitude of the wave in a direction perpendicular to the direction of propagation. As the base moved, a pen attached to the immobile pendulum drew a line that not only recorded the amplitude of the wave but was oriented perpendicular to the direction of the line of propagation. Note that the wave could have approached in one of two directions. Unfortunately, the instrument is incapable of distinguishing between the two possibilities.*

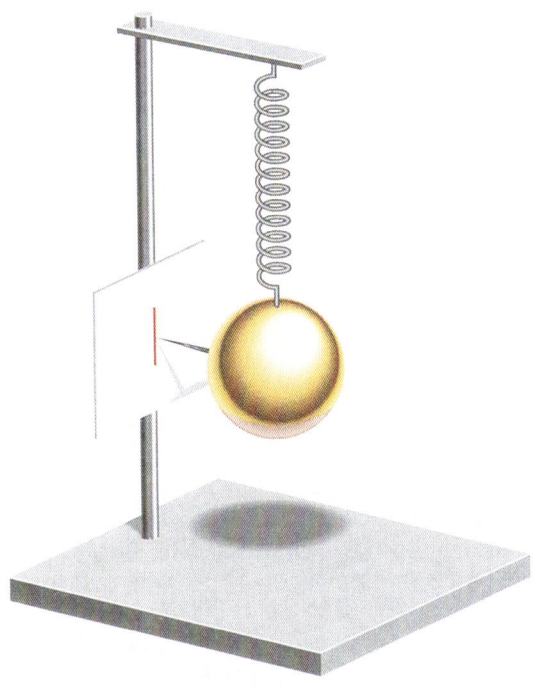

FIGURE 18.16 *The vertical motion imposed by the Rayleigh waves is recorded by a similar instrument where the pendulum is suspended by a spring. As the instrument is displaced vertically by the wave, a pen attached to the immobile pendulum draws a line whose length records the wave amplitude.*

wanted to make a detailed study of the large number of earthquakes to which Japan is subjected each year but recognized that the hanging pendulum instruments could not record any time-related parameters. He therefore set out to find a way to modify the instrument so it would record temporal data.

Milne's modifications were simple and few. His major design change was to reorient the pendulum horizontally. He placed the pendulum mass at one end of the rigid, horizontal bar that pivoted at the other end, allowing the bar to move in a plane parallel to the base of the instrument. A cylinder was mounted on the instrument base just forward of the pendulum mass with its axis perpendicular to the pendulum bar (Figure 18.17). The cylinder was driven by a clock mechanism that simultaneously rotated the cylinder and moved it parallel to its axis (translation). With the rotation of the drum set at some convenient speed, for example, once per hour, the time at the beginning of each line could be noted and the length of each line subdivided into minutes and seconds.

As a seismic wave moves the base of the instrument and the attached rotating and translating cylinder moves back and forth, the trace of the pen generates a **seismogram** that records (1) the arrival of the individual types of seismic waves, (2) the change in both wavelength and amplitude of the seismic waves with time, and (3) the end of the quake (Figure 18.18). Although modern instruments are significantly more sensitive and have incorporated all the benefits of modern electronic and computer technologies, the basic instrument design devised by John Milne more than a hundred years ago still underlies each modern seismograph.

Note that a seismic wave approaching the instrument perpendicular to the pendulum bar causes the pendulum to move with the instrument base. Because the pendulum does not remain stationary as the instrument base moves, the instrument cannot

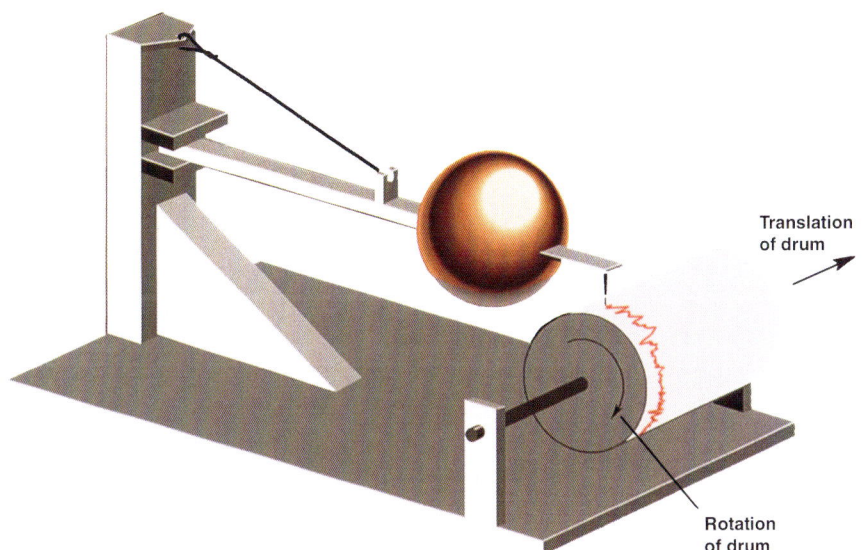

FIGURE 18.17 *John Milne made two major design changes to the seismograph: (a) he mounted the pendulum at the end of a horizontal bar, and (b) he mounted a rotating, translating drum to the base of the instrument on which the pendulum pen rested. Milne's instrument is the basis for all modern seismographs. Subsequent improvements increased the sensitivity of the instrument to allow the recording of all seismic waves including both the low-amplitude body waves and the higher-amplitude surface waves.*

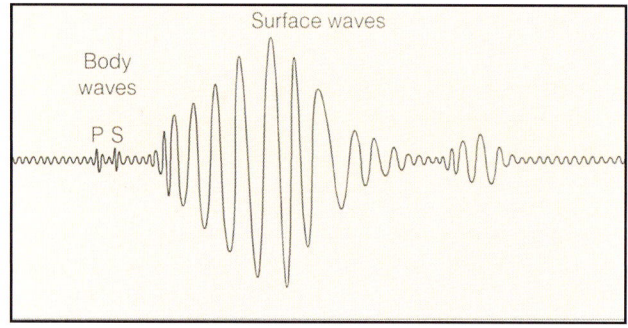

FIGURE 18.18 *Perhaps the most important new capability provided by Milne's improved design was the recording of time-related parameters. Because the seismogram is recorded on the surface of each wave, the change of both wavelength and amplitude of each wave with time, and the end of the seismic event.*

record the seismic wave (Figure 18.19). As a result, in order for a seismic station to be able to record seismic waves approaching from all directions, it must have two horizontal pendulum instruments oriented at right angles to each other.

The vertical movements of Earth are still detected and recorded by a spring-mounted vertical pendulum. The only basic difference in design from the earlier seismograph is the inclusion of a rotating, translating drum with its associated electronics to record the earth movement.

LOCATING EARTHQUAKES FROM SEISMIC DATA

Figure 18.20 is an actual seismogram. The seismogram shows the arrival of the body waves with the higher-velocity compressional P wave being recorded first, followed by the shear-type S wave. Note that the amplitudes of both body waves are small. Following the arrival of both body waves, the high-amplitude surface waves arrive with the faster Love waves arriving first, followed by the slower Rayleigh waves.

Assuming this is a typical seismogram, we might ask several questions: (1) Where did the earthquake occur? (2) What was the magnitude of the earthquake? (3) How deep was the focus? Remember that the information recorded on a seismogram describes the seismic waves as they arrive at the seismic station, *not* at the point of origin.

First, let us consider the problem of determining the distance from the seismograph station to the

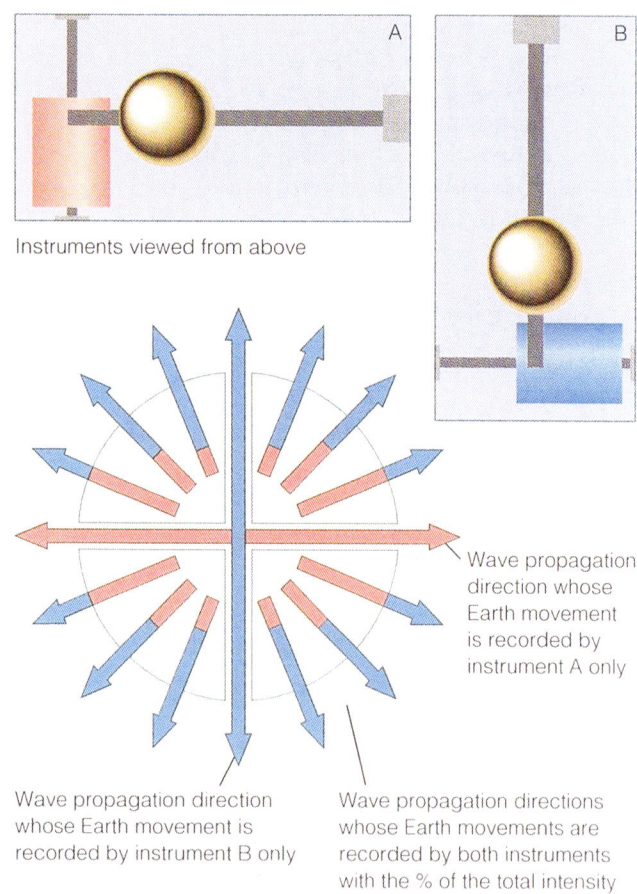

FIGURE 18.19 *A seismic wave approaching a horizontal-pendulum instrument perpendicular to the length of the pendulum bar cannot be detected because the entire instrument, including the pendulum, will be set in motion. Remember that in order for a pendulum instrument to detect movement, the pendulum must remain immobile. As a result, in order to ensure the recording of seismic waves approaching from all directions, two horizontal instruments oriented perpendicular to each other must be employed. Except for waves approaching perpendicular to one of the instruments, waves will be recorded by* both *instruments, with the amplitude recorded for a particular seismic wave depending on the angle of approach.*

earthquake epicenter. You will remember from Figure 18.4 that both of the body waves follow the same pathways. Consider the specific pathway taken by the seismic waves from the earthquake focus to the seismic station that generated our seismogram. The different velocities with which the two kinds of body waves travel through the rocks of Earth's interior are known with great precision. Knowing that they follow the same path, we can plot time versus

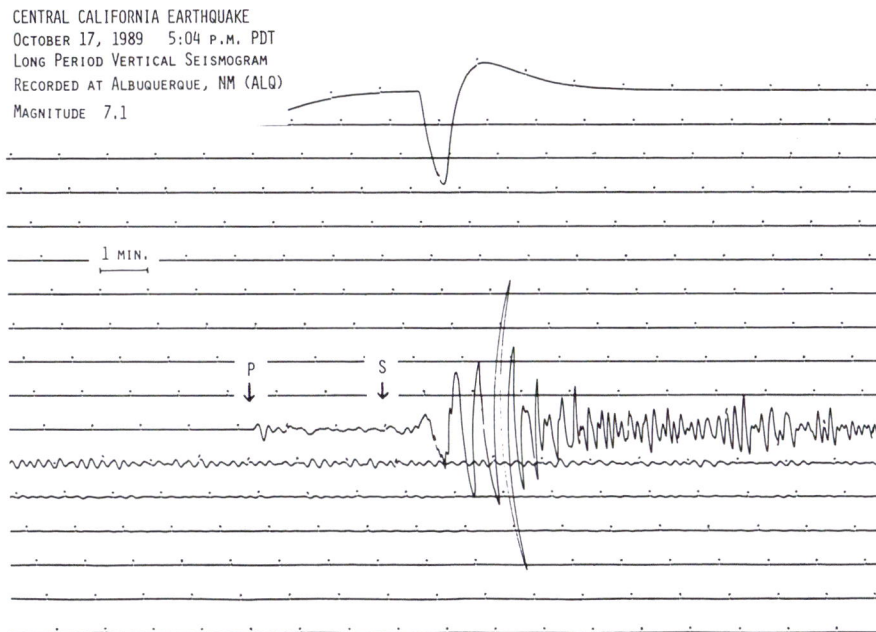

FIGURE 18.20 *An actual seismogram recorded at the Albuquerque, New Mexico seismic station illustrating the arrival of the body and surface waves during the Loma Prieta Earthquake on October 17. 1989. (Courtesy of USGS)*

distance traveled by the two body waves to produce a **time-distance** (or **travel-time**) curve. A typical time-distance curve, shown in Figure 18.21, graphically illustrates the relationship between the time and distance of travel. In other words, it is a graphic portrayal of the "race" between the P wave and the S wave. As time passes (increases), the distance between the advancing fronts of the two body waves will increase correspondingly. Because every seismogram is a timed recording, it accurately records the arrival times of the P wave and S wave. Once the difference in arrival times of the body waves is known, the time-distance curve provides the distance from the seismic station to the earthquake epicenter (Figure 18.22a).

The next step in locating the epicenter of an earthquake is to draw a circle on a map with the center at the location of the station and the radius equal to the distance determined to the earthquake focus. We now know that the earthquake occurred somewhere on the circle. But where? The exact location is most easily determined by using distance information from two other seismic stations. When the distance circles determined by three different stations for a specific earthquake are plotted on the same map, the three circles will overlap (refer to Figure 18.22b). The three circles intersect at a single, discrete point on the map, the earthquake focus is located on the surface at the point of intersection (Figure 18.23).

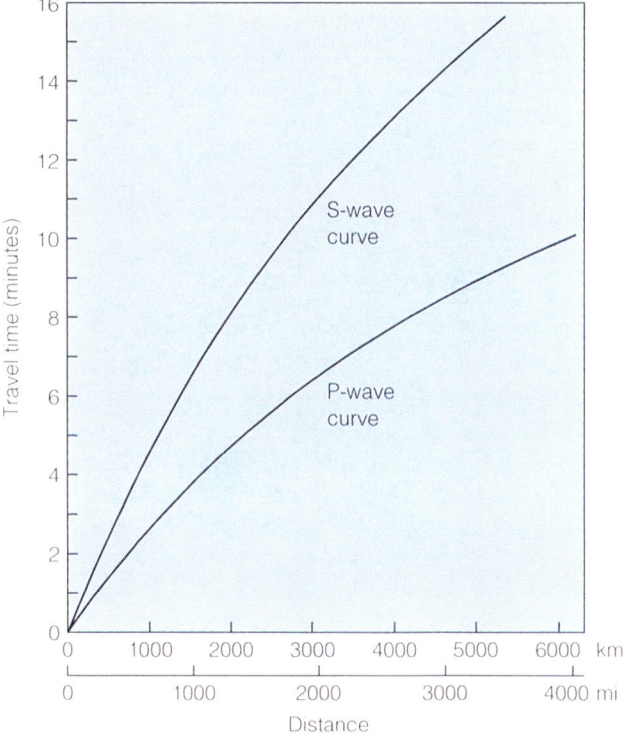

FIGURE 18.21 *Because the P and S body waves travel at different, but known velocities, a graph can be made that portrays the time-distance relationship for each. Called a time-distance or travel-time curve, the plot is used to determine the distance between the focus of an earthquake and the station recording the event.*

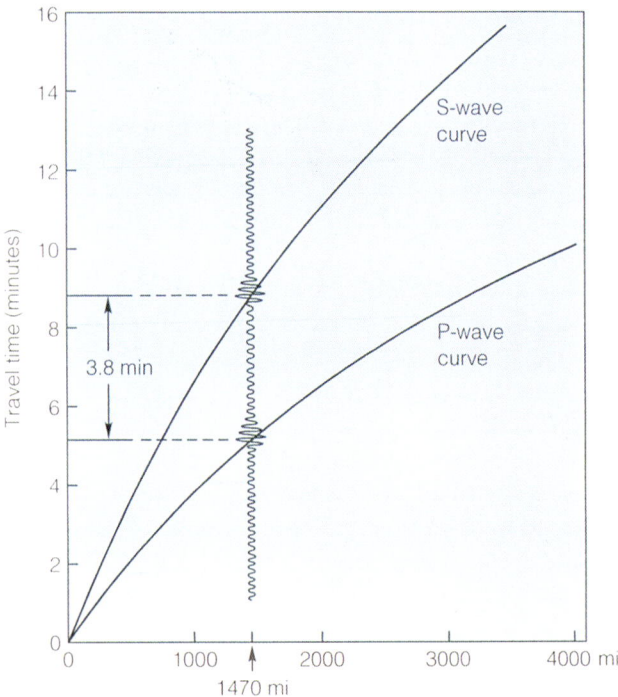

FIGURE 18.22 *The location of an earthquake epicenter is accomplished in two steps. (a) First, using the difference in arrival times of the P and S body waves at three different seismic stations, the distance to the epicenter from each station is determined using the time-distance curve shown in Figure 20.21. With the three distances known, a circle of appropriate radius is then drawn around the map location of each of the three seismic stations (b). The epicenter of the earthquake is located where all three circles intersect.*

(a)

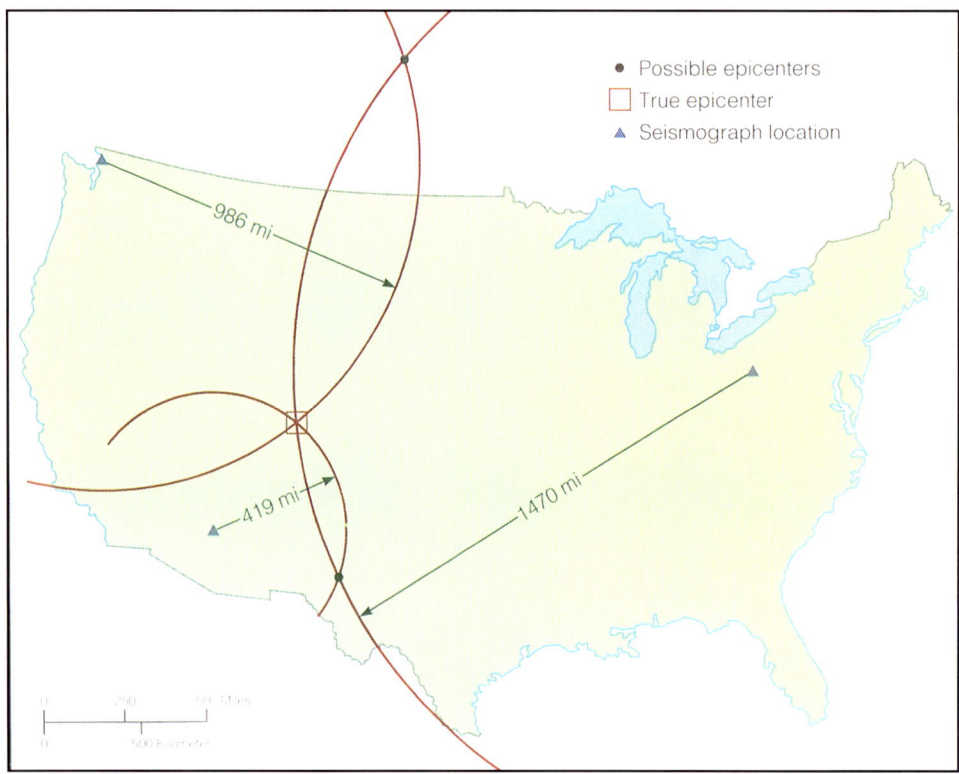

(b) Location of earthquake epicenter by triangulation

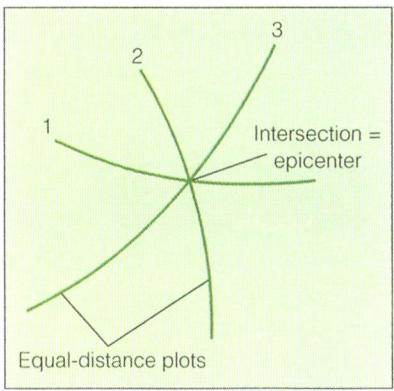

FIGURE 18.23 *The intersection of the three distance circles at a common point indicates that the earthquake focus was located at the surface.*

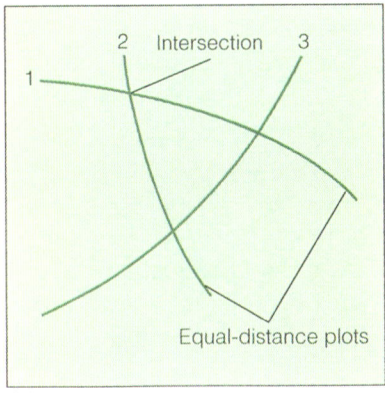

FIGURE 18.24 *The location of an earthquake focus below Earth's surface is indicated when the distance circles from three seismic stations intersect in the form of a spherical triangle.*

Should the intersection of the three circles produce a spherical triangle as illustrated in Figure 18.24, the epicenter is located at the center of the triangle, and the focus is located some distance below the surface. The depth to the focus is then determined by drawing a large number of station-focus distance data as hemispheres with centers located at the corresponding stations. The point of intersection of all the hemispheres then gives the location of the earthquake focus (Figure 18.25).

Because the amplitudes of the surface waves decrease away from the epicenter, the Richter Scale magnitude of an earthquake at its epicenter can be calculated using the distance data determined from the body wave arrivals and the maximum amplitude of the surface waves recorded on the seismogram. Richter

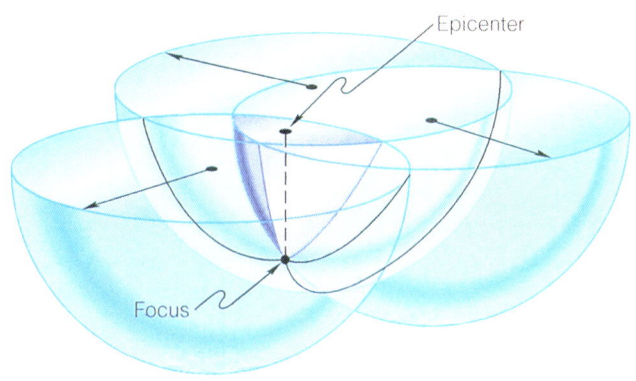

FIGURE 18.25 *In Figure 18.22, the distance was taken as the radius of a circle centered at the seismic station when, in reality, the distance determination provided by the time-distance graph locates the earthquake focus somewhere on the surface of a hemisphere centered at the seismic station. The location of the earthquake focus is actually the point of intersection of the three hemispheres, a determination that allows an earthquake to be categorized as shallow, intermediate, or deep focus.*

provided equations for the calculation of magnitude, which take into consideration (1) the distance from the earthquake to the seismic station, (2) the kind of rock through which the seismic waves pass, and (3) the physical parameters of the specific seismograph employed at the seismic station. A nomograph used to determine the magnitude of an earthquake based upon distance and maximum surface wave amplitude is shown in Figure 20.26.

SPOT REVIEW

1. What are the differences between shear- and compression-type waves?
2. Which seismic waves are responsible for most of the damage resulting from an earthquake?
3. How does the seismograph satisfy the requirements for any instrument designed to measure the movement of another object?
4. How was the seismic instrument designed by John Milne able to record time-related seismic parameters?
5. How are the data recorded on a seismogram used to locate the epicenter, focus, and magnitude of the earthquake?

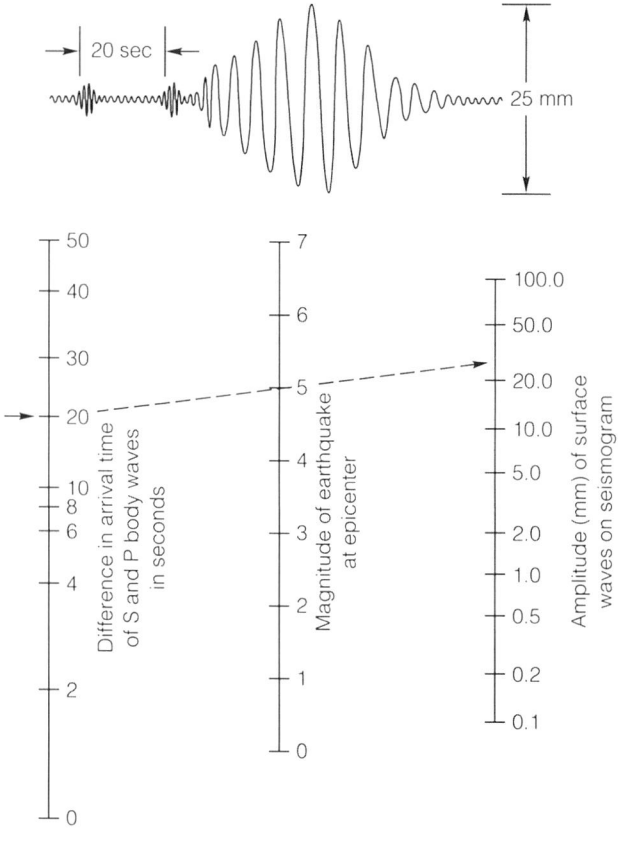

FIGURE 18.26 *Because the* amplitude *of the surface waves at an epicenter is determined by the* magnitude *of the earthquake and* decreases *with* distance *from the epicenter in a regular fashion, a nomograph can be used to determine the* magnitude *of the earthquake at the epicenter based on the data recorded on an individual seismogram. A nomograph consists of three parallel scales, each graduated for a different variable such that a straight line cutting all three scales intersects at the related values for each variable. Note that the three scales in the nomograph are (1) the difference in arrival times for the body waves (a measure of the* distance *between the epicenter and the seismic station), (2) the* magnitude *of the earthquake at the epicenter, and (3) the* amplitude *of the surface waves on the siesmogram. In our example, by drawing a line that connects the difference in arrival times of the P and S waves (20 seconds) with the amplitude of the surface waves as recorded by our seismograph (25 mm), the nomograph indicates that the magnitude of the earthquake at the epicenter was about 5 on the Richter Scale. (Note that both the arrival times scale and the amplitude scale are logarithmic).*

WORLDWIDE SEISMIC NETWORK

The stresses that build along each plate margin are, at least in part, the result of the combined movement of all the plates. The data amassed by seismologists so far clearly indicate that the full understanding of earthquakes will require research based upon a worldwide database rather than upon detailed studies of specific earthquake-prone areas. Potentially, the knowledge gained by the monitoring and study of seismic activity on a global scale could spare many lives and untold amounts of property loss by providing accurate predictions of the place, time, and magnitude of future earthquakes.

To provide such a database, in 1966 the U.S. Department of the Interior established the **National Earthquake Information Center (NEIC)** at Golden, Colorado, which collects seismic data from all over the world. More than 3,300 seismic stations from 80 foreign countries plus 120 stations of the **U.S. National Seismograph Network (USNSN)**, originally established to monitor Soviet underground nuclear testing, send data to the center in Golden (Figure 18.27).

The center's first mission is to obtain estimates of the magnitude and intensity of an earthquake occurring anywhere in the world and to disseminate the information as quickly as possible to the appropriate federal and state agencies such as the *Federal Emergency Management Agency (FEMA)* and the tsunami early warning system. The center also notifies agencies responsible for providing disaster relief assistance throughout the world.

The second mission of the NEIC is to establish and maintain the national and global network of seismic stations and to serve as a data collection and dissemination center. The data collected by the NEIC are available to anyone.

The third mission is to conduct research that will not only improve the ability to detect and locate earthquakes but will also allow the development of a potential to forecast the occurrence of earthquakes worldwide. Ultimately, the goal of the NEIC is to reduce the incidence of earthquake hazards to all humans.

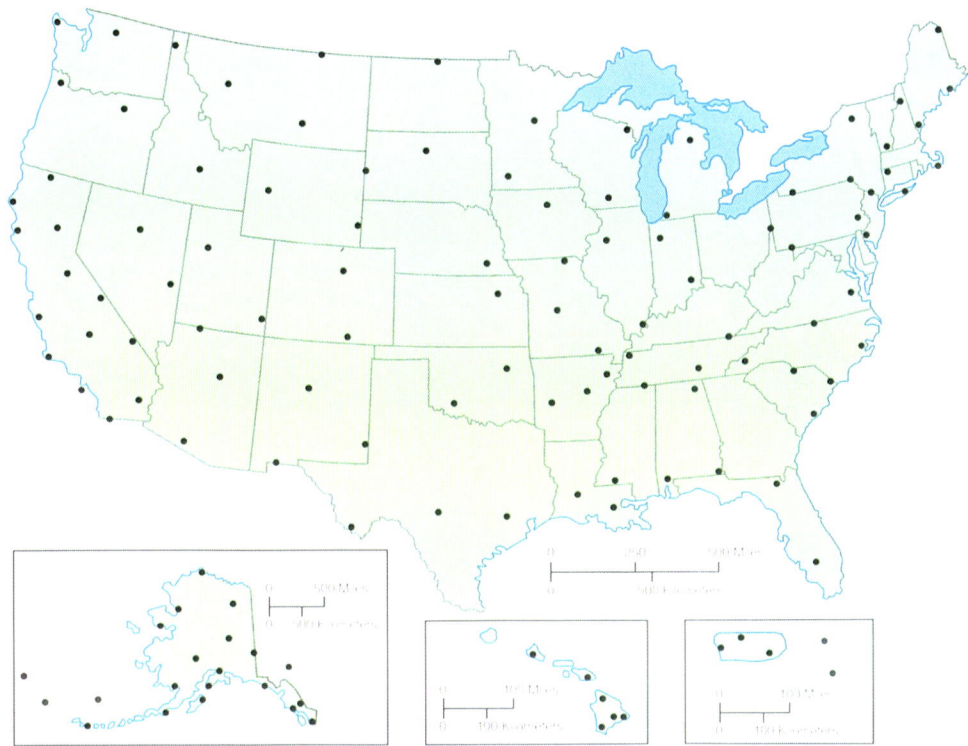

FIGURE 18.27 *The United States National Seismograph Network (USNSN) consists of 120 cooperating seismic stations that provide seismic data to the* National Earthquake Information Center *(NEIC) at Golden, Colorado, where they are combined with data from more than 3,300 stations worldwide. The center not only conducts research to improve our ability to detect and locate earthquakes worldwide, but also is closely associated with organizations whose mission is to provide disaster relief to areas struck by major earthquakes anywhere in the world.*

EARTHQUAKE PREDICTION

Since a major earthquake is capable of destroying an entire city and killing hundreds of thousands of people within a matter of minutes, the ability to accurately predict the time, location, and magnitude of an impending earthquake would be of immense value. Unfortunately, the prediction of earthquakes suffers the same shortcomings as those previously outlined for the prediction of volcanic eruptions. Historic and prehistoric data can be assembled to allow **long-term prediction**; that is, that a major earthquake will take place sometime within the next hundred to a few *hundred thousand* years. Most people responsible for disaster planning and most private individuals find long-range predictions to be of limited use. At the other extreme is **short-term prediction** where the earthquake can be accurately predicted to within days or hours of its actual time of occurrence. In 1975, Chinese scientists used a series of low-magnitude foreshocks (up to a magnitude of 3) to make an accurate short-range prediction of a 7.5 magnitude earthquake that enabled thousands of people to be saved in a city of 90,000 that was nearly destroyed. When the same set of criteria were applied a year later, they failed to predict the Tangshan earthquake that killed several hundred thousand people. It appears that accurate short-range earthquake prediction is still a long way off.

Seismologists have had some success in **medium-term prediction**, which establishes the time of occurrence of an earthquake within a window of a few years to a few months. Medium-range prediction is based on identifying a series of precursory events that have been shown to precede a major earthquake. For example, several major earthquakes have been preceded by a series of low-magnitude foreshocks. Such relationships are then used to devise an empirical model that evaluates the predictive potential of each parameter. The sum total of all events is then applied to a fault zone to predict future fault movements.

One criterion is the historical earthquake data for the fault zone, in particular, the identification of **seismic gaps** or segments of the fault where no recent earthquake activity has occurred. Because these zones have not experienced recent fault movement, stress is still accumulating within the rocks, and, as a result, they are considered to be the most likely sites of future earthquakes. Another observation shown to precede major earthquakes is an as yet unexplained increase in radon gas within deep water wells in or near the fault zone.

One of the most widespread events observed to precede some major fault movements (and other natural disasters) is the erratic behavior of animals. Animals have been observed to run in random patterns, dogs to howl, and burrowing animals to emerge from underground. According to some accounts, the reported use of foreshocks by the Chinese to predict the 1975 earthquake was coincidental, and the prediction was actually based on the behavior of hibernating animals who left their burrows in December (the month the earthquake occurred). Although scientists have had some successes in medium-term prediction, the existing models are of limited use.

With no accurate predictive model in the offing, much effort has been spent in attempts to minimize the destruction of life and property. One area of active research is the design of buildings and other structures to withstand moderate-magnitude earthquake shocks, in particular, the horizontal motions that have been shown to be especially destructive. Based on the findings associated with the Mexico City earthquake, engineers and architects are attempting to design buildings whose inherent vibrational frequency will fall outside the normal range of vibrational frequencies induced by seismic waves.

The basic concepts of land-use planning have been applied to restrict the construction of certain kinds of structures such as homes, schools, and hospitals within active fault zones to minimize injury and loss of life. Storage facilities for dangerous materials such as oil, caustic chemicals, and liquefied gas are excluded from zones of potentially high earth movement. In planning future construction, geologic maps are used extensively to delineate areas underlain by unconsolidated materials as well as areas underlain by bedrock. Many cities located in earthquake-prone areas have established disaster relief systems to alert citizens to potential dangers as well as to be responsible for implementing various procedures such as evacuation, fire fighting, and the distribution of medical help and food supplies.

Through it all, the inhabitants of an earthquake-prone area must be educated to the potential dangers. Some people are not aware of the potential threat. Many individuals are aware but not impressed with a model that predicts that a major earthquake will occur within the next half-century. Such people apparently do not consider an event that may occur only once within two or three generations to be a real threat. Also, there will always be a few individuals who *are* aware of the dangers, *know* that a major earthquake will most likely occur within their lifetime, and yet either choose to stay or, unfortunately, have no choice but to stay.

SPOT REVIEW

1. What parameters are required for the true prediction of any future natural disaster? With our present understanding of earthquakes, how precisely can these parameters be evaluated?
2. What kinds of efforts are currently being taken to minimize the death and destruction that may arise from future earthquakes?

CONCEPTS AND TERMS TO REMEMBER

Distribution of earthquakes
  Plate margins
Frequency and location of earthquakes
  seismic waves
  subduction zones
    Benioff Zone
  Focus or hypocenter
    shallow, intermediate, and deep

epicenter
Measurement of earthquakes
  intensity (damage)
    Modified Mercalli Scale
  magnitude
    earth movements and energy released
    Richter Scale
    seismograph

Earthquake damage
  tsunami
    Seismic Sea Wave Warning System (SSWWS)
Seismic waves
  kinds of waves
    shear
    compressional-extensional (compression)

body waves
   S and P waves
surface waves
   Love (LQ) waves
   Rayleigh (LR) waves
Seismology
seismograph
seismogram
   time-distance (travel-time) curve
World seismic network
National Earthquake Information Center (NEIC)
U.S. National Seismograph Network (USNSN)
Prediction
   long-, medium-, and short-term prediction
   seismic gaps

REVIEW QUESTIONS

1. Most earthquakes and nearly all major earthquakes are located
   a. along the oceanic ridges.
   b. along the zones of subduction.
   c. at points where major masses of magma are moving within the underlying crust.
   d. in association with oceanic hot spots.
2. The most intense earthquake in North America within historic time occurred at
   a. San Francisco, California, in 1906.
   b. Los Angeles, California, in 1847.
   c. Boston, Massachusetts, in 1636.
   d. New Madrid, Missouri, in 1811.
3. Each higher step in the Richter Scale represents a _____-fold increase in energy released.
   a. 10   b. 50
   c. 30   d. 100
4. A tsunami is a(an)
   a. surface shock wave large enough in amplitude to result in an earthquake of at least 8.0 on the Richter Scale.
   b. earthquake body wave transmitted totally within the crust.
   c. seawave generated by an earthquake located within or near an ocean basin.
   d. major earth movement such as a landslide or rock fall resulting from an earthquake.
5. The earthquake shock wave that causes the most damage is the
   a. Love surface wave.
   b. Rayleigh surface wave.
   c. P body wave.
   d. S body wave.
6. The distance from an earthquake focus to a seismic station is determined by the
   a. difference in the arrival times of the body waves.
   b. arrival time of the Love surface wave.
   c. difference in the arrival times of the body and surface waves.
   d. arrival time of the P body wave.
7. Why is the amount of earthquake energy released from zones of subduction much greater than that released from oceanic spreading centers (oceanic ridges)?
8. Why don't body waves contribute to the damage caused by an earthquake?
9. An earthquake is the result of what kind of strain?
10. Why can compression-type shock waves be transmitted through any medium while shear-type shock waves can only be conducted through solids?
11. Why must seismic stations have at least two horizontal pendulum instruments to record the arrival of earthquake shock waves?
12. How are the times of arrival of earthquake body waves used to determine the distance from the earthquake focus to the seismic station?

THOUGHT PROBLEMS

1. If any instrument designed to measure the movement of another object requires a part that either remains immobile while the rest of the instrument moves or that is moving in a known direction at a known speed, how does the speedometer in an automobile work? How is the airspeed of an airplane determined by an on-board instrument?

2. Individual faults experience successive movements as the rocks on opposite sides of the fault surface "lock up," allowing stresses to rebuild. If the rocks are prevented from locking up, successive faulting, and subsequent earthquakes, will not occur. Is it feasible to prevent lockup and, if so, how could it be done?

FOR YOUR NOTEBOOK

Since most of us do not live in active earthquake areas, our information about recent earthquakes is obtained from the media. Using information reported in newspapers and weekly periodicals, compile information about the magnitude and intensity of some of the major earthquakes that have occurred during the past 10 or so years. Compare the geological settings, impact on the inhabitants of the region, and extent of damage. For sequences of earthquakes in a given area, establish the frequency of earth movement. Note in particular instances where earthquake prevention methods, or their lack, were involved and especially where particular structures either survived or failed to survive the earthquake.

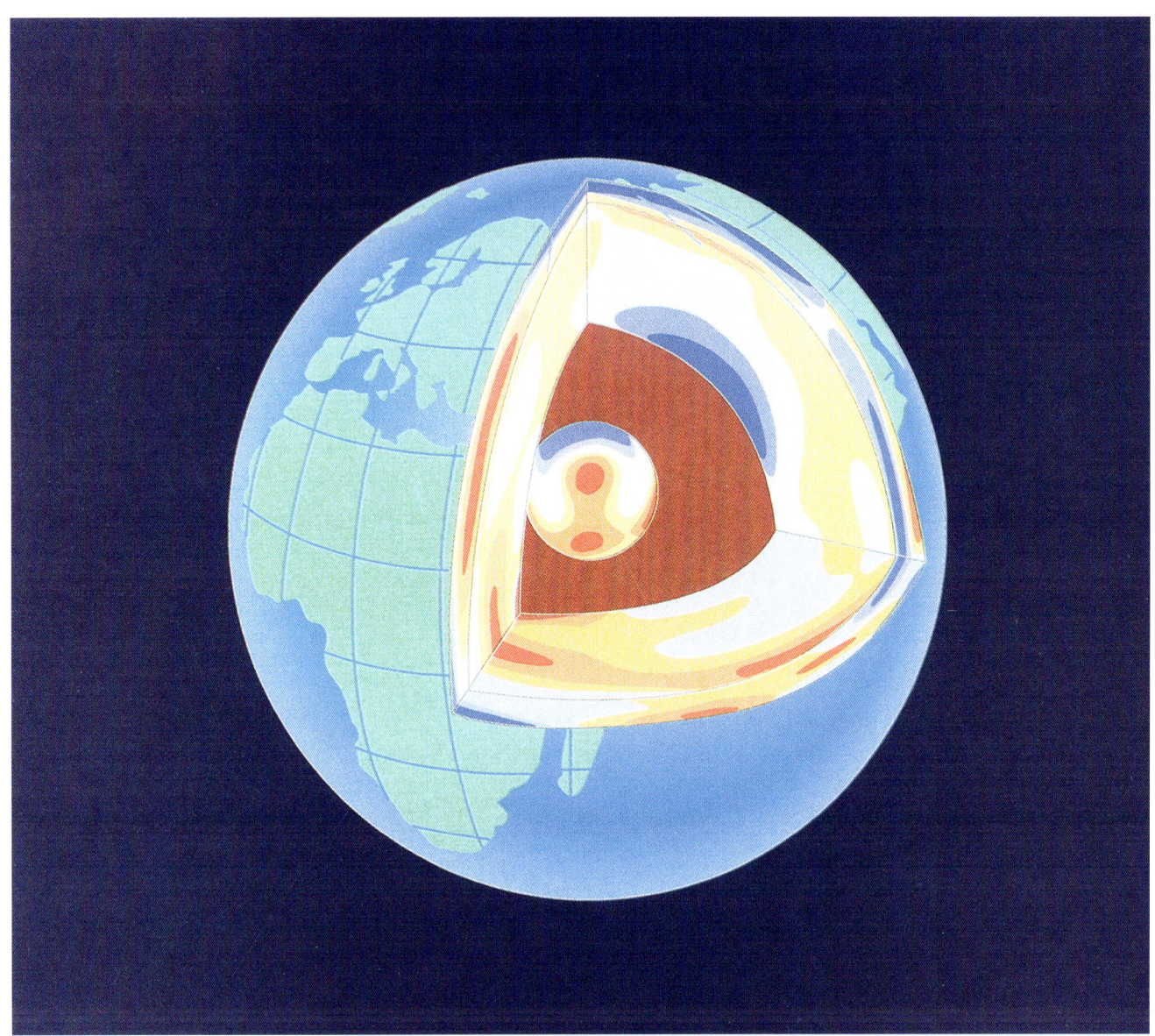

CHAPTER OUTLINE

INTRODUCTION

SEISMIC WAVE VELOCITIES

THE CRUST-MANTLE BOUNDARY

THE MANTLE-CORE BOUNDARY

ISOSTASY
   Isostatic Balance

CRUSTAL LOADING AND UNLOADING

CHAPTER 19

Earth's Interior

INTRODUCTION   With this chapter, we have come full circle in our discussion of the structure of Earth. Chapter 1 introduced a cross section of Earth showing the basic subdivision into the **crust, mantle,** and **core** (Figure 19.1). The **lithosphere** was defined as the combination of the crust and the outer brittle portion of the mantle. Chapter 2 introduced the topic of plate tectonics and discussed how the lithosphere is broken into plates that move over Earth's surface. We also introduced the **asthenosphere**, the plastically behaving layer underlying the lithosphere and a source of the convection cells that drive the plate movements (refer to Figure 19.1). In those early chapters, we did not discuss how scientists came to propose such an internal anatomy but deferred that topic until other information needed to understand the model had been presented.

Our present understanding of Earth's interior is the product of seismological studies that began in the early part of this century. Seismologists studying seismograms for individual earthquakes recorded worldwide observed systematic variations in the arrival times of body waves that eventually led them to develop the model we now accept for the interior of Earth. In this chapter, we will discuss the data and how they were interpreted. In preparation for our discussion, you may want to review the basic information about seismic waves presented in Chapter 18.

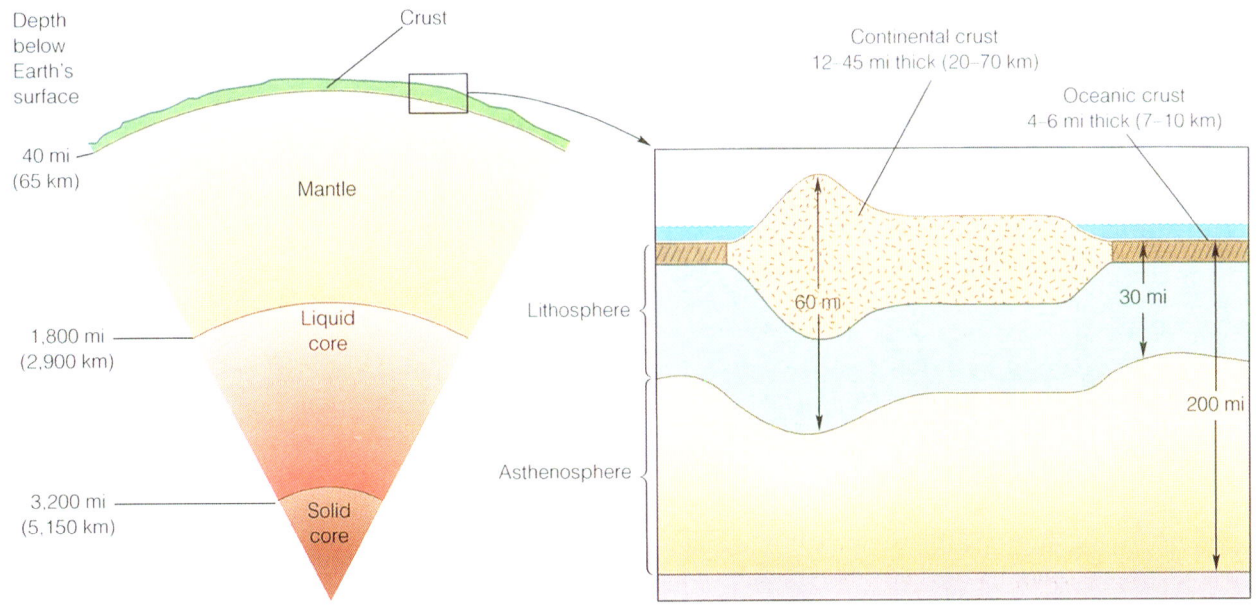

FIGURE 19.1   *The three major subdivisions of Earth are the* core, mantle, *and* crust. *The crust and the outer brittle portion of the mantle are combined in the* lithosphere, *which in turn overlies a plastic portion of the mantle called the* asthenosphere.

SEISMIC WAVE VELOCITIES

To understand how seismologists used seismic data to determine Earth's internal structure, we must discuss the influence of the physical state of the transmitting medium on seismic wave velocity. Seismic velocity is commonly described as being proportional to density whereas, in fact, the main factors determining the velocity of seismic wave transmission are the elastic properties of the material and its **rigidity**. Rigidity is the property of a material to resist an applied stress that would tend to distort it. Fluids (gases and liquids) have no rigidity.

The proportionality that seems to exist between seismic wave velocity and density occurs because both rigidity and density increase with the increased confining pressure of burial. The increase in seismic wave velocity with increasing rigidity also explains why a compression wave is transmitted at a higher velocity through the solid inner core than through the liquid outer core.

Another important aspect of propagating seismic waves is the directional response of the waves as they cross boundaries between media of different transmission velocities. Figure 19.2 shows the two possible scenarios. In one case, the seismic wave passes from a medium of lower velocity into one of higher velocity while in the other scenario, the relationship is reversed. In both cases, if the seismic wave approaches the contact between the two media at a right angle, the wave will cross the boundary and undergo the appropriate change in velocity but will *not* change direction.

If, however, the seismic wave approaches the interface between the two media at an angle less than 90°, the wave will change both velocity and direction as it crosses the interface. The rule is that if the seismic wave passes from a medium of *lower* velocity to one of *higher* velocity, its path will turn *away* from a perpendicular erected to the plane of contact at the point of crossing. On the other hand, if the seismic wave passes from a medium of *higher* velocity to one of *lower* velocity, the change in direction will be *toward* the perpendicular. The angle of refraction will be determined by the angle between the direction of approach and the perpendicular and the difference in the velocities on opposite sides of the interface between the two media.

These same rules hold for light waves crossing the contact between water and air. You see the effect as the apparent "bending" of a pencil placed in a glass

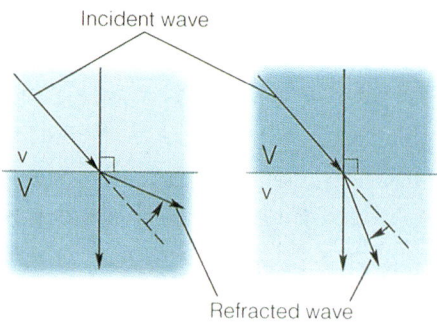

FIGURE 19.2 *Except when approaching at right angles, both the* velocity *and* direction *of waves will change as they cross the contact between layers of different wave velocities. The rule for directional changes is that the wave will turn* away *from a perpendicular constructed at the crossing point when passing from a* lower-velocity layer *to a* higher-velocity layer *and* toward *the perpendicular when passing from a* higher-velocity *to a* lower-velocity layer.

of water as the light changes direction upon passing from the higher-density water to the lower-density atmosphere (Figure 19.3).

The change in direction with changing rigidity, and therefore velocity, explains the curvature of the raypaths of the body waves as they propagate through Earth (Figure 19.4). As the waves penetrate deeper into Earth's interior and meet rocks of progressively higher rigidity, the direction of propagation continuously turns *away* from a line drawn from Earth's center as the seismic waves encounter an infinite number of layers of *increasing* rigidity. At the point where the seismic waves reach the maximum depth of penetration and begin to move toward Earth's surface, the angle of directional change progressively *decreases* as the raypaths begin to encounter layers of *decreasing* rigidities and turn *toward* a line drawn from Earth's center. The total effect is a smooth and symmetrical, upwardly concave raypath.

SPOT REVIEW

1. What rock property or properties determine the transmission velocity of seismic waves?
2. Why is the transmission velocity of seismic waves always greater in solids than in liquids?
3. What is the rule describing directional changes in waves as they pass across the boundary between materials of different transmission velocities?

FIGURE 19.3 *The apparent bending of a pencil in a glass of water is due to the change in the direction of light waves as they pass from air to water. Note that the portion of the pencil in the water appears to bend* away *from the perpendicular erected at the surface of the water. What does that say about the relative velocities of light in air and water? (VU/© SIU)*

THE CRUST-MANTLE BOUNDARY

Even before the contact between the crust and the mantle had been located, geologists intuitively believed that the crust and the underlying mantle differ in composition. They were aware of the granitic composition of the crust from studies of continental rocks and the basaltic character of oceanic rocks from studies of oceanic shield volcanoes such as the Hawaiian Islands. They were aware of intrusions into the crust of ultramafic rocks such as kimberlite, which were totally different in composition from any crustal rock; such intrusions suggested very strongly that there was a more mafic layer below. It remained for seismologists, however, to prove that such a compositional transition exists.

In 1909, a Yugoslavian seismologist, *Andrija Mohorovičić,* first identified the contact between the crust and the mantle by analyzing worldwide seismic data. As we discuss how Mohorovičić was able to recognize the crust-mantle boundary, keep in mind that he did not have our advantage of hind-sight. Mohorovičić's accomplishment was to devise a model that fitted and explained the data he had collected.

In our discussions of earthquakes, we compared the passage of the body waves through Earth to a race. Following that same approach, let us describe several races between two runners labeled B_1 and B_2 on a unique race course that is delineated by two parallel, curved boundaries (Figure 19.5). Each successive race will be longer than the preceding one with the start and the finish located on the upper boundary.

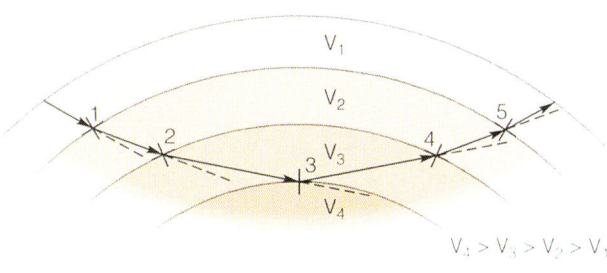

FIGURE 19.4 *The curved raypaths of body waves are due to the application of the directional-change rule. Until body waves reach their greatest depth, they are deflected away from the perpendicular as they penetrate layers of* increasing *velocity. As they return to the surface and encounter layers of* decreasing *velocity, they are deflected* toward *the perpendicular.*

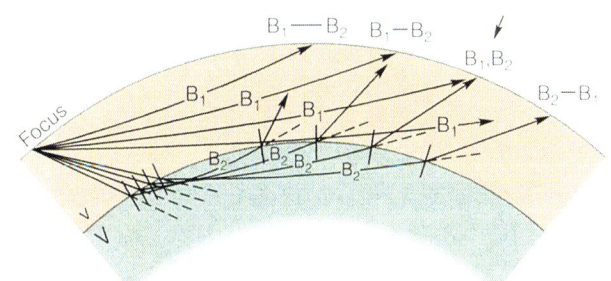

FIGURE 19.5 *The multiplicity of body waves seen on some seismograms is due to the presence of two sets of body waves: one set is propagated totally within the lower-velocity crustal rocks, while a second set passes through the crust-mantle boundary and travels for a period of time through the higher-velocity rocks of the mantle. At a point determined by the thickness of the mantle, the two sets of body waves arrive simultaneously at seismic stations and appear on seismograms as a single set of body waves.*

The races must conform to several rules: (1) neither racer may step across the upper boundary; (2) between the two boundaries, the racers must walk; (3) in each race, B_1 must stay between the two boundary lines and is therefore limited to walking; and (4) B_2 has the option of crossing over the lower boundary and running.

Although B_1 must walk from the starting point to the finish line in each race, B_2 must make some fundamental decisions. One decision is the optimum path that will allow B_2 to take the fullest advantage of the running portion of the course while at the same time keeping the total distance covered to a minimum. Obviously, crossing the lower boundary directly opposite the starting point, running to a point directly opposite the finish line, and walking across the track to the finish point would *maximize* the running portion of the race but would also represent the *longest* possible distance. The *optimum* path would entail walking across the track at a certain *angle* to the racecourse, crossing the lower boundary into the running area, running more or less parallel to the course a specified distance, recrossing the lower boundary at the same angle to the course, and walking to the finish point. Note that as B_2 crosses the lower boundary, the path taken turns *away* from a line drawn perpendicular to the course but when B_2 recrosses from the running area to the walking area, the path runs *toward* a line drawn perpendicular to the direction of the course. In the first race, even though B_2 is allowed to run once the lower boundary is reached, the running portion of the first race is not long enough to allow a win.

Race 2 follows all the same rules: but as you might predict, the time interval between B_1 and B_2 at the finish line decreases as the running portion of B_2's race increases. Eventually, a race will be long enough that B_1 and B_2 will tie. Once the tie occurs, B_2 will win all subsequent races with the winning margin increasing with each race.

Note also that changing the distance between the two boundary lines would change the distance required for the race to end in a tie. As the distance between the two lines *increases,* the race ending in a tie will *lengthen*, and as the distance *narrows*, the race will become *shorter.*

The data Mohorovičić recorded from the seismograms resembled the data one would record for the sequence of races we have just described. While studying suites of seismograms recorded worldwide for individual earthquakes (each representing a separate race), he observed that certain seismograms recorded two sets of **body wave arrivals** (B_1 and B_2 in our hypothetical race). He noted further that the difference in the arrival times of the two sets of body waves seemed to depend upon the distance of the recording seismic station from the focus of the earthquake. The interval between arrival times of the sets of body waves progressively decreased away from the focus to a location where there was a single set of body waves. Beyond this point, two sets of body waves again arrived with the time interval between them increasing with increasing distance from the earthquake focus. His eventual solution was the scenario illustrated by our hypothetical race.

Mohorovičić suggested that one set of body waves was being transmitted totally within a lower-velocity layer we now call the crust while the second set of body waves had crossed a boundary between the lower-velocity upper layer and a higher-velocity layer below. He interpreted the difference in velocities as being due to a difference in rock composition and concluded that a layer consisting of a basaltic oceanic crust or a granitic continental crust overlay a peridotitic mantle. This interpretation fit the structure most geologists accepted for the outer portion of Earth.

From the variation in distance from the earthquake foci to the point where the two sets of body wave arrivals were superimposed, Mohorovičić deduced that the thickness of the crust varied. He showed that the oceanic crust as thin and relatively uniform in thickness. He showed further that not only was the continental crust significantly thicker than the oceanic crust, but that it varied, being thickest under mountains, intermediate beneath the continental interiors, and thinnest at the continental margins. The cross section in Figure 19.1 showing the lithosphere is a graphic portrayal of Mohorovičić's interpretation. The contact between the crust and mantle was later named the **Mohorovičić Discontinuity** (or **Moho**) in recognition of his contributions to our understanding of Earth.

THE MANTLE-CORE BOUNDARY

In 1914, a German seismologist, *Beno Gutenberg,* identified the contact between the mantle and the core. Like Mohorovičić, Gutenberg was studying worldwide body wave arrival data for individual earthquakes. He noticed a band around the Earth from about 7,000 to 10,000 miles (11,260–16,090

Earth's Interior 553

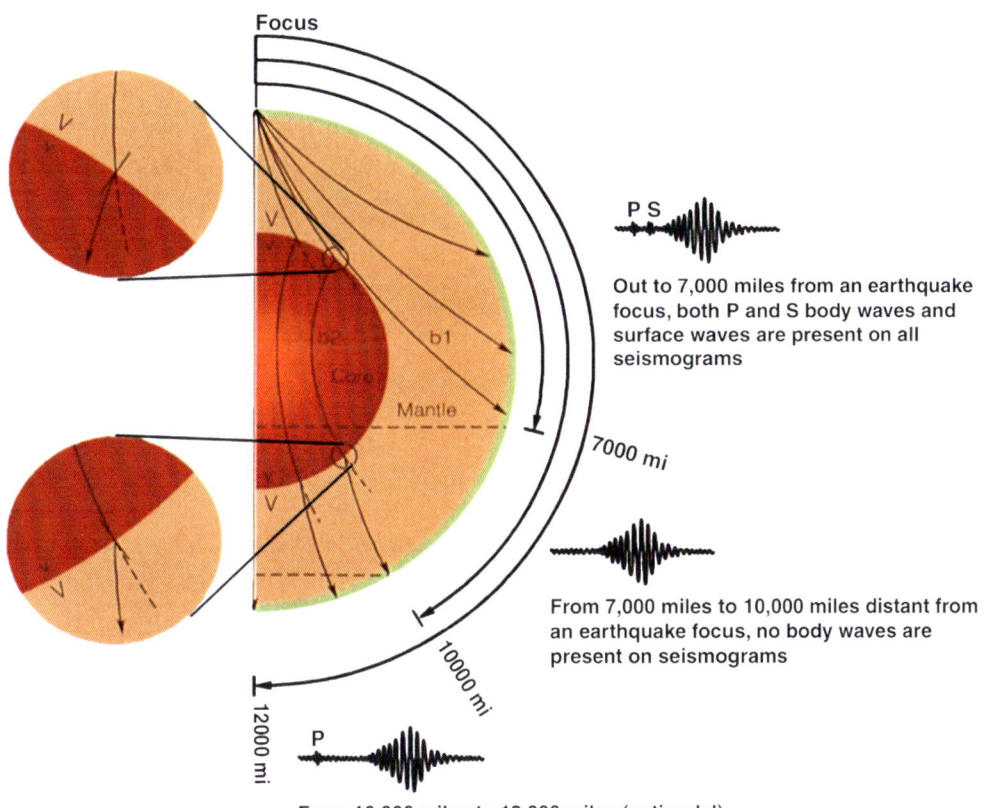

FIGURE 19.6 *The absence of strong body wave arrivals on seismograms located from 7,000 to 10,000 miles (11,260–16,090 km) distant from an earthquake focus is due to the low rigidity of the core. The fact that no S body waves appear on seismograms recorded beyond 10,000 miles from an earthquake focus proves that at least the outer portion of the core is molten.*

km) distant from the earthquake focus within which no strong body waves were recorded (Figure 19.6). He also noted that beyond 10,000 miles distant from the earthquake focus, P waves reappeared on the seismograms, whereas S waves did not.

How did Gutenberg interpret these data? The body wave raypath labeled b_1 in Figure 19.6 just misses the core and was recorded on seismograms with both P and S wave arrivals. The very next deeper body wave raypath, labeled b_2, intersects the core-mantle boundary and is sharply refracted through the core to the opposite side of the Earth where it was recorded *only* with a P wave arrival. Note that the next lower raypath is deflected into the core and that the change in direction of the body wave raypath was *toward* a perpendicular drawn at the point of entry into the core. According to our previous discussions, this change indicates a passage from a medium of *higher* velocity to one of *lower* velocity. Secondly, note that the angle of the directional change indicated that the P wave velocity below the contact between the two layers was *significantly lower* than above the contact. Gutenberg interpreted the data as meaning that the upper layer, called the mantle, was composed of a material *significantly more rigid* than that of the underlying layer, called the core. The explanation for the sharp decrease in velocity was that the seismic waves had passed from a rigid, *solid* mantle into either a *plastic* or a *liquid* core. Note that if the contact was between a *solid* mantle and a *plastic* core, both P and S waves would appear on seismograms recorded on the side of Earth directly opposite the earthquake foci since S waves can be transmitted through plastic solids. If, on the other hand, the mantle-core boundary was a contact between a *solid* mantle and a *liquid* core, only the P wave could penetrate Earth's center. Gutenberg's data showed an absence of S waves in the region on the opposite side of Earth from the focus of an earthquake. This suggested that the outer core was liquid rather than plastic. In recognition of his discovery, the mantle-core contact is called the **Gutenberg Discontinuity**.

Inge Lehmann later found evidence for a solid center for the core in faint P wave arrivals in the region from 7,000 to 10,000 miles (11,260–16,090 km) (Figure 19.7). She showed that, upon passing from the liquid outer core into the solid inner core,

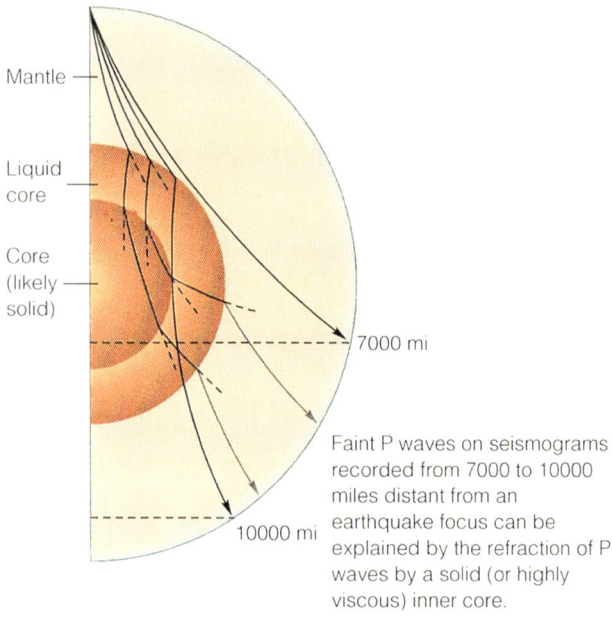

FIGURE 19.7 *The identification of faint P body waves on seismograms recorded from 7,000 to 10,000 miles (11,260–16,090 km) distant from an earthquake focus indicates that the center of the core is either solid or a highly viscous liquid.*

the P waves would both increase in velocity and change direction *away* from the perpendicular, placing them on a trajectory that would intersect Earth's surface within the band where strong P waves are excluded.

Since the pioneering discoveries of Mohorovičić, Gutenberg, and Lehmann, other important seismic studies have laid the foundations for our present concept of plate tectonics. You will remember from our early discussions of continental drift that Alfred Wegener could not satisfactorily explain the source of energy that transported the continents. At that time, Earth was thought to be a solid mass of rock surrounding a molten core, with the outermost layer, the crust, being a brittle layer of rock lying upon a solid, and presumably rigid, mantle.

Since that time, the determination of body wave velocities from a large number of earthquakes has shown that P and S wave velocities vary with depth (Figure 19.8). The increase in velocity with depth within the crust is explained by an increase in rigidity with increased depth. As the S and P waves pass from the crust into the underlying mantle, the velocities of both continue to increase. However, there is a zone extending from about 60 to about 120 miles (90–195 km) within which the S wave velocities decrease. The zone of decreased S wave velocity is referred to as the **low-velocity zone**. Similar to the contact between the mantle and the core, the rocks within the zone are *significantly less rigid* than the rocks both above and below. The fact that S waves propagate through the zone indicates that it is *not liquid*. If the layer were *plastic*, however, the decrease in rigidity would explain the decrease in velocity.

You will remember from our discussion of plate tectonics in Chapter 2 that Holmes proposed that convection cells provide the energy needed to move the lithospheric plates, a proposal implying a liquid-like response of that portion of the mantle. The existence of a low-velocity layer within the upper portion of the mantle supports his idea.

From the base of the asthenosphere down to a depth of about 550 miles (880 km), the S wave velocities undergo a number of marked increases. These velocity increases are explained by changes in rock rigidity resulting from modifications in the crystal structure of the main mineral component of peridotite, olivine. Using data from laboratory experiments as well as seismic data, scientists theorize that as pressures increase with depth, the crystal structure of the mantle minerals undergoes phase changes. The initial phase change takes place at a depth of about 250 miles (400 km) where the structure of olivine changes to a more compact structure similar to that of the mineral spinel. At about 400 miles (640 km), the pressures and temperatures become so high that the ferromagnesian minerals begin to decompose into basic oxides. From a depth of 550 miles (885 km) to the contact with the molten core, the body wave velocities increase only slowly, indicating that little further change occurs in the makeup of the mantle rocks.

SPOT REVIEW

1. What seismic evidence did Mohorovičić use to determine that the crust and mantle have different compositions and that the crust varies in thickness?
2. What evidence did Gutenberg use to prove that the outer portion of the core is liquid?
3. What evidence did Lehman present to indicate that the center of the core is either solid or a very dense liquid?

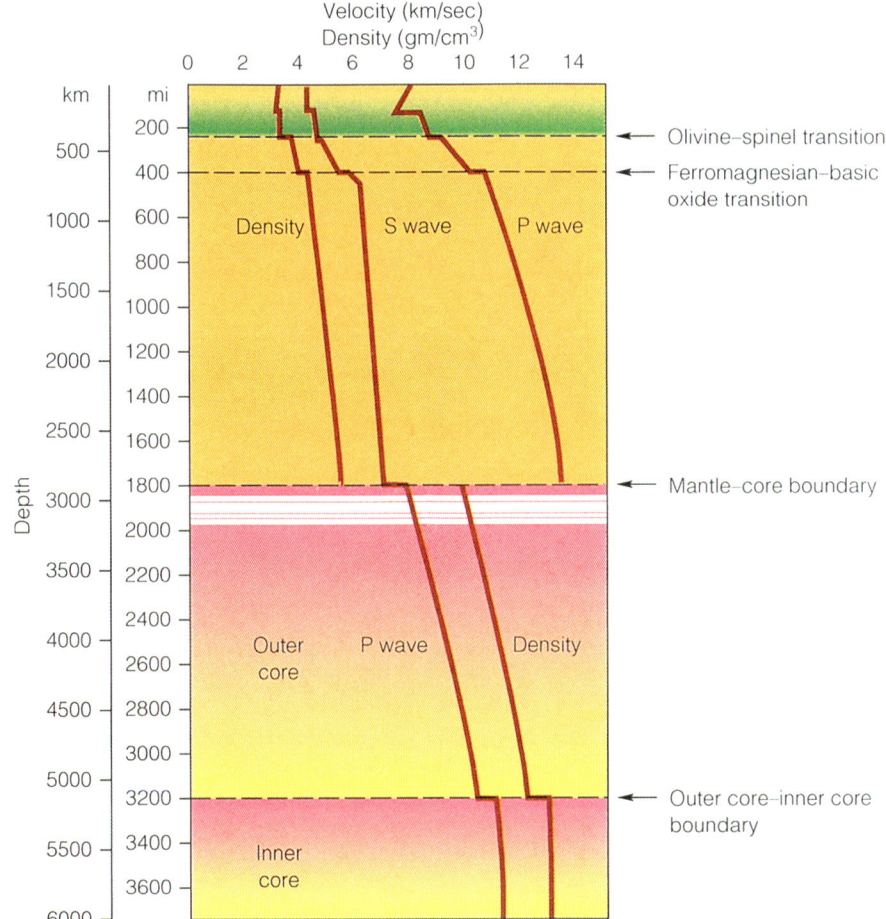

FIGURE 19.8 *The changes detected in body wave velocities between Earth's surface and the core-mantle contact are due to changes in the physical and chemical properties of the rocks encountered with depth. The exclusion of S body waves from the core is due to the fact that shear-type waves will only propagate through rigid materials.*

ISOSTASY

Throughout the text, we have stressed the *horizontal* movements of the lithosphere plates observed at divergent plate boundaries, transform faults, and convergent margins. Some horizontal plate movements, however, result in the *vertical* movement of the crust. For example, horizontal *extensional* forces produce downfaulted blocks such as those along the oceanic ridges and both downfaulted and upfaulted blocks in block-fault mountains; horizontal *compressional* forces generated at the convergent plate margins are responsible for the uplift of most of Earth's major mountain ranges. Other vertical displacements of the continental crust are induced by epeirogenic forces that do not seem to be directly related to plate movements. It is believed that the crustal movement involved in epeirogenic uplift is driven by the buoyancy of hot mantle rocks concentrated beneath the lithosphere.

In addition to the heat-driven movements of Earth's crust, there are other, more subtle, vertical movements that are driven entirely by the force of gravity. These vertical movements involve the gravity-driven plastic flow of asthenospheric rocks. The movements are summarized in the concept of **isostasy** (from the Greek meaning "equal standing").

Mohorovičić showed that Earth's crust varies in both thickness and density. Notwithstanding the broad oceanic ridges and the hot spot volcanoes that dot the ocean floor, the oceanic crust is relatively uniform in composition, density, and thickness. In contrast, the continental crust varies significantly in composition, density, and thickness. The oceanic crust is a relatively uniform 3 miles (5 km) thick and is uniformly composed of gabbro-basalt with a density of about 3 gm/cm^3. The continental crust, on the other hand, ranges in thickness from about 3 miles (5 km) at the margins through an average of about 22 miles (35 km) and up to 45 miles (70 km) beneath foldbelts (Figure 19.9). Continental crust is composed of a mixture of igneous, metamorphic, and sedimentary rocks with an average density of about 2.7 gm/cm^3. The mass (volume times density)

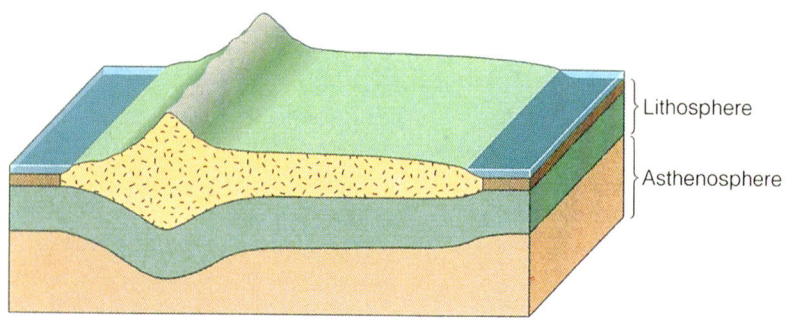

FIGURE 19.9 *The lower-density rocks of the lithosphere literally* float *on the higher-density rocks of the asthenosphere. Following the rules that apply to all floating bodies, the continental lithosphere rises higher and penetrates deeper into the underlying asthenosphere because of its greater mass.*

When the mass of the liquid is displaced, represented by the shaded area, is equal to the mass of the solid, equilibrium will be established and the solid will float.

(a)

FIGURE 19.10 *A solid will sink into a liquid of higher density until the mass of liquid displaced is equal to the mass of the solid, at which point, a* gravitational balance *is achieved* (a). *If the density of the solid is greater than that of the liquid, gravitational balance cannot be achieved, and the solid will sink to the bottom. If the solid is removed from the liquid, the liquid will be out of gravitational balance. Gravitational forces will cause the liquid to replace the mass of liquid originally displaced, and gravitational balance will be reestablished* (b).

Solid in gravitational balance with liquid.

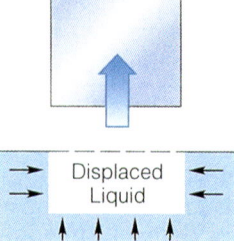

Solid removed. Liquid out of gravitational balance. Forces will replace displaced liquid.

Displaced liquid replaced. Gravitational balance re-established.

of the continental crust is several times greater than that of the oceanic crust. The thickness of the rigid portion of the mantle contained within the lithosphere is relatively uniform (about 25–30 miles or 40–48 km), and its density is bout 3.3 gm/cm³. The variation in the lithosphere's thickness and mass is therefore largely the result of the variation in the thickness and mass of the crust, in particular, the continental crust.

Gravity will cause any object to settle into a liquid medium until the mass of the liquid displaced is equal to the mass of the object (Figure 19.10). At that depth, a **gravitational balance** is achieved between the mass of the object and the mass of the displaced liquid. If the density of the object is *less* than the density of the liquid medium, a gravitational balance will be reached *before* the object is completely immersed, and it will *float*. If the density of the object is *greater* than that of the liquid, the object will *sink* because it is never able to displace a mass of liquid equal to its own mass. This was the situation that existed during the initial differentiation of Earth when molten iron, with a density of about 15 gm/cm³, began to settle to the center of protoplanet Earth to form the core while the lower-density silicates rose toward the surface.

You have seen examples of the principles behind the gravitational balance of floating objects. Because the density of ice is 0.9 gm/cm³ and the density of water is 1.0 gm/cm³, a block of ice floats in a glass of water or on a pond with 90% of the mass under water

(Figure 19.11). The mass of ice "settles" into the water until the mass of the displaced water is equal to the mass of the ice and a gravitational balance has been attained. For the same reason, a piece of wood with a density of 0.5 gm/cm^3 will float half out of water, and a cork with a density of 0.25 gm/cm^3 will float with 75% of its mass out of the water (Figure 19.12).

The depth of penetration of a floating object can be determined from the ratio of the densities of the two materials, *regardless* of the dimensions of the object. Consider, for example, two blocks of ice of different dimensions (Figure 19.13). Assume that one block measures 10 cm on a side and has a mass of 900 grams while the other measures 100 cm by 100 cm by 1 cm thick with a mass of 9,000 grams. The smaller of the two blocks will sink 9 cm into the water to displace 900 grams of water and will have 1 cm exposed. The larger block will settle only 0.9 cm into the water to displace 9,000 grams of water and will have 0.1 cm exposed. In both cases, the same percentage of the thickness is submerged even though the dimensions and masses of the two blocks are different. Note that the *deeper* the block penetrates *below* the surface of the underlying fluid, the *higher* the upper surface of the block will ride *above* the surface of the fluid.

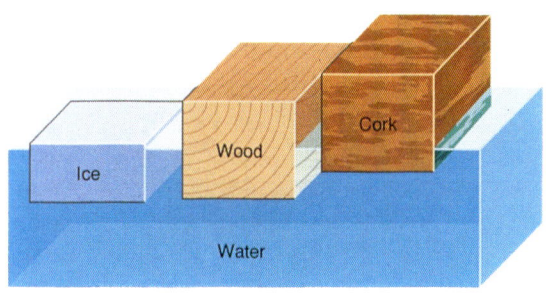

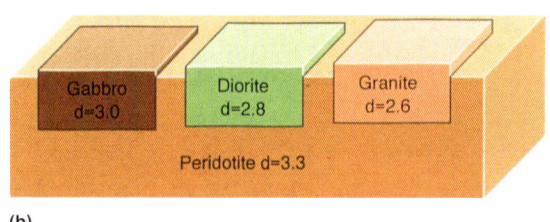

FIGURE 19.12 *To further illustrate the relationship between density and the depth to which solids will float in a denser liquid, cubes of ice, with a density of 0.9, wood, with a density of 0.5, and cork, with a density of 0.25, will float in water with 90%, 50%, and 25% of their masses submerged respectively* (a). *The comparable relationship between gabbro (d=3.0), diorite (d=2.8), and granite (d=2.6) in gravitational balance with mantle peridotite (d=3.3) is shown in* (b).

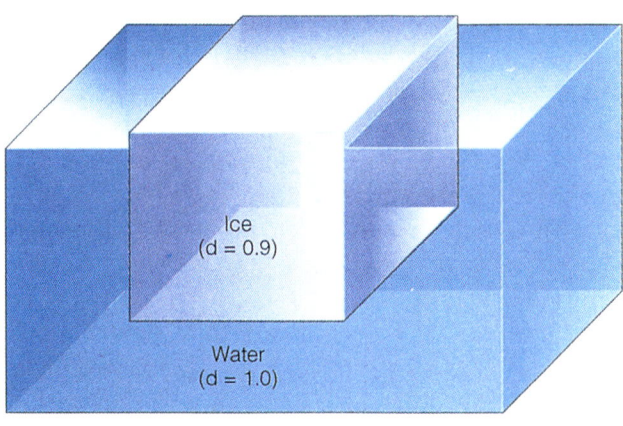

FIGURE 19.11 *When gravitational balance is achieved between water and a solid with a density less than 1.0, the solid will float with a percentage of its volume under water equal to its density. For example, ice with a density of 0.9 will float with 90% of its volume submerged.*

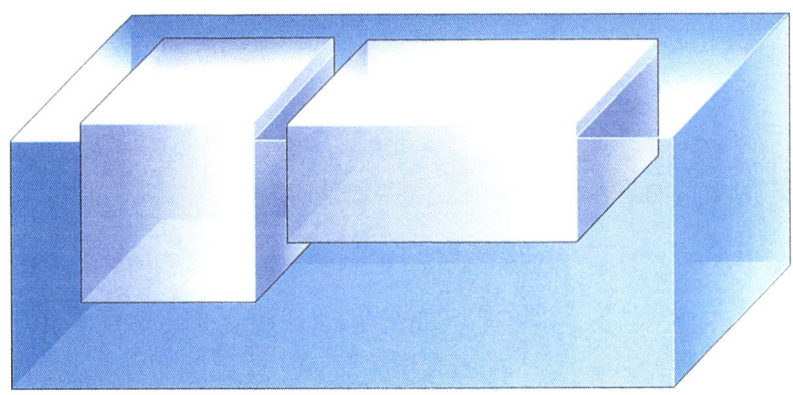

FIGURE 19.13 *Gravitational balance is achieved when the mass of liquid displaced is equal to the mass of the floating solid. How deep a floating object will sink below or rise above the surface of the liquid depends on its mass and shape. In this case, a smaller cube of ice (900 grams) will sink deeper below and rise higher above the surface of water than a larger slab of ice (9,000 grams) because of the difference in thickness (10 cm versus 1 cm respectively). Note, however, that both masses of ice will achieve gravitational balance when they have displaced a mass of water equal to their mass and both will float with 90% of their thickness submerged.*

Isostatic Balance

Isostasy is the gravitational balance between the lithosphere and the underlying asthenosphere. The lithosphere has a density of about 3 gm/cm^3 while the density of the asthenosphere is 3.3 gm/cm^3. **Isostatic balance** is achieved when the mass from any point on Earth's surface to the center of Earth is constant. Remembering that mass is volume times density, consider the drawings in Figure 19.14. Figure 19.14a represents a portion of Earth's lithosphere where the volume and density of the rocks are constant. Notwithstanding the presence of broad oceanic ridges and shield volcanoes over hot spots, Figure 19.14a is a reasonable representation of the oceanic lithosphere. Consider the hypothetical situation illustrated in Figure 19.14b where a segment of the oceanic crust, with a density of about 3 gm/cm^3, has been replaced by an *equal thickness* of continental crust with a density of about 2.7 gm/cm^3. Because of the *decreased* density of the continental crust, the mass measured from any point on the surface of the segment of continental crust to the center of the Earth will be *less* than from any point on the surface of the adjoining oceanic segments. As a result, the lithospheric segment bearing the continental crust would be *out* of isostatic balance. In order to reestablish isostatic balance, *additional mass* must move into the area under the segment of continental lithosphere. The only portion of Earth that is capable of such plastic movement is the asthenosphere. As asthenospheric rock flows laterally from beneath the adjacent oceanic crust into the area below the segment of continental crust, the lithosphere bearing the continental crust will be displaced *upward* to allow the denser asthenospheric rock to be introduced below. The surface of the continental crust will therefore *rise* above the upper surface of the oceanic crust until the mass measured from any point on the surface of the continental crust to the center of the Earth is *equal* to that measured from anywhere on the surface of the adjacent portions of oceanic crust. A simple comparison would be a block of cork (density of 0.25 gm/cm^3) floating on water between two ice cubes (density 1.0 gm/cm$_3$), all of equal thickness. The cork will float higher than the ice cubes because less of it will extend into the water.

This demonstration has been hypothetical because, everywhere except at its margins, the continental crust is thicker than the adjoining oceanic crust. Let us now consider a more accurate model that takes into account the difference in the relative thickness of the oceanic and continental crust (Figure 19.15). Because of the greater thickness of the continental crust, the mass of the segment of continental crust is *greater* than the mass of a segment of oceanic crust of equal area. In Figure 19.15a, the mass from any point on the surface of the continental crust to Earth's center is *greater* than from any point on the surface of adjoining oceanic segments, rendering the segment of continental crust out of isostatic balance. To reestablish isostatic balance, the mass is *reduced* by having asthenospheric rock flow out from under the segment of continental lithosphere. As the asthenospheric rock is removed, the segment of lithosphere and the overlying continental crust will *sink* into the asthenosphere until the mass of the removed asthenospheric rock is equal to the additional mass of continental crust; at this point, isostatic balance is reestablished (Figure 19.15b.) The thickest block represents the rocks involved in a foldbelt mountain range such as the Rocky Mountains, the block with the intermediate thickness represents the vast continental interior, and the thinnest block represents a

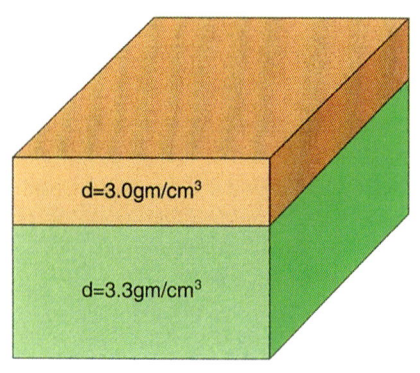
(a)

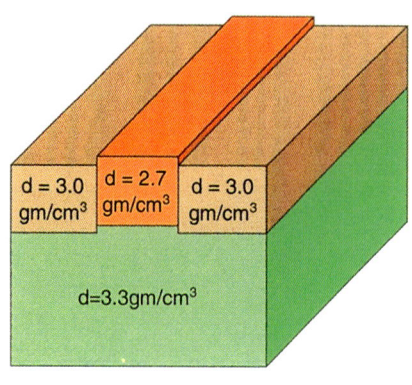
(b)

FIGURE 19.14 *In this hypothetical case, the lower-density continental rocks (d=2.7 gm/cm^3) are lifted above the surface of the oceanic rocks (d=3.0 gm/cm^3) as higher-density asthenospheric rocks (d=3.3 gm/cm^3) flow under the continental rocks in order to maintain a constant amount of mass between Earth's surface and the center of the core.*

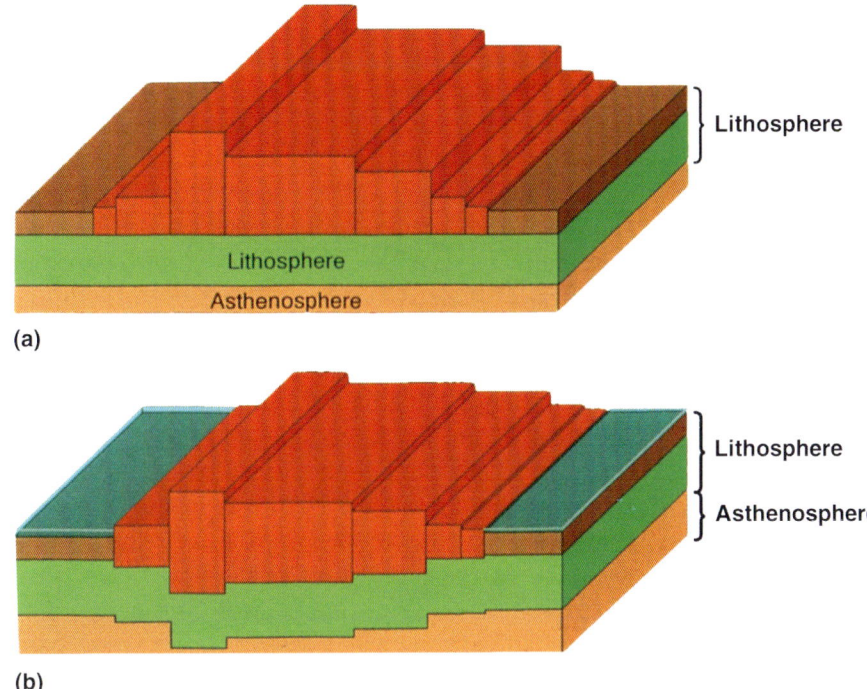

FIGURE 19.15 *In this more realistic demonstration, each block of continental crust and the associated portion of the lithosphere sink into the asthenosphere until the mass of asthenospheric rocks displaced is equal to the mass of the overlying segment of the lithosphere. When gravitational balance is achieved, the amount of mass from Earth's surface to the center of the core is everywhere equal. The thickest portion of the lithosphere penetrates most deeply into the underlying asthenosphere and rises to the highest elevation above Earth's surface.*

continental margin. Following the principles just described, the depth to which each block will sink into the underlying asthenosphere and the elevation of the upper surface of each block after isostatic balance is achieved will depend upon the thickness of the block. The *thickest* segment, representing the mountain range, will penetrate *deepest* into the asthenosphere and will stand at the *highest* elevation above the adjoining oceanic crust. The *block* representing the continental interior will penetrate into the asthenosphere *more* than the block representing the continental margin but *less* than that representing the mountain range. As a result, the elevation of the block representing the continental interior will be *below* that of the mountain range but, above that of the continental margin. This simple experiment demonstrates that the topography of the continental surface is largely dependent upon isostatic adjustments resulting from lateral variations in the thickness of continental crust. Note, however, that upon attaining isostatic balance, the elevation of the upper surface of the continental crust will be *higher* than the surface of the adjoining oceanic crust. The difference in thickness and mass of oceanic and crustal rocks, combined with the effects of gravity, results in the average elevation of the continents being nearly 3 miles (5 km) above the level of the ocean floor (refer to Figure 19.9).

The mass of continental rock penetrating into the asthenosphere beneath mountain ranges is referred to as the **roots** of the mountain range, and we speak of the roots *supporting* the elevations of the mountains. This is, in fact, an accurate statement. According to Archimedes' principle, "when a body is immersed in a fluid, the fluid exerts an upward force on the body equal to the weight of the fluid displaced." Isostatic balance is the balance between the upward force defined by Archimedes and the force of gravity.

As new mountains such as Himalayas rise and increase in mass, more asthenospheric rock is displaced to maintain isostatic balance between the two rock masses, and the roots penetrate deeper into the asthenosphere. On the other hand, as mountain summits are reduced in elevation by erosion and the mass of rock decreases, a mass of asthenospheric rock equal to the mass of rock removed will flow back under the roots of the mountains. Eventually, when the mountains have been removed by erosion and the area is reduced to a peneplane, the roots will have disappeared (Figure 19.16).

The only type of mountain that does *not* have roots, and where the summit elevations are *not* maintained by isostatic balance, are block-fault mountains such as those of the Basin and Range Province. In such mountains, the elevations are being maintained by the buoyancy of an underlying mass of hot, low-density mantle rocks.

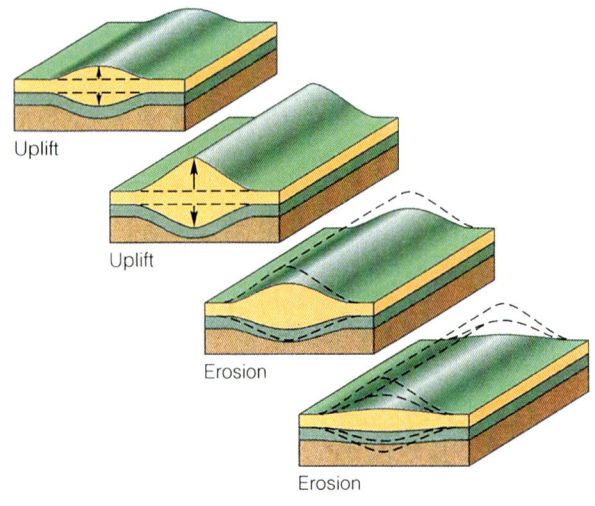

FIGURE 19.16 *The highest elevations of the continental crust, the mountains, are maintained by the gravitational balance achieved with the underlying rocks of the asthenosphere. As the mountains undergo erosion, the elevations of the mountains are reduced and the "roots" simultaneously rise as denser asthenospheric rocks flow under the continent to compensate for the lost mass. Should the mountains be completely removed by erosion, the "roots" of the mountains will disappear.*

SPOT REVIEW

1. What is isostasy?
2. How would you go about calculating the depth to which a cube of a material with a density of less than 1.0 would sink into water?
3. What is meant by the statement that mountains have "roots"?
4. Why does the continental crust sink deeper into the underlying mantle than the oceanic crust does even though continental crust is lower in density?
5. Consider a cubic iceberg 1,000 feet thick. How many feet will be exposed above sea level? How much ice must melt to lower the summit elevation by 50 feet?

CRUSTAL LOADING AND UNLOADING

Vertical crustal movements can also be a response to **crustal loading**. An example is the response of Earth's crust to the mass of the continental ice sheets. As the ice formed and imposed additional mass onto the surface of the continental crust, the surface of the crust was bowed down as a volume of asthenospheric rocks equal in mass to the ice cap flowed out from under the point of loading. At the present time, portions of the bedrock surfaces of both Antarctica and Greenland are depressed below sea level because of the mass of their respective ice caps (Figure 19.17). At the climax of each advance of the Great Ice Ages, the continental surfaces of Canada and the Scandinavian Peninsula were similarly depressed. At the close of the Great Ice Age when the ice melted, the crust began to rebound as isostatic readjustments were initiated by the loss of the ice loading (refer to Figure 19.17c). Ice can melt at a faster rate than the rocks within the asthenosphere can flow, however, so the elastic rebound of the original continental surfaces is not yet complete. The areas of depressed continental crust are still occupied by oceanic waters, with Hudson Bay in central Canada and the Baltic Sea in northern Europe being the aftermaths of the Laurentian and Scandinavian ice sheets, respectively. The North American crust is rising at a rate of about 3 feet (90 cm) per century, while the rocks of the Scandinavian Peninsula are rising at a rate of about 1 foot (30 cm) per century. With time, these bodies of water will continuously decrease in size, and eventually, when isostatic rebound is complete, both areas will once again become dry land.

Another possible example of isostatic adjustments, which we have previously discussed, is the loading of the oceanic crust under shield volcanoes such as the Hawaiian Islands. Some geologists are of the opinion that most of the sinking experienced by the islands is the result of the islands moving into progressively deeper water as they are carried off the lithospheric bulge that formed over the mantle plume. Other geologists believe the sinking is due to isostatic adjustments in which asthenospheric rock flows out from beneath the islands in compensation for the weight of the island. Still others feel that the sinking of the islands is primarily due to an increase in density due to cooling. Any one of these mechanisms, combined with erosion and the slumping of the outer slopes of the cones, would eventually result in the island disappearing from view and becoming seamounts and guyots.

The amount of loading of the crust necessary to induce isostatic adjustments does not need to be of

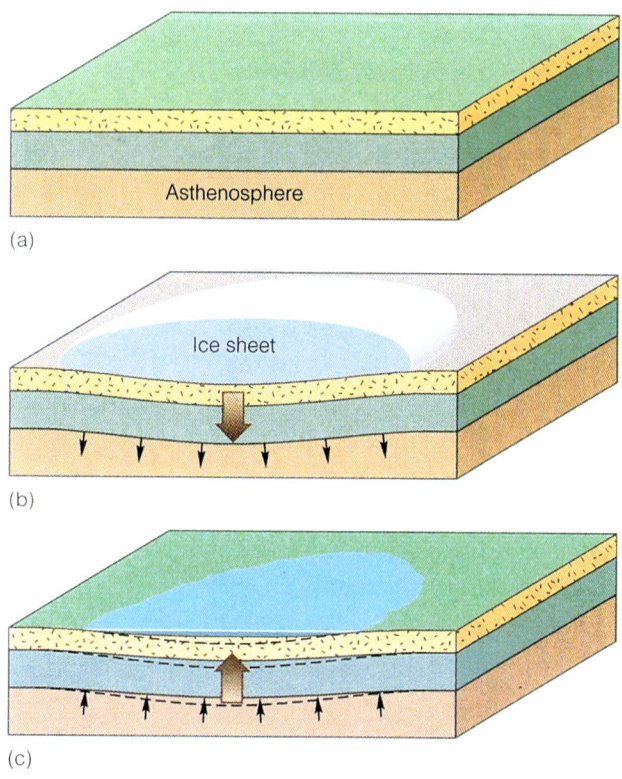

FIGURE 19.17 Crustal loading *refers to the accumulation of mass on Earth's surface. An example that we have discussed is the Pleistocene ice sheets. The accumulation of glacial ice over Canada and Scandinavia caused the lithosphere to be depressed as asthenospheric rocks were displaced to accommodate the added mass of ice. Because ice melts faster than asthenospheric rocks can flow, portions of Canada and Scandinavia are still depressed and out of gravitational balance with the underlying asthenosphere. Asthenospheric rocks are flowing back under the depressions, however, and the lithophere is rising. In time, gravitational balance will be reestablished, and the surface of the land will return to the preglacial elevations.*

FIGURE 19.18 *Not all examples of crustal loading are of the scale of continental ice sheets. When Hoover Dam was filled with 12 billion tons of water, the crust beneath the lake was depressed about 6 inches (15 cm). Based on the incidence of low-magnitude earthquakes recorded in the vicinity of the lake, isostatic adjustment seems to be largely accomplished by fracturing of the underlying rocks. (Courtesy of the Las Vegas Convention and Visitors Authority)*

such massive proportions as those of continental ice sheets, volcanoes, or mountain ranges. When the waters of the Colorado River began to collect to form Lake Mead behind Hoover Dam (Figure 19.18), minor earthquakes in the area signaled the brittle response of the underlying rocks to the loading of the crust. When filled, the mass of approximately 12 billion tons of water was sufficient to cause a regional downwarping of Earth's crust of approximately 6 inches (15 cm) surrounding the lake. The continued incidence of low-magnitude earthquakes in the area indicates that isostatic adjustments to the mass of overlying water are still taking place.

SPOT REVIEW

Why have the areas around Hudson Bay and the Scandinavian Peninsula not yet returned to their pre-Ice Age elevations?

CONCEPTS AND TERMS TO REMEMBER

Basic subdivision of Earth
  crust
  mantle
  core
  lithosphere
  asthenosphere
Seismic wave velocities
  effect of rigidity

directional response
Crust-Mantle boundary
Mohorovičić Discontinuity
  (Moho)
body wave arrivals
Mantle-core boundary
Gutenberg Discontinuity
body wave extinctions

low-velocity zone
Isostasy
  gravitational balance
  isostatic balance
    roots of mountains
Crustal loading

REVIEW QUESTIONS

1. The Mohorovičić Discontinuity is the contact between the
   a. crust and the mantle.
   b. lithosphere and the asthenosphere.
   c. asthenosphere and the underlying mantle.
   d. mantle and the core.

2. Between 7,000 and 10,000 miles distant from an earthquake focus, seismograms show
   a. all body and surface waves.
   b. body waves but no surface waves.
   c. surface waves but no direct body waves.
   d. no S body waves.

3. The proof that the core is partially molten is the absence of _____ from 10,000 to 12,000 miles distant from the earthquake focus.
   a. P body waves.
   b. surface waves.
   c. S body waves.
   d. all body waves.

4. Isostasy is the gravitational balance between the
   a. crust and the mantle.
   b. lithosphere and the asthenosphere.
   c. asthenosphere and the underlying mantle.
   d. mantle and the core.

5. The lithosphere penetrates most deeply into the asthenosphere beneath
   a. continental cratons.
   b. ocean basins.
   c. mountains.
   d. deep-sea trenches.

6. The distance that a solid will sink into a liquid is proportional to the
   a. ratio of the density of the solid to that of the liquid.
   b. product of the densities of the solid and the liquid.
   c. difference between the densities of the solid and the liquid.
   d. the sum of the densities of the solid and the liquid.

7. How are body waves able to appear, though faintly, in the zone between 7,000 and 10,000 miles distant from an earthquake focus, the so-called "body wave shadow zone"?

8. Why are the raypaths of body waves curved rather than straight?

9. What determines how deeply a floating solid will sink into a liquid? Why do some solids sink?

THOUGHT PROBLEMS

1. What effects would be experienced by Earth's surface if the plasticity of the asthenosphere were to decrease? If it became rigid?

2. How is isostasy used to explain the elevations of mountains?

THOUGHT PROBLEMS

Many students have difficulty visualizing the concept of isostasy. Because the major topics of this chapter are impossible to view directly, perhaps a few experiments to firm the idea of gravitational balance are in order. Assemble a selection of solid materials and determine their densities by observing how far they submerge in water. Exchange the water with clear syrup and repeat the experiments. Explain the difference in the outcome of the experiments.

Using wood blocks to represent continental lithosphere and syrup to represent the asthenosphere, design and conduct a series of experiments that will (1) illustrate the growth of a mountain chain and the development of mountain "roots," and (2) show the effects of erosion on the lithosphere-asthenosphere relationship.

CHAPTER OUTLINE

INTRODUCTION
CLASSIFICATION OF NATURAL RESOURCES
    Ore, Protore, and Gangue
    Reserves and Resources
METALLIFEROUS DEPOSITS
    Concentrations of Metals
    Enrichment of Ore Bodies by Weathering
    Banded Iron Formations
    Alluvial and Eolian Concentrations
NONMETALLIC NONFUELS
FUELS
    Coal
    Petroleum
    Oil Shales
THE FUTURE OF FOSSIL FUELS
NUCLEAR POWER
    Fission Reactors
    Fusion
ALTERNATIVE SOURCES OF ENERGY
    Geothermal Power
    Wind
    The Tides
    Hydropower
    Biomass
    Conservation
    Solar Energy
ENVIRONMENTAL CONCERNS
    Coal Mining
    Environmental Problems
    Treating the Acid Mine Drainage Problem

CHAPTER 20

Economic Geology and Energy

INTRODUCTION Economic geology involves the study and exploitation of materials at or near the surface of Earth for the use of society. The first human use of Earth materials probably involved the selection of a stream pebble of the proper shape to make a hand ax, the earliest of tools. *Homo habilis* discovered that certain rock materials could be shaped to a sharp cutting edge better than others. Eventually, fine-grained chert, commonly called *flint,* and *obsidian* came to be preferred because of the ease with which they could be chipped and flaked to produce sharp edges (Figure 20.1).

By the culmination of the Neolithic or Late Stone Age, humans were making tools by grinding the polishing stone and other materials using pieces of sandstone, the predecessor to the modern whetstone. During this same period, humans also developed the technology of smelting copper and firing clay to make pottery.

Four millenia before the birth of Christ, the discovery of **native metals**, elements such as copper, gold, and platinum that occur naturally in metallic form, was to change society. The first native metal to be used, *copper,* signaled the end of the Stone Age. The Age of Copper was short-lived, however. Too soft to be effectively used for tool making, copper soon gave way to *bronze.* Consisting of copper and tin, bronze was the first human-made alloy and soon became the primary metal for the manufacture of tools and weapons, first in Europe and later in western Asia and Egypt.

FIGURE 20.1 *The two most common materials chosen for the manufacture of early cutting tools and points were* flint *and* obsidian. *(VU/© John D. Cunningham)*

Bronze remained the metal of choice until the discovery and use of *iron*. The earliest reference to iron tools dates to about 2500 B.C. in Antolia, the region between the Black Sea and the Mediterranean Sea now known as Turkey. The technology spread to Italy by 1200–1000 B.C. and into the rest of Europe around 1000 B.C. The discovery of iron, and later the manufacture of its by-product, *steel,* was perhaps the most important single human achievement in the use of metals. Today, iron is the most widely used of all the metals. You need only look about to see how pervasive iron is in our daily lives.

CLASSIFICATION OF NATURAL RESOURCES

The importance of metals in society is reflected in the classification of natural resources into **metals** and **nonmetals,** and the specific importance of iron can be seen in the subdivision of metals into **ferrous** and **nonferrous** categories (Figure 20.2). Similarly, the subdivision of nonmetals into **fuels** and **nonfuels** clearly reflects the importance of fuels in modern society. Included in the fuel category are *petroleum* (gas and oil), *coal,* and *uranium.* Petroleum and coal are commonly referred to as **fossil fuels** because they are both made from the converted remains of organisms. Of the present U.S. energy budget, 70% is provided by petroleum, 20% by coal, and 5% by uranium.

Nonmetallic, nonfuel resources include a wide variety of materials ranging from crushed and cut rock to sand and gravel.

Ore, Protore, and Gangue

Although the term **ore** can be used to describe any deposit that can be exploited *at a profit,* the term is usually restricted to metalliferous deposits. Uneconomical deposits are referred to as **protore** or **tailings.** In most deposits, the ore is mixed with the enclosing rock matrix, commonly referred to as **gangue.** After the gangue is removed from the ore and discarded, the ore is usually smelted to extract the metal. It should be noted that as the supplies of certain metals decline and the value increases, discarded gangue is sometimes mined and reprocessed as ore to extract the small amount of metal remaining within the waste rock.

Whether a metalliferous deposit is considered economical or not depends on the market value of the metal per unit mass (dollars per ounce, pound, or ton) relative to the cost per unit mass of producing the metal for sale. The market value of any commod-

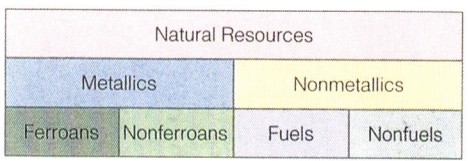

FIGURE 20.2 *The basic subdivision of natural resources is based upon distinguishing metals from nonmetals, with metals being further subdivided into ferrous and nonferrous, indicating the importance of iron in society. The subdivision of nonmetals into fuels and nonfuels shows the importance of energy in our society.*

ity is a function of two factors: (1) *availability* and (2) *demand.* In general, low availability and high demand tend to increase market value. The values of gold, diamonds, and platinum are examples. On the other hand, high availability and low demand result in a low market value. Grass clippings come to mind. Availability and demand can change over time as the value of antiques illustrates. Common items discarded by one generation as having little or no value become sought-after treasures to the next.

The cost of production includes all expenses incurred from prospecting to the production of the final product. Like market value, the production cost changes constantly and is one of the reasons why a marginally profitable ore can change to protore, or vice versa, literally overnight.

Reserves and Resources

Two terms that are crucial to all discussions of natural materials are *reserves* and *resources.* **Reserves** are those deposits (1) that have already been located, (2) whose extent has been determined, and (3) that can be worked at a profit using existing technology. **Resources** include (1) reserves, (2) all other known but currently uneconomical deposits, plus (3) deposits that have not yet been found but, based on geological evidence, are considered likely to be present. In financial terms, your reserves consist of all your cash, all money in your bank accounts, plus any stocks or bonds that you could sell for cash. Your resources include these assets *plus* all worldly possessions that you could sell for some as yet unknown sum, any loans that you could acquire using some possession for collateral, any money you could borrow on your signature (all of which would be of unknown value until the transaction was attempted), plus all future earnings and your long-shot winning of the state lottery. In summary, reserves are part, possibly only a *small* part, of the total of a given resource.

METALLIFEROUS DEPOSITS

You will recall from our discussions of minerals in Chapter 3 that only 13 elements constitute all but a tiny fraction of 1 percent of the rocks of Earth's crust. Of those, the only metallic elements used by modern society that have crustal abundances above 1 weight percent (wt%) are *aluminum* (8 wt%), *iron* (6 wt%), and *magnesium* (3 wt%). All other metals considered to be of value are present at concentrations of less than 1 wt%. For example, the crustal concentration of copper, an indispensable metal in a variety of electrical applications, is only about 0.0058 wt%. Gold is present at only 0.0000004 wt%! Table 20.1 lists the major metals and their average crustal abundances.

TABLE 20.1
CRUSTAL ABUNDANCE OF SOME ORE-FORMING ELEMENTS

Element	Abundance in Weight Percent
Aluminum	8
Iron	6
Magnesium	3
Zinc	0.008
Nickel	0.007
Copper	0.0058
Lead	0.0015
Uranium	0.0002
Mercury	0.00001
Gold	0.0000004

Concentrations of Metals

The distribution of metalliferous deposits leaves little doubt that magma is the ultimate source of most metals. Metalliferous deposits are usually contained either within the larger igneous rock bodies such as stocks and batholiths or within the surrounding country rocks. Most metalliferous deposits are found associated with the host rocks, indicating that most ore deposits result from the reaction of the host rock with fluids originally contained within the crystallizing magma or with circulating hydrothermal solutions.

Concentrations within Igneous Rock Bodies

With the exception of iron, aluminum, sodium, potassium, calcium, and magnesium, few metallic elements contained within a magma are needed to produce the common rock-forming silicate minerals. As the magma cools and the silicate minerals crystallize, the charge and/or ionic radii of rare and economically interesting elements such as platinum, copper, and nickel are incompatible with the crystal structures of the common igneous minerals. As a result, these metals begin to concentrate in the remaining melt. Some of these metals may co-crystallize with the silicate minerals, however; then, due to their higher density, the metals separate from the surrounding mush by a process known as **magmatic segregation** and settle to the bottom of the magma chamber (Figure 20.3). Concentrations of

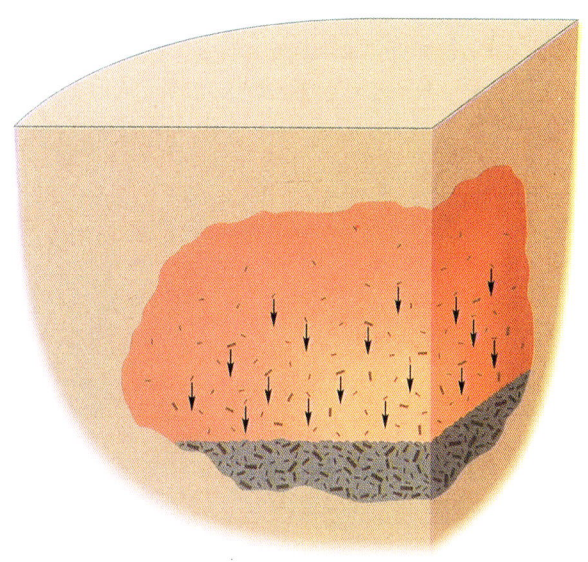

FIGURE 20.3 *When the density of the early-formed minerals is higher than that of the surrounding magma, the minerals settle to the bottom of the magma chamber, a process called* magmatic segregation.

platinum, chromium, nickel, and iron minerals are commonly created by this process. An example of such a deposit is found at Sudbury in Ontario, Canada, where magmatic segregation has resulted in the accumulation of layers of nickel, copper, and iron sulfides within a lopolith (a bowl-shaped igneous pluton) (Figure 20.4). However, intrusive rock bodies, except for those enriched in iron and magnesium, are generally devoid of major concentrations of metalliferous minerals.

Concentrations in Intruded Country Rocks As noted earlier, most metalliferous deposits are found in the rocks surrounding major igneous intrusions.

From our discussions of the crystallization of magmas in Chapter 5, you will remember that as the magma nears the end of the crystallization process, a hydrothermal solution is left over containing concentrations of the elements that have not been "accepted" in the common silicate minerals because of their ionic size or charge restrictions. These solutions are injected into fractures in the surrounding rocks in the form of superheated gaseous or liquid fluids. In some cases, the fluids cool, and the last remaining minerals precipitate in cracks and fissures within the host rock to form **veins** or within pores to produce **disseminated deposits**. Veins are commonly filled with **pegmatite**, an extremely coarse-grained rock with a composition often similar to that of granite. Because pegmatites form from the final portion of the magma and often contain high concentrations of relatively rare elements (Figure 20.5), they are eagerly sought as

FIGURE 20.4 *The sulfide minerals mined at Sudbury, Ontario, Canada, were concentrated by magmatic segregation within a lopolithic intrusion. (Landsat imagery courtesy of the Canada Centre for Remote Sensing, Energy, Mines, and Resources, Canada)*

FIGURE 20.5 *Because the rarer elements are not used in the formation of the common silicate minerals, they become concentrated in the final magma fluids. When these fluids are injected into the surrounding rocks, the veins of pegmatite that subsequently form may contain economic concentrations of valuable elements such as gold. (© Peter Kresan)*

possible economic deposits of elements such as lithium, boron, niobium, tantalum, and the rare earth elements—elements with atomic numbers between 57 (lanthanum) and 71 (lutecium).

Most concentrations of metalliferous minerals within the host rock are produced as the injected solutions react with the surrounding rocks. In some cases, new minerals form from the reaction between the existing minerals and the elements borne in solution. In other cases, the original rock mass is replaced with new minerals by the process of **metasomatic replacement** (refer to Chapter 16). Because of the prevalence of sulfur in magmas (consider all the sulfurous gas emitted during volcanic eruptions), most of the metals are precipitated with sulfur as **sulfide** minerals. Hydrothermal sulfide deposition is responsible for much of the ore mined throughout the world including copper (CuS, Cu_2S, $CuFeS_2$, Cu_5FeS_4), lead (PbS), zinc (ZnS), silver (Ag_2S), nickel (NiS), iron (FeS, FeS_2), and many others.

Enrichment of Ore Bodies by Weathering

As in the case with any mineral assemblage, sulfide ore bodies readily undergo chemical weathering by oxidation and solution when exposed at Earth's surface. An important concentration reaction resulting from the weathering of sulfide mineral deposits is called **supergene enrichment**. Pyrite (FeS_2), one of the common sulfide minerals present in most sulfide ore bodies, is critical to the process. A typical situation is a noneconomic copper deposit where the mineral assemblage of the ore body is a mixture of chalcocite (Cu_2S) and pyrite (FeS_2) disseminated throughout the host rock. As precipitation infiltrates the zone of aeration and moves toward the water table, pyrite oxidizes and hydrolyzes to produce a mixture of iron hydroxide and sulfuric acid (H_2SO_4):

$$FeS_2 + {}^{15}/_2 O_2 + {}^7/_2 H_2O \rightarrow$$
pyrite + oxygen + water →
$$Fe(OH)_3 + 4H^{1+} + 2SO_4^{2-}$$
iron + sulfuric acid
hydroxide

The iron hydroxide precipitates and stains the surface of the regolith, producing the telltale orange-yellow **gossan** or **iron hat** sought by exploration geologists to locate a potentially enriched underlying sulfide deposit (Figure 20.6).

The acidic solutions dissolve the chalcocite present within the rocks above the water table and

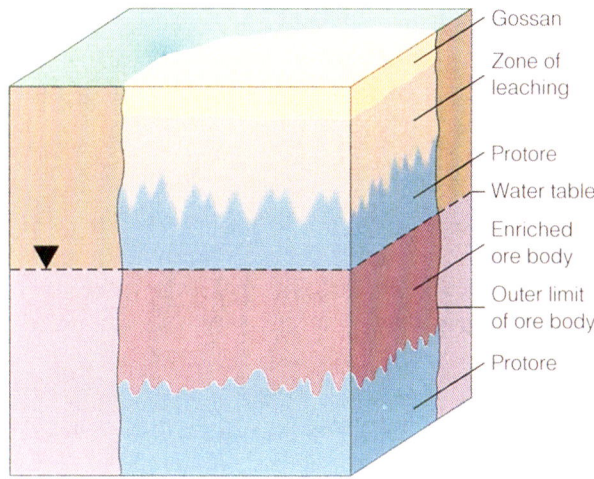

Above the water table the two major reactions are:

(1) Oxidation of pyrite produces acid solutions and forms the iron-rich gossan:
$$FeS_2 + \tfrac{15}{2}O_2 + \tfrac{7}{2}H_2O \longrightarrow Fe(OH)_3 + 2SO_4^{2-} + 4H^+$$

(2) Leaching of copper from protore to below the water table:
$$Cu_2S + 5O_2 + 4H^+ \longrightarrow 4Cu^{2+} + 2SO_4^{2-} + 2H_2O$$

Below the water table, the Cu^{+2} ions removed from the protore above the water table displace other metal ions from the protore to create a zone of copper enrichment:
$$Cu^{2+} + ZnS \longrightarrow CuS + Zn^{2+}$$
$$14Cu^{2+} + 5FeS_2 + 12H_2O \longrightarrow 7Cu_2S + 5Fe^{2+} + 3SO_4^{2-} + 24H^+$$
The displaced metals are leached from the system by groundwater movement.

FIGURE 20.6 *Exploration geologists look for iron-stained* gossans *because surface deposits indicate that a subsurface chemical reaction may have upgraded a noneconomic sulfide protore to an ore body.*

produce solutions of copper sulfate ($CuSO_4$) that percolate downward:

$$Cu_2S + 5O_2 + 4H^{1+} \rightarrow$$
Chalcocite + oxygen + hydronium →
ion
$$4Cu^{2+} + 2SO_4^{2-} + 2H_2O$$
copper sulfate + water

As the solutions move below the water table and the oxygen-rich environment of the zone of aeration gives way to the reducing (oxygen-poor) zone of saturation, the copper ion displaces metals from other sulfide minerals, producing covellite (CuS) and chalcocite (Cu_2S), which precipitate within the ore body and the adjacent rocks:

$$Cu^{2+} + ZnS \rightarrow CuS + Zn^{2+}$$
copper ion + zincite covellite + zinc ion

$$14Cu^{2+} + 5FeS_2 + 12H_2O \rightarrow$$
$$\text{copper} + \text{pyrite} + \text{water} \rightarrow$$
$$\text{ion}$$

$$7Cu_2S + 5Fe^{2+} + SO_4^{2-} + 24H^{1+}$$
$$\text{chalcocite} + \text{iron} + \text{sulfate} + \text{hydronium}$$
$$\qquad\qquad\text{ion}\quad\ \text{ion}\qquad\text{ion}$$

The result is a zone of copper enrichment beneath the water table that may be sufficient to convert a protore into an ore.

One of the most impressive examples of a copper deposit whose original copper concentration was increased by supergene enrichment is *Bingham Canyon,* just west of Salt Lake City, Utah (Figure 20.7). Bingham Canyon was the world's largest working open pit mine and second largest producer of copper when it was closed a few years ago because the base of the mine had extended below the zone of concentration. The mine has since reopened after a $400 million modernization.

Other examples of ore bodies produced by weathering are the lateritic accumulations of hematite (Fe_2O_3) and bauxite, a mixture of hydrous aluminum oxides and hydroxides. You will recall from our discussions of tropical soils in Chapter 7 that due to the extreme intensity of chemical weathering in the tropics, few silicate minerals survive the process of dissolution. Once released into solution, iron and aluminum oxidize in the leached zone and may accumulate in concentrations sufficient to produce commercial deposits such as lateritic bauxites, which are the world's major sources of aluminum.

Banded Iron Formations

Most of the iron ore mined in the United States comes from the iron ore district of the Lake Superior *Mesabi Range* in Minnesota and the iron ranges of northern Michigan, where the iron ore exists as interlayered iron oxide and chert (Figure 20.8). This and similar deposits around the world formed at a time in Earth's history when photosynthetic organisms had recently evolved. Before the evolution of green plants, the atmosphere was devoid of free oxygen. A large portion of the primeval atmosphere was composed of volcanically derived gases such as carbon dioxide (CO_2), hydrogen sulfide (H_2S), and sulfur dioxide (SO_2), which reacted with water to form strong acids that bathed Earth's surface in super "acid rain." The attack of the strong acidic solutions removed iron from the land, where it is primarily contained within igneous rocks in the more soluble ferrous form, and carried it to the oceans where it remained in solution due to the acidity of the water. As the green plants evolved and oxygen produced by photosynthesis began to saturate the ocean waters, the dissolved iron oxidized and precipitated onto the ocean floor as hematite (Fe_2O_3). In time, the layers of hematite were buried under fine-grained siliceous marine sediments. Both were eventually lithified to produce the **banded iron deposits** of the Mesabi Range of Minnesota and the iron ranges of Michigan.

A much later deposit of sedimentary iron oxide ore provides an example of a different mode of origin for iron oxide. During the Silurian Period of Earth

FIGURE 20.7 *Recently reopened, the world's largest open pit mine,* Bingham Canyon *near* Salt Lake City, Utah, *produces copper minerals that were concentrated by supergene enrichment. (VU/© Victor H. Hutchinson)*

FIGURE 20.8 *The iron ore mined at the* Mesabi Range of Minnesota *and the equivalent deposits in northern* Michigan *are examples of* banded iron formations *that formed billions of years ago before oxygen was an abundant component of Earth's atmosphere.* (Courtesy of W. B. Hamilton/USGS)

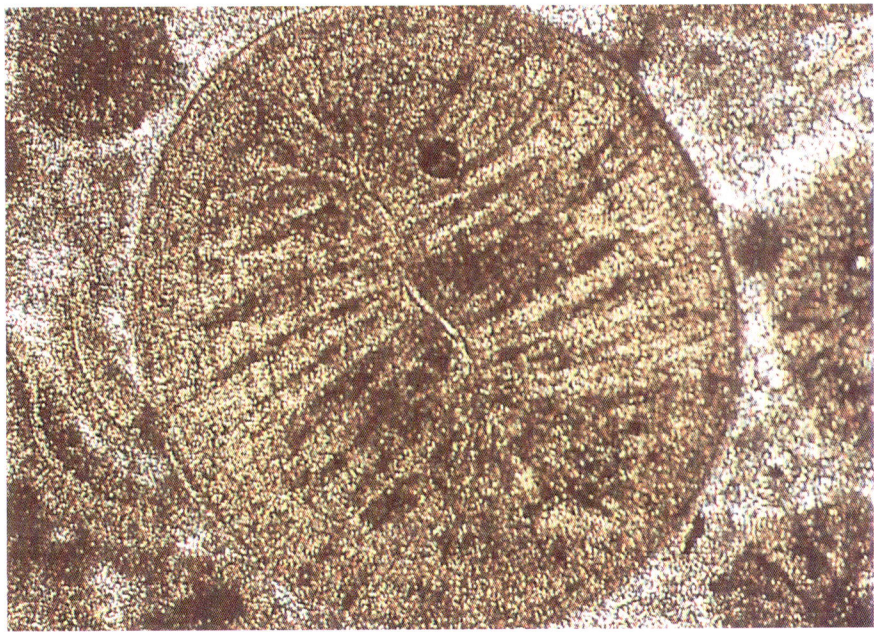

FIGURE 20.9 *The ooids of iron oxides found in the Clinton Iron Ore formed during Silurian time on the shallow ocean bottom in much the same fashion as calcareous ooids can be observed forming today off the Bahamas.* (Photo Researchers/© Ray Simons, 1974)

history about 430 million years ago, chemical weathering of the silicate minerals removed iron from the surrounding highlands as soluble bicarbonates. The dissolved bicarbonate was carried into a shallow portion of the sea occupying the Appalachian geosyncline where the iron oxidized and was deposited as a layer of hematite (Fe_2O_3). In some localities where the water was sufficiently shallow to allow the bottom sediments to be agitated by wave action, the iron oxide precipitated as **ooids** similar to the calcium carbonate ooids now forming in parts of the shallow water off the Bahamas (Figure 20.9). Subsequently buried and lithified, the layer of hematite was incorporated into the sedimentary rock record and exists today as the *Clinton Iron Ore*. Extending from New York to Alabama, the Clinton Iron Ore supplied much of the ore for the iron industries during the early days of settlement of the eastern United States. The remains of small iron furnaces that used the Clinton Iron Ore as feedstock can still be found throughout the Appalachians (Figure 20.10).

In more recent times, the Clinton Iron Ore was the basis for the steel industry centered at Birmingham, Alabama. The immediate proximity of all three of the resources essential for steel making—iron ore, limestone to remove impurities from the molten iron ore, and coal both to produce the coke needed to reduce the iron oxide to metallic iron

FIGURE 20.10 *Throughout the 1800s, small furnaces were used to produce the iron needed by the early inhabitants of Appalachia. This particular furnace, last worked in 1881, produced about 8,500 tons of iron per year. (Courtesy of Robert Behling)*

FIGURE 20.11 *Although most gold is not extracted from its ore by panning, the procedure is still used to remove gold from stream sediments. (Courtesy of Montana Historical Society)*

and to generate energy—resulted in Birmingham becoming known as the "Pittsburgh of the South" in the days when Pittsburgh, Pennsylvania, was a major producer of steel.

Alluvial and Eolian Concentrations

Minerals such as gold, platinum, cassiterite (SnO_2), zircons, and diamonds, which are characterized by either exceptionally high densities or resistance to chemical weathering, commonly concentrate in alluvial and eolian accumulations called **placer deposits**.

Placer deposits are examples of natural sorting and concentration. Within the original rock, placer minerals are usually present at trace to subtrace concentrations. The combined effects of weathering over millions of years, coupled with the downhill movement of mass wasting and the sorting power of streams and the wind, result in higher concentrations of the mineral.

Although some placer accumulations are found in beach sands, sand dunes, and soil, most are associated with stream sediments. Once transported to the streambed, these dense, chemically resistant minerals become components of the bed load and suspended load, accumulate in cracks and crevices within the bedrock, or are transported during episodes of high flow to the mouth of the stream where they may become incorporated in deltaic deposits. If accumulated at the shore, these minerals may be redistributed throughout beach sands.

The mineral most often associated with alluvial placer deposits is *gold*. In the United States, the best-known placer gold deposits are those romanticized in portrayals of the gold panners of the California Gold Rush of 1849 (Figure 20.11). First discovered at Sutter's Mill, near Coloma, California, in 1848, the major finds of placer gold were in the streams draining westward from the Sierra Nevada. Evidence of these early workings can still be seen throughout the area.

Soon after the initial discoveries of gold in California, an environmentally destructive mining process called **hydraulic mining** replaced panning as

FIGURE 20.12 Hydraulic mining, *such as seen here at Bonanza Creek, Yukon, Canada, uses high-pressure water to excavate unconsolidated sediments that contain placer deposits.* (VU/© Steve McCutcheon)

the major technique for extracting placer gold. Hydraulic mining utilizes high-pressure water to excavate large volumes of unconsolidated deposits in search of the gold (Figure 20.12). The resultant slurries are then moved across shaking tables to separate the gold. Hydraulic mining was outlawed by the California legislature in 1884 as being environmentally destructive. Elsewhere, however, it is still widely used. The most recent publicized use of hydraulic gold mining is in Brazil. There, after the gold is excavated and concentrated by settling, it is panned and subjected to a dangerous extraction technique using mercury amalgamation. Where sediment accumulation has resulted in more disseminated deposits such as beach sands, valley fills, and stream channels, placer deposits are commonly worked by the use of dredges (Figure 20.13).

The California gold seekers realized that the placer deposits must have come from a primary deposit upslope. Consequently, they began to search for the primary source of the mineral soon after the original placer deposits were discovered in the streams draining the western slopes of the Sierra Nevada. They soon found the *Mother Lode* in the Mariposa slate exposed in the foothills of the Sierra Nevada (Figure 20.14). The name *Mother Lode* derives from the fact that ore deposits are sometimes called **lodes**. The rocks that make up the Mother Lode consisted originally of sedimentary rocks that were subsequently metamorphosed by the crystallization of the massive intrusions that make up the Sierra Nevada. Most of the gold is contained within quartz veins as pockets or as small veins of native gold with pyrite. Although the lure of placer gold still attracts a

FIGURE 20.13 Dredging *is commonly used to mine unconsolidated stream and beach deposits in preparation for the recovery of concentrations of placer minerals.* (Courtesy of E. E. Brabb/USGS)

few fortune seekers, all of the major gold production in California today is from the Mother Lode.

Another example of a search for the mother lode was the attempt to locate the source of the placer gold found along the Gold Coast of Africa (now Ghana). It was eventually found in northern Brazil! Use what you now know about plate tectonics to understand why.

Not all placers are associated with alluvial deposits. The tiny diamonds found in the sands of the Namib Desert of southwestern Africa are an example of an eolian placer deposit. (See the discussion of deserts in Chapter 11).

SPOT REVIEW

1. How do ores and protores differ? How can one be transformed into the other?
2. Why are the resources of any economic material always greater than the reserves?
3. How are metalliferous deposits created by magmatic segregation, dissemination, supergene enrichment, and density separation?
4. How do the modes of formation of the deposits of the Mesabi Range and the Clinton Iron Ore differ?
5. What properties do all placer deposits have in common?

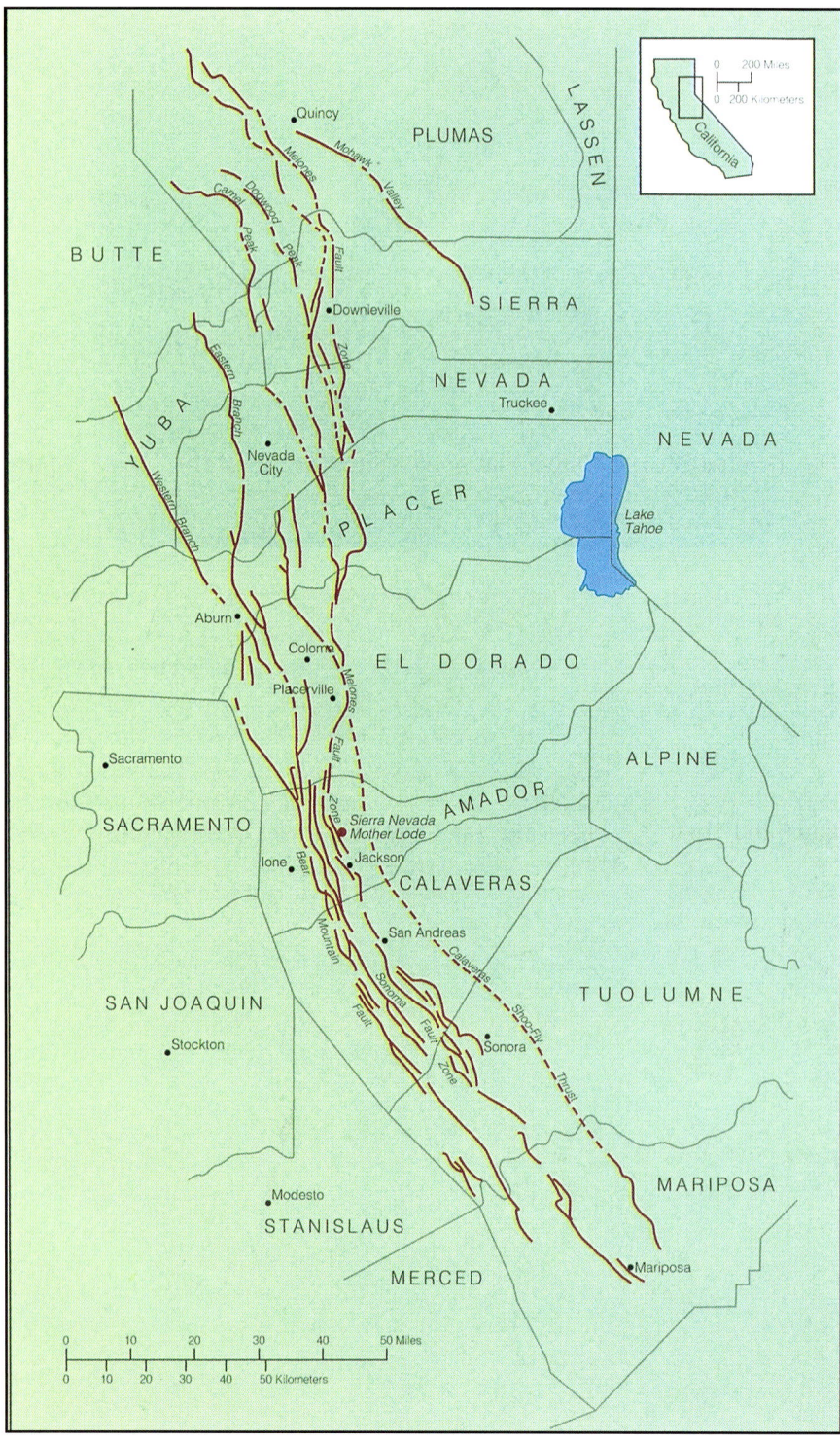

FIGURE 20.14 *All of the present gold production in California comes from the Mother Lode. Located along the foothills of the Sierra Nevada, the Mother Lode follows the trend of the Melones fault within the Mariposa Slate. The gold appears as pockets in quartz veins or as veins of gold and pyrite.*

NONMETALLIC NONFUELS

Although nonmetallic nonfuel deposits lack the romantic appeal and colorful history of metals exploration and mining, they are the most abundant economic deposits. Nonmetallic nonfuels include crushed and cut stone, sand and gravel, and a variety of evaporite materials such as salt, borax, and gypsum. The dependence of modern society upon nonmetallic nonfuels is obvious if only for one material, concrete. Concrete incorporates sand and gravel held together with Portland cement, a product made by roasting a mixture of limestone and shale.

Other examples of the utilization of nonmetallic nonfuels are the use of phosphates, nitrates, and lime as agricultural soil additives, the use of clay materials to produce bricks and a wide variety of ceramics from drain tile to the finest china, and the use of quartz in the manufacture of glass.

FIGURE 20.15 *Improvements made by James Watt on Thomas Newcomen's original steam engine ushered in the Industrial Revolution and forever changed how work is accomplished. (Reproduced by courtesy of the Trustees of the Science Museum, London)*

FUELS

An essential need of modern society is energy. Throughout most of human history, the main source of energy has been human beings themselves. In many developing countries, the human body is still the major source of energy. At some time and place lost in antiquity, humans began to domesticate animals both for food and to provide the energy needed to help perform tasks such as moving goods, plowing fields, and lifting water. Although other sources of energy such as the wind and running water were utilized, the main sources of energy until the eighteenth or nineteenth century remained humans and their domesticated animals.

The entire energy picture was revolutionized by the invention of the steam engine by *Thomas Newcomen* in 1705. At the end of the eighteenth century, the Scottish inventor *James Watt* improved the design of the steam engine to the point where it could drive industrial machinery (Figure 20.15). The ability to provide enormous amounts of power that could be harnessed to perform a number of tasks via belts, gears, and chains revolutionized civilization.

The introduction of the steam engine brought with it a need for a source of heat to generate the steam. Although this energy was initially supplied by burning wood, it soon became evident that wood could not provide the amount of heat necessary to supply the required volume of steam. Coal then became the major source of energy and not only fueled the *Industrial Revolution*, but also fired the boilers that provided the steam for the engines of the first intercontinental locomotives and transoceanic liners. The rise of coal to the status of preferred fuel came none too soon. At the rate trees were being felled, the forests of Europe would have been depleted in a short time.

The supremacy of coal as the major source of energy lasted until just after the beginning of the twentieth century. The first commercial oil well was drilled and completed in 1858 in Oil Springs, Ontario, Canada, by James M. Williams. The following year—the same year Charles Darwin published *The Origin of Species*—E. L. Drake completed the first well drilled in the United States solely for the production of petroleum near *Titusville, Pennsylvania* (Figure 20.16). Although it cannot be verified, Drake reportedly learned how to drill for oil by observing the Williams operation.

Easier to transport and cleaner burning than coal and leaving no ash to be disposed of, petroleum took over as the preferred fuel within less than half a century. By about 1910, oil was the leading fuel in the United States as coal fell from providing 90% of the energy budget to second place. Today, oil and gas provide about 70% of the U.S. energy budget with coal providing 20%.

On December 2, 1944, the age of nuclear power was born with the achievement of the first sustained chain reaction in a laboratory beneath the stadium at

FIGURE 20.16 *The first well in the United States drilled solely for the production of oil was completed in 1859 by* E. L. Drake *at* Titusville, Pennsylvania. *Drake may have seen the process applied a year earlier to bring in the world's first oil well at* Oil Springs, Ontario, Canada. *(VU© Science VU-API)*

the *University of Chicago*. In 1947, utilizing a nuclear reactor salvaged from a decommissioned naval ship, the world's first nuclear power plant was brought on-line at *Shippensport, Pennsylvania.*

Although it promised to produce power so cheaply that it would be virtually free, the nuclear power industry has never lived up to expectations. Enormous building costs, cost overruns, problems with complex safety systems, and waste disposal problems, as well as the reluctance of the public to accept a source of power they did not understand and in many cases feared, have limited nuclear power in the United States to about 5% of the total energy demand. Other countries, such as France, however, which are less fortunate than we in reserves of petroleum and coal, are becoming increasingly dependent on nuclear power.

Coal

Coal is a fossil fuel because it is made from preserved plant tissues, in particular, wood. The life expectancy of most plant debris is short. Upon falling to the ground, plant remains are quickly attacked by a combination of microbial and inorganic processes that convert them to humus and soluble compounds that return to the soil and provide the food for existing plants. If plant debris is to be preserved in sufficient quantities to produce coal, it must accumulate in a special environment that will deter decomposition. The environment that serves that purpose, the **swamp,** is an acidic, oxygen-deficient wetland dominated by woody plants. (Figure 20.17).

FIGURE 20.17 *The* swamp *is a* paludal environment *where the debris from wood-rich plants may accumulate and be partially preserved as peat. (Courtesy of J. F. Elder/USGS)*

FIGURE 20.18 *Although limited in its size, the* Okefenokee Swamp *on the Georgia-Florida border is a good chemical model for a coal-forming swamp. Because the swamp is isolated from the sea and, except for a small portion along its southern edge, completely underlain by noncalcareous sediments, the pH of the swamp water is less than 3. With the water pH below 3, microbial degradation of the plant debris is so depressed that appreciable amounts of peat can accumulate. (VU/© John D. Cunningham)*

Note the contrast with a **marsh**, which is a wetland dominated by nonwoody plants. In the United States, the best models for coal-forming swamps are the Okefenokee Swamp on the Georgia-Florida border and the Great Dismal Swamp of Virginia (Figure 20.18). The coastal swamps of the Atlantic seaboard and the Gulf of Mexico, although frequently cited as examples of coal-forming swamps, will not accumulate sufficient organic material to allow the formation of coal for reasons that will be discussed shortly.

Microbes—fungi and bacteria—are mainly responsible for the decomposition of plant materials. Microbes are most active at or near neutral pH (pH about 7). As the pH of the depositional environment drops, microbial activity decreases, becoming essentially nonexistent below about pH3. The swamp environment retards microbial activity by maintaining a low pH in the swamp water. Naturally acidified rainwater combined with organic acids produced by the decomposing plant debris will maintain pH values as low as 2 if sources of alkaline water are not present.

There are three natural sources of alkaline water: (1) the weathering of exposures of carbonate rocks, (2) alkaline groundwater, and (3) the introduction of seawater. Swamps such as those of the south Florida Everglades will not preserve sufficient plant debris to be a precursor to coal because the dissolution of carbonate bedrock maintains the swamp water at a neutral pH, which enhances microbial activity and therefore significantly limits the amount of preserved plant tissue.

Most of the coastal swamps along the southern margin of the Mississippi delta or along the Atlantic and Gulf coastlines that are commonly cited as coal-forming swamps are partially affected by the tides. Because the average pH of seawater is 8.4, the pH values of the water throughout most of the swamps are too high to preserve the volumes of plant debris needed to produce coal.

In the continental United States, only the Okefenokee and Great Dismal swamps have the correct swamp water chemistry to adequately preserve organic debris and therefore are the only swamps that can be considered models of coal-forming swamps. Because these swamps are developed on noncarbonate siliceous sediments and are isolated from the ocean, the pH values of their waters are maintained low enough to inhibit microbial degradation of the plant debris. Even under ideal conditions of preservation, however, more than half of the organic material represented by the plant debris is decomposed and either returns to the ground as food for the existing plant population or passes into the atmosphere as carbon dioxide (CO_2).

Once accumulated, the plant debris goes through a partial decomposition-conversion process that changes the material into **peat**. Coal-forming peat differs from most peat used in gardens or potted plants in that it contains more wood products. The peat purchased for gardening is usually derived from sphagnum moss and other nonwoody plants.

Coal Rank Eventually, the peat is buried either by floodplain deposits as the sediments subside and streams meander across the swamp or by marine deposits as sea level rises and the sea transgresses onto the land. As the thickness of the overlying sediments increases, the peat is subjected to compaction and slightly elevated temperatures and undergoes a series of processes that eventually convert it into other forms of coal (some coal geologists consider peat to be a form of coal). During the compaction and converstion processes, the thickness of the peat is reduced by about 90%. In other words, 1 foot of coal requires the accumulation of about 10 feet of peat.

Heated to no more than 300°F (150°C), the compacted peat slowly cooks and undergoes a process called **coalification**, the chemical details of which are not precisely understood. The major change that takes place is the progressive transformation of the fleshy materials that make up the plant parts to the black, solid material that constitutes most of the mass of coal. Evidence for the plant origin of coal includes the fossilized remains of the more resistant plant parts such as cell walls, spores, and, in more modern coals, pollen that are revealed by microscopic examination of coal (Figure 20.19).

What takes place following coalification can best be illustrated by considering the two fundamental components of any organic material: (1) *volatiles* and (2) *fixed carbon*. **Volatiles** are the portion of an organic material that can be driven off by heating; **fixed carbon** is the carbonaceous residue that remains behind. As the newly formed materials respond to the temperatures of burial, the volatiles are driven off and the carbon content increases. Peat, with an average carbon content of about 55%, changes into a material with an average carbon content of about 65% called **brown coal** or **lignite**. If the baking continues and the average carbon content rises to about 75%, lignite is converted into **subbituminous coal**. With an increase in average carbon content to about 85%, subbituminous coal changes to **bituminous coal**.

The removal of the last 10% of the least volatile materials usually requires more heat than can be provided by burial alone. However, if the coal is part of the sedimentary sequence contained within a passive continental margin, back-arc basin, or marginal sea that becomes involved in a mountain-building episode, the added energy of metamorphism may provide the heat necessary to increase the carbon content to 95% or more and convert the bituminous coal into **anthracite**. If the conditions of metamorphism are sufficiently severe, all of the volatiles may be driven off to form **graphite**, which is pure carbon. (Diamonds will not form by this process. Diamonds are formed within the mantle and are brought to the surface as component of an ultramafic rock called kimberlite.) This series from peat through anthracite coal is called the **coal rank series** with peat being the lowest rank coal and anthracite the highest (Figure 20.20).

FIGURE 20.19 *Microscopic examination of a polished coal block reveals the fragmentary remains of various kinds of plant tissue. The bright objects are pyrite crystals that have grown within a folded spore. The field of view is approximately 100 microns (1/10 of a millimeter).*

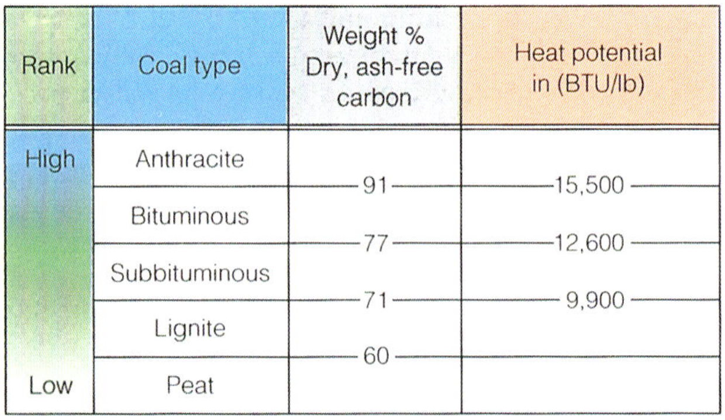

Rank	Coal type	Weight % Dry, ash-free carbon	Heat potential in (BTU/lb)
High	Anthracite		
		91	15,500
	Bituminous		
		77	12,600
	Subbituminous		
		71	9,900
	Lignite		
		60	
Low	Peat		

FIGURE 20.20 *The various members of the* coal rank series *form as metamorphism of the organic portion of peat following burial progressively changes the relative concentration of carbon and volatiles.*

Heat Content of Coal The amount of energy released from any mass of coal is primarily a function of the carbon content. The energy present in all fossil fuels is solar energy originally absorbed and stored in the carbon atoms of living plants. As the plant tissues are converted, the stored solar energy is transferred to the coal, to be released when the coal is burned.

The heat potential of coal is measured in terms of **BTUs per pound**. (The BTU is a *B*ritish *T*hermal *U*nit and is equal to 251.98 calories.) In general, heat content rises with increasing rank (refer to Figure 20.20). The slightly lower BTU value of anthracite coal as compared to bituminous coal is due to a basic change in the composition of the organic portion of the coal as bituminous coal undergoes the final stages of metamorphism.

Coal Quality Coal quality is determined by (1) the amount of **ash** (mineral matter) and (2) the **sulfur** content.

All plants contain mineral matter. The ash remaining after burning a log in a fireplace or campfire is the mineral matter originally contained in the wood. As woody tissue is transformed into coal, some of the original mineral matter is incorporated into the coal and, when burned, remains behind as ash. Increased mineral matter content reduces the energy potential of coal in two says: (1) the mineral matter will not burn and contribute to the production of heat, and (2) it accumulates as ash, which must be removed from the boiler and disposed of.

In unweathered coal, the sulfur is found in two forms: (1) **organic sulfur** (the sulfur that is part of the organic components) and (2) **pyritic sulfur** (the sulfur in the mineral pyrite, FeS_2). As a rule of thumb, organic sulfur is the dominant form of sulfur in coals with less than 1 wt% total sulfur content; when the total sulfur content of coal exceeds 1 wt%, most of the increase is due to pyrite.

In general, the sulfur and ash contents increase and decrease together. Based on ash and sulfur content, coals are arbitrarily classified as *high, medium,* or *low quality*. High-quality coals are those with less than 10 wt% ash and 1 wt% sulfur. The division between medium-quality and low-quality coal is less definite and depends more upon the intended mode of utilization. In general, however, coals with more than 20 to 25 wt% ash and 3 to 4 wt% sulfur would be considered too poor in quality for most uses today.

Coal Utilization The way coal is used is largely determined by its quality. Historically, coal has been used primarily for two purposes: (1) the manufacture of *coke* and (2) the production of *steam*. Coke is carbon produced by roasting coal in an oxygen-starved oven. The primary use of coke is in the steel industry where it is used to reduce iron ore (Fe_2O_3) to free iron. The quality requirements for coking coal are very stringent, less than 6 wt% ash and 0.5 wt% sulfur. Although there is no rank requirement, coke makers usually prefer bituminous coal.

In most countries, and in the United States before 1967, any coal that did not fit the requirements for coke was considered steam coal. In 1967, the U.S. Congress passed the Environmental Protection Act (EPA), which limits the amount of sulfur that may be introduced into the atmosphere from coal-burning power plants to 1.2 pounds of sulfur in the form of sulfur dioxide (SO_2) per million BTUs generated. In terms of the sulfur content of the coal, this amounts to about 1.2 wt% sulfur. Unfortunately, much of the coal in the United States has a sulfur content in excess of 1.2 wt%. In order for coal with a sulfur content in excess of 1.2 wt% to meet the EPA standards and be considered a **compliance coal**, either the coal or the flue gases produced during combustion must be *treated*.

Four basic treatments can be used to bring a coal into compliance: (1) *coal cleaning,* (2) *coal blending,* and the use of (3) *getters* or (4) *scrubbers*. Cleaning refers to the physical removal of pyrite using various density separation processes (Figure 20.21). Organic sulfur, however, cannot be removed by any preburning treatment.

In blending, noncompliance coal is mixed with high-quality compliance coal to produce a mixture that falls within compliance standards for sulfur. Most coal-fired power plants in the eastern United States use blended coals as feedstock.

Getters are chemicals that are added to the coal and react with the SO_2 as it is produced in the boiler and thereby prevent it from being vented into the atmosphere. In modern steam boilers, powdered limestone is mixed with the powdered coal before it is introduced into the firebox. As the coal burns, the limestone decomposes producing calcium oxide (CaO), which in turn reacts with (gets) the SO_2 released from the combustion of the sulfur-bearing materials to produce calcium sulfate ($CaSO_4$), which is removed with the ash.

FIGURE 20.21 *The physical removal of pyrite-rich coal and coal-associated rock at* coal preparation plants *is a means of producing* compliance coal.

Scrubbers are mechanical devices placed at the exhaust of the firebox to wash or "scrub" the emerging gases with water. The water reacts with the SO_2 to produce hydrogen sulfite (H_2SO_3), which is drained from the scrubber and disposed of.

By using combinations of these treatments, many medium-quality coals can be brought into compliance with EPA requirements. The upper limit of sulfur content of coal that can be successfully treated and brought into compliance is about 3 to 4 wt%.

The greatest potential for poor-quality coals that cannot be brought into compliance by the various treatments is as feedstock for coal **liquefaction**, which converts the coal into petroleum-like liquids or **synfuel**. For reasons that have not yet been fully explained, the coals with the highest ash and sulfur contents convert most efficiently into synfuel. Once the coal is liquefied, the ash can be physically removed by filtration while the sulfur is removed chemically. Coal liquefaction was developed in Germany in the late 1930s because of the lack of petroleum resources.

Geographic Distribution of U.S. Coal The coals in the United States are divided into **western** and **eastern** coal-producing areas. The western coals occur in many relatively small basins scattered throughout the Rocky Mountains and western Great Plains while the eastern coals occur in four relatively large basins (Figure 20.22). The western coals are mostly low in rank (subbituminous and lignite) but are all high quality (less than 10% ash and 1% sulfur). As a result, *all* western coals meet EPA compliance standards without treatment. The major use of western coals is to produce steam for the generation of electricity. Because of the distances involved, the western power-generating industry builds *mine-mouth* power plants where the coal is taken from the mine, pulverized, and fed directly into the boiler (Figure 20.23). Rather than selling and shipping coal, the western utilities sell power. In contrast, eastern power-generating plants are usually supplied with coal shipped from several distant sources by rail or a combination of rail, barge, and lake freighter. In addition to being consumed at mine-mouth power plants, western coal is also shipped east in large amounts to serve as a compliance coal or for blending.

With the exception of a small area of anthracite in eastern Pennsylvania, the eastern coals are all high-rank bituminous, which makes them better heat producers than the western coals. Unfortunately, they suffer from the presence of sulfur. The quality of the eastern coals varies from poor in the Midcontinent Basin to high in the southern Appalachian Basin. The major coal producer in the east is the *Appalachian Basin,* which is divided by coal quality into the

Economic Geology and Energy 583

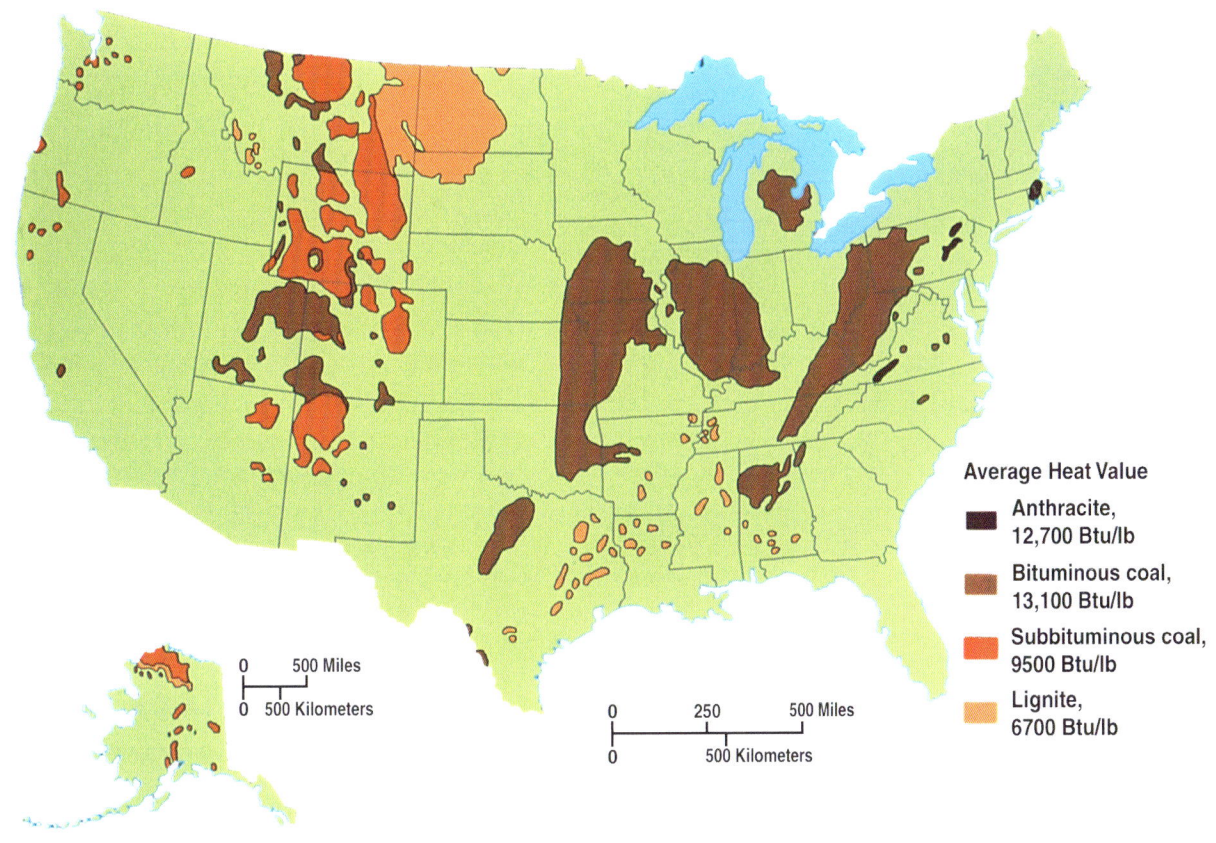

FIGURE 20.22 *The coals in the continental United States are distributed among the eastern and western coal fields. In general, the eastern coal fields contain low- to high-quality, high-rank, bituminous coal while the western fields contain high-quality coal, mostly lignite and subbituminous in rank.*

Average Heat Value
- Anthracite, 12,700 Btu/lb
- Bituminous coal, 13,100 Btu/lb
- Subbituminous coal, 9500 Btu/lb
- Lignite, 6700 Btu/lb

FIGURE 20.23 *To minimize the overall costs of power production, most western coals are burned in nearby power plants, such as this plant in Rosebud County, Montana, and the power is shipped by way of the national electrical grid. (Courtesy of E. N. Hinrichs/USGS)*

FIGURE 20.24 *The coal within the* Appalachian Basin *is clearly divided in quality between the high-quality coals of the southern* Pocahontas Basin *and the lower-quality coals of the northern* Dunkard Basin; *the rank throughout both basins is bituminous. Historically, the Pocahontas Basin has been the source of high-quality coking coals while the coals of the Dunkard Basin have largely been used to produce steam in coal-fired power plants.*

northern *Dunkard Basin*, characterized by medium-quality coals, and the southern *Pocahontas Basin* with its high-quality coals (Figure 20.24). The Dunkard Basin (northern West Virginia, western Pennsylvania, and eastern Ohio) produces primarily steam coals while the coals in the Pocahontas Basin (southern West Virginia and eastern Kentucky) are premier coking and compliance steam coals. Except for the coals of the Pocahontas Basin, all eastern coals must be treated before or during combustion to bring them into compliance.

In general, most Midcontinent Basin coals are too low in quality to have any current market value. They should make excellent feedstock for coal liquefaction, however. Except for a few individual coal beds, most of the coals in the *Illinois Basin* are noncompliance, even with treatment. The *Michigan Basin* is at present not a commercial producer of coal.

SPOT REVIEW

1. What conditions are required within a swamp to preserve sufficient wood to produce a precoal peat?
2. What coal compositional parameters change with rank?
3. What determines coal quality?
4. What processes can be employed to bring noncompliance coal into compliance with EPA requirements?
5. What differences exist between the coal deposits of the eastern and western United States?

Petroleum

Whereas coal is made from the preserved remains of *land* plants, **petroleum** forms from the preserved remains of *marine* microorganisms, mostly plants. As with coal, the energy extracted by burning oil and gas is solar energy originally sent up the food web. The debris of microorganisms accumulates and is preserved on the ocean bottom where special conditions restrict water movement and exclude oxygen and potential microbial scavengers in much the same way as the low pH of swamp waters promotes the preservation of land plant debris or where organic residues have survived bacterial degradation. With time, the preserved plant remains are buried and undergo a conversion process similar to coalification. As with coalification, the actual chemical reactions that transform marine organisms into oil and gas are not fully understood.

The difference between the conversion products of land plants (coal) and marine plants (oil and gas) is primarily the result of basic differences in plant composition. The main constituent of coal forms by the conversion of the high-molecular-weight cellulose and lignin used by land plants to strengthen their stems so that they may grow upward against the pull of gravity. Marine plants, being supported by water and needing no such strengthening, do not contain cellulose and lignin, but rather are dominated by lower-molecular-weight materials such as those that constitute the fleshy portion of leaves. These lower-molecular-weight materials give rise to relatively low-molecular-weight liquid and gaseous products such as oil and gas rather than solids.

Reservoirs Once produced, the buoyancy of the relatively volatile and highly fluid oil and gas causes them to migrate into rocks called **reservoirs** that have sufficient porosity and permeability to contain and transmit them. The discussion of porosity and permeability in relation to groundwater in Chapter 14 is directly applicable here as well. Like aquifers, most petroleum reservoirs are sedimentary rocks; and of the sedimentary rocks, sandstones make the best reservoirs. Shales are usually poor reservoirs. Although some shales may contain petroleum because of their relatively high porosity, the low permeabilities characteristic of shales do not allow the transmission of the oil or gas. Fine-grained limestones with low permeabilities are very poor reservoirs, but limestones such as reef limestones that possess cavernous openings can contain enormous amounts of oil and gas.

Cap Rock In order for the petroleum to remain in the reservoir, the rock must be overlain by a low-permeability rock layer, appropriately called a **cap rock**. It should come as no surprise to find that the most common cap rock is shale. Note that the combination of the reservoir and the cap rock is the situation previously described in Chapter 14 for a confined aquifer. Like water in a confined aquifer, petroleum contained within a reservoir is under pressure.

Traps Because of the pressures that exist within the reservoir, the oil is intimately suspended in the water, and the gas is in solution. A geologic structure called a **trap** restricts the movement of oil and gas within the reservoir and produces economic concentrations of oil or gas within smaller reservoir volumes. The first geologic structure suggested as a trap was the anticline (Figure 20.25). During the early years of the oil industry when new oil wells were located without the benefit of science, *I. C. White*, state geologist and founder of the West Virginia Geological and Economic Survey published a paper in the June 26, 1885, issue of *Science Magazine* that included a discussion of what has come to be called the **anticlinal theory**. According to White's theory, if gas and oil are present in a sequence of sedimentary rocks, the most likely place for accumulation is in the axial regions of anticlines. This theory represented the first application of geology in the oil and gas industry and demonstrated that science could be successful in the search for oil and gas.

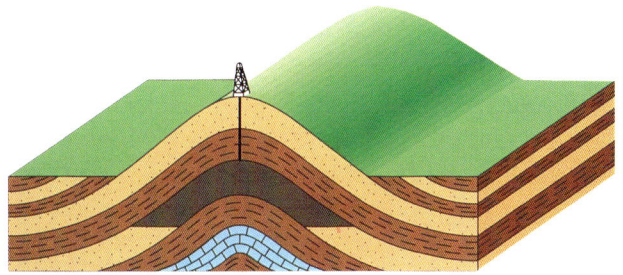

FIGURE 20.25 *In 1885*, I. C. White, *suggested that if oil or gas were present in rocks, they would accumulate in the crestal region of anticlines. His theory ushered in the modern era of petroleum exploration.*

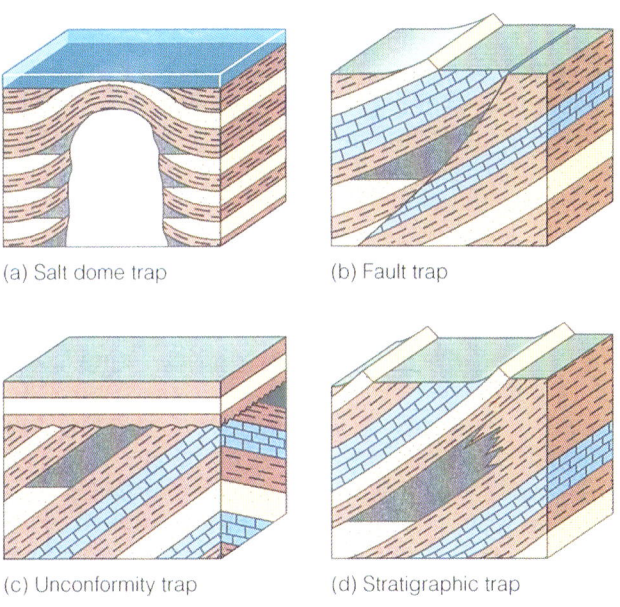

FIGURE 20.26 *A number of different geologic situations may result in the entrapment of oil or gas. The salt dome is the major type of trap in the Gulf coastal areas of the United States while the most difficult, the most expensive, type of trap to find is the stratigraphic trap.*

Over the years, the anticlinal trap has produced more gas and oil than any other. Unfortunately, since most anticlinal accumulations have now been found, the remaining petroleum is contained within other types of traps that are generally more difficult and, therefore, more expensive to find. Figure 20.26 shows examples of some other kinds of traps, including the salt dome trap, which produces most of the oil and gas from the Gulf coast of the United States.

Oil Shales

Another hydrocarbon source, and one that is relatively untapped, is **oil shale**. The name is something of a misnomer in that many of the rocks involved are not shales but shaley limestones, and the organic materials in them are not oil but rather solid kerogen. The richest oil shale deposits in the United States are in the *Green River Formation* located in the tristate area of Utah, Colorado, and Wyoming (Figure 20.27). In the eastern United States, organic-rich rocks are found in shales and siltstones of lower Devonian age but with lower kerogen concentrations than those of the Green River Formation, which, in some cases, are so high that the rock actually smells oily.

During the OPEC oil embargo of 1973 when the $30 per barrel price of oil was comparable to the cost of producing the oil from the shales, preparations were made to exploit the Green River Shale as a source of oil. When the price of oil dropped to less than $20 per barrel, however, the development of the resource was put on hold.

The exploitation of oil shales is not without potentially severe environmental problems. The oil is extracted by roasting crushed rock to volatize the organic material. The gases are then condensed to produce oil, a process requiring large volumes of water in an area that is already water-poor. In addition, the volume of the rock increases during roasting, creating both disposal and reclamation problems. In spite of these environmental problems, oil shales are a resource that will undoubtedly be tapped in the future.

THE FUTURE OF FOSSIL FUELS

The future of fossil fuels is clouded by the same limitation that faces all nonrenewable resources, that is, depletion. A typical production-time curve for any nonrenewable resource is shown in Figure 20.28. The initial fast rise in the curve represents growing demand and production from deposits that are large and easy to find. As the remaining deposits become smaller, poorer in quality, and more difficult to find, production peaks and then begins to decline, even though demand may still remain high. Eventually, the resource will be exhausted. Important dates for any nonrenewable resource are the years in which production will peak and in which production will fall to the point where the commodity is economically depleted, usually considered to be about 80% depletion.

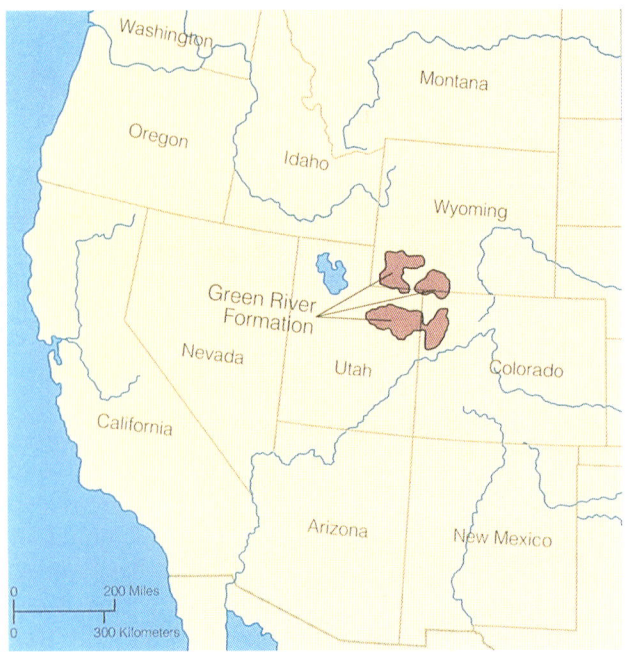

FIGURE 20.27 *The rocks of the* Green River Formation *are found in four major basins in contiguous parts of Utah, Colorado, and Wyoming. Generally found in flat-lying beds, the petroleum content of some of the rock layers is so high that it can be smelled. Although not economically feasible at the present time, oil produced by roasting the rocks will become a significant portion of the U.S. domestic oil production when worldwide supplies begin to dwindle. (Photo courtesy of R. L. Elderkin/USGS)*

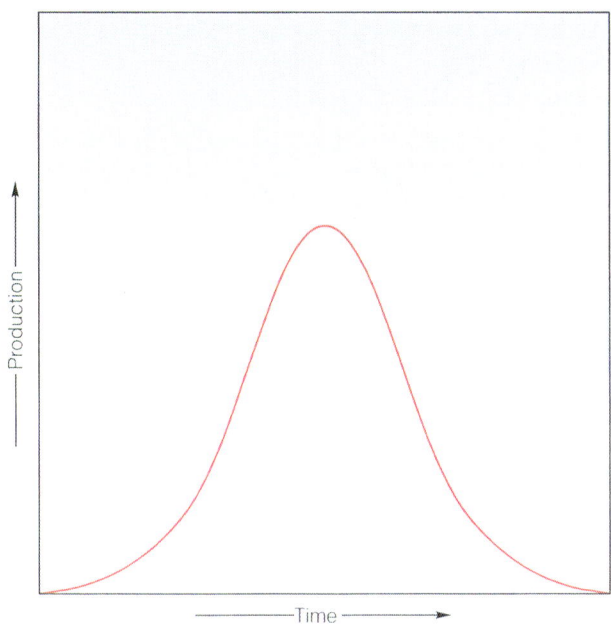

FIGURE 20.28 *The production-time plot for any nonrenewable resource is a* Gaussian curve. *For oil, the curve begins in 1858 with the well at Oil Creek, Canada. According to one prediction, world production will peak by the year 2000, and by 2100, all recoverable oil will have been produced. It is significant to note that should the prediction be accurate, in fewer than 250 years, we will have consumed all the oil that was produced over a period of half a billion years.*

Country		Percentage of world's proven reserves of oil
Middle East	Saudi Arabia	25.5
	Iraq	9.9
	Kuwait	9.4
	Iran	9.2
	Abu Dhabi	9.1
	Others	2.5
	Venezuela	5.8
	Former USSR	5.6
	Mexico	5.2
	United States	3.4
	China	2.4
	Libya	2.3
	Nigeria	1.7
	Indonesia	1.1
	Algeria	0.9
	Canada	0.8
	India	0.8
	Norway	0.8
	Egypt	0.4
	United Kingdom	0.4
	Others	2.8

FIGURE 20.29 *The Middle East holds by far the world's largest proven reserves of oil, with more than 60%. By comparison, the United States and Canada together have approximately 4% of the world's oil reserves, which explains why a large amount of oil must be imported every year for domestic consumption.*

For the United States, the future for coal looks good. The U.S. reserves of coal are estimated at about 200 billion tons, about one-third of the world's reserves. The U.S. resources of coal are probably in the order of 2 trillion tons. With only a few percent of the total amount of available coal having been produced to date, it is estimated that the U.S. production of coal will not peak for more than a century, and projections indicate that coal will be a major energy source for several centuries.

When it comes to petroleum resources, however, the United States is not as fortunate. The distribution of the world's remaining reserves of petroleum is shown in Figure 20.29. The United States, which has only 3% of the world's proven reserves of oil and yet consumes about 30% of the entire world production, must import about 50% of domestic consumption. It is important to note that the imported oil is primarily used to satisfy our demand for cheap transportation fuels. Scientists estimate that if the average fuel consumption of U.S. automobiles were increased to 35 miles per gallon, there would be no need to import oil.

The real question is how long will our petroleum resources last? In the early 1950s, the late *M. King Hubbert,* a world authority on energy resources, predicted that U.S. domestic production of oil would peak in the late 1960s or early 1970s. As close as can be determined, U.S. production peaked about 1968. Unfortunately, demand continued to rise, and we became increasingly dependent upon imported oil. Although discoveries continue to be made, domestic production of oil is declining. Discoveries such as the huge Prudhoe Bay field on the north slope of Alaska will not reverse the trend. Even in the Prudhoe Bay field, production has passed its peak. The fact is that the United States will never be self-sufficient in oil again. Our demand is too large.

The ultimate question is, when will *world* oil production peak? In the same paper where Hubbert predicted the peak of U.S. Production, he predicted that world oil production would peak at about the turn of the century. At the time of his prediction, the turn of the century was still half a century away. Today, the new century will be here in less than a decade. According to Hubbert's estimates, by the year 2100 the world production of oil will have declined to the point where it will no longer be considered a major energy source. Although that event is beyond the life expectancy of anyone reading this text, it is within the expected lifetime of the generation to follow. What will replace oil as the world's major energy source? An obvious possibility is coal until it is exhausted. Despite all of the admitted environmental problems involved in both the mining and utilization of coal, it will be called on to fill the need. It is hoped that by then major advances in mining, reclamation, and clean coal combustion, combined with an increased reliance upon the conversion of coal to sulfur-free liquids, will have solved most of the problems involved in the utilization of coal. The other source of energy that is technologically ready to fill the void left by oil is uranium.

SPOT REVIEW

1. What are the source materials for petroleum?
2. What distinguishes a good petroleum reservoir from a bad reservoir?
3. What is the function of a petroleum trap?
4. According to M. King Hubbert, what is the outlook for future U.S. domestic production of petroleum? For world production?

NUCLEAR POWER

Nuclear power can be produced by two mechanisms: (1) **fission,** which splits large atomic nuclei such as uranium into smaller nuclei, with energy released in the form of heat, and (2) **fusion,** which combines the smallest of atomic nuclei, hydrogen, to form a larger atom, helium, with the liberation of energy (heat). Of the two, only fission is now available.

Fission Reactors

The **fission reactor** is employed in all present-day nuclear power plants. Most fission reactors use uranium 235 (^{235}U) as fuel. The reaction converts ^{235}U into lighter elements including a number of highly radioactive by-products, such as cesium 137 (^{137}Cs) and strontium 90 (^{90}Sr), and neutron radiation.

When mined, more than 99% of the uranium is nonfissioning ^{238}U with less than 1% being ^{235}U fuel. After processing the ore to increase the ^{235}U content to about 3%, the fuel is fabricated into *fuel pellets* about as large as the end of your little finger and assembled into *fuel rods* about 15 feet (5 m) long. With a half-life of 713 million years, ^{235}U undergoes fission at a *very* slow, which means that uranium is not very radioactive; you could hold ^{235}U fuel pellets in your hand with no danger from radiation. If the mass of uranium reaches the *critical mass,* however, the intensity of the neutron flux being generated is strong enough to initiate nuclear fission and produce a *chain reaction.* The critical mass is achieved by stacking about 100 tons of fuel rods together in the *core* of a reactor. The chain reaction is controlled by absorbing the neutrons with graphite rods called *control rods,* which are inserted between the fuel rods.

Reactor Design The design of a fission reactor facility is illustrated in Figure 20.30. The core of fuel rods and graphite control rods is placed within a thick-walled stainless steel *reaction vessel*. The rate of the chain reaction and the amount of heat generated are controlled by the positioning of the control rods. The heat is removed from the core by the primary cooling system, which pumps water under high pressure through the reactor vessel to a *heat exchanger* located outside the core. Within the heat exchanger, heat is transferred from the *primary,* high-pressure cooling system to a *secondary,* low-pressure water system where the water is converted to steam. The steam is used to drive a turbine, which in turn drives a generator to produce electricity. Upon emerging from the turbines, the steam is condensed in the cooling towers and pumped back into the secondary cooling system.

Waste Disposal Every year, about a third of the fuel rods are replaced with rods containing fresh fuel. The disposal of the spent fuel and the highly radioactive decay products is a major problem facing the industry. Because most of the radioactive products of the ^{235}U fission have half-lives measured in tens of years, they are *highly* radioactive but relatively short-lived. Even with such short half-lives, however, about 10,000 years must pass before the fission products are no longer life threatening. Furthermore, one of the highly toxic products of the fission reaction,

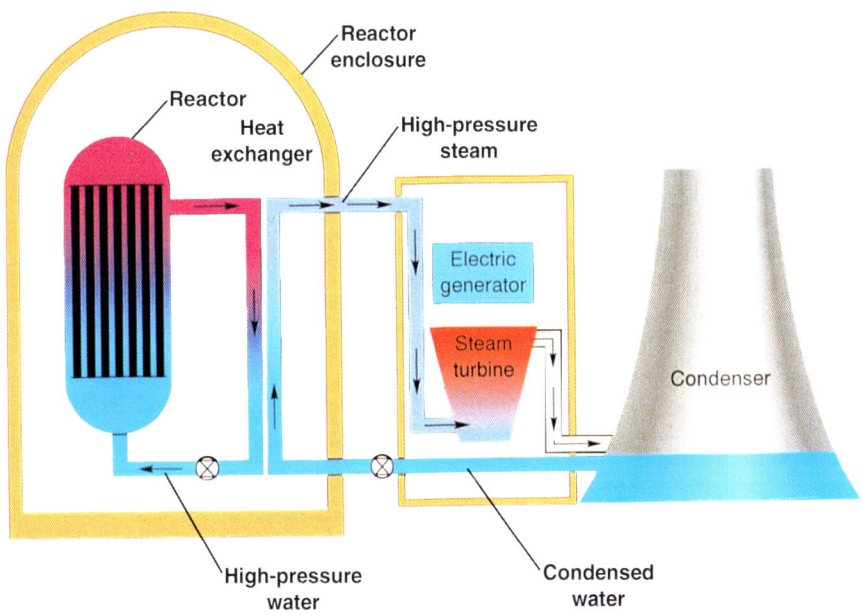

FIGURE 20.30 *In the modern fission reactor, heat generated by the radioactive breakdown of the uranium fuel is used to produce the steam that drives a turbine-powered generator. The absence of smoke stacks identifies a nuclear power station. The round-topped building houses the reactor. (Photo VU/© Albert J. Copley)*

plutonium 239 (^{239}Pu), has a half-life of 24,000 years. Depending on the ^{239}Pu content, the waste may need to be sequestered for up to 200,000 years!

Suggestions for disposal have included sending the materials into deep space or into the Sun, disposing of them within zones of subduction, and burying them in holes drilled deep into batholiths. To date, however, the waste has been contained in tanks that are either stored at the surface or buried at shallow depths. In the United States, low-level radioactive wastes, such as the residues from chemical processing, and slightly contaminated materials, such as hospital gowns and surgical gloves, are buried in near-surface sites where they can be monitored. High-level radioactive wastes such as spent reactor fuel are buried in underground containers, which present the potential problems of container deterioration and leakage of the radioactive materials into the groundwater supply. At the present time, high-level liquid and solid radioactive waste is stored at U.S. Department of Energy repositories at Savannah River, Georgia; Idaho Falls, Idaho; and Hanford, Washington.

Attempts to establish a federal underground disposal site by utilizing abandoned salt mines at Lyons, Kansas, failed because of public objections to the site. The reluctance of the residents of the proposed site to have high-level radioactive waste brought into their area was understandable. Salt seems to be an ideal burial medium in that the heat generated by the decomposition of the radioactive waste products will cause the salt to flow plastically and encapsulate the containers in an environment totally devoid of water. Opponents of salt encapsulation, however, argue that future changes in the movement of groundwater may subject the salt to dissolution.

The most recent site selected to be the federal repository for high-level radioactive waste is Yucca Mountain, Nevada, where testing to establish the site's suitability is still underway. Although the site is isolated from large population centers, some geologists feel that the tectonic setting and geologic history of the region render the site unacceptable for the disposal of high-level radioactive waste.

The other major problem facing the nuclear power industry is the public fear of a reactor accident such as those that occurred at Three Mile Island, Pennsylvania, and, more recently, at Chernobyl in Ukraine (Figure 20.31). Although the commonly expressed fear of a nuclear explosion is unwarranted, the loss of cooling and subsequent rupture of the core containment vessel, such as occurred at the Chernobyle site, demonstrate that a disaster can occur without a nuclear explosion. Although public rejection of nuclear power has brought the construction and planning of new nuclear power plants in the United States to a halt, other countries are becoming increasingly dependent upon nuclear power, especially in Europe where few other sources of energy exist.

Breeder Reactors One of the fundamental problems that faces the nuclear power industry if ^{235}U continues to be used as the fuel is that uranium is a very scarce element. The average concentration of uranium in Earth's crust is only 0.0002 wt%. The United States has as much uranium as any country, but no nation has very much. It is estimated that if we were to depend upon the typical water-cooled, ^{235}U-fueled fission reactor to supply 100% of the U.S. energy needs, the entire U.S. reserves of uranium would be exhausted in fewer than 10 years! If fission is to be a major energy source for a significant portion of the future, the industry must convert to **breeder reactors.**

The breeder reactor, developed by the U.S. nuclear research effort, is commonly said to produce more energy than it uses, a statement that needs some explanation. When the spent fuel rods are removed from a ^{235}U-fueled fission reactor core, the 97% of the original fuel represented by nonfissionable ^{238}U still remains. The design of the breeder reactor allows the surplus neutrons generated within the core to convert the nonfissionable ^{238}U to fissionable ^{239}Pu, which can then be used as a fuel. For every atom of ^{235}U that was originally placed into the core and used as fuel, 60 atoms of ^{239}Pu fuel can be made from the ^{238}U. In other words, more fuel will be made than was (originally)

FIGURE 20.31 *The explosion at Chernobyl released highly toxic materials into the atmosphere. Most of those involved in the initial attempts to contain the radiation have died of radiation poisoning. The final toll in terms of lives lost, lives shortened and debilitating effects of radiation overdose will not be known for decades. (Science VU/Visuals Unlimited)*

used, and the life of the available uranium is extended by a factor of at least 60. Rather than having the potential to meet 100% of the U.S. demand for fewer than 10 years, the available U.S. uranium reserves could provide 100% of all of our power demands for more than 500 years, assuming constant consumption rates.

The problem with dependence upon breeder reactors for the commercial production of electricity is that not only is ^{239}Pu highly toxic, but it is used to make nuclear weapons. Any nation possessing a breeder reactor supposedly for the production of electrical power could, if it so desired, become a manufacturer of atomic weapons. The Carter administration terminated the U.S. breeder research program because of concerns about the potential proliferation of nuclear weapons. To date, only two commercial breeder reactors are in operation, both in France.

Fusion

Nuclear **fusion** promises unlimited power from an inexhaustible supply of fuel. The concept is quite simple. It combines four hydrogen atoms to form a helium atom and heat. Rather than producing highly radioactive by-products, the fusion process generates only tritium (an isotope of hydrogen), heat, and harmless helium gas. Because hydrogen atoms are available in inexhaustible supplies in Earth's hydrosphere, exhausting the fuel supply is not a problem. It all sounds too good to be true. Unfortunately, as you will recognize from our discussion of the origin of the Universe in Chapter 1, the reaction converting hydrogen to helium is the process that fuels the stars. Although the United States, Russia, and a European consortium have active fusion research programs, creating and containing a star with an estimated start-up temperature of 10^8 kelvins poses certain fundamental problems that researchers have not yet been able to solve. How close any country is to initiating and maintaining a successful fusion reaction is unknown.

SPOT REVIEW

1. What is the difference between nuclear fission and nuclear fusion?
2. What fuel is used in a fission reactor? In a fusion reactor?
3. Why is the disposal of spent fuel from fission reactors such an environmental problem?
4. What is a breeder reactor and how is it able to "produce more fuel than it uses"?

ALTERNATIVE SOURCES OF ENERGY

Faced with the predicted demise of oil as the major energy source within the twenty-first century and wanting neither the environmental problems commonly associated with coal nor the potential disasters of nuclear power, we must look for other sources of energy. What are these alternatives?

Geothermal Power

Geothermal power utilizes the natural heat flow from Earth, usually in the form of steam. The operational requirements for the standard steam turbine, steam at a temperature of 930°F (500°C) and at a pressure of 2,400 to 3,500 lb/in^2, are met by some fumaroles. The best-known locality utilizing natural steam is Iceland where the steam is used to produce electrical power and to provide hot water to heat buildings. The only commercial natural steam power plant in the United States is located at Geysers, California, just north of San Francisco (Figure 20.32). Rated at about 850 megawatts, the power-generating capacity of the Geysers facility is smaller than the average (1,000 megawatt) coal-fired power plant.

Utilizing natural steam presents several problems. First, sufficient steam is available only at limited localities. Secondly, the life expectancy of any natural steam field is limited. A proposed solution to both problems is the *Dry Hot Rocks Geothermal Project (DHRGP)*, a federally funded research effort of the national laboratory at Los Alamos, New Mexico. The DHRGP technology employs heat contained within the rocks surrounding magma chambers (Figure 20.33). Two wells are drilled into the hot rocks (not in the magma), and the rocks between the bottoms of the two holes are then shattered by injecting a sand-water slurry under high pressure. Water is then pumped down one well into the shattered rock where it is converted to steam and withdrawn through the second well. The steam is then used to drive a turbine-generator system. At the present time, the project has had a 10-megawatt power plant operating in northern New Mexico for about 10 years and just completed a 100-megawatt station.

One aim of the DHRGP was to locate areas where magmas were close enough to the surface for the heated rocks surrounding them to be accessible with a drill. Since modern technology enables us to drill to depths of nearly 30,000 feet (9,000 m), it is estimated that if all of the available sites were fully utilized, geothermal power using the DHRGP technique could

FIGURE 20.32 *The only natural steam power plant in the United States is at* Geysers, California. *With a power rating about that of an average coal-fired power plant, Geysers provides a significant portion of the electrical demands of San Francisco.* (© Science VU/Visuals Unlimited)

provide about 6% of U.S. energy needs, not in itself an impressive number. Then, too, since pumping cold water into the dry, hot rocks and withdrawing the heat as steam will in time render the rocks wet and cold, the life expectancy of any DHRGP site is only about 50 years. Because rocks are excellent insulators and, once cooled, will not reheat within the duration of human existence, geothermal power must be considered a nonrenewable resource.

Wind

Another area of potential power that has been investigated by the federal government is **wind**. Wind has been used for power for millennia. In many parts of the country, windmill-driven pumps bringing groundwater to the surface are a frequent sight (Figure 20.34). More romantic are the windmills of Holland where, as elsewhere in the world, the wind has been used for centuries to provide power for grinding grain (Figure 20.35).

As part of the project investigating the potential of the wind, two large wind-powered stations were built, one in the Smoky Mountains of the southern Appalachians and the other in New England. Both sites are now defunct. The government also located sites where the wind could potentially be a significant source of power. For anyone who has been to the coast, it should come as no surprise to find that both the Atlantic and Pacific coastal areas are potential sites for wind-powered generating stations. But interestingly enough, the areas that have the strongest sustained winds are located in the Great Plains, especially along the Front Range of the Rocky Mountains.

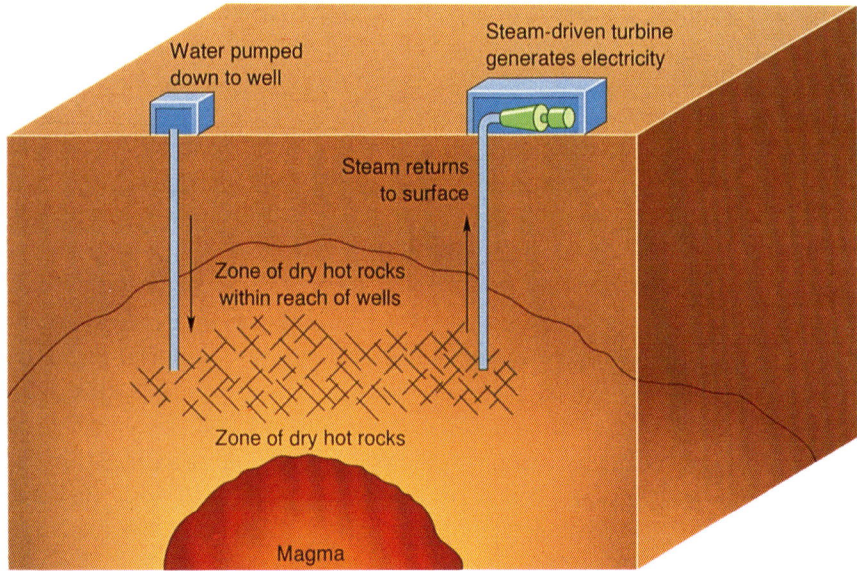

FIGURE 20.33 *The* Dry Hot Rocks Geothermal Project *is a federally funded project designed to tap the heat of subterranean magma by utilizing the heat stored in the rocks that surround the magma chamber.*

Economic Geology and Energy   593

The technology would require the construction of large numbers of interconnected windmills, each capable of about 1 megawatt of power. An example of such an array is located in Tehachapi Pass in California where hundreds of windmills of various designs cover the hilltops. They were constructed during a period when the federal government provided funds and tax incentives to develop wind power (Figure 20.36). Few have been constructed since, an

FIGURE 20.34   *Throughout the country,* windmill-driven pumps *used to raise water to the surface for domestic needs, for irrigation, or for watering livestock are common sights.* (VU/© LINK)

FIGURE 20.35   *Most people are familiar with the windmills of Holland that have been used for centuries to grind grain. Today networks of windmills take advantage of the seaside winds to generate power.* (Photos courtesy of Netherlands Board of Tourism)

FIGURE 20.36   *Windmills of several different designs line the slopes of the Sierra Nevada at Tehachapi Pass east of San Francisco, California.*

indication that the economic advantages of wind as a means of generating electricity on a large scale are currently marginal.

If all of the potential sites were fully utilized, the wind would only generate an estimated 6% of our total energy needs. Even in areas of strong winds, generation will still be intermittent, and power will have to be stored for times of calm. In defense of using the wind as an energy source, environmentalists point out that the environmental impact is small or nonexistent, although residents in the vicinity of large windmills have complained about the near-continuous swishing sound made by the rotors.

Although windmills are usually welcomed by environmental groups as sources of clean energy, there is concern that large numbers of birds are being killed each year as they fly into the rotors. Currently California produces about 80% of the world's wind energy, about 1% of its own energy demand. Although the largest windmill farms have about 6,000 units, Altamont Pass, about an hour east of San Francisco, has about 8,000 windmills covering the hills. These windmills, standing about 150 feet high, with rotor blades 40 or more feet long, each generate the same amount of power every year as burning 100 tons of coal or 360 barrels of oil. Windmill manufacturers are testing various methods to scare the birds away from the spinning rotors including painting the rotors and hubs bright colors, painting eyes on the hubs, and using sound as a warning.

The Tides

The **tides** are another potential energy source that is often suggested. Harnessing the tides requires the building of a gated dam at the mouth of a major estuary (Figure 23.37). During the flood tide, the gates are opened and the water is allowed to enter. The gates are closed at peak high tide, and the water is forced to exit the reservoir through a port equipped with a turbine similar to those in hydroelectric plants. At the present time, only two commercial tidal power stations are in operation. One, a 250–megawatt station is located on the estuary of the Rance River along the Atlantic coast of France (Figure 20.38). The other, a smaller station, is located in the Bay of Fundy, Nova Scotia, Canada. Like many of the alternate energy sources, the tide will not be a major source of electrical power. It is estimated, for example, that if all of the potential bays and estuaries available along the U.S. coastline were utilized, tidal power would produce only about 3% of the total energy demand while at the same time disrupting the intertidal/estuarine environment.

Hydropower

Another source of energy that has been used for centuries is **hydropower** (Figure 20.39). Water contained behind dams has been released over waterwheels for centuries and through ports to drive turbines for decades. In 1948, hydropower accounted for nearly a third of the entire U.S. energy budget. Compared to today, however, the demands for electricity in the years following World War II were considerably less, with

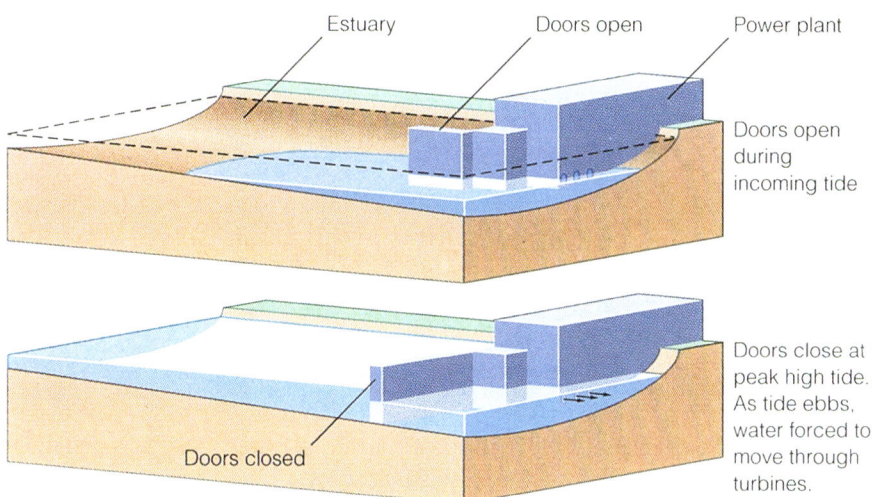

FIGURE 20.37 *It has been suggested that the energy potential of the tides be harnessed by allowing the rising tide to freely enter an enclosure through opened doors that are closed at peak high tide. The entrapped water is then forced to drain through turbine-fitted passageways as the tide ebbs.*

FIGURE 20.38 *A commercial tidal power plant takes advantage of a 44-foot (13.5 m) tidal range at the entrance to the Rance River along the Breton coast of France. (Courtesy of Institute Geographique National)*

FIGURE 20.39 Hoover Dam, *near* Las Vegas, Nevada, *was built to utilize the potential energy of the Colorado River to generate electricity for much of the southwestern United States. (Courtesy of Las Vegas Convention and Visitor's Authority)*

little use of air-conditioning and few "labor-saving" electrical appliances. Today, hydropower accounts for about 4% of our total energy budget and is not expected to grow any larger. Most dammable streams are already being utilized to capacity, and the life span of hydropower systems is limited by the infilling of the reservoirs with sediment. Although major hydropower projects are underway in the rest of the world, the contribution of hydropower in the United States may be expected to decline into the next century.

Biomass

Biomass includes any organic material that can be burned or converted to a combustible gas or liquid that can be burned. It includes a wide variety of materials such as refuse from sawmills, leftovers from the harvesting of agricultural products, waste paper, methane generated in solid waste disposal sites and sewage treatment plants, and garbage.

Before the advent of coal, oil, and gas, *wood* was the main source of energy. Over the past few decades, the use of wood as an energy source has increased, primarily in wood-burning home heating systems. Wood could be called on to provide a greater portion of the U.S. energy budget. However, its use as a major source of power would either require the use of large tracts of existing forests or the rededication of extensive tracts of land for the planting of new forests allocated to power generation. Since the present demand for wood products is expected to double by the turn of the century and available arable land is needed for the production of food, it seems unlikely that the United States would be able to generate the large volumes of wood needed to fuel a wood-based energy system. However, wood waste from sawmills and wood products facilities that is now disposed of could be utilized as a fuel and contribute significantly to the national energy budget.

Another potential source of energy is the *methane* generated during the decomposition of organic wastes in sewage treatment plants and sanitary landfills (Figure 20.40). In a number of large sewage treatment facilities, methane is collected and used on-site as a source of heat. Although technology to collect methane generated within landfills has been developed, its widespread use will depend on economics. As long as the supply of natural gas remains high and the unit cost low, the utilization of methane generated from such systems will be limited.

The disposal of *garbage* is a major problem in the United States. Not only do we generate more garbage per capita than any other nation, but its disposal is fast becoming a major national problem. Existing landfills are becoming full while other landfills are being closed because of their potential to contaminate groundwater supplies. At the same time, permits to open new sites are being refused because of the potential environmental problems. Everyone realizes that garbage must be disposed of, but no one wants the disposal site to be in their area, an example of the *NIMBY* (*n*ot *i*n *m*y *b*ack *y*ard) syndrome. With land disposal becoming more difficult, pressure to permit expanded ocean disposal is increasing while at the same time it is becoming increasingly obvious to many that the oceans can no longer be used as dumping grounds. Recent accounts of human waste and hospital wastes such as used syringes being washed up onto beaches have made people aware of the absurdity and dangers of ocean dumping.

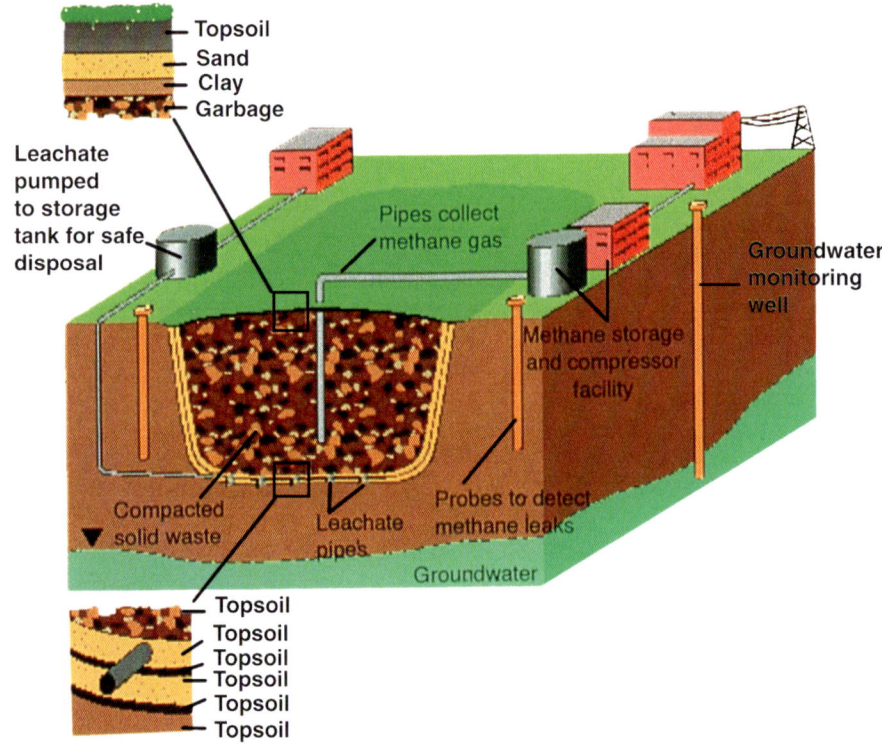

Figure 20.40 *Modern solid waste disposal sites often have facilities to collect methane gas generated by the decomposition of organic wastes and utilize it to produce electricity.*

With the exception of metals and glass that can be removed for recycling, nearly all of the materials now disposed of in landfills and by ocean dumping can be *incinerated* to generate steam and power. Unfortunately, at the present time, incineration is only marginally profitable. Furthermore, the public is wary of incineration because of the potential generation of toxic air pollutants from materials such as plastics. Due to this combination of marginal economics and public reluctance, incineration of combustible garbage has not been widely utilized.

A material that has the potential to be a major source of biomass energy is *alcohol* produced from grains such as corn. The alcohol can either be used directly as a fuel or mixed with gasoline. In Brazil, for example, a mixture of alcohol and gasoline (gasohol) is the major transportation fuel. In the United States, alcohol-gasoline mixtures are widely used by commercial fleet vehicles. Nevertheless, except in a few states such as California where government-funded projects have made alcohol-gasoline available at public service stations, the use of alcohol-gasoline mixtures for domestic use is limited. Most of the current research involving the use of alcohol as an alternative vehicular fuel is directed to solving the emission problems. Although most of the emissions from the burning of alcohol-gasoline mixtures are less toxic than those from gasoline or diesel fuel, the formation of formaldehyde is especially troublesome because of its reaction with ozone. Further reduction in the protective atmospheric ozone layer may result in an increase in ultraviolet-induced skin cancer. Studies are underway to devise methods of combining optimum alcohol-gasoline mixtures and combustion parameters with various additives that will reduce the production of formaldehyde to acceptable limits.

Each of the alternative energy sources discussed so far could make only a relatively small potential contribution to the U.S. energy budget by current estimates. Biomass, however, has the potential to supply an estimated 12% of the total energy demands of the United States. Certainly, another advantage of using biomass as a fuel is that it is constantly being generated and therefore has the added attraction of being a *renewable* energy source.

Conservation

One way out of the energy problem is to *conserve* energy and simply not use any more than is absolutely needed. We have made great strides toward conservation since the OPEC oil embargo of 1973 resulted in long lines at gasoline stations and created real concerns that insufficient heating oil would be available to heat homes during the winter. Although the energy consumption of the United States has been reduced by 25%, we still have a lot to learn about real conservation. Life-styles in other parts of the world reflect the fact that most other countries have always been energy-poor. In Europe, for example, coal deposits are largely exhausted, and the combined resources of oil and gas only amount to a few percent of the world's remaining supplies. In Europe, energy conservation takes many forms including a strong dependence on public transportation, the use of small automobiles, the use of motor scooters and bicycles for personal transportation, the sparse use of air-conditioning, and the space heating of individual rooms rather than whole-house heating.

A number of conservation practices could be applied in our own homes with little effort and inconvenience. These include turning off lights that are not needed, setting lower temperatures for space heating and higher temperatures for air-conditioning, operating washing machines only when fully loaded, and reducing the time spent taking showers. Another domestic activity that consumes a disproportionate portion of the energy budget in the average U.S. household is the heating of water. It is estimated that maintaining a constant supply of hot water may cost the average U.S. household from 10% to 25% of the total household energy budget. The reason for the high cost is that water heaters operate constantly in most U.S. homes. In Europe and elsewhere in the world, hot water heaters are operated with timers that shut the power off overnight and during the day when there is little demand for hot water. By using timers, water is heated only when needed.

We have already mentioned that most imported oil is used to produce transportation fuels, mostly gasoline. In 1989, the United States consumed 2.7 billion barrels of gasoline, a level of consumption equal to 484 gallons per person per year. In comparison, the next largest consumer of gasoline, Sweden, consumes only 221 gallons per capita per year. Our consumption of oil, and especially our dependence on foreign oil, could be significantly reduced by the use of alternate transportation fuels. We have already mentioned the use of an alcohol-gasoline mixture. The most promising potential fuel, however, is natural gas, which is plentiful, inexpensive, and clean burning. Most gasoline-powered engines can be converted to natural gas at a cost of about $1,000. Immediate benefits of using natural gas could be realized by converting fleet vehicles such as those owned by federal and state agencies. Homes supplied with natural gas can be equipped with a fixture that would allow the gas supply of family vehicles to be recharged overnight. The use of natural gas as a replacement for gasoline would save significant supplies of oil for other needs such as home heating fuels and petrochemicals.

Undertaking more extensive conservation of energy does not mean returning to the horse and buggy days—that is not what most people want nor is it necessary. If we simply employed the levels of energy conservation that the rest of the world has been using for years, it is estimated that we could reduce our total energy demand by as much as 40%! With demand reduced to that level, even the seemingly small contributions of some of the alternate energy sources could have considerable impact on reducing the total demand for whatever energy source will eventually replace oil.

Solar Energy

There is no doubt that the ultimate source of energy is the *Sun*. In fact, many of the sources of energy we have already discussed, from fossil fuels to the wind, are in reality forms of **solar energy**. The difficulty in using the Sun as a major source of national and world energy is harnessing its energy in sufficient quantities to replace oil.

At the present time, two methods of converting solar radiation into electricity are being considered: (1) *photovoltaic cells* and (2) *mirrors*. Photovoltaic cells convert solar radiation directly into electricity. The most widespread use of photovoltaic cells is to provide power for satellites (Figure 20.41), although you may have a hand calculator that employs a solar-powered battery for power. The problems with using photovoltaic cells as a major method of converting solar radiation are their low efficiency, high cost, and low manufacturing capacity. Their efficiency of conversion in only 10 to 20%. Assuming 5 hours of direct sunlight for each of 300 days per year, it would take a collecting surface equal to about one-third the area of the state of Arizona to produce 100% of the U.S. energy demand. With a cost of more than $2,000 per square meter and with only several thousand square meters being produced per year, the use of photovoltaic cells as a major source of electricity now or in the foreseeable future is unlikely. However, one rather ambitious proposal to utilize photovoltaic cells would reduce the total collection area needed to an estimated few hundred square miles. The scheme involves placing photovoltaic collectors in geosynchronous orbits above the United States. At elevations of 24,000 miles (39,000 km), the incoming solar

FIGURE 20.41 *One of the major uses of the photovoltaic cells manufactured in the United States is to provide electricity for satellites. (Courtesy of NASA)*

radiation intercepted by the cells would be much more intense because it would not have passed through Earth's atmosphere. According to the proposal, the electricity generated by the photovoltaic cells would be converted to microwaves, beamed down to Earth, collected by stationary antennae, and reconverted into electricity. At the present time, the cost of such a venture would be prohibitive.

The major area of research in solar power generation is the use of mirrors. The technique is similar to the use of a magnifying glass to concentrate the image of the Sun on a spot on a piece of paper. An array of concave mirrors focuses the solar image onto a steam or heat generator (Figure 20.42).

About a dozen mirror-type power-generating plants exist around the world, most of them in the Alps. A few years ago, the U.S. Department of Energy completed a 100-megawatt power station utilizing computer-controlled mirrors in the Mojave Desert in southern California (Figure 20.43). Unfortunately, the plant is no longer in operation. Assuming a conversion efficiency of about 25% and the same availability of direct solar radiation as for the photovoltaic cells, mirrors with a total collecting area of approximately one-fourth the area of Arizona would be required to provide 100% of all the power we need.

Both solar techniques present some fundamental problems that may be insurmountable. First, the only part of the United States where sufficient, direct, cloudless sunlight would be available is the Southwest. Appropriating the extensive areas of land that would be needed would certainly be met with resistance by the involved landowners. The surface of the land below the collectors would be unusable for any other purpose. Deprived of sunlight, it would support no life-forms. In addition, some individuals contend that, international relationships being what they are, concentrating a national energy source in such a small, exposed area would be, strategically, a poor decision.

Because solar power plants would only be operable during the day, continuous generation of power would require the storage of either heat or electricity during the daylight hours. Heat can be stored in liquids such as water or oil or in solids such as crushed rock, but the efficiency of retrieval is not impressively high. Electricity can be stored in batteries. Although the efficiency of modern batteries is more than 90%, they are very expensive.

At the present time, solar power technology is incapable of providing the amounts of power needed to allow the Sun to be considered a major energy source. When the Sun does become the ultimate

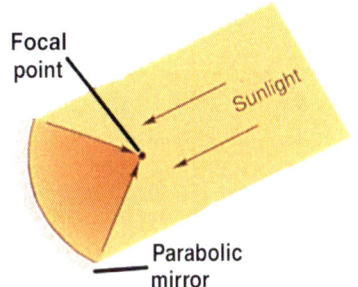

(a) The concept

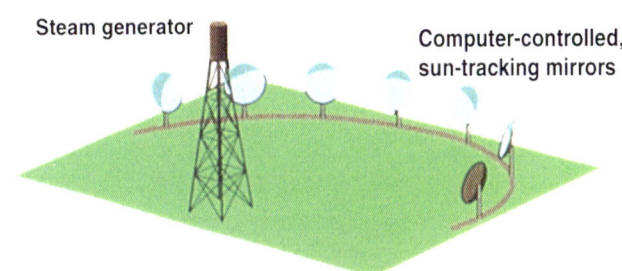

(b) Multi-mirror, single focal spot array

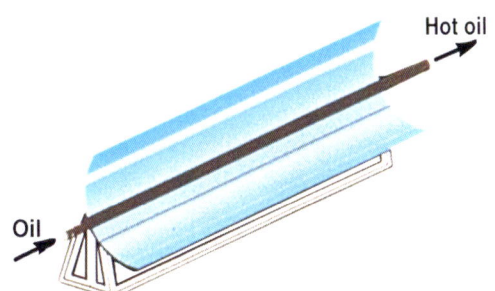

(c) Linear Sun-tracking mirror with heat transfer liquid

FIGURE 20.42 *Most solar power stations employ a convex mirror to focus the rays of the Sun to a point where the heat is concentrated. The designs differ in shape of the reflector and in the way the heat is utilized. In one design, hundreds of computer-controlled mirrors are arranged in a circular pattern. Depending on the time of day and the orientation of the individual mirror, each mirror tracks the Sun and directs its rays to a common point where they fall on a steam generator. In another design, computer-controlled, linear reflectors focus the Sun's rays along the length of a pipe through which a fluid such as oil is passed. The fluid collects the heat and transfers it to an adjoining power station.*

source of domestic and world energy, the technique employed will undoubtedly be one that has not yet been suggested. Despite all the problems that remain to be solved, using the Sun as a source of power offers one great advantage: as long as it shines, it can be counted upon for unlimited energy. According to the best estimates, the Sun is going to shine for another 5 billion years.

FIGURE 20.43 *With few days of cloudy skies, the Mojave Desert is an ideal site for a solar power station. The original U.S. Department of Energy solar facility utilized a circular array of individual mirrors to direct the Sun's rays to a steam generator located atop a centrally located tower. The DOE facility is no longer operating. A nearby commercial generating station utilizes computer-controlled linear-reflectors and oil to conduct the heat to the generating plant. (Photo a © Science VU/Visuals Unlimited)*

In closing, two important points should be made concerning the use of alternate energy sources: (1) although individual contributions to the total energy demand may be relatively small, the *combined* contribution of all the alternate sources can be considerable, and (2) on a local scale, the contribution of any one technique can be significant. For example, although the geothermal power station at Geysers, California, is smaller than an average coal-fired power plant, it supplies a significant portion of San Francisco's electricity demand. Widespread use of *passive* solar heating could significantly reduce the amount of nonalternate power now being consumed to heat buildings and water while the energy of the wind can be used to lift water from the bottom of a well, with electricity only being used during times of calm.

SPOT REVIEW

1. What is the concept behind the Dry Hot Rocks Geothermal Project?
2. What is the estimated future potential of wind power?
3. How are the tides harnessed to produce commercial quantities of power?
4. Why is hydropower not likely to contribute a larger portion of the future U.S. energy budget?
5. What sources of biomass have the greatest potential to provide significant portions of the total U.S. energy demand?
6. Discuss several examples where conversion can significantly reduce the U.S. energy demand.
7. How is solar power being harnessed as a source of electrical power?

ENVIRONMENTAL CONCERNS

Society depends on the products and services provided by the mining industry. We have already mentioned the widespread use of iron in modern society. To see other such products, you need only look about your home. Your coffee cup and the glaze on fine paper are made from clay minerals produced by quarrying residual soils. The concrete that was used in the construction of your home was made from sand and gravel quarried from sites such as glacial moraines or dredged from streambeds. The lead in your car's battery, the zinc coating that protects the surface of a steel bucket from oxidation, and the copper wiring that conducts electricity throughout your home were most likely obtained by mining sulfide minerals originally emplaced by hydrothermal metamorphism. The electricity used to power most of your household appliances is most likely provided by the burning of coal mined throughout the country.

Aquisition of the resources that made these products and services available requires the disruption of Earth's surface. As mining exposes additional rock and mineral surfaces to the atmosphere in quarries, surface mines, tailings from deep mines, and the refuse from cleaning and processing plants, chemical weathering introduces a variety of dissolved products into the environment. In addition, the fine sediments released from many mining operations infiltrate streams and lakes and may physically disrupt the life cycles of plants and animals. In addition to the *chemical* and *physical* pollution of the environment, various aspects of mining contribute to *visual pollution* in the form of gob piles and moonscape-type land surfaces.

Nevertheless, it is quite obvious that we cannot do without the products and services provided by the mining industry. While we accept the potential environmental impact of mining, both society and the mining industry have begun to take steps to minimize the environmental impact of resource exploitation by initiating various reclamation procedures. Although individual mining industries have unique environmental problems and solutions to those problems, the kinds of problems faced by the coal industry are common to many mining industries, as are the advances in mining and reclamation techniques the coal industry has initiated to minimize the environmental impact of its operations. Therefore, it will be used here to illustrate developments throughout the mining industry.

Coal Mining

Large-scale coal mining began in the eighteenth century with the beginning of the Industrial Revolution. Until the beginning of the twentieth century, coal was the main source of energy for the industrial nations. Shortly after the turn of the century, coal was replaced by oil as the major energy source. Today, coal represents only about 20% of our energy budget.

According to 1991 figures, Wyoming is the leading producer of coal in the United States, producing nearly 20% of the total U.S. output of 993,486,000 short tons. Kentucky and West Virginia are tied for second, each producing about 17% of the total.

Originally, most coal was produced from deep mines. With the advent of large earth-moving equipment capable of cost-effectively removing thick sections of overburden, the volume of surface-mined coal increased. At the present time, 59% of U.S. coal comes from surface mines, largely from states such as Wyoming where 99% of the production is from surface mines. Surface mining is not employed so extensively in the coal-producing regions east of the Mississippi; West Virginia where surface mining accounts for only 25% of the coal production is typical.

Environmental Problems

Until the mid-1960s, the environmental problems associated with coal mining were more or less accepted as the necessary price of producing coal. In the mid-1960s, however, state and federal legislation began to be enacted that dictated strict reclamation requirements. In the years that followed, the environmental problems of coal mining, although certainly not eliminated, have been significantly reduced by the cooperative efforts of government agencies, researchers, and the coal industry.

Revegetation   The kind of environmental problem resulting from coal mining depends largely on the geographic location and whether the coal is produced from deep or surface mines. In the western states, for example, where most coal is surface mined, the major environmental problem is reestablishing vegetative cover on the disturbed lands. Soils in arid to semiarid regions are normally very thin and require very long periods of time to develop. Because they are so fragile, these soils are usually destroyed during the initial mining operations. Reclamation legislation requires the reestablishment of grass cover, which, unfortunately, requires extensive irrigation in areas that are already water-poor. In most western states, reclamation legislation dictates a specified period of irrigation, during which the sites become covered with dense stands of grass. Although much of the grass may die when the irrigation is terminated, whatever remains serves to protect the surface from wind and water erosion and helps to reestablish a soil profile.

Acid Mine Drainage   Acid mine drainage (AMD) is not a serious problem in the western mining areas because the *pyrite* content of most western coal is low. It is, however, the major environmental problem in the northern Appalachian bituminous coal basin (northern West Virginia, Ohio, western Pennsylvania) and the anthracite basin of eastern Pennsylvania.

The acid is formed when pyrite undergoes chemical weathering; the acid-producing process is described by the following reactions:

$$FeS_2 + 7/2 O_2 + H_2O \rightarrow Fe^{2+} + 2SO_4^{2-} + 2H^{1+} \quad (1)$$

$$Fe^{2+} + 1/4 O_2 + H^{1+} \rightarrow Fe^{3+} + 1/2 H_2O \quad (2)$$

$$FeS_2 + 14Fe^{3+} + 8H_2O \rightarrow$$
$$15Fe^{2+} + 2SO_4^{2-} + 16H^{1+} \quad (3)$$

$$Fe^{3+} + 3H_2O \rightarrow Fe(OH)_3 + 3H^{1+} \quad (4)$$

The acid (H^{1+}) enters streams and lakes and may seriously affect the aquatic biota. The iron hydroxide, $Fe(OH)_3$, precipitates and coats the bed loads of the affected streams, producing the characteristic "yellow boy" stain seen commonly throughout the coal-producing areas of Appalachia. We will discuss some of the treatments that are being used to solve the acid drainage problem below.

Slope Stability   Acid generation is not a serious problem in the southern Appalachian coal basin (southern West Virginia, eastern Kentucky, and northern Alabama) because the coal and coal associated rocks contain little pyrite.

In this region, most of the coal is produced by *contour mining,* a process where the coal is mined by excavating the outcrop of the coal around the contour of a slope. Because of the rugged topography that characterizes much of the southern Appalachians, the major environmental problem is maintaining slope stability once the sites are reclaimed and the steep slopes are returned to the original topography. The planting of fast-growing grasses and trees serves to stabilize the slopes and reduce erosion.

Treating the Acid Mine Drainage Problem

The law now stipulates that AMD must be treated and the acid neutralized before the water is released into surface streams. Treatment facilities are required at all active mine sites to treat mine waters before they are released into the environment.

Unfortunately, much of the acid polluting the streams throughout Appalachia is being generated in abandoned deep mines. In cases where the acid emerges at a point source, it can be collected and neutralized by one or more of the treatments described below before being discharged into the streams. A major problem exists, however, when the acid emerges at an inaccessible site or seeps to the surface at many diffuse sites, making the collection and treatment of the waters difficult or impossible. One solution that has been suggested for such sites is to utilize the fly ash generated by coal-fired power plants. Rather than treating the ash as refuse that must be disposed of, it might be used to make concrete. This, in turn, would be injected into the abandoned workings, sealing the mine from the groundwater and thereby preventing additional production of acid.

Chemical Treatments to Neutralize Acidity
The major chemical treatments used today to neutralize acid mine waters are sodium hydroxide, NaOH, sodium carbonate, Na_2CO_3, calcium hydroxide, $Ca(OH)_2$, and ammonium hydroxide, NH_4OH. Each chemical reacts with water to form an alkaline solution. Which of the chemicals is selected for a specific site depends largely on the volume of water to be treated and the level of acid to be neutralized.

Sodium hydroxide, which is added as an aqueous solution, is commonly chosen where the treatment site does not have access to electricity to power an automatic dispensing system. Being a strong base, sodium hydroxide quickly raises the pH of the treated solution.

Sodium carbonate is limited in the amount of acid it can neutralize. For this reason, it is usually used when both the volume of water and the level of acidity are low. Because it may be purchased in solid briquettes, it is convenient to apply as either a water treatment or a surface application.

Calcium hydroxide is selected where large volumes of highly acidic water are to be treated over extended periods of time. Calcium hydroxide is prepared by mixing calcium oxide (CaO) with water. Because a power blender is needed for the mixing, calcium hydroxide is only used in areas where electricity is available.

Ammonium hydroxide is a commonly used treatment due to its relatively low cost. It also has a high neutralization potential, making it capable of neutralizing large volumes of highly acidic water.

It is important to note that these treatments are intended to neutralize acid that has *already been produced;* they are not meant to *prevent the formation* of acid.

Treatments to Prevent the Formation of Acid
Much of the research now underway is designed to develop treatment procedures that will prevent the formation of acid by eliminating the oxidizing agents responsible for the decomposition of pyrite.

You will note that in the equations describing the formation of AMD, the initial oxidizing agent is dissolved oxygen ($O_2 + H_2O$ in Equation 1). Attempts to eliminate oxygen from the water percolating through reclamation sites have not been very successful. Note, however, that the oxidation of pyrite by dissolved oxygen is *not* responsible for most of the generated acid. The ferric ion (Fe^{3+}) generated in Equation 2 not only produces *eight times* more acid than does dissolved oxygen (refer to Equation 3), but produces it at a *faster* rate. Eliminating dissolved iron from the percolating water would, therefore, be much more effective at reducing the production of acid than would the elimination of dissolved oxygen. A number of procedures have shown promise.

Apatite One treatment that has proved effective in reducing the concentration of dissolved iron in leachates is to mix the mineral *apatite* ($CaPO_4$) with the potentially toxic rock material. As apatite dissolves under acid conditions, the phosphate ion (PO_4^{2-}) is released into solution and combines with the dissolved iron to produce a highly insoluble precipitate of iron phosphate. The process not only removes iron from the system, but the precipitate coats the surfaces of the pyrite grains and seals them from any additional reactions.

With the *major* oxidizing agent eliminated, the rate of acid production is significantly reduced. Any hydrogen ion (acid) produced by the initial reaction of dissolved oxygen and pyrite (refer to Equation 1) can be removed from solution by cation exchange onto the surfaces of the clay minerals contained within the rocks.

Unfortunately, the cost of apatite is a limiting factor in some parts of the country. In the Appalachian region, for example, the nearest source of apatite is easternmost South Carolina. Because the cost of delivering apatite to the Appalachian coal fields is high, its use as an AMD treatment is not cost-effective. In the western states, however, where deposits of apatite are more common, its use can be considered economically feasible.

Fluidized-Bed Combustor Ash Another material that has shown promise as an AMD treatment is the ash from the fluidized-bed combustors employed in modern coal-burning power plants. The fluidized-bed combustor differs from conventional systems in that hot air is forced upward through the materials (making it a "fluidized bed") to promote complete combustion.

The fluidized-bed combustor utilizes a combination of calcium oxide, CaO, and calcium hydroxide, $Ca(OH)_2$, as *getters* to remove the sulfur dioxide, SO_2, produced by the combustion of the sulfur contained in the fuel. You will remember from our previous discussion of coal utilization that getters are employed to remove sulfur dioxide from the flue gases to prevent it from entering the atmosphere and contributing to the acid rain problem. The oxides and hydroxides of calcium are produced by the thermal disintegration of powdered limestone introduced with the fuel.

Because more limestone is usually added to the original charge of fuel than is needed to react with all of the generated sulfur dioxide, the ash from the combustor commonly contains unreacted calcium carbonate, $CaCO_3$, and both unused CaO and $Ca(OH)_2$. The CaO and $Ca(OH)_2$ readily dissolve in water to produce a highly alkaline solution that neutralizes the hydrogen ion (acid) produced in Equation 1. Note that with the hydrogen ion from Equation 1 eliminated, the reaction cannot go on to Equation 2. The acid-producing reaction therefore comes to a stop.

It should be noted that many of these treatments of coal-generated acid are equally applicable to the metals mining industry where, among the other sulfide minerals, pyrite is a major constituent and produces acid in the same way.

Coal Refuse: Environmental Problem or Wasted Fuel?
One of the advantages of a fluidized-bed combustor is that the combustion process is so efficient that coal *refuse,* the by-product of coal-cleaning plants, can be used as a significant part of the fuel. In a local coal-refuse-fired power plant, for example, coal refuse represents 60% of the fuel.

Because it is pyrite-rich, coal refuse invariably produces high levels of acid when exposed to the atmosphere and must be treated before disposal to prevent the generation of acid. The disposal of these materials is one of the major environmental problems faced by the coal industry.

It should be noted, however, that while coal refuse represents an environmental problem, it commonly contains as much as 45% organic material by weight. This is enough to provide the energy-producing potential of lignite. In other words, not only may the disposal of these materials generate a potential environmental problem, but a valuable source of energy is being discarded at a time when we should be conserving all sources of energy. In this regard, it is important to note that the use of coal refuse as a fuel for fluidized-bed combustors *simultaneously eliminates* the toxic potential of the material by thermally decomposing the pyrite and utilizes its energy potential.

Certainly, the mining industry still has many environmental problems to be solved. What is important, however, is that the modern mining community realizes that its survival depends on solutions being found for the problems that remain. As a researcher in environmental geology, the cooperation and support that I and other researchers receive from the mining industry indicates that they want the problems solved as much as we do.

CONCEPTS AND TERMS TO REMEMBER

Native metals
Classification of natural resources
    metalliferous deposits
        ferrous deposits
        nonferrous deposits
        ore
    protore or tailings
        gangue
    nonmettaliferous deposits
        fuels
            fossil fuels
        nonfuels
    reserves
    resources

Metals
    magmatic segregation
    disseminated deposits
    veins
        pegmatite
    metasomatic replacement
    sulfide minerals
    supergene enrichment
        gossan or iron hat
    banded iron deposits
    ooids
    placer deposits
        hydraulic mining
    lodes

Fuels
    coal
        swamp versus marsh
        peat
            coalification
        rank series
            volatiles and fixed carbon
            brown coal (lignite)
            subbituminous coal
            bituminous coal
            anthracite
            graphite
        heat content (BUTs per pound)
        quality

ash
sulfur
   organic and pyritic
utilization
  compliance coal
  liquefaction (synfuel)
distribution
  western versus eastern United States

petroleum
reservoir
cap rock
trap
  anticlinal theory
oil shale
Nuclear power
fission and fission reactors
breeder reactors

fusion
Alternate energy sources
  geothermal
  wind
  tides
  hydropower
  biomass
  solar

REVIEW QUESTIONS

1. The most widely used metal in the history of the human race has been
   a. gold.
   b. copper.
   c. iron.
   d. tin.
2. Most metalliferous deposits are associated with
   a. granitic batholiths.
   b. the host rock surrounding major igneous intrusions.
   c. deep-sea sedimentary rocks.
   d. ultramafic intrusions.
3. A physical characteristic of most placer materials is
   a. low density.
   b. high density.
   c. high hardness.
   d. high melting point.
4. The highest rank of coal is
   a. bituminous.
   b. lignite.
   c. anthracite.
   d. subbituminous.
5. The major source of energy in the United States is
   a. coal.
   b. oil.
   c. uranium.
   d. hydropower.
6. Most coal is used for
   a. the production of electricity.
   b. the production of petroleum-type products.
   c. the production of coke for the steel industry.
   d. home heating.
7. Excluding solar power, the alternate energy source that represents the greatest potential to replace oil as a major energy source is the
   a. tides
   b. burning of biomass.
   c. use of geothermal steam.
   d. use of the wind.
8. What is the difference between resources and reserves, ores and protores?
9. What evidence can be cited to uphold the statement that the source of the metals in most metalliferous deposits is magma?
10. What chemical reactions are involved in the concentration of certain metals by the process of supergene enrichment?
11. How do coal, oil, and uranium compare as potential sources of energy for the next two centuries?
12. What currently are the limitations to the use of the Sun as a major source of energy?

THOUGHT PROBLEMS

1. At the end of World War II when the other nations of the world decided to build their infrastructures on expensive transportation fuels, the U.S. government made the decision that ours would be based on cheap transportation fuels. What differences in the life-styles of Europe and the United States can be directly attributed to these energy policies? What would be the economic and social impacts if the United States were to convert overnight to a policy of expensive transportation fuels? What would be the effects if we were to change over a period of a decade or two?

2. In what ways will society adjust, change, or react as supplies of nonrenewable resources such as oil, gas, coal, and essential metals become exhausted?

FOR YOUR NOTEBOOK

Make a detailed analysis of the kinds of economic deposits in your immediate area and in your state. For each material, include the kinds of mining or exploitation techniques that are used to acquire the material. Investigate and compare potential or existing environmental impacts that result from each kind of operation as well as the procedures that are being used by each industry to eliminate or minimize the environmental impact of their operations. Determine the contributions that each industry makes to the economy of your area and state. Include in your analysis other industries that depend on or are served by the various minerals industries.

CHAPTER OUTLINE

INTRODUCTION

EARLY ESTIMATES OF EARTH'S AGE

DATING METHODS
   Relative dating
   Absolute Dating
   Dating by Sediment Accumulation, Ocean Salinity, and Earth's Heat
   Radiometric Dating

ROCK TIME SCALES
   The Modern Time Scale
   Periods of the Paleozoic Era
   Periods of the Mesozoic Era
   Periods and Epochs of the Cenozoic Era

CHAPTER 21

The Age of Earth

INTRODUCTION The construction of monuments such as *Stonehenge* indicates that the builders were aware of time. Some historians believe that the builders of Stonehenge not only kept records of passing events but used them to predict future occurrences of certain celestial events such as eclipses. These civilizations measured the passage of time with the same kind of celestial observations we use today. The rising and setting of the Sun defined the day, the phases of the Moon determined the month, and the passage of the seasons established the year. To these natural cycles, we have added the artificial subdivisions of weeks, hours, minutes, and seconds in an attempt to make our record keeping more precise.

The idea of an ancient Earth is not a recent concept. The *Brahmins* of India considered the world to be eternal. The early Greek philosophers were also aware of the antiquity of Earth. The Greek philosopher *Xenophanes* (c. 570–470 B.C.) correctly concluded that areas where fossiliferous rocks were exposed had once been covered by the sea and that significant amounts of time must have passed since the land was a part of the ocean floor. In c. 450 B.C., the Greek historian *Herodotus* watched the Nile delta slowly build up with each yearly flood and realized the enormity of time needed to amass the entire structure. Unfortunately, with the fall of the Roman Empire, most of the peoples of Europe were more concerned with protecting themselves from roving marauders than with intellectual pursuits. Questions about topics such as the age of Earth and the mysteries of the heavens were seldom raised until the end of the Middle Ages.

EARLY ESTIMATES OF EARTH'S AGE

The first real attempt to establish the age of Earth was made in 1644 by *John Lightfoot,* the vice chancellor of Cambridge University, who claimed that Earth was created at exactly 9:00 A.M. on October 26 in the year 3926 B.C. In 1658, *James Ussher,* the archbishop of Armagh, Ireland, determined that Earth was created on October 23 in the year 4004 B.C. The two churchmen proposed such similar dates because both used the Bible as their source, in particular, the Old Testament Book of Numbers, which relates the genealogy of the tribes of Israel. Both men started with Adam and Eve and then attempted to calculate how long it would take to evolve all the tribes of Israel. Although today some may scoff at such attempts to establish the age of Earth, one must keep in mind that both men were serious scholars of theology who used what they considered to be the most reliable source of information, namely, the Bible. Unfortunately, in 1701, *Bishop Lloyd* inserted the 4004 B.C. date for the creation of Earth into a footnote to the Great Edition of the English Bible, and for nearly a hundred years thereafter, to deny a 6,000-year age for Earth was tantamount to heresy.

Limited to a short 6,000 years, many early geologists were forced to use extraordinary means to explain the origin of Earth's rocks and landforms. Of these theories, the most widely accepted was **catastrophism**. According to catastrophists, most features of Earth's crust were created by short-lived, violent processes. For example, catastrophists claimed that all mountains were born when violent, convulsive eruptions lifted huge masses of rock out of the ocean depths. Although some volcanoes do arise in such spectacular fashion, we now know that most of Earth's mountains rise slowly over geologic periods of time, driven by forces generated by the convergence of the lithospheric plates. Similarly, catastrophists would have explained features like the Grand Canyon as having been born as the result of a cataclysmic wrenching and tearing of Earth's crust rather than by millions of years of stream erosion during and after slow uplift. But with only 6,000 years at their disposal, early students of Earth had little choice but to propose such origins, especially since they also wanted to avoid contradicting the teachings of the church.

The first real challenge to catastrophism and to the short 6,000-year history of Earth came in 1785 with the efforts of *James Hutton*. Regarded as the founder of modern geology, Hutton was a wealthy physician and landowner who spent much of his time roaming the countryside studying the rocks of his native Scotland. Hutton saw rocks being slowly reduced to rock fragments and minerals, shifted downslope by almost imperceptible movements, and eventually carried off by streams. He watched the streams deposit the materials into the ocean and correctly surmised that the materials were picked up by ocean currents, spread out on the ocean floor, and eventually transformed into sedimentary rocks. He visualized that the newly formed sedimentary rocks were then uplifted from the ocean floor to create bedded rocks similar to those he observed on land. Most importantly, it was obvious to Hutton that given the extremely slow rates of the processes he saw going on about him, the rocks and landforms could not possibly have been created in only 6,000 years.

Hutton's opinion of the extreme age of Earth was reinforced in 1788 when he first saw the rock exposure at *Siccar Point* along the southeastern coast of Scotland (Figure 21.1). As Hutton viewed the near-horizontal sedimentary layers of the Old Red Sandstone resting on the vertical beds of the underlying sedimentary rocks, he immediately grasped the significance of the formation. Hutton recognized that before the sediments that eventually formed into the upper rocks were deposited, the lower rocks had been uplifted, tilted, subjected to a long period of erosion, and resubmerged beneath the sea. As he looked at this single outcrop, he understood that it had required many millions of years to accomplish. *John Playfair,* a friend who accompanied Hutton on the historic visit to Siccar Point, was to write a few years later, "The mind seemed to grow giddy by looking so far back into the abyss of time...."

Hutton published his ideas in *Theory of the Earth.* His concepts eventually gave rise to one of the most important tenets of geology, the concept of **uniformitarianism.** In direct contradiction to the then-popular tenets of catastrophism, uniformitarianism states that the processes that shape Earth's surface today are the same as those that acted in the past. The concept is commonly summarized in the statement "the present is the key to the past." In support of uniformitarianism, Hutton argued that although catastrophic events were not excluded as natural

FIGURE 21.1 *It was the exposure at* Siccar Point, *along the North Sea coast of Scotland, that convinced James Hutton of the great age of Earth. The Devonian Old Red Sandstone resting on the upturned edges of the folded Silurian sedimentary rocks below is an example of an angular unconformity.*

processes, they were not required to produce Earth's landforms. It was Hutton's view that given sufficient time, very slow-acting, seemingly insignificant processes could create all the landforms attributed to catastrophism. One of the most important results of Hutton's work was the general acceptance of a great age for Earth. Hutton summed up his view of Earth's antiquity in a paper published by the Royal Society of Edinburgh in 1788: "The results, therefore, of our present inquiry is that we find no vestige of a beginning no prospect of an end." With Hutton's ideas firmly entrenched, others soon attempted to determine the true age of Earth.

Because Hutton's writings were very difficult to read, James Playfair published *Illustrations of the Huttonian Theory* in 1802 in which he championed Hutton's ideas in a much more readable form. In 1830, *Sir Charles Lyell* published his *Principles of Geology,* which expanded and built on Hutton's basic concepts. Lyell's book not only laid the groundwork for the early development of the science of geology but also established the basis for modern geology.

DATING METHODS

The age of an object or the timing of an event can be determined by either (1) *relative dating* or (2) *absolute dating.* **Relative** (or **sequential**) **dating** only requires that one object or event be determined to be younger or older than another. For example, based upon physical appearance alone, one could conclude with a degree of certainty that the average high school student is younger than the average retiree. **Absolute** (or **chronologic**) **dating** of these two individuals would entail determining how many years have passed since each was born. The availability of other information, such as the number of months, weeks, days, hours, and minutes that have passed, would allow the absolute age to be determined with increasing precision. Although absolute dating has obvious advantages, most geologists employ relative dating techniques to establish the ages of rocks and events because absolute dating techniques are quite expensive and cannot be applied to all rocks.

Relative Dating

The procedures used to determine the relative age of rocks and events date back to 1669 when *Nicolas Steno,* a Danish physician, published the results of studies that he had made of rocks in the vicinity of

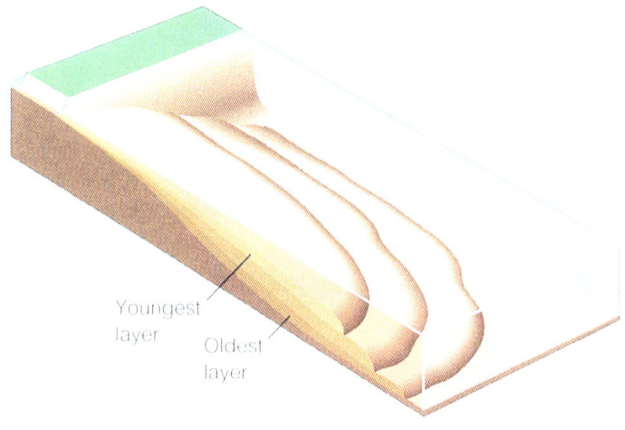

FIGURE 21.2 *This schematic drawing of a delta illustrates the relative ages of the individual layers of the* principle of superposition.

Tuscany, Italy. In his publication, Steno introduced two of the most important concepts upon which relative dating of rocks is based. The first is the **principle of superposition,** which states that in a sequence of layered rocks, unless it has been overturned by folding or faulting, younger layers are positioned on top of older layers (Figure 21.2). Although this may seem obvious to us now, keep in mind that Steno made this deduction nearly a hundred years before the monumental works of Hutton.

Geologists employ the principle of superposition every time they observe an outcrop of rocks for the first time. The first determination that must be made is whether the sequence of rocks is "right side up" or "overturned." According to Steno's principle of superposition, if the rock layers are right side up, the layers will become progressively *younger upward* while if they are overturned, they will become progressively *older upward.*

In Chapter 13, you learned a number of features that can be used to establish the original position of sedimentary layers. For example, *graded bedding* is a common sedimentary feature that forms when coarse particles accumulate, followed by particles of progressively smaller sizes (Figure 21.3). To find a bed with coarse grains at the top and progressively finer particles toward the bottom could be evidence of overturning.

The *cross-bedding* that exists within some sedimentary layers is another sedimentary feature used to determine the orientation of beds. During the formation of cross-bedded layers, newly deposited layers commonly erode the uppermost portions of the

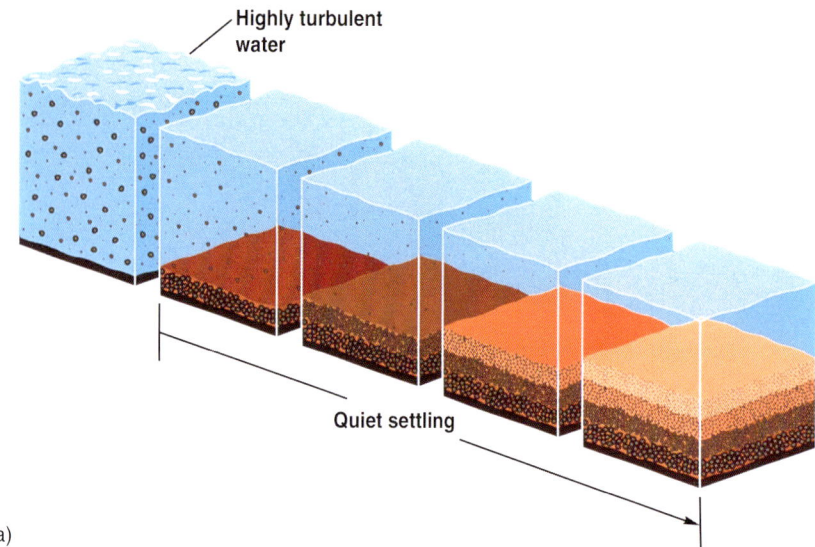

FIGURE 21.3 a and b Graded beds *form as particles settle out of water suspension by size; the largest particles settle first, and the smallest particles settle last. (VU/© 1993 Martin G. Miller)*

underlying bed as they are laid down, producing a truncation of the cross-beds at the top of the layer (Figure 21.4). Truncated cross-beds located at the tops of sedimentary beds are therefore evidence of a "right side up" orientation.

Ripple marks are formed on the upper surface of layers of sand-sized deposits as they are deposited by water or wind. When originally formed, the peaks of the waveforms point upward (Figure 21.5). To find a bed in a sequence of rocks with the peaks of the ripples pointing downward would be evidence that the sequence was overturned.

Steno's second major contribution was the **principle of original horizontality**. Steno correctly surmised that sediment layers are generally horizontal or near-horizontal when first deposited. To observe layered rocks tilted from the horizontal means that some force moved the rocks from their original horizontal state (Figure 21.6).

As is often the case with radical new ideas, Steno's colleagues rejected his work. The significance of Steno's contributions to the development of the science of geology would not be recognized and appreciated until the work of Hutton nearly a century later.

In the mid-eighteenth century, the work of *William Smith,* an English engineer, resulted in the formulation of a concept called **correlation** that allowed the age equivalency of distant rock layers to be established. Smith was involved in the construction of waterways throughout England where most of the sedimentary beds are relatively horizontal. He

The Age of Earth 613

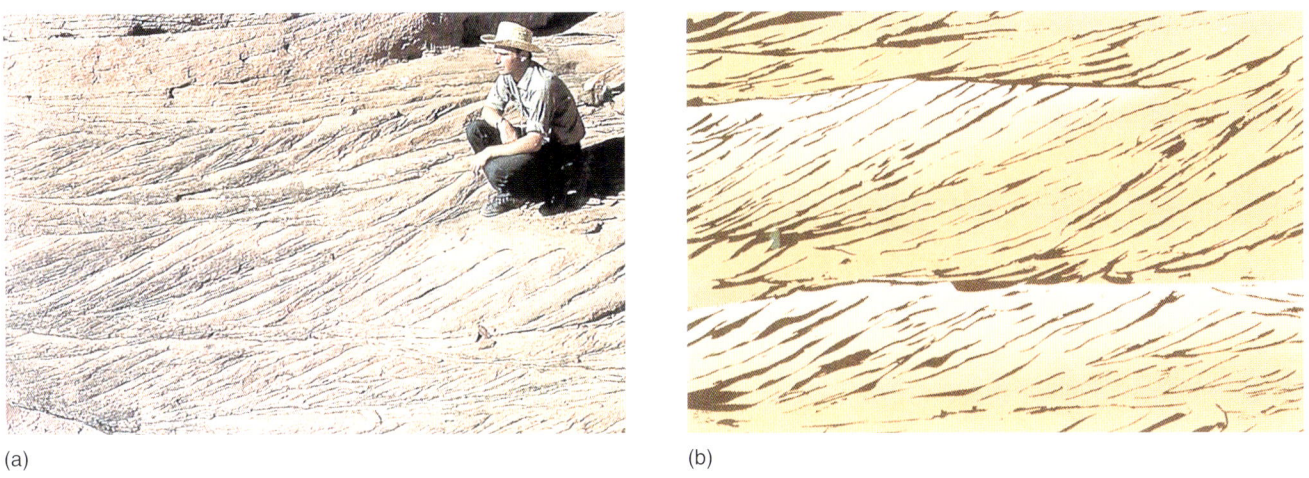

FIGURE 21.4 *The lateral movement of each new bed of cross-bedded sand removes the upper portion of the underlying bed. As a result, the cross-beds at the top of the underlying bed are truncated. (Courtesy of E. D. McKee/USGS)*

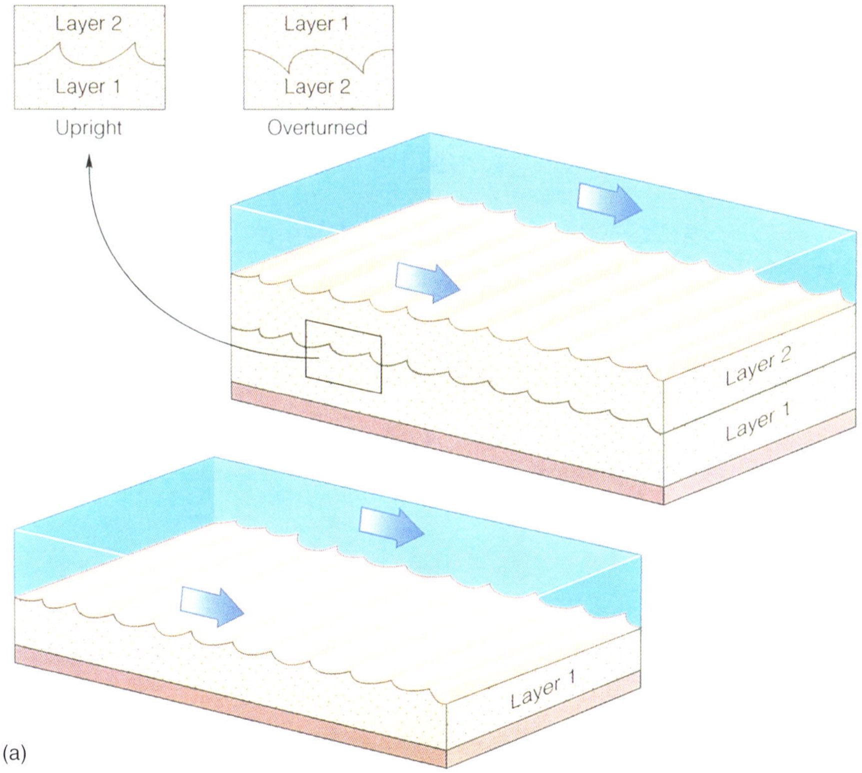

FIGURE 21.5 *(a) The presence of truncated cross-beds at the top of a bed indicates that the bed is right side up. (b) Symmetrical ripple marks commonly form on the surface of layers of sand subjected to oscillatory water movement such as in the shallow water between low and high tide. (Courtesy of Joe Donovan) (c) Asymmetrical ripple marks form where the surface of a layer of sand is being subjected to directional water currents such as on the beach at Vancouver Bay, British Columbia, Canada. (Courtesy of Joe Donovan) (d) If a ripple-marked surface is subsequently covered by sediments and the entire sequence is buried and undergoes lithification, the ripple marked surface may be preserved on the bedding surface of a sandstone. (Courtesy of E. D. McKee/USGS)*

FIGURE 21.6 *When folded rocks are visible in an outcrop, one must keep in mind that folding can only occur deep within Earth. Only after the overlying rocks are removed by millions of years of erosion are the folded rocks exposed at the surface.*

soon was able to recognize certain rock layers based upon their physical appearance and composition and noted further that they always appeared in a definite vertical (stratigraphic) sequence. Once familiar with the rock sequence in a given area, he was able to predict which rock layers would be found from one site to the next. Having identified the rock unit exposed at the surface, Smith was also able to predict the sequence of rock layers that would be found below.

Because most of the rocks Smith encountered in his work were very fossiliferous, he was also able to identify individual rock layers by their diagnostic fossil contents. He demonstrated that over relatively short distances, the fossil content of individual rock layers remained reasonably constant. This allowed Smith to correlate the rock layers from one locale to another. Over long distances, he observed that similar fossil assemblages were found in different lithologies (kinds of sedimentary rocks). At the time, Smith could not explain the cause of these observations. We now know that the rocks Smith was observing formed from sediments that accumulated in a nearshore marine environment, which consists of different subenvironments usually depending on the distance from the shore. Individual rock types represent different depositional environments. Each environment is inhabited by a certain assemblage of plants and animals that is favored by the conditions that exist within that environment. The result is a particular lithology with a diagnostic fossil content.

However, because the individual species and communities of plants and animals evolve over time, rocks deposited during any given period of time will contain a fossil assemblage characteristic of that time interval and different from the assemblages in rocks deposited earlier or later (Figure 21.7). As a result, certain fossil assemblages can be found in rocks of different lithologies.

Smith's inability to understand the reasons for the fossil distributions he observed was primarily due to the fact that the basis for our understanding of the spatial and temporal variation in fossils was not to come until more than a half century later with the publication of Charles Darwin's *On The Origin of Species*.

As an understanding of plant and animal evolution developed, Smith's work was eventually formulated in the **principle of faunal succession**. This concept states that the fossil assemblage found in rocks accumulated during any particular period of geologic time is unique to that time interval and is fundamentally different from the assemblages found in rocks that accumulated earlier or later. More than any other single tool at the disposal of the geologist, fossils are used to correlate rocks or similar or dissimilar lithologies and to establish the age equivalence of isolated exposures anywhere on Earth.

Another important tool that allows geologists to establish the relative ages of rocks was proposed by Charles Lyell in 1830 with the publication of his

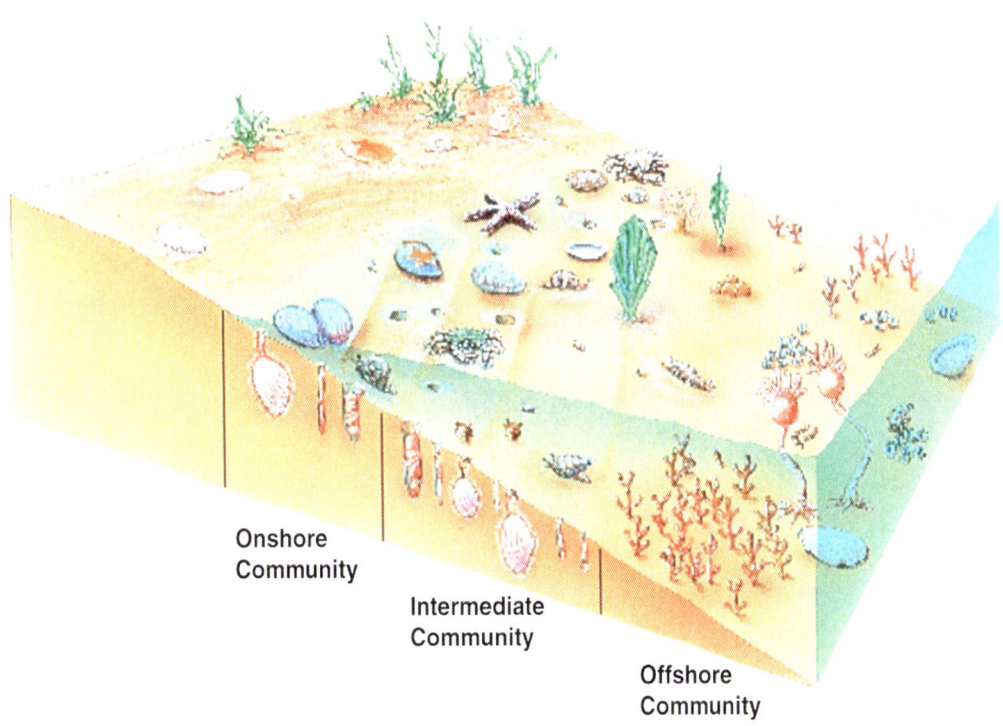

FIGURE 21.7 *Modern plant and animal assemblages change with the depth of water. Recognizing such changes in the fossil record allows geologists to determine the relative depth of water, or distance from shore, in which the sediments forming a particular sedimentary rock were deposited.*

Principles of Geology. Lyell introduced a concept that has become known as the **principle of cross-cutting relations**, which states that a geologic feature or rock is younger than any rock or geologic feature that it cuts across (Figure 21.8). An excellent example is the relationship that Hutton observed at Siccar Point, Scotland, where near-horizontal beds of the Old Red Sandstone overlie the truncated vertical beds of the sedimentary rocks below. According to the principle, the Old Red Sandstone is younger than the truncated rocks below because the sandstone layers cut across the layers of the underlying beds.

The Siccar Point exposure is an example of a relationship known as an **unconformity**. (Figure 21.9) The term was first used in 1805 by *Robert Jameson* to describe surfaces of nondeposition and/or erosion that represent breaks in the geologic record. The time required to develop a surface of unconformity is not recorded either because the sediments needed to record time were not accumulating or because rocks that recorded the passage of geologic time have been stripped away by erosion. The first case is comparable to not recording several days' events in a personal diary while the latter case is comparable to tearing out several pages from the completed portion of the diary. In both cases, a gap exists in the record. The interval of unrecorded time between the accumulation of the older sequence of rocks and that of the younger sequence is called a **hiatus**.

The most difficult type of unconformity to detect is a **disconformity**, which is an unconformable surface between parallel bedded sedimentary rocks (refer to Figure 21.9b). The parallelism of the beds indicates that the older rocks were uplifted without deformation or that there was a period of nondeposition (a hiatus) after which sedimentation was resumed. Disconformities are usually identified by recognizing significant differences in the fossil assemblages contained in the rocks on either side of the surface of disconformity. There are several major disconformities in the upper sedimentary sequence within the Grand Canyon (Figure 21.10).

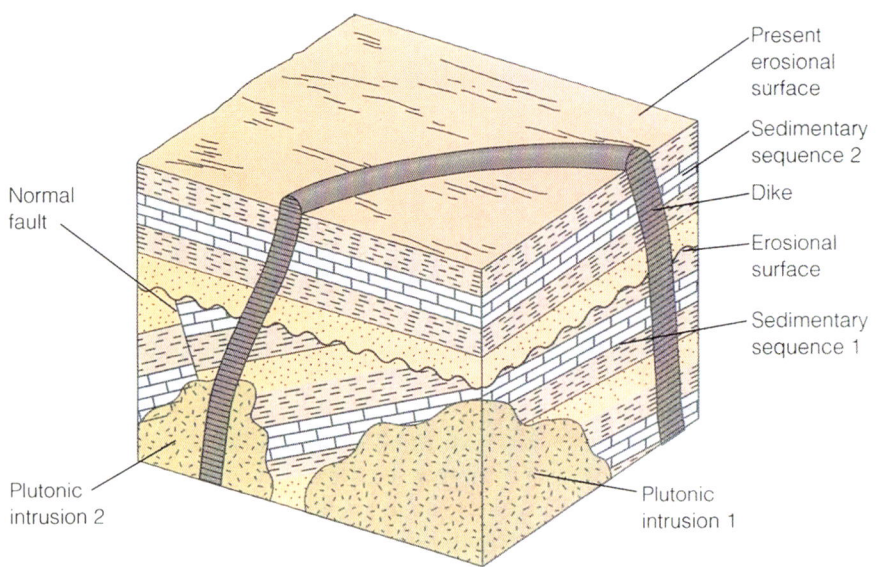

FIGURE 21.8 *The relationship between the boundary of one rock body or geologic feature and another allows a geologist to determine which of the two is the younger or older by applying the* principle of cross-cutting relations.

Suggested order of features or events:
(1) Formation of sedimentary sequence 1
(2) Deformation with formation of normal fault
(3) Uplift and formation of erosional surface 1
(4) Intrusion of plutonic intrusion 2
(5) Subsidence and accumulation of sediments
(6) Formation of sedimentary sequence 2
(7) Intrusion of dike
(8) Uplift and formation of present erosional surface

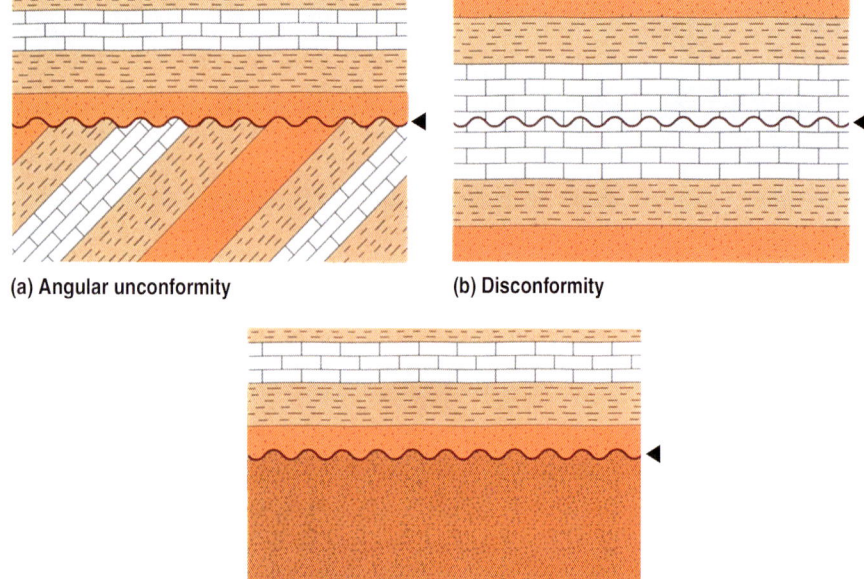

FIGURE 21.9 *The various types of* unconformities *are determined by the combination of rocks and the attitude of the sedimentary beds involved. The most difficult type of unconformity to recognize is the* disconformity *that exists between parallel layers of sedimentary rocks.*

Usually, an unconformity implies deformation, uplift, and erosion of the older rocks before the accumulation of the younger rocks. If the older rocks are sedimentary and underwent deformation by folding or faulting during uplift, the beds of the younger and older rocks will usually meet at an angle, giving rise to an **angular unconformity** such as the exposure at Siccar Point (refer to Figure 21.9a and Figure 21.11).

An unconformity where sedimentary rocks are deposited on the surface of an eroded plutonic

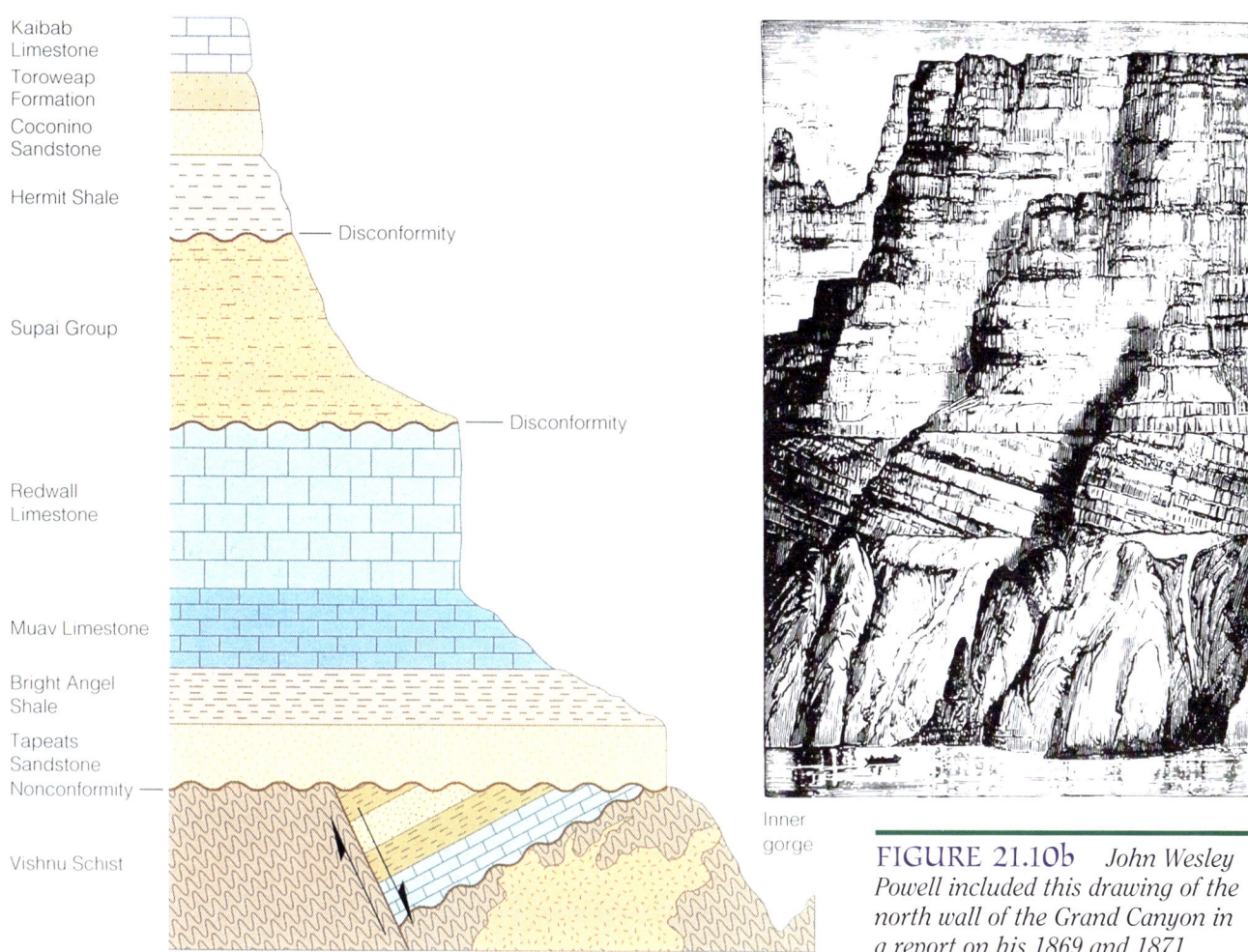

FIGURE 21.10a *The rocks exposed in the Grand Canyon represent a tremendous span of geologic time, but this record is not continuous. Unconformities within the rocks represent vast amounts of time that have been lost to the geologic record.*

FIGURE 21.10b *John Wesley Powell included this drawing of the north wall of the Grand Canyon in a report on his 1869 and 1871 expeditions. (From* The Colorado River Region, *John Wesley Powell, 1969. U. S. Geological Survey Professional Paper 669.)*

igneous or metamorphic rock mass is called a **nonconformity** (refer to Figure 21.9c). A well-known example of a nonconformity can be seen in the Grand Canyon where the Tapeats Sandstone overlies the erosional surface developed on the Vishnu Schist and associated intruded granites and pegmatites (Figure 21.10). It is important to note that the type of unconformity may change from place to place as the kinds of geologic structures and rock types under the erosional surface change.

A number of other cross-cutting relationships between rock bodies can also be observed in the Grand Canyon (refer to Figure 21.12). The continuity of the contacts of the granitic and pegmatitic bodies within the Vishnu Schist indicates that the igneous bodies are younger than the Vishnu Schist. Similar relationships exist where dikes and sills intrude and cut across host rocks.

The principle of cross-cutting relationships can also be applied to determine the relative ages of geologic features other than rock bodies. Note in Figure 21.10 that the unconformity below the Tapeats Sandstone cuts across major fractures, called normal faults, in the underlying rocks. The principle of cross-cutting relationships indicates that the faults are older than both the erosional surface and the Tapeats

(a)

(b)

FIGURE 21.11 *Angular unconformities, such as those seen at (a) Siccar Point and (b) within the deep gorge of the Grand Canyon, represent a great interval of unrecorded time. Before the sediments from which the younger rocks formed had even accumulated, the older rocks had been deformed, uplifted, eroded, and resubmerged beneath the sea. Several cross-cutting relations exist within the rocks of the Grand Canyon, each of which can be used to determine the relative ages of rocks and geologic events. For example, the unconformity at the base of the Tapeats Sandstone cuts across the metamorphic structures within the Visnu Schist, normal faults, the sequence of sedimentary rocks titled by the faulting, and the unconformity that exists below the titled sedimentary rocks. Thus, according to the principle of cross-cutting relations, the unconformity at the base of the Tapeats Sandstone, and all the rocks above it, are younger than any of the features it cuts across. Can you find other examples where the principle of cross cutting relations can be used to establish relative ages of rocks or events within the rocks of the Grand Canyon? (Courtesy of E. D. McKee/USGS)*

(a)
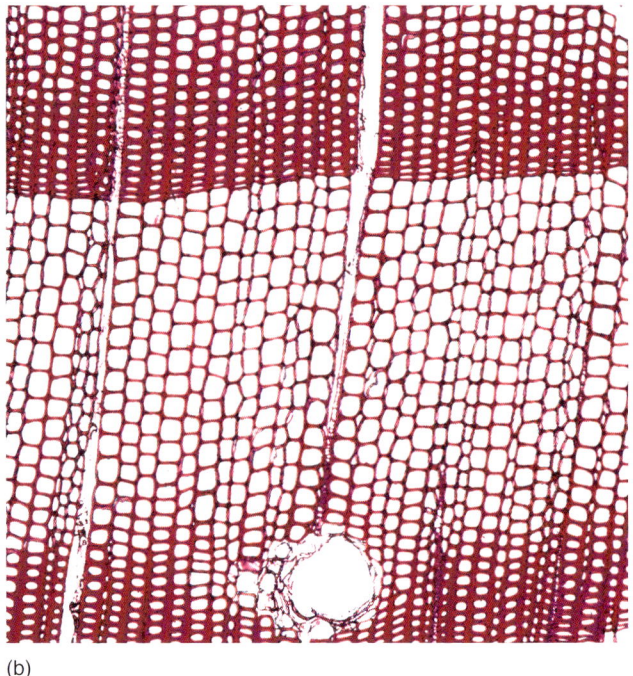
(b)

FIGURE 21.12 *Each year, trees add another layer of wood cells to the outer surface of their stems. The cells are large during the summer when growth is most rapid, but become smaller during the winter when growth slows. This difference in cell size produces the annual rings that are counted to give the age of the tree. (a VU/© H. A. Miller; b VU/© Bruce Iverson)*

Sandstone but younger than both the Vishnu Schist and the sediments of the Grand Canyon Series that they displace.

Absolute Dating

The objective of absolute (chronologic) dating is to determine the age of an event or an object in years. For example, the absolute age of a relatively young object such as a tree can be determined by counting growth rings, each of which represents a year's addition of new wood to the outer surface of the stems (Figure 3.12). Because the growth rate of plants is influenced by the prevailing climatic conditions including temperature, moisture, and available light, growth rings can be used to correlate the ages of individual trees growing within a specific region by identifying similar patterns of growth ring thickness. In the southwestern United States, a procedure called **dendrochronology** (Greek *dendros* = tree and Greek *khronos* = time), which uses wood growth rings to date events of the recent past, has enabled archaeologists to date human habitation sites back about 6,000 years.

Another example of a feature that can be used to date relatively young materials is the **varve**. Varves are very thin (1–2 mm) sedimentary beds or laminations that represent yearly cycles of deposition in lakes located in temperate-humid regions (Figure 21.13). Some of the best-developed varves have been found in sediments of the lakes that formed as the ice caps of the Great Ice Age waned. If a sequence of varves includes the currently forming layer, the varves can be used for absolute dating. However, without the present layer to establish the absolute age of the youngest varve, the varves can only be used for relative dating.

As with tree growth rings, regional climate affects the thickness of varves that accumulate within a particular year, allowing the correlation of varves from different lakes that have been subjected to common climatic influences. Periods of high rainfall, for example, will result in increased runoff and erosion and, subsequently, a thicker yearly varve. A period of drought will have the opposite effect.

Dating by Sediment Accumulation, Ocean Salinity, and Earth's Heat

Dating events and rocks with ages that span geologic time requires the use of more complex methods. In the late 1800s and early 1900s, some scientists attempted to determine the age of Earth based upon the rate of accumulation of sedimentary rocks. The

FIGURE 21.13 *In temperate climates, lakes commonly deposit a yearly graded bed. Called* varves, *these graded beds can be used to determine information such as lake age, relative age of the sediments, and climatic changes that have taken place over the lifetime of the lake. If the same event can be recognized in the varves of two lakes, the lake sediments can be correlated. (Courtesy of P. Carrara/USGS)*

researchers assumed that if they could determine the average rate at which sediments accumulate and accurately estimate the total thickness of the sedimentary rocks accumulated since the creation of Earth, the age of Earth could be calculated by simple division. Obviously, such an approach entails numerous pitfalls. For one thing, the rates at which sediments accumulate vary tremendously from the thousands of years required to accumulate to a few millimeters of deep-sea sediments to the hours or minutes needed to amass several feet of postflood debris. Secondly, there is no single place where sediments have been accumulating continuously since the formation of Earth. Therefore, estimating the total thickness of sedimentary rocks that have accumulated since the formation of Earth would require summing thicknesses of individual sedimentary rock sequences from all over the world that were assumed not only to have accumulated sequentially at different sites but to have done so with no interruption of time. Finding such rock sequences is highly improbable. To complicate the situation further, even if geologists found sequences of sedimentary rocks that recorded an unbroken sedimentary history, they would have no way to estimate either the amount of time that had lapsed between the accumulation of the individual sedimentary beds or the length of time represented by any unconformities that might exist within the individual rock sequences. Nevertheless, despite these potential problems, researchers estimated that 100,000 to 300,000 feet (30,000–90,000 m) of sedimentary rock had accumulated since the formation of Earth at an average rate of accumulation of about 1 foot (30 cm) per 1,000 years. Based on these data, the age of Earth was estimated at about 500 million years. As we will see later, although the estimate of 500 million years falls far short of the actual age of Earth, it was useful in convincing geologists that 6,000 years was not enough time to accumulate the thick layers of sedimentary rocks that exist at Earth's surface.

In 1899, an Irish chemist and geologist, *John Joly* (perhaps the first *geochemist*), suggested that the age of Earth could be determined based upon the salinity (total of all dissolved solids) of the ocean. Joly assumed that the oceans were originally filled with fresh water and became salty as a result of water-soluble materials being washed from the land. Joly made his calculations based upon estimates of (1) the volume of the oceans, (2) the yearly volume of water added to the oceans from the land, and (3) an average value for the salt content of oceanic and stream waters. Estimating that the oceans contained 16 quadrillion tons of salt (16 followed by 12 zeros) and that salt was being added at a rate of 160 million tons per year, Joly calculated the age of Earth at about 100 million years.

Like the researchers trying to determine the accumulation of sedimentary rocks, Joly made some erroneous assumptions and consequently significantly underestimated the age of Earth. For one thing, there is no reason to believe that the oceans were ever fresh. Salts began to be delivered to the ocean basin as soon as the rocks were subjected to weathering. In addition, some salt is contributed as a result of sea-floor spreading. Secondly, accurate estimates of ocean volumes did not exist in Joly's time, nor could he obtain much data on the world-wide salt content of either oceanic or stream waters. A large portion of Joly's error lay in his assumption that salt, once deposited in the ocean, stays in solution; in fact, large volumes of salt, including the sources of most of our domestic and industrial salt, are periodically removed from the ocean and deposited within sedimentary rock sequences.

At about the same time Joly was attempting to determine the age of Earth, *Lord Kelvin,* one of the leading physicists of the day, suggested a procedure that he believed would solve the dilemma once and for all. An expert in heat and heat flow, Lord Kelvin assumed that, in the beginning, Earth was a completely molten sphere of rock. He then calculated how long it would take for a mass of molten rock with the diameter of Earth to cool to the point where the release of heat from its surface equaled the measured heat flow at Earth's surface. Kelvin determined that Earth was no older than 70 million years, an estimate significantly shorter than the ages calculated from sedimentation rates or the accumulation of oceanic salt. When geologists suggested that his estimate might be too low, Kelvin intimidated them with mathematics that "proved" that he was correct and that they must be mistaken in thinking that Earth was any older.

But as with the other attempts at age estimation, Kelvin's error lay in his basic assumptions. For one thing, there is no reason to believe that Earth was ever a completely molten blob of rock. Most damaging, however, was his assumption that all of the heat radiating from Earth's surface was heat remaining from its initial fromation. We now know that most of the heat radiating from Earth's surface is generated within the crust by the disintegration of radioactive elements. In all fairness to Kelvin, he could not have been aware of the radioactive source of heat inasmuch as radioactivity was not discovered until 1896 when *Henri Becquerel* observed that photographic plates darkened when exposed to uranium-bearing minerals.

In 1903, *Maria Sklodowska Curi* and her husband, *Pierre,* became the first to isolate a measurable quantity of a radioactive element, radium—an achievement for which they received the Nobel Prize. Within a few years, *Lord Rutherford,* another well-known scientist determined that the rate at which radioactive **isotopes** disintegrate is a physical constant measured by the **half-life** of the isotope. Isotopes are atoms whose nuclei contain the same number of protons but different numbers of neutrons. As a result, the atoms have the same **atomic number** (determined by the number of protons) but different **atomic mass** (the sum of the number of protons and neutrons in the nucleus). We will discuss the significance of the isotopic half-life in the next section.

In 1905, Lord Rutherford was the first to determine the age of a rock by using the rate of radioactive breakdown of uranium. With that experiment, modern age-dating procedures were born. For the first time, it was possible to determine the age of a rock with reasonable accuracy and, eventually, to determine the age of Earth.

Radiometric Dating

The use of radioactive isotopes to determine the age of rocks is based upon the disintegration of radioactive **(parent)** isotopes into stable **(daughter)** isotopes. (Should you not be familiar with the structure of the atom, it might be well for you to read the appropriate section in Chapter 3 before continuing with the following discussion.) The disintegration of radioactive isotopes involves the release of one or both of two atomic particles from the nucleus: (1) *beta particles* or (2) *alpha particles* (Figure 21.14). **A beta particle** is an electron lost from a neutron. The emission of a beta particle therefore converts a neutron into a proton and increases the atomic number of the daughter isotope by one unit over that of the parent isotope. Since the sum of protons and neutrons has not changed, however, the atomic mass is unaffected. An example would be the disintegration of thorium (atomic number 90, atomic mass 234) into protactinium (atomic number 91, atomic mass 234).

An **alpha particle** is the combination of two protons and two neutrons. With each alpha particle released from the nucleus, the atomic number of the daughter isotope will decrease by two, and the atomic mass will decrease by four relative to the parent isotope. An example would be the disintegration of uranium (atomic number 92, atomic mass 238) into thorium (atomic number 90, atomic mass 234).

Another pair of isotopes used in radiometric dating, potassium 40 (^{40}K) and argon 40 (^{40}Ar), involves a different process. The change of ^{40}K to ^{40}Ar involves the capture of an electron from the innermost electron shell by a proton in the nucleus of the potassium atom. As a result, the atomic number of the daughter isotope (^{40}Ar) is one unit less than that of the parent isotope (^{40}K). Potassium 40 may also disintegrate to calcium 40 (^{40}Ca) by the process of beta decay. The radiation released from the nucleus of the parent isotope during disintegration affects neither the atomic number nor the atomic mass.

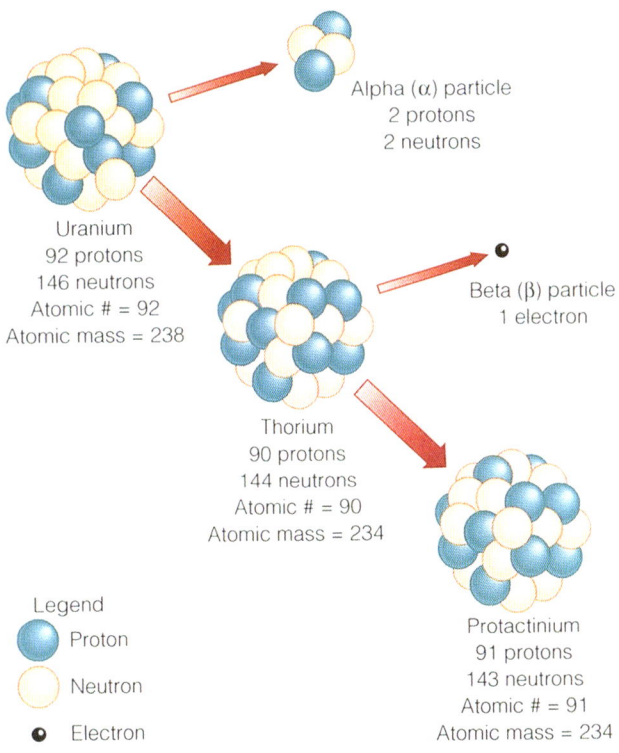

FIGURE 21.14 *Radioactive isotopes disintegrate by releasing subatomic particles from the nucleus.*

The rate of disintegration of a parent isotope is measured by the physical constant called the half-life. Defined by Lord Rutherford, the half-life of the parent isotope is the amount of time required for one-half of the parent atoms of the isotope to disintegrate into atoms of the daughter isotope. For example, the half-life of the isotope uranium 235 (^{235}U) is 713 million years. Every 713 million years, one-half of any number of ^{235}U atoms will disintegrate into atoms of the daughter isotope, lead 207 (^{207}Pb). The disintegration of a radioactive isotope is graphically portrayed in Figure 21.15, which relates the atomic fraction of the radioactive isotope to the number of expired half-lives.

To date a rock by a radiometric technique, the number of atoms of parent and daughter isotopes in the rock are counted using an instrument called a **mass spectrometer** (Figure 21.16). The sum of the number of parent and daughter isotopes in the rock equals the number of parent isotopes originally present in the rock. The atomic fraction of the original parent isotope concentration represented by the remaining parent isotopes is calculated by dividing

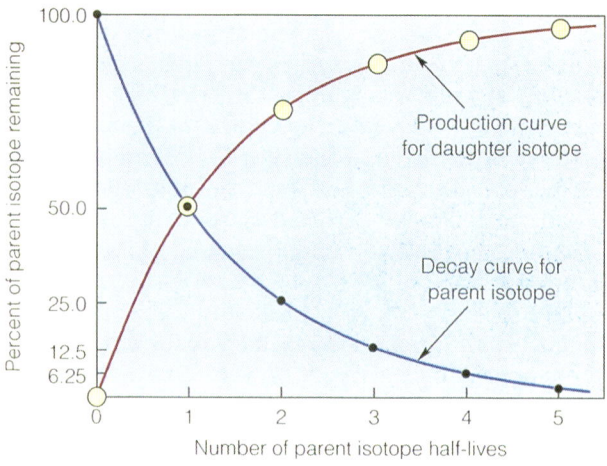

FIGURE 21.15 *The rate of disintegration of radioactive isotopes is measured by the isotope's half-life. During one half-life, one half of the existing mass of the* parent *isotope disintegrates into the* daughter *element. Therefore, after one half-life, 50% of the original mass of the parent will remain; after two half-lives, 25% will remain, and so forth.*

the number of parent isotopes remaining by the sum of the existing parent and daughter isotopes. The number of half-lives required to reduce the parent isotope concentration to the determined level is then calculated by using a plot such as the one shown in Figure 21.15. If the calculations show, for example, that only 25% of the original concentration of parent isotopes remains, two half-lives have passed since the rock formed. If the parent isotope was ^{235}U with a half life of 713 million years, the age of the rock is 2×713 million years or 1.4 billion years.

As with any age-dating technique, the accuracy depends on the validity of certain basic assumptions. One assumption is that no atoms of the parent or daughter isotopes have been removed from or added to the rock by any process other than radioactive disintegration. If the isotopic concentrations have been modified by any outside process such as weathering, the calculated ratio of parent to daughter isotopes and any subsequent age determination will be in error. To minimize such errors, geologists analyze specific minerals within the rock that are known to be especially resistant to external modification. For example, when the ^{235}U/^{207}Pb technique is used, the dating is usually based upon the analysis of *zircon* crystals ($ZrSiO_4$) painstakingly removed from the rock. Because zirconium and uranium have similar atomic radii (0.74Å for Zr and 0.97 Å for U), uranium atoms commonly substitute for zirconium atoms within the zircon crystal structure during its formation. Atoms of lead, on the other hand, with atomic radii of 1.20Å cannot enter the growing crystal structure of the zircon grain because of the size restrictions discussed in Chapter 3. Any lead atoms found in the zircon crystal can therefore be assumed to have formed from the radioactive disintegration of uranium.

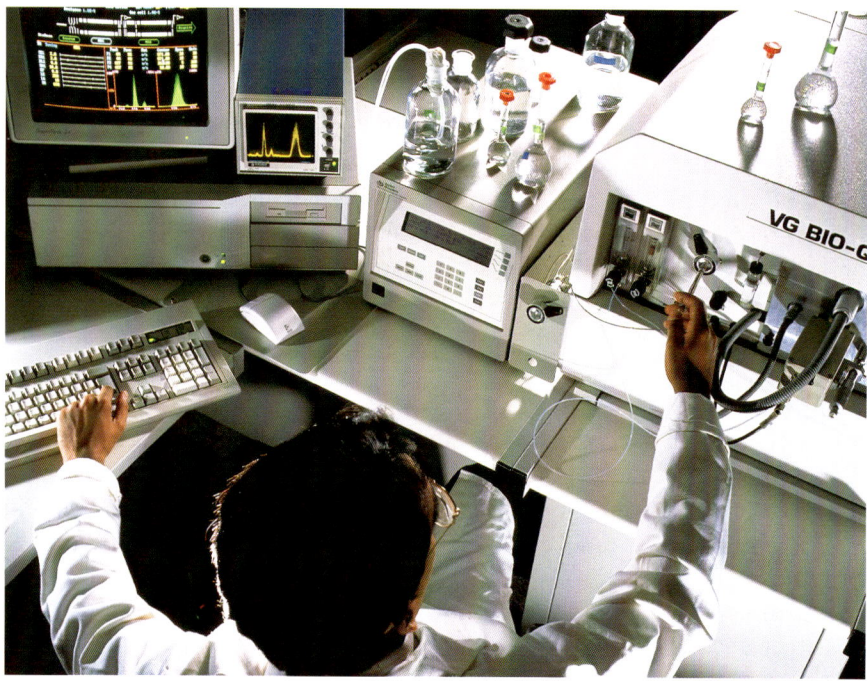

FIGURE 21.16 *The* mass spectrometer *is an instrument that allows the number of atoms of a particular element in a sample to be counted. (Geoff Tompkinson/Science Photo Library/Photo Researchers, Inc.)*

The half-life of the parent isotope limits the use of any particular isotopic parent/daughter pair for dating. If, for example, a parent isotope with a very long half-life is used to date a relatively young rock, the number of atoms of daughter isotopes created since the formation of the rock may be too few to be counted even with the highly sensitive mass spectrometer. Because the concentration of the daughter isotope will be evaluated as zero, the atomic fraction of the parent isotope will be calculated as 1.0, and according to the chart in Figure 21.15, no half-lives will have passed since the formation of the rock.

If, on the other hand, a parent isotope with a very short half-life is used to date a very old rock, the number of parent isotopes remaining within the rock will be too few to count. In this case, the atomic fraction of the remaining parent isotope will be vanishingly small, and the curve showing the half-lives in Figure 21.15 will be so close to the horizontal axis that the number of half-lives cannot be calculated, again precluding an age determination.

To allow the dating of rocks with a wide range of ages, a number of isotopic techniques have been devised that utilize parent isotopes with differing half-lives (Table 21.1). The older the rock, the longer the parent isotope half-life must be to determine the age accurately. For example, because the isotope rubidium 87 (^{87}Rb) has a half-life of 47 billion years, the isotopic combination used to determine the ages of the very oldest rocks is rubidium/strontium (^{87}Rb/^{87}Sr). The most widely used isotopic parent/daughter pair is uranium/lead (^{235}U/^{207}Pb).

TABLE 21.1

ISOTOPES USED FOR RADIOMETRIC DATING

PARENT ISOTOPE	DAUGHTER ISOTOPE	HALF-LIFE (IN YEARS)
Carbon 14	Nitrogen 14	5,730
Uranium 235	Lead 207	713 million
Potassium 40	Argon 40	1.3 billion
Uranium 238	Lead 206	4.5 billion
Thorium 232	Lead 208	14 billion
Rubidium 87	Strontium 87	47 billion

Because of the abundance of potassium in Earth's crust (2.6% by weight), another isotopic pair commonly used to date rocks is potassium/argon (^{40}K/^{40}Ar). A potential problem that must be recognized in using the ^{40}K/^{40}Ar pair, however, is that argon is a gas. Consequently, some of the argon may have escaped from the rock, thereby introducing the possibility of error into the age determination. This is especially true if the rock has been heated by volcanic or metamorphic activity. If fact, the disintegration of radioactive potassium in the crust and the subsequent "leaking" of the argon from the rocks exposed at the surface are responsible for argon being the third most abundant gas in Earth's atmosphere.

The age of human remains and artifacts is determined by using carbon 14 (^{14}C) or **radiocarbon dating**. Carbon 14 is generated in the atmosphere when cosmic rays impact nitrogen atoms (^{14}N). The ratio of ^{14}C to ^{12}C isotopes in the atmosphere is constant with ^{12}C being the more abundant. Because all isotopes of an element have identical chemical responses, both the ^{12}C and the ^{14}C atoms react with oxygen to form carbon dioxide, which mixes throughout the atmosphere. The carbon then enters the tissues of plants during photosynthesis and the tissues of animals when they eat the plants. As long as the organism is alive, the ratio of ^{14}C to ^{12}C within the plant and animal tissues is maintained at a constant value. When the organism dies, however, the ^{14}C with a half-life of 5,730 years disintegrates by giving off beta particles and reverts back to nitrogen, thereby changing ^{14}C/^{12}C ratio in a predictable way. Because ^{14}C has a short half-life, radiocarbon dating is restricted to materials with maximum ages of about 50,000 to 60,000 years.

With the exception of the radiocarbon techniques used to date plant or animal remains, most radioactive dating techniques are used to date igneous rocks. It is assumed that all the minerals contained within an igneous rock formed in a relatively short period of time. If the assumption that the abundances of parent and daughter isotopes have not been altered by any outside process is valid, the radioactive date will give the time the rock crystalized. If an igneous rock is metamorphosed at a grade sufficiently high to form new minerals, however, the atomic "clock" may be reset, and rather than dating the age of the original rock, the radioactive procedure will date the time of metamorphism. Because sedimentary rocks are made up of the remains of other rocks, dating a sedimentary rock will not give the time the rock formed but rather the time of formation of the rocks from which the fragments were derived.

Since Lord Rutherford's first successful dating of rocks in 1905, a large number of crustal rocks have been dated. Combining these results with ages obtained by dating meteorites and rocks brought back from the Moon, we have now arrived at an age for Earth of 4.6 billion years.

With radioactive techniques available to determine the absolute ages of rocks, why do most geologists use relative dating techniques? As previously discussed, not all rocks can be successfully dated using radiometric techniques. One of the major reasons, however, why most geologists do not use radiometric dating is because the mass spectrometer used to determine isotopic abundances is an expensive instrument, both to purchase and to maintain on a day-to-day basis, and highly trained individuals are required to prepare the samples, operate the instrument, and interpret the data. For these reasons, the absolute dating of rocks is very expensive and is usually far beyond the financial capabilities of most geologists. As a result, although they would prefer absolute dating, most geologists, taking advantage of the ages of rocks that *have* been dated by radiometric methods via the methods of correlation, usually establish the *relative* ages of rocks utilizing the principles set forth by Steno, Hutton, and Lyell more than a hundred years ago.

SPOT REVIEW

1. Why are estimates of Earth's age based on criteria such as the total accumulation of sedimentary rocks and the salinity of the oceans since Earth's creation always in error?
2. Why does the particular isotopic parent/daughter pair used for dating a rock depend on the age of the rock?
3. Why are most radioactive dating techniques used only to determine the age of igneous rocks?

ROCK TIME SCALES

Utilizing relative dating techniques, geologists attempt to place rocks in their proper chronological order. One of the earliest attempts to establish the chronology of Earth's rocks was a subdivision proposed by *Giovanni Arduino* in 1759 based on his studies in the Alps. Arduino subdivided all rocks into three groups. He assigned the oldest rocks to the *primitive group,* which included all the igneous and metamorphic rocks that he observed in the core of the Alps. Younger than the primitive group was the *secondary group,* which consisted of the sedimentary rocks that he observed overlying the crystalline core of the Alps and making up most of the landscape. Arduino then assigned surface accumulations of unconsolidated materials such as regolith and soil to the *tertiary group.* Later, in 1830, a fourth group, the *quaternary,* was added to include the most recent materials such as glacial and lake deposits.

The Modern Time Scale

The modern time scale was first developed during the nineteenth century as the knowledge and understanding of both organic evolution and the record of evolution represented by fossils expanded. The geologic time scale that we use today was created by assembling data from rock sequences from localities around the world that contained fossil assemblages that were both distinctly different from those in the adjacent rock sequences and showed evolutionary changes (Figure 21.17).

Within limited continental areas, geologists can often determine the relative age of a rock sequence simply by recognizing certain rock layers or sequences of lithologies based on their physical appearance. Within the Appalachian Mountains, for example, the Tuscarora Sandstone, which is responsible for most of the topographic relief of the area, can be easily recognized by its white quartzose composition and its location above the red sandstones and shales of the Juniata Formation. Other rock units with similar lithologies may be identified and distinguished by diagnostic fossils or sedimentary features. The identification of a particular rock unit such as the Tuscarora Sandstone allows a geologist to immediately establish the position of the rocks in the relative age scale.

On the other hand, between distant regions within a continent such as between the eastern and western United States and certainly between continents, the lithologies of the rocks that accumulated during any interval of time will be different, and lithologic similarities cannot be used to establish the relative ages of rocks. Nevertheless, even though the lithologies of the rocks exposed in different regions of a continent or between continents may vary, the similarity of the fossil assemblages will usually allow the age equivalency to be established.

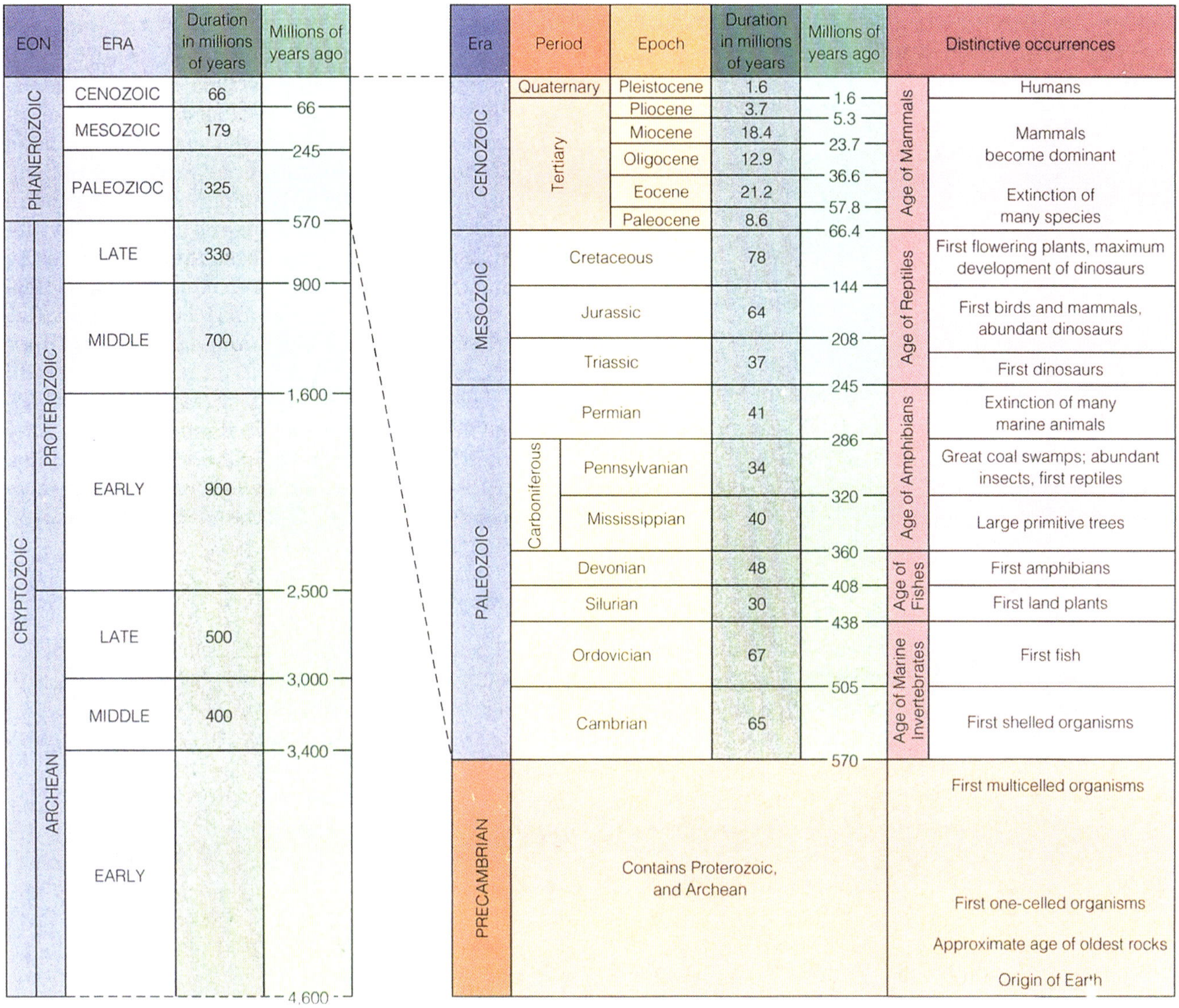

EON	ERA	Duration in millions of years	Millions of years ago
PHANEROZOIC	CENOZOIC	66	66
	MESOZOIC	179	245
	PALEOZOIC	325	570
CRYPTOZOIC — PROTEROZOIC	LATE	330	900
	MIDDLE	700	1,600
	EARLY	900	2,500
CRYPTOZOIC — ARCHEAN	LATE	500	3,000
	MIDDLE	400	3,400
	EARLY		4,600

Era	Period	Epoch	Duration in millions of years	Millions of years ago		Distinctive occurrences
CENOZOIC	Quaternary	Pleistocene	1.6	1.6	Age of Mammals	Humans
	Tertiary	Pliocene	3.7	5.3		
		Miocene	18.4	23.7		Mammals become dominant
		Oligocene	12.9	36.6		
		Eocene	21.2	57.8		Extinction of many species
		Paleocene	8.6	66.4		
MESOZOIC	Cretaceous		78	144	Age of Reptiles	First flowering plants, maximum development of dinosaurs
	Jurassic		64	208		First birds and mammals, abundant dinosaurs
	Triassic		37	245		First dinosaurs
PALEOZOIC	Permian		41	286	Age of Amphibians	Extinction of many marine animals
	Carboniferous — Pennsylvanian		34	320		Great coal swamps; abundant insects, first reptiles
	Carboniferous — Mississippian		40	360		Large primitive trees
	Devonian		48	408	Age of Fishes	First amphibians
	Silurian		30	438		First land plants
	Ordovician		67	505	Age of Marine Invertebrates	First fish
	Cambrian		65	570		First shelled organisms
PRECAMBRIAN	Contains Proterozoic, and Archean					First multicelled organisms
						First one-celled organisms
						Approximate age of oldest rocks
						Origin of Earth

FIGURE 21.17 *The geologic time scale is largely based on the chronological evolution of life as recorded in the fossils contained in sedimentary rocks.*

Eons of Time The longest period of geologic time is the *eon*. All geologic time is subdivided into two eons, the **Cryptozoic Eon** and the **Phanerozoic Eon** (Figure 21.17). The *Cryptozoic Eon* is subdivided into the **Archean** and **Proterozoic**. The subdivision is primarily based on the formation of large, granitic continents; during the older *Archean,* there were not well-established granitic cratons. Another important development that accompanied the formation of continents in the *Proterozoic* was the development of the shallow water marine environments that, in turn, allowed the development of shallow-water life forms. The importance of fossils in establishing relative age is clearly illustrated by the names given to the various subdivisions of geologic time. The stem *-zoic* is from the Greek word meaning "life," in recognition of the importance of life as a basis for subdivision. The prefix *crypto-* comes from the Greek word meaning "hidden" and refers to the fact that although the most ancient rocks may be devoid of fossils, we realize that life did exist. Because hard body components such as shells, bones, and teeth that are the basis of nearly all fossil remains had not yet widely evolved, indications of their presence are *hidden* from observation. In some cases, a wispy film of carbon on the bedding surface of a sedimentary rock or on the cleavage surface of a metamorphic rock may be all that remains to indicate the presence of former life.

The Cryptozoic Eon includes the vast interval of time from the formation of Earth 4.5 to 5.0 billion years ago until recognizable fossils began to appear consistently in the sediments about 600 million years ago. Attempts to subdivide the Cryptozoic Eon have met with little success due to the low abundance and diversity (the number of different kinds of organisms in an assemblage) of fossils and the deformation to which most of Earth's most ancient rocks have been subjected over the billions of years since their formation.

The appearance of fossils in sedimentary rocks about 600 million years ago is used to subdivide Earth history to Cryptozoic and Phanerozoic Eons in much the same fashion as the appearance of writing about 5,000 years ago is used to establish the "prehistoric" and "historic" portions of human existence. The Phanerozoic Eon beings the "historic" portion of Earth's existence. The prefix *phanero-*, from the Greek work meaning "apparent," refers to the presence of diverse and abundant identifiable fossil remains. A span of about 200 million years prior to the transition and the 20 to 40 million years between the two eons of time is the period during which animals (and to a lesser degree plants) evolved body parts that were sufficiently resistant to decomposition so as to be capable of entering the fossil record.

Eras of Time   Eons of time are subdivided into *eras*. Based upon the overall characteristics of the lifeforms, the Phanerozoic Eon has been subdivided into three eras: (1) the *Paleoszoic*, (2) the *Mesozoic*, and (3) the *Cenozoic*.

The **Paleozoic Era** extends from approximately 570 million to about 245 million years ago. The prefix *paleo-*, from the Greek word *palaios* meaning "ancient" or "old," refers to the dominance of marine invertebrates. Although the life of the Paleozoic was dominated by the invertebrates, the era was a time of spectacular evolutionary change. The Paleozoic had not been long underway when all of the major animal phyla had evolved. Familiar forms of fish were present in the lakes and streams, and simple plants carpeted the land. During the late Paleozoic, dragonflies with three-foot wingspreads buzzed through the extensive swamps that characterized the time. The peat that accumulated in these swamps gave rise to a large percentage of Earth's coal deposits.

The swamps of the early Devonian were the sites where amphibians evolved from air-breathing fish that struggled for life in the oxygen-poor waters. Although the amphibians were the first animals to adapt to the land, they never made a complete break with the marine environment. Because of their fragile, unprotected eggs, they had to return to the water to spawn. As a result, the amphibians were never able to completely colonize the land. Always having to remain near the water, they were unable to venture into the drier or more upland parts of the land.

By the end of the Pennsylvanian, one group of amphibians had evolved into the reptiles by evolving the shelled egg. The new, protected egg not only provided a better chance of survival, but it also allowed the reptiles to more successfully and completely invade the land.

The shrinking of the vast wetlands and the onset of the drier climate toward the end of the Paleozoic spurred rapid reptilian evolution. At the close of the Pennsylvanian and into the Permian, a "mammal-like" reptile evolved that was to foretell the coming of the mammals. The animal had changed its stance from the characteristic reptilian sprawl by developing strong, muscular legs that lifted and supported it's body off the ground. The shape of the teeth had changed from the spikelike teeth of the reptile to more diversified shapes better designed to eat a variety of food. An enlarged skull indicated a larger and better developed brain.

The **Mesozoic Era** extends from 245 million years ago to 65 million years ago. The prefix *meso-*, from the Greek word *mesos* meaning "middle," refers to the middle position the dominant Mesozoic reptilian lifeforms represent between the marine invertebrates of the Paleozoic and the mammalian species of the Cenozoic Era. During the Mesozoic, the reptiles occupied all of Earth's major environments as they spread across the land, returned to the water, and took to the air (Figure 21.18). Huge marine reptiles such as *Elasmosaurus* (whose descendants are still thought by devout Scots to live in the black depths of Loch Ness) dominated the sea (Figure 21.19). *Pteranodon*, the great "flying" reptile, gained access to the air by taking advantage of folds of skin that stretched between its fingers and rear limbs not unlike those possessed by the modern bat (Figure 21.20).

Best known, however, was a new class of animal called the *dinosaur* that evolved early in the Mesozoic and was destined to dominate the land during that

FIGURE 21.18   The dinosaurs, which dominated every ecological niche during the Mesozoic Era, went to extinction at the close of the Cretaceous Period. Exactly what caused the worldwide extinction is still the topic of debate.

FIGURE 21.19   *Elasmosaurus,* a well-known Cretaceous marine reptile, was over 36 feet (12 m) long. Devout Scots believe that the creature supposedly living in the depths of Loch Ness may be an elasmosaur.

FIGURE 21.20   Not all Mesozoic reptiles were tied to the land or dwelt in the sea. *Pteranodon* is an example of a reptile that was able to overcome the force of gravity and achieve soaring flight. Although its body was only about the size of a goose, *Pteranodon* had a wingspan of over 20 feet (7 m).

era. The name means "terrible lizard," and although it was coined with the large carnivorous kinds in mind, most dinosaurs did not fit the description.

As the era closed, birds evolved from the biped dinosaurs, which, according to some, had already developed internal body thermostats that allowed their blood to warm. The end of the Mesozoic Era saw a mass extinction of marine and flying reptiles, flightless dinosaurs, and many marine invertebrates. The scene was set for the mammals.

Although mammals had evolved early in the Mesozoic, their small size indicates that they were unable to compete with the reptiles and dinosaurs. Most early mammals were the size of mice with the largest being no bigger than a cat. They probably survived by keeping out of the way of the larger and more numerous reptiles and dinosaurs and subsisted on seeds an insects. As the Mesozoic came to a close and the numbers of reptiles declined and the dinosaurs vanished, the mammals underwent rapid evolution and soon filled the vacated ecological niches.

The **Cenozoic Era** extends from 65 million years ago to the present. The name *Cenozoic* is derived from the Greek word *kainos* meaning "recent." At the beginning of the Cenozoic Era, all of the dinosaurs had become extinct. With both the land and the oceans free of the "terrible lizards," the mammals were able to come out of hiding and undergo a meteoric evolution. The Cenozoic is definitely the "Age of the Mammals."

All of the major features of the modern landscape also evolved during the Cenozoic. Continents moved to their present positions and took on their present sizes and shapes. The great mountains (the Rockies, the Alps, the Himalayas, the Andes, and the Appalachians) were sculpted to their present appearance by the processes of erosion during the Cenozoic. In fact, most of the landforms we see today are the result of the activity of geological processes during the last 2 million years.

Periods of Time An era is subdivided into **periods** of time with the sequence of rocks deposited during a period being referred to as a **system** of rocks. The subdivision of the rock column into periods and systems is based upon both fossil and rock contents and includes the missing time represented by the unconformities.

The present geologic time scale was largely established by European geologists of the eighteenth and nineteenth centuries who began to recognize the repetition of particular rock types from place to place and the sequences of life's evolution represented by their fossil contents. They began to assign the time of formation of the rocks to periods and the rocks themselves to systems. The geographic locality where the rocks of a particular system were first described is called the **type locality**. Whoever first described the rocks was afforded the privilege of assigning the name, usually of local origin. Rocks of similar age are recognized throughout the world by referring back to the rock sequence exposed in the type locality and in particular to the diagnostic fossil assemblage that the sequence contains.

Periods of the Paleozoic Era

The oldest rocks of the Paleozoic Era belong to the **Cambrian System**, first described in 1831 by *Adam Sedgwick,* one of the foremost geologists of the day. At the type locality of the Cambrian System in northern Wales, the rocks are a sequence of sparsely fossiliferous sedimentary and volcanic rocks. Sedgwick called the system the Cambrian after *Cambria,* the Roman name for Wales. Because the Cambrian is the first period of the Phanerozoic Eon, the rocks belonging to the Cryptozoic Eon are usually referred to as **Precambrian**.

While Sedgwick was studying and describing the rocks of northern Wales, a colleague and friend, *Roderick Murchison*, was studying a sequence of fossiliferous rocks in southern Wales that he named the **Silurian System** after an ancient Welsh tribe, the *Silures.* In 1835, Sedgwick and Murchison presented a joint paper describing the Cambrian and Silurian Systems in England and Wales.

In later years, a dispute broke out between the two friends that alienated them for the rest of their lives. As they continued to study the rocks of England and Wales, it became evident that the upper portion of Sedgwick's Cambrian System overlapped with the lower portion of Murchison's Silurian. Because Sedgwick did not provide the detailed description of the fossil evidence needed to prove that the fossils were of Cambrian age, Murchison suggested that only Sedgwick's lower unfossiliferous rocks were Cambrian in age and that the rocks of Sedgwick's upper Cambrian belonged in his own Silurian System. Needless to say, Sedgwick did not agree.

The dispute that subsequently developed between Sedgwick and Murchison not only divided the two men but also split the geologists of Europe into two camps until a solution to the problem was suggested in 1879 by another geologist, *Charles Lapworth.* Lapworth proposed that the disputed sequences of rock be combined into a new system that he proposed to name the **Ordovician**, after another ancient Welsh tribe, the *Ordovices.* Although generally accepted by most geologists as a period of geologic time, the Ordovician Period is not universally recognized because it is not based on rocks exposed in a type locality.

In 1839, before the beginning of their feud, Sedgwick and Murchison had studied a section of rock in Devonshire in southwestern England and tentatively identified it as either Cambrian or what was later to be described as Ordovician. However, based on fossil evidence, *William Londsdale* showed the rocks were intermediate in life-forms between the Silurian and the Carboniferous Systems and named the system **Devonian.**

In 1822, *William Coneybeare* and *William Phillips* described a section of rocks in northern England that they named the **Carboniferous System** because of its coal content. In North America, because the coals are largely restricted to the upper portion of the system, the Carboniferous System is subdivided into two systems, the older **Mississippian** and a younger **Pennsylvanian**, which have a widespread unconformity between them. Although the Mississippian System in North America does contain some coal, the dominant lithology of the Mississippian is limestone. The Mississippian System was named after rock exposures along the Mississippi River that were first described and subdivided as the Lower Carboniferous by *Alexander Winchell* in 1870 about the time that he became the first chancellor of Syracuse University. In 1891, *Henry Shaler* proposed that the rocks represented a separate system and assigned the name Mississippian. The upper portion of the Carboniferous in North America was named the Pennsylvanian System in 1891 by *H. S. Williams* after excellent exposures in Pennsylvania where the rock sequence contains rich deposits of coal.

In 1841, the tsar of Russia invited Murchison to study a sequence of rocks west of the Ural Mountains. Murchison found the rocks contained fossils of a younger stage of biological succession than those of the Carboniferous. On that basis, he designated the sequence as a separate system that he called the **Permian** after the Perm Province in which the rocks were described.

Periods of the Mesozoic Era

The Mesozoic Era is subdivided into three periods of time with their respective systems of rock, the *Triassic, Jurassic,* and *Cretaceous.*

The **Triassic System** was named by a German geologist, *Frederich von Alberti,* in 1834 after a sequence of rocks in Germany. The name refers to the threefold lithologic subdivision of the sequence in the type locality (not characteristic of the system elsewhere) where a marine rock sequence is sandwiched between two continental sequences. In eastern North America, some of the best known Triassic rocks come from the down-faulted troughs called the Triassic Basins. Extending from North Carolina to Nova Scotia, the Triassic rocks of the eastern United States are known for their content of fossil dinosaur tracks (Figure 21.21). In the western United States,

FIGURE 21.21   *The study of dinosaur tracks has allowed paleontologists to determine much more than just the size of the animal. The running stride and the distances covered while running prompted some paleontologists to suggest that many of the biped dinosaurs were actually warm-blooded. Since then, other information, such as the internal structure of bones, has supported the idea of warm-blooded dinosaurs. (VU/© Scott Berner)*

Triassic rocks are responsible for much of the beauty of the southwestern desert including the Painted Desert, Zion National Park, and the Petrified Forest (Figure 21.22).

The name **Jurassic** was originally applied in 1795 by a German geologist, *Alexander von Humboldt,* to a sequence of rocks exposed in the Jura Mountains, located between France and Switzerland. At the time

FIGURE 21.22 *The rocks of the Triassic provide the Southwest with some of its most spectacular scenery and information about the environment. (a) The sandstones of the Wingate and Navajo Formations exposed in the cliffs of* Zion National Park *are fossil dunes that record a vast desert. (b) The colored rocks of the* Painted Desert *record the presence of streams and lakes. (Courtesy of National Park Service) (c) The silicified logs that characterize the* Petrified Forest National Park *were trees similar to modern redwoods that became trapped in logjams during floods, were buried by stream deposits, and eventually were preserved as silica replaced the wood. (Courtesy of E. D. McKee/USGS)*

Humboldt described the rocks, the concept of periods of time and systems of rocks had not yet been developed. Later in 1839, the rocks were redefined as a system by another German geologist, *Leopold von Buch.*

The richly fossiliferous rocks of England that inspired William Smith were Jurassic in age. The Jurassic rocks are known worldwide for their abundance of dinosaur remains. In the United States, for example, the fossil content of the *Morrison Formation* in Colorado and Wyoming has provided a rich picture of the life of Jurassic time (Figure 21.23). One of the single most famous fossil species, the remains of the first bird, *Archaeopteryx,* was found in the Jurassic-age rocks within the lithographic stone quarries near Solenhofen, Bavaria (Figure 21.24).

The **Cretaceous System** was established based upon studies of the rocks in the chalk cliffs along the Strait of Dover, England, and within the structural basins surrounding London and Paris. The name, which was first applied by a Belgian geologist, *Omalius d'Halloy,* in 1822, is derived from the Latin word *creta* for "chalk," a common lithology within the Cretaceous System around the world. The rocks of the Cretaceous record a burst of evolution among the mammals as the Age of the Reptiles began to decline, but not before the evolution of the most fearsome of the "terrible lizards," *Tyrannosaurus rex.*

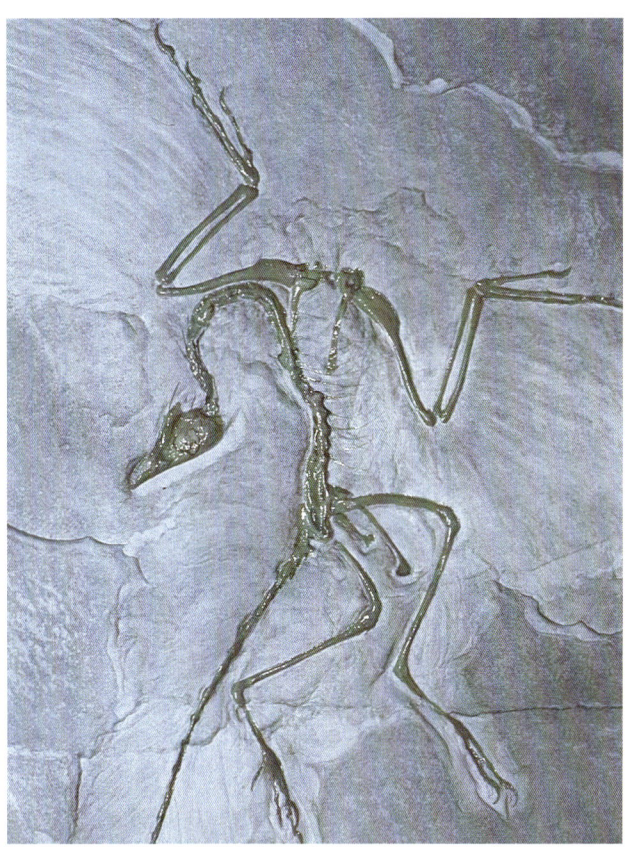

FIGURE 21.24 *Archaeopteryx is considered the first bird. About the size of a crow, the fossil remains first found in Bavaria show the similarity between birds and reptiles that had been observed by naturalists since the time of Darwin. (VU/© John D. Cunningham)*

In addition to its spectacular fossil content, the Cretaceous is important as a source of economic deposits of coal throughout the world, including most of the major coal deposits of the western United States.

Periods and Epochs of the Cenozoic Era

The Cenozoic Era is subdivided into two periods, the older **Tertiary** and the younger **Quanternary**. Although the names are carryovers from Arduino's original classification of rocks, they no longer have their original connotations.

A period of time is subdivided into *epochs.* In some cases, periods are subdivided into *lower, middle,* and *upper* time units while in other cases, such as the Tertiary Period, specific names are applied to the individual epochs.

The subdivision of the Tertiary Period into *epochs* of time represents the first application of statistics to geology (Table 21.2). The names applied to each of the epochs of the Tertiary were based upon

FIGURE 21.23 *The* Morrison Formation *has preserved a rich record of Jurassic dinosaur remains. A well-known locality for observing dinosaur remains is at Dinosaur National Park. (Courtesy of National Park Service)*

TABLE 21.2
THE SUBDIVISION OF THE TERTIARY BASED ON THE PERCENTAGE OF MODERN SPECIES REPRESENTED IN THE FOSSIL ASSEMBLAGE OF THE PARIS BASIN

Period	Epoch		Percentage of Modern Species
Quarternary	Pleistocene	(Greek *pleistos*, most + *kainos*, recent)	90–100
	Pliocene	(Greek *pleios*, more + *kainos*)	50–90
	Miocene	(Greek *meios*, less + *kainos*)	20–40
Tertiary	Oligocene	(Greek *oligos*, little + *kainos*)	10–15
	Eocene	(Greek *eos*, dawn + *kainos*)	1–5
	Paleocene	(Greek *palaios*, ancient + *kainos*)	0

detailed studies of the fossil assemblages in the rocks of the Paris Basin. Originally studied by the French geologist *Gérard-Paul Deshayes,* the assemblages attracted the attention of Lyell who noticed that as the individual beds became younger, they showed a greater percentage of living species. Lyell then proposed a classification scheme, based upon the percentage of still-living shelled invertebrates, which subdivided the Tertiary Period into the **Eocene, Miocene,** and **Pliocene Epochs** (refer to Figure 21.17). The stem *-cene* is from the Greek *kainos* meaning "recent" while the prefixes *eo-, mio-,* and *plio-,* referring to the fossil content, are from the Greek words for "dawn," "less," and "most," respectively. The other epochs, the **Paleocene** (Greek *palaios* = ancient) and the **Oligocene** (Greek *oligos* = little), were added later.

The *Quaternary Period* is the shortest of the periods of time and consists of a single epoch, the **Pleistocene**. Although it represents only the last 2 million years of Earth history, it was during the Pleistocene that much of the modern world developed. During this short period of time, the continental glaciers gripped the northern continents, the continents took on their modern shapes and sizes, and the finishing touches were applied to the modern landscape.

Not only was the land taking on its modern appearance, but so were the life-forms. Forests were essentially modern by the beginning of the Pleistocene. Grasslands carpeted the plains, giving rise to herds of grazing animals including elephants, camels, and horses. Familiar (toothless) birds dominated the skies while a few large flightless birds served as predecessors to modern flightless birds such as the ostrich.

Perhaps the most significant evolutionary development of the Pleistocene was the arrival and meteoric evolution of *Homo sapiens.* Beginning with *Australophithecus* in the Pliocene and culminating with *Homo sapiens,* the human was to become the life-form that has had the greatest impact upon Earth since the blue-green algae changed the composition of the atmosphere from predominantly carbon dioxide to 21% oxygen and 78% nitrogen.

SPOT REVIEW

What is the basis for subdividing geologic time into eons, eras, periods, and epochs?

CONCEPTS AND TERMS TO REMEMBER

Early estimates of Earth's age
   catastrophism
   uniformitarianism
Dating methods
   relative dating
      principle of superposition
      principle of original horizontality

correlation
principle of faunal succession
principle of cross-cutting relations
unconformity
hiatus
disconformity

angular unconformity
nonconformity
absolute dating
   dendrochronology
   varves
radiometric dating
   isotope
   half-life

atomic number
atomic mass
parent isotope
daughter isotope
beta particle
alpha particle
mass spectrometer
radiocarbon dating
Geologic time scales
eons of time
Cryptozoic
Archean
Proterozoic
Phanerozoic

eras of time
Paleozoic
Mesozoic
Cenozoic
periods of time
system of rocks
type locality
Cambrian
Precambrian
Silurian
Ordovician
Devonian
Carboniferous
Mississippian

Pennsylvanian
Permian
Triassic
Jurassic
Cretaceous
Tertiary
Quaternary
epochs of time
Eocene
Miocene
Pliocene
Paleocene
Oligocene
Pleistocene

REVIEW QUESTIONS

1. The age of Earth is about _____ years.
   a. 500 million
   b. 1 billion
   c. 5 billion
   d. 10 billion
2. A surface of unconformity between parallel sedimentary beds is called a (an)
   a. angular unconformity.
   b. nonconformity.
   c. hiatus.
   d. disconformity.
3. The person credited with originating the principles of superposition and original horizontality is
   a. Lord Kelvin.
   b. Nicolas Steno.
   c. James Hutton.
   d. Charles Lyell.
4. The percentage of radioactive isotopes that disintegrate during each half-life period
   a. increases at a constant rate.
   b. decreases at a constant rate.
   c. depends on the atomic number of the isotope.
   d. is constant.
5. When 75% of a parent isotope has disintegrated to the daughter isotope, _____ half-lives have passed.
   a. 1   c. 3
   b. 2   d. 4
6. The radioactive isotope used to date human remains and artifacts is
   a. carbon 14.
   b. uranium 235.
   c. potassium 40.
   d. rubidium 87.
7. The greatest portion of Earth history is included in the
   a. Paleozoic Era.
   b. Phanerozoic Eon.
   c. Cryptozoic Eon.
   d. Mesozoic Era.
8. The appearance of the present surface of Earth evolved during the
   a. Miocene Epoch.
   b. Pleistocene Epoch.
   c. Pliocene Epoch.
   d. Eocene Epoch.
9. What assumptions are made in using (a) sediment thickness, (b ocean water salinity, and (c) radioactivity to determine the age of rocks?

REVIEW QUESTIONS *continued*

10. How are the principles of superposition, original horizontality, and cross-cutting relations used to determine the relative age of rocks?
11. Why can't a radioactive isotope with a short half-life be used to date very old rocks?
12. What is the basis behind the prefixes *paleo-, meso-,* and *ceno-* used with the various geologic eras of time?

THOUGHT PROBLEMS

1. The concept of uniformatarianism seems to preclude the existence of castrophism, yet there are ample examples of natural catastrophes such as floods, earthquakes, and volcanic eruptions. How can this dilemma be resolved? Can there be "catastrophic uniformitarianism"?
2. The oldest rocks yet found are only about 4.1 billion years old, although the age of Earth is estimated at 4.5 to 5.0 billion years. How can this discrepancy be explained? Do you think older rocks exist? If so, where do you think they are most likely to be found, and how old do you think they will be?
3. One of the justifications given for the space program and visits to distant celestial bodies is that the information obtained will shed light on the age and origin of Earth. How valid do you think these arguments are?

FOR YOUR NOTEBOOK

Geologic maps give information about the kind, distribution, and age of the rocks in any area. A geologic map of the United States has been included in the appendix of this text. Acquire a geologic map of your area from your instructor or from the state geological survey, and investigate the ages of rocks that occur in your area in more detail. Most people are intrigued by fossils. Investigate the fossil assemblage that are found in the rocks in your area, and attempt to determine the kind of environment that existed when the rocks were formed or accumulated.

APPENDIX A:

Glossary

A

A horizon The dark-colored soil horizon located below the O horizon that is a mixture of humus and decomposed regolith.

aa The Hawaiian term for lava characterized by jagged, rough surfaces and sharp edges. See also *Pahoehoe*.

abrade To wear away by physical or mechanical means.

abrasion The process by which rock surfaces are worn away by the frictional contact or impact of rock particles transported by wind, running water, waves, glacial ice, or gravity.

absolute dating Any of a number of procedures that give the definite age of a rock, fossil, mineral, or geologic event in units of time, usually years. See also *relative dating*.

abyssal hill A dome-shaped feature up to several hundred feet (meters) in height and several miles (kilometers) in diameter found on the ocean floor. Although found in all deep-ocean basins, abyssal hills are most prevalent in the Pacific Ocean basin.

abyssal plain The perfectly flat portion of the ocean floor, located beyond the base of the continental rise with a slope of less than 1:1,000; most commonly found in opening oceans. The abyssal plain consists of a layer of sediments provided by turbidity currents from the continental margin and fine-grained materials settling from the ocean surface that covers the pre-existing topography (the abyssal hills) of the ocean floor.

accretionary wedge A mixture of rock materials stripped from the descending lithospheric plate in zones of subduction that is accreted to the edge of the overlying plate landward of the deep-sea trench. See also *melange*.

acidic (*Igneous*) A term synonymous with *felsic* and referring to rocks that contain more than 60% SiO_2. (*Chemistry*) Refers to solutions with pH values less than 7.0.

activation energy The energy required to initiate a chemical reaction or process.

active continental margin The continental margin adjacent to a convergent plate boundary. Also called a leading edge.

alkali elements The monovalent elements of sodium (Na) and potassium (K).

alkaline earth elements The divalent elements of calcium (Ca), magnesium (Mg), and barium (Ba).

alluvial fan A gently sloping, fan-shaped deposit of unconsolidated materials usually deposited where a mountain stream flows out upon an adjoining plain, especially in semiarid or arid regions.

alluvium A general term for unconsolidated detrital materials deposited in relatively recent geologic time by running water.

alpha particle A subatomic particle containing 2 protons and 2 neutrons emitted from the nucleus of an atom during a type of radioactive decay.

alpine or valley glacier Any glacier found in mountainous terrains except for ice caps or ice sheets. Alpine glaciers usually originate in a cirque and flow down a former stream valley. See also *continental glacier*.

altitude The vertical height of an object above a given reference datum, in particular, the angle of the Sun above the horizon.

Alvin A manned submersible for deep-sea exploration.

amorphous Any solid material that lacks a definite crystalline structure. An example is obsidian.

amphibole group A common group of dark-colored, rock-forming, ferromagnesian, silicate minerals that occur most frequently in igneous and metamorphic rocks and are constructed of a cross-linked double chain of silicon-oxygen tetrahedra with a silicon:oxygen ratio of 4:11. An example is hornblende.

amplitude (or wave height) The distance from the crest of a wave to the adjoining trough. See also *frequency* and *wavelength*.

andesite A fine-grained igneous rock composed mostly of plagioclase feldspar with 25% to 40% amphibole, pyroxene, and biotite, but containing no quartz or orthoclase. Andesite is the extrusive equivalent of diorite and is believed to be formed by the partial melting of basalt within the zone of subduction. The name andesite is taken from the Andes Mountains where it is abundantly found.

angle of repose The angle up to which loose unconsolidated materials will be at rest; when the angle is exceeded, the materials will begin to slide and roll downslope.

angular unconformity An unconformity where the younger sediments rest on the eroded edges of older tilted or folded rocks.

animal trails Trace fossils left by ancient organisms as they carried on their daily lives. Dinosaur tracks are good examples.

anion An ion with a net negative charge due to the gaining of electrons. See also *cation*.

annular stream pattern A drainage pattern characterized by streams following a circular or concentric path along a belt of weak rocks. Best displayed by streams draining domes.

Antarctic Circle The latitude at approximately 66.5° south of the equator.

anthracite The highest rank of coal with a fixed carbon content in excess of 95% on a dry, mineral-matter-free basis.

anticlinal theory The theory of petroleum entrapment first proposed by I.C. White that states that oil and gas will accumulate in the axial region of an anticline.

anticline A fold, usually convex upward in which the oldest rocks are found in the center. See also *monocline* and *syncline*.

aphanitic A textural term that describes an igneous rock with individual mineral grains too small to be seen by the unaided eye. Aphanitic texture indicates that rapid cooling of the molten rock limited the size of crystal growth. See also, *glassy, phaneritic,* and *porphyritic*.

aquiclude An impermeable rock unit or unconsolidated material that effectively prevents the flow of groundwater. See also *aquifer* and *aquitard*.

aquifer A rock unit or unconsolidated deposit with sufficient porosity and permeability to conduct groundwater and to provide economically significant volumes of water to a well or spring. See also *aquiclude* and *aquitard*.

aquitard A semipermeable rock unit or unconsolidated deposit that does not readily allow the flow of water to a well or spring but may serve as a storage unit for groundwater. See also *aquiclude* and *aquifer*.

Arctic Circle The latitude at about 66.5° north of the equator.

arête A knife-edged mountain ridge formed by alpine glacial erosion.

Argo (and ***Jason***) Robot vehicles launched from mother ships that are used to explore the deep ocean.

argon The third most abundant gas in the atmosphere (0.9%); argon is an inert element produced by the decomposition of radioactive potassium.

aridosols Alkaline or saline soils that develop under arid and semiarid conditions.

arkose A current-deposited sandstone of continental origin containing at least 25% feldspar, chiefly microcline, that gives the rock a pink or red color.

ash Pyroclastic material with diameters less than 2.0 mm produced during volcanic eruptions.

ash flow A density current consisting of a highly heated mixture of gases, ash, and unsorted pyroclastic materials produced by the explosive eruption of viscous lavas.

ash flow tuff A pyroclastic igneous rock formed by the consolidation of materials deposited by an ash flow.

asteroid Any of a large number of small celestial bodies that orbit the Sun, mostly between the orbits of Mars and Jupiter.

asthenosphere Part of the upper mantle, directly below the lithosphere, that is characterized by its plastic response to stress and strong attenuation of seismic waves.

asymmetric A descriptive term applied to objects whose shapes change from one side of the object to the other.

asymmetrical ripple marks Ripple marks that have one side steeper than the other when viewed in cross section.

atmosphere The layer of gas that surrounds Earth and is composed primarily of nitrogen (78%) and oxygen (21%) with trace amounts of other gases.

atoll A ring-shaped group of coral islands that encloses a shallow interior lagoon with the deep ocean on the seaward side.

atom The smallest unit of an element that retains all of the unique physical and chemical properties of the element.

atomic mass The mass of an atomic constituent expressed in atomic mass units (amu or u) with one amu (u) equal to $1/12$ the mass of the ^{12}C isotope.

atomic number The number of protons in the nucleus of an atom. The atomic number determines the identity of an atom.

aureole The zone of metamorphosed rock that surrounds an igneous intrusion.

axial plane An imaginary plane that parallels the length of a fold and divides the cross section of a fold into halves.

axis The line of intersection between the axial plane and the limbs of a fold.

B

B horizon The portion of the soil profile below the A, O, and E horizons where materials leached from the overlying horizons are deposited, giving rise to its designation as the *zone of deposition*.

back-arc basin The depositional basin between an island arc and the continental landmass that accumulates sediments from both the island arc and the continental landmass.

backwash The flow of water down a beach face after being driven onto the beach by the surf.

bajada A gently inclined, detrital surface extending outward from the base of a mountain range and formed by the coalescence of alluvial fans. See also *alluvian fan* and *bolson*.

barchan dune A crescent-shaped dune, formed in areas of limited sand supply, with the "horns" of the crescent pointing downwind.

barrier islands Elongated sand islands oriented parallel to the shore.

barrier reef A long, narrow reef that parallels the shore, commonly separated from the land by a lagoon of substantial width and depth.

basal sliding A method of glacial ice movement in which the glacier slides over its bed. The fact that it is more prevalent in temperate climates indicates that subglacial meltwater may play an important role in the process.

basalt A dark-colored, fine-grained, extrusive igneous rock made up chiefly of calcic plagioclase and pyroxenes. The fine-grained equivalent of gabbro.

baselevel The theoretical elevation below which a stream cannot erode its channel. See also *temporary baselevel* and *ultimate baselevel*.

Basin and Range topography A topography characterized by a series of tilted fault blocks forming parallel mountain ranges and intermontane basins.

batholith A massive, intrusive igneous body with an exposure of 40 square miles (100 km^2) or more. See also *stock*.

bathymetry The measurement of the depth of the ocean and the charting of the topography of the ocean floor.

bay barrier (or baymouth bars) A deposit that begins as a spit and grows to enclose an inlet and separates it from the main body of water.

beach The unconsolidated material that covers a gently sloping surface extending from the low-water line landward to a place where there is a definite change in either material or topography (such as a cliff).

bed A laterally continuous layer of sedimentary rock or unconsolidated sediments that is easily distinguished from the layers above and below.

bed load That part of the stream load moved along the channel bottom. See also *dissolved load* and *suspended load*.

Benguela Current The cold ocean current responsible for the Namib fog desert of South Africa.

Benioff Zone The plane beneath the trenches of the circum-Pacific belt that dips toward the continents at about 45, and along which earthquake foci cluster.

beta particle An electron released from the nucleus of an atom during a type of radioactive decay.

Big Bang The theoretical explosion that initiated the formation and expansion of the Universe.

biochemical Refers to nondetrital sedimentary rocks that are composed of materials generated by living organisms. Reef limestones are an example.

biomass (*Biology*) The total mass of living material in a particular area. (*Energy*) Any organic material that can be used to produce energy.

bituminous coal The coal rank between subbituminous and anthracite.

block An angular rock fragment in excess of 64 mm in diameter formed during a volcanic eruption. See also *bomb*.

block-fault mountains Linear mountain ranges formed under extensional forces and bounded on each side by normal faults. See *Basin and Range topography*.

blowout A cup- or saucer-shaped depression formed on the surface of a sand deposit by wind erosion.

body wave A seismic wave that travels through Earth's interior.

bolson An alluvium-covered basin into which drainage from surrounding mountains flows. See also *alluvial fan* and *bajada*.

bomb An aerodynamically shaped, smooth, pyroclastic rock with a diameter greater than 64 mm formed as blobs of molten lava, ejected during a volcanic eruption, solidify in midair. See also *block*.

bottomset bed The horizontal layer of fine-grained material deposited at the terminus of a prograding delta associated with a bed load-dominated stream. See also *foreset beds* and *topset beds*.

braided stream A stream characterized by an interlacing network of small branching and reuniting streams, separated by sand and gravel bars, that develops in response to being overloaded with sediment that it is unable to transport.

breccia A clastic rock characterized by angular, gravel-sized and larger particles showing little evidence of transport, held together with fine-grained materials or cement. Breccia can be sedimentary in origin or formed by igneous processes. See *volcanic breccia*.

breeder reactor A type of nuclear reactor where nonfissionable ^{238}U is converted to fissionable ^{239}Pu that is used as a fuel.

brittle deformation Deformation that results in failure after a short period of elastic deformation. See also *elastic deformation* and *plastic deformation*.

bronze An alloy of copper and tin.

BTU A *B*ritish *T*hermal *U*nit. One BTU is the energy required to raise one pound of water one degree Fahrenheit.

butte An isolated, steep-sided, flat-topped hill or small mountain often capped with a resistant layer of rock and surrounded by talus. A butte is smaller than a mesa. See also *mesa* and *needle*.

C

C horizon   The lowest horizon of the soil profile; the parent material from which the soil forms.

caldera   A large, basin-shaped, volcanic depression with a diameter many times greater than the associated vent. Calderas form by collapse of the rocks overlying a magma chamber following the eruption of large volumes of magma.

calving   The process whereby blocks of ice break off the front of a glacier, usually where it enters a body of water.

Cambrian Period   The earliest period of the Paleozoic Era; lasting from about 570 to 505 million years ago.

cap rock   A relatively impermeable layer of sedimentary rock, usually shale, that immediately overlies a petroleum reservoir.

capacity   The total amount of load that a stream can move. See also *load*.

carbon dioxide (CO_2)   A requirement for photosynthesis and, when dissolved in water, forms carbonic acid, the major agent of the weathering process of carbonation/hydrolysis. Largely the product of volcanic activity, carbon dioxide is also an important agent in the greenhouse effect.

carbonation   The process of chemical weathering whereby rocks and minerals containing calcium, magnesium, potassium, and iron are transformed into carbonates or bicarbonates by reacting with carbonic acid (dissolved CO_2).

carbonic acid   A weak acid (H_2CO_3) formed by the reaction of carbon dioxide and water.

Carboniferous Period   Named for the large amounts of contained coal, the Carboniferous Period is that portion of the Paleozoic Era lasting from 360 to 286 million years ago. In North America, the Carboniferous Period is subdivided into the older Mississippian Period and younger Pennsylvanian Period.

catastrophism   The hypothesis that proposes changes in living forms and modification of Earth's crust were brought about by recurrent catastrophic events throughout Earth's history.

cation   An ion with a net positive charge brought about by the loss of electrons.

cation adsorption   The affixing of cations to the surface of certain mineral grains, in particular, the clay minerals, as a means of neutralizing a deficiency of positive charge.

cation exchange capacity   The ability of a material to exchange cations held on its surface for cations present in surrounding solutions.

cementation   The process by which unconsolidated sediments are converted to rock as minerals precipitate in the pore space between the grains.

Cenozoic Era   Meaning *recent life*, the Cenozoic Era covers the last 65 to 70 million years of the Earth's history and is divided into the Tertiary and Quaternary Periods.

Channeled Scablands   Deeply eroded lands of eastern Washington that were formed by the catastrophic release of water from glacial Lake Missoula upon retreat of the Pleistocene ice sheet.

chemical sedimentary rock   A sedimentary rock composed of nondetrital materials that were not generated by living organisms. An example is chert.

chemical weathering   The decomposition of rocks and minerals by the processes of dissolution, oxidation, and carbonation/hydrolysis.

chert   A chemical sedimentary rock composed of microcrystalline or cryptocrystalline quartz.

cinder cone   A cone-shaped hill composed of loose volcanic fragments that accumulated around a volcanic fissure or vent.

cirque   A bowl-shaped mountain depression formed by erosion at the source of an alpine glacier.

clay mineral   One of a complex group of fine-grained, hydrous silicate minerals. Formed from the weathering of most silicate minerals, crystalline clay minerals have a sheet structure similar to the micas. Clay minerals are a major component of soil.

climate   The characteristic weather of a region in terms of temperature and precipitation, averaged over an extended period of time.

climatic maxima   A climatic condition that exists when the tropical Hadley atmospheric circulation cell is at its maximum width.

climatic minima   A climatic condition that exists when the tropical Hadley atmospheric circulation cell is at its minimum width.

coal   A readily combustible rock containing more than 50% by weight and 70% by volume carbonaceous material, formed by the compaction and transformation of plant remains, mostly wood, similar to those found in peat.

coal blending   The mixing of low- and high-quality coals in order to comply with environmental standards.

coal cleaning   The physical removal of pyrite-rich materials from a coal in order to comply with environmental standards.

coal liquefaction   The process by which coal is converted into petroleum-like liquids.

coal rank   The degree of metamorphism of coal ranging from peat to anthracite as measured by carbon content.

coalification The process by which buried peat is transformed into higher ranks.

coastal wetlands An area of low hydraulic gradient and high water table found along flat areas adjacent to the coast occupied by marsh-swamp complexes that may or may not be protected by barrier islands.

cockpit karst A type of karst landscape characterized by isolated rounded hills separated by bowl-shaped valleys.

cohesion The strength of a material derived from properties other than intergranular friction.

coke A combustible material produced by fusing the mineral and fixed carbon of coal in an oxygen-starved oven after the volatile matter is driven off. Coke is used in the steel industry to reduce iron ore to free iron.

col A mountain pass formed by the intersection of cirques being eroded from opposite sides of a ridge.

columnar jointing Parallel, prismatic columns, polygonal in cross section, commonly seen in basalt flows and other igneous rock bodies, both intrusive and extrusive.

comet A celestial body, thought to be composed largely of water ice, that orbits the Sun in very large, highly elliptical patterns. Ionization of the comet's surface by the solar wind produces the distinctive tails seen as it swings around the Sun.

compass A devise used to locate the north magnetic pole.

competence The largest particle a stream can move.

compliance coal Coal that meets or exceeds the minimum EPA (Environmental Protection Agency) emission standards to qualify as a fuel for coal-fired power plants.

compound A substance containing two or more chemically bound elements.

compression Stress that acts toward a body and tends to reduce the volume and dimensions of the body. See also *tension*.

concordant Said of an igneous intrusion whose surfaces are parallel to the layering of the intruded rock. See also *discordant*.

cone of depression The cone-shaped depression of the water table that forms around a well as water is withdrawn.

confined aquifer An aquifer bounded both above and below by impermeable beds or by beds of distinctly lower permeability than the aquifer itself. The water contained in a confined aquifer is under pressure.

conglomerate A sedimentary rock made up of rounded particles, granule size and larger, that show evidence of transport, held together by cement or a finer-grained matrix.

contact metamorphism The type of metamorphism resulting from the direct heating of the country rock by an intruding igneous body.

continental arc The chain of volcanic mountains that form near the edge of the overriding continental plate at a convergent plate boundary. The Cascade Mountains are an example. See also *island arc*.

continental climate The climate of a continental interior, characterized by large daily and annual temperatures ranges, one-month seasonal heating and cooling lags, and a tendency to have low amounts of precipitation and humidity. See also *maritime climate*.

continental crust The granitic crust that underlies the continents. Ranging in thickness up to 45 miles (70 km) thick under mountain ranges, the continental crust is richer in silica and alumina than the oceanic crust and has an average density of 2.7 gm/cm^3. See also *oceanic crust*.

continental divide The drainage divide that separates streams flowing toward opposite sides of a continent. In North America, continental divides separate the water-sheds of the Pacific Ocean, the Gulf of Mexico, the Atlantic Ocean, and the Arctic Ocean.

continental glacier A glacier of considerable thickness that covers a large part of a continent, or an area of at least 20,000 square miles (50,000 km^2), and obscures the topography of the underlying surface.

continental margin That part of the continent between the shoreline and the abyssal ocean floor. It typically includes the *continental shelf, continental slope,* and *continental rise.*

continental rise That portion of the continental margin located between the continental slope and the abyssal plain. The lsope is typically gentle, with slopes of 1:2,000 to 1:40.

continental rocks Rocks that make up the continental crust. They are typically less dense (2.7 gm/cm^3) than those of the oceanic crust 3.0 gm/cm^3) and richer in silica and alumina.

continental shelf The shallow-water portion of the continental margin that extends from the shore to the continental slope; the seaward extension of the coastal plain.

continental slope The relatively steep portion of the continental margin between the continental shelf and the continental rise.

convection The lateral or vertical movement of subcrustal or mantle materials as the result of variations in heat.

convection cell The pattern of heat-driven movement of mantle material wherein the central heated portion rises and the cooled outer portion sinks.

convergent plate boundary A boundary at which two plates are moving toward each other. See also *subduction zone*.

coquina A detrital limestone composed wholly or chiefly of weakly to moderately cemented shells and shell fragments.

coral The general name for any of a large group of bottom-dwelling, sessile, marine invertebrates of the class *Anthozoa*, phylum *Coelenterata* that produce external skeletons of calcium carbonate and live individually or in colonies. Colonial corals make up the framework of most modern reefs.

core (*Earth structure*) The innermost part of the Earth, believed to be composed of an iron-nickel mixture. Seismic data suggest that the outer core is liquid while the inner core is either solid or a highly viscous liquid. (*Exploration*) A cylindrical portion of bedrock or sediment recovered by drilling. (*Structure*) The rocks located at the center of a fold.

Coriolis effect The tendency for particles in motion on Earth's surface to be deflected to the right in the Northern Hemisphere and to the left in the Southern Hemisphere to an extent dependent on the velocity of the particles and the latitude.

correlation The establishment of age equivalence for two separate geologic phenomena or objects in different areas.

country rock The rock enclosing or traversed by a mineral deposit. See also *host rock*.

covalent bonding A type of chemical bonding where the electrons are shared between atoms.

crater A bowl-shaped depression formed by volcanic eruptions or meteorite impact.

craton A part of Earth's crust that has not been subjected to mountain building or tectonic activities for prolonged periods of time. See also *shield*.

creep The slow, continuous movement of regolith downslope under the influence of gravity.

Cretaceous Period The final period of the Mesozioc Era, the Cretaceous Period lasted from about 144 to 66 million years ago.

crevasse (*Stream*) A breach in the bank of a river or canal. (*Glacier*) A deep, nearly vertical fissure in glacial ice formed by stresses resulting from differential movement of the ice.

cross-bed A single bed inclined at an angle to the main layers of a sedimentary deposit or rock.

crust The outermost layer of Earth, making up less than 1% of the total volume of Earth, that includes the granitic continental crust and the basaltic oceanic crust.

Cryptozoic Eon Literally meaning "hidden life," the Cryptozoic Eon is the oldest and longest portion of geologic time; the little evidence of life that was preserved consists of the remains of primitive life-forms.

crystal form The geometric shape of a crystal as defined by the faces and angles between the faces.

crystal lattice or structure The three-dimensional, systematically repeated network of atoms within a crystal.

crystal settling A process of magmatic differentiation whereby the first crystals to form in a cooling magma settle to the bottom of the magma chamber.

crystalline Any substance possessing a crystal lattice.

Curie point The temperature above which spontaneous magnetic ordering cannot occur.

current ripple mark The asymmetric ripple mark formed by the flow of water or air across the surface of a layer of sand.

cutbank The outer bank in the bend of a meandering stream that is the site of maximum erosion. See also *point bar*.

cyclic stratigraphy The sequence of rocks resulting from a circuitous sequence of conditions, such as recurring climatic conditions or sea level changes, that affect the weathering and erosion of bedrock and the transportation and deposition of sediments.

D

daughter element The element produced by the radioactive decay of a parent element.

deadmen Horizontal extensions of a retaining wall that use the weight and friction of the overlying rock materials to counter the downslope movement of the retained rock materials.

debris avalanche The movement and flowage of soils and loose bedrock.

debris flow A moving body of unconsolidated material where more than 50% of the particles are sand size and larger.

debris slide The slow or fast slide of predominantly dry, unconsolidated rock and soil.

decomposition Any chemical process whereby the original composition of the material is changed. See also *chemical weathering*.

deep-focus earthquake An earthquake whose focus is located between 185 and 450 miles (300–700 km) below the surface.

deep-sea fan A fan-shaped deep-sea deposit found at the mouth of a submarine canyon or in deep water opposite the site of a major delta.

deep-sea trench A deep, linear depression that forms at a convergent plate boundary and marks the location of the zone of subduction.

deep-water mass A mass of dense ocean water that sinks to the ocean bottom as a thermohaline current and moves along the ocean floor.

deflation A process of wind erosion whereby clay- and silt-sized particles are preferentially removed from surface deposits.

deformation The change in shape, volume, or orientation of rock bodies by the application of various stresses.

delta The deposit that forms where a stream enters a larger body of water. The deposit is named after the Greek letter delta (Δ) in reference to the triangular shape of the deposit at the mouth of the Nile River.

dendritic A stream pattern resembling the veins in a leaf.

dendrochronology The determination and dating of past events using tree growth rings.

density Mass per unit volume, usually measured in grams per cubic centimeter (gm/cm^3). See also *specific gravity*.

density current A gravity-induced flow of dense water where the density has been increased by changes in temperature, salinity, and suspended solids.

deposition The process by which transported particles are laid down from an agent of erosion.

depositional environment Any environment in which sediments can accumulate.

desert An area that receives less than 10 inches (25 cm) of rain per year and is so devoid of vegetation as to be incapable of supporting any considerable population.

desert pavement The mantle of granule-sized and larger particles that covers the surface of some deserts after the finer materials have been removed by wind and water.

detrital Refers to materials eroded, transported, and deposited at a location remote from the point of origin.

detritus Any material that has been derived from the weathering of rock or minerals and transported from its place of origin.

Devonian Period A time interval in the Paleozoic Era spanning from 400 to 360 million years ago. The period is named after Devonshire, England, where rocks of this age were first studied.

dew point The temperature at which air, cooled at constant pressure and water-vapor content, becomes saturated with water. See also *relative humidity*.

dewatering Any process by which water is removed from an aquifer.

diagonal slip fault A type of fault where the offset has both a horizontal and a vertical component.

diastrophism A general term for all movements of Earth's crust produced by tectonic processes.

dike A tabular, discordant, intrusive, igneous body.

dip The angle that a structural surface, such as a bed or the limb of a fold, makes with the horizontal.

dipmeter An instrument used to measure the angle between the magnetic field and Earth's surface.

direct lifting The process by which forces generated by running water, glacial ice, or wind lift particles from the surface.

disappearing stream A surface stream that is diverted into a subterranean fracture or cave.

discharge The volume of water passing a given point within a fixed interval of time, usually measured in cubic feet per minute in the case of streams or gallons per minute for springs and wells. The stream parameter that determines the amount of energy available to the stream for erosion and transportation.

disconformity A type of unconformity where the erosional surface is located between parallel sedimentary beds.

discordant Said of an intrusive igneous body that cuts across the bedding, foliation, or structure of the county rock into which the rock intrudes. See also *concordant*.

disintegration Any process by which a rock is broken into smaller pieces with no change in composition. See *mechanical weathering*.

displacement The actual amount of movement along a fault.

disseminated deposits Mineral or ore deposits where the desired minerals are scattered throughout the county rock.

dissolution The dissolving of a solid by a solvent.

dissolved load The load carried by a stream in solution. See also *bed load* and *suspended load*.

distributary A small channel that carries water and load away from a main channel within a delta. See also *tributary*.

divergent plate boundary A plate boundary at which two plates are moving away from each other. See also *oceanic ridge*.

doldrums A term given by mariners in the days of sailing ships to the zone between 5° north and south latitude characterized by calm to still wind conditions due to the vertical motions of the air masses.

dolomite A carbonate mineral consisting of equal proportions of calcium and magnesium, $CaMg(CO_3)_2$.

dolostone A sedimentary rock composed of at least 50% dolomite, $CaMg(CO_3)_2$.

domal mountains Mountains formed by localized, epeirogenic upwarping of continental crust.

domed swamp The type of swamp that forms under ever-wet conditions. Because the water is totally provided as rainfall, the peat accumulates vertically into a mound. See also *planar swamps*.

dormant A volcano that is not currently active but has been active in historic time and is expected to become active in the future.

double-chain structure A type of silicate structure where two parallel chains of silicon-oxygen tetrahedra are joined along their lengths.

downcutting The process by which streams and alpine glaciers erode their channels.

downstream flood A flood that affects the lower reaches of a stream system and occurs after long periods of heavy rainfall.

drag fold A minor fold, usually formed in incompetent beds on opposite sides of a fault as a result of the movement of the rocks.

drainage basin A region or area bounded by a drainage divide and occupied by a drainage system.

drainage divide The boundary between adjacent drainage basins.

drainage system The total network of impounded water, streams, and tributaries that removes water from a given area.

dripstone A general term applied to rock formed from calcite or other minerals that precipitate from water as it flows across the surface. See also *stalactites* and *stalagmites*.

drumlin A teardrop-shaped landform composed of till that forms beneath continental ice sheets. Drumlins align with the direction of ice flow with the blunt end pointing in the direction from which the ice approached.

dry-based glacier A glacier that is frozen to the underlying bedrock surface. See also *wet-based glacier*.

dug well A water well that has been dug by hand.

dune A low mound-shaped deposit composed of granular, windblown material, usually sand. Dunes are named by the appearance that they maintain as they move. The movement of dunes is controlled by the amount of vegetation, the nature of the substrate, the quantity of available sand, and the strength of the wind.

dunite An ultramafic igneous rock composed almost exclusively of olivine.

dynamo-thermal metamorphism A type of regional metamorphism involving direct pressures and shear stresses as well as a wide range of confining pressures and temperatures.

E

E horizon The soil layer consisting primarily of quartz sand found below the A and O horizons; referred to as the zone of leaching.

Earth The third planet from the Sun.

earthflow A type of mass wasting in which soil and loose rock material move over a laterally confined, basal shear zone oriented roughly parallel to the ground surface, with little rotation of the sliding materials.

ebb tide The tidal current associated with the retreating tide, generally flowing seaward.

ecliptic The imaginary plane within which, or near to which, all of the planets in the Solar System except Pluto orbit the Sun.

effluent stream A stream whose channel is below the water table that receives water from the zone of saturation. See also *gaining stream*.

elastic deformation Deformation that disappears after the deforming stresses are removed.

electron A subatomic particle with negligible mass and a negative charge.

electron capture A mode of radioactive decay in which an orbital electron is captured by a proton within the nucleus of an atom.

emergent (or high-energy coastline) A coastline that is rising or has risen in relation to sea level as a result of tectonic uplift of the land or a decrease in the elevation of sea level.

Eocene Epoch The epoch of the Tertiary Period between the Paleocene and Oligocene Epochs lasting from 58 to 37 million years ago.

eolian Pertaining to the wind, especially in reference to deposits, structures, and processes of wind erosion and deposition.

eon The longest interval of formal geologic time, next in order above an *era*.

epeirogeny Primarily vertical crustal movements, either upward or downward, affecting large parts of a continent. See also *orogeny*.

epicenter The point on Earth's surface directly above the focus of an earthquake.

epoch A division of geologic time longer than an *age* but shorter than a *period*.

equatorial countercurrent A narrow, surface ocean current near the equator that flows eastward between the westward-flowing equatorial currents.

equatorial currents Currents just north and south of the equator that are driven by the trade winds, southwestward or westward in the Northern Hemisphere and northwestward or westward in the Southern Hemisphere.

equilibrium line The line of a glacier where the mass of ice lost by ablation is equal to the mass of ice gained by ice accumulation and movement.

equinox The biannual point in time when the Sun is directly over the equator and the number of hours of daylight and dark are equal at all latitudes.

era The geologic time unit shorter than an eon.

erosion The wearing away of any part of Earth's surface by natural processes.

erratic A rock carried by a glacier and deposited a significant distance from its point of origin. The lithology of the rock is usually, but not always, different from the rocks of the area in which it is found.

esker A long, sinuous landform composed of sorted gravels and sands that are believed to have been deposited by a subglacial river or stream upon melting of the ice.

evaporite A nondetrital sedimentary rock or mineral formed from the extensive evaporation of saline solutions. Examples are halite and gypsum.

exfoliation A mechanical weathering process by which concentric layers of rock of various thicknesses are removed from a rock mass (like the removal of layers from an onion).

external drainage Drainage in which the water directly or indirectly reaches the ocean.

extinct Refers to a volcano that is not active and is not likely to become active in the future.

extrusive Refers to igneous rocks that are formed from molten rock that has erupted onto Earth's surface or to the processes by which extrusion takes place.

F

facet A nearly flat, smooth area on a rock formed by the abrasion of wind-driven sand.

facies An assemblage of mineral, rock, or fossil features that reflects a particular environment of deposition.

failed arm The triple junction fracture that failed to open beyond the formation of a rift valley.

fault Any fracture along which there has been movement or displacement.

feldspar group Silicate minerals containing aluminum and one or more of the metals sodium, calcium, or potassium. The most abundant of all the mineral groups, the feldspar minerals constitute 60% of Earth's crust and are found in all kinds of rocks.

felsic A term derived from *f*eldspar + and *s*ilica + *c* referring to igneous rocks having a high percentage of light-colored minerals (quartz, feldspars, feldspathoids, muscovite), to those minerals as a group, and to the magmas from which they form. See also *mafic*.

felsic eruptions Violent volcanic eruptions due to the highly viscous nature of felsic magmas.

Ferrel (temperate Cell) The air circulation cell located between the subtropical high-pressure and polar low-pressure zones.

ferrous Refers to any material containing appreciable amounts of iron.

filter pressing A method of magmatic differentiation wherein a "mush" of precipitated crystals is separated from the magma by Earth movements.

firn A transitional material between snow and glacial ice. See also *névé*.

firn line The line on the surface of a glacier to which snow, accumulated during the winter, retreats during the summer season.

fissile The ability to be split along closely spaced, parallel planes.

fission reactor A nuclear reactor that creates energy using radioactive isotopes that split to form atoms of smaller atomic mass.

fissure eruption A volcanic eruption that takes place along the length of a fissure rather than from a central vent.

fjord A deep, narrow, steeply sided, U-shaped embayment, usually the seaward end of a glacial U-shaped valley or trough.

flood basalts Horizontal to subhorizontal flows of basaltic lavas that form by the simultaneous extrusion of fluid basaltic lava from many fissures over a vast area.

flood delta A tidal delta developed on the landward side of an inlet from materials deposited by the incoming (flood) tide.

flood frequency The average interval at which a flood of given height and/or discharge can be expected to recur.

flood tide A rising tide.

floodplain The flat, alluvium-covered area adjacent to a stream channel that is covered by water when the stream is in a state of flood.

floodplain management The controlling of human activities and construction on floodplains in order to reduce the loss of materials and lives during major flood events.

floodway A large-capacity drainage way to divert floodwaters from flood-prone areas.

fluvial A term referring to any aspect of a stream.

fluvial landform Any landform that has been created by the direct action of streams.

focus The point at which rocks rupture and release the energy of an earthquake. See also *hypocenter*.

fog desert A desert formed on the west coast of a warm continental landmass due to the presence of a cold, offshore surface current.

fold A warp or bend in Earth's crust resulting from plastic deformation.

foldbelt mountains Mountains typically formed by compressive forces at convergent plate boundaries. Most of Earth's major mountains are foldbelt mountains.

foliation A general term referring to the planar or layered arrangement of textural or structural features in any type of rock.

foot wall The rock mass located beneath an inclined fault. See also *hanging wall*.

force Any quantity capable of producing motion.

foreset beds The inclined layers on the front of a prograding delta formed from materials deposited from a bed load-dominated stream.

fracture (*Structural geology*) Any break in a rock due to mechanical failure. (*Mineral*) The breaking of a mineral other than along planes of cleavage.

fracture porosity Porosity resulting from fractures within a rock.

framework structure The silicate structure characterized by a three-dimensional arrangement of tetrahedra in which each oxygen is shared by an adjoining tetrahedron, giving a silicon:oxygen ratio of 1:2.

free-fall Unhindered fall of any material responding solely to the effects of gravity.

freeze-thaw cycle The recurrent frost action due to daily or seasonal variations in temperature.

frequency (*General*) The time interval that designates when an event will or might recur. (*Wave*) The number of wavelengths that will pass a point in space per unit of time.

friction The force that resists the motion or tendency of motion between two bodies or between a body and a medium.

fringing reef A reef that is attached to the border of an island or continent on one side and slopes steeply to the ocean on the other. There may be a shallow channel on the landward side. See also *barrier reef* and *atoll*.

frost action A mechanical weathering process caused by the alternate freezing and thawing of water in rock fractures. The expansion of the freezing water breaks the rock apart, and the thawing allows additional water to enter the widened cracks as the next cycle begins.

frost heaving The lifting of the top layers of soil or regolith by the subsurface freezing of water and subsequent growth of ice crystals.

fumarole A volcanic vent that emits hot gasses, usually associated with late stage volcanic activity.

fusion A nuclear reaction during which atoms of hydrogen combine to form helium and release energy; the reaction that goes on within the cores of main-line stars.

G

gabion Wire mesh baskets filled with crushed rock that are used to provide protection from water erosion and various processes of mass wasting.

gaining stream A stream that receives water from groundwater discharge within its channel and banks. See also *effluent stream*.

galaxy A large celestial collection of stars, planets, and other bodies in the Universe.

gangue The waste rock component of an ore.

geocline The wedge of sediment that accumulates at the margin of a continental trailing edge.

geosyncline A regional, elongate or basinlike downwarping of the continental margin. The concept was originally presented by Hall in 1859 and named by Dana in 1873 to explain the great thickness of sedimentary rocks found in foldbelt mountains. Some geologists have suggested that the term should be superseded by the term *geocline*.

geothermal power A source of power utilizing Earth's internal heat.

getters Chemicals that are mixed with coal in modern coal-fired power plants to remove SO_2 from the hot flue gases, thereby preventing it from being vented to

geyser A type of hot spring that, through the interaction of ground water and an underlying heat source, intermittently ejects steam and hot water.

glacial trough A U-shaped valley cut by an alpine glacier.

glass An amorphous material commonly considered a solid, that is technically considered to be supercooled liquid.

glassy A term applied to the texture of an igneous rock formed by the very rapid cooling of lava that lacks any crystallinity.

Glomar Challenger A specially equipped ship for deep-sea drilling. The first research ship for studying the deep-ocean basins, the *Glomar Challenger* has since been replaced by more modern platforms.

gneiss A foliated metamorphic rock formed by regional metamorphism in which light- and dark-colored minerals have segregated into layers.

gneissic banding The distinctive texture associated with gneiss, consisting of alternating light and dark layers of felsic and mafic minerals.

Gondwana The Late Paleozoic continent of the Southern Hemisphere. Originally combined with Laurasia as the supercontinent of Pangea, Gondwana broke up in the early Mesozoic to form the southern continents and India.

gossan The iron-bearing weathered material overlying a sulfide deposit. Oxidation of pyrite and the subsequent production of acid leached out most metals and left behind the hydrated iron oxides characteristic of the gossan.

graben The downthrown block associated with block-fault mountains. See also *horst*.

gradation The leveling of a land area to a gentle, continuous slope by the erosion of bedrock and the subsequent transportation and deposition of sediments.

grade The state of equilibrium that exists among a stream's gradient, sediment supply, sediment load, channel characteristics, and capacity to carry its load.

graded bedding Bedding showing a progressive change in grain size from the bottom to the top, usually from course grained at the base of the bed to fine grained at the top, that results from the settling of particles in quiet water.

granite A coarse-grained igneous rock composed mostly of potassium feldspar, plagioclase, and quartz with small amounts of mica and amphiboles. A major component of the continental crust.

granodiorite A coarse-grained igneous rock composed mainly of quartz, potassium feldspar, and plagioclase but containing less potassium feldspar than granite, Mafic constituents are often biotite and hornblende. With granite, a major component of the continental crust.

grassland soil A mollisol. An alkaline, organic-rich, very fertile soil especially suited to the growth of grass. These soils produce most of the world's grain, except for rice.

gravitational sliding The gravity-induced downward movement of soil, rock, and vegetation on a slope.

gravity The force of attraction between two bodies. The force of gravity is proportional to the product of the masses of the two bodies and inversely proportional to their distance of separation ($F = f\frac{m_1 m_2}{d^2}$).

graywacke A dark gray, coarse-grained sandstone made up of poorly sorted, angular quartz and feldspar particles with a variety of rock fragments embedded in a clayey matrix.

Great Ice Age The period of time during the Pleistocene Epoch when much of the Northern Hemisphere was covered by continental ice sheets.

greenhouse effect The interception of long-wavelength terrestrial radiation by carbon dioxide and water vapor in the lower portion of the atmosphere that is primarily responsible for the temperature of the lower portion of the atmosphere and Earth's surface.

groins Artificial structures built perpendicular to the shoreline to protect the beach from erosion by waves or tides or to trap sand for the purpose of building a beach.

ground moraine Till deposited during the retreat of a glacier. Ground moraine can cover large areas and generally produces an undulating surface.

groundmass (*Igneous*) The material between the phenocrysts of a porphyritic igneous rock. (*Sedimentary*) Synonymous with the *matrix* of a sedimentary rock.

groundwater The portion of subsurface water contained in the zone of saturation.

Gutenberg discontinuity The seismic-wave velocity discontinuity at about 1,800 miles (2,900 km) that marks the contact between the mantle and the core.

guyot A flat-topped seamount named after Arnold Guyot, a nineteenth-century Swiss-American geologist.

gypsum A common evaporite mineral composed of hydrous calcium sulfate ($CaSO_4 \cdot 2H_2O$) or the sedimentary rock formed primarily from the mineral.

gyre A large, oceanic water-circulation system rotating clockwise in the Northern Hemisphere and counterclockwise in the Southern Hemisphere.

H

Hadley (tropical) Cell The air circulation cell located between the subtropical high-pressure zone and the Intertropical Convergence Zone.

half-life The amount of time required for half of the mass of a radioactive isotope to be converted to the daughter element.

halite A common evaporite mineral composed of sodium chloride (NaCl); halite is commonly known as table salt or rock salt.

halo A circular distribution pattern that forms around the source of a mineral, an ore body, or a petrographic feature.

hanging valley A U-shaped valley located on the wall of a deeper U-shaped valley.

hanging wall The rock mass located above an inclined fault plane. See also *foot wall*.

hanging water table The water table associated with a mass of groundwater isolated from the main body of groundwater by an impermeable layer of rock.

hardness (*Mineral*) The ability of mineral to resist being scratched by another mineral. A diagnostic physical property of minerals. See also *Mohs's Scale of Hardness*. (*Water*) A property of water containing high concentrations of magnesium and calcium ions to precipitate soap.

hardpan A general term for a relatively hard, impervious layer just below the surface of soil or regolith in arid and semiarid regions.

Hawaiian-type eruptions A nonexplosive type of volcanic eruption characterized by the relatively quiet production of large quantities of fluid, basaltic lava.

headward erosion The lengthening of a youthful valley beyond the original source of the stream by erosion of the upland at the head of the valley.

headwater The source area of a stream where the stream first becomes identifiable as a continuous, perennial, fluvial body.

heat exchanger A device used to transfer the heat generated in the core of a nuclear reactor to water in order to produce the steam needed to drive the turbines of a nuclear power station.

heave The horizontal component of a fault displacement. See also *displacement* and *throw*.

helium The inert gaseous element formed within mainline stars by a thermonuclear reaction involving hydrogen.

hiatus A break or interruption in the continuity of the geologic record represented by an unconformity.

high-grade metamorphism A type of metamorphism brought about by high pressure and temperatures. See also *low-grade* and *intermediate-grade metamorphism*.

hill A natural rise in the land surface that stands out from the surrounding landscape. Generally, hills are less than 1,000 feet (300 m) from base to summit. Whether a rise is a hill or a mountain depends on local terminology.

honeycomb weathering A type of chemical weathering where the rock surface is covered by numerous pits.

horn A sharp-pointed, mountain peak, carved by the combined effects of several alpine glaciers, bounded by the walls of three or more cirques. Examples are the Matterhorn in the Alps and the Grant Tetons west of Jackson's Hole, Wyoming.

hornfels A fine-grained metamorphic rock formed by contact metamorphism and typically composed of unoriented, equidimensional grains.

horse latitude The area of the ocean located below the subtropical high-pressure zones and characterized by long periods of calm winds.

horst The upthrown block of block-fault mountains. See also *graben*.

host rock A rock body that serves as the host for other rocks or for mineral deposits. See also *country rock*.

hot spot A volcanic center, usually located within a lithospheric plate, that persists for at least a few tens of millions of years and is thought to be associated with a rising mantle plume. Hot spots are not associated with convergent plate boundaries but may be associated with oceanic ridges.

hot spring Surface emissions of heated groundwater.

Humboldt Current The cold offshore ocean current responsible for the fog deserts of Chile.

hydraulic mining A mining technique utilizing high-pressure water streams to liberate desired ores from deposits.

hydrogen The most abundant element in the Universe; the element with the lowest atomic number and mass.

hydrologic cycle The circulation pattern of water from the ocean, throughout the atmosphere, to the land, and eventually return to the ocean.

hydrolysis Any chemical reaction involving water.

hydropower Electrical energy generated by the passage of water through a turbine, usually involving the gravitational fall of water.

hydrosphere The water that exists on Earth. See also *atmosphere* and *lithosphere*.

hydrothermal metamorphism A type of metamorphism where the parent rocks are altered by reaction with hot water or gases derived from a magmatic source.

hypocenter The point of release of earthquake energy; also called the focus. See also *focus*.

I

igneous The type of rock formed by the solidification of magma or lava.

ignimbrite The rock type formed by the consolidation of ash flows and nuées ardentes.

incised meander An older stream meander that has become deepened by rejuvenation.

index minerals Minerals formed under specific conditions of pressure and temperature that can be used to determine the grade of a metamorphic rock or a metamorphic rock assemblage.

inert gas Any of the six elements, helium, neon, argon, xenon, krypton, and radon, that have no tendency to react with other elements.

inertia The property of matter that tends to resist any change in motion.

influent stream A stream where there is a flow of water into the zone of saturation. See also *losing stream*.

insolation The amount of solar and sky radiation that reaches Earth and the rate at which it is received.

intergranular porosity The void spaces between the individual particles of a rock or sediment. See also *fracture porosity*.

intermediate-focus earthquake An earthquake with a focus between 40 and 185 miles (65–300 km) below the surface.

intermediate-grade metamorphism A type of metamorphism brought about by moderate pressures and temperatures.

internal drainage Surface drainage where the water flows into an interior basin. Most prevalent in arid and semiarid regions.

Intertropical Convergence Zone (ITCZ) The zone extending slightly north and south of the equator where subtropical air masses converge, rise, and form the tropical low-pressure zone.

intrusive Refers to igneous rocks that were formed by the subsurface cooling of magma. Intrusive rocks are typically coarse grained because of their slow rates of cooling.

iron hat See *gossan*.

iron meteorite A kind of meteorite that is composed dominantly of nickeliferous iron.

island arc A linear belt of volcanic islands constructed on the ocean floor above a zone of subduction. The Aleutian Islands are an example.

isoclinal Refers to a fold with parallel limbs.

isolated tetrahedral structure A silicate structure within which the silicon-oxygen tetrahedra are connected by other metal ions. See also *double-chain structure, single-chain,* and *sheet structures.*

isotasy The condition of gravitational balance, comparable to floating, between the lithosphere and the underlying asthenosphere.

isostatic balance The gravitational balance that exists between the continental crust and the underlying mantle.

isotope An atom of an element that has the same number of protons in the nucleus but a different number of neutrons and thus a different atomic mass than another atom of the same element.

J

jetty A structure, often built in pairs at harbor entrances, designed to maintain water currents of sufficient strength to prohibit the deposition of sediment within the harbor entrance. See also *groins.*

joint A fracture in a rock along which there has been no displacement.

Jovian planet Any of four very large planets of the Solar System, including Jupiter, Saturn, Uranus, and Neptune, that are made up largely of frozen gas.

Jupiter The largest planet of the Solar System. Jupiter lies between Mars and Saturn.

Jurassic Period The period of geologic time extending from 208 to 144 million years ago, during which the dinosaurs were the dominant vertebrate animals and reached their maximum size.

K

kame A deposit of stratified sand or gravel that originated as a delta at the margin of a melting glacier or in a depression on the ice surface that was deposited as a mount-like landform as the ice melted.

kame terrace A terracelike feature composed of stratified deposits found along a glacial valley wall that are made from material carried along the edges of the glacier and laid down along the valley walls as the glacier melted and receded.

karst A term used to designate the topography, landform features, and deposits formed in an area from the dissolution of soluble bedrock. Most karst is developed in areas underlain by limestone and is typified by caves, sinkholes, and an absence of surface drainage.

karst tower A residual pillar of limestone in a karst landscape surrounded by an alluviated plain.

kettle A depression left in the surface of ground marine that formed as a block of ice, buried by a retreating glacier, melted and caused the overlying ground to subside into the new void. See also *kettle lake.*

kettle lake A lake formed by the filling of a kettle.

komatiite A suite of rare igneous rocks noted by the presence of ultramafic lavas.

L

laccolith A concordant, massive intrusive igneous body. The overlying country rock is arched upward, and the underlying rock remains relatively horizontal.

lacustrine A term that applies to any process or deposit associated with lakes.

lagoon A shallow body of water open to the sea but protected from the energy of the open sea by organic reefs or sand barriers.

lahar A mudflow composed chiefly of volcanic materials originally accumulated on the flank of a volcano.

laminar flow A type of flow where the individual flow paths of the particles are parallel to one another and do not cross the laminar sublayer. See also *turbulent flow.*

laminar layer A thin layer of water, up to 10 mm thick, at the contact with a stream channel within which water moves by laminar flow. See also *turbulent flow.*

landslide A general term for a variety of fast, gravity-induced, downslope mass movements of rock and soil.

lateral moraine A low, ridge-like moraine carried on or deposited at or near the side of an alpine glacier. Unlike kame terraces, lateral moraines are not fluvial in nature and show no stratification.

lava Extruded molten rock or the rock that forms by the solidification of extruded molten rock. See also *magma.*

left-lateral Refers to a strike-slip fault where, when facing the trend of the fault, the side opposite the observer has moved to the left.

levee A natural or artificial embankment along the bank of a stream that serves to confine the stream flow to the channel.

light-year A unit of celestial measurement equal to the distance of a photon of light will travel in one year, equal to 5.9 trillion miles.

lignite (or brown coal) The coal rank between peat and subbituminous coal.

limb The part of a fold between the axes of an anticline and the adjacent syncline.

limestone A sedimentary rock consisting of more than 50 weight percent calcium carbonate, $CaCO_3$, primarily in the form of the mineral calcite.

linear ocean An intermediate stage between a rift valley and an ocean basin; a flooded rift valley. An example is the Red Sea.

listric fault An upward concave fault.

lithification Any process by which unconsolidated sediments are converted to a coherent, solid rock.

lithosphere The layer of Earth consisting of the crust and the outer brittle portion of the mantle.

load The material being moved by a stream, river, wind, or glacier. Depending on the transporting mediums, loads can be carried in solution, suspended, or as bed loads. See also *capacity*.

lode A mineral deposit consisting of a zone of veins, disseminations, or planar breccias.

loess Wind-derived deposits of silt- and clay-sized particles.

longitudinal dune A long, narrow sand dune, normally symmetrical in profile, whose long dimension is oriented parallel to the prevailing wind direction. These dunes often form where sand is plentiful and winds are strong. See also *seif dune*.

longshore current An ocean current, generated by waves approaching the coast at an angle, that flows parallel and near to the shore.

longshore transport The transport of sediment along a shoreline by longshore currents.

lopolith A concordant, massive, intrusive igneous rock body with a concave upper surface due to the sagging of the underlying country rock in response to the weight of the igneous intrusion.

losing stream A stream that is losing water from its channel to the zone of saturation. See also *influent stream*.

Love wave A seismic surface shear wave that moves the surface horizontally in the direction perpendicular to the direction of wave propagation.

low-grade metamorphism A grade of metamorphism caused by low or moderate temperatures and pressures.

low-velocity zone The zone in the upper mantle where seismic velocities are slower than in the outermost mantle. It is located from approximately 35 to 155 miles (60–250 km) below the surface.

luster The appearance of a mineral under reflected light. An example is metallic versus nonmetallic.

M

mafic A term derived from *magnesium* + *ferric* + *ic* that is used to refer to rocks or minerals high in magnesium and iron and the magmas from which they form. See also *felsic*.

magma Molten rock that has not reached Earth's surface. See also *lava*.

magmatic segregation The concentration of crystals of a particular mineral in portions of magma as it cools. Some valuable ore deposits form by this process.

magnetic inclination The angle between Earth's magnetic field and the horizontal.

magnetic reversal A change in the polarity of Earth's magnetic field, that is, a switching of the north and south magnetic poles.

magnetometer An instrument used to measure Earth's magnetic field.

magnitude A measure of the strength of an earthquake or the strain released during an earthquake based on seismographic observations. Reported relative to the Richter Scale.

mantle The part of Earth between the base of the crust and the top of the core.

mantle drag The force generated on the base of the lithosphere by the motion of the underlying asthenosphere. See also *ridge push* and *slab pull*.

marble A metamorphic rock formed by the recrystallization of limestone or dolomite by heat and pressure.

marginal sea A semi-enclosed sea adjacent to a continent, in particular, between an island arc and the mainland. An example is the Sea of Japan.

marine rock A sedimentary rock formed from sediments accumulated in an oceanic environment.

maritime climate The climate of islands and landmasses bordering the ocean that experience only mild diurnal and annual temperature ranges and where maximum and minimum temperatures occur longer after the solstice than in areas affected by continental climates. See also *continental climate*.

Mars The fourth planet from the Sun.

marsh A wetland dominated by grasses.

marsh-swamp complex A wetland within which areas of both marsh and swamp are present along with streams and areas of open water.

mass spectrometer A device used to determine the relative abundance of isotopes within a sample.

mass wasting The downslope movement of rock materials by gravitational forces without being carried within, on, or under any other medium.

massive (*Sedimentary*) A term used to describe rocks that occur in thick, homogeneous beds.

(*Igneous*) Said of an intrusive, igneous body where the largest dimension is less than 10 times the smallest dimension.

maturity Pertaining to the relative age of a landscape as proposed by Davis. Landscapes that have reached maturity have low-gradient streams with developed meanders and wide floodplains and experience more deposition than active erosion. Oxbow lakes may be present.

maximum rock transport The dominant direction in which rock has been displaced by compressive forces.

meander The sinuous pattern characteristic of a stream whose valley has progressed to the mature stage of development.

mechanical weathering Any process whereby rocks are broken and reduced in particle size with no change in composition. See also *disintegration*.

medial moraine A moraine carried upon or in the middle of a glacier parallel to the valley walls formed by the coalescing of two inner lateral moraines below the junction of two alpine glaciers.

melange A mass of folded, faulted, and metamorphosed rocks formed at convergent plate boundaries. See also *accretionary wedge*.

Mercury The planet closest to the Sun.

mesa An isolated, flat-topped landmass standing above the surrounding terrain with steep, shear walls bounded by talus deposits. See also *butte* and *needle*.

Mesozoic Era Literally meaning "middle life," the Mesozoic Era lasted from 245 to 66 million years ago. Dinosaurs and other reptiles flourished during the Mesozoic, which was known as the "Age of Reptiles," and were extinct by the beginning of Cenozoic.

metal An element such as iron or gold that has a metallic luster, conducts heat and electricity well, and is opaque, fusible, and generally malleable and ductile.

metallic bonding The kind of bonding characteristic of metals where valence electrons are free to roam through the entire mass of the metal.

metamorphic Refers to rocks or minerals that have been formed by the physical or chemical alteration of a parent material by the application of heat, pressure, and/or chemically active fluids without melting, or to the processes by which the change takes place.

metamorphic facies A mineral assemblage that reflects the metamorphic conditions under which a metamorphic rock formed and is different from the mineral assemblage in an adjacent rock known to have formed under different metamorphic conditions.

metamorphic grade Refers to the intensity or rank of metamorphism; that is, the extent to which a rock or mineral has been physically or chemically altered from its original state by metamorphic processes.

metamorphism The combined mineralogical, chemical, and structural changes that take place within a rock mass in response to conditions of temperature, pressure, and chemically active fluids different from those under which it formed. Metamorphic conditions are only found at depth, usually within a zone of subduction.

metasomatic replacement. The hydrothermal metamorphic process whereby simultaneous capillary dissolution and deposition partially or totally replace the original mineral assemblage of the host rock with a new mineral or assemblage of minerals.

meteor The visible streak of light resulting from the incineration of a meteoroid as it enters the atmosphere.

meteorite A meteoroid that has impacted the Earth.

meteoroid A fragment of rock or iron found in interstellar space, distinguishable from asteroids or planets by its smaller size.

mica group A group of silicate minerals exhibiting sheet structures and perfect basal cleavage. Micas are prominant minerals in igneous and metamorphic rocks. Muscovite and biotite are the common forms.

migmatite A rock made up of both metamorphic and igneous minerals that forms under such intense metamorphic conditions that some melting takes place.

Milankovitch cycles (*climate*) Cyclic variations in the intensity of solar radiation reaching Earth's surface due to the "wobble" of Earth's rotational axis and the ellipticity of Earth's orbit around the Sun.

Milky Way Galaxy The galaxy in which our Solar System resides.

mineral A naturally occurring, solid, inorganic element or compound having an orderly internal structure and characteristic chemical composition, crystal form, and physical properties.

mineral cleavage The breaking of a mineral along planes of weakness within its crystal structure. See also *rock cleavage*.

mineraloid A material that satisfies all the requirements of a mineral except that it lacks a crystalline structure.

Miocene Epoch The epoch that preceded the Pliocene and followed the Oligocene in the Tertiary Period, lasting from 24 to 5 million years ago.

Mississippian Period The lower division of the Carboniferous Period in North America lasting from 360 to 320 million years ago. This subdivision of the Carboniferous Period is not recognized in Europe.

Modified Mercalli Scale A scale used to designate the intensity of earthquakes based on the effect the earthquake has on people and structures, i.e., a measure of damage.

Moho An abbreviation for the Mohorovičić discontinuity.

Mohorovičić discontinuity The sharp seismic-velocity discontinuity occurring from 6 to 22 miles (10–35 km) below Earth's surface that separates the crust from the mantle.

Moh's Scale of Hardness A standard scale of 10 minerals that is used to determine the hardness of an unknown mineral.

monocline A local steepening in an otherwise uniform dip. See also *anticline* and *syncline*.

moraine A mound, ridge, or other distinctive landform made up of dominantly unstratified, poorly sorted deposits that were transported and deposited by the direct action of glacial ice.

mother lode A main mineralized zone that may not itself be economical to mine, but is associated with local concentrations of minerals that are.

mountain Any part of Earth's surface that is elevated above the surrounding land higher than a hill (1,000 feet or 300 m) and is considered to be worthy of a distinctive name.

mouth The point at which a stream enters a larger body of water. The place of termination of a stream as opposed to its headwaters.

mud cracks Shrinkage cracks that generate from roughly poloygonal-shaped plates in fine-grained unconsolidated deposits as a result of drying or freezing.

mudflat A shallow, flat area of fine-grained sediments found within a lagoon or in the shallow water around an island or shoreline that is alternately submerged and exposed during tidal cycles.

mudflow A process of mass wasting characterized by a flowing mixture of water and fine-grained materials possessing a high degree of fluidity during movement. With increased fluidity, mudflows grade into turbid and clear streams. With decreased fluidity, they grade into earthflows. The term is also used to describe the deposit that forms following the deposition of the materials. See also *lahar*.

mudstone A sedimentary rock similar to shale but lacking the fine laminations characteristic of shale.

N

National Earthquake Information Center (NEIC) A federal agency established in 1966 to collect worldwide seismic data to aid in earthquake research.

National Seismograph Network (USNSN) A federal agency that was originally founded to monitor Soviet nuclear testing, but now gathers seismic data.

native metal A metal found chemically uncombined in nature. An example is gold.

neap tides Bimonthly tides that occur when the Moon is in the first and third quarters during which the tidal range is substantially less than the average tidal range.

nebulae Intersellar dust clouds faintly visible from Earth.

needle (*rock*) A pointed, elevated, detached mass of rock formed by erosion in arid and semiarid regions. The end product of the mesa-butte-needle rock sequence. See also *butte* and *mesa*.

Neptune The outermost of the Jovian planets. Because the highly elliptical orbit of Pluto sometimes brings it inside the orbit of Neptune, Neptune is at times the most distant planet from the Sun.

neutron A subatomic particle located in the nucleus of an atom that has no charge and a mass of 1 amu (u); a combination of a proton and an electron.

névé Originally used in English as synonymous with firn, and still so used, although some geologists have suggested restricting the definition to the hardened snow at the source of a glacier that remains throughout the period of melting. See also *firn*.

nitrogen A nonreactive element that is the most abundant gas in the atmosphere.

nonconformity An unconformity where younger sedimentary rocks lie atop a surface of erosion developed on older igneous or metamorphic rocks.

nuée ardente Literally meaning "glowing cloud," a nuée ardente is a fast-moving cloud of hot, turbulent, sometimes incandescent, gas containing ash and other pyroclastic materials in its lower part as an ash flow.

O

O horizon The surface layer of some soils consisting of organic debris derived from the decay of accumulated vegetation.

oblique-slip A type of fault movement consisting of both vertical and horizontal components of offset.

obsidian A dark-colored, amorphous material similar in composition to rhyolite that forms by the superfast cooling of lava. See also *tachylyte*.

ocean basin The area between continental margins or between a continental margin and an oceanic ridge.

oceanic crust The crust that underlies the ocean basins. Oceanic crust (3.0 gm/cm^3) is denser than continental crust (2.7 gm/cm^3). See also *continental crust*.

oceanic ridge A continuous mountain range rising from the abyssal ocean floor at the divergent plate boundaries and interconnected between all ocean basins. The oceanic ridge is the site of oceanic spreading, upwelling of basaltic magma, and seismic activity.

oceanography The study of the ocean including the ocean basin and all physical, chemical, biological, and geological processes acting within the basin.

off-road vehicles (ORVs) Vehicles from mountain bikes to four-wheel trucks that, when operated with disregard, can have detrimental effects on environmentally and biologically sensitive areas.

oil shale A type of sedimentary rock that yields gaseous or liquid hydrocarbons upon distillation.

old age The third and final stage of Davis's threefold landscape evolution theory. An old age landscape has very low relief; its streams occupy floodplains many times wider than the channel width and show extreme sinuous meander patterns; and the valley contains many meander scars and oxbow lakes.

Oligocene Epoch The epoch of the early Tertiary Period that preceded the Miocene and followed the Eocene, lasting from 37 to 24 million years ago.

olivine group A group of ferromagnesian silicate minerals with independent tetrahedral structures forming an isomorphous series of minerals with fayalite (Fe_2SiO_4) and forsterite (Mg_2SiO_4) as end members.

Olympus Mons A volcano on Mars whose base would cover an area equal to the combined areas of Pennsylvania and New York with a summit elevation of 80,000 feet (24,000 m).

ooid An individual spherical component of an oolitic rock, created by the chemical precipitation of calcite around a nucleus such as a sand grain.

Oort Cloud The hypothetical cloud of comets that surrounds our Solar System. According to the hypothesis, individual comets are dislodged by the gravitational pull of a passing celestial body and sent into a highly elliptical orbit around the Sun.

ophiolite suite A sequence of mafic rocks that makes up the oceanic crust; consists of gabbro at the base, followed upward by sheeted dikes, basaltic pillow lavas and lava flows, and deep-sea sediments.

Ordovician Period A period of the early Paleozoic Era spanning the time from 500 to 438 million years ago.

ore A naturally occurring material from which a mineral or minerals can be extracted at a reasonable profit.

organic sulfur The organically bound sulfur in coal. See also *pyritic sulfur*.

orogeny The process by which the structures seen in foldbelt mountains were formed including folding and faulting of the upper layers and plastic deformation, metamorphism, and plutonism found in the deeper layers. Orogenic forces are usually associated with the horizontal forces associated with convergent plate boundaries as opposed to the vertical forces associated with epeirogenic processes.

oscillation ripple marks Symmetrical ripple marks formed by the alternating movement of water.

outwash plain A broad, flat deposit of stratified sand, gravels, and cobbles eroded and transported from the terminal moraine of a continental glacier by meltwater streams and deposited beyond the terminus of the glacier.

oversteepening Any process by which slopes are rendered unstable by being steepened beyond the angle of repose.

overturned Said of a fold or the limb of a fold that has been tilted beyond the perpendicular.

oxbow lakes Lakes formed from an isolated stream meander resembling the U-shaped frame used to attach implements to the neck of draft animals.

oxidation A chemical reaction by which compounds combine with oxygen.

oxisols A deeply weathered soil formed on stable surfaces in tropical and subtropical regions consisting of quartz, free oxides, and organic material and lacking clearly marked horizonation.

oxygen The most abundant element in Earth's crust and the second most abundant constituent (28%) of Earth's atmosphere.

P

P waves Body waves of the compression type; also called *Primary* waves. See also *S waves*.

pahoehoe A Hawaiian term for a type of basaltic lava characterized by a smooth, ropy surface. See also *aa*.

paired terrace One of two terraces located on opposite sides of a stream valley at the same elevation that are the remnants of the same floodplain. See also *unpaired terrace*.

Paleocene Epoch The oldest epoch of the Tertiary Period, lasting from 66 to 58 million years ago.

paleosol A fossil soil.

Paleozoic Era The era following the Precambrian, preceding the Mesozoic, and lasting from about 570 to 245 million years ago.

paludal Refers to an environment characterized by swampy or marshy conditions.

Pangaea The supercontinent that existed from 300 to 200 million years ago. Pangaea represented most of the continental crust in existence at the time and was the crustal mass from which the present continents were derived by rifting.

parabolic dune A crescent-shaped dune similar in shape to a brachan dune except that it is convex downwind with the horns pointing upwind. Parabolic dunes are often sparsely covered with vegetation and occur in coastal areas where there are strong onshore currents and a good supply of sand. See also *barchan dune*.

parallel drainage   A drainage pattern where the streams are regularly spaced and flow parallel to each other over considerable distances. Parallel drainage is characteristic of areas with uniformly sloping topography underlain by relatively homogeneous bedrock.

partial melting   The process of rock melting in which minerals with low melting points melt before minerals with higher melting points as a rock mass is heated and/or subjected to a decrease in pressure.

particle   A general term applied to any piece of rock or mineral regardless of origin, shape, composition, or internal structure.

patch reef   A moundlike flat-topped reef usually not more than 0.5 miles (1 km) across. Patch reefs may be part of a larger reef-forming complex, but they also exist as solitary formations.

paternoster lake   One of a chain of small, interconnected lakes occupying rock basins within a glacial valley.

patterned ground   A surface feature found in polar and subpolar regions where intensive frost action has heaved rocks into circular, polygonal, step, or stripe patterns.

peat   An accumulation of unconsolidated, semicarbonized plant debris in a water-saturated environment such as a marsh, swamp, or bog. Under certain conditions, peat is the precursor to higher rank coal.

pedestal rock   A rock formation, resulting from differential weathering, that is characterized by a large rock supported atop a relatively slender column of rock.

pediment   A broad, gently sloping, erosional surface typically found in semiarid to arid regions that extends outward from the base of a receding mountain range. A pediment is underlain by bedrock but may be covered with a thin veneer of sediment.

pedology   The scientific study of soils.

pegmatite   A very coarse-grained igneous rock, usually found in fractures, lenses, or veins at the margins of batholiths, that forms from the final hydrous portion of a cooling magma. Most pegmatites are granitic in composition and often contain high concentrations of rare elements.

Pelean-type eruption   A violent volcanic eruption that includes a large amount of ejected pyroclastic material as well as a nuée ardente.

peneplane   A term suggested by Davis for a broad, low, gently undulating, flat surface that forms at or near baselevel as the result of long-term subaerial erosion; the end product of erosion.

Pennsylvanian Period   Known as the Upper Carboniferous in Europe, in North American terminology, the Pennsylvanian Period lasted from 320 to 286 million years ago, follows the Mississippian Period, and precedes the Permian Period.

perched water table   See *hanging water table*.

peridotite   A coarse-grained, ultramafic, plutonic, igneous rock composed mostly of olivine with or without other mafic minerals such as amphiboles or pyroxenes; the major component of the upper mantle.

periglacial   Said of processes, conditions, regions, climates, and topographic features that exist at or near the margins of a past or present glacier and are influenced by the cold environment caused by the glacier; an area where frost action is an important factor.

period   A unit of geologic time longer than an epoch and shorter than an era; the fundamental unit of geologic time worldwide.

permeability   The ability of rock or unconsolidated rock materials to transmit a fluid.

Permian Period   The last period of the Paleozoic Era, lasting from 286 to 245 million years ago.

petroleum   A general term for all naturally occurring hydrocarbons, whether gaseous, liquid, or solid.

phaneritic   A textural term applied to igneous rocks in which the individual mineral crystals can be seen with the unaided eye; indicates that the rock formed by the slow cooling of magma.

Phanerozoic Era   That part of geologic time where evidence of life is abundant in the rock record. The interval of time from the Cambrian Period to the present.

phenocrysts   Large, readily seen crystals set in a fine-grained groundmass. See also *porphyry*.

photic zone   The part of the water column in a lake or ocean where the penetration of light is sufficient to support photosynthesis; usually the upper 260 feet (80 m).

Photovoltaic cell   A manufactured cell capable of producing a voltage when exposed to light.

phyllite   A foliated metamorphic rock intermediate in grade between slate and schist. Minute crystals of chorlite or sericite impart a characteristic silky sheen to the cleavage surfaces.

phyllitic cleavage   The smooth, undulating rock cleavage surfaces typically exhibited by phyllite.

physical property   Any property of a mineral that can be determined with the senses and used as an aid to identification including color, hardness, luster, cleavage, form, taste, and smell.

pillow lava   A type of lava, characterized by discontinuous pillow-shaped masses ranging in size from inches to several feet in longest dimension, that forms by the subaqueous cooling of lava.

placer An accumulation of dense or resistant mineral grains such as gold, usually stream, beach, or dune deposits.

plagioclase A group of feldspars that, at high temperatures, forms a complete solid-solution series from calcium-rich *anorthite* to sodium-rich *albite*.

planar swamps Swamps that develop in areas subjected to seasonable wet-dry conditions where nutrients for the vegetation are largely provided by ground and surface water. See also *domed swamp*.

planet One of nine celestial bodies that revolve around the Sun in elliptical orbits and in the same direction.

planetesimal In the planetesimal hypotheses, the fragments of rock and ice that originally orbited the new-formed Sun and that accreted to form the protoplanets. Some believe that the asteroids are planetesimals that were prevented by the gravitational pull of the Sun and Jupiter from accreting to form a planet between Mars and Jupiter.

plant wedging A process of mechanical weathering whereby plant roots enter fractures within coherent rock masses, grow in diameter, and break the rock into smaller rock fragments.

plastic deformation Permanent deformation of the shape or volume of a substance without rupture.

plate tectonics The concept that Earth's lithosphere is divided into about a dozen large, rigid plates and a few smaller plates, all of which are moving relative to each other in response to convection cells within the mantle.

plateau Any comparatively flat area of great extent and elevation formed by erosional, tectonic, or volcanic processes.

plates Pieces of the lithosphere involved in plate tectonics that move horizontally and adjoin other plates at seismically active convergent or divergent boundaries.

playa The lowest part of an undrained desert basin; usually underlain by sediments and commonly by soluble salts.

playa lake A shallow lake found in arid and semiarid regions that occupies a playa during the wet season but evaporates during the dry season.

Pleistocene Epoch The last 2 million years of geologic time. The only epoch of the Quaternary Period; the interval of time during which the Northern Hemisphere was subjected to the Great Ice Age.

Plinian-type eruption A very violent style of volcanic eruption during which a highly turbulent, high-velocity stream of intermixed fragmented magma and superheated gas is released from a vent and driven upward to form an eruption column of great height.

Pliocene Epoch The uppermost division of the Tertiary Period, lasting from 5 to 2 million years ago.

plunge The angle between a fold axis and the horizontal.

Pluto The "maverick" planet, not thought to be one of the originally formed planets, that is located farthest from the Sun when its highly elliptical orbit takes it beyond the orbit of Neptune.

pluton Any intrusive igneous rock body.

pluvial lake A lake formed during periods of high rainfall; specifically, a lake formed in the Pleistocene Epoch during a time of glacial advance and now either gone or existing in reduced size. An example is glacial Lake Bonneville of which Great Salt Lake is a remnant.

point bar The deposit of sand and gravel that accumulates on the inside of a growing meander as the stream channel migrates in the direction of the outer bank.

point source A source of groundwater contamination that can be traced to a discrete discharge point.

polar circulation cell The atmospheric circulation cell located between the polar low-pressure zone and the polar high-pressure zone.

polar front The contact between the warm, lower latitude air masses and the cold polar air masses.

polar high-pressure zone The high-pressure zone over the North and South poles created by descending masses of cold, polar air.

polar wandering The apparent movement of the magnetic poles over time with respect to the continents.

pores The space that exists between grains in rocks or unconsolidated sediments.

porosity The percentage of the total volume of a rock or unconsolidated material represented by void space.

porphyritic Said of the texture of an igneous rock where large crystals (phenocrysts) are enclosed by a finer-grained matrix (ground mass), either crystalline and/or glassy. Porphyritic textures implies a two-stage, interrupted cooling history.

porphry An igneous rock of any composition consisting of conspicuous phenocrysts contained within a finer-grained matrix.

Precambrian The total interval of time and the rocks that formed before the beginning of the Paleozoic Era, representing about 90% of all geologic time.

pressure melting (*Igneous*) The melting of hot rocks by a reduction of pressure such as occurs at the top of the asthenosphere below the oceanic ridges. (*Glacial*) The melting of ice due to increased pressure favoring the higher-density phase. An example is the trail of water left behind the blade of an ice skate.

pressure surface The surface to which water under pressure will rise when freed from its confinement; also called the *potentiometric surface*. An example of water rising to a pressure surface is an artesian well producing water from a confined aquifer.

pressure zones Zones of low or high air pressure circumscribing the globe that are created by the heating and rotation of Earth.

principle of cross-cutting relations The principle that a rock body or geologic feature is younger than any rock body or geologic feature that it cuts across.

principle of faunal succession The principle that fossils in a stratigraphic sequence succeed one another in a definite, recognizable order.

principle of original horizontality The principle that all sedimentary rocks form from sediments that were originally laid down as horizontal beds.

principle of superposition The principle that in a sequence of sedimentary rocks, unless overturned, the oldest beds are at the bottom and the youngest beds are at the top.

progradation The seaward building of a shoreline or coastline by the deposition of stream-borne or wave-transported materials. An example is the building out of a delta.

prograde metamorphism Metamorphism that increases in grade with increasing heat and pressure.

proton A subatomic particle in the nucleus of an atom that has a positive charge and a mass of one amu (u).

protoplanets In the planetesimal hypothesis of planet formation, an intermediate stage between the planetsimal and the planet; essentially, an orbiting body that has not yet developed the size or internal stratification required of a planet. See also *planetesimal*.

protore The rock below the sulfide zone of a supergene enrichment deposit.

pyritic sulfur The component of sulfur contained within coal that is combined with iron in the mineral pyrite (or marcasite). See also *organic sulfur*.

pyroclastic Refers to clastic rock material ejected during a volcanic eruption, the rock that forms from those materials, and the texture of the rock.

pyroxene group A group of dark-colored, ferromagnesian, silicate, rock-forming minerals that are made up of single-chain silicon-oxygen tetrahedra.

Q

quarrying (or plucking) The process of glacial erosion whereby rock fragments are loosened, detached, and removed from bedrock as the ice advances.

Quaternary Period The late period of the Cenozoic Era, extending from 2 million years ago to the present.

R

radial stream pattern A drainage pattern developed on domal structures and volcanic cones where streams diverge outward from a central point much like the spokes of a wheel.

radioactive Said of elements whose nuclei spontaneously disintegrate to form elements of smaller atomic mass.

radiocarbon dating The dating technique that uses the ratio of ^{14}C and ^{12}C to obtain the absolute age of the remains of once-living organisms. Generally restricted to materials no older than 60,000 years.

radiometric dating Dating techniques that use the ratio of parent and daughter isotopes of radioactive elements to obtain the absolute age of rocks and minerals. The procedure is primarily used to date igneous rocks.

rainshadow The dry area that exists on the leeward side of a topographic obstacle, usually, a mountain range.

rainshadow desert A desert located on the leeward side of a mountain or mountain range. An example is Death Valley, California.

Rayleigh waves A type of seismic surface wave that has a retrograde elliptical motion relative to the direction of propagation.

reach The straight stream segment between the bends of a meandering stream.

recessional moraine An end or lateral moraine deposited during a significantly long stillstand in the final retreat of a glacier.

recharge The process, either natural or artificial, whereby water is returned to the zone of saturation.

rectangular stream pattern A drainage pattern characterized by many segments, often of nearly the same length and oriented at right angles to each other. This type of drainage pattern is typical of streams that follow faults or fractures in the underlying bedrock.

recumbent Refers to folds whose axial planes are horizontal or near-horizontal.

reef A rigid, moundlike structure built by sedentary calcareous organisms, especially coral, and their remains.

regelation The process whereby glacial ice pressure melted on the upstream side of an obstacle, generating water that flows around the object and refreezes on the downstream side as the pressures are relieved.

regelation slip The process by which glacial ice moves downslope by having the mass of ice transferred from the upstream side of an obstacle to the downstream side by regelation.

regional metamorphism A general term for metamorphism affecting a large area.

regional water table The water table and groundwater flow system underlying a region.

regolith The layer of unconsolidated material accumulated above bedrock.

rejuvenation The process of renewed erosion by a stream in response to an increase in the distance between the stream channel and the baselevel resulting from an uplift of the land or the depression of the baselevel; the restoration of more youthful qualities to a stream whose valley has attained maturity or old age.

relative dating The establishment of the proper chronological position of an object, feature, or event within the framework of geologic time without establishing the absolute age. See also *absolute dating*.

relative humidity The ratio of the amount of water contained within a mass of air to the amount of water it is capable of containing at a given temperature. See also *dew point*.

relief The vertical difference in elevation between the hilltops or mountain summits and the lowlands or valleys of a region.

reserves Identified deposits of minerals or fuels that can be extracted profitably with existing knowledge and technology. See also *resource*.

reservoir A subsurface rock body that has sufficient porosity and permeability to allow the accumulation of oil or gas within a trap, usually a sedimentary rock.

resources All mineral deposits, including reserves, known and unknown, whether they may be profitably exploited at this time or not, that may become available sometime in the future. See also *reserves*.

retrograde metamorphism A type of metamorphism where lower-grade minerals are formed at the expense of higher-grade minerals as the rock readjusts to less severe conditions of temperature and pressure.

reverse fault A fault formed under compressive stresses where the hanging wall has moved up in relation to the foot wall.

reversed radial A drainage pattern where streams flow into a central depression such as a sinkhole.

Richter Scale A scale that measures the intensity of earthquakes based on the amount of energy released.

ridge push A gravity-induced force generated at an oceanic ridge that results in the lateral movement of the lithosphere over the asthenosphere away from the oceanic ridge. The force originates because of the elevated position of the oceanic ridge due to the buoyancy of the underlying magma. See also *mantle drag* and *slab pull*.

rift valley A valley that has developed along a rift.

rift zone A system of crustal features associated with tensional forces that signals the development of a potential divergent plate boundary.

right-lateral A type of strike-slip fault where the block on the opposite side of the fault from an observer has moved to the right.

rigidity The property of a material to resist stress that normally would result in deformation.

ripple marks A series of parallel or subparallel, small-scale ridges and valleys that form as currents of water or air move over the surface of a sand deposit.

roche moutonnée A relatively small, elongate, protruding hill of bedrock that has been scupted by a glacier. The hill is oriented with its long dimension parallel to the direction of ice movement with a gentle slope on the upstream side and a steeper slope on the downstream side.

rock A coherent aggregate of minerals.

rock avalanche A rapid flowage of rock fragments during which rocks may be further reduced in size and pulverized.

rock cleavage The ability of a rock to split along planes of weakness.

rock cycle The orderly sequence of events that describes the formation, modification, destruction, and re-formation of rocks as a result of processes acting on and within Earth.

rock fall The free-fall of rock fragments newly detached from a very steep surface.

rock glacier A mass of poorly sorted rock fragments, cemented by ice a few feet (meters) below the surface, and moving slowly like a small alpine glacier.

rockslide A sudden and rapid movement of rock down a preexisting, inclined surface such as a bedding plane, joint, or fault.

rock stream A layer of angular blocks of rock, often found at the head of a ravine, accumulated on the surface of solid or weathered bedrock, colluvium, or alluvium.

rock transport The lateral movement of rock masses by the forces involved in deformational events such as the formation of foldbelt mountains.

roundness A characteristic of sand-sized or larger grains that involves the removal of corners and sharp edges by abrasion during transport. A perfectly rounded particle such as a sphere has a roundness of 1.0 with less well rounded particles having roundness values less than 1.0. Not to be confused with sphericity. See also *sphericity*.

S

S waves Shear-type, seismic body waves, also called secondary waves. See also *P waves*.

salinity current An ocean density current formed or maintained because of a salt content higher than the surrounding ocean water.

saltation The movement of particles by water or wind in intermittant bounces, leaps, or jumps.

saltwater encroachment The displacement of low-density fresh groundwater by higher-density saline water due to the withdrawal of fresh water from near-shore aquifers or those located near playa lakes.

sandstone A sedimentary rock composed dominantly of sand-sized particles.

sanitary landfill A site where municipal waste is deposited, compacted, and buried in a fashion that minimizes the potential environmental impact of effluents.

Saturn A Jovian planet that is the sixth planet from the Sun.

scarp A line of cliffs formed by fault action or erosion.

schist A metamorphic rock of intermediate to high grade characterized by strong foliation.

schistosity The foliation in schist or other coarse-grained metamorphic rocks due to the parallel alignment of platy minerals.

scrubber A mechanical device attached to the exhaust of the firebox of a coal-fired power plant that removes SO_2 from the flue gases by causing it to react with water.

sea arch A bridge of rock formed when wave action along a high-energy coastline erodes a channel through a headland.

sea cave A cave or cavity formed along a high-energy coastline at or near sea level by the preferential erosion of weaker rocks in the base of a sea cliff.

sea cliff A cliff formed by wave erosion along high-energy coastlines that represents the inner limit of beach erosion.

sea stacks Freestanding rock masses along high-energy coastlines that have been totally detached from the headland by wave erosion.

sea-floor spreading The movement of newly formed oceanic lithosphere away from an oceanic ridge as a result of convection cells within the upper mantle.

seamount A peak, usually a shield volcano, that rises at least 3,000 feet (1,000 m) above the sea floor. Seamounts can be peaked or flat topped. See also *guyot*.

secondary porosity The porosity that develops in a rock by dissolution or by fracturing subsequent to its formation.

sediment A general term applied to any unconsolidated material that accumulates in layers on Earth's surface. The materials can be detritus generated by the weathering of rocks and subsequently transported and deposited from water, wind, or ice, or materials precipitated from solution by chemical or biochemical processes.

sedimentary Pertaining to or containing sediment.

seif dune A longitudinal dune or a chain of dunes that may reach a height of 650 feet (200 m) and a length of 60 miles (100 km).

seismic activity Pertaining to Earth vibrations or earthquakes of natural or artificial origin.

seismic gap A portion of a fault that has not experienced major movement during a period of time when other parts of the same fault have been active.

Seismic Sea Wave Warning System A system established in 1946 to warn inhabitants of Pacific coastal areas of a potentially destructive tsunami.

seismic wave A general term used to describe all elastic waves generated naturally by earthquakes or artificially by explosions.

seismogram A record of seismic activity generated by a seismograph.

seismograph An instrument that detects, magnifies, and records Earth vibrations.

shale A fine-grained sedimentary rock formed by the lithification of silt- or clay-sized sediments showing distinct fissle character; the most abundant sedimentary rock.

shallow-focus earthquake An earthquake with a focus less than 40 miles (70 km) below the surface; the most frequent type of earthquake.

shear The type of deformation resulting from stresses that cause contiguous parts of a body to be displaced parallel to their plane of contact.

shear joint A fracture formed by shear forces along which no movement has taken place.

shear wave A seismic wave that imparts an oscillating motion in the transporting medium perpendicular to the direction of propagation.

sheet dike A vertical basaltic dike that occurs as part of the oceanic crust. See also *ophiolite suite*.

sheeting A series of joints oriented parallel to the surface of exposed granitic plutons produced by pressure release. See also *exfoliation*.

shield A large expanse of exposed basement rock within a craton, commonly with a gentle convex upward surface, that is surrounded by a sediment-covered platform. See also *stable platform*.

shield volcano A broad, low volcanic cone built by repeated flows of very fluid basaltic lava. See also *stratovolcano*.

silicate A mineral whose crystal structures contain silicon-oxygen tetrahedra as basic building blocks.

siliciclastic Refers to clastic, non-carbonate sediments and rocks that are composed almost exclusively of quartz or other silicate minerals.

silicon-oxygen tetrahedron An ion formed by four oxygen ions surrounding a silicon ion in a tetrahedral configuration, SiO_4^{4-}, with a negative charge of 4; the basic structural unit of all silicate minerals.

sill   A tabular, concordant, intrusive igneous body.

siltstone   A sedimentary rock composed of lithified silts and fine sand.

Silurian Period   A period in the lower Paleozoic Era lasting from 438 to 408 million years ago.

single-chain structure   A silicate structure formed by single chains of silicon-oxygen tetrahedra. See also *pyroxene group*.

sinkhole   A circular or elliptical surface depression of variable size characteristic of karst topography.

sinuosity   The ratio of the length of a stream channel between two points to the down-valley distance between the same two points.

slab pull   The force generated by the sinking portion of a mantle convection cell that is believed by some geologists to be primarily responsible for the movement of the lithospheric plates. See also *ridge push* and *mantle drag*.

slate   A low-grade metamorphic rock with distinct rock cleavage formed almost exclusively from shale.

slaty cleavage   A form of rock cleavage that develops in slate and other rocks formed by low-grade metamorphism due to the parallel alignment of fine-grained platy minerals.

slickenside   A polished, smoothly striated surface resulting from the mutual abrasion of rocks on opposite sides of a fault.

slope failure   Any process by which soil, regolith, or rock moves downslope under the force of gravity.

slope stability   The disposition of a slope to resist failure by mass wasting.

slump   A type of mass wasting characterized by a shearing and rotation of regolith or rock along curved, concave-upward, slip surfaces with an axis of motion parallel to the slope face.

snowline   The line separating areas where snow, deposited during the winter, disappears during the summer from those areas where the snow remains throughout the year. See also *firn line*.

soil   That portion of the regolith that supports plant life out of doors.

solar nebula   The cloud of gas and solar debris from which the Sun and other bodies within the Solar System formed.

solar power   Energy sources generated from the conversion of the Sun's radiant energy.

solar wind   The stream of ionized particles moving away from the Sun.

solid-solution series   Two or more minerals with physical properties that vary uniformly. Examples are the olivine and plagioclase groups of silicate minerals.

solifluction   The slow, downslope motion of water-saturated regolith that occurs in regions underlain by frozen ground that serves as a barrier to the downward percolation of meltwater generated by the thawing of snow and ground ice.

solstice   The time of the year when the Sun is either directly overhead at the Tropic of Cancer (about June 22) or the Tropic of Capricorn (about December 22).

solution channel   A channel opened in soluble bedrock by dissolution.

sonar   A technique that detects sound waves reflected from the ocean bottom and used extensively by scientists to study sea-floor topography; an acronym of *so*und *na*vigation and *r*anging.

sorting   The separation of particles by size during transportation and deposition by water or wind.

spalling   A weathering process whereby rock fragments, usually relatively thin and curved, are removed from the surface of a rock body. See also *exfoliation*.

spatter cone   A low, steep-sided cone developed around a fissure or vent by the accumulated spatter of basaltic lava.

specific gravity   The ratio of the weight of an object to the weight of an equal volume of water; equal numerically to density but without units of measurement.

specific heat   The energy required to raise the temperature of one gram of a substance by one degree Celsius.

speleothem   Any mineral deposit, usually calcium carbonate, that is formed in a cave by the deposition of soluble materials from percolating water.

sphericity   The ratio of the true nominal diameter of a particle to the diameter of a circumscribing sphere (usually the longest diameter). The nominal diameter of a particle is the calculated diameter of a sphere having the same volume as that of a particle. See also *roundness*.

spheroidal weathering   A form of weathering where concentric or spheroidal layers of rock are removed. See also *exfoliation*.

spreading rate   The rate, usually measured in millimeters per year, at which new oceanic crust moves away from an oceanic ridge.

spring tides   Bimonthly tides when the tidal range is at a maximum; occur at the times of the new and full moon phase when the gravitational pull of the Sun and Moon are additive.

stable platform   The part of the continent, generally surrounding the shield, that is covered by flat-lying or gently tilted sedimentary rocks and underlain by igneous and metamorphic rocks of the basement that have not been subjected to extensive crustal deformation for a long period of geologic time.

stalactite A conical formation of dripstone that hangs down from the ceiling of limestone caves or caverns. See also *speleothem*.

stalagmite A conical formation of dripstone that protrudes up from the floor of limestone caves or caverns. See also *speleothem*.

stock A massive, intrusive, igneous body with an exposure less than 40 square miles (100 km^2). See also *batholith*.

stony meteorite A meteorite composed predominantly or entirely of mafic silicate minerals such as olivine and pyroxene.

stoping The process by which plutons are emplaced by detaching and engulfing pieces of host rock that sink into the magma or are assimilated into the melt.

strain The change in the shape or volume of a body as a result of stress.

strain ellipsoid An ellipsoid in the deformed state that is derived from a sphere in the undeformed state.

stratovolcano or composite volcano A volcanic cone composed of alternating layers of pyroclastic materials and lava, typically associated with zones of subduction. See also *shield volcano*.

streak The color of a powdered mineral.

stream Any body of running water that moves under gravity to progressively lower levels in a relatively narrow, defined channel on the surface of the ground, within a cave or cavern, or on or under a glacier.

stream gradient The angle between the water surface (of a large stream) or the channel floor (of a small stream) and the horizontal.

stream order The classification system used to designate the relative position of a stream in a drainage system based on the pattern of tributaries.

stream pattern The arrangement in plan view of the stream courses in an area.

stream piracy The natural diversion of the headwaters of one stream into the channel of another having greater erosive power and flowing at a lower elevation.

stream-worn pebble A rock particle that has been rounded and smoothed by abrasion during stream transport.

strength The ability of a material to withstand differential stress.

stress The force per unit area working on or within a body.

striation Scratches or minute grooves, generally parallel, on a rock surface produced mostly by the abrasion of particles being transported by streams or glaciers.

strike The direction of the line of intersection between a plane and the horizontal.

strike-slip A type of fault where there is horizontal offset with little or no vertical offset.

Strombolian-type eruption A volcanic eruption characterized by frequent, relatively mild eruptions of basaltic magma from a center crater.

subbituminous coal A coal rank between lignite and bituminous coal.

subduction The process by which one lithospheric plate moves beneath another. See also *convergent plate boundary*.

subduction zone The long, narrow belt parallel to convergent boundaries along which subduction takes place.

sublimation The process in which a solid vaporizes without going through a liquid state. An example is the vaporization of dry ice (solid CO_2).

submarine canyon A steep-sided, V-shaped valley that crosses the continental shelf or slope. Similar in appearance to a youthful stream valley.

submergent (or low-energy) coastline A coastline currently being inundated due either to the sinking of the land or to a rise in sea level and usually characterized by relatively gentle offshore slopes and low-energy surfs.

subpolar low-pressure zone A low-pressure zone that circumscribes the globe and is associated with rising air masses over the polar front.

subsidence The sudden or gentle sinking of Earth's surface with little or no horizontal motion.

subtropical desert A desert formed on landmasses located under the descending masses of warm, dry air of the subtropical high-pressure zones, approximately the latitudes of the Tropic of Cancer in the Northern Hemisphere and the Tropic of Capricorn in the Southern Hemisphere.

subtropical high-pressure zone A global high-pressure zone between 30° and 35° north and south latitude associated with the downward movement of air masses form high altitudes.

sulfide A mineral formed by the combination of sulfur and a metal. Examples are galena (PbS) and pyrite (FeS_2).

sulfur An orthorhombic mineral and native nonmetallic element.

supergene enrichment A process of sulfide mineral concentration whereby the oxidation of near-surface sulfide minerals produces acid that dissolves and carries metals downward where they are reprecipitated to enrich sulfide minerals already present.

surf The wave activity in the surf zone.

surf zone The area between the landward limit of wave onlap and the most seaward breaker.

surface wave A seismic wave that travels along Earth's surface, the Rayleigh and Love waves.

surge A period of rapid glacial advance.

suspended load That portion of the total stream load, consisting primarily of silt- and clay-sized

Glossary 659

particles, that is transported for a considerable period of time within the mass of moving water. See also *bed load* and *dissolved load*.

swallow hole   The point in a stream channel where a stream disappears into a subterranean cave or fracture system.

swamp   A wetland whose vegetation is dominated by woody-tissued plants, i.e., trees. See also *marsh*.

swell   A long-wavelength, flat-crested wave that has moved out of its area of formation.

symmetric   The quality of having symmetry; the repeat of a similar pattern from one side of an object to the other.

symmetrical ripple mark   A ripple mark that has a symmetrical cross section.

syncline   A concave-upward fold in which the core contains stratigraphically younger rocks. See also *anticline* and *monocline*.

synfuel   *Synthetic fuel*, usually produced by the liquefaciton of coal.

system (of rocks)   The rocks that accumulate during a *period* of geologic time.

T

tabular (*Three dimensional*)   A descriptive term applied to a rock formation or feature that has two dimensions substantially greater than the third. (*Two dimensional*) Said of a body where the largest dimension is more than 10 times the smallest dimension. An example is a dike or a sill. See also *massive*.

tachylyte   A volcanic glass of various colors commonly found in the chilled margins of intrusive igneous bodies formed from basaltic magma. See also *obsidian*.

talus   An accumulation of rock fragments at the base of a steep slope such as a cliff or road cut.

tarn   A glacial lake located within a cirque.

temporary baselevel   Any baselevel, other than sea level, below which, for a limited period of time, a land area cannot be reduced by stream erosion. See also *ultimate baselevel*.

tension   Stresses that act away from a body and lend to lengthen or increase the volume of the body.

tensional joint   A fracture within a rock body along which there has been no displacement, formed under tensional stress.

tephra   A general term for all pyroclastic materials produced by the eruption of a volcano.

terminal moraine   The moraine that marks the furthest extent of a glacial advance.

terrestrial planet   One of the four planets of the Solar System nearest the Sun; Mercury, Venus, Earth, or Mars. See also *Jovian planet*.

Tertiary Period   The first period of the Cenozoic Era spanning the time from 66 to 2 million years ago.

texture   The general appearance of a rock in terms of the size, shape, and arrangements of its constituents.

thermohaline currents   Vertical movements of oceanic water due to density differences resulting from variations in temperature and salt content.

throw   The vertical component of fault displacement.

thrust fault   A type of reverse fault with a fault plane that dips 45° or less along most of its extent.

tidal inlet   Any inlet through which seawater flows alternately during the rising and falling of the tides.

tide   The rhythmic rising and lowering of sea level under the combined gravitational pull of the Sun and the Moon.

till   Unsorted, unstratified, usually unconsolidated material deposited by and/or beneath a glacier without being reworked by meltwater.

tiltmeter   An instrument commonly used in volcanology and seismology to detect and measure slight changes in the tilt (slope) of Earth's surface.

time-distance curve   In seismology, a curve relating the distance traveled by the s and p body waves to the time elapsed since the rupture of the rocks that is used to calculate the distance from a seismic station to the earthquake focus.

topography   The general configuration of Earth's surface in terms of relief and the location of both natural features and those constructed by humans.

topset beds   The uppermost, horizontal beds of a prograding delta formed by the deposition of the load transported by a bed load-dominated stream.

traction   The mode of transport by which the bed load of a stream is carried along the channel bottom by bouncing, rolling, sliding, pushing, or saltating.

trade winds   The system of tropical air currents that move from the subtropical highs toward the equatorial lows, from northeast to southwest in the Northern Hemisphere and from southeast to northwest in the Southern Hemisphere.

transform fault   A special type of strike-slip fault that offsets the oceanic ridges and allows the lithospheric plates to move on Earth's spherical surface.

transpiration   The process by which the water absorbed by plants by way of the roots is released to the atmosphere as water vapor.

transportation   Any process by which the products of weathering, both solid and dissolved, are moved from one location to another.

transverse dune   An asymmetric dune, elongated perpendicular to the direction of the prevailing wind, having a gentle windwarp slope and a steep leeward slope.

trap Any geologic structure that results in the accumulation of economic deposits of oil and/or gas.

trellis stream pattern A stream pattern characterized by parallel main streams, intersected by their tributaries at or nearly at right angles, which are in turn fed by tributaries flowing parallel to the main streams. The pattern typically develops in areas characterized by parallel belts of resistant and nonresistant rocks such as the rejuvenated fold-belt mountains of the Appalachian Valley and Ridge Province.

Triassic Period The first period of the Mesozoic Era spanning the time from 245 to 208 million years ago.

tributary A stream of smaller order that enters a larger trunk stream and contributes to its total flow. See also *distributary*.

triple junction The point where three lithospheric plates meet.

Tropic of Cancer The latitude at about 23.5° north of the equator where the Sun's rays are vertical to Earth's surface at the summer solstice.

Tropic of Capricorn The latitude at about 23.5° south of the equator where the Sun's rays are vertical to the Earth's surface at the winter solstice.

tropical low-pressure zone Also referred to as the Intertropical Convergence Zone, the tropical low-pressure zone is produced by the upward movement of air masses over the warm equatorial regions.

tsunami A sea wave produced by any large-scale, short-duration disturbance of the ocean floor, usually a shallow submarine earthquake or volcanic eruption.

tuff A general term for any rock composed of pyroclastic materials.

turbidite The sediment or the rock formed from the sediments deposited from a turbidity current.

turbidity current A density current in any fluid formed by different amounts of suspended material that travels quickly downslope under the influence of gravity. Examples are the movement of materials down the continental slope onto the abyssal plain and the nuée ardente resulting from a violent volcanic eruption.

turbulent flow A type of fluid flow where the flow paths cross one another. See also *laminar flow*.

type locality The locality at which the rock body defining a system of rocks (the *stratotype*) is first recognized and described.

U

ultimate baselevel The lowest possible baselevel. For an external stream, the ultimate baselevel is sea level. For an internal stream, it is the elevation of the lowest basin into which the water flows. See also *temporary baselevel*.

ultramafic Refers to igneous rocks composed almost entirely of mafic minerals or the magma from which they form. See also *felsic* and *mafic*.

unconfined aquifer An aquifer having a water table. See also *confined aquifer*.

unconformity A surface of erosion and/or nondeposition representing a substantial time break in the geologic record. See also *hiatus*.

uniformitarianism The concept that the geologic processes now operating on Earth operated throughout geologic time in the same way and with the same intensity. The concept is summarized by the saying "the present is the key to the past."

unpaired terrace A stream terrace with no corresponding terrace on the opposite side of the valley. Usually produced by a meandering stream. See also *paired terrace*.

Uranus The Jovian planet that is seventh from the Sun.

V

valley Any low-lying ground bordered by higher ground.

valley flat The bedrock surface produced by stream erosion in mature and old age valleys that is covered with alluvium to form the floodplain.

valley train The materials deposited by meltwater streams beyond the terminal moraine of an alpine glacier.

van der Waals bonding The weakest of the four types of chemical bonds.

varve A sedimentary bed or lamina deposited in a body of still water within one year's time.

vein A mineral-filled fracture.

ventifact Any rock fragment that has been shaped, worn, faceted, or polished by the abrasive action of windblown sand.

Venus The terrestrial planet that is second from the Sun.

vertisol A soil containing expandable clay minerals that forms in areas where the soil is subjected to seasonal periods of drying and wetting that causes the clay minerals to alternately shrink and swell.

vesicular An igneous rock texture describing lavas that contain many small holes created by the release and expansion of gas while the lava was still molten.

viscosity The resistance of a fluid to internal deformation or flow.

volcanic breccia A pyroclastic rock composed of angular volcanic materials larger than 64 mm in diameter.

volcanic neck The exposed core of an extinct volcano. An example is Ship Rock, New Mexico.

volcanic rocks Any extrusive rock resulting from volcanic action.

volcanism The process by which magmas move to the surface and are extruded as lava.

Vulcanian-type eruption A volcanic eruption characterized by infrequent but severe explosions accompanied by the explusion of large quantities of pyroclastic material.

W

Walther's law The concept that contiguous marine depositional environments will be represented by superposed sedimentary beds within a vertical sequence of sedimentary beds.

water gap A narrow passage cut perpendicular to a mountain ridge through which a stream flows.

water table The contact between the zone of aeration and the zone of saturation, where the hydrostatic pressure is equal to atmospheric pressure.

wave An oscillatory movement of water characterized by a rise and fall of the surface.

wave-built platform A gently sloping surface built by the deposition of sediment on the seaward side of a wave-cut platform.

wave-cut cliff A cliff formed by the undercutting and collapse of the headland along a high-energy coastline.

wave-cut platform A gently sloping surface produced by wave erosion that extends from the base of a wave-cut cliff seaward to the wave-built platform.

wavelength The distance between wave crests or any other equivalent points of adjacent waves.

weather The condition of Earth's atmosphere in a particular place and time in relation to winds, humidity, barometric pressure, temperature, precipitation, and clouds.

weathering Any process by which rocks disintegrate or decompose.

westerlies Winds within the Ferrel Cell that move from the subtropical high to the subpolar low.

wet-based glacier A glacier underlain by a thin layer of water. See also *dry-based glacier*.

windshadow The area on the leeward side of an obstacle where the velocity of the wind is sharply reduced.

wind-driven currents Ocean surface currents driven by the wind. An example is the Gulf Stream. See also *gyre*.

wind-induced currents Vertical currents within the upper 3,000 feet (900 m) of the ocean that are generated to replace masses of surface water displaced by wind-driven currents and not replenished by other wind-driven currents.

X

xenolith A fragment of country rock that sank into the magma chamber during magma emplacement but was not assimilated before the magma solidified.

Y

youth The first of three stages of landform developed proposed by Davis. A youthful landform is characterized by streams that are actively downcutting and flow straight for relatively long distances with numerous waterfalls and rapids. Youthful valleys have little or no floodplain and are typically V-shaped.

Z

zone of ablation The glacial zone within which there is a net loss of snow and ice by melting, sublimation, and/or calving.

zone of accumulation The glacial zone within which there is a net gain of snow and ice.

zone of aeration The zone above a water table where, except for absorbed water, the pores are empty of water.

zone of saturation The zone below the water table where all pore space is filled with water.

zone of subduction The long, narrow belt where one lithospheric plate descends beneath another.

APPENDIX B:

Tables

I. Earth Data
II. The Periodic Table
III. Average Composition of Earth's Crust
IV. Water Composition
   1. Composition of Seawater: Elements with Concentrations over .001 ppm
   2. Major Ions in River Water
V. The Geologic Time Scale
VI. Conversion Factors
VII. Prefixes for Metric System Units
VIII. Scientific Notation
IX. Common Rock-Forming and Accessory Minerals

I. EARTH DATA

DIMENSIONS
Mean radius:	3,960 mi	(6,371 km)
Equatorial radius:	3,964 mi	(6,378 km)
Polar radius:	3,951 mi	(6,357 km)
Polar circumference:	24,866 mi	(40,009 km)
Equatorial circumference:	24,908 mi	(40,077 km)

ROCK VOLUMES
Total volume:	2.6×10^{11} mi^3	(1.08×10^{12} km^3)
Volume of continental crust:	1.49×10^8 mi^3	(6.21×10^8 km^3)
Volume of oceanic crust:	0.64×10^8 mi^3	(2.66×10^8 km^3)
Volume of mantle:	2.16×10^5 mi^3	(8.99×10^5 km^3)
Volume of core:	4.31×10^4 mi^3	(1.75×10^5 km^3)

DENSITIES
Mean density of Earth:	5.5 gm/cm^3
Mean density of the crust:	2.8 gm/cm^3
Density of the continental crust:	2.7 gm/cm^3
Density of the oceanic crust:	3.0 gm/cm^3
Density of the mantle:	4.5 gm/cm^3
Density of the core:	10.7 gm/cm^3

AREAS
Total area:	1.97×10^8 mi^2	(5.10×10^8 km^2)	
Ocean area:	1.39×10^8 mi^2	(3.61×10^8 km^2)	70.6% of total
Land area:	5.75×10^7 mi^2	(1.49×10^8 km^2)	29.3% of total
Asia:	1.80×10^7 mi^2	(4.66×10^8 km^2)	
Africa:	1.15×10^7 mi^2	(2.98×10^7 km^2)	
North America:	8.2×10^6 mi^2	(2.1×10^7 km^2)	
South America:	7.6×10^6 mi^2	(1.9×10^7 km^2)	
Europe:	4.2×10^6 mi^2	(1.1×10^7 km^2)	
Australia:	3.0×10^6 mi^2	(7.8×10^6 km^2)	

WATER VOLUMES
Total volume:	3.39×10^8 mi^3	(1.41×10^9 km^3)
Volume of oceans and seas:	3.3×10^8 mi^3	(1.37×10^9 km^3)
Volume of glaciers:	7×10^6 mi^2	(2.5×10^7 km^3)
Volume of groundwater:	2×10^6 mi^3	(8.4×10^6 km^3)
Volume of lakes:	3×10^4 mi^3	(1.25×10^5 km^3)
Volume of rivers:	3×10^2 mi^3	(1.25×10^3 km^3)

RELIEF
Mean height of land above sea level:	2,871 ft	(875 m)
Mean depth of ocean:	12,467 ft	(3,800 m)

II. THE PERIODIC TABLE

III. AVERAGE COMPOSITION OF EARTH'S CRUST[a]

	AVERAGE IGNEOUS ROCK, %	AVERAGE SHALE, %	AVERAGE SANDSTONE, %	AVERAGE LIMESTONE, %	WEIGHTED-AVERAGE CRUST,[b] %
SiO_2	59.12	58.11	78.31	5.19	59.07
TiO_2	1.05	0.65	0.25	0.06	1.03
Al_2O_3	15.34	15.40	4.76	0.81	15.22
Fe_2O_3	3.08	4.02	1.08	0.54	3.10
FeO	3.80	2.45	0.30		3.71
MgO	3.49	2.44	1.16	7.89	3.45
CaO	5.08	3.10	5.50	42.57	5.10
Na_2O	3.84	1.30	0.45	0.05	3.71
K_2O	3.13	3.24	1.32	0.33	3.11
H_2O	1.15	4.99	1.63	0.77	1.30
CO_2	0.10	2.63	5.04	41.54	0.35
ZrO_2	0.04	—	—	—	0.04
P_2O_5	0.30	0.17	0.08	0.04	0.30
Cl	0.05	—	Tr^c	0.02	0.05
F	0.03	—	—	—	0.03
SO_3	—	0.65	0.07	0.05	—
S	0.05	—	—	0.09	0.06
$(Ce,Y)_2O_3$	0.02	—	—	—	0.02
Cr_2O_3	0.06	—	—	—	0.05
V_2O_3	0.03	—	—	—	0.03
MnO	0.12	Tr^c	Tr^c	0.05	0.11
NiO	0.03	—	—	—	0.03
BaO	0.05	0.05	0.05	0.00	0.05
SrO	0.02	0.00	0.00	0.00	0.02
Li_2O	0.01	Tr^c	Tr^c	Tr^c	0.01
Cu	0.01	—	—	—	0.01
C	0.00	0.80	—	—	0.04
Total	100.00	100.00	100.00	100.00	100.00

[a] After F.W. Clarke and H.S. Washington, "The Composition of the Earth's Crust," *U.S. Geological Survey Professional Paper* 127 (1924), p. 32. As printed in Judson and Kauffman, *Physical Geology* 8/e (Englewood, N.J.: Prentice-Hall, 1990), page 483. Used with permission.

[b] Weighted average: igneous rock, 95% shale, 4%; sandstone, 0.75%; limestone, 0.25%.

[c] Trace.

IV. WATER COMPOSITION

1. Composition of Seawater: Elements with Concentrations over .001 ppm

ELEMENT	CONCENTRATION IN PARTS PER MILLION (PPM)
Oxygen (as H_2O)	857,000
Hydrogen	108,000
Chlorine	19,400
Sodium	10,800
Magnesium	1,350
Sulfur	904
Calcium	400
Potassium	380
Carbon (inorganic)	28
Strontium	8
Bromine	6.7
Oxygen (dissolved)	6.0
Boron	4.6
Silicon	2.9
Carbon (dissolved organic)	2.0
Fluorine	1.3
Nitrogen (as NO_3^{1-}, NO_2^{1-}, NH_4^{4+})	0.6
Nitrogen (dissolved N_2)	0.5
Argon	0.45
Lithium	0.17
Rubidium	0.12
Phosphorous	0.08
Iodine	0.06
Barium	0.03
Aluminum	0.01
Iron	0.01
Molybdenum	0.01
Zinc	0.01
Arsenic	0.003
Copper	0.003
Uranium	0.003
Nickel	0.002
Manganese	0.002
Vanadium	0.002
Titanium	0.001

IV. WATER COMPOSITION

2. Major Ions in River Water

ION	CONCENTRATION IN PARTS PER MILLION (PPM)
HCO_3^{1-}	58.4
Ca^{2+}	15
SO_4^{2-}	11.2
Cl^{1-}	7.8
Si^{4+}	6.5
Na^{1+}	6.3
Mg^{2+}	4.1
K^{1+}	2.3
NO_3^{1-}	1
Fe_T	0.004

V. THE GEOLOGIC TIME SCALE

ERA	PERIOD	EPOCH	AGE (10⁶ YR)	LIFE-FORMS	GEOLOGIC EVENTS
Cenozoic	Quaternary	Recent	0.01	Age of humans	Glacial advances and retreats
		Pleistocene	1.6	Appearance of hominids	Gulf of California opens
		Pliocene	5		Rocky Mountains rise
	Tertiary	Miocene	24	Whales, grazing animals	
		Oligocene	37	Appearance of apes	Rise of Alps and Himalayas
		Eocene	58		
		Paleocene	66	First primates	
Mesozoic	Cretaceous		144	Dinosaur extinctions	
	Jurassic		208	Dinosaur zenith	South Atlantic begins to open
	Triassic		245	First dinosaurs	Gondwana rifts and North Atlantic begins to open
Paleozoic	Permian		286	Primitive mammals Reptiles evolve	Appalachians formed
	Pennsylvanian		320	Frist reptiles	Glaciation in Gondwana
	Mississippian		360		
	Devonian		408	First amphibians and forests	
	Silurian		438	First land plants	Mountain building in Europe and eastern North America
	Ordovician		505	First fishes and vertebrates	
	Cambrian		600	Age of marine invertebrates	
Proterozoic			2,500	First life	Extensive mountain building
Cryptozoic			4,500	Oceans form Earth's crust forms	

Source: Paul Pinet, *Oceanography* (St. Paul, Minn.: West, 1992). Used with permission.

VI. CONVERSION FACTORS

TO CONVERT FROM	TO	MULTIPLY BY
Centimeters	Feet	0.0328
	Inches	0.394
Meters	Inches	39.37
	Feet	3.2808
	Yards	1.0936
	Miles	0.0006214
Kilometers	Miles	0.621
Square kilometers	Square miles	0.386
Inches	Centimeters	2.54
	Meters	0.0254
Feet	Centimeters	30.48
	Meters	0.3048
Yards	Meters	0.9144
Miles	Kilometers	1.609
	Feet	5280
	Yards	1760
Square miles	Square kilometers	2.59
	Acres	640
Acres	Hectares	0.4047
Hectares	Acres	2.47
Grams	Ounces (avdp)	0.03527
	Pounds (avdp)	0.002205
Kilograms	Ounces (avdp)	35.27
	Pounds (avdp)	2.2046
Ounces	Grams	28.35
Pounds (avdp)	Grams	453.6
	Kilograms	0.4536

VII. PREFIXES FOR SYSTEM UNITS

PREFIX	POWER		EQUIVALENT
Tera	10^{12} =	1,000,000,000,000	Trillion
Giga	10^{9} =	1,000,000,000	Billion
Mega	10^{6} =	1,000,000	Million
Kilo	10^{3} =	1,000	Thousand
Hecto	10^{2} =	100	Hundred
Deca	10^{1} =	10	Ten
	10^{0} =	1	One
Deci	10^{-1} =	.1	Tenth
Centi	10^{-2} =	.01	Hundredth
Milli	10^{-3} =	.001	Thousandth
Micro	10^{-6} =	.000001	Millionth
Nano	10^{-9} =	.000000001	Billionth
Pico	10^{-12} =	.000000000001	Trillionth

VIII. SCIENTIFIC NOTATION

Scientific notation is a shorthand way of designating numbers and is especially useful when dealing with very large or very small numbers, both of which are difficult to read and equally difficult to notate. For example, the distance from the Sun to Pluto is three billion, six hundred and sixty-seven million miles (five billion, nine hundred million kilometers). Even if we reduce the distances to numerical notation, one still has to write 3,667,000,000 miles (5,900,000,000 km). Scientific notation reduces such numbers to a number *greater than 1* and *less than 10*, called the *real constant*, times 10 to an exponent. When the original number is larger than 1, the exponent of 10 is *positive*; when it is smaller than 1, the exponent is *negative*. The exponent tells how many *powers of ten* are needed to convert the real constant to the original number.

Let us first consider a simple case. The real constant for the number 100 would be 1. (Remember, the real constant is a number between 1 and 10.) The real constant, 1, must be multiplied by *two powers of ten* (10×10) to equal 100. The scientific notation for 100 would therefore be 1 (the real constant) $\times$ 10 raised to a power of two (the exponent). The scientific notation for 100 would therefore be 1×10^2. Similarly, the real constant for the number 1,000 is also 1, but it would have to be multiplied by *three powers of ten* ($10 \times 10 \times 10$) to equal 1,000. The scientific notation for 1,000 would therefore be 1×10^3.

Now let's convert the distance to Pluto in miles, 3,667,000,000, to scientific notation. The real constant, which would be 3.667, needs to be multiplied by *nine powers of ten* or *1 billion* to equal the original number. The scientific notation would therefore be 3.667×10^9) miles. The scientific notation is a much easier way to both record and convey the distance from the Sun to Pluto.

SUMMARY OF THE SCIENTIFIC NOTATION METHOD

10^6	=	1,000,000
10^5	=	100,000
10^4	=	10,000
10^3	=	1,000
10^2	=	100
10^1	=	10
10^0	=	1
10^{-1}	=	.1
10^{-2}	=	.01
10^{-3}	=	.001
10^{-4}	=	.0001
10^{-5}	=	.00001
10^{-6}	=	.000001

Note that for numbers greater than 1, the exponent of 10 is *positive* and is *equal* to the number of zeros *before* the decimal. For numbers smaller than 1, the exponent of 10 is *negative* and is equal to *one more* than the number of zeros *after* the decimal point. For the number 1, the exponent of 10 is zero.

The same procedure is applied to determine the scientific notation of very small numbers. For example, the diameter of the carbon atom is 0.000,000,000,154 meters. The real constant would be 1.54. Because the number is *less* than 1, the exponent of 10 will be *negative* and will be equal to the number of digits the decimal point was shifted to the right to arrive at the real constant. Note that to arrive at 1.54, the decimal point is shifted *10 places*. The scientific notation for the diameter of a carbon atom in meters would be 1.54×10^{-10}.

IX. COMMON ROCK-FORMING AND ACCESORY MINERALS

Table follows on pages 672 to 683.

MINERAL GROUP OR MINERAL NAME	FORMULA	CRYSTAL SYSTEM	DENSITY	HARDNESS
Olivine Group				
Fayalite	Fe_2SiO_4	Orthorhombic	4.39	6.5
Forsterite	Mg_2SiO_4	Orthorhombic	3.2	6.5
Pyroxene Group				
Augite	$Ca(Mg,Fe,Al)(Al,Si)_2O_6$	Monoclinic	3.2–3.4	5–6
Diopside	$CaMgSi_2O_6$	Monoclinic	3.2	5–6
Enstatite	$MgSiO_3$	Orthorhombic	3.2–3.5	5.5
Hedenbergite	$CaFeSi_2O_6$	Monoclinic	3.5	5–6
Hypersthene	$(Mg,Fe)SiO_3$	Orthorhomibc	3.4–3.5	5–6
Amphibole Group				
Actinolite	$Ca_2(Mg,Fe)_5Si_8O_{22}(OH)_2$	Monoclinic	3.1–3.3	5–6
Hornblende	$(Ca,Na)_{2-3}(Mg,Fe,Al)_5Si_6(Si,Al)_2O_{22}(OH)_2$	Monoclinic	3.0–3.4	5–6
Tremolite	$Ca_2Mg_5Si_8O_{22}(OH)_2$	Monoclinic	3.0–3.2	5–6
Feldspar Group				
Potassium Feldspars				
Microcline	$KAlSi_3O_8$	Triclinic	2.5–2.6	6
Orthoclase	$KAlSi_3O_8$	Monoclinic	2.57	6
Plagioclase Feldspars				
Albite	$NaAl\,Si_3O_8$	Triclinic	2.62	6
Anorthite	$CaAl_2Si_2O_8$	Triclinic	2.76	6

COLOR	STREAK	LUSTER	OTHER PROPERTIES
Dark green	Pale Green	Glassy	Minerals of the olivine group are commonly found in mafic rocks.
Yellow-green	White	Glassy	The magnesium-rich variety of olivine, forsterite, is the most common variety.
Green-black	Greenish gray	Glassy	The most common pyroxene; found chiefly in mafic igneous rocks where it is a major component.
White-green		Glassy	Forms as the result of the thermal metamorphism of siliceous, magnesium-rich limestones and dolomites. Transparent forms of diopside have been used as gemstones.
Yellow-green to brown		Glassy	Along with hypersthene, enstatite is commonly found in mafic igneous rocks and in both stony and metallic meteorites.
Dark green		Glassy	Occurs in more iron-rich metamorphic rocks and in the late stages of igneous crystallization.
Green-brown		Glassy	Found in iron-rich metamorphic rocks and in both stony and metallic meteorites.
White-green	Colorless	Glassy	A common metamorphic mineral commonly found in greenschists. The fibrous forms of both actinolite and tremolite are used as asbestos.
Green-black	Colorless	Glassy-silky	A common mineral found in both igneous and metamorphic rocks.
White-green	White	Glassy	Found in metamorphosed, impure dolomitic limestones and in talc-rich schists.
White-pink	White	Glassy	Common in felsic igneous rocks that cooled slowly at considerable depth, arkoses, and gneisses. It is used in the manufacture of porcelain.
White-pink	White	Glassy	A major constituent of granites, granodiorites, and syenites that have cooled relatively fast at moderate depths.
White-gray White-green to yellow-red	White White	Glassy-pearly Glassy-pearly	The plagioclase feldspars are more widely distributed in rocks and are more abundant than the potassium feldspars; commonly found in igneous and metamorphic rocks, more rarely in sedimentary rocks.

MINERAL GROUP OR MINERAL NAME	FORMULA	CRYSTAL SYSTEM	DENSITY	HARDNESS
Mica Group				
Biotite	$K(Mg,Fe)_3AlSi_3O_{10}(OH)_2$	Monoclinic	2.8–3.2	2.5–3
Muscovite	$KAl_2Si_3O_{10}(OH)_2$	Monoclinic	2.7–2.9	2–2.5
Clay Minerals				
Chlorite	$(Mg,Fe)_3(Si,Al)_4O_{10}(Mg,Fe)_3(OH)_6$	Monoclinic	2.6–3.3	2–2.5
Illite	$(Ca,Na,K)(Al,Fe^{3+},Fe^{2+},Mg)_2(Si,Al)_4O_{10}(OH)_2$	An earthy substance resembling mica		
Kaolinite	$Al_2Si_2O_5(OH)_4$	Triclinic	2.6	2
Montmorillonite	$(Al,Mg)_8(Si_4O_{10})_3(OH)_{10}12H_2O$	Monoclinic	2.5	1–1.5
Vermiculite	$(Mg,Fe^{3+})_3(Si,Al)_4O_{10}(OH)_2$	Monoclinic	2.4	1.5
Quartz	SiO_2	Hexagonal	2.65	7

COLOR	STREAK	LUSTER	OTHER PROPERTIES
Black-brown to dark green	Colorless	Glassy-pearly	Forms under a wide variety of conditions. It is found in both felsic and mafic igneous rocks, in metamorphoc rocks, and in clay-rich sedimentary rocks.
Colorless, yellow-green, or red	Colorless	Glassy-pearly or silky	Commonly found in felsic igneous rocks, metamorphic rocks, and, as illite, in sedimentary rocks. Because of its high dielectric and heat-resisting properties, muscovite is used as an insulator in electrical appliances. Sheets of muscovite, called isinglass, are used as windows in furnaces.
Green	Colorless	Glassy-pearly	A common mineral found in low-grade metamorphic rocks.
Earthy clay mineral resembling mica			A clayey material that forms by the chemical decomposition of muscovite. Commonly found in soils and clay-rich sedimentary rocks.
White	Colorless	Dull-earthy	A common mineral formed by the chemical weathering of aluminum-rich silicate minerals. The main ingredient of kaolin clay, which is widely used for the manufacture of ceramics.
Earthy clay mineral			A clay mineral with the unique property of swelling and contracting as it absorbs and releases water.
Mineral formed by the alteration of mica			A platy mineral that forms by the chemical alteration of biotite.
Colorless-white to varioius colors	Colorless	Glassy	A common and abundant mineral in a great variety of geologic environments including igneous, metamorphic, and sedimentary rocks. A major constituent of most vein deposits where it occurs in the form of chert and flint. Sandstones and quartzites are often pure quartz. Occurs as rock crystal, chalcedony, and agate.

MINERAL GROUP OR MINERAL NAME	FORMULA	CRYSTAL SYSTEM	DENSITY	HARDNESS
Carbonates				
Ankerite	$CaFe(CO_3)_2$	Rhombic	2.95–3	3.5
Aragonite	$CaCO_3$ (orthorhombic)	Orthorhombic	2.95	3.5–4
Calcite	$CaCO_3$ (hexagonal)	Rhombohedral	2.71	3
Dolomite	$CaMg(CO_3)_2$	Rhombic	2.85	3.5–4
Siderite	$FeCO_3$	Rhombic	3.96	3.5–4
Sulfates				
Anhydrite	$CaSO_4$	Orthorhombic	2.9–3	3–3.5
Barite	$BaSO_4$	Orthorhombic	4.5	3–3.5
Gypsum	$CaSO_4 \cdot 2H_2O$	Monoclinic	2.32	2

COLOR	STREAK	LUSTER	OTHER PROPERTIES
White-yellow	Colorless	Glassy	Forms by the low-grade metamorphism of calcite or dolomite.
Colorless or white-tinted	Colorless	Glassy	The pearly portion of shells and the major constituent of pearls; also forms in association with hot springs.
Colorless or white-tinted	Colorless	Glassy	A very common mineral; the major constituent of limestones and marble and also associated with dolomite in dolostone.
White-pink to brown-black	Colorless	Glassy-pearly	The main constituent of dolostone and also found in association with calcite in limestones and marble. Dolomite is used for the manufacture of certain cements and for the manufacture of the magnesia (MgO) used in the preparation of refractory liners in furnaces.
Light to dark	Colorless	Glassy	Commonly found in clay ironstone, in concentric concretions, and in the rocks associated with coal. Siderite has been used as an ore of iron, but only in Great Britain and Austria.
White-tinted	Colorless	Glassy-pearly	Found in limestones, in evaporite deposits, and in the cap rocks of salt domes. Ground anhydrite is used as a soil conditioner.
White, shades of blue, or red	Colorless	Glassy-pearly	Commonly found in hydrothermal veins in ores of a variety of metals, in limestones, and in the deposits of hot springs.
Colorless or white-tinted	Colorless	Glassy-pearly to silky	The most common sulfate mineral; widely found in sedimentary rocks and in association with evaporite deposits where it is one of the first minerals to crystallize from evaporating salt waters. Gypsum is used for the manufacture of plaster of Paris and cements.

MINERAL GROUP OR MINERAL NAME	FORMULA	CRYSTAL SYSTEM	DENSITY	HARDNESS
Sulfides				
Chalcocite	Cu_2S	Orthorhombic	5.5–5.8	2.5–3
Chalcopyrite	$CuFeS_2$	Tetragonal	4.1–4.3	3.5–4
Cinnabar	HgS	Rhombohedral	8.1	2.5
Galena	PbS	Isometric	7.4–7.6	2.5
Sphalerite	ZnS	Isometric	3.9–4.1	3.5–4
Molybdenite	MoS_2	Hexagonal	4.6–4.7	1–1.5
Pyrite	FeS_2	Isometric	5.0	6–6.5
Realgar	AsS	Monoclinic	3.5	1.52

COLOR	STREAK	LUSTER	OTHER PROPERTIES
Shiny lead or gray	Gray-black	Metallic	One of the major ores of copper, primarily being formed as the result of secondary enrichment.
Brassy yellow	Green-black	Metallic	The most widely occurring copper mineral and one of the major sources for copper. Occurs most commonly in hydrothermal veins.
Red	Red	Adamantine (brilliant)	The most important ore of mercury but not widely found. Occurs as vein fillings near recent volcanic rocks and hot springs.
Lead gray	Lead gray	Metallic	A very common metallic sulfide found in hydrothermal veins, commonly with silver minerals where it often contains commercial amounts of silver.
			The most important ore mineral of zinc. Commonly found with galena, having formed under similar conditions.
Lead gray	Gray-black	Metallic	The principal ore of molybdenum, the mineral forms as an accessory mineral in certain granites and pegmatites.
Brass yellow	Green-black	Metallic	The most common and widespread of the sulfide minerals. It occurs as magmatic segregations, in hydrothermal veins, and in sedimentary rocks, especially those associated with coal.
Red-orange	Red-orange	Resinous	Found in veins of lead, gold, and silver ores and in hot spring deposits.

MINERAL GROUP OR MINERAL NAME	FORMULA	CRYSTAL SYSTEM	DENSITY	HARDINESS
Oxides				
Cassiterite	ZnO_2	Tetragonal	6.8-7.1	6-7
Corundum	Al_2O_3	Rhombohedral	4.0	9
Cuprite	Cu_2O	Isometric	6.1	3.5-4
Hematite	Fe_2O_3	Rhombohedral	4.8-5.3	5.5-6
Ilmenite	$FeTiO_3$	Rhombohedral	4.7	5.5-6
Magnetite	Fe_3O_4	Isometric	5.2	6
Pyrolusite	MnO_2	Tetragonal	4.8	1-2
Rutile	TiO_2	Tetragonal	4.2-4.3	6-6.5
Spinel	$MgAl_2O_4$	Isometric	3.5-4.1	8
Zincite	ZnO	Hexagonal	5.7	4
Nitrates				
Niter	KNO_3	Orthorhombic	2.1-2.2	2
Soda niter	$NaNO_3$	Rhombohedral	2.3	1-2
Phosphates				
Apatite	$Ca_5(PO_4)_3(F,Cl,OH)$	Hexagonal	3.1-3.2	5
Turquoise	$CuAl_6(PO_4)_4(OH)_8 \cdot 4H_2O$	Triclinic	2.6-2.7	6

COLOR	STREAK	LUSTER	OTHER PROPERTIES
			The principal ore of tin. Most of the world's supply comes from alluvial deposits where it accumulates because of its relatively high density.
			Common as an accessory mineral in some metamorphic rocks and in silica-deficient igneous rocks. Artificial corundum is manufactured from bauxite.
			Forms by supergene enrichment and, in places, is an important ore of copper.
			Widely distributed in rocks of all kinds and of all ages, hematite is the most abundant and important ore of iron.
			A common accessory mineral in igneous rocks, it is the main source of titanium.
			Found as an accessory mineral in many igneous rocks and in metamorphic rocks. Magnetite is a major source of iron in the United States where it makes up about 25% of the iron formations in the Lake Superior region.
			The most common ore of manganese. Nodular deposits are found on the bottoms of lakes and bogs and on the ocean floor.
			Found in granite, pegmatites, gneisses, and schists and in carbonate rocks. Most rutile is used as a coating for welding rods.
White-red, green-brown, or black	White	Glassy	When pure and of good color, it is a gem. Usually red, it is called ruby spinel. It is a high-temperature metamorphic mineral in association with the contact metamorphism of limestones and clay-rich sedimentary rocks.
Yellow-orange to deep red	Orange-yellow	Subadamantine	Found almost exclusively at the zinc deposits at Franklin, New Jersey. It is primarily used for the production of zinc oxide.
White	White	Glassy	Found as crusts in soils and rock surfaces.
White-gray to yellow-brown	White	Glassy	Because of its solubility, niter is only found in arid regions. It is used primarily in the manufacture of explosives and fertilizers. The world's largest deposits are in northern Chile.
Green-brown or red	White	Glassy	An accessory mineral in all kinds of rocks. The major constituent of bone and tooth dentine, it is also used extensively as a source of phosphorous in fertilizers.
Green-blue		Waxy	Found as a secondary mineral in small veins within decomposed volcanic rocks in arid regions. It is primarily used as a gemstone.

MINERAL GROUP OR MINERAL NAME	FORMULA	CRYSTAL SYSTEM	DENSITY	HARDNESS
Chlorides				
Halite	NaCl	Isometric	2.16	2.5
Sylvite	KCl	Isometric	2	2

COLOR	STREAK	LUSTER	OTHER PROPERTIES
Colorless to white-tinted	Colorless	Glassy-dull	Formed primarily as the result of the evaporation of salt water, its greatest use is in the chemical industry where it is the source of chlorine for the manufacture of hydrochloric acid and a wide variety of sodium compounds; also used as a familiar seasoning for food.
Colorless to various tinted		Glassy	Has the same origin as halite but is far less abundant. Sylvite is one of the last salts to precipitate as salt water evaporites. Its chief use is as a source of potassium for fertilizer.

ANSWERS TO MULTIPLE CHOICE QUESTIONS

CHAPTER 1
1. b; 2. b; 3. d; 4. c; 5. b; 6. b; 7. b; 8. d; 9. a; 10. a; 11. a; 12. b; 13. c; 14. d

CHAPTER 2
1. a; 2. b; 3. c; 4. a; 5. d; 6. c

CHAPTER 3
1. a; 2. a; 3. c; 4. b; 5. c; 6. a; 7. a; 8. b; 9. b; 10. b

CHAPTER 4
1. b; 2. a; 3. d; 4. d; 5. b; 6. c; 7. c; 8. c; 9. b; 10. d

CHAPTER 5
1. b; 2. b; 3. b; 4. c; 5. b; 6. a; 7. a; 8. b; 9. d; 10. d

CHAPTER 6
1. c; 2. b; 3. b; 4. c; 5. a; 6. c; 7. c; 8. b

CHAPTER 7
1. c; 2. c; 3. a; 4. b; 5. b; 6. a; 7. b; 8. d; 9. b

CHAPTER 8
1. a; 2. b; 3. b; 4. c; 5. a; 6. c; 7. a

CHAPTER 9
1. c; 2. b; 3. c; 4. b; 5. b; 6. c; 7. d; 8. a; 9. a; 10. d

CHAPTER 10
1. d; 2. c; 3. b; 4. d; 5. c; 6. c

CHAPTER 11
1. b; 2. a; 3. d; 4. b; 5. b; 6. d

CHAPTER 12
1. b; 2. a; 3. b; 4. c; 5. c; 6. a

CHAPTER 13
1. b; 2. b; 3. a; 4. c; 5. c; 6. a; 7. d

CHAPTER 14
1. d; 2. a; 3. d; 4. b; 5. c; 6. d.; 7. a

CHAPTER 15
1. c; 2. a; 3. d; 4. c; 5. b; 6. a; 7. d

CHAPTER 16
1. a; 2. b; 3. a; 4. b

CHAPTER 17
1. b; 2. c; 3. b; 4. c; 5. b

CHAPTER 18
1. b; 2. d; 3. b; 4. c; 5. a; 6. a

CHAPTER 19
1. a; 2. c; 3. c; 4. b; 5. c; 6 a

CHAPTER 20
1. c; 2. b; 3. b; 4. c; 5. b; 6. a; 7. b

CHAPTER 21
1. c; 2. d; 3. b; 4. d; 5. b; 6. a; 7. c; 8. b

INDEX

Note: Page numbers followed by f indicate figures; those followed by t indicate tables.

A

Aa lava, 104, 105f
Ablation zone, in glaciation, 288-289, 289f
Abrasion
    by glaciers, 290, 290f
    by streams, 249, 251
    by wind, 338-339, 339f, 340f
Absolute dating, 611, 619
Abyssal floor, deposition on, 408
Abyssal hills, 15, 56f, 358-359
Abyssal plains, 13-15, 56f, 359, 361
Accretionary wedge, 512f, 513
Accumulation zone, in glaciation, 288-289, 289f
Acid(s)
    in mineral identification, 84
    in ore enrichment, 571-572
    properties of, 190
Acid mine drainage, 170, 180, 602-604
Acid rain, 191f, 194
    dissolution and, 172, 173f
Acid soil, 190-192
Actinolite, 672t-673t
Activation energies, 176
Active margin, 9-10, 9f, 10f, 14f, 66, 356, 359, 360f, 363-364, 363f
Agassiz, Louis, 306, 528f
Agate, 401
Age of Earth, 609-632
    dating methods for, 611-624. See also Dating
    early estimates of, 609-611
Agriculture
    conservation techniques in, 345
    desertification and, 344-345
    groundwater pollution from, 443
    irrigation in, 196, 345-346
    soil erosion and, 198, 199f
    water pollution and, 178-179, 276, 276f
A horizon, 192, 194
Alberti, Frederich von, 629
Albite, 90, 672t-673t
Alcohol, as fuel, 596
Alkali, in soil, 194
Alkali feldspar, 90, 91f
Alkaline earth, 194
Alkaline soil, 190-192
Alkaline water
    in peat formation, 579
    sources of, 579
Alluvial deposits, 261-274
    channel, 261, 262f

deltas and, 264-272
floodplain, 261-264, 262f-264f
point bar, 261, 262f
Alluvial fans, 272-274, 273f
    in deserts, 329-330, 330f, 335-336, 337f
Alluvial mineral concentrations, 574-575, 574f
Alpha particle, 74, 621
Alpine glaciers, 283, 284f, 285f. See also Glacier(s)
    erosion by, 290-294, 291f-294f
    formation of, 313
Alumina, in soil, 196
Aluminum, abundance of, 569t
Ammonium hydroxide, for acid mine drainage, 603
Amorphous materials, 80
Amphibians, Paleozoic, 626
Amphibole minerals, 86t, 141, 672t-673t
Amplitude
    of folds, 462
    of seismic waves, 541, 542f
Ancient Greeks
    astronomical observations of, 27-28
    hydrologic theories of, 425
Andalusite, 488
Andean-type (continental) arcs, 11f, 99, 101f, 363, 363f, 504
Andesite, 148t, 149
Andesitic magma, 114, 150, 151-152, 151f, 153
    eruptive intensity and, 114
Angle of repose, 208, 210, 211
Angular particles, 395
Angular unconformity, 616, 616f, 618f
Anhydrite, 676t-677t
Animals
    desert, 321-323, 323f
    in earthquake prediction, 544
    in soil formation, 186-187
Animal trails, 413-414, 415f
Anions, 75, 76, 77t
Ankerite, 676t-677t
Annular stream pattern, 240, 242f
Anorthite, 90, 672t-673t
Anthracite, 580
Anticlinal theory, 585, 585f
Anticlinal traps, 585, 585f
Anticlines, 462, 462f
Apatite, 82, 91, 680t-681t
    for acid mine drainage, 603
Aphanitic texture, 146, 146f-148f
Aqueducts, 425f
Aquicludes, 428-429
Aquifers, 428-429
    confined, 434-437, 435f-437f
    dewatering of, 445-446, 446f
    flow rate through, 429, 429i
    pressure surface of, 435f, 436

recharge from, 434
saltwater encroachment into, 446-447, 447f
unconfined, 435-436
Aquitards, 428-429
Aragonite, 399, 676t-677t
    in cementation, 409
Archaeopteryx, 631, 631f
Archean Eon, 625, 625f
Arduino, Giovanni, 624
Arête, 290, 291f
Argillaceous rock, 488
Argon 40, in radiometric dating, 621, 623t
Arid erosion, 206f, 331-336
Aristarchus, 28
Aristotle, 425
Arkose, 397
Arroyo, 329-330, 330f
Arsenic, 91
Artesian well, 436
Ash
    coal quality and, 581
    fluidized-bed combuster, 603-604
    volcanic, 107, 107f
Ash flows, 108
Ash-flow tuffs, 108, 108f
Asteroids, 4
Asthenosphere, 8f, 9, 61, 61f
    convection cells and, 57-58, 60-64, 61f-63f
    definition of, 549
    lithosphere and, 12, 556-559, 556f-560f
    movement of, 61-63, 61f-63f, 558. See also Plate tectonics
Astronomy
    history of, 27-28
    modern, evolution of, 29-30
Asymmetrical folds, 463, 463f, 464-465, 464f-465f
Atmosphere, 21
    in ecosphere, 185, 185f
    gases in, 36
Atmospheric temperatures
    cyclic variations in, 313, 315
    El Niño and, 372-373
    glaciation and, 313-314
    rising. See Greenhouse effect
    volcanic eruptions and, 97, 106-107, 128, 314
    weathering and, 168, 174
Atolls, 383, 383f, 384f
Atom(s), 73-75, 74f, 75f
    components of, 73, 74f
    definition of, 73
    electron shell of, 73, 74f-76f, 75-76
    identity of, 74, 75f
Atomic mass, 74
    of isotopes, 621
Atomic number, 73-74
    of isotopes, 621

685

Augite, 83, 88, 672t-673t
Aureole, 488, 488f
Avalanche
	rock (debris), 216-218, 217f, 218f
	snow, 211
Axial plane, of fold, 462
Axis of maximum extension, 457
Axis of maximum shortening, 457

B

Back-arc basin, 510, 511f, 514
Backwash, 365
Bacon, Francis, 49
Bacon rinds, 443
Bajada, 272-274, 273f, 336, 337f
Baked apple theory, 509-510
Baker, H.B., 50
Banded iron formations, 572-574, 573f
Barchan dunes, 342, 343f
Barite, 676t-677t
Barrier islands, 379-382, 380f, 381f
	as depositional environments, 406, 407f, 408f
	erosion of, 271-272, 272f
	formation of, 271, 271f, 406, 407f, 408f
Barrier reefs, 382-383, 382f-384f, 384f, 415, 416f
	destruction of, 386
Barrington meteorite, 42, 42f
Basal sliding, 287
Basalt, 5, 5f, 9, 35, 36, 148t, 149
Basaltic lava/magma, 103-104, 105f, 113, 150-152, 150f-152f, 153
	eruptive intensity and, 112-113
	in flood basalts, 117-118
Baselevel
	incised meander and, 254, 255f
	tectonic influences on, 260
	temporary, 249, 249f
	terracing and, 254, 255f
	ultimate, 249, 249f
Basin, drainage, 236-238, 236f, 237f
Basin and Range topography, 335-336, 336f, 337f, 507, 508f
Basins, ocean, 10, 10f, 12-16. *See also* Ocean basins
Batholith, 154, 155
Bathymetry, 54
Bauxite, 91, 196, 572
Bay barriers, 378, 379f
Baymouth bars, 378, 379f
Bays, filling of, 268, 269f
Beaches, 406, 406f, 407f. *See also under* Coastal; Coastline
Becquerel, Henri, 620
Bedded chert, 400
Bed load, 247-248, 269f
Bedrock, in slump, 214, 214f
Beds/bedding, 410, 410f
	bottomset, 264
	cross-bedding, 261, 262f, 410-411, 412f, 416, 417f, 611-612, 613f
	foreset, 264, 266f
	graded, 411-413, 413f, 414f, 611, 612f
	overturning and, 611-612
	in relative dating, 611-614, 612f, 613f
	ripple marks on, 410, 411f, 612, 613f

Benching, 224, 225f
Benguela current, fog deserts and, 327, 327f
Benioff Zone, 525, 525f
Bernhardi, 306
Beta particle, 74, 621
B horizon, 193, 194-195
Big Bang, 31
Biochemical limestone, 401-402, 401f
Biochemical sedimentary rock, 400-402
Biomass, as power source, 595-597, 596f
Biosphere, in ecosphere, 185, 185f
Biotite, 89, 674t-675t
Bismuth, 91
Bituminous coal, 580
Block-fault mountains, 504, 507, 507f, 559
Blocks, 106, 107f
Blowouts, 338, 339f
Body wave, 534, 534f, 550, 551f
Body wave arrivals, 552
Bolson, 273f, 274, 336, 337f
Bombs, 106, 107f
Bonds, 78-80, 79f
	covalent, 79, 79f
	ionic, 76f, 78-79
	metallic, 79
	Van der Waals, 79-80
Bottomset beds, 264
Bowen, N.L., 144-145, 145t
Brahmins, 609
Braided streams, 242, 243f
	in periglacial regions, 305
Breadbasket soil, 195
Breccia, 397, 397f
Breeder reactors, 590-591
Bretz, Harlan, 311
Bridal veils, 443
Brittle deformation, 455-456, 455f
	in rock, 460-461, 461f
Brittle fracture, 287, 288f
Bronze, 569-570
Brown coal, 402, 580
Brucite, 490
BTUs, 581
Buch, Leopold von, 631
Buffon, Compte de, 49
Bulge, volcanic, 123
Buttes, 18, 20f, 333, 335, 335f

C

Calcite, 83, 91, 399, 484, 490, 490f, 676t-677t
	in cementation, 409
	dissolution of, 171-172, 172f, 173f
	karst topography and, 438
Calcium carbonate, in limestone, 399
Calcium hydroxide
	for acid mine drainage, 603
	in fluidized-bed combuster, 603-604
Calcium oxide, in fluidized-bed combuster, 603-604
Calderas, 118-121, 118f-120f
Callisto, 38, 38f
Calving, 289, 289f
Cambrian System, 628-629
Canyons, 18, 20f
	in caves, 440
	submarine, 361

Capillary action, 194
Cap rock, 585
Carbon, 91
	fixed, 580
Carbonate minerals, 676t-677t
Carbonate muds, 394, 394f, 416f
	of ocean bottom, 400-401, 400f
Carbonate rocks, karst topography and, 438
Carbonates, 86t, 91
Carbonation, 171
	with hydrolysis, 171
Carbon dioxide
	greenhouse effect and, 103
	in karst development, 438
	from volcanoes, 103
Carbonic acid, 171
Carboniferous System, 625f, 629
Cassiterite, 680t-681t
Catastrophism, 609-610
Cation(s), 75, 76, 77t
	clay mineral adsorption of, 189, 189f
	in soil, 188-189, 189f
Cation adsorption, 189, 189f
	soil acidity and, 191-192
Cation exchange, soil acidity and, 191-192
Cation exchange capacity, 189, 189f
Caverns, 440-443, 441f-443f
	collapse of, 226
Caves, 440-443, 441f-443f
	collapse of, 226, 419, 440-441
	construction phase of, 441-443
	development phase of, 440
	entrances to, 441, 441f
	sea, 375, 376f
Cementation, of sedimentary deposits, 409
Cenozoic Era, 625f, 628
	periods and epochs of, 625f, 631-632, 632t
Cerebrovascular accident, water quality and, 180
Chalcocite, 571-572, 678t-679t
Chalcopyrite, 678t-679t
Challenger expedition, 351
Change-in-direction rule, for seismic waves, 550, 550f, 551f
Chang Heng, 535
Channel creation. *See also* Stream valleys
	by abrasion, 249, 251
	in arid areas, 256, 256f
	baselevel and, 249, 249f
	in deserts, 329-330, 329f
	by downcutting, 251, 252, 252f, 254, 255f
	fluvial landforms and, 249-258, 250f-258f. *See also* Fluvial landforms
	by headward erosion, 238, 239f, 251, 251f
	by lateral cutting, 251, 252
	by mass wasting, 250f-258f, 251-258
	by mass weathering, 251-258
Channel deposits, 261, 262f
Channels, on Mars, 36, 37f
Channel switching, 270, 270f, 271
Channel texture, 244
Charon, 40
Chemical bonding, 78-80, 79f
	covalent, 79, 79f
	ionic, 76f, 78-79
	metallic, 79
	Van der Waals, 79-80

Chemical compounds. *See* Compounds
Chemical limestone, 399
Chemically active fluids, in metamorphism, 483, 484, 489-490
Chemical sedimentary rock, 398-399, 399f, 400f
Chemical weathering, 170-176, 332-333. *See also* Weathering
   in deserts, 333
   in humid areas, 332-333
   rate of, 174-176
Chernozems, 195, 195f
Chert, 400-401, 401f
Chimney, cave, 441, 441f
Chloride minerals, 682t-683t
Chlorite, 491, 674t-675t
C horizon, 194
Chronologic dating, 611, 619
Cinder cone, 109, 110f
Cinders, volcanic, 106-107, 107f
Cinnabar, 678t-679t
Cirque, 290, 291f
Clay minerals, 90, 173, 176, 188-189, 674t-675t, 674t-676t. *See also* Soil
   cation adsorption by, 189, 189f
   cation exchange capacity of, 189, 189f
   electrical charge of, 188-189
   expandable, 197, 198t
   hydrogenation of, 190, 192
   recrystallization of. *See* Metamorphism
   in sedimentary rock, 394
   structure of, 188
Clay-sized material, 394, 394f
Cleavage, 484-485
   gneissic banding and, 486, 487f
   mineral, 83, 84f
   phyllitic, 486, 486f
   schistose, 486, 487f
   slaty, 484-485
Cliffs. *See* Sea cliffs
Climate. *See also* Atmospheric temperatures
   soil formation and, 187, 188
   stream grade and, 260
Climate change
   deforestation and, 196
   greenhouse effect and, 103
   ice cap melting and, 307, 315
Closing oceans, 356
Coal, 577, 578-584
   anthracite, 580
   ash and, 581
   bituminous, 580
   blending of, 581
   brown, 402, 580
   cleaning of, 581
   compliance, 581, 582f
   distribution of, 582-583, 583f
   eastern, 582, 583f
   formation of, 402-403, 402f, 416-417, 578-579, 578f, 584
   future of, 586-588
   heat content of, 581
   liquefaction of, 582
   quality of, 581
   steam, 581
   subbituminous, 580
   sulfur content of, 581
   treatment of, 581, 582f
   utilization of, 581-582
   western, 582, 583f

Coalification, of peat, 580
Coal mining, 601-604. *See also* Mining
   acid drainage and, 170, 180, 602-604
   contour, 602
   deep, 601
   environmental consequences of, 601-604
   revegetation and, 602
   slope stability and, 602
   surface, 601
Coal rank, 580, 580f
Coal refuse, 604
Coastal processes, 374-387
   at high-energy coastlines, 374-378, 375f-377f
   at low-energy coastlines, 378-382, 378f-381f
Coastal wetlands, 382
   as depositional environments, 407, 408f
   elimination of, 386
   erosion of, 272
Coastline
   emergent (high-energy), 374-378, 375f-377f
   submergent (low-energy), 378-382, 378f-381f
Coastline processes
   environmental aspects of, 385-386
   modification of, 385
Cockpit karst, 439f
Cohesion
   factors affecting, 208-211
   slope stability and, 208
Coke, 581
Col, 290, 291f
Cold glaciers, 287
Collapse sinkholes, 226, 419, 440-441, 441, 441f
Color, of minerals, 80, 81f
Columnar joints, 470-471, 470f
Columns
   vs. arches, 458, 459f
   dripstone, 442, 442f
Comets, 42-43
Compaction, of sedimentary deposits, 409
Compliance coal, 581, 582f
Composite volcanoes, 111, 112f, 504
Compounds, 75-78
   electric neutrality of, 76, 78
   formation of, 75-77
   precipitation of, 76-77
Compression, 454, 454f
   in dynamothermal metamorphism, 484, 485, 485f
Compression waves, 533-534, 534f
Conchoidal fracture, 83-84, 84f
Concordant intrusive rock bodies, 154, 155f
Cone of depression, 434, 434f
Coneybeare, William, 629
Confined aquifers, 434-436, 435f, 436f
   artificial, 436-437, 436f, 437f
Conglomerate, 397, 397f
Construction projects
   in earthquake-prone areas, 544
   mass wasting and, 222-223, 223f, 224, 225f
   sedimentary rock and, 419-420
Contact metamorphism, 488, 488f
Continent(s), 17-18
   area of, 664
   composition of, 17
   elevation of, 12, 13f
   landforms of, 17-18, 19f-21f

Continental arcs, 11f, 99, 101f, 363, 363f, 504
Continental crust, 8f, 9, 12, 13f, 35-36. *See also* Crust
   composition of, 9, 555
   density of, 8, 8f, 664
   depth of, 555
   loading/unloading of, 560-561, 561f
   mass of, 555-556
   triple junction in, 354, 354f
   volume of, 664
Continental deposits, 403-405, 403f-405f
Continental divides, 236-238, 237f
Continental drift, 10-12. *See also* Plate tectonics
   energy source for, 57-58, 57f, 60-64, 61f-65f
   fossil evidence for, 50, 51f
   magnetic reversal and, 53, 59, 60f
   mantle convection and, 57-58, 60-64, 61f-65f
   modern investigations of, 52-64
   in ocean basin formation, 353-356, 354f, 355f
   rock magnetism and, 52-54, 53f-55f, 59, 60f
   seismic data and, 60-64, 61f-65f
   triple junction and, 354, 354f
   Vine-Matthews-Morely hypothesis of, 59, 59f
   Wegener's theory of, 50-52
Continental glaciers, 283, 285f, 305-314. *See also* Glacier(s)
   crustal loading by, 560
   early theories of, 305-306
   erosion by, 294-297, 296f, 297f
   formation of, 313-314
   melting of, 307, 315
   Pleistocene, 307-312
Continental leading edge, 9-10, 9f, 10f, 14f, 66, 356, 359, 360f, 363-364, 363f
Continental margins
   convergent, 9, 10-12, 11f, 12f, 13, 14f, 66, 99, 100f, 356, 359-364, 360f
   divergent, 9-10, 9f, 10f, 13, 14f, 66, 99, 99f, 356, 359, 360f, 363-364, 363f
Continental rifts. *See also* Rifting/rift zone
   magma formation and, 150, 150f
Continental rise, 13, 14f, 361, 361f
Continental rocks, 7
Continental shelf, 13, 14f, 56f, 359-360, 361f, 407-408, 409f
Continental slope, 13, 14f, 361, 361f
   deposition on, 408
Continental trailing edge, 9, 10-12, 11f, 12f, 14f, 66, 99, 100f, 359-363, 360f-362f
   ocean closing and, 356, 359, 360f
Continent-continent orogenesis, 515-518, 516f, 517f
Contour mining, 602
Convection cell theory, 57-58, 60-64, 61f-63f
Convergent plate margins, 9, 10-12, 11f, 12f, 66, 99, 100f, 360f. *See also* Plate tectonics
   ocean closing and, 356, 359, 360f
Conversion factors, 670t
Copernicus, Nicolaus, 29
Copper, 91, 569
   abundance of, 569t
   enrichment of, 571-572

Copper sulfate, in ore enrichment, 571-572
Coquina, 402, 402f
Coral reefs, 382-383, 382f-384f, 415, 416f
   destruction of, 386
Core, 8-9, 8f, 35, 35f
   composition of, 9
   density of, 8, 8f, 664
   liquid, 549f
   solid, 549f
   volume of, 664
Core-mantle boundary, 552-554, 553f-555f
Corundum, 83, 680t-681t
Covalent bonds, 79, 79f
Covellite, 571
Crab nebula, 31, 32f
Craters
   meteoroid, 42, 42f
   volcanic, 118, 118f
Craton, 17
Creep, 206, 211, 212f
Cretaceous System, 631, 631f
Crevasses, glacial, 287
Crevasse splay, 268, 271f
Cross-bedding, 261, 262f, 410-411, 412f, 416, 417f, 611-612, 613f
   in relative dating, 611-612, 613f
Cross-cutting relationships, 615-619, 616f-618f
Crust, 8, 8f, 9, 35-36, 35f, 549f
   composition of, 9, 666
   continental, 8f, 9, 12, 13f, 35-36. See also Continental crust
   density of, 8, 8f, 664
   in gravitational balance, 556-557, 556f, 557f
   isostasy and, 12, 556-559, 558f-560f
   oceanic, 8f, 9, 12, 13f, 35, 57, 358, 358f. See also Oceanic crust
   orogenesis and, 513
   vertical movements of, 12, 555-559, 556f-560f
   volume of, 664
Crustal loading/unloading, 560-561, 561f
Crust-mantle boundary, 12, 13f, 551-552, 551f
Cryptozoic Eon, 625, 625f, 626
Crystal form, 83, 83f
Crystal lattice, 77, 78f, 80, 80f. See also Crystal structure
   disruption of, 141. See also Melting
   formation of, 80f, 141. See also Crystallization
Crystalline materials, 80
Crystallization
   lattice formation in, 80f, 141
   of lava, 143-145
   order of, 143-145, 145f, 145t
   temperature and, 141-142
Crystallization temperature, 141-142
Crystal settling, 143
Crystal structure, 77, 78f, 80, 80f, 83, 83f, 86-91, 87f-91f
   double-chain, 88, 88f
   framework, 90-91, 90f, 91f
   isolated tetrahedral, 87, 87f
   lattice, 77, 78f, 80, 80f
   sheet, 88-90, 89f
   single-chain, 88, 88f
   weathering and, 175-176
Cuprite, 680t-681t
Curi, Maria Sklodowska, 620

Curi, Pierre, 620
Curie point, 52
Current ripple marks, 410, 411f
Currents
   longshore, 367, 367f, 378
   ocean, 367-374. See also Ocean currents
Cutbank, 257, 257f, 258

D

Dams, siting of, 494, 494f, 495f
Dana, James D., 509
Darcy, Henry, 429
Darwin, Charles, 383, 577, 614
Dating
   absolute (chronological), 611, 619
   bedding in, 611-614, 612f, 613f
   correlation and, 612-614
   by Earth's heat, 620
   fossils in, 614, 615f
   by ocean salinity, 620
   principle of cross-cutting relations and, 615-619, 616f-618f
   principle of faunal succession and, 614
   principle of original horizontality and, 612
   principle of superposition and, 611-612, 611f
   radiometric, 621-624, 621f, 622f, 623t
   relative (sequential), 611-619
   by sediment accumulation, 619-620
   varves in, 619, 619f
Davis's theory of fluvial landforms, 249-258, 250f-258f
Deadman, 226, 227f
Debris avalanche, 216-217, 217f, 218f
Debris flows, 218-219
Debris slides, 217-218, 219f
Decomposition, 164
Deep-focus earthquake, 525, 525f
Deep-sea drilling, 353, 354f
Deep-sea fans, 361-363, 362f
Deep-sea trenches, 16, 17f, 56-58, 56f, 58f, 63, 359
Deflation, 338, 339f
Deforestation, 196
   desertification and, 345
   slope instability and, 222
Deformation
   brittle, 455-456, 455f
   definition of, 454
   elastic, 455, 455f
   plastic, 455, 455f
   pressure and, 458
   rate of, 456-457
   rock, 453, 458-461. See also Rock deformation
   strain ellipsoid and, 456-457, 456f-457f
   stress and, 454-455, 454f
   temperature and, 458
Delta(s), 264-272
   bay filling and, 268, 269f
   channel switching and, 270, 270f, 271
   coastal erosion and, 271-272
   definition of, 264
   depositional processes in, 266-270, 267f-271f, 405, 406f
   ebb, 381
   evolution of, 266-272, 267f-271f

flood, 381, 381f
loss of land in, 271
shape of, 264, 265f
Delta front, 264, 267f
Delta lobes, 27, 266, 268f, 270f
   destruction/abandonment of, 270-271, 270f, 271f
Delta plain, 264, 267, 267f
Dendritic stream pattern, 238-240, 240f
Dendrochronology, 619
Density, 8
   gravitational balance and, 556-557, 556f, 557f
   seismic waves and, 550, 550f, 551f
Density currents, 361, 361f, 364, 369-374
   salinity-induced, 370-374
   thermohaline, 367, 367f, 368f, 369-370, 371f
   turbidity, 361, 361f, 374, 412
Deposition, 5, 7f
Depositional environments, 403-408. See also under Sedimentary
   continental, 403-405, 403f-405f
   environmental concerns and, 419-420
   interpretation of, 414-418
   marine, 407-408, 409f
   transitional, 405-407, 406f-408f
   Walther's law and, 417-418, 418f
Desert(s)
   animal life in, 321-323, 323f
   basin and range topography in, 335-336, 336f, 337f
   chemical weathering in, 333-335, 334f, 335f
   deforestation and, 345
   environmental concerns and, 344-346
   expansion of, 344-345, 344f
   fog, 326-327, 327f
   frost wedging in, 333
   isolation, 327-329, 328f
   lakes in, 330, 331f
   mass wasting in, 333-335, 334f, 335f
   overcultivation and, 345
   overgrazing and, 344-345
   overview of, 321-324
   plant life in, 321, 322f, 342, 342f
   rainshadow, 324-325, 325f, 326f
   relative humidity in, 324
   salinization and, 345-346
   stream erosion in, 329-331, 330f, 332f, 333-335, 334f, 335f
   subtropical, 325-326, 326f, 327f
   temperatures in, 321
   water erosion in, 329-337, 330f-337f
   wind deposition in, 339-343, 340f-343f
   wind erosion in, 337-339, 338f-340f
Desert, wind deposition in, 394, 403-405, 404f
Desertification, 344-345, 344f
Desert pavement, 338, 339f
Deshayes, Gérard-Paul, 632
Detrital sedimentary rocks, 394, 397-398, 397f, 398f
Dewatering, 445-446
Dew point, 324
Diagonal-slip faults, 468, 468f
Diamonds, 82, 83
Diastrophism, 502
Diatoms, 400
Differential weathering, 20f
Diffractogram, 85, 86f

Index 689

Dike, 154, 155f
Dinosaurs, 626-631, 627f, 629f, 631f
Dione, 39f
Diopside, 672t-673t
Dip, 461, 461f
Directional response, of seismic waves, 550, 550f, 551f
Disappearing streams, 242, 243f, 439
Disconformity, 615, 616f
Discordant intrusive rock bodies, 154
Disintegration, 164
Disseminated deposits, metal, 570-571
Dissolution
 in channel cutting, 249
 weathering and, 171-172, 173f
Dissolved load, 247
Distributaries, switching among, 270, 270f, 271
Distributary front, 267
Distributary mouth bar, 267, 269f
Distributary streams, 264, 266f
Disulfides, 86t, 91
Ditches, drainage, 224, 225f
Divergent plate margins, 9-10, 9f, 10f, 13, 66, 99, 99f, 356, 360f, 363-364, 363f. *See also* Plate tectonics
 ocean spreading and, 356
Diversion drains, 224, 225f
Divides, drainage, 236-238, 237f
Dolomite, 399, 676t-677t
 as carbonate rock, 438
 in cementation, 409
Dolostone, 399
 as carbonate rock, 438
Domal mountains, 504-507, 505f-507f
Doming
 local, 504-505, 505f
 regional, 505-507, 505f-507f
Downcutting, 251, 251f, 252
 rejuvenation and, 254, 254f
Downdropping, 474
Downstream floods, 274
Drag folds, 475, 476f
Drainage. *See also* Stream patterns
 in karst areas, 439-440
 parallel, 240, 242f
Drainage basins, 236-238, 236f, 237f
Drainage ditches, 224, 225f
Drainage divides, 236-238, 237f
Drainage systems, 236-238, 236f, 237f. *See also* Streams
 divides in, 236-238, 237f
 external, 236
 internal, 236, 330-331, 337f
 stream order in, 238, 238f
Drake, E.L., 577
Dredging, 271, 575, 575f
Drift, 297
Drilling
 deep-sea, 353, 354f
 oil, 226, 229f
Drip irrigation, 346
Dripstone, 442, 442f
Drumlins, 301-302, 302f
Dry-based glaciers, 287
Dry Hot Rocks Geothermal Project, 591-592
Dug wells, 433, 434f
Dunes, 340-342, 341f-343f, 394, 403-405, 404f
 in deserts, 340-342, 341f-343f, 394, 403-405, 404f

 as eolian deposits, 403-405, 404f, 405f
 in periglacial regions, 305
 transverse, 342, 343f
Dunite, 149
Dust
 soil erosion and, 198, 199f
 volcanic, 107-108, 107f
Dust Bowl, 198
Dynamo-thermal metamorphism, 485, 488-489, 489f

E

Earth
 age of. *See* Age of Earth
 atmosphere of, 36
 circumference of, 664
 density of, 35, 664
 diameter of, 28f, 35
 dimensions of, 664
 early history of, 36-37
 internal structure of, 8-12, 8f-12f, 549-559. *See also* Internal structure of Earth
 mantle of, 8, 8f, 9, 35, 35f
 measurement of, 28f
 orbit of, 33, 34f, 40, 40f
 properties of, 35-36
 radius of, 664
 surface of, 12-18, 36
 as terrestrial planet, 35-36
Earthflow, 211-213, 213f
Earth history, depositional environments and, 414-418
Earth moving, slope instability and, 222-223, 223f
Earthquake(s), 523-544
 causes of, 524
 construction siting and methods and, 544
 convection cell theory and, 60-64, 61f-65f
 damage from, 526, 528f, 530-533, 530f-533f
 depth of, 525, 525f
 distribution of, 523-524, 524f
 early records of, 523
 epicenter of, 525-526, 526f, 538-541
 faults and, 464-469
 focus (hypocenter) of, 525, 525f, 526f, 539-541, 541f
 frequency of, 524
 intensity of, measurement of, 526-527
 intraplate, 523-524
 location of, 538-541, 538f-542f
 magnitude of, 527, 527t, 541, 542f
 prediction of, 543-544, 543f
 preparedness for, 544
 seismic waves in, 524, 533-535, 534f, 535f. *See also under* Seismic
 sites of, 524-525
 volcanoes and, 97
Ebb delta, 381
Ebb tide, 381
Echo-sounding studies, of ocean bottom, 54, 55f, 56
Ecliptic, 33, 34f
Economic geology, 567-604. *See also* Fuel(s)

Ecosphere, 185f
Effluent streams, 432
E horizon, 192-193, 194
Einstein, Albert, 29
Elasmosaurus, 626, 627f
Elastic deformation, 455, 455f
 in rock, 460
Electric neutrality, 76, 78
Electron(s), 73, 74f
 gain/loss of, 75-76, 76f
Electron capture, 75
Electron shells, 73, 74f-76f, 75-76
Elements
 abundance of, 77t
 in functional groups, 76
 heavy, 177
 identity of, 73-74, 75f
 ion affinities of, 77t
 isotopes of, 74. *See also* Radioactive isotopes
 light, 177
 trace, 177-179
El Niño, 372-373
Emergent coastline, 374-378, 375f-377f
End moraines, 297, 298f
Energy conservation, 597-598
Energy sources
 nonrenewable, 577-590
 renewable, 591-600
Engineering projects
 in earthquake-prone areas, 544
 mass wasting and, 222-223, 223f, 224, 225f
 sedimentary rock and, 419-420
Enstatite, 672t-673t
Environmental quality
 coal mining and, 601-604
 deserts and, 344-346
 expandable soils and, 197
 mass wasting and, 220-229
 nuclear power and, 588-590
 oceans and coastlines and, 385-387
 renewable resources and, 591-600
 resource conservation and, 597-598, 601
 sedimentary deposits/rocks and, 419-420
 streams and, 274-276
 urbanization and, 199-200
 weathering and, 177-180
Eocene Epoch, 625f, 632, 632t
Eolian deposits, 403-405, 404f, 405f, 575
Eons of time, 625-626, 625f
Epeirogenic forces, 502
 in block-fault mountain formation, 507, 508f
 in domal mountain formation, 504-507, 505f-507f
Epicenter, 525-526, 526f
 location of, 538-541, 538f-542f
Equatorial countercurrents, 367, 368f
Equatorial currents, 367, 368f
Equilibrium line, 389, 389f
Eratosthenes, 28
Erosion, 5, 7f, 163. *See also* Mass wasting; Weathering
 arid, 206f
 arid cycle of, 331-336
 coastal, 271-272, 272f
 of delta lobes, 271, 271f
 in deserts, 329-343. *See also* Desert(s)
 by glaciers, 289-297, 290f-298f
 headward, 238, 239f, 251, 251f

in periglacial regions, 303-305, 304f, 305f
soil, 198-199, 199f
by streams, 246-249, 246f, 248f, 250-258, 251f-258f, 329-336, 329f-340f. *See also* Channel creation
by waves, 375-382
weathering and, 163. *See also* Weathering
by wind, 305, 337-339, 338f-340f
Erratics, 300-301, 301f
Eskers, 302, 303f
Europa, 38, 38f
Evaporation, exfoliation and, 165-167, 165f-168f
Evaporitic sedimentary rocks, 402
Exfoliation, 165-167, 165f-168f
Expandable soils, 197
External drainage systems, 236
Extrusive igneous rock bodies, 4, 141, 154

F

Facets, 338, 339f
Facies, 417
   metamorphic, 491, 492f
Failed arm, 354, 354f
Fans
   alluvial, 272-274, 273f, 329-330, 330f, 335-336, 337f
   deep-sea, 361-363, 362f
Farming. *See* Agriculture
Fault(s), 464-469
   definition of, 464
   diagonal-slip, 468, 468f
   distribution of, 464-465
   earthquakes and, 464-469
   with folding, 468-469, 469f
   formation of, 465-466, 468-469
   listric, 467, 467f
   movement of, 473-475, 473f-477f
   normal, 467, 467f, 471, 472f
   oblique-slip, 468, 468f
   reverse, 466, 466f, 467f, 471, 472f
   strain ellipsoid and, 471-472, 472f
   strike-slip, 467-468, 468f, 471, 472f
   thrust, 466, 466f
   transform, 66, 66f
Fault-block mountains, 504, 507, 507f, 559
Fault trap, 585, 585f
Faunal succession, principle of, 614
Fayalite, 87, 672t-673t
Feldspar/feldspar group, 83, 86t, 90, 91, 141t, 149, 672t-673t
   carbonation/hydrolysis of, 173
   in sedimentary rock, 394
   structure of, 90, 91f
Felsic eruptions, 118
Felsic rock, 148f, 149, 150
Ferromagnesian minerals, 81f, 85, 86t. *See also* Mineral(s)
   weathering of, 175. *See also* Weathering
Ferrous metals, 568, 568f
Fields, drumlin, 301-302, 302f
Filter pressing, 143
Firn, 286, 286f
Firn line, 389, 389f
Fishing, 386

Fission, 590
Fission reactors, 588-590, 589f, 590f
Fissure eruptions, 116-118, 117f
Fixed carbon, 580
Fjords, 293-294
Flint, 401, 401f, 567, 567f
Floating objects, gravitational balance of, 556-557, 557f
Flood basalts, 117-118, 117f
Flood delta, 381, 381f
Flood/flooding, 274-275, 274f
   in deserts, 330, 331f
   downstream, 274
   of rift valleys, 356
Floodplain, 252, 252f, 275
   deposition in, 261-264, 262f-264f
   features of, 261-263
   management of, 275
   stream gradient and, 260
Flood tide, 381
Floodway, 275
Floodway fringe, 275
Fluidized-bed combuster ash, 603-604
Fluorine, 178
Fluorite, 83
Fluvial deposits, 403, 403f
Fluvial landforms, 249-257
   Davis's theory of, 249-258
   evolution of, 249-258, 250f-258f
Focus, of earthquake, 525, 525f, 526f, 539-541, 541f
Fog deserts, 326-327, 327f
Fold(s), 453f, 461-465, 462f-465f
   classification of, 462-463, 463f
   drag, 475, 476f
   with faulting, 468-469, 469f
   features of, 462
   formation of, 461, 464-465, 464f-465f, 468-469, 473-474, 473f, 474f
   strain ellipsoid and, 464-465, 464f-465f
Foldbelt mountains
   epeirogenic forces and, 507-511, 508f-511f
   orogenic forces and, 511-518, 512f-517f
Foliated metamorphic rocks, 484-486, 485f-487f
Foot wall, 466
Forel, François, 526
Foreset beds, 264, 266f
Forests. *See* Deforestation; Plants
Form, crystal, 83, 83f
Forsterite, 87, 672t-673t
Fossil(s), 415
   Cambrian, 628
   in dating of Earth, 614, 615f, 624
   disconformities and, 615, 616f
   in geologic time scale, 624, 625-626, 625f
   of Jurassic Period, 631, 631f
   principle of faunal succession and, 614
   Triassic, 629-630, 639f
Fossil fuels
   coal as, 578-584
   future of, 586-588
   natural gas, 597-598
   oil shale as, 586
   petroleum as, 584-585
Fracture(s), 83-84, 84f, 426, 426f
   brittle, 287, 288f
   conchoidal, 83-84, 84f
   sinkholes and, 439, 439f

Fracture porosity, 426, 426f
Framework structure, 90-91, 90f, 91f
Free-fall, 206
Freeze-thaw cycles, 164, 209, 209f
   creep and, 211, 212f
   in periglacial regions, 303-305, 304f
Freezing temperature, 141-142
Friction
   factors affecting, 208-211
   slope stability and, 208
Fringing reef, 383, 383f, 384f
Frost action
   in deserts, 333, 333f
   rock glacier and, 215, 216f
   weathering and, 164-165, 164f, 165f, 209-210, 210f, 333, 333f
Frost heaving, 209, 303
   creep and, 211, 212f
Frost wedging, 303
   in deserts, 333, 333f
Fuel(s), 577-588
   alcohol as, 597
   biomass as, 595-597, 596f
   coal as, 577, 578-584
   conservation of, 597-598
   development of, 577-578
   future of, 586-588
   methane as, 596
   natural gas as, 597-598
   nuclear, 588-591
   petroleum as, 584-586, 585f
   steam as, 577, 591-592
   volatiles in, 580, 580f
   wood as, 596
Fuel vs. nonfuel nonmetals, 568, 568f
Fumaroles, 122, 122f, 443, 444f
Functional groups, 76
Fusion, nuclear, 590

G

Gabbro, 149
Gabbro-basaltic melt, 150-151
Gabions, 226, 227f
Gaining streams, 432
Galaxies, 30, 30f
   formation of, 31
Galena, 84, 678t-679t
Galileo, 29
Gangue, 568
Garbage, as fuel source, 596-597
Garnet, 87, 491
Gas, 584-585
   conservation of, 597-598
   future of, 586-588
Gases
   atmospheric, 36
   inert, 74
Geocline, 510, 510f
Geologic features, size of, 154
Geologic structures
   dip in, 461, 461f
   faults, 464-469
   folds, 461-465, 462f-465f
   joints, 469-471, 469f-471f
   strike in, 461, 461f
Geologic time scale, 669
Geologic time scales, 624-632
   eons in, 625-626, 625f

eras in, 625f, 626-628
periods in, 625f, 626-628
Geology, economic, 567-604
Geomorphology, 357
Geosyncline, 509, 510
Geothermal phenomena, 121-122, 121f, 122f
Geothermal power, 122, 591-592, 592f
Getters, 581, 603
Geysers, 121, 122f, 443, 444f
Glacial budget, 288-289
Glacial deposits, 297-302, 298f-303f, 403, 404f
Glacial erosion, 289-297, 290f-298f
by abrasion, 290, 290f
by Alpine glaciers, 290-294, 291f-294f
by plucking, 290, 290f
by quarrying, 290, 290f
Glacial ice
accumulation/ablation of, 288-289, 289f
origin of, 283-286, 283f
plastic flow of, 286-287, 287f, 288f
in sorting, 395
temperature of, 287
thickness of, 286-287, 287f
Glacial lakes, 291-293, 293f, 294f, 295, 296f
Glacial surges, 288
Glacial trough, 290
Glacier(s), volume of, 664
Glaciers, 283-316. *See also* Glacial ice
advance/retreat of, 288-289, 289f
Alpine (valley), 283, 284f, 285f, 290-294, 291f-294f, 313
cold, 287
continental, 283, 284f, 305-314. *See also* Continental glaciers
crustal loading by, 560
distribution of, 283, 283f
dry-based, 287
formation of, 283, 283f
of Great Ice Age, 307-312
melting of, 307, 315
movement of, 286-288, 287f
periglacial regions and, 303-305, 304f, 305f
piedmont, 283, 286f
Pleistocene, 307-312
rock, 215, 216f, 304f, 305
wet-based, 287
zone of ablation in, 288-289, 289f
zone of accumulation in, 288-289, 289f
Glass, 80, 144
felsic, 149
mafic, 149
Glassy texture, 146, 146f-148f
Global warming
deforestation and, 196
greenhouse effect and, 103
ice cap melting and, 307, 315
Glomar Challenger, 353, 354f
Gneissic banding, 486, 487f
Goethite, 170
Goiter, 179, 315-316
Gold, 84, 91
abundance of, 569t
alluvial deposits of, 574-575, 574f
mining of, 574-575, 575f
panning for, 574, 574f
Goldrich weatherability series, 175t
Gossan, 571, 571f
Graben, 507

Gradation, 205
Graded bedding, 411-413, 413f, 414f
in relative dating, 611, 612f
Gradient, stream, 244, 244f, 247, 258-261
Granite, 3, 148t, 149
durability of, 163
weathering of, 169f, 174
Granitic magma, 150, 151f, 153-154
Granitization, 490
Granodiorite, 148t, 149
Graphite, 580
Grassland soils, 195
Gravitational balance, 556-557, 557f
lithosphere-asthenosphere, 12, 556-559, 556f-560f
Gravitational sliding, 64
Gravity, in mass wasting, 208, 208f
Graywacke, 397-398
Great Ice Age, 307-312, 313
Greenhouse effect, 103
deforestation and, 196
ice cap melting and, 307
on Venus, 34
Groins, 385, 386f
Ground covers, 222, 222f
Groundmass, 148, 148f
Ground moraines, 297, 298f, 305-306, 306f
Groundwater, 21, 425-447. *See also* Water
aquicludes and, 428-429
aquifers and, 428-429. *See also* Aquifers
aquitards and, 428-429
conduits for, 428-429
dewatering and, 445-446, 446f
early concepts of, 425-426
environmental concerns and, 443-447
flow rates of, 429, 429f
in hydrologic cycle, 235-236, 236f
karst topography and, 437-443
magma-heated, 121-122, 121f, 122f, 443, 444f
movement of, 194
municipal water supplies and, 436-437, 436f, 437f
pollution of, 443, 443f
in regolith, 194
saltwater encroachment in, 446-447, 447f
in soil, 194-196
sources of, 425-426
subsurface drainage of, 439-440
surface drainage of, 439
trace elements in, 177
volume of, 235, 664
water table and, 430-432, 430f-433f
water wells and, 433-437
Growth rings, 618f, 619
Gutenberg, Beno, 552-553
Gutenberg Discontinuity, 553
Guyot, 359
Gypsum, 82, 91, 166, 402, 676t-677t
in cementation, 409
formation of, 419
Gyres, 367, 368, 368f

H

Half-life, of radioactive isotopes, 620-623, 622f, 623t
Halides, 86t, 91

Halite, 76, 83, 85, 402
dissolution of, 171
halite, 682t-683t
Hall, James, 508-509
Halley's comet, 42, 43
Halloy, Omaliu d,' 631
Halo, 488, 488f
Hamatite, 680t-681t
Hanging valley, 291
Hanging wall, 466
Hanging water table, 431-432, 431f, 432f
dewatering of, 445-446, 446f
Hardness, of minerals, 82-83, 82t
Hardpan, 195
Harmful pollutants, 275
Hawaiian-type eruptions, 114, 114f
Headward erosion, 238, 239f, 251, 251f
Heart disease, water quality and, 179
Heat, in metamorphism, 483, 484, 488-490
Heat of Earth, dating by, 620
Heavy elements, 177
Hedenbergite, 672t-673t
Heliopause, 40
Hematite, 82, 91, 170, 572, 573
in soil, 196
Herodotus, 609
Hess, Harry H., 57
Hiatus, 615
High-energy coastline, 374-378, 375f-377f
High-grade metamorphism, 491
Hills, 17
abyssal, 15, 56f, 358-359
Holmes, Arthur, 57
Honeycomb weathering, 166, 167f
Horizons, soil, 192-196, 193f
Horn, 290, 292f
Hornblende, 83, 88, 672t-673t
Hornfels, 488
Horst, 507
Hot spots, 100, 101f, 132, 132f
magma formation and, 150, 151
Hot spot volcanoes, 503-504, 503f
Hot springs, 121, 121f, 443, 444f
Hubbert, M. King, 587, 588
Hubble Space Telescope, 30, 31f
Humboldt, Alexander von, 630
Humboldt current, fog deserts and, 327, 327f
Humidity, relative, 324
Humus, 192
Hutton, James, 610-611
Hutton's theory of uniformitarianism, 414-415
Hydraulic gradient, 429, 429f
Hydraulic head, 429, 429f
Hydraulic mining, 574-575, 575f
Hydrogenation, of clay minerals, 190, 192
Hydrologic cycle, 235-236, 236f
Hydrolysis, 171
Hydropower, 594-595
Hydrosphere, 21, 235
definition of, 351
in ecosphere, 185, 185f
source of water in, 102-103
Hydrothermal metamorphism, 489-490
Hypersthene, 672t-673t
Hypocenter, of earthquake, 525, 525f, 526f, 539-541, 541f
Hypsographic curve, 12, 13f

I

Ice, glacial. *See* Glacial ice
Ice Age
    Little, 315
    Pleistocene, 307-312, 313
Ice caps, 21
    melting of, 307
Ice formation, mass wasting and, 209-210
Ice sheets, 21
Igneous rocks, 3-5, 3f-5f, 141-157, 154-157
    aphanitic, 146, 146f-148f
    classification of, 148-150, 148f, 154-157, 155f-157f
    composition of, 149, 666
    cooling of, 146-148, 148f
    extrusive, 4, 141, 154
    formation of, 143-145, 145f, 145t
    glassy, 146, 146f-148f
    intrusive, 3, 3f, 141, 154-157, 155f-157f. *See also* Intrusive igneous rock bodies
    joints in, 469-470, 470f
    metal concentrations in, 569-570
    phaneritic, 146, 146f-148f
    porphyritic, 146, 146f
    pyroclastic, 146, 146f
    in rock cycle, 21-22, 22f
    texture of, 146-148, 146f-148f
Ignimbrite, 108
Illite, 674t-675t
Illustrations of the Huttonian Theory (Playfair), 611
Ilmenite, 680t-681t
Incised meanders, 254, 255f
Index minerals, 491
Industrial Revolution, 577
Inert gases, 74
Influent streams, 432
Interfluves, 236, 237f
Intergranular porosity, 426, 426f
Intermediate axis, of strain ellipsoid, 457
Intermediate-focus earthquake, 525, 525f
Intermediate-grade metamorphism, 491
Internal drainage systems, 236, 330-331, 337f
Internal structure of Earth. *See also* Core; Crust; Mantle
    crust-mantle boundary and, 551-552, 551f
    gravitational balance in, 556-557, 556f, 557f
    isostasy and, 12, 555-559, 556f-560f
    isostatic balance in, 558-559, 558f-560f
    mantle-core boundary and, 552-554, 553f-555f
    seismic wave velocities and, 550-554, 550f, 551f, 553f, 554f
Intraplate volcanism, 100, 101f
Intrusive igneous rock bodies, 3, 3f, 141, 154-157, 155f-157f
    massive, 155-157, 156f, 157f
    tabular, 154, 155f
Iodine, 179
    goiter and, 179, 315-316
Ionic bonds, 76f, 78-79
Ions, 75-76, 76f, 77t
    charge of, 75-76
    formation of, 76
    in mineral formation, 77-78, 78f
Iron, 85
    abundance of, 569t
    oxidation of, 170
    in tool making, 570
Iron hat, 571, 571f
Iron meteorites, 41
Iron ore, banded deposits of, 572-574, 573f
Iron oxide, in cementation, 409
Irrigation, 196
    drip, 346
    salinization and, 345
Island(s)
    barrier, 271-272, 271f, 272f, 379-382, 380f, 381f, 406, 407f, 408f. *See also* Barrier islands
    volcanic, 16, 16f, 131-136, 131f-136f, 359, 383, 384f
Island arcs, 11f, 99, 101f, 363, 363f, 504
Isoclinal folds, 463, 463f
Isolated tetrahedral structure, 87, 87f
Isolation deserts, 327-329, 328f
Isostasy, 12, 555-559, 556f-560f
Isostatic balance, 558-559, 558f-560f
Isotopes, 74
    radioactive. *See* Radioactive isotopes

J

Jameson, Robert, 615
Jasper, 401
Jetties, 385, 385f
Johnston, David, 128-129
Joints, 469-471, 469f-471f
Joly, John, 620
Jovian planets, 33, 37-40, 38f, 39f
Jupiter, 37-38, 38f, 41
Jurassic Period, 630-631, 631f

K

Kame, 300
Kame terrace, 300
Kaolinite, 173, 674t-675t
Karst
    cockpit, 439f
    definition of, 437
    development of, 438
Karst areas
    caves and caverns in, 440-443, 441f-443f
    drainage in, 439-440
    sinkholes in, 438-439, 438f, 439f
    water table in, 439-440
Karst towers, 439, 440f
Kelvin, Lord, 620
Kettle, 300, 300f
Kettle lakes, 300, 300f
Keystone, 458, 459f
Komatiite, 149-150
Kyanite, 491

L

Laccolith, 155, 156f
Lacustrine deposits, 405
Lagoons, 380, 381f
Lahars, 108, 109f, 220, 221f
Lake(s), volume of, 664
Lakes, 21
    glacial, 291-293, 293f, 294f, 295, 296f
    graded bedding in, 412-413, 413f
    kettle, 300, 300f
    oxbow, 252-253, 253f
    paternoster, 291-293, 294f
    playa, 330, 331f
    pluvial, 310-311, 311f
    trough, 293, 294f
Laminar sublayer, 245
Land, mean height of, 664
Landfills, sanitary, 445, 446f
Landslides, 213-220
    slump and, 214, 214f, 215f
La Niña, 372
Lapworth, Charles, 629
Lateral cutting, in channel creation, 251, 252
Lateral moraines, 297, 299f
Lattice, crystal, 77, 78f, 80, 80f, 141
Lava, 3, 4, 103-106, 104f-106f. *See also* Magma; Volcanic eruptions
    aa, 104, 105f
    basaltic, 103-104, 105f
    composition of, 103-104
    crystallization of, 143-145
    definition of, 141
    extrusive igneous rock from, 4, 141, 154
    felsic, 149
    mafic, 149
    pahoehoe, 104, 105f
    pillow, 106, 106f, 116, 116f, 357-358
    sampling of, 144, 144f, 145f
Lava tubes, 428, 428f
Leachates, acid neutralization for, 602-604
Lead, abundance of, 569t
Lead 207, in radiometric dating, 621, 622
Leading edge, 9-10, 9f, 10f, 14f, 66, 356, 359, 360f, 363-364, 363f
LeChatelier's rule, 142
Left-lateral faults, 467-468, 468f, 471, 472f
Lehmann, Inge, 553-554
Lemaître, Canon Georges Henri, 31
LePichon, X., 65
Levees, 261, 262f
    artificial, 261, 271
Light elements, 177
Lightfoot, John, 609
Light microscopy, 85, 85f
Light waves, 550, 551f
Light-years, 30
Lignite, 402, 580
Lime, agricultural, 191-192
Limestone, 7, 398-399, 399f
    biochemical, 401-402, 401f
    as carbonate rock, 438
    composition of, 666
    coral and, 382f
    disappearing streams and, 243f
    dissolution of, 171-172, 172f, 173f, 438
    engineering problems with, 419

oolitic, 399, 399f
  as petroleum reservoir, 585
  sinkholes and, 226, 228f, 243f, 419, 438-439, 438f-440f
Limonite, 170
Linear oceans, 10
Listric faults, 467, 467f
Lithification, 7, 409
  porosity and, 427
Lithologies, 614
Lithosphere, 8f, 9, 61, 61f. *See also* Plate(s)
  asthenosphere and, 12, 555-559, 556f-560f
  composition of, 61, 61f
  convection cells and, 57-58, 60-64, 61f-63f
  definition of, 61, 549
  in ecosphere, 185, 185f
  as floating object, 556, 556f
  fracture of, 57f, 65-66, 65f
  in gravitational balance, 556-557, 556f, 557f
  in isostatic balance, 558-559, 558f-560f
  mass of, 556
  movement of, 61-63, 61f-63f. *See also* Plate tectonics
  thickness of, 556
Little Ice Age, 315
Lloyd, Bishop, 609
Load, stream, 247-248, 248f
Local doming, 504-505, 505f
Local metamorphism, 489
Lodes, 575
Loess, 305, 339-340, 340f, 394, 403-405, 405f
Longitudinal dunes, 342, 343f
Longshore currents, 364, 367, 367f, 378
Longshore transport, 367
Lopolith, 155, 156f, 570, 570f
Losing streams, 432
Love Canal, 199-200
Love waves, 534-535, 535f
Low-energy coastline, 378-382, 378f-381f
Low-grade metamorphism, 491
Low-velocity zone, 554
LQ waves, 534-535, 535f
LR waves, 534-535, 535f
Luster, of minerals, 84
Lyell, Charles, 611, 614-615

M

Mafic rock, 148f, 149, 150
Magma, 3, 5. *See also* Lava; Volcanic eruptions
  andesitic, 114, 150, 151-152, 151f, 153
  basaltic, 103, 112-113, 113, 150-151, 150-152, 150f-152f, 153
  composition of, 150
  crustal areas and, 150
  crystallization of, 143-145
  definition of, 141
  eruptive intensity and, 112-114
  felsic, 149, 150
  formation of, 150-154
  gas content of, 112-113
  granitic, 150, 151f, 153-154
  groundwater heating by, 121-122, 121f, 122f, 443, 444f
  intrusive igneous rock from, 4, 141, 154
  mafic, 149
  metals in, 569-571, 569f
  rhyolitic, 114, 118
  in sheet dikes, 116, 117f
  silicone content of, 149
  subduction zone, 114
  synthetic, 144, 145f
  viscosity of, 5, 113, 149
  water content of, 149
Magma oceans, 103
Magmatic fractionation, 143
Magmatic segregation, 569-570, 569f
Magnesium, 85
  abundance of, 569t
Magnetic field
  geologic record of, 52-53
  reversal of, 53, 59, 60f
Magnetic inclination, 52, 53f
Magnetic intensity, of rocks, 59
Magnetic poles, wandering, 54, 55f
Magnetic rocks, plate tectonics and, 52-54, 53f-55f
Magnetic zonation, of oceanic crust, 59, 59f, 60f
Magnetism
  intensity of, 59
  plate tectonics and, 52-54, 53f-55f
  polar wandering and, 54, 55f
Magnetite, 52, 680t-681t
Magnetometer, 53
Magnitude, earthquake, 527, 527t, 541, 542f
Mantle, 8, 8f, 9, 35, 35f, 549f
  composition of, 9
  convection cells of, 57-58, 60-64, 61f-65f
  density of, 8, 8f, 556, 664
  isostasy and, 12
  thickness of, 556
  volume of, 664
Mantle-core boundary, 552-554, 553f-555f
Mantle-crust boundary, 12, 13f, 551-552, 551f
Mantle drag, 63
Marble, 7, 486, 487, 487f
Marginal sea, 511
Marine environment. *See* Coastline; Ocean(s)
Marine plants
  in biochemical limestone, 401-402, 401f
  in petroleum formation, 584
Marine rocks, 7
  sedimentary, 407
Mars, 36-37, 36f, 41
Marshes, 263, 263f, 268. *See also* Wetlands
  coastal erosion and, 272
  definition of, 579
  paludal deposits in, 405, 405f
  saltwater, 381f, 382
  vs. swamps, 578-579
Marsh-swamp complex, 381f
Massive intrusive rock bodies, 155-157, 156f, 157f
Mass spectrometer, 621-622, 622f
Mass wasting, 5, 6f, 187, 187f, 205-229. *See also* Erosion
  creep and, 211, 212f
  dangerous site identification and, 223, 224f
  debris flows and, 218-219
  debris slides and, 217-218, 219f
  deforestation and, 222, 223f
  in deserts, 333-335, 334f, 335f
  earthflow and, 211-213, 213f
  earth moving projects and, 222, 223f
  economic impact of, 220
  environmental concerns and, 220-229
  freeze-thaw cycle and, 164, 209, 209f
  frost heaving and, 209
  gravity in, 208, 208f
  ice formation and, 209-210, 209f
  landslides and, 213-220
  mudflows and, 220, 221f
  overview of, 205-206
  in periglacial regions, 303
  processes of, 206, 206f, 207f, 211-220
  rainfall and, 187, 187f, 329-330
  rock falls and, 215, 215f, 216f
  rock slides and, 216-217, 216-218, 218f, 219f
  slope instability and, 210-211, 210f, 211f, 220-222
  slope stability and, 208-211
  slump and, 214, 214f, 215f
  solifluction and, 213
  vegetation and, 222, 223f
  water saturation and, 208-209, 221-222
Matthews, D.H., 59
Maximum rock transport, 473-474, 473f
McKenzie, D.P., 65
Meandering streams, 210, 210f, 239f, 252, 252f, 254, 255f, 257f-258f
  navigation of, 259
  sinuosity of, 252, 252f, 258
  water velocity in, 257-258, 258f
Measuring units
  conversion factors for, 670t
  prefixes for, 670t
Mechanical weathering, 164-169. *See also* Weathering
Medial moraines, 297, 299f
Mélange, 513
Melting
  of ice caps, 307
  of minerals, 142-143, 142f
  pressure, 286, 287
  of snow, 286
Melting temperature, 141-142, 142f
Mercury, 91
  abundance of, 569t
Mercury (planet), 33-34, 34f
Mesas, 18, 20f, 333, 335, 335f
Mesozoic Era, 626-628
  periods of, 625f, 629-631
Mestable minerals, 176
Metallic bonds, 79
Metals, 569-575. *See also* Mineral(s); Ore
  abundance of, 569t
  disseminated deposits of, 570-571
  distribution of, 569-570
  ferrous, 568, 568f
  in igneous rocks, 569-570
  in intruded country rocks, 570-571
  native, 569
  nonferrous, 568, 568f
  placer deposits of, 574
  veins of, 570, 570f
Metamorphic facies, 491, 492f
Metamorphic grade, 491, 491f, 492f

Metamorphic rock, index minerals in, 491
Metamorphic rocks, 7-8, 483-494
    definition of, 483
    distribution of, 492-493, 493f
    foliated, 484-486, 485f-487f
    formation of, 492-493, 493f
    grade of, 491, 491f, 492f
    nonfoliated, 484, 486 487, 487f
    in rock cycle, 21-22, 22f
    weathering of, 494
Metamorphism
    chemically active fluids in, 483, 484, 489-490
    in coal formation, 403
    contact, 488, 488f
    definition of, 484
    dynamo-thermal, 485, 488-489, 489f
    environmental concerns and, 494, 494f, 495f
    heat in, 483, 484, 488
    hydrothermal, 489-490
    local, 489
    plate tectonics and, 492-493, 493f
    pressure in, 483, 484, 488-489
    prograde, 491
    recrystallization in, 484
    regional, 488-489, 491, 492f
    retrograde, 491
Metaquartzite, 487
Metasediment, 484
Metasomatic replacement, 571
Metasomatism, 484, 490
Meteorites, 41-42
Meteoroids, 41-42, 41f, 42f
Meteors, 41-42
Methane, as fuel, 596
Mica/mica group, 83, 86t, 141t, 488, 491, 674t-675t
Micocline feldspar, 90
Microcline, 672t-673t
Microscopy, light, 85, 85f
Midstone, 398
Migmatite, 8, 484
Milankovitch, Milutin, 313
Milky Way Galaxy, 30, 30f
Milne, John, 536-537
Mine-mouth power plants, 582, 583f
Mineral(s), 73-91. See also Metals
    abundance of, 77t
    alluvial concentrations of, 574-575, 574f
    amphibole, 86t, 141, 672t-673t
    carbonate, 676t-677t
    chemical composition of, 73-80
    chloride, 682t-683t
    classification of, 81f, 85, 86t
    clay, 90, 173, 176, 188-189, 674t-676t-675t. See also Clay minerals
    crystal structure of, 77, 78f, 80, 80f, 83, 83f, 86-91, 87f-91f. See also Crystallization
    definition of, 73
    dredging for, 575, 575f
    eolian concentrations of, 403-405, 404f, 405f, 574-575, 574f
    feldspar. See Feldspar/feldspar group
    ferromagnesian, 81f, 85, 86t, 141t
    formation of, 77-78
    identification of, 80-85
    index, 491
    melting of, 141-143, 142f
    mestable, 176

    mica. See Mica/mica group
    mining of. See Mining
    nitrate, 680t-681t
    nonferromagnesian, 81f, 85, 86t, 141t
    nonsilicate, 85, 86, 86t, 91
    olivine, 35, 87, 141, 672t-673t
    oxide, 680t-681t
    phosphate, 680t-681t
    physical properties of, 80-85
    platy, 484, 485f
    pyroxene, 35, 86t, 141t, 672t-673t
    silicate, 77, 85, 86-91, 86t, 87t-91t, 141t
    sulfate, 676t-677t
    sulfide, 571, 678t-679t
Mineral cleavage, 83, 84f
Mineral fracture, 83-84, 84f
Mineraloids, 80
Mining
    hydraulic, 574-575, 575f
    pollution from, 170, 180, 601-604
    stream gradient and, 260
    subsidence and, 228, 229f
    water quality and, 170, 180
Minoans, 128
Miocene Epoch, 625f, 632, 632t
Mirrors, for solar power, 599, 599f
Mississippian System, 625f, 629
Modern time scale, 624-628
Modified Mercalli Scale, 526-527, 527t
Moho, 552
Mohorovicic, Andrija, 551
Mohorovicic Discontinuity, 552
Mohs's Scale of Hardness, 82t, 8282
Mollisols, 195, 195f
Molybdenite, 678t-679t
Monoclines, 461, 462f
Monsoon, ocean currents and, 368
Montmorillonite, 674t-675t
Moon, tides and, 365-366, 365f, 366f
Moraines, 297, 298f, 299f, 305-306
    ground, 297, 298f, 305-306, 306f
    lateral, 297, 299f
    medial, 297, 299f
    recessional, 297, 298f
    terminal (end), 297, 298f, 306, 306f
Morely, L., 59
Morgan, W.J., 65
Mother lode, 575, 576f
Mountain(s), 17, 501-518
    in Basin and Range topography, 335-336, 336f, 337f, 507, 508f
    block-fault, 504, 507, 507f, 559
    definition of, 501-502
    domal, 504-507, 505f-507f
    foldbelt, 507-518, 508f-517f. See also Foldbelt mountains
    formation of. See Mountain formation
    isostatic balance in, 559, 560f
    oceanic ridges as, 502-503, 503f
    rainshadow deserts and, 324-325, 325f, 326f
    roots of, 559
    types of, 502-511
    volcanic, 4, 4f, 363, 363f, 502-504, 502f-504f
Mountain formation, 17, 19f, 21, 22f
    diastrophic, 502
    epeirogenic forces in, 502, 504-507, 504f-507f
    orogenic forces in, 502, 511-518, 512f-518f

    plate tectonics and, 507-518
    volcanism in, 502-504, 502f-504f
Mountain ridges, 17-18, 19f
Mountain valleys, 17-18, 19f
Mud
    carbonate, 394, 394f, 416f
    of ocean bottom, 400-401, 400f
Mud cracks, 413, 414f
Mudflats, 381-382, 381f, 407, 408f
Mudflows, 220, 221f
    volcanic, 108, 109f
Mudstone, formation of, 409
Municipal water supplies, 436-437, 436f, 437f
Murchison, Roderick, 628-629
Muscovite, 89, 674t-675t
Mylonite, 489

N

National Earthquake Information Center (NEIC), 542, 543f
National Seismograph Network (USNSN), 542
Native elements, 86t, 91
Native metals, 569
Natural gas, 597-598
Natural resources
    classification of, 568-569
    conservation of, 597-598
    nonrenewable, 577-588
    production-time curve for, 586-587, 587f
    renewable, 591-600
    vs. reserves, 568-569
Navigation, of meandering streams, 259
Neap tides, 366, 366f
Nebulas, 30, 30f
    solar, 32
Needle rocks, 333, 335, 335f
Neptune, 40, 40f
Neumann, F., 526
Neutrons, 73, 74f
Névé, 286, 286f
Newcomen, Thomas, 577
Newton, Isaac, 29
Newton's laws of motion, 29
Nickel, abundance of, 569t
NIMBY syndrome, 596
Niter, 680t-681t
Nitrate minerals, 680t-681t
Nitrates, 86t, 91
Nondetrital sedimentary rocks, 398-403, 399f-402f
Nonferromagnesian minerals, 81f, 85, 86t. See also Mineral(s)
    weathering of, 175. See also Weathering
Nonferrous metals, 568, 568f
Nonfoliated metamorphic rocks, 484, 486-487, 487f
Nonfuels, nonmetallic, 568, 568f, 577
Nonmetals
    fuel, 568, 568f, 577-588. See also Fuel(s)
    nonfuel, 568, 568f, 577ls
Nonpoint sources, for water pollutants, 276, 276f
Nonrenewable resources, 577-588. See also Fuel(s)
    future of, 586-588
    production-time curve for, 586-587, 587f

Index   695

Nonsilicate minerals, 85, 86, 86t. *See also* Mineral(s)
Normal faults, 467, 467f, 471, 472f
Nuclear power, 588-591
Nuclear reactors
   breeder, 589-590
   fission, 588-590, 589f, 590f
   fusion, 590
Nueé ardente, 108, 115-116, 115f

O

Oblique-slip faults, 468, 468f
Obsidian, 149, 567, 567f
Ocean(s)
   area of, 664
   closing of, 356, 359
   depth of, 351-352, 352f, 664
   in hydrologic cycle, 235-236, 236f
   linear, 10
   magma, 103
   opening of, 356, 359, 512f
   volcanic source of, 102-103
   volume of, 664
   warming of, 315
Ocean basins, 12-16
   closing, 359
   earthquakes in, 532-533
   features of, 13-16
   formation of, 10, 10f, 56f, 57, 58f, 61, 62f, 353-357, 354f, 355f, 360f
   location of, 14f
   opening, 359
   spreading of, 359, 360f
Ocean bottom
   as depositional environment, 407-408, 409f
   depth of, 12, 13f
   echo-sounding studies of, 54, 55f, 56
   exploration of, 351-353
   landforms of, 357-359, 357f, 358f
   muds (oozes) of, 400-401, 400f
   sediment on, 152-153, 153f
   spreading of, 356
Ocean bottom topography, 357-359, 357f, 358f
   continental drift and, 54-60, 55f-60f
Ocean-continent orogenesis, 511-513, 512f, 513f
Ocean currents, 367-374
   density, 361, 361f, 369-374
   El Niño and, 372-373
   equatorial, 367, 368f
   monsoons and, 368
   thermohaline, 367-368, 367f, 368f, 369-370
   turbidity, 361, 361f, 374, 412
   wind-driven, 367-368, 367f, 368f
   wind-induced, 369, 370f
Ocean dumping, 386-387
Ocean floor. *See* Ocean bottom
Oceanic crust, 8f, 9, 12, 13f, 35, 57, 358, 358f. *See also* Crust
   composition of, 9, 117f, 555
   in continental drift, 57-58, 58f
   creation/destruction of, 57-58
   density of, 8, 8f, 664
   depth of, 555

   magnetic zonation of, 59, 59f, 60f
   volume of, 664
Oceanic ridges, 15, 15f, 357-358, 357f
   continental drift and, 56, 56f
   fissure eruptions at, 116, 116f
   magma formation and, 150, 15015f
   magnetic bands of, 59, 60f
   as mountains, 502-503, 503f
   volcanism and, 98f, 99, 116, 116f, 150
Ocean–island arc–continent orogenesis, 514, 514f, 515f
Oceanography, 353
Ocean reefs, 382-383, 382f-384f, 415, 416f
   destruction of, 386
Ocean salinity, dating of Earth by, 620
Ocean water
   elemental concentrations in, 667t
   movement of, 364-374
Ocean waves, 364-365, 364f, 365f. *See also under* Wave
   refraction of, 367, 367f
   tsunami, 532-533, 533f
Octet rule, 74
Off-the-road vehicles, 199, 200f
O horizon, 192, 194
Oil, 584-585
   conservation of, 597-598
   future of, 586-588
Oil drilling, subsidence and, 226, 229f
Oil shale, 586, 586f
   future of, 586-588
Oligocene Epoch, 625f, 632, 632t
Olivine minerals, 35, 87, 141, 173, 672t-673t
On the Origin of Species (Darwin), 577, 614
Ooids, 399, 573, 573f
Oolitic limestone, 399, 399f
Oort cloud, 43
Oozes, 400-401, 400f
Opal, 80
Opening oceans, 356, 359, 512f
Ophiolite suite, 358, 358f
Orbits
   of comets, 42, 43f
   of planets, 33, 34f, 40, 40f
Ordovician Period, 625f, 629
   glaciation in, 313
Ore, 568. *See also* Metals
   enriched, 571-572
   weathering of, 571
Organic sedimentary rock, 402-403, 403f
Organic sulfur, in coal, 581
Original horizontality, principle of, 612
The Origin of Continents and Oceans (Wegener), 50-52
Orogenesis
   continent-continent, 515-518, 516f, 517f
   ocean-continent, 511-513, 512f, 513f
   ocean–island arc–continent, 514, 514f, 515f
Orogenic forces, 502
   in foldbelt mountain formation, 511-518, 512f-517f
Orthoclase, 82, 672t-673t
Orthoclase feldspar, 90, 91f, 149
Oscillation ripple marks, 410
Outwash plains, 302, 302f
Overcultivation, 345
Overfishing, 386
Overgrazing, 344-345

Overturned folds, 463, 463f
Overturning, 611-612
Oxbow lake, 252-253, 253f
Oxidation, weathering and, 170
Oxide minerals, 680t-681t
Oxides, 86t, 91

P

Pahoehoe lava, 104, 105f
Paired terraces, 254, 255f
Paleocene Epoch, 625f, 632, 632t
Paleozoic Era, 625t, 626-628
   periods of, 625f, 628-629
Paludal deposits, 405, 405f
   in coal formation, 578-579, 578f
Panning, for gold, 574, 574f
Parabolic dunes, 342, 343f
Parallel drainage, 240
Parker, R.L., 65
Partial melt, 143
Particles, in sedimentary rock, 394
Particle shape, 395-396, 395f, 396f
Particle size
   classification of, 394, 394f
   porosity and, 426-427, 426f, 427f
   stream erosion and, 246-247, 246f
   weathering and, 174, 175f
Particle sorting, 394-395
   porosity and, 426-427, 427f
Passive margin, 9, 10-12, 11f, 12f, 14f, 66, 99, 100f, 359-363, 360f-362f
   ocean closing and, 356, 359, 360f
Patch reef, 382-383, 383f
Paternoster lakes, 291-293, 294f
Patterned ground, 303, 304f
Peat
   coal-forming, 579-580
   formation of, 402, 417, 578, 578f, 579
Pebbles, stream-worn, 248, 248f
Pedestal rocks, 209, 210f, 339, 340f
Pediment, 336
Pediment slope, 336
Pedology, 185
Pegmatite, veins of, 570, 570f
Pelean-type eruptions, 115-116, 115f
Peneplanes, 253, 257
Penicillus, 401-402, 401f
Pennsylvanian Period, 625f, 629
   glaciation in, 313
Perched water tables, 431-432, 431f, 432f
   dewatering of, 445-446, 446f
Peridotite, 35, 148t, 149
   magma from, 150-151
Periglacial regions, 303-305, 304f, 305f
   mass wasting in, 303
   streams in, 305
   weathering in, 204f, 303-305
Periodic table, 665
Permeability, 428
Permian Period, 625f, 629
   glaciation in, 313
Perrault, Pierre, 425
Petroleum, 584-586, 585f
   conservation of, 597-598
   future of, 586-588
   production peak for, 587-588
   production-time curve for, 586-587, 587f

Petrologists, 144
Phaneritic texture, 146, 146f-148f
Phanerozoic Eon, 625, 625f, 626
Phenocrysts, 108
Phillips, William, 629
Pholtovoltaic cells, 598-599
Phosphate minerals, 680t-681t
Phosphates, 86t, 91
Photic zone, 382, 415
Phyllitic cleavage, 486, 486f
Piedmont glaciers, 283, 286f
Pilings, 226, 228f
Pillars, dripstone, 442, 442f
Pillow lava, 106, 106f, 116, 116f, 357-358
Pingo, 303, 304f
Placer deposits, 574
Placet, François, 49
Plagioclase feldspar, 90, 91f, 141t
Plains, 18
   abyssal, 13-15, 56f, 359, 361
   delta, 264, 267, 267f
   outwash, 302, 302f
Planetesimals, 32
Planets, 33-41. *See also specific planets*
   formation of, 32, 33f
   Jovian, 33, 37-40, 38f, 39f
   orbits of, 33, 34f
   terrestrial, 33-37, 34f-37f
Plants
   in coal formation, 579-580, 580f, 584
   desert, 321, 322f
   in dune stabilization, 342, 342f
   growth rings in, 618f, 619
   marine, 401-402, 401f, 584
   in petroleum formation, 584
   in slope stability, 222, 222f
   soil acidity and, 190-192
   in soil formation, 186
Plant wedging, 169, 169f
Plastic deformation, 455, 455f
   in rock, 460, 460f
Plastic flow, of glacial ice, 286-287, 287f, 288f
Plate(s)
   convergent, 9, 10-12, 11f, 12f, 66, 99, 100f, 356, 359, 360f
   divergent, 9-10, 9f, 10f, 13, 66, 356, 360f, 363-364, 363f
   movement of, 10-12, 11f, 12f
   volcanoes and, 97, 99-100, 99f, 101f
Plateaus, 17
Plate fracture
   sites of, 354, 354f, 355f
   steps in, 354
Plate margins
   active, 9-10, 9f, 10f, 14f, 66, 356, 359, 360f, 363-364, 363f
   convergent, 9, 10-12, 11f, 12f, 66, 99, 100f, 356, 359, 360f
   divergent, 9-10, 9f, 10f, 13, 66, 99, 99f, 356, 360f, 363-364, 363f
   earthquakes at, 523-524, 524f
   volcanic eruptions at, 98f
Plate tectonics, 12, 49-67
   continental drift and, 50-52. *See also* Continental drift
   continental glaciation and, 314
   convection cell theory and, 57-58, 60-64, 61f-63f
   early theorists of, 49-52

   fault movement and, 473-475, 473f-476f
   foldbelt mountain formation and, 507-518
   historical development of, 49-52
   magma formation and, 150-154
   metamorphism and, 492-493, 493f
   ocean basin formation and, 353-356
   ocean bottom topography and, 54-59
   principles of, 65-67
   rock magnetism and, 52-54, 53f-55f, 59, 60f
   seismic data and, 60-64, 61f-65f, 554
   stream gradient and, 260
   supercontinent cycles and, 356-357, 357f
   triple junction and, 354, 354f, 355f
   as unifying theory, 49
Plato, 28
Platy minerals, 484, 485f
Playa lakes, 330, 331f
Playfair, John, 610, 611
Pleistocene Epoch, 625f, 632, 632t
Pleistocene Ice Age, 307-312, 313
Plinian eruptions, 115f, 116
Pliocene Epoch, 625f, 632, 632t
Plucking, 290, 290f
Plunge, of folds, 462, 463f
Pluto, 40-41, 40f
Plutonium, as nuclear waste, 589
Plutons, 3, 156f. *See also* Intrusive igneous rock bodies
Pluvial lakes, 310-311, 311f
Point, of meandering stream, 257, 257f
Point bar, 257f, 258, 261, 403, 403f
Point sources, for water pollutants, 275, 276f
Polar ice caps. *See also* Continental glaciers
   melting of, 307
Polar wandering, 54
Pollutants
   harmful, 275
   nonpoint sources for, 276, 276f
   point sources for, 275-276, 276f
   undesirable, 275
Pollution. *See also* Environmental quality
   acid rain and, 172, 173f
   from mining, 170, 180, 601-604
   soil contamination and, 199-200
   trace elements and, 177-179
   water, 275-276, 443, 443f. *See also* Water quality
Pores, 426
   volume of, 427f
Porosity, 426-427, 426f, 427f
   fracture, 426, 426f
   intergranular, 426, 426f
Porphyritic texture, 146, 146f, 148, 148f
Porphyry, 148, 148f
Potassium 40, in radiometric dating, 621, 623t
Powell, John Wesley, 617f
Power plants
   mine-mouth, 582, 583f
   nuclear, 588-591, 589f, 590f
   solar, 598-599, 599f, 600f
Precambrian Period, 628-629
   glaciation in, 313
Precipitation, 77-78
   in hydrologic cycle, 235-236, 236f
Pressure, in metamorphism, 483, 484, 488-489

Pressure melting, 286, 287
Pressure surface, of aquifer, 435f, 436
Principles. *See specific principles, e.g.,* Superposition, principle of
Principles of Geology (Lyell), 611, 614-615
Prodelta, 264, 267, 267f
Production-time curve, 586-587, 587f
Prograde metamorphism, 491
Proterozoic Eon, 625, 625f
Protons, 73, 74f
Protoplanets, 32, 33f
Protore, 568
Pteranodon, 627f, m626
Ptolemy, 28
P waves, 534, 534f, 538-539, 539f, 540f
   density and, 553-554, 553f-555f
Pyrite, 84, 571, 678t-679t
Pyritic sulfur, in coal, 581
Pyroclastic materials, 4, 106
Pyroclastic texture, 146, 146f
Pyrolusite, 680t-681t
Pyroxene minerals, 35, 86t, 141t, 672t-673t
Pythagoras, 27-28

Q

Quarrying, 290, 290f
Quartz, 83, 84, 149, 173, 674t-675t
   in cementation, 409
   chemical stability of, 175, 176
   in sedimentary rock, 394
Quartz sandstone, 397-398, 398f
Quaternary Period, 625f, 632, 632t

R

Radial stream pattern, 240, 242f
Radioactive isotopes, 75, 620-621
   atomic number and mass of, 620-621
   daughter, 621, 623, 623t
   decay of, 75, 620-621, 621, 622f
   half-life of, 620-623, 622f, 623t
   parent, 621, 623, 623t
   in radiometric dating, 621-624, 621f, 622f, 623t
Radiocarbon dating, 623, 623t
Radiolaria, 400
Radiometric dating, 621-624, 621f, 622f, 623t
   radiocarbon, 623, 623t
Rainfall. *See also* Water
   acidity of, 172, 173f, 191f, 194
   amounts of, 194-196
   containments for, 447, 447f
   mass wasting and, 187, 187f, 329-330
   soil acidity and, 190-191
   soil horizons and, 194
Rainforest, destruction of, 196
Rainshadow, 324, 324f
Rainshadow deserts, 324-325, 325f, 326f
Rapids, 251, 251f
Rayleigh waves, 534-535, 535f
Reach, of meandering stream, 257, 257f
Realgar, 678t-679t
Recessional moraines, 297

Recharge, from aquifer, 434, 435f
Recrystallization, in metamorphism, 484
Rectangular stream pattern, 240, 240f
Recumbent folds, 463, 463f, 464-465, 464f-465f
Red giants, 31, 32f
Reefs, 382-383, 382f-384f
   barrier, 382-383, 382f-384f, 415, 416f
   destruction of, 386
Regelation, 287
Regelation slip, 287
Regional doming, 505-507, 505f-507f
Regional metamorphism, 488-489
   grade and, 491, 492f
   metamorphic facies distribution in, 492-493
Regional water table, 431-432, 431f, 432f
Regolith, 5, 176, 176f, 177f, 185, 186f. *See also* Soil
   composition of, 176
   in creep, 201, 211
   definition of, 176
   in earthflow, 211-213, 213f
   formation of, 205
   in slump, 211f, 214, 214f
   stability of, 208
   thickness of, 186, 187
   water movement in, 194
Rejuvenation, stream, 254, 255f
Relative dating, 611-619
Relative humidity, 324
Renewable resources, 591-600
Reptiles
   Mesozoic, 626, 627f
   Paleozoic, 626
Reserves, vs. resources, 568-569
Reservoirs, petroleum, 585
Resources
   classification of, 568-569
   conservation of, 597-598
   nonrenewable, 577-588
   production-time curve for, 586-587, 587f
   renewable, 591-600
   vs. reserves, 568-569
Retaining walls, 226
Retrograde metamorphism, 491
Reversed radial streams, 242, 243f, 439
Reverse faults, 466, 466f, 467f, 471, 472f
Rhyolitic magma, 114
   eruptive intensity and, 114
   in felsic eruptions, 118
   weathering of, 174
Richter Scale, 524, 527, 527t
Ridge(s), 17-18, 19f
   oceanic, 15, 15f, 357-358, 357f. *See also* Oceanic ridges
Ridge push, 64
Rift, 9, 10f
Rifting/rift zone, 9, 10f, 63. *See also* Plate tectonics
   formation of, 57, 57f, 61, 62f, 354, 354f, 355f
   magma formation and, 150, 150f
   mechanisms of, 354, 354f, 355f
   ocean basin formation and, 354-356, 354f, 355f, 360f
   at triple junction, 354, 354f, 355f
   volcanism and, 9, 10f, 57, 57f, 356, 502
Rift valleys, 9, 10f, 99, 99f, 354
   flooding of, 356, 360f
   formation of, 57, 57f, 61, 62f, 354, 354f, 355f, 360f
   stream, 238, 239f, 249-257, 250f-256f
   at triple junction, 354, 354f, 355f
Right-lateral faults, 467-468, 468f, 471, 472f
Rigidity, seismic wave velocities and, 550
Ripple marks, 410, 411f, 612, 613f
Rivers. *See* Streams
Road construction
   mass wasting and, 222-223, 223f, 224, 225f
   sedimentary rock and, 419-420
Roche moutonnée, 295-297, 297f
Rock(s)
   argillaceous, 488
   cap, 585
   carbonate, 438
   continental, 7
   detrital, 394
   igneous, 3-5, 3f-5f, 141-157. *See also* Igneous rocks
   marine, 7, 407
   metamorphic, 7-8. *See also* Metamorphic rocks
   nondeformational uplifting of, 453, 453f
   nondetrital, 394
   overturned, 611
   pedestal, 209, 210f, 339, 340f
   permeability of, 428
   porosity of, 426-427, 426f, 427f
   pyroclastic, 4, 106
   radiometric dating of, 621-624, 621f, 622f, 623t
   rounding of, 167-168, 168f, 248f, 256, 395, 396f
   sedimentary, 5-7, 6f, 7f, 393-420. *See also* Sedimentary rocks
   size of, 154
   stratigraphic sequence of, 417-418, 418f
   stream-worn, 248, 248f
   strength of, 458
   system of, 628
   types of, 3-8
   vertical sequence of, 417-418, 418f
   volcanic, 4, 154
   volumes of, 664
   weathering of, 5, 6f, 163-180. *See also* Weathering
Rock avalanche, 216-217, 217f, 218f
Rock cleavage, 484-485
Rock cycle, 21-22, 22f
Rock deformation, 453, 458-461
   brittle, 455-456, 455f, 460-461, 461f
   direction of movement in, 473-475, 473f-477f
   elastic, 460
   faults and, 465-469
   folds and, 461-465
   joints and, 469-471
   plastic, 460, 460f
   rock composition and, 458
   stress and, 454-455, 454f
Rock falls, 215, 215f, 216f
   in deserts, 333
Rock glaciers, 215, 216f, 303-305, 304f
Rock magnetism. *See also under* Magnetic
   intensity of, 59
   plate tectonics and, 52-54, 53f-55f
   polar wandering and, 54, 55f

Rock movement, direction of, 473-475, 473f-477f
Rock particles. *See also under* Particle
   in sedimentary rock, 394
   shape of, 395-396, 395f, 396f
   size of, 174, 175f, 246-247, 246f, 394, 394f
   sorting of, 394-395
Rock salt, 76, 83, 85, 402
   dissolution of, 171
Rock slides, 216-217, 217f
   in deserts, 333
Rock streams, 305, 305f
Rock time scale, 624-632. *See also* Geologic time scale
Rock transport, maximum, 473-474, 473f
Roman arch, 458, 459f
Roots, of mountain range, 559
Rossi, M.S. de, 526
Rounding, 167-168, 168f, 248f, 256, 395, 396f
Roundness, of particles, 395, 395f, 396f
Rutherford, Lord, 620-621
Rutile, 680t-681t

S

Sabkhas, 402
Salinity, dating of Earth by, 620
Salinity-induced density currents, 370-374
Salinization, 345-346
Saltation, 248, 338, 338f
Salt dome trap, 585, 585f
Saltwater encroachment, 446-447, 447f
Saltwater marsh, 381f, 382. *See also* Wetlands
   coastal erosion and, 272
Sand
   of beaches, 406, 406f, 407f
   compaction of, 409
   longshore transport of, 367
   wind transport of, 338, 338f
Sand dunes. *See* Dunes
Sandstone, 7, 397-398, 398f
   composition of, 666
   honeycomb weathering of, 166, 167f
   as petroleum reservoir, 585
Sand storm, 338, 338f
Sanidine feldspar, 90
Sanitary landfills, groundwater pollution from, 443, 443f, 446f
Saturn, 37, 38, 39f
Scarp, 507
Schist, 489, 489f
Schistose cleavage, 486, 487f
Schistosity, 486
Scientific notation, 671, 671t
Scouring rushes, 175
Scrubbers, 582
Sea(s). *See also* Ocean(s)
   marginal, 511
Sea arches, 376-378, 377f
Sea caves, 375, 376f
Sea cliffs
   of emergent coastlines, 375-378, 375f-377f
   oversteepening of, 211f, 214
   wave-cut, 375, 376f, 377f

Sea level, glacial melting and, 307, 315
Seamounts, 16, 16f, 100, 110, 131-136, 131f-136f, 359
   formation of, 132
Sea stacks, 376-378, 377f
Sedgwick, Adam, 628-629
Sediment, definition of, 393
Sedimentary deposits
   as aquifers and aquitards, 428-429
   in back-arc basin formation, 510-511, 511f
   bedding of. *See* Beds/bedding
   cementation of, 409
   compaction of, 409
   continental, 403-405, 404f, 406f
   control of, 419
   dating of Earth by, 619-620
   eolian, 403-405, 404f, 405f
   fluvial, 403, 403f
   in foldbelt mountain formation, 509-510, 510f
   glacial, 297-302, 298f-303f, 403, 404f
   interpretation of, 414-418
   lacustrine, 405
   layering of, 410, 410f, 614. *See also* Beds/bedding
   Lithification of, 409
   marine, 407-408, 409f
   orientation of, 611-612
   paludal, 405, 405f
   porosity of, 427, 427f
   sequence of, 417-418, 418f
   in transitional environments, 405-407, 406f-408f
   Walther's law and, 417-418, 418f
   as water conduits, 420
Sedimentary rocks, 5-7, 6f, 7f, 393-420
   abundance of, 393
   animal trails in, 413-414, 415f
   as aquifers and aquitards, 428-429
   beds (strata) of, 410, 410f. *See also* Beds/bedding
   biochemical, 400-402
   chemical, 398-399, 399f, 400f
   classification of, 394-396
   components of, 393
   composition of, 394
   cross-bedding in, 261, 262f, 410-411, 412f, 416, 417f, 611-612, 613f
   dating of, 414-417
   dating of Earth by, 619-620
   definition of, 393
   deformation of, 453-477. *See also* Deformation
   detrital, 394, 397-398, 397f, 398f
   distribution of, 393
   economic aspects of, 393
   engineering problems with, 419
   environmental concerns and, 419-420
   evaporitic, 402
   formation of, 409, 510-511, 511f
   importance of, 393
   internal features of, 410-414, 410f-415f
   layering of. *See* Beds/bedding
   marine, 407
   mud cracks in, 413, 414f
   nondetrital, 398-403, 399f-402f
   organic, 402-403, 403f
   overturned, 611-612
   porosity of, 427, 427f
   ripple marks in, 410, 411f, 612, 613f
   in rock cycle, 21-22, 22f
   sorting of, 394-395
   stratification of, 410, 410f, 614
   Walther's law and, 417-418, 418f
   as water conduits, 420
   weathering of, 5, 6f, 20f
Sediment erosion and transport, 245-248
Seif dunes, 342, 343f
Seismic activity, 123-124
Seismic data
   for convection cell theory, 60-64, 61f-65f
   in crust-mantle boundary identification, 551-552, 551f
   in earthquake location, 538-541, 538f-542f
   in earthquake prediction, 543-544, 543f
   earth's interior and, 550
   in mantle-core boundary identification, 552-554, 553f-555f
   plotting of, 538-541, 539f-541f
   time-distance curve for, 539, 539f
   worldwide network for, 542
Seismic gaps, 543
Seismic Sea Wave Warning System, 533
Seismic tomography, 64, 65f
Seismic waves, 524, 533-535, 534f
   amplitude of, 541-542
   body, 534, 534f, 550, 551f, 552
   compression, 533-534, 534f
   density and, 550, 550f, 551f, 553-554, 553f, 554f
   directional changes in, 550, 550f, 551f
   Love, 534-535, 535f
   P, 534, 534f, 538-539, 539f, 540f, 553-554, 553f-555f
   plate tectonics and, 554
   Rayleigh, 534-535, 535f
   rigidity and, 550
   S, 534, 534f, 538-539, 539f, 540f, 553-554, 553f-555f
   shear, 533-534, 534f
   surface, 534-535, 534f, 535f
   velocity of, 550, 550f
Seismogram, 537, 537f, 538f
Seismograph, 527, 535-538
   development of, 535-538, 535f-537f
Selenium, 178-179
Septic systems, pollution from, 443, 443f
Sequential dating, 611-619
Serpentine, 490, 490f
Serpentinite, 487
Sewage, water pollution and, 275, 276f
Shafts, cave, 441, 441f
Shale, 7, 398
   composition of, 666
   engineering problems with, 419
   formation of, 409
   oil, 586, 586f
   as petroleum reservoir, 585
Shaler, Henry, 629
Shallow-focus earthquake, 525, 525f
Shear, 454, 454f
   dynamothermal metamorphism and, 485, 485f
   faults and, 464-465
Shear joints, 469, 469f
Shear waves, 533-534, 534f
Sheet dikes, 116, 117f
Sheeted dikes, 358
Sheeting, 167, 168f
Sheet structure, 88-90, 89f
Shells, in biochemical limestone, 401, 402, 402f
Shields, 17, 19f
Shield volcanoes, 16f, 110-111, 111f, 359
   hot spots and, 503-504, 503f
Shoreline. *See* Coastline
Shrinking Earth theory, 509-510
Siderite, 676t-677t
Silica
   chert forms of, 400-401
   in soil, 196
Silicate minerals, 77, 85, 86t, 141t. *See also* Mineral(s)
   crystallization of, 143-145, 144f, 145f, 145t
   ferromagnesian, 81f, 85, 86t, 141t
   nonferromagnesian, 81f, 85, 86t, 141t
   structure of, 86-91, 87f-91f
Silicon, magma viscosity and, 149
Silicon-oxygen tetrahedron, 86, 87f
Sill, 154, 155f
Sillimanite, 488, 491
Siltstone, 398
Silver, 91
Sinkholes, 226, 228f, 419, 438-439, 438f-440f
   collapse, 226, 419, 440-441, 441f
   solution, 438-439
   stream patterns and, 242, 243f
   surface/subsurface drainage and, 439-440
Sinuosity, of meandering stream, 252, 252f, 258
Size, of geologic features, 154
Slab pull, 64
Slate, 7
Slaty cleavage, 484-485
Slickenside, 475, 477f
Slope instability
   coal mining and, 602
   cohesion and, 208-211
   earth moving projects and, 222-223, 223f
   friction and, 208-211
   gravity and, 208, 208f
   identification of, 223, 224f
   mass wasting and, 208-211, 220-222. *See also* Mass wasting
   prevention/remediation of, 224-229
   vegetation and, 222, 222f
   water saturation and, 208-209, 221-222
Slope orientation, soil formation and, 187, 188f
Slope oversteepening, 210-211, 210f, 211f
Slope stabilization, 224-229, 225f-228f
Slope steepness, stream order and, 238
Slope wash, 187
Slump, 214, 214f, 215f
Smith, William, 612-614, 631
Snider-Pellegrini, Antonio, 49
Snow
   melting of, 286
   sublimation of, 286
   transformation of into glacial ice, 286, 286f

Snow avalanche, 211
Snow line, 283
Soda niter, 91, 680t-681t
Sodium carbonate, for acid mine drainage, 603
Sodium hydroxide, for acid mine drainage, 603
Soil, 176, 177f, 185-200. *See also* Regolith
    acidity/alkalinity of, 190-192
    breadbasket, 195
    cation adsorption by, 189, 189f, 191
    cation exchange in, 189, 189f, 191-192
    chemistry of, 189-192
    clay minerals in, 188-189. *See also* Clay minerals
    contamination of, 199-200
    definition of, 176, 185
    erosion of, 198-199, 199f
    expandable, 197
    fertility of, 194-196
    formation of, 163, 185-188
    grassland, 195
    irrigation of, 196
    plants and, 190-191
    rainfall and, 194-196
    regolith and, 176, 177f, 178f, 186f
    superalkaline, 195
    thickness of, 188
    trace elements in, 177-179
    in tropical regions, 196
    types of, 196-197, 198t
    water volume of, 194-196
Soil horizons, 192-196
    development of, 194-196
    types of, 192-194, 193f
Soil taxonomy, 196-197, 198t
Solar energy, 598-600, 598f, 600f
Solar nebula, 32
Solar System
    Copernican, 29
    formation of, 32, 33f
    Ptolemic, 28
Solar wind, 32, 33f
Solfateras, 122
Solids, melting and crystallization of, 141-142
Solid-solution series, 87
Solid waste, as fuel, 596-597
Solid waste disposal, 387
    via ocean dumping, 386-387
Solifluction, 213, 303
Solution sinkholes, 438-439
Sonar, in ocean exploration, 351-352, 352f
Sorting, 394-395
    porosity and, 426-427, 427f
Southwestern monsoon, 368
Space. *See also* Planets
    distances in, 30-31
Spalling, 166, 167f
Spatter cone, 109-110, 110f
Specific gravity, of minerals, 84
Speliothems, 400f
Sphalerite, 678t-679t
Sphericity, of particles, 395-396, 395f, 396f
Spheroidal weathering, 167-168, 168f
Spinel, 680t-681t
Spits, 378, 379f
Sponges, 400
Spreading centers, 66
Spring tides, 366

Stable platform, 17
Stalactites, 442, 442f
Stalagmites, 442, 442f
Stars
    death of, 31, 32, 32f
    formation of, 31, 32
Staurolite, 491
Steam, in geothermal power, 591-592, 592f
Steam coal, 581
Steam engine, 577, 577f
Steno, Nicolas, 611-612
Stock, 154, 155
Stony-iron meteorites, 41
Stony meteorites, 41
Stoping, 156, 156f
Strain, definition of, 454
Strain ellipsoid, 456-457, 456f-457f
    faults and, 471-472, 472f
    folds and, 464-465, 464f-465f
Stratigraphic sequence of rocks, 417-418, 418f
Stratigraphic trap, 585, 585f
Stratovolcanoes, 111, 112f, 504
Streak, 82, 82f
Stream(s), 21, 235-276
    alluvial deposits of, 261-274. *See also* Alluvial deposits
    alluvial fans and, 272-274, 273f, 329-330, 330f, 335-336, 337f
    banks of. *See under* Slope
    baselevel of, 249, 249f
    braided, 242, 243f, 305
    capacity of, 247
    channel cutting by, 249-258. *See also* Channel creation
    channel texture of, 244
    competence of, 246, 247, 337-338
    deltas and, 264-272
    desert, 329-331, 330f, 332f
    disappearing, 242, 243f, 439
    distributary, 264, 266f
    in drainage systems, 236-238, 236f, 237f
    effluent, 432
    energy of, 242-244. *See also* Water flow
    environmental concerns and, 274-276
    erosion by, 245-249, 246f, 248f, 250-258, 251f-258f, 329-336, 329f-340f. *See also* Channel creation
    evolution of, 248-258, 250f-258f
    first-order, 238, 239f
    flooding of, 274-275, 274f
    gaining, 432
    grade of. *See* Stream grade
    gradient of, 244, 244f
    headward erosion and, 238, 239f
    hydrologic cycle and, 235-236, 236f
    influent, 432
    losing, 432
    meandering, 210, 210f, 239, 252, 252f, 254, 255f, 257-258, 257f-258f, 259
    oversteepening and, 210-211, 210f, 211f
    in periglacial regions, 305
    rapids in, 251, 251f
    rejuvenation of, 254, 255f
    reversed radial, 242, 243f, 439
    rock, 305, 305f
    second-order, 238, 239f
    sediment erosion and transport by, 245-248
    topographic relief and, 249-258

    tributary, 264
    volume of, 664
    waterfalls in, 251, 251f
    water flow in, 242-245, 244f, 245f. *See also* Water flow
    water volume in, 244-245
Stream channels, cutting of, 249-258
Stream discharge, 242-244
Stream grade, 247
    climatic influences on, 260
    development of, 258-261
    human influences on, 260
    tectonic influences on, 260
Stream gradient, 244, 244f
Stream load, 247-248, 248f
    bed, 247-248
    dissolved, 247
    suspended, 247, 248f
Stream order, 238, 238f, 239f
Stream patterns, 238-240
    annular, 240, 242f
    dendritic, 238-240, 240f
    parallel, 240, 242f
    radial, 240, 242f
    rectangular, 240, 240f
    reversed radial, 242, 243f
    trellis, 240, 241f
Stream piracy, 238
Stream systems, 236-238, 236f, 237f
    divides in, 236-238, 237f
    external, 236
    internal, 236, 330-331, 337f
    stream order in, 238, 238f
Stream valleys, 238, 239f
    evolution of, 238, 239f, 249-258, 250f-256f, 250f-258f. *See also* Channel creation
    rejuvenation of, 254, 254f
    V-shaped, 251, 251f
Stream water. *See also* Water
    elemental concentrations in, 668t
Stream-worn pebbles, 248, 248f
Strength
    definition of, 454
    of rocks, 458
Stress
    compressive, 454, 454f
    deformation and, 454-455, 454f
    folds and, 461-465, 462f-465f
    shear and, 454, 454f
    strain and, 454
    tensional, 454-455, 454f, 458
Striations, in slickenslides, 475, 477f
Strike, 461, 461f
Strike-slip faults, 467-468, 468f, 471, 472f
Strip mining. *See* Mining
Stroke, water quality and, 180
Strombolian eruptions, 114, 114f
Subbituminous coal, 580
Subdelta, 268, 271f
Subduction, 10
    zone of. *See* Zone of subduction
Subdued topography, 294
Sublimation, of snow, 286
Submarine canyons, 361
Submergent coastline, 378-382, 378f-381f
Submersibles, in ocean exploration, 352-353, 353f
Subsidence
    dredging and, 271

mining and, 228, 229f
natural, 268
oil drilling and, 226, 229f
Subsurface drainage, 439
Subtropical deserts, 325-326, 326f, 327f
Suess, Edward, 49-50
Sulfate minerals, 676t-677t
Sulfates, 86t, 91
Sulfide deposits, identification of, 571, 571f
Sulfide minerals, 571, 678t-679t
Sulfides, 86t, 91
  weathering of, 170
Sulfur, 91
  in coal, 581, 603-604
Sulfur dioxide
  getters for, 603-604
  from volcanoes, 103
Sun
  distance to, 30
  formation of, 32
  location of, 30
Superalkaline soils, 195
Supercontinent cycles, 356-357
Supergene enrichment, 571
Supergiants, 31
Supernova, 31, 32f
Superposition, principle of, 267, 611-612, 611f
Surf, 365
Surface drainage, 439
Surface mining. *See* Coal mining
Surges, glacial, 288
Suspended load, 247, 248f
Swallow holes, 439
Swamps, 263, 268. *See also* Wetlands
  coastal, 381f
  definition of, 578
  vs. marshes, 578-579
  paludal deposits in, 405, 405f, 578-579, 578f
  peat formation in, 405, 405f, 578-579, 578f
S waves, 534, 534f, 538-539, 539f, 540f
  density and, 553-554, 553f-555f
Swells, 364
Sylvite, 85, 682t-683t
Symmetrical folds, 463, 463f, 464-465, 464f-465f
Synclines, 462, 462f
Synfuel, 582
System of rocks, 628

T

Tabular intrusive rock bodies, 154, 155f
Tachylyte, 149
Tailings, 568
Talc, 82
Talus, 164, 215, 216f
  in deserts, 333, 333f
Tarn, 291, 293f
Taste, of minerals, 85
Taylor, F.B., 50
Temperature, melting-crystallization, 141-142
Temperatures, atmospheric
  cyclic variations in, 313, 315
  El Niño and, 372-373

glaciation and, 313-314
rising. *See* Greenhouse effect
volcanic eruptions and, 97, 106-107, 128, 314
weathering and, 168, 174
Temporary baselevel, 249, 249f
Tension, 454-455, 454f
Tensional joints, 469, 470f
Tensional stress, 454-455, 454f, 458
Tephra, 4, 106-108
Terminal moraines, 297, 298f, 306, 306f
Terraces, 254, 255f
  kame, 300
Terrestrial planets, 33-37, 34f-37f
Tertiary Period, 625f, 631-632, 632t
Tethys, 39f
Thalweg, 258
Thea, 39f
Theory of the Earth (Hutton), 610-611
Thermohaline circulation, 369
Thermohaline currents, 367, 367f, 368f, 369-370
Thrust faults, 466, 466f
Tidal inlets, 381, 381f
Tides, 365-366, 365f, 366f
  ebb, 381
  flood, 381
  neap, 366, 366f
  as power source, 594, 595f
  spring, 366
Till, 297
Tilt, volcanic, 123
Tiltmeters, 123
Time-distance (time-travel) curve, 539, 539f
Time scales, rock, 624-632. *See also* Geologic time scale
Tomography, seismic, 64, 65f
Tools, early, 401, 401f, 567, 567f
Topaz, 87
Topographic relief, development of, 249-258
Topset beds, 264
Topsoil, 192
Trace elements, 177-179
Traction, in water flow, 246
Trade winds, 367, 368f
  El Niño and, 372-373
Trailing edge, 9, 10-12, 11f, 12f, 14f, 66, 99, 100f, 359-363, 360f-362f
  ocean closing and, 356, 359, 360f
Transform faults, 66, 66f, 357, 357f
Transitional depositional environments, 405-407, 406f-408f
Transverse dunes, 342, 343f
Traps, 585, 585f
Travertine, 399, 400f
Trees, growth rings in, 618f, 619
Trellis stream pattern, 240, 241f
Tremolite, 672t-673t
Trenches, deep-sea, 16, 17f, 56-58, 56f, 58f, 63, 359
Triassic System, 625f, 629-631
Tributaries, 236, 238
Triple junction, 354, 354f
Tropical regions, soil in, 196
Trough, glacial, 290, 293f
Trough lakes, 293, 294f
Tsunami, from earthquake, 532-533, 533f
Tubes
  in caves, 440
  lava, 428, 428f

Tuff, 108, 108f
  ash-flow, 108, 108f
  welded, 108
Turbidites, 361, 361f, 412
Turbidity currents, 361, 361f, 374, 412. *See also* Density currents
Turquiose, 680t-681t
Type locality, 628

U

Ultimate baselevel, 249, 249f
Ultramafic rock, 149
Unconfined aquifers, 435-436
Unconformity, 615, 616-617
  angular, 616, 616f, 618f
Unconformity trap, 585, 585f
Underground mining. *See* Mining
Underthrusting, 474
Undesirable pollutants, 275
Uniformitarianism, 414-415
Unifying theory of geology, 49
Unit cell, 80f
United States National Seismograph Network (USNSN), 542
Units of measure
  conversion factors for, 670t
  prefixes for, 670t
Universe, origin and evolution of, 31
Unpaired terraces, 254, 255f
Uplifting, nondeformational, 453, 453f
Upstream floods, 274
Uranium
  abundance of, 569t
  as fuel, 588, 590-591
Uranium 235, in radiometric dating, 621-622, 623, 623t
Uranus, 37, 38-40, 39f, 40f
Urbanization
  environmental problems of, 199-200
  flooding and, 274-275
U-shaped valleys, 290-291, 293f
Ussher, James, 609

V

Valley(s), 17-18, 19f
  desert, 334f, 335, 335f
  glacial, 290, 293f
  hanging, 291
  rift. *See* Rift valleys
  U-shaped, 290-291, 293f
Valley flat, 252, 252f
Valley glaciers, 283, 284f, 285f. *See also* Glacier(s)
  erosion by, 290-294, 291f-294f
  formation of, 313
Valley train, 302
Van der Waals bonding, 79-80
Varve, 413, 414f, 619, 619f
Vegetation, desert, 321, 322f, 342, 342f
Veins, 570, 570f
Venetz-Sitten, Ignatz, 305-306
Ventifacts, 338, 339f
Venus, 34-35, 35f
Verde antique, 490, 490f

Vermiculite, 674t-675t
Vertisols, 197, 198t
Vine, F., 59
Vine-Matthews-Morely hypothesis, 59, 59f
Viscosity, definition of, 149
Viscous sublayer, 245
Volatiles, 580
Volatile substances, melting-crystallization temperature and, 142
Volcanic breccia, 108, 108f
Volcanic eruptions
    atmospheric temperatures and, 97, 106-107, 128, 314
    felsic, 118
    fissure, 116-118, 117f
    gases emitted in, 102-103
    Hawaiian-type, 114, 114f
    intensity of, 112-116, 113f
    lahars and, 108, 109f, 220, 221f
    noteworthy examples of, 125-136
    Pelean-type, 114, 114f
    Plinian, 115f, 116
    prediction of, 123-124
    seismic activity and, 123-124
    Strombolian, 114
    Vulcanian, 114
Volcanic islands, 16, 16f, 131-136, 131f-136f, 359
    reefs and, 383, 384f
Volcanic mountains, 4, 4f, 363, 363f, 502-504, 502f-504f
Volcanic neck, 156-157, 157f
Volcanic rocks, 4, 154
Volcanoes/volcanism, 97-136
    active, 97
    calderas of, 118-121, 118f-120f
    cinder cone, 109, 110f
    composite, 111, 112f, 504
    continental arc, 11f, 99, 101f, 363, 363f, 504
    continental glaciation and, 313-314
    convergent margin, 99, 100f, 101f
    craters of, 118, 118f
    definition of, 109
    distribution of, 99-100
    divergent margin, 99, 99f
    dormant, 97
    earthquakes and, 97
    eruption of. See Volcanic eruptions
    extinct, 97
    geothermal phenomena and, 121-122, 121f, 122f
    hot spot, 100, 101f, 132, 132f, 150, 151, 503-504, 503f
    intraplate, 100, 101f
    island arc, 11f, 99, 101f, 363, 363f, 504
    lava from, 3, 4, 103-106, 104f-106f. See also Lava
    magma from. See Magma
    on Mars, 36-37, 37f
    monitoring of, 123-124
    mountain formation and, 502-504, 502f-504f
    noteworthy examples of, 125-136
    oceanic ridge, 98f, 99, 116, 116f, 150
    plate boundaries and, 97, 99-100, 99f-101f
    rifting and, 9, 10f, 57, 57f, 356, 502
    seamounts and, 16, 16f, 100, 110, 131-136, 131f-136f
    shield, 16f, 359, 503-504, 503f
    spatter cone, 109-110, 110f
    subduction zone, 4f, 10, 11f, 100, 104f, 111, 150, 151, 504, 504f
    submerged, 16, 16f
    tephra from, 106-108
    types of, 109-111, 110f-112f
    on Venus, 35, 35f
    as water source, 102-103
Vulcanian eruptions, 114, 114f

W

Walther's law, 267, 417-418, 418f
Waste
    as fuel, 596-597
    nuclear, 588-590
Waste disposal, 387
    via ocean dumping, 386-387
Water. See also Groundwater; Rainfall
    alkaline, 579
    containments of, 21, 235-236
    desert erosion from, 329-337, 330f-337f
    diversion of, 224, 225f
    elemental concentrations in, 667t-668t
    evaporation of, exfoliation and, 165-167, 165f-168f
    expansion of with freezing, 164
    hardness of, 179
    in hydrologic cycle, 235-236, 236f
    melting temperature and, 142, 142f
    in metamorphism, 483, 484, 489-490
    municipal, 436-437, 436f, 437f
    ocean, movement of, 364-374
    pollution of. See Water quality
    in regolith, 194
    in sedimentary deposits/rocks, 420
    in soil, 194-196
    solutes in, 171
    in sorting, 394
    trace elements in, 177
    weathering and, 174
Waterfalls, 251, 251f
Water flow
    channel texture and, 244
    direct particle lifting by, 246
    dynamics of, 245
    force generation by, 246
    gradient and, 244, 244f
    laminar, 245, 245f
    in meandering stream, 257-258
    sediment erosion and transport and, 245-248
    turbulent, 245, 245f
    velocity of. See Water velocity
    water volume and, 244-245
Watergaps, 240, 241f
Water quality, 275-276
    agriculture and, 178-179
    coastal, 386
    disease and, 179-180
    mining and, 170, 180
    trace elements and, 177-179
Water saturation, mass wasting and, 208-209, 221-222
Water spouts, 376f
Water tables, 430-432, 430f-433f
    depth to, 430, 430f
    dewatering of, 445-446, 446f
    in karst areas, 439-440
    perched (hanging), 431-432, 431f, 432f
    regional, 431-432, 431f, 432f
Water vapor
    relative humidity and, 324
    from volcanoes, 102-103
Water velocity. See also Water flow
    channel texture and, 244
    gradient and, 244, 244f
    horizontal, 246
    in meandering stream, 257-258, 258f
    sediment erosion and transport and, 246
    water volume and, 244-246
Water volume, in streams, 244-246
Water volumes, 664
Water wells, 433-437
Watt, James, 577, 577f
Wave(s)
    light, 550, 551f
    ocean. See Ocean waves
    seismic. See Seismic waves
Wave-built platform, 376, 377f
Wave-cut cliffs, 375, 376f, 377f
Wave-cut platform, 376
Weathering, 5, 6f, 163-180. See also Erosion
    carbonation in, 173
    in channel cutting, 251
    chemical, 170-176, 332-333
    crystal structure and, 175-176
    decomposition in, 164
    definition of, 164
    in deserts, 333-335, 334f, 335f
    differential, 20f
    disintegration in, 164
    dissolution and, 171-172, 173f
    environmental quality and, 177-180
    exfoliation and, 165-167, 165f-168f
    frost action and, 164-165, 164f, 165f, 209-210, 210f
    honeycomb, 166, 167f
    hydrolysis in, 173
    mechanical, 164-169
    of metamorphic rocks, 494
    mineral composition and, 175
    ore enrichment by, 571-572
    oxidation and, 170
    particle size and, 174, 175f, 394, 394t
    in periglacial regions, 303-305, 304f
    plant wedging and, 169, 169f
    rate of, 174-176
    rock rounding and, 167-168, 168f, 248f, 256
    rock slides and, 216
    sheeting and, 167, 168f
    soil formation and, 163
    solid products of, 173, 175-176, 176, 185, 186f. See also Regolith; Sediment
    spalling and, 166, 167f
    spheroidal, 167-168, 168f
    temperature and, 168, 174
    trace elements and, 177-179
    water and, 174
Wegener, Alfred, 50-52, 313, 554
Weighted lines, in ocean exploration, 351
Welded tuff, 108
Wells, 433-437
    artesian, 436
    dewatering of, 445-446
    saltwater encroachment into, 446-447, 447f
Westerly winds, 367, 368f

Wet-based glacier, 287
Wetlands, 263, 263f, 268
    coastal, 272, 381f, 382, 385, 407, 408f
    paludal deposits in, 405, 405f
White, I.C., 585
White dwarfs, 31
William, H.S., 629
Williams, James, 577
Winchell, Alexander, 629
Wind(s)
    competence of, 337-338
    solar, 32, 33f
    in sorting, 394
    trade, 367, 368f, 372-373
    westerly, 367, 368f
Wind deposition, 339-343
    in deserts, 339-343, 340f343f, 394, 403-405, 404f
    dunes and, 340-342, 341f-343f
    loess and, 339-340, 340f
Wind-driven ocean currents, 367-368, 367f, 368f

Wind erosion
    abrasion in, 338-339, 339f, 340f
    deflation in, 338, 339f
    in deserts, 337-339, 338f-340f
    in periglacial regions, 305
    saltation in, 338, 338f
Wind-induced currents, 369, 370f
Wind power, 592-594, 593f, 594f
Windshadow, 341, 342f
Winter monsoon, 368
Wood, as fuel, 596
Wood, H.O., 526

X

Xenolith, 156, 157f
Xenophanes, 609
X-ray diffraction, 85, 86f

Z

Zinc, 179
    abundance of, 569t
Zincite, 680t-681t
Zircon, 622
Zone of ablation, 288-289, 289f
Zone of accumulation, 288-289, 289f
Zone of aeration, 430, 430f
Zone of deposition, 193
Zone of leaching, 192-193
Zone of rifting. *See* Rifting/rift zone
Zone of saturation, 430, 430f
Zone of subduction, 4f, 10, 11f, 66
    earthquakes and, 525, 525f
    formation of, 58f, 63, 63f
    leading edge and, 14f, 359, 360
    in magma formation, 150, 151-153, 151f, 153f
    in mountain formation, 512, 512f
    in orogenesis, 512, 512f, 513, 514
    volcanism and, 4f, 10, 11f, 100, 104f, 111, 504, 504f

LOCATION INDEX

Note: Pages followed by f indicate figures.

A

Adirondack Mountains
    erosion of, 294, 295f
    stream patterns in, 240
Africa
    collision of with Europe, 518
    deserts of, 321, 326-327, 328f
    drifting of, 49, 50, 511, 518
    movement of, 10, 11f
    overgrazing in, 344-345
    soils of, 195
Alaska
    oil reserves in, 587
    valley glaciers in, 287, 288
Aleutian Islands
    as island arc, 99, 101f, 363
    subduction zone of, 131
    as volcanic mountains, 66, 151, 153, 504
    zones of subduction in, 151-152
Alexandria (Egypt), 28
Alps, 10, 17, 284, 290, 292f
    as foldbelt mountains, 508
    formation of, 10, 518
Altamont Pass, 594
Amazon River, order of, 238
Anak Krakatoa, 128
Andes Mountains
    as continental arc, 363, 504
    deep-sea trenches and, 56
    as foldbelt mountains, 508
    formation of, 512f
    as volcanic mountains, 66, 153, 513, 513f
    zones of subduction in, 151
Antarctica
    crustal loading in, 560, 561
    deserts of, 328
    ice sheet of, 283, 285f
    rock samples in, 174
Appalachian Basin, coal in, 582, 584f
Appalachian Mountains, 17
    as foldbelt mountains, 460f, 463f, 508-510, 509f
    formation of, 466, 518
    mass wasting in, 207f
    mining in, 602
    Oriskany Sandstone of, 426
    rock movement direction in, 473-474, 473f
    stream patterns in, 240, 241f
    Valley and Ridge Province of, 240, 241f

Appalachian Plateau, 453, 518
Arctic Ocean, 14f
    cooling of, 314
Asia, movement of, 11f, 12, 12f
Atacama Desert, 327, 328f
Atchafalaya distributary, 270, 270f
Atlantic Ocean, 14f
    barrier islands of, 379-382, 379f, 380f, 407f, 408f
    bathymetry of, 54, 55f
    creation of, 10
    currents in, 369, 370
    formation of, 10, 10f, 357, 358f, 518
    low-energy coastline of, 378-382, 378f-381f
    surface water exchange in, 368
Atlantic Oceanic Ridge, 56, 59, 99, 357, 358f
Atlantis, 127f, 128
Australia
    deserts of, 326
    Great Barrier Reef of, 383, 383f
    soils of, 195
Austral Islands, 100, 102f

B

Bahamas, 416, 573
Balize delta lobe, 268f, 269f, 270, 406f
Baltic Sea, crustal loading and, 560
Barringer Crater, 42, 42f
Basin and Range Province, 17-18, 19f, 310, 335, 336f, 507, 508f
Bau of Fundy, tides of, 366, 366f
Bengal fan, 362, 362f
Benguela Current, 327, 327f, 369
Bingham Canyon, 572, 572f
Birmingham (Alabama), steel industry in, 573-574
Black Bay (Canada), 297f
Black Hills
    aquifers of, 436
    local doming in, 504-505, 505f
    stream patterns in, 240, 241f
Black Sea, 10, 356
Blue Ridge Mountains, 466, 518
    stream patterns in, 241f
Bonanza Creek, 575f
Brazil
    mining in, 575
    Mother Lode in, 575
Breed's Hill, 302
Bryce Canyon, 18, 20f, 506, 506f
Bryd Station (Antarctica), 287
Bunker Hill, 302

C

California
    Gold Rush in, 574, 574f
    slope instability in, 220, 222, 222f
    subsidence in, 226, 229f
California Coastal Range, 492, 494f, 513, 513f
Canada
    crustal loading in, 560
    roche moutonnées in, 297f
Canadian Shield, 17, 19f, 20f
    as peneplane, 257
    regional metamorphism in, 491, 491f
    stream patterns in, 240
Carlsbad Caverns, 443, 443f
Cascade Mountains, 10, 11f, 325
    as continental arc, 99, 101f, 363, 363f
    deep-sea trenches and, 56
    formation of, 10, 11f
    rainshadow of, 325
    as volcanic mountains, 66, 131, 151, 153, 220
Caspian Sea, 10
    as Tethys Sea remnant, 10, 356
Channeled Scablands, 311-312, 312f
Chernobyl, 590, 590f
China
    earthquakes in, 530
    karst towers in, 440f
    loess deposits in, 340, 341f
Clark Fork River, 311
Clinton Iron Ore, 573, 573f
Coast Ranges, deserts of, 324
Cockpit Region, 439f
Colorado Plateau, 18, 20f-21f, 21f, 453, 505-507, 506f
Colorado River, 20f, 21f, 330
    damming of, 561, 561f
    as rejuvenated stream, 254
Columbia Plateau, 54f
    Channeled Scablands of, 311-312, 312f
    flood basalts of, 117, 117f
    formation of, 152, 152f
Columbia River, 220, 312
Crater Lake, 120, 120f
Crater Lake National Park, 108f

D

Dakota Sandstone, 436
Dead Horse Point (Utah), 255f, 453f
Death Valley, 171f, 249, 321, 325, 326f
Devil's Post Pile, 470f
Diamond Head, 384f

704 Location Index

Donner Pass, 507
Dunkard Basin, 584, 584f
Dust Bowl, 198

E

Earthquake Lake, 217, 219f
East African Rift Valley, 9, 10, 10f, 99, 99f, 336, 354, 355f, 502
    fault movements in, 474-475
Egypt, 28
El Capitan, 157, 157f
Emperor Seamounts, 131, 134, 137
England, 631, 631f
Entrada Sandstones, 405
ET's Finger, 335f
Europe, drifting of, 518
Everest, Mount, 516

F

Finger Lakes
    formation of, 295f, 296
    as trough lakes, 293, 294f
Florida Bay, 401, 401f, 416, 416f
Florida Keys, 383, 386, 416
France, 594, 595f
Franciscan Formation, 492, 494f, 513
Fujiyama, Mount, 111, 112f

G

Gandy River, 243
Geysers (California), 122, 591, 592f, 600
Ghana, Mother Lode in, 575
Gilbert Islands, 100, 102f
Glacier Bay National Monument, 280
Gobi Desert, 327-329.327f, 340
Gondwana, 11f, 49, 51f, 257, 356
    glaciation of, 314
    mountain building on, 515-516, 516f
Goosenecks, 254, 255f
Grand Canyon, 18, 20f-21f, 331, 332f, 453, 506, 506f
    discontinuities in, 615, 617f
    nonconformities in, 617
    as youthful valley, 254
Grand Tetons, 290, 292f
Great Barrier Reef, 383, 383f
Great Dismal Swamp, 579
Great Lakes, formation of, 295, 296f, 307-308, 308f, 309f
Great Plains, 18
    aquifers of, 436
    Dust Bowl of, 198
    kettle lakes in, 300, 300f
    loess deposits in, 341f
    overcultivation in, 345
    soils of, 178, 195
    wind power and, 592
Great Salt Lake, 249, 311, 311f

Great Valley (California)
    deserts of, 324
    subsidence in, 226
Greece, Ancient
    architecture in, 458, 459f
    astronomy in, 27-28
Greenbrier Limestone, 243f
Greenland, 315
    crustal loading in, 560, 561
    ice sheet of, 283, 285f
Greenland Gap, 241f
Green River Formation, 586f
Gros Ventre River, 217, 218f
Gulf of Aden, 35f5f, 354, 356
Gulf of Mexico
    barrier islands of, 379-382, 379f, 380f
    coastal erosion in, 272, 272f
    low-energy coastline of, 378-382, 378f-381f

H

Harper's Ferry, 241f
Hawaiian Island–Emperor Seamount chain, 100, 102f
Hawaiian Islands
    calderas of, 118-120, 119f
    formation of, 133
    rock formations of, 133-134
    as seamounts, 16, 16f, 503f
    shield volcanoes of, 110, 111f, 503-504, 503f
    volcanic eruptions in, 103, 104f, 106f, 112, 113f, 114, 131-136, 131f-136f
    as volcanic islands, 5, 5f, 16, 16f, 151, 151f
    volcano observatory on, 144, 144f, 145f
Henry Mountains, 155, 156f
Herculaneum, 126, 126f
Hickory Run (Pennsylvania), 305, 305f
Himalaya Mountains, 12, 12f, 17, 328, 493
    as foldbelt mountains, 508
    formation of, 12, 12f, 516f-518, 517f, 559
    rock movement direction in, 473, 473f
Hoover Dam, 561, 561f, 595f
Hot Springs (Virginia), 443, 444f
Hualalai volcano, 132, 134f
Huascarán, Mount, debris flow on, 218-219
Hudson Bay, crustal loading and, 560
Hudson River, 361
Humboldt Current, 327, 327f, 369
    El Niño and, 372-373

I

Iceland, 15, 15f, 56, 315
    Atlantic Oceanic Ridge and, 99, 357, 358f
    fissure eruption on, 117
    geothermal phenomena in, 122
Icelandic Plateau, 357, 358f
Illinois Basin, 584
Imperial Valley, 196
India, drifting of, 11f, 12, 12f

Indian Ocean, 14f, 362-363, 362f
    formation of, 10, 10f, 357
    gyre of, 367, 368f
    winter monsoon and, 368
Indus fan, 362, 362f
Isles Dernieres, 271-272, 272f

J

Jamaica, Cockpit Region of, 439f
Japanese Islands, 10, 11f, 514f
    deep-sea trenches and, 56
    formation of, 10, 11f
    pollution-related disease in, 180
    tsunami in, 532
    as volcanic mountains, 66, 151, 153
    volcanoes of, 111, 112f
    zones of subduction in, 152
John Pennecamp Coral Reef State Park, 386

K

Kalahari Desert, 326
Kesterson Reservoir, 178
Kilauea volcano, 5f, 112, 113f, 134, 134f, 136
    eruption of, 5f
Kilimanjaro, Mount, 356, 502
Kohala volcano, 132, 134f
Komati River, 149
Krakatoa volcano, 116, 118, 128, 129f, 131
Kuei-lin (China), 440f

L

Lake Athabasca, 297f
Lake Bonneville, 311, 311f
Lake Missoula, 311-312, 312f
Lassen National Park, 121
Laurasia, 11f, 51f, 356
Lockport Dolostone, 308
Loihi volcano, 132, 134f
Loma Prieta Earthquake, 528, 528f, 530-532, 531f
Long Beach (California), subsidence in, 226, 229f
Long Valley Caldera, 123-124, 123f, 131
Los Alamos, 591
Louisiana
    barrier islands of, 271-272, 272f
    coastal erosion in, 380f
Love Canal, 199-200
Lower Snake River, 312
Lyons (Kansas), 590

M

Madison River, 217, 219f
Mammoth Caves, 443f
Mammoth Hot Springs, 121, 121f

Mammoth Lake, 123-124, 123f
Maringouin/Sale delta lobe, 270
Marshall Islands, 100, 102f
Martinique, volcanic eruption on, 115-116, 115f
Matterhorn, 290, 292f
Mauna Kea volcano, 132, 134f
Mauritanide Mountains, 518
Maxwell, Monte, 35, 36
Mayon volcano, 314
Mazama, Mount, 120, 120f
McKenzie River, 304f
Mediterranean Sea, 10, 518
    currents in, 370-372, 371f
    as Tethys Sea remnant, 10, 356
Mekong delta, 265f
Mesabi Range, 572, 573f
Meteor Crater, 42, 42f
Mexico City earthquake, 532, 532f
Michigan Basin, 584
Midcontinent Basin, 582
Middle East, oil reserves in, 587f
Mississippi delta, 7f, 265-272, 265f-271f
    formation of, 260
Mississippi fan, 361-362
Mississippi River
    levee systems of, 260, 263f
    navigation of, 259
    order of, 238
Mississippi River Valley, loess deposits in, 339, 340f
Missouri River, navigation of, 259
Mohawk River, 307
Mojave Desert, solar power generation in, 599, 600f
Monroe County (Illinois), flooding of, 262f
Monte Maxwell, 35, 36
Monument Valley, 18, 20f, 506, 506f
    erosion in, 206f, 331, 333-335, 334f
Morrison Formation, 631, 631f
Mother Lode, 575, 576f

N

Namib Desert, 327, 328f, 575
Navajo Sandstones, 405
Nepal, 502
New Madrid earthquake, 524, 527
New Orleans, flood prevention in, 271
New Zealand, geysers in, 121
Niagara Escarpment, 307, 308
Niagara Falls, 308, 309f
    Love Canal and, 199-200
Niagara River, 308
Niger delta, 265f
Nile River, 330
Nile River delta, 264, 265f
North America
    divides in, 237, 237f
    ice sheet of, 30f, 310
    rainfall distribution in, 190f
    as trailing edge, 359
North American Cordilleria, 514, 515f
North Atlantic gyre, 368
North Atlantic Oceanic Ridge, 15, 15f

O

Ohio River, navigation of, 259
Oil Creek (Canada), 587f
Oil Springs (Ontario), 577, 578f
Okefenokee Swamp, 579, 579f
Old Faithful, 121, 122f
Old Red Sandstone, 615
Olympic Mountains, deserts of, 324
Olympus Mons, 36, 37f
Oriskany Sandstone, 426

P

Pacific Coastal Range, fog in, 325f
Pacific Ocean, 14f
    currents in, 369, 370, 372-373
    deep-sea trenches of, 16, 16f, 17f, 359
    earthquakes in, 532
    El Niño and, 372-373
    high-energy coastline of, 375-378, 375f-377f
    Ring of Fire in, 4f, 98f, 100f
    sea-floor spreading in, 356
    shield volcanoes of, 111f
    surface water exchange in, 368
Painted Desert, 630, 630f
Pangaea, 10, 11f, 50, 51f, 52, 357
Pelée, Mont, 115-116, 115f
Peru, debris flow in, 218-219
Petrified Forest, 630, 630f
Philippines, volcanoes of, 103f, 314
Piedmont, 466, 518
Pinatubo, Mount, 4f, 103f
Pinnacles, 108f
Pocahontas Basin, 584, 584f
Point Lobos, 375, 375f
Pompeii, 126, 126f
Potomac River, 241f
Pozzuoli (Italy), 124
Prudhoe Bay oil field, 587
Pu'u O'o volcano, 134, 135f

R

Rance River, 594, 595f
Ranier, Mount, 220
Red Sea, 10, 10f, 354, 355f, 356
    formation of, 10, 10f
Rift Valley, 10, 10f
Ring of Fire, 4f, 98f, 100f
Rio Grande Rift, 9, 99, 100f, 502
Rocky Mountains
    divide in, 237, 237f
    erosion of, 294, 295f
    as foldbelt mountains, 508
    formation of, 514, 515f
    horns in, 290, 292f
    ice caps of, 310
Rome, Ancient, architecture in, 458, 459f
Rosebud County (Montana), 583f

S

Sahara Desert, 321, 326, 330
St. Francis Dam, 494, 494f, 495f
St. Helens, Mount, eruption of, 104f, 108, 109f, 112, 113, 128-131, 130f, 220, 221f
St. Lawrence fan, 362
St. Lawrence River, 308
St. Lawrence Valley, 307
St. Pierre, 115-116, 115f
Sakurajima volcano, 136, 136f4
San Andreas Fault, 467-468, 468f, 475
    Loma Prieta Earthquake and, 528, 528f
    San Francisco Earthquake and, 528, 528f
San Francisco earthquake, 528, 528f
San Francisquito Canyon, 494, 494f, 495f
San Joaquin Valley, selenium pollution in, 178
San Juan River, Goosenecks of, 254, 255f
San Rafael Swell, 462f
Santoria, 126-128, 127f
Santorini volcano, 126-128, 127f
Scandinavian Peninsula, crustal loading in, 560, 561f
Scotland, 610, 610f
Sea of Japan, as marginal sea, 511
Sevier Lake, 311
Seyne (Sudan), 28
Shasta, Mount, 220
Shippensport (Pennsylvania), 578
Siccar Point, 610, 610f, 615, 616, 618f
Sierra Nevada, 3, 3f, 325
    Mother Lode in, 575, 576f
Sinks of Gandy, 243f
Snake River, 428, 428f
South Africa, komatites in, 149-150
South America
    deserts of, 326-327, 328f
    drifting of, 49, 50, 511
    pampas of, 195
    as trailing edge, 359
South Pacific Equatorial Current, 372
South Pole, ice sheet of, 288
Spirit Lake, 129
Stonehenge, 27, 27f, 609
Strait of Dover, 631, 631f
Strait of Gibraltar, 371, 371f
Stromboli, 114
Sudan, 28
Sudbury (Ontario), 570, 570f
Sumbawa, 97
Sutter's Mill (California), 574, 574f
Switzerland, mountains of, 10, 17, 284, 290, 292f

T

Tambora volcano, 97, 116, 314
Tapeats Sandstone, 617, 618f
Tehachapi Pass, 592, 593f
Tethys Sea, 10, 11f, 356, 516
    remnants of, 10, 356
Thousand Springs, 428, 428f

706 Location Index

Three Mile Island, 590
Tigris-Euphrates Valley, salinization in, 345
Titusville (Pennsylvania), 577, 578f
Tonga Islands, as island arc, 363, 504
Toutle River, 129, 220, 221f
Tuamotu Archipelago–Line Island chain, 100, 102f
Tuscany, 611
Tuscarrora Sandstone, 624

U

United States
    overcultivation in, 345
    rainfall distribution in, 190f
    soil orders in, 197f
University of Chicago, 578
Ural Mountains, 518
Ukraine, soils of, 195
Utah Lake, 311

V

Valles Merineris, 36f, 37
Valley and Ridge Province, stream patterns in, 240, 241f
Vancouver Bay, 613f
Variegated Glacier, 288
Vesuvius, Mount, 124, 125-126, 125f, 126f
Vishnu Schist, 617, 618f
Volcano (island), 114

W

Warm Springs (Virginia), 443, 444f
Wasatch Formation, 20f
Willamette Valley, deserts of, 324
Wills Mountain Anticline, 241f
Winter Park (Florida), sinkhole in, 226, 228f

Y

Yellow River, 340
Yellowstone Lake, 121
Yellowstone National Park
    formation of, 114
    geothermal phenomena in, 121, 121f, 443, 444f
    hot spot under, 100, 102f, 117
    volcanic eruptions in, 118, 121, 154
Yosemite National Park, sheeting in, 167, 168f, 470f
Yosemite Valley, glacial trough in, 291
Yucca Mountain, 590
Yungay (Peru), destruction of, 219

Z

Zion National Park, 18, 20f, 405f, 416, 417f, 506, 507f, 630, 630f